# Dynamics, Information and Complexity in Quantum Systems

# Theoretical and Mathematical Physics

The series founded in 1975 and formerly (until 2005) entitled *Texts and Monographs in Physics* (TMP) publishes high-level monographs in theoretical and mathematical physics. The change of title to *Theoretical and Mathematical Physics* (TMP) signals that the series is a suitable publication platform for both the mathematical and the theoretical physicist. The wider scope of the series is reflected by the composition of the editorial board, comprising both physicists and mathematicians.

The books, written in a didactic style and containing a certain amount of elementary background material, bridge the gap between advanced textbooks and research monographs. They can thus serve as basis for advanced studies, not only for lectures and seminars at graduate level, but also for scientists entering a field of research.



For further volumes:
http://www.springer.com/series/720

Fabio Benatti

# Dynamics, Information and Complexity in Quantum Systems

Springer

Dr. Fabio Benatti
Università Trieste
Dipto. Fisica Teorica
Strada Costiera, 11
34014 Trieste
Miramare
Italy
benatti@ts.infn.it

ISBN 978-1-4020-9305-0 e-ISBN 978-1-4020-9306-7

DOI 10.1007/978-1-4020-9306-7

Library of Congress Control Number: 2008937916



Printed on acid-free paper

9 8 7 6 5 4 3 2 1

springer.com

To Heide, for her scholarship, for her friendship

# Preface

The aim of this book is to offer a self-consistent overview of a series of issues relating entropy, information and dynamics in classical and quantum physics. My personal point of view regarding these matters is the result of what I had the good fortune to learn in the course of the years from various scientists: Heide Narnhofer in the first place, who introduced me to quantum dynamical entropies and was a precious guide ever since, then Robert Alicki, Mark Fannes, Giancarlo Ghirardi, Andreas Knauf, John Lewis, Geoffrey Sewell, Franco Strocchi, Walter Thirring, Armin Uhlmann. To me, all of them have been a constant example of rigorous mathematics and physical intuition jointly at work.

Last but not least, my deep gratitude goes to my family and to the many friends on whom I could always count for support and encouragement with a special thought for Traude and Wolfgang Georgiades.

Trieste, 6 August 2008 Fabio Benatti

# Contents

## Part II Quantum Dynamical Systems

## Part III Quantum Dynamical Entropies and Complexities

# 1 Introduction

This book focusses upon quantum dynamics from various points of view which are connected by the notion of dynamical entropy as a measure of information production during the course of time.

For classical dynamical systems, the notion of dynamical entropy was introduced by Kolmogorov and developed by Sinai (*KS* entropy) and provided a link among different fields of mathematics and physics. In fact, in the light of the first theorem of Shannon, the *KS* entropy gives the *maximal compression rate* of the information emitted by ergodic information sources. A theorem of Pesin relates it to the *positive Lyapounov exponents* and thus to the exponential amplification of initial small errors, in a word to classical chaos. Finally, a theorem of Brudno links the *KS* entropy to the compressibility of classical trajectories by means of computer programs, namely to their *algorithmic complexity*, a notion introduced, independently and almost simultaneously by Kolmogorov, Solomonoff and Chaitin.

In a previous book by the author, the notion of quantum dynamical entropy elaborated by A. Connes, H. Narnhofer and W. Thirring (*CNT* entropy) was presented within the context of quantum ergodicity and chaos. The *CNT* entropy is a particular proposal of how the *KS* entropy might be extended from classical to quantum dynamical systems.

After the appearance of the *CNT* entropy, other proposals of quantum dynamical entropies appeared which in general assign different entropy productions to the same quantum dynamics. The basic reason is that each proposal is built according to a different view about what information in quantum systems should mean. Concretely, it is a general fact that, in order to gain information about a system and its time-evolution, one has to observe it and a quantum fact that observations may be invasive and perturbing. Should this fact be considered inescapable and thus incorporated in any good quantum dynamical entropy or, rather, should it be avoided as a source of spurious effects that have nothing to do with the actual quantum dynamics?

This is an unavoidable question and, based on the possible answers, one is led to different notions of quantum dynamical entropies. These will be sensitive to different aspects of the quantum dynamics and thus, not unexpectedly, not equivalent: the real issue is which these aspects are and what kind of informational meaning they do posses.

F. Benatti, *Dynamics, Information and Complexity in Quantum Systems*, Theoretical and Mathematical Physics, DOI 10.1007/978-1-4020-9306-7_1,

In view of the role of the *KS* entropy in classical chaos, one of the principal applications of the quantum dynamical entropies has been to the phenomenology of quantum chaos. The scope has now become wider: quantum compression theorems and recent attempts at formulating a non-commutative algorithmic complexity theory motivate the study of whether and how the different quantum dynamical entropies are related to these new concepts. In particular, a better understanding of the many facets of information in quantum systems may come from clarifying the relations of the various quantum dynamical entropies among themselves and their bearing on quantum compression schemes and the algorithmic reproducibility of quantum dynamics.

The issue at stake can be conveniently conveyed by an example: the simplest classical ergodic information source emits bits independently of each other with probabilities 1/2 for both 0 and 1. The *KS* entropy is $\log 2$ and represents

1. the information rate of a classical source emitting independent bits;
2. the Lyapounov exponent of the classical dynamical system consisting in throwing a fair coin;
3. the algorithmic complexity of almost every resulting sequence of tails and heads.

The quantum counterpart of such an information source is a so-called quantum spin chain, that is a one-dimensional lattice carrying a $2 \times 2$ matrix algebra at each of its infinitely many sites: each site carries a so-called *qubit* . The dynamics of such a system is just the shift from one site to the other and the infinite dimensional algebra of operators is equipped with a translation-invariant state. These non-commutative structures have recently become of primary importance in the boosting field of quantum information. What is relevant is that one can construct subalgebras of quantum spin chains characterized by varying degrees of non-commutativity between their operators. Depending on that degree, the *CNT* entropy, varies between zero and $\log 2$, while another quantum dynamical entropy, the *AFL* entropy of Alicki, Fannes and Lindblad, is always $\log 2$. The *CNT* entropy thus appears to be sensitive to the amount of non-commutativity between operators, whereas the *AFL* entropy is apparently independent of that structural algebraic property.

Because of its unifying properties, the *KS* entropy can be taken as a good indicator of classical randomness and complexity; one would then like to assign a similar role to the quantum dynamical entropies. Does this mean that, in accordance with the *CNT* entropy behavior, quantum dynamical systems have varying degrees of complexity or randomness depending on the degree of non-commutativity ? Or, according to the *AFL* entropy, the algebraic structural properties have no bearing on dynamical randomness or complexity, which are rather related to the statistics of such systems, namely to their shift-invariant state?

More concretely, one may ask which one of the two quantum dynamical entropies is closer to the actual quantum informational structure of these quantum sources. Regarding this issue, of particular interest are the yet unexplored relations of the quantum dynamical entropies to the quantum algorithmic complexities.

Indeed, as there inequivalent generalizations of the *KS* entropy, so there are different extensions of the classical algorithmic complexity. These extensions have been motivated by the possibility of a model of computation based on the laws of quantum mechanics and on the theoretical formulation of the notion of *Quantum Turing Machines* (*QTMs* ). Like *Classical Turing Machines* (*TMs* ), *QTMs* consist of a read/write head moving on tapes with, say, binary programs written on them. Only, the tapes of *QTMs* can occur in linear superpositions of the classical configurations of 0's and 1's. In a word, inputs and outputs of *QTMs* are *qubits*.

Since the various quantum dynamical entropies were proposed, independently of quantum information, as tools to better study the long-time dynamical features of infinite quantum systems, one may doubt that relations should exist between them and quantum information. One notices, however, that the *CNT* entropy was developed using the notion of *entropy of a subalgebra* which, years later, independently appeared in quantum information theory as a measure of *entanglement* known as *entanglement of formation.* Also, the *AFL* entropy is based on techniques that in quantum information theory are fundamental tools to describe quantum channels and, more in general, all quantum operations that may affect quantum systems.

The book is organized in three parts.

In the first part, the first chapter presents basic notions of ergodic theory, the second gives an overview of entropy in information theory, the third addresses the notion of *KS* entropy and the classical compression theorems, while algorithmic complexity is the subject of the fourth chapter.

The second part consists of three chapters; the first offers an overview of algebraic quantum mechanics with particular emphasis on the notions of positivity and complete positivity of quantum maps and quantum time-evolutions, both reversible and irreversible. The second chapter introduces the fundamentals of quantum information, the relations between positive and completely positive maps and quantum entanglement, the entropy of a subalgebra, the entanglement of formation and the accessible information of a quantum channel. The third concerns infinite quantum dynamical systems and quantum ergodicity, quantum chains as quantum sources and the quantum counterparts to Shannon's theorems.

In the first chapter of the third part, a detailed introduction is given to the *CNT* and *AFL* entropies and to their use in the study of dynamical information production in quantum systems. Finally, the second and last chapter of the book focusses on some recent extensions of algorithmic complexity to quantum systems, starting with a discussion of quantum Turing machines and quantum computers and concluding with an exploration of the possible role played in this context by the quantum dynamical entropies.

The topics addressed come from rather different fields that only recently, because of the birth and rapid development of quantum information, quantum communication and computation have started to overlap. This book has been written not as an introduction to any of these topics (of which exhaustive presentations do exist in plenty), rather as an attempt to provide readers with expertise in some, but not in all of the topics, with a self-consistent overview of these many subjects. Therefore, care has been taken to give proofs of almost all of the results that have been used, apart from basic and standard facts, and to illustrate them by means of selected examples.

# Part I

# Classical Dynamical Systems

In the first part of the book classical dynamical systems are presented from the points of view of ergodic, information and algorithmic complexity theory.

Ergodic theory studies the clustering properties of equilibrium states; in information theory the central notion of entropy is used to quantify the degree of predictability of phase-space trajectories, while algorithmic complexity theory quantifies their randomness in terms of how easily they can be described by algorithms.

The purpose of this presentation is to set up a suitable algebraic framework that makes easier the extension of these three points of view to quantum dynamical systems.

# 2 Classical Dynamics and Ergodic Theory

In this chapter the term *classical dynamical system* will broadly refer to one-parameter families of transformations, or *dynamical maps*, $T_t$ acting on a phase space $\mathcal{X}$ whose points $x$ describe the system degrees of freedom. In physical applications, $x$ identifies an initial state, or configuration, $T_t x$ the resulting state or configuration after a span of time of length $t$. If $t$ is discrete, $t \in \mathbb{Z}$, one speaks of a *reversible time-evolution* through discrete time steps with trajectories $\{T_t x\}_{t \in \mathbb{Z}}$ consisting of countably many configurations at negative and positive integer times. If $t \in \mathbb{N}$, this means that the dynamics can only develop forward in time and is thus irreversible. In the case of a continuous-time dynamics, trajectories through $x \in \mathcal{X}$ at $t = 0$ are continuous sets $\{T_t x\}_{t \in \mathbb{R}}$ of configurations if the dynamics is *reversible*, otherwise trajectories are only forward in time, $\{T_t x\}_{t \in \mathbb{R}^+}$.

Once the description of a system by means of a phase-space $\mathcal{X}$ has been chosen, any phase-point $x \in \mathcal{X}$ contains all possible information about the system state. When all this information is not available, the state of a system amounts to a normalized positive measure on $\mathcal{X}$, a probability distribution, such that the volume of a measurable subset gives the probability that $x$ belong to it. *Entropy* quantifies the amount of information corresponding to such probability distribution, that is how informative the measure is about the actual state of the system.

Beside the knowledge of the state of classical systems, information can also concern how states change in time, in particular, as regards foreseeing their behavior; the degree of predictability of dynamical systems is measured by *dynamical entropies*. Intuitively, regular time-evolutions should allow for reliable predictions, which are instead hardly possible for irregular dynamics; roughly speaking, irregularity is expected to correspond to the fact that the past does not completely contain the future.

Information about the state or the time-evolution of physical systems can be obtained by measuring suitable quantities accessible to experiments. These quantities, called observables for short, correspond to functions on $\mathcal{X}$. Unlike for quantum dynamical systems, for classical ones any measuring protocol can in principle be assumed not to interfere with the system observed, the basic reason being that classical descriptions involve commuting objects, as functions on the phase-space $\mathcal{X}$ indeed are.

F. Benatti, *Dynamics, Information and Complexity in Quantum Systems*, Theoretical and Mathematical Physics, DOI 10.1007/978-1-4020-9306-7_2,

Which observables are appropriate to describe a dynamical system depends on the structure of the chosen phase-space $\mathcal{X}$; for instance, statistical descriptions require that $\mathcal{X}$ be endowed with a measure-structure, whereby measurable functions constitute appropriate observables. On the other hand, $\mathcal{X}$ might be provided with a topology and typical observables would then correspond to continuous functions.

## 2.1 Classical Dynamical Systems

In this section we review some basic facts relative to classical dynamical systems mainly adopting a measure-theoretic point of view; in this way a minimum of constraints is put on the mathematical properties of states, observables and dynamical maps and the emerging technical context is broad enough to describe a large variety of physical phenomena, from those typical of Hamiltonian mechanics to those better understood in terms of discrete dynamical systems.

**Definition 2.1.1.** *Classical dynamical systems are triplets* $(\mathcal{X}, T, \mu)$, *where*

1. $\mathcal{X}$ *is a measure space with an assigned* $\sigma$*-algebra* $\Sigma$ *of measurable sets;*
2. $T$ *is measurable, that is* $A \in \Sigma \Rightarrow T^{-1}(A) \in \Sigma$*;*
3. $\mathcal{X}$ *is endowed with a* $T$*-invariant, positive normalized measure* $\mu$*, such that* $\mu(\mathcal{X}) = 1$ *and* $\mu \circ T^{-1} = \mu$.

**Remarks 2.1.1.**

1. A collection $\Sigma$ of subsets $S \subseteq \mathcal{X}$ is called a *measure-algebra* if 1) $\mathcal{X} \in \Sigma$, 2) $S \in \Sigma$ implies $\mathcal{X} \setminus S \in \Sigma$, where $S_1 \setminus S_2$ denotes the complement of the subset $S_2$ relative to the subset $S_1$, and 3) $S_i \in \Sigma$ for $i = 1, 2, \ldots, n$, implies $\bigcup_{i=1}^n S_i \in \Sigma$. A measure-algebra $\Sigma$ is a *measure* $\sigma$*-algebra* if it is closed not only with respect to finite unions of its elements, but also with respect to countable unions, that is if $\bigcup_{n=1}^\infty S_n \in \Sigma$ for all $\{S_n\}_{n=1}^\infty$, $S_n \in \Sigma$. Since the complements of unions of sets are the intersections of the complements of the sets, namely $\mathcal{X} \setminus (A \cup B) = (\mathcal{X} \setminus A) \cup (\mathcal{X} \setminus B)$, $\sigma$-algebras contains infinite intersections of their elements, too.
2. Let $\Sigma_0$ be a measure-algebra, by adding to $\Sigma_0$ infinite unions and intersections of elements of $\Sigma_0$ one obtains a $\sigma$-algebra $\Sigma$ which is the smallest one containing $\Sigma_0$; such $\Sigma$ is called the $\sigma$-algebra generated by $\Sigma_0$. If the measure space $\mathcal{X}$ is endowed with a topology, then, the $\sigma$-algebra generated by the open subsets is known as *Borel* $\sigma$*-algebra* and its elements as *Borel sets* [258].
3. A positive function $\mu : \Sigma \mapsto \mathbb{R}^+$, such that $\mu(\mathcal{X}) = 1$ is a probability measure on $\mathcal{X}$ relative to a $\sigma$-algebra $\Sigma$ if it is $\sigma$*-additive*, namely if

$$\Sigma \supset \{S_n\}_{n=1}^{\infty}\ ,\ S_i \cap S_j = \emptyset \Longrightarrow \mu\left(\bigcup_{n=1}^{\infty} S_n\right) = \sum_{n=1}^{\infty} \mu(S_n)\ .$$

Notice that $\mu$ is automatically monotone under inclusion, namely

$$A \subseteq B \Longrightarrow A = (A \setminus B) \cup B \Longrightarrow \mu(A) = \mu(A \setminus B) + \mu(B) \geq \mu(B)\ .$$

4. The following criterion is rather useful: an additive positive finite map $\mu : \Sigma \mapsto \mathbb{R}^+$ is $\sigma$-additive if and only if $\lim_n \mu(B_n) = 0$ for any collection $\{B_n\}_{n=1}^{\infty}$ of sets $B_n \in \Sigma$ such that $B_{n+1} \subseteq B_n$ and $\bigcap_n B_n = \emptyset$. Indeed, suppose $\mu$ is $\sigma$-additive and $\{B_n\}_{n=1}^{\infty}$ has decreasing properties and empty intersection; then, the sets $C_n := B_n \setminus B_{n+1}$ are disjoint and $B_n = \bigcup_{k \geq n} C_k$. It thus follows that $\mu(B_1) = \sum_{k=1}^{\infty} \mu(C_k)$, whence

$$\lim_{n\to\infty} \mu(B_n) = \lim_{n\to\infty} \sum_{k=n}^{\infty} \mu(C_k) = 0\ .$$

Vice versa, let $\mu$ be positive, finite and additive on $\Sigma$ and take any collection $\{C_n\}_{n=1}^{\infty}$ of disjoint subsets of $\Sigma$; because of additivity

$$\mu\Big(\bigcup_{k=1}^{\infty} C_k\Big) = \sum_{k=1}^{n} \mu(C_k) + \mu\Big(\bigcup_{k=n+1}^{\infty} C_k\Big)\ .$$

Since $B_n := \bigcup_{k=n+1}^{\infty} C_k \subseteq B_{n-1}$ and $\bigcap_n B_n = \emptyset$, $\sigma$-additivity follows. If $\mu$ is $\sigma$-additive over a measure algebra $\Sigma_0$ it can be extended in an unique way to the $\sigma$-algebra $\Sigma$ generated by $\Sigma_0$. In other words, given a $S \in \Sigma$, for any $\varepsilon > 0$, there exists $S' \in \Sigma_0$ such that $\mu(S \,\Delta\, S') < \varepsilon$, where

$$S \,\Delta\, S' = (S \setminus S') \cup (S' \setminus S) = (S \cup S') \setminus (S \cap S')\ . \tag{2.1}$$

5. A *regular Borel measure* on $\mathcal{X}$ is a measure on the Borel $\sigma$-algebra such that, for any measurable subset $B$ and $\varepsilon > 0$, there exists an open, $U_\varepsilon$, and a closed subset, $C_\varepsilon$, with $C_\varepsilon \subseteq B \subseteq U_\varepsilon$ such that $\mu(U_\varepsilon \setminus C_\varepsilon) < \varepsilon$ [258, 313].

Definition 2.1.1 provides an appropriate framework for irreversible dynamical systems in discrete time whereby the time-evolution of *phase-points* $x \in \mathcal{X}$ consists in successively applying the dynamical map $T$ to $x$ so that trajectories are given by countable sets $\{T^n x\}_{n\in\mathbb{N}}$. For reversible, discrete-time dynamics, also $T^{-1}$ is assumed measurable, that is $T(A) \in \Sigma$ if $A \in \Sigma$ with $\mu \circ T = \mu$; trajectories are then of the form $\{T^n x\}_{n\in\mathbb{Z}}$.

The measure $\mu$ defines a *probability distribution* over $\mathcal{X}$: if $f : \mathcal{X} \mapsto \mathbb{R}$ is a measurable function (an observable of the system), its *mean value* is

$$\mu(f) := \int_{\mathcal{X}} \mathrm{d}\mu(x) f(x)\ . \tag{2.2}$$

In particular, if $A \subseteq \mathcal{X}$ is a measurable subset and $\mathbf{1}_A(x)$ its *characteristic function* [1], the volume

$$\mu(A) := \int_{\mathcal{X}} \mathrm{d}\mu(x)\, \chi_A(x) , \tag{2.3}$$

has a natural interpretation as the probability that $x \in \mathcal{X}$ belong to $A$. We shall as well refer to these probability distributions as to the states of a classical dynamical system. In fact, in the case of a continuous phase-space, access to phase-points is practically never achievable; thus, one has to content oneself with the knowledge of how phase-points are distributed over $\mathcal{X}$. From a physical point of view, the fact that states $\mu$ are assumed to be $T$-invariant means that the statistical description of dynamical systems refers to *equilibrium states.* Interestingly, a measure-theoretical dynamical triplet can be represented in terms of a unitary operator on a Hilbert space [17, 61].

**Example 2.1.1 (Koopmann-von Neumann Formalism).** [175]

Let $(\mathcal{X}, T, \mu)$ be a measure-theoretic dynamical triplet. Finite additions and multiplications of characteristic functions of measurable subsets $A_i \subseteq \mathcal{X}$ give the algebra $\mathfrak{S}(\mathcal{X})$ of *simple functions* $s = \sum_i c_i \mathbf{1}_{A_i}$ over $\mathcal{X}$. Lebesgue-integration with respect to $\mu$ defines a scalar product $\langle s_1 \,|\, s_2 \rangle_\mu$ over $\mathfrak{S}(\mathcal{X})$,

$$\langle s_1 \,|\, s_2 \rangle_\mu := \sum_{i,j} (c_i^1)^* c_j^2 \int_{\mathcal{X}} \mathrm{d}\mu(x) \mathbf{1}_{A_i^1}(x) \mathbf{1}_{A_j^2}(x) = \sum_{i,j} (c_i^1)^* c_j^2\, \mu(A_i^2 \cap A_j^2) ,$$

for $\mathbf{1}_A(x)\mathbf{1}_B(x) = \mathbf{1}_{A\cap B}(x)$. Further, by linearly extending the map defined by $\mathbf{1}_A \mapsto U_T\, \mathbf{1}_A := \mathbf{1}_A \circ T = \mathbf{1}_{T^{-1}(A)}$, one gets a linear operator $U_T$ on $\mathfrak{S}(\mathcal{X})$. Since $\mu \circ T^{-1} = \mu$, $U_T$ preserves scalar products

$$\langle U_T s_1 \,|\, U_T s_2 \rangle_\mu := \sum_{i,j} (c_i^1)^* c_j^2\, \mu\Big(T^{-1}(A_i^1 \cap A_j^2)\Big) = \langle s_1 \,|\, s_2 \rangle_\mu .$$

Therefore, the *Koopman operator* $U_T$ can be extended to an isometric implementation of the dynamics (invertible and thus unitary in the reversible case) on the Hilbert space $\mathbb{L}^2_\mu(\mathcal{X})$ of square-summable functions on $\mathcal{X}$,

$$(U_T \psi)(x) = \psi(Tx) \qquad \forall \psi \in \mathbb{L}^2_\mu(\mathcal{X}) \; , \; \forall x \in \mathcal{X} . \tag{2.4}$$

The spectral properties of $U_T$ will turn out to be of particular relevance for ergodic theory (see Section 2.3). Using a bra-ket quantum like notation, we observe that:

1. the identity function $\mathbb{1}(x) = 1$ almost everywhere with respect to $\mu$, is always an eigenvector of $U_T$ with eigenvalue 1, $U_T |\, \mathbb{1} \rangle = |\, \mathbb{1} \circ T \rangle = |\, T \rangle$;

[1] $\mathbf{1}_A(x) = 1$ if $x \in A$, $\mathbf{1}_A(x) = 0$ otherwise

2. if 1 is a degenerate eigenvalue, then there exist constants of the motion $\mathbb{1} \neq \psi \in \mathbb{L}^2_\mu(\mathcal{X})$, $U_T|\psi\rangle = |\psi \circ T\rangle = |\psi\rangle$;
3. mean values are scalar products, $\mu(\psi) = \langle \mathbb{1} | \psi \rangle$, for all $\psi \in \mathbb{L}^2_\mu(\mathcal{X})$;
4. products of mean values amount to the matrix elements of the orthogonal projection $|\mathbb{1}\rangle\langle\mathbb{1}|$

$$\mu(\psi)\,\mu(\phi) = \langle \psi^* \,|\, \mathbb{1}\rangle\langle \mathbb{1} \,|\, \phi \rangle \ , \quad \forall \psi, \phi \in \mathbb{L}^2_\mu(\mathcal{X}) \ , \tag{2.5}$$

where $\psi^*$ is the complex conjugate of $\psi$.

## Hamiltonian Mechanics

Hamiltonian mechanics is an important source of classical dynamical systems [16, 17, 299]. Systems with $f$ degrees of freedom are described by a phase-space which is a $2f$-dimensional manifold $M_f \subseteq \mathbb{R}^f \times \mathbb{R}^f$ whose points $\boldsymbol{r} = (\boldsymbol{q}, \boldsymbol{p})$ consist of positions $\boldsymbol{q} = (q_1, \ldots, q_f) \in \mathbb{R}^f$ and momenta $\boldsymbol{p} = (p_1, \ldots, p_f) \in \mathbb{R}^f$. The phase-space inherits a *symplectic geometry* from the *symplectic matrix* $\mathbb{J} := [J_{ij}] = \begin{pmatrix} \mathbb{O}_f & \mathbb{1}_f \\ -\mathbb{1}_f & \mathbb{O}_f \end{pmatrix}$ where $\mathbb{O}_f$ and $\mathbb{1}_f$ are the $f \times f$ zero and identity matrices, respectively. Via the symplectic matrix one defines the *Poisson brackets* of two (differentiable) functions $F, G : M_f \mapsto \mathbb{R}$,

$$\{F\,,\,G\}(\boldsymbol{r}) := \sum_{i,j=1}^{2f} \frac{\partial F(\boldsymbol{r})}{\partial r_i} J_{ij} \frac{\partial G(\boldsymbol{r})}{\partial r_j} \ . \tag{2.6}$$

With respect to them, $\boldsymbol{q}$ and $\boldsymbol{p}$ are *canonical coordinates*: $\{q_i\,,\,p_j\} = \delta_{ij}$ and the time-evolution is generated by the *Hamilton equations*

$$\frac{\mathrm{d}\boldsymbol{q}}{\mathrm{d}t} = \partial_{\boldsymbol{p}} H(\boldsymbol{r}) \ , \quad \frac{\mathrm{d}\boldsymbol{p}}{\mathrm{d}t} = -\partial_{\boldsymbol{q}} H(\boldsymbol{r}) \ ,$$

where $H = H(\boldsymbol{r})$ is a (time-independent) Hamiltonian or energy function of the system. They are solved by the *Hamiltonian flux* $\boldsymbol{r} \mapsto \boldsymbol{r}(t) = \Phi^H_t(\boldsymbol{r})$, $t \in \mathbb{R}$ [2]. The time-evolution of functions $F$ on $M_f$ then amounts to a group of dynamical maps $F \mapsto F_t := F \circ \Phi^H_t$ that solves the time-evolution equation

$$\frac{\mathrm{d}F_t(\boldsymbol{r})}{\mathrm{d}t} = \{F_t\,,\,H\}(\boldsymbol{r}) \ . \tag{2.7}$$

Suppose $M_f = \mathbb{R}^{2f}$; then, a natural $\sigma$-algebra for the phase-space $M_f$ is the Borel $\sigma$-algebra (see Remark 2.1.1.2) containing all open subsets of

---

[2] One can always extract a discrete time-evolution $\{T^n\}_{n\in\mathbb{Z}}$ from it by fixing $t = 1$ and setting $T := \Phi^H_1$.

the topology of $M_f$ given by the Euclidean distance. The *Liouville measure* $\mathrm{d}\boldsymbol{r} = \prod_{i=1}^{f} \mathrm{d}q_i \mathrm{d}p_i$ is invariant under the Hamiltonian flux $\Phi_t^H$; however, $\int_{M_f} \mathrm{d}\boldsymbol{r}$ diverges and cannot be normalized to a probability distribution. A way out typically occurs when there are constants of the motion, that is functions $F$ on $M_f$, like the Hamiltonian itself, such that $\{F\,,\,H\} = 0$. By fixing their values, the dynamics is restricted to time-invariant submanifolds that usually have finite volumes. Instances of equilibrium states leading to descriptions of Hamiltonian systems as measure-theoretical triplets $(M_f, \Phi_1^H, \mu_H)$ (discrete time), or $(M_f, \{\Phi_t^H\}_{t\in\mathbb{R}}, \mu_H)$ (continuous time), are in general provided by probability distributions $\mathrm{d}\mu_H(\boldsymbol{r}) = f(\boldsymbol{r})\mathrm{d}\boldsymbol{r}$, where $f : \mathcal{M}_f \mapsto \mathbb{R}^+$ is a normalized, positive functions such that $\{f\,,\,H\} = 0$. Prominent instances of such probability measures are the micro-canonical, canonical and grand-canonical states of classical statistical mechanics [300].

The time-invariance of states as the previous ones deserves to be examined in some more detail as it follows from a *duality* argument which we shall frequently encounter in the following. Duality is essentially the observation that the mean value of a function $F$ at time $t$, $F_t$, with respect to a state $\mu$ equals the mean value of $F$ with respect to the state $\mu_t$ at time $t$, $\mu(F_t) = \mu_t(F)$. This defines the time-evolution of states as the dual of the time-evolution of observables (functions); indeed, from time-invariance of the Liouville measure it follows that

$$\mu(F_t) = \int_{M_f} \mathrm{d}\boldsymbol{r}\; \mu(\boldsymbol{r})\, F(\Phi_t^H(\boldsymbol{r})) = \int_{M_f} \mathrm{d}\boldsymbol{r}\; \mu(\Phi_{-t}^H \boldsymbol{r})\, F(\boldsymbol{r}) =: \mu_t(F)\;, \tag{2.8}$$

whence $\mu_t := \mu \circ \Phi_{-t}^H$ solves the time-evolution equation

$$\frac{\partial \mu_t(\boldsymbol{r})}{\partial t} = \{H\,,\,\mu_t\}(\boldsymbol{r})\;. \tag{2.9}$$

**Example 2.1.2 (Regular Motion).** Consider two uncoupled harmonic one-dimensional oscillators described by $\boldsymbol{r} = (q_1, q_2, p_1, p_2) \in M_2 = \mathbb{R}^4$ and by the Hamiltonian

$$H(\boldsymbol{r}) = \underbrace{\frac{p_1^2}{2m_1} + \frac{m_1\omega_1^2}{2} q_1^2}_{H_1(\boldsymbol{r})} + \underbrace{\frac{p_2^2}{2m_2} + \frac{m_2\omega_2^2}{2} q_2^2}_{H_2(\boldsymbol{r})}\;.$$

By fixing the single oscillator energies $H_i(\boldsymbol{r}) = E_i$, $i = 1, 2$, the motion develops on the 2-torus $\mathbb{T}^2 := \{\boldsymbol{\theta} = (\theta_1, \theta_2) : \theta_i \in [0, 2\pi)\}$, where it amounts to a two-dimensional rotation. Indeed, setting

$$J_i(\boldsymbol{r}) := H_i(\boldsymbol{r})/\omega_i = \frac{p_i^2}{2m_i\omega_i} + \frac{m_i\omega_i}{2} q_i^2\;, \quad \tan\theta_i := \frac{p_i}{m_i\omega_i q} \qquad \text{whence}$$

$$q_i = \sqrt{\frac{2J_i}{m_i\omega_i}}\,\cos\theta_i\;, \quad p_i = \sqrt{2m_i\omega_i J_i}\,\sin\theta_i\;,$$

one gets *angle-action variables* $(\boldsymbol{\theta}, \boldsymbol{I})$, $\boldsymbol{\theta} = (\theta_1, \theta_2)$, $\boldsymbol{I} := (I_1, I_2)$.

These are canonical coordinates, with Poisson brackets $\{\theta_i\,,\,J_k\} = \delta_{ik}$; moreover, $H(\boldsymbol{r}) = K(\boldsymbol{J}) = \omega_1 J_1 + \omega_2 J_2$. Thus, the corresponding Hamilton equations,

$$\frac{\mathrm{d}\boldsymbol{\theta}}{\mathrm{d}t} = \boldsymbol{\omega}\ , \qquad \frac{\mathrm{d}\boldsymbol{J}}{\mathrm{d}t} = 0\ , \qquad \boldsymbol{\omega} = (\omega_1, \omega_2)\ ,$$

are solved by the Hamiltonian flux

$$T_t : \boldsymbol{\theta} \mapsto \boldsymbol{\theta}(t) := T_t(\boldsymbol{\theta}) = \boldsymbol{\theta} + \boldsymbol{\omega}\, t\ . \tag{2.10}$$

By varying $E_1, E_2$ and thus $\boldsymbol{I}$, the phase-space $\mathbb{R}^4$ is covered by non-intersecting 2-dimensional tori. On each fixed torus, $\mathrm{d}\boldsymbol{\theta}/(2\pi)^2$ gives a probability measure which is invariant under the Hamiltonian flux. The triplet $(\mathbb{T}^2, T := T_1, \mathrm{d}\boldsymbol{\theta}/(2\pi)^2)$ fulfils the requirements in Definition 2.1.1.

In the Koopman-von Neumann formalism, the unitary operator $U_T$ implementing $T$ on $\mathbb{H} := \mathbb{L}^2(\mathbb{T}^2, \mathrm{d}\boldsymbol{\theta}/(2\pi)^2)$ has the exponential functions $e_{\boldsymbol{n}}(\boldsymbol{\theta}) = \exp(i\boldsymbol{n}\cdot\boldsymbol{\theta})$, $\boldsymbol{n} \in \mathbb{Z}^2$, as eigenfunctions ,

$$(U_T e_{\boldsymbol{n}})(\boldsymbol{\theta}) = e_{\boldsymbol{n}}(\boldsymbol{\theta} + \boldsymbol{\omega}) = \mathrm{e}^{i\sum_{j=1}^2 n_j(\theta_j + \omega_j)} = \mathrm{e}^{i\sum_{j=1}^2 n_j\omega_j}\, e_{\boldsymbol{n}}(\boldsymbol{\theta})\ . \tag{2.11}$$

Therefore, the time-evolution of $\mathbb{H} \ni |\,\psi\,\rangle = \sum_{\boldsymbol{n}\in\mathbb{Z}^2} \widehat{\psi}(\boldsymbol{n})|\,e_{\boldsymbol{n}}\,\rangle$ is given by

$$|\,\psi\,\rangle \mapsto U_T^k|\,\psi\,\rangle = \sum_{\boldsymbol{n}\in\mathbb{Z}^2} \widehat{\psi}(\boldsymbol{n})\,\mathrm{e}^{ik\sum_{j=1}^2 n_j\omega_j}\,|\,e_{\boldsymbol{n}}\,\rangle\ , \quad k \in \mathbb{Z}\ . \tag{2.12}$$

**Remarks 2.1.2.**

1. If $\dfrac{\omega_2}{\omega_1} = \dfrac{p}{q}$, $p, q \in \mathbb{N}$, trajectories close since $\boldsymbol{\theta}(2q\pi/\omega_1) = \boldsymbol{\theta} \bmod 2\pi$.
2. If there are no $0 \neq n_{1,2} \in \mathbb{Z}$ such that $n_1\omega_1 + n_2\omega_2 = 0$, then, every trajectory $\{\boldsymbol{\theta}(t)\}_{t\in\mathbb{R}}$ fills the 2-torus $\mathbb{T}^2$ densely. Namely, for any $\varepsilon \geq 0$, $\boldsymbol{\phi}, \boldsymbol{\theta} \in \mathbb{T}^2$, there is $t \in \mathbb{R}$ such that $\|\boldsymbol{\theta}(t) - \boldsymbol{\phi})\| \leq \varepsilon$, where the norm is the Euclidean norm computed $\bmod 2\pi$. Indeed, using (2.10),

$$t_* := (\phi_1 - \theta_1)/\omega_1 \Longrightarrow \theta_1(t_* + 2n\pi/\omega_1) = \phi_1 \bmod 2\pi\ ,$$

for all $n \in \mathbb{Z}$. Since $\mathbb{T}$ is compact, the sequence $\{\theta_2(t_* + 2n\pi/\omega_1)\}_{n\in\mathbb{Z}}$ has accumulation points; thus, for any $\epsilon \geq 0$ there exist $n, p \in \mathbb{N}$ such that

$$\Big|\theta_2(t_* + 2(n+p)\pi/\omega_1) - \theta_2(t_* + 2n\pi/\omega_1)\Big| = 2p\pi\frac{\omega_2}{\omega_1} \bmod 2\pi \leq \epsilon\ ,$$

whence the sequence $\{\theta_2(t_* + 2np\pi/\omega_1)\}_{n\in\mathbb{N}}$ subdivides the circle into disjoint intervals $\Delta_n$ of length

$$\Big|\theta_2(t_* + 2(n+1)p\pi/\omega_1) - \theta_2(t_* + 2np\pi/\omega_1)\Big| \leq \epsilon\ .$$

Therefore,

$$\phi_2 \in \Delta_m \Longrightarrow \|\boldsymbol{\theta}(t_* + 2mp\pi/\omega_1) - \boldsymbol{\phi}\| = \Big|\theta_2(t_* + 2pm\pi/\omega_1) - \phi_2\Big| \leq \epsilon\ .$$

3. A similar argument as before shows that, in discrete time, trajectories $\{\boldsymbol{\theta}(n)\}_{n\in\mathbb{Z}}$ fill $\mathbb{T}^2$ densely if and only if there are no integers $n_{1,2} \neq 0$ such that $n_1\omega_1 + n_2\omega_2 = 2\pi p$ with $\mathbb{Z} \ni p \neq 0$ [91].
4. Example 2.1.2 is a particular instance of the Liouville-Arnold theorem [16, 17, 299] on integrable Hamiltonian systems. Suppose a canonical system with $f$ degrees of freedom possesses $f$ global constants of the motion $K_i$, $K_1 := H$ in involution, that is $\{K_i\,,\,K_j\} = 0$, $i,j = 1,2,\ldots,f$. If the subset $N_{\boldsymbol{k}} := \{K_i(\boldsymbol{r}) = k_i : i = 1,2,\ldots,f\} \subseteq M_f$ is compact and connected and the differential 1-forms $\mathrm{d}K_i$ are linearly independent on it, then $N_{\boldsymbol{k}}$ is isomorphic to the $f$-torus $\mathbb{T}^f$. Moreover, there exists a canonical transformation from $\boldsymbol{r} \in N_{\boldsymbol{k}}$ to angle-action variables $(\boldsymbol{\theta}\,,\,\boldsymbol{J})$ such that the Hamiltonian flux $\Phi_t^H$ is isomorphic to an $f$-dimensional rotation on $\mathbb{T}^f$ with $\boldsymbol{J}$-dependent frequencies: $\boldsymbol{\theta}(t) = \boldsymbol{\theta} + \boldsymbol{\omega}(\boldsymbol{J})t$. Accordingly, the phase-space $M_f$ foliates into disjoint $f$-tori which are filled densely by the trajectories $\{\boldsymbol{\theta}(t)\}_{t\in\mathbb{R}}$ when $\sum_{i=1}^f n_i\omega_i(\boldsymbol{J}) = 0$, $n_i \in \mathbb{Z}$, only if all $n_i = 0$. Tori such that $\sum_{i=1}^f n_i\omega_i(\boldsymbol{J}) = 0$ for $0 \neq n_i \in \mathbb{Z}$ are called *resonant* and on them trajectories close. The independence of the oscillation frequencies $\boldsymbol{\omega}$ from the actions $\boldsymbol{J}$ in Example 2.1.2 is an exception due to the linearity of the Hamilton equations.

Integrable Hamiltonian systems cannot behave too irregularly as their motion amounts to a multi-dimensional rotation over invariant tori. In order to increase the degree of irregularity, some constants of the motion must disappear in order to let the trajectories wander around according to less predictable patterns. In the following example, a constant of the motion is eliminated by means of a *folding* condition.

**Example 2.1.3 (Hyperbolic Behavior).** [17, 271]
Let $\delta_p(t)$ denote the periodic delta function $\sum_{n\in\mathbb{Z}}\delta(n-t)$ with unit period and consider a free one-dimensional motion with periodic quadratic kicks, occurring with strength $\beta \in \mathbb{R}$, according to the pulsed Hamiltonian

$$H = \frac{1}{2}\left(p^2 + \delta_p(t)\beta q^2\right) .$$

A natural dynamical map $T$ consists in updating the vector $\boldsymbol{r} = (q,p)$ on phase space from immediately after the $n$-th kick to immediately after the $n-i$-th one; namely $T : \boldsymbol{r}_n \to \boldsymbol{r}_{n+1}$, where $\boldsymbol{r}_n := (q_n, p_n)$ and

$$q_n := \lim_{\varepsilon\to 0^+} q(n+\varepsilon) \;, \quad p_n := \lim_{\varepsilon\to 0^+} q(n+\varepsilon) \;.$$

Integrating the Hamilton equations

$$\frac{\mathrm{d}q}{\mathrm{d}t} = p \;, \quad \frac{\mathrm{d}p}{\mathrm{d}t} = -\delta_p(t)\,\beta\,q \;,$$

first between $Tn+\varepsilon$ and $T(n+1)-\varepsilon$ and then between $T(n+1)-\varepsilon$ and $T(n+1)+\varepsilon$, yields

$$\begin{aligned} q(n+1+\varepsilon)-q(n+\varepsilon) &= p(n+\varepsilon)+\int_{n+1-\varepsilon}^{n+1+\varepsilon} \mathrm{d}s\, p(s) \\ p(n+1+\varepsilon)-p(n+\varepsilon) &= -\beta\, q(n+1)\ . \end{aligned}$$

By letting $\varepsilon \to 0^+$, the integral is of order $\varepsilon$ and vanishes; thus, the dynamical map $T$ reduces to a $2\times 2$ matrix acting on $\mathbb{R}^2$:

$$\boldsymbol{r} = \begin{pmatrix} q \\ p \end{pmatrix} \mapsto \mathbb{A}\boldsymbol{r}\ ,\quad \mathbb{A} = \begin{pmatrix} 1 & 1 \\ -\beta & 1-\beta \end{pmatrix}\ . \tag{2.13}$$

Since $\det(\mathbb{A}) = 1$, the Liouville measure $\mathrm{d}\boldsymbol{r} = \mathrm{d}q\,\mathrm{d}p$ is $T$-invariant. The eigenvalues of $\mathbb{A}$,

$$\alpha^{\pm 1} = \frac{2-\beta \pm \sqrt{\beta(\beta-4)}}{2} = \frac{2-\beta \pm \sqrt{(\beta-2)^2-4}}{2}\ ,$$

are real with $|\alpha| > 1$ when $\beta < 0$ or $\beta > 4$. The corresponding eigenvector $|\,a_+\,\rangle$ identifies a direction in $\mathbb{R}^2$ along which lengths increase exponentially for $n \geq 0$, while they contract exponentially along the direction of the eigenvector $|\,a_-\,\rangle$ relative to the other eigenvalue $|\alpha|^{-1} < 1$. This motion is called *hyperbolic*.

For $\beta = -1$, $\mathbb{A} = \begin{pmatrix} 1 & 1 \\ 1 & 2 \end{pmatrix}$ is symmetric thus $\langle\, a_- \,|\, a_+\,\rangle = 0$ and, writing $|\,\boldsymbol{r}\,\rangle = \gamma|\,a_+\,\rangle + \delta|\,a_-\,\rangle$,

$$\|\boldsymbol{r}_n\|^2 = |\gamma|^2 \mathrm{e}^{2n\log\alpha} + |\delta|^2 \mathrm{e}^{-2n\log\alpha}\ , \tag{2.14}$$

where $\boldsymbol{r}_n := \mathbb{A}^n\boldsymbol{r}$. Therefore, the norms of all vectors $\boldsymbol{r} \neq 0$ increase exponentially while remaining on the hyperbolae selected by fixing a value of $F(\boldsymbol{r}) := q^2 - p^2 + qp$. Indeed, one can directly check that $F(\boldsymbol{r}_{n+1}) = F(\boldsymbol{r}_n)$, whence this function is a constant of the motion [118]. This is no longer true if one imposes a *folding condition* that forces the dynamics to develop on the two-dimensional torus $\mathbb{T}^2 := \{\mathbb{R}^2 \ni \boldsymbol{r} = (q,p) \bmod (1)\}$, namely if one defines the dynamical map

$$T_{\mathbb{A}} : \mathbb{T}^2 \ni \boldsymbol{r} \mapsto \boldsymbol{r}_n := (\mathbb{A}^n \boldsymbol{r} \bmod 1) \in \mathbb{T}^2\ . \tag{2.15}$$

Then, the resulting triplet $(\mathbb{T}^2, T_{\mathbb{A}}, \mathrm{d}\boldsymbol{r})$ is as in Definition 2.1.1 and the map $T$ is known as *Arnold Cat Map* [17].
More in general, one may consider the dynamics on the 2-dimensional torus $\mathbb{T}^2$ generated as in (2.15) by a matrix

$$\mathbb{A} = \begin{pmatrix} a & b \\ c & d \end{pmatrix}\ ,\quad a,b,c,d \in \mathbb{Z} \,:\, ad-bc = 1\ ,\ |a+d| > 2\ , \tag{2.16}$$

with eigenvalues $\alpha^{\pm 1} \in \mathbb{R}$. Since $\mathbb{A}$ need not be Hermitian, its normalized eigenvectors $|\,a_\pm\,\rangle = \begin{pmatrix} a_{1\pm} \\ a_{2\pm} \end{pmatrix}$ are in general only linearly independent; one explicitly computes

$$a_{1\pm} = b\Delta_\pm \ , \ a_{2\pm} = (\alpha^{\pm 1} - a)\Delta_\pm \quad \text{where} \quad \Delta_\pm := \frac{1}{\sqrt{b^2 + (\alpha^{\pm 1} - a)^2}} \ , \tag{2.17}$$

and expands $\mathbb{R}^2 \ni |\,\boldsymbol{r}\,\rangle = \begin{pmatrix} x \\ y \end{pmatrix} = C_+(\boldsymbol{r})|\,a_+\,\rangle + C_-(\boldsymbol{r})|\,a_-\,\rangle$ with

$$C_+(\boldsymbol{r}) := \frac{x a_{2-} - y a_{1-}}{\Delta} \ , \quad C_-(\boldsymbol{r}) := \frac{y a_{1+} - x a_{2+}}{\Delta} \tag{2.18}$$

where

$$\Delta := \mathrm{Det}\begin{pmatrix} a_{1+} & a_{1-} \\ a_{2+} & a_{2-} \end{pmatrix} = b(1 - \alpha^2)\,\Delta_+\Delta_- \ . \tag{2.19}$$

Then, the hyperbolic behavior shows up since

$$\mathbb{A}^k|\,\boldsymbol{r}\,\rangle = \alpha^k\,C_+(\boldsymbol{r})\,|\,a_+\,\rangle \,+\, \alpha^{-k}\,C_-(\boldsymbol{r})\,|\,a_-\,\rangle \tag{2.20}$$

and the absolute value of one of the eigenvalues $\alpha_\pm = \dfrac{a + d \pm \sqrt{(a+d)^2 - 4}}{2}$ is larger than 1.

Consider now the Koopman operator $U_\mathbb{A}$ on $\mathbb{H} := \mathbb{L}^2_{\mathrm{d}\boldsymbol{r}}(\mathbb{T}^2)$; the orthogonal exponential functions

$$e_{\boldsymbol{n}}(\boldsymbol{r}) := \exp(2\pi i\,\boldsymbol{n}\cdot\boldsymbol{r}) \tag{2.21}$$

are such that ($\mathbb{A}^T$ denotes the transposed of $\mathbb{A}$)

$$(U_\mathbb{A}\,e_{\boldsymbol{n}})(\boldsymbol{r}) = \mathrm{e}^{2\pi i\,\boldsymbol{n}\cdot(\mathbb{A}\boldsymbol{r})} = \mathrm{e}^{2\pi i\,(\mathbb{A}^T\boldsymbol{n})\cdot\boldsymbol{r}} = e_{\mathbb{A}^T\boldsymbol{n}}(\boldsymbol{r}) \ , \tag{2.22}$$

whence, setting $\psi(\boldsymbol{n}) := \langle\,e_{\boldsymbol{n}}\,|\,\psi\,\rangle$ for all $\psi \in \mathbb{H}$, it turns out that

$$(U_\mathbb{A}\psi)(\boldsymbol{n}) = \langle\,e_{\boldsymbol{n}}\,|\,U_\mathbb{A}\,|\psi\,\rangle = \langle\,e_{\mathbb{A}^{-T}\boldsymbol{n}}\,|\,\psi\,\rangle = \psi(\mathbb{A}^{-T}\boldsymbol{n}) \ .$$

Therefore, $U_\mathbb{A}$ has no other eigenvector but $e_{\boldsymbol{0}} = 1\!\!1$: if $U_\mathbb{A}|\,\psi\,\rangle = \mu|\,\psi\,\rangle$ for $\psi \in \mathbb{H}$ with $|\mu| = 1$; then, with $|\,\psi\,\rangle = \sum_{\boldsymbol{n}\in\mathbb{Z}^2}\widehat{\psi}(\boldsymbol{n})|\,e_{\boldsymbol{n}}\,\rangle$,

$$\langle\,e_{\boldsymbol{m}}\,|\,U_A^p\psi\,\rangle = \sum_{\boldsymbol{n}\in\mathbb{Z}^2}\widehat{\psi}(\boldsymbol{n})\,\langle\,e_{\boldsymbol{m}}\,|\,e_{\mathbb{A}^p\boldsymbol{n}}\,\rangle = \widehat{\psi}(\mathbb{A}^{-p}\boldsymbol{m}) = \mu^p\widehat{\psi}(\boldsymbol{m}) \ ,$$

for any fixed $\boldsymbol{m} \in \mathbb{Z}^2$. Since $\widehat{\psi}(\boldsymbol{n}) \to 0$ with $\|\boldsymbol{n}\| \to \infty$, if $\widehat{\psi}(\boldsymbol{m}) \neq 0$, then $\lim_p \widehat{\psi}(A^{-p}\boldsymbol{m}) = 0$ because of hyperbolicity, while $\mu^p\widehat{\psi}(\boldsymbol{m})$ oscillates.

The exponential amplification of small errors that results from (2.14) (or form (2.20)) cannot hold for arbitrarily large $n$: In fact, $\|\boldsymbol{r}_n\| \leq \sqrt{2}$

so that the expansion is eventually counteracted by the folding condition in (2.15). Suppose $|\, \boldsymbol{r}\,\rangle = \varepsilon|\, a_+\,\rangle$, then $\|\boldsymbol{r}_n\| = \varepsilon \mathrm{e}^{n \log \alpha} \leq \sqrt{2}$ increases until $n \leq \log(\varepsilon^{-1}\sqrt{2})/(\log \alpha)$.

This argument applies to any pair of initial conditions $\boldsymbol{r}^{1,2}$; their distance $\|\boldsymbol{r}^1 - \boldsymbol{r}^2\|$ increases exponentially due to the expanding contribution from the component of $\boldsymbol{r}^1 - \boldsymbol{r}^2$ along $|\, a_+\,\rangle$ until the folding condition affects one of the two cartesian components of $\boldsymbol{r}^1 - \boldsymbol{r}^2$. Notice however that the smaller is $\|\boldsymbol{r}^1 - \boldsymbol{r}^2\|$, the longer the amplification lasts. This observation allows the introduction of the notion of asymptotic divergence rate of initially close trajectories even when they develop on compact phase-spaces: these rates are known as *Lyapounov exponents* and are a measure of dynamical instability.

**Definition 2.1.2 (*Maximal Lyapounov Exponent*).** *[199, 106] The maximal positive Lyapounov exponent of a dynamical triplet $(\mathcal{X}, T, \mu)$ equipped with a distance $d(x, y)$ is defined by*

$$\lambda_M(x) = \lim_{n\to\infty} \frac{1}{n} \lim_{d(x,y)\to 0} \log \frac{d(T^n x\,,\, T^n y)}{d(x, y)} \;.$$

Of course $\mathcal{X}$ may be a multi-dimensional space and thus there might be more directions along which distances expand exponentially fast with exponents $\lambda(x) > 0$; the intuitive picture behind the definition is that, for sufficiently small $d(x, y)$, the distance at time $n$ is such that

$$d(T^n x, T^n y) \simeq \mathrm{e}^{n\lambda_M(x)}\, d(x, y)\left(1 + O\Big(\mathrm{e}^{-n(\lambda_M(x) - \lambda(x))}\Big)\right) ,$$

where $\lambda(x) < \lambda_M(x)$ [62].

**Remarks 2.1.3.**

1. A rigorous approach to Lyapounov exponents can be found in [199]; here, we sketch a few basic facts (see [106, 313]). Assume the phase-space $\mathcal{X}$ to be a compact manifold with a $C_\infty$ differentiable structure, a Borel $\sigma$-algebra and a Riemannian metric such that the tangent spaces $\tau_x(\mathcal{X})$ at $x \in \mathcal{X}$ are isomorphic to $\mathbb{R}^k$ equipped with an Euclidean structure. The dynamics $T : \mathcal{X} \mapsto \mathcal{X}$ is assumed to be continuous with continuous first derivatives, so that one can focus upon its linearization $\tau_x(T)$ that maps the tangent space $\tau_x(\mathcal{X})$ into the tangent space $\tau_{Tx}(\mathcal{X})$. In particular, one is interested in the asymptotic behavior of $\|\tau_x(T^n)\|$ where, by the chain rule,
$$\tau_x(T^n) = \tau_{T^{n-1}x}(T) \circ \tau_{T^{n-2}x} \circ \cdots \tau_x(T) \;.$$
Let $\mathcal{X}$ be equipped with a $T$-invariant regular Borel measure $\mu$ (see Remark 2.1.1.5); then, there exists a measurable subset $B \subseteq \mathcal{X}$ with $\mu(B) = 1$ and a positive measurable function $s : B \mapsto \mathbb{R}_+$ such that, given

$x \in B$, there are real numbers $\{\lambda^{(j)}(x)\}_{j=1}^{s(x)}$, $\lambda^{(j)}(x) < \lambda^{(j+1)}(x)$, and linear subspaces of $\mathbb{R}^k$, $\{V^{(j)}(x)\}_{j=0}^{s(x)}$, $V^{(0)} = \{0\}$, $V^{(j)}(x) \subset V^{(j+1)}(x)$, $V^{(s(x))} = \mathbb{R}^k$, for which

*a)* $\lim_{n\to\infty} \frac{1}{n} \log \|\tau_x(T^n)\boldsymbol{r}\| = \lambda^{(j)}(x)$ for all $\boldsymbol{r} \in W_j(x) := V^{(j)}(x) \ominus V^{(j-1)}(x)$;

*b)* $\lambda^{(j)}(x)$ is defined, measurable and $T$-invariant on the subset of $x \in B$ such that $s(x) \geq j$, that is $\lambda^{(j)}(Tx) = \lambda^{(j)}(x)$;

*c)* $\tau_x(T)V^{(j)}(x) \subset V^{(j)}(Tx)$ for all $j \leq s(x)$.

It thus follows that, if $\lambda^{(j)}(x) < 0$, the norms of all $\boldsymbol{r} \in V^{(j)}(x)$ go to 0 exponentially fast with $n \to +\infty$. On the other hand, if $\lambda^{(j)}(x) > 0$, the norms of all vectors $\boldsymbol{r} \in V^{(j)}(x) \ominus V^{(j-1)}(x)$ diverge exponentially fast.

2. There can be more than one positive Lyapounov exponent thus more than one amplifying direction in space. In volume-preserving dynamical systems to any amplifying direction there corresponds a shrinking direction (amplifying in the past).
3. On compact manifolds, the two limits in Definition 2.1.2 do not commute: the numerator is limited by compactness, whence the $1/n$ limit vanishes if performed before letting $d(x, y) \to 0$.
4. If there is an intrinsic smallest distance $\delta > 0$ between points $x, y \in \mathcal{X}$ and the largest possible distance $\Delta$ is finite, then the Lyapounov exponent is zero. This means that exponential separation or amplification cannot be extended beyond the *logarithmic time-scale* set by $\delta\, \mathrm{e}^{\lambda t} \leq \Delta$. This gives a so-called *breaking-time* [118] $t_B := \frac{1}{\lambda} \log \frac{\Delta}{\delta}$.
5. When the motion develops on a compact phase-space, the existence of positive Lyapounov exponents is known as *extreme sensitivity to initial conditions* and provides a widely accepted definition of *classical chaotic motion* [271, 228]. Notice that without the folding condition, also an inverted harmonic oscillator with Hamiltonian $H(\boldsymbol{r}) = p^2/(2m) - m\omega q^2/2$ would show an exponentially fast separation of initial conditions, though far less irregular and interesting than one on a compact manifold.

### 2.1.1 Shift Dynamical Systems

Phase-spaces with a finite or a countable number of states are typical either of systems which arise from suitable discretizations of otherwise continuous phase-spaces or of intrinsically discrete systems as cellular automata [62]. The first possibility arises in particular when the observations aimed at identifying the system state as a point of phase-space have a finite accuracy; then, one performs a *coarse-graining* of phase space into a certain number of regions whose volume is determined by the given accuracy and whose interior points are accessible only through observations of higher accuracy. As we shall see in later sections, in such a case, the system states are identifiable with the labels of the regions where the system state is localized and the dynamics corresponds to jumping from label to label rather than from point to point of the phase-space.

Instead, the phase-space of cellular automata [62, 18] is discrete from the start as they consist of copies of a same system (automaton) described by a $d$-valued function $i$, for instance, in the binary case $i = 0$ may be used to signal when an automaton is deactivated, $i = 1$ when it is activated.

The phase-space $\mathcal{X}$ of a cellular automaton with $N$ systems comprises $d^N$ configurations (states) corresponding to finite strings $\boldsymbol{i}^{(N)} = (i_1, i_2, \ldots, i_N) \in \Omega_d^{(N)} := \{1, 2, \ldots, d\}^N$. The dynamics is given in discrete time by a map $T : \Omega_d^{(N)} \mapsto \Omega_d^{(N)}$ that updates the configurations from time $n$ to time $n+1$: $\boldsymbol{i}^{(N)}(n) \mapsto \boldsymbol{i}^{(N)}(n+1)$. The state $i_k(n+1)$ of the $k$-th automaton at time $n+1$ in general depends on the states of some or all other automata at time $n$. In the following, we shall focus upon a most simple class of cellular automata, that is *shift dynamical systems* [17, 61, 164, 313].

Let the space $\mathcal{X}$ be the collection $\Omega_d := \{0, 1, \ldots, d\}^{\mathbb{N}}$ of all sequences $\boldsymbol{i} = \{i_j\}_{j\in\mathbb{N}}$ of symbols from a finite alphabet $i_j = 1, 2, \ldots, d$. Each $\boldsymbol{i}$ can be interpreted as a configuration of a countable network of cellular automata, each of them being indexed by an integer $j \in \mathbb{N}$, with $\boldsymbol{i}_j$ denoting its actual state among the $d$ possible ones. Let $T_\sigma : \Omega_d \mapsto \Omega_d$ be the *left shift* along sequences,

$$(T_\sigma \boldsymbol{i})_j = \boldsymbol{i}_{j+1} \ , \tag{2.23}$$

and set $\boldsymbol{i}(n) := T_\sigma^n \boldsymbol{i}$: $T_\sigma$ amounts to a rather trivial dynamics, namely to a deterministic updating whereby the state $\boldsymbol{i}_j(n+1)$ of the $j$-th automaton at time $n+1$ depends only on (is equal to) that of its right nearest neighbor at time $n$:

$$\boldsymbol{i}_j(n+1) := (T_\sigma^{n+1} \boldsymbol{i})_j = (T_\sigma \boldsymbol{i})_j(n) = \boldsymbol{i}_{j+1}(n) \ .$$

From the point of view of a fixed automaton, say the 0-th one, this kind of dynamics is typically like tossing a coin. Indeed, suppose the initial configuration $\boldsymbol{i}(0)$ of the network is to be chosen randomly, according to a probability distribution where all automaton states occur with the same probability $2^{-N}$. Because of the dynamics, this property is then inherited by the sequence $\{i_0(n)\}_{n\in\mathbb{N}}$ of successive states of the 0-th automaton.

In order to provide the shift along binary sequences with a measure-theoretic formulation as in Definition 2.1.1, the set $\Omega_d$ of infinite sequences has to be equipped with a $\sigma$-algebra of measurable sets. The standard way to do this is by means of the so-called *cylinders* [61, 164, 91, 17]; they consist of all sequences whose entries have fixed values within chosen intervals:

$$\underbrace{C^{[j,k]}_{i_j i_{j+1} \cdots i_k}}_{\boldsymbol{i}^{(k-j+1)}} := \Big\{ \boldsymbol{i} \in \Omega_2 \, : \, \boldsymbol{i}_{j+\ell} = i_{j+\ell} \, , \, \ell = 0, 1, \ldots, k-j \Big\} \ . \tag{2.24}$$

They are labeled by the interval $[j, k]$ and by the binary string $\boldsymbol{i}^{(k-j+1)}$ of length $k - j + 1$ of assigned digits within that interval; each one of them can be obtained as a finite intersection of *simple cylinders* $C^{\{\ell\}}_{i_\ell}$,

$$C^{[j,k]}_{\boldsymbol{i}^{(k-j+1)}} = \bigcap_{\ell=0}^{k-j} C^{\{j+\ell\}}_{i_{j+\ell}} \ , \quad C^{\{\ell\}}_{i_\ell} := \left\{ \boldsymbol{i} \in \Omega_2 \, : \, \boldsymbol{i}_\ell = i_\ell \right\} . \tag{2.25}$$

We shall denote by $\mathcal{C}_{[j,k]}$ the sets consisting of the $2^{(k-j+1)}$ cylinders $C^{[j,k]}_{\boldsymbol{i}^{(k-j+1)}}$. The $\sigma$-algebra $\Sigma$ is obtained from all possible unions and intersections of simple cylinders. Further, pre-images of cylinders under $T^{-1}$ remain cylinders: in fact

$$\begin{aligned} T_\sigma^{-1}\Big(C^{\{\ell\}}_{i_\ell}\Big) &:= \left\{ \boldsymbol{i} \in \Omega_2 \, : \, T_\sigma \boldsymbol{i} \in C^{\{\ell\}}_{i_\ell} \right\} = \left\{ \boldsymbol{i} \in \Omega_2 \, : \, \boldsymbol{i}_\ell(1) = \boldsymbol{i}_{\ell+1} = i_\ell \right\} \\ &= C^{\{\ell+1\}}_{i_\ell} \ , \end{aligned} \tag{2.26}$$

whence $T_\sigma$ is measurable with respect to $\Sigma$.

**Remark 2.1.4.** The left shift on unilateral sequences is not invertible; it becomes so by choosing instead of $\Omega_d$ the set $\Omega_d^{\mathbb{Z}}$ of all doubly infinite sequences $\boldsymbol{i} = \{i_j\}_{j\in\mathbb{Z}}$. Then, the same result as in (2.26) holds for the pre-images of cylinders under $T_\sigma$, $T_\sigma(C^{\{\ell\}}_{i_\ell}) = C^{\{\ell-1\}}_{i_\ell}$, whence $T_\sigma^{-1}$ is also measurable.

We shall refer to any probability measure $\mu$ on $\Sigma$ as to a global state on $\Omega_d$; to any such $\mu$ there correspond *local states* $\mu_{[j,k]}$ on the cylinder sets $\mathcal{C}_{[j,k]}$. As cylinders in $\mathcal{C}_{[j,k]}$ are in one-to-one correspondence with strings $\boldsymbol{i}^{(k-j+1)} \in \Omega_d^{(k-j+1)}$ of length $k-j+1$, these local states are probability distributions on $\Omega_d^{(k-j+1)}$:

$$\begin{aligned} &\mu_{[j,k]} = \left\{ p_{[j,k]}(\boldsymbol{i}^{(k-j+1)}) \right\}_{\boldsymbol{i}^{(k-j+1)} \in \Omega_d^{(k-j+1)}} \\ &p_{[j,k]}(\boldsymbol{i}^{(k-j+1)}) \geq 0 \ , \qquad \sum_{\boldsymbol{i}^{(k-j+1)} \in \Omega_d^{(k-j+1)}} p_{[j,k]}(\boldsymbol{i}^{(k-j+1)}) = 1 \ . \end{aligned} \tag{2.27}$$

Consider the sequence of local states $\{\mu^{(n)}\}_{n\in\mathbb{N}}$,

$$\mu^{(n)} := \{p^{(n)}(\boldsymbol{i}^n)\}_{\boldsymbol{i}^{(n)} \in \Omega_d^{(n)}} \ , \quad p^{(n)}(\boldsymbol{i}^n) = p_{[1,n]}(\boldsymbol{i}^n) \ , \tag{2.28}$$

on the cylinder sets $\mathcal{C}_{[1,n]}$; since $C^{[1,n]}_{i_1 i_2 \cdots i_n} = \bigcup_{i=1}^d C^{[1,n+1]}_{i_1 i_2 \cdots i_n i}$, from the additivity of the measure the following compatibility condition follows

$$p^{(n)}(i_1 i_2 \ldots i_n) = \mu\Big(C^{[1,n]}_{i_1 i_2 \ldots i_n}\Big) = \sum_{i=1}^d p^{(n+1)}(i_1 i_2 \ldots i_n i) \ . \tag{2.29}$$

Particularly important global states over $\Omega_d$ correspond to shift-invariant probability measures $\mu$, $\mu \circ T_\sigma^{-1} = \mu$. From

$$\mu\Big(C^{[2,n+1]}_{i_1 i_2 \dots i_n}\Big) = \mu\Big(T_\sigma^{-1}(C^{[1,n]}_{i_1 i_2 \dots i_n})\Big) = \mu\Big(C^{[1,n]}_{i_1 i_2 \dots i_n}\Big)$$

it then follows that

$$\begin{aligned} \sum_{i=1}^{d} p^{(n)}(i i_2 \cdots i_n) &= \sum_{i=1}^{d} \mu\Big(C^{[1,n]}_{i i_2 \dots i_n}\Big) = \mu(C^{[2,n]}_{i_2 \dots i_n}) = \mu\Big(T_\sigma^{-1}(C^{[1,n-1]}_{i_2 \dots i_n})\Big) \\ &= \mu\Big(C^{[1,n-1]}_{i_2 \dots i_n}\Big) = p^{(n-1)}(i_2 \cdots i_n) \ . \end{aligned} \tag{2.30}$$

As a consequence, if $\mu$ is shift-invariant the probabilities assigned to cylinders $C^{[j,k]}_{i_j i_{j+1} \dots i_k}$ depend only on the values $i_j i_{j+1} \dots i_k$ defining the cylinder, but not on the interval $[j,k]$.

**Remark 2.1.5.** Interestingly, the conditions (2.29) and (2.30) defines a dynamical triplet $(\Omega_d, T_\sigma, \mu)$ in the sense of Definition 2.1.1. This is the content of *Kolmogorov representation theorem* [266]: if $\mathcal{X} = \{1, 2, \dots, d\}$, the set $\Omega_d$, as the infinite Cartesian product $\mathcal{X}^{\times\infty}$ of countably many copies of $\mathcal{X}$ can be equipped with the *product topology* which is the coarsest one with respect to which the projection maps $\pi_j : \boldsymbol{i} \mapsto \boldsymbol{i}_j$ are continuous, namely the one generated by union and intersections of preimages $\pi_j^{-1}(B)$ of sets $B \in \mathcal{X}$ that are open with respect to the discrete topology of $\mathcal{X}$. Then, $\Omega_d$ is a compact set by Tychonoff theorem [251]. Namely, any open cover of $\Omega$ also contains a finite subcover, whence in any collection of closed sets in $\Omega$ with empty intersection there also exists a finite sub-collection with empty intersection [251].

Suppose one is given a collection of numbers $p^{(n)}(\boldsymbol{i}^{(n)})$ as in (2.28) satisfying (2.27); they assign volumes $\mu\Big(C^{[1,n]}_{\boldsymbol{i}^{(n)}}\Big) := p^{(n)}(\boldsymbol{i}^{(n)})$, and define local states on the measure algebras generated by these cylinders. If the quantities $p^{(n)}(\boldsymbol{i}^{(n)})$ fulfil (2.29), the local states extend to a positive, finite and additive function $\mu$ on the $\sigma$-algebra $\Sigma$ generated by cylinders. In order to show that $\mu$ is also $\sigma$-additive and thus a measure, one uses Remark 2.1.1.4 and that each set in $\Sigma$ the $\sigma$-algebra is closed in the product topology. Therefore, given any decreasing sequence $\Sigma \supset \{C_n\}_{n=1}^{\infty}$ with empty intersection, compactness ensures that there exists a finite sub-collection $\{C_{n_j}\}_{j=1}^{k}$ such that $\bigcap_{j=1}^{k} C_{n_j} = \emptyset$, whence $\lim_{n\to\infty}(C_n) = 0$.

Further, suppose that the quantities $p^{(n)}(\boldsymbol{i}^{(n)})$ also fulfil (2.30), then it turns out that

$$\begin{aligned} \mu\Big(C^{[j,k]}_{i_j \dots i_k}\Big) &= \sum_{i_1 i_2 \dots i_{j-1}} p^{(k)}(i_1 i_2 \dots i_{j-1} i_j \dots i_k) \\ &= \sum_{i_\ell \dots i_{j-1}} p^{(k-\ell+1)}(i_\ell \dots i_{j-1} i_j \dots i_k) \ , \end{aligned}$$

for all $\ell = 1, 2, \ldots, j-1$. Therefore, $\mu\Big(C^{[j,k]}_{i_j \ldots i_k}\Big) = p^{(k-j+1)}(i_j \ldots i_k)$ whence $\mu\Big(T_\sigma^{-1}(C^{[j,k]}_{i_j \ldots i_k})\Big) = \mu\Big(C^{[j,k]}_{i_j \ldots i_k}\Big)$ and the measure $\mu$ is shift-invariant.

**Example 2.1.4 (Bernoulli shifts).** Consider a shift dynamical system $(\Omega_d, T_\sigma, \mu)$; the simplest choice of local states $\mu^{(n)}$ corresponds to *product measures* on $\Omega_d^{(n)}$:

$$p^{(n)}(i_1 \cdots i_n) = \prod_{j=1}^{n} p(i_j) \ , \quad p(i) \geq 0 \ , \ \sum_{i=1}^{d} p(i) = 1 \ . \tag{2.31}$$

These dynamical triplets are known as *Bernoulli-shifts*; if $d = 2$ and $\mathcal{X} = \{0, 1\}$, $(\Omega_2, T_\sigma, \mu)$ amounts to repeatedly tossing a coin, possibly biased if the probabilities of head (0) and tail (1) are different.

**Example 2.1.5 (Markov Chains).** Shift dynamical systems slightly more correlated than Bernoulli shifts are the so-called *Markov shifts.* Given the local states $\mu^{(n)} = \{p^{(n)}(\boldsymbol{i}^{(n)}\}_{\boldsymbol{i}^{(n)} \in \Omega_d^{n)}}$, the ratios

$$p(i_n | i_1 i_2 \cdots i_{n-1}) := \frac{p^{(n)}(i_1 i_2 \cdots i_n)}{p^{(n-1)}(i_1 i_2 \cdots i_{n-1})} \tag{2.32}$$

define *conditional probabilities* for the $n$-th symbol to be $i_n$ if the previous $n-1$ ones are $i_1 \cdots i_{n-1}$. The global state $\mu$ is said to possess the Markov property if and only if the following conditions occur:

$$p(i_n | i_1 i_2 \cdots i_{n-1}) = p(i_n | i_{n-1}) \tag{2.33}$$

$$\sum_{i=1}^{d} p(i|j) = 1 \tag{2.34}$$

$$\sum_{j=1}^{d} p(i|j)\, p(j) = p(i) \ . \tag{2.35}$$

Condition (2.33) means that the conditional probabilities (2.32) depend only on $i_n$ and on $i_{n-1}$ and not on the previous symbols, so that

$$p^{(n)}(i_1 i_2 \cdots i_n) = \Big(\prod_{\ell=1}^{n-1} p(i_{\ell+1} | i_\ell)\Big)\, p(i_1) \ . \tag{2.36}$$

Therefore, local states $\mu^{(n)}$ are completely specified by the $d \times d$ matrix $P = [p(i_n | i_{n-1})]$ and the probability vector $|\, p\rangle = \{p(j)\}_{j=1}^{d}$.

Because of (2.34), the matrix $P$ is a *stochastic matrix*, namely its entries $p(i|j)$ are positive and qualify as *transition probabilities* as they express the fact that the system cannot but remain in the same state or change into another one. It follows that condition (2.29) is satisfied, indeed from (2.36)

$$\begin{aligned}\sum_{i=1}^{d} p^{(n)}(i_1 \cdots i_{n-1} i) &= \sum_{i=1}^{d} p(i|i_{n-1}) \Big(\prod_{\ell=1}^{n-2} p(i_{\ell+1}|i_\ell)\Big) p(i_1) \\ &= \Big(\prod_{\ell=1}^{n-2} p(i_{\ell+1}|i_\ell)\Big) p(i_1) = p^{(n-1)}(i_1 i_2 \cdots i_{n-1}) \ .\end{aligned}$$

Further, because of (2.35), the probability vector is an eigenvector with eigenvalue 1 of the matrix $P$ and (2.30) is also satisfied, whence the local states $\mu^{(n)}$ generate a global shift-invariant states on $\Omega_d$. In fact,

$$\begin{aligned}\sum_{i=1}^{d} p^{(n)}(i i_2 \cdots i_n) &= \Big(\prod_{\ell=2}^{n-1} p(i_{\ell+1}|i_\ell)\Big) \sum_{i=1}^{d} p(i_2|i) p(i) \\ &= \Big(\prod_{\ell=2}^{n-1} p(i_{\ell+1}|i_\ell)\Big)\, p(i_2) = p^{(n-1)}(i_2 i_3 \cdots i_n) \ .\end{aligned}$$

Notice that Bernoulli shifts are particular instances of Markov chains with transition probabilities $p(i|j) = p(i)$ for all $j = 1, 2, \ldots, d$.

## 2.2 Symbolic Dynamics

As already remarked, states corresponding to a continuous phase-space can only be identified with finite precision that is they can be located within subsets of small, but finite size, and cannot be further resolved. A typical case is when the finite accuracy available corresponds to the subdivision of the phase-space in a finite number of non-overlapping measurable subsets, namely to a *coarse-graining* of the phase-space $\mathcal{X}$ by means of a so-called *finite partition* [7, 167].

**Definition 2.2.1 (Partitions).**

1. *A finite, measurable partition (partition for short) $\mathcal{P}$ of $(\mathcal{X}, T, \mu)$ is any collection of measurable subsets $P_i \subseteq \mathcal{X}$, $i \in I_\mathcal{P}$, $I_\mathcal{P}$ an index set of finite cardinality, such that $P_i \cap P_j = \emptyset$ for $i \neq j$ and $\bigcup_{i \in I_\mathcal{P}} P_i = \mathcal{X}$ The subsets $P_j$ are called atoms.*
2. *A partition $\mathcal{P} = \{P_i\}_{i \in I_\mathcal{P}}^{p}$ is finer than a partition $\mathcal{Q} = \{Q_j\}_{j \in I_\mathcal{Q}}^{q}$ ($\mathcal{Q}$ coarser than $\mathcal{P}$), symbolically $\mathcal{Q} \preceq \mathcal{P}$, if the atoms of $\mathcal{Q}$ are unions of atoms of $\mathcal{P}$: $Q_j = \bigcup_{i \in I_j \subseteq I_\mathcal{P}} P_i$, for all $j \in I_\mathcal{Q}$.*

*3. Given two partitions* $\mathcal{P} = \{P_i\}_{i\in I_{\mathcal{P}}}$ *and* $\mathcal{Q} = \{Q_j\}_{j\in I_{\mathcal{Q}}}$*, the partition* $\mathcal{P} \vee \mathcal{Q} = \{P_i \cap Q_j\}_{i\in I_{\mathcal{P}}, j\in I_{\mathcal{Q}}}$ *is the coarsest refinement of* $\mathcal{P}$ *and* $\mathcal{Q}$*.*

**Example 2.2.1.** [61] Quite often, $\mathcal{X}$ is endowed with a $\sigma$-algebra $\Sigma$ which is generated by a measure-algebra $\Sigma_0$; it is then possible to approximate within $\varepsilon$ any finite $\Sigma$-measurable partition $\mathcal{P} = \{P_i\}_{i=1}^d$ by a finite $\Sigma$-measurable partition $\mathcal{Q} = \{Q_i\}_{i=1}^d$ with atoms $Q_i \in \Sigma_0$, in the sense that (see (2.1)) $\mu(P_i \,\Delta\, Q_i) < \varepsilon$, $i = 1, 2, \ldots, d$. Indeed, because of Remark 2.1.1.4, given $\delta > 0$, for any $P_i \in \mathcal{P}$ one can find $Q'_i \in \Sigma_0$ such that $\mu(P_i \,\Delta\, Q'_i) < \delta$; notice that $P_i \cap P_j = \emptyset$, thus $x \in Q'_i \cap Q'_j$ and $x \notin P_i$ yield $x \in Q'_i \,\Delta\, P_i$, whence

$$Q'_i \cap Q'_j \subseteq Q'_i \,\Delta\, P_i \cup Q'_j \,\Delta\, P_j \Longrightarrow \mu(Q'_i \cap Q'_j) \leq 2\delta \ .$$

The sets $Q'_i$ need not form a partition; however, let $Q' := \bigcup_{i,j=1}^{d-1} Q'_i \cap Q'_j$, which is such that $\mu(Q') \leq d(d-1)\,\delta$ and set

$$Q_i := Q'_i \setminus Q' \ , \ i = 1, 2, \ldots, d-1 \ , \quad Q_d := \mathcal{X} \setminus \bigcup_{j=1}^{d-1} Q_j \ .$$

These are atoms of a partition $\mathcal{Q} \subset \Sigma_0$. Consider first the symmetric differences $Q_i \,\Delta\, P_i$, $i = 1, 2, \ldots, d-1$; one has that, if $x \in Q_i$ and $x \notin P_i$, then $x \in Q'_i \,\Delta\, P_i$, while, if $x \in P_i$ and $x \notin Q_i$, then $x \in Q'_i \,\Delta\, P_i$ or $x \in P_i \cap Q'$, whence

$$Q_i \,\Delta\, P_i \subseteq Q' \cup (Q'_i \,\Delta\, P_i) \Longrightarrow \mu(Q_i \,\Delta\, P_i) \leq (d(d-1)+1)\,\delta \ .$$

Since $P_d = \mathcal{X} \setminus \bigcup_{j=1}^{d-1} P_j$ and $(\mathcal{X} \setminus A) \,\Delta\, (\mathcal{X} \setminus B) = A \,\Delta\, B$,

$$Q_d \,\Delta\, P_d = \left(\bigcup_{j=1}^{d-1} Q_j\right) \Delta \left(\bigcup_{j=1}^{d-1} P_j\right) \subseteq \bigcup_{j=1}^{d-1} (Q_j \,\Delta\, P_j)$$

yields $\mu(Q_d \,\Delta\, P_d) \leq (d-1)(d(d-1)+1)\,\delta$, whence the result follows by choosing $\delta = (d-1)^{-1}(d(d-1)+1)^{-1}\varepsilon$.

The volumes $\mu(P_i) =: p(i)$ of the atoms of any partition $\mathcal{P}$ provide a discrete probability measure $\mu_{\mathcal{P}} := \{\mu(P_i)\}_{i\in I_{\mathcal{P}}}$ on $\mathcal{P}$. While atoms in general change under the dynamics $T$,

$$P_i \mapsto T^{-j}(P_i) := \{x \in \mathcal{X} : T^j x \in P_i\} \qquad \forall j \geq 0 \ , \tag{2.37}$$

their volumes do not for $T$ is assumed to preserve $\mu$.

Further, if $P_{i_1} \cap P_{i_2} = \emptyset$, then $T^{-j}(P_{i_1}) \cap T^{-j}(P_{i_2}) = \emptyset$. Therefore, for all $j \in \mathbb{N}$ ($j \in \mathbb{Z}$ if $T$ has a measurable inverse)

$$\mathcal{P}^j := T^{-j}(\mathcal{P}) = \{T^{-j}(P_i)\}_{i \in I_\mathcal{P}} \tag{2.38}$$

are partitions with the same probability distribution of $\mathcal{P}$: $\mu_{\mathcal{P}^j} = \mu_\mathcal{P}$. Further, partitions at successive times are all refined by the partition

$$\mathcal{P}^{(n)} := \bigvee_{j=0}^{n-1} \mathcal{P}^j = \mathcal{P} \vee T^{-1}(\mathcal{P}) \vee \cdots \vee T^{1-n}(\mathcal{P}) \; . \tag{2.39}$$

If $p := \mathrm{card}(I_\mathcal{P})$, the atoms

$$P^{(n)}_{\boldsymbol{i}^{(n)}} := P_{i_0} \cap T^{-1}(P_{i_1}) \cap \cdots T^{-n+1}(P_{i_{n-1}}) \tag{2.40}$$

of $\mathcal{P}^{(n)}$ are labeled by strings $\boldsymbol{i}^{(n)} := i_0 i_1 \cdots i_{n-1} \in \Omega_p^{(n)}$. We shall denote by

$$\mu^{(n)}_\mathcal{P} = \left\{ p^{(n)}(\boldsymbol{i}^{(n)}) := \mu(P^{(n)}_{\boldsymbol{i}^{(n)}}) \right\}_{\boldsymbol{i}^{(n)} \in \Omega_p^{(n)}} \; , \tag{2.41}$$

the probability distribution associated with $\mathcal{P}^{(n)}$ and consisting of the volumes of its atoms with respect to the given probability measure $\mu$.

**Remark 2.2.1.** Notice that a phase-point $x \in \mathcal{X}$ belongs to the atom $P_{\boldsymbol{i}^{(n)}}$ of $\mathcal{P}^{(n)}$ if and only if $T^j x \in P_{i_j}$ for all $0 \leq j \leq n-1$. As a consequence, the atoms of $\mathcal{P}^{(n)}$ contain all phase-points $x \in \mathcal{X}$ whose trajectories $\{T^j x\}_{j \in \mathbb{Z}}$ successively intercept the atoms $P_{i_j}$ of $\mathcal{P}$ identified by the string $\boldsymbol{i}^{(n)} = i_0 i_1 \cdots i_{n-1} \in \Omega_p^{(n)}$. As an effect of the coarse-graining, segments of different trajectories $\{T^j x\}_{j=0}^{n-1}$ may correspond to a same $\boldsymbol{i}^{(n)} \in \Omega_p^{(n)}$; thus, as normalized volumes, the probabilities $p^{(n)}(\boldsymbol{i}^{(n)})$ quantify how likely is it that different initial conditions give rise to a same segment of trajectory between (discrete) time $j = 0$ and $j = n-1$.

**Lemma 2.2.1.** *Given a reversible dynamical system $(\mathcal{X}, T, \mu)$ and a partition $\mathcal{P} = \{P_i\}_{i \in I_\mathcal{P}}$, card$(I_\mathcal{P}) = p$, the dynamics $T : \mathcal{X} \mapsto \mathcal{X}$ corresponds to the left-shift (2.23) on sequences $\boldsymbol{i} \in \Omega_p^\mathbb{Z}$ (see Remark 2.1.4).*

**Proof:** Let $\boldsymbol{i}(x) \in \Omega_p^\mathbb{Z}$ be the sequence of atom labels corresponding to the trajectory $\{T^j(x)\}_{j \in \mathbb{Z}}$ with initial point $x \in \mathcal{X}$. According to section 2.1.1 and to (2.37), $\boldsymbol{i}_j(x) = i_j$ if and only if $T^j x \in P_{i_j}$. Then, from (2.23),

$$\boldsymbol{i}_j(Tx) = i_j \Leftrightarrow T^{j+1} x \in P_{i_j} \Leftrightarrow \boldsymbol{i}_{j+1}(x) = (T_\sigma \boldsymbol{i})_j(x) = i_j \; .$$

□

Therefore, any coarse-graining of $\mathcal{X}$ by means of a partition $\mathcal{P}$ provides a description of the dynamical triplet $(\mathcal{X}, T, \mu)$ in terms of the left shift on the sequences in $\Omega_p^{\mathbb{Z}}$. The segments of trajectories up to time $n-1$ are in one-to-one correspondence with the sets of strings $\boldsymbol{i}^{(n)} \in \Omega_p^{(n)}$ and the probability distribution $\mu_{\mathcal{P}}^{(n)}$ provides states over the cylinder sets $\mathcal{C}^{[0,n-1]}$. By means of (2.40) and of the $T$-invariance of $\mu$, one shows that conditions (2.29) and (2.30) are fulfilled, whence the local states $\mu_{\mathcal{P}}^{(n)}$ define a global shift-invariant state $\mu_{\mathcal{P}}$ over $\Omega_p^{\mathbb{Z}}$. By varying $x \in \mathcal{X}$, the trajectories $\{T^j x\}_{j\in\mathbb{Z}}$ gets in general encoded by a subset $\widetilde{\Omega}_p^{\mathbb{Z}} \subset \Omega_p^{\mathbb{Z}}$.

**Definition 2.2.2 (Symbolic Models).** *Given a partition $\mathcal{P}$ of $\mathcal{X}$, the triplet $(\widetilde{\Omega}_p^{\mathbb{Z}}, T_\sigma, \mu_{\mathcal{P}})$ provides a symbolic model for the dynamical system $(\mathcal{X}, T, \mu)$.*

**Example 2.2.2 (Baker Map).** The Baker map (see Figure 2.1) is the invertible map of the two-dimensional torus $\mathbb{T}^2 = \{\boldsymbol{x} = (x_1, x_2)\ ,\ \mathrm{mod}\,1\}$ into itself given by

$$T_B\boldsymbol{x} = \begin{cases} \left(2x_1, \dfrac{x_2}{2}\right) & 0 \le x_1 < \dfrac{1}{2} \\ \left(2x_1 - 1, \dfrac{1+x_2}{2}\right) & \dfrac{1}{2} \le x_1 < 1 \end{cases}$$

$$T_B^{-1}\boldsymbol{x} = \begin{cases} \left(\dfrac{x_1}{2}, 2x_2\right) & 0 \le x_2 < \dfrac{1}{2} \\ \left(\dfrac{1+x_1}{2}, 2x_2 - 1\right) & \dfrac{1}{2} \le x_2 < 1 \end{cases} .$$

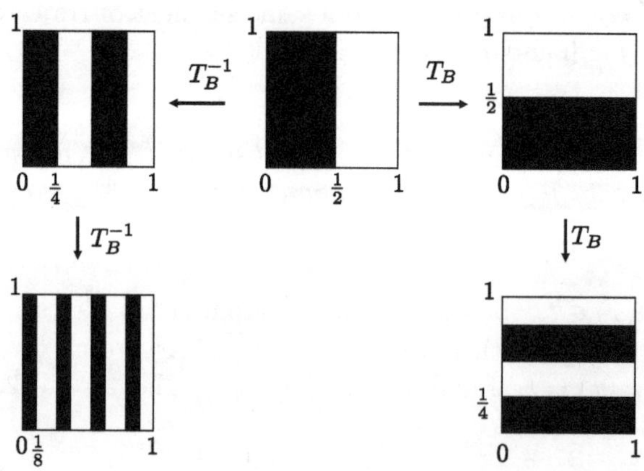


**Fig. 2.1.** Baker Map

The map $T_B$ is measurable with respect to the Borel $\sigma$-algebra of $\mathbb{T}^2$ and preserves the Lebesgue measure $\mathrm{d}\mu(\boldsymbol{x}) = \mathrm{d}x_1\mathrm{d}x_2$: altogether, one has a dynamical system described by the measure-theoretic triplet $(\mathbb{T}^2, T_B, \mathrm{d}\boldsymbol{x})$. It is evident that, when $n \to +\infty$, a sufficiently small distance between two points $\boldsymbol{x}$ and $\boldsymbol{x}+\boldsymbol{\delta}$ increases as $2^n$ along the horizontal direction until it gets of order 1. Therefore, Definition 2.1.2 gives $\log 2 > 0$ as maximal Lyapounov exponent of the Baker map; instead, small distances decrease exponentially along the vertical direction with the same speed so that volumes are conserved.

Let $\omega_+(x_1) = \{\omega_i\}_{i\geq 0}$ and $\omega_-(x_2) = \{\omega_{-j}\}_{j\geq 1}$ be the half-sequences consisting of the coefficients of the binary expansions of $x_1$, respectively $x_2$:

$$x_1 = \sum_{j=0} \frac{\omega_j}{2^{j+1}} \;, \qquad x_2 = \sum_{j=1} \frac{\omega_{-j}}{2^j} \;.$$

Setting $\omega(\boldsymbol{x}) := (\omega_-(x_1), \omega_+(x_2)) = \{\omega_j(\boldsymbol{x})\}_{j\in\mathbb{Z}} \in \Omega_2$ and using the mod 1 folding condition defining $\mathbb{T}^2$, it turns out that $T_B$ is isomorphic to the left shift $T_\sigma$ on $\Omega_2$, namely $\omega_j(T_B\boldsymbol{x}) = \omega_{j+1}(\boldsymbol{x})$.

Further, the Lebesgue measure on $\mathbb{T}^2$ corresponds to the uniform product measure (2.31) on the $\sigma$-algebra generated by cylinders. This can be seen as follows. According to Remark 2.25, cylinders are intersections of simple cylinders as $C_0^{\{0\}}$ and $C_1^{\{0\}}$ that correspond to the vertical rectangles $P_0 = \{\boldsymbol{x} : 0 \leq x_1 < 1/2\}$ and $P_1 = \{\boldsymbol{x} : 1/2 \leq x_1 < 1\}$ and their images $C_{i_j}^{\{j\}} = T_B^{-j}(C_{i_j}^{\{0\}})$, $j \in \mathbb{Z}$ (see (2.26)). Under $T_B^{-1}$ they get rotated into horizontal rectangles; successive applications of the Baker's map split them into horizontal rectangles of half height, each one of them having as neighbors halved rectangles coming from the other initial rectangle.

It turns out that $C_{i_1,\ldots,i_{n-1}}^{[1,n-1]}$ is a horizontal rectangle of width 1 and height $2^{-n+1}$; a further intersection with $C_{i_0}^{\{0\}}$ provides the cylinder $C_{i_0,i_1,\ldots,i_{n-1}}^{([0,n-1]}$ corresponding to a horizontal rectangle of width $1/2$ and height $2^{-n+1}$ whose area is $2^{-n}$. These areas may only come from a product measure,

$$\mu_B^{(n)}(C_{\boldsymbol{i}^{(n)}}^{([0,n-1]}) := \prod_{j=0}^{n-1} \mu_B^{(1)}(C_{i_j}^{\{j\}}) \;, \quad \mu_B^{(1)}(C_{i_j}^{\{j\}}) = \frac{1}{2} \qquad \forall j \;.$$

Therefore, the coarse-graining of $\mathbb{T}^2$ given by $\mathcal{P} = \{P_0, P_1\}$ provides the symbolic model $(\Omega_2, T_\sigma, \mu_B)$ for $(\mathbb{T}^2, T_B, \mathrm{d}\boldsymbol{x})$.

### 2.2.1 Algebraic Formulations

In this section, instead of referring to phase-space trajectories, we shall consider classical dynamical systems from the point of view of their observables and of their time-evolution. By observables we mean suitable functions over the phase-space.

It is convenient to consider complex-valued functions $f : \mathcal{X} \mapsto \mathbb{C}$; their values $f(x)$ can be inferred by measuring real, $\Re(f)$, and imaginary parts, $\Im(f)$. Further, it is reasonable to assume that functions $f, g$ in a suitably chosen class of observables give observables in the same class under addition, $(f,g) \mapsto (f+g)(x) = f(x)+g(x)$, and multiplication either by scalars $\alpha \in \mathbb{C}$, $(\alpha, f) \mapsto (\alpha f)(x)\alpha f(x)$, or by another observable, $(f,g) \mapsto fg(x) = f(x)g(x)$.

In other words, it is a reasonable physical assumption to require that observables constitute algebras of functions on $\mathcal{X}$: these algebras are *commutative* for $fg = gf$. Physically speaking, there are no fundamental obstructions to the fact that classical measuring processes can, in line of principle, be performed without effects on the state of the measured system. As the measured values depend on the system state, it follows that measuring $g$ and then $f$ yields the same results as measuring $f$ and then $g$.

Also, it is practically convenient to approximate certain observables in the algebra by means of other observables that are in a certain sense close to them; we shall thus assume these commutative algebras of observables to be endowed with topologies and to be closed with respect to them; in particular, we shall consider algebras of observables where converging sequences of functions do converge to observables in the algebra.

Like in Examples 2.1.2, 2.1.3, in the following we shall assume $\mathcal{X}$ to be compact in a metric topology and measurable with respect to the Borel $\sigma$-algebra that contains all its open and closed sets. Then, a natural algebra of observables is provided by the continuous functions on $\mathcal{X}$ [258].

**Definition 2.2.3.** *Let $\mathcal{X}$ be a compact metric space; $C(\mathcal{X})$ will denote the Banach $^*$-algebra (with identity) of continuous functions $f : \mathcal{X} \mapsto \mathbb{C}$ endowed with the uniform topology given by the norm*

$$C(\mathcal{X}) \ni f \mapsto \|f\| = \sup\{|f(x)| \, : \, x \in \mathcal{X}\} \, . \tag{2.42}$$

**Remarks 2.2.2.**

1. If $f, g \in C(\mathcal{X})$ and $\alpha \in \mathbb{C}$ then $f + \alpha g \in C(\mathcal{X})$ as well as $fg \in C(\mathcal{X})$. Sums, multiplications by complex scalars, by continuous functions and complex conjugation $* : f(x) \mapsto f^*(x)$ all map $C(\mathcal{X})$ into itself. These facts make $C(\mathcal{X})$ a $^*$-algebra with a norm $f \mapsto \|f\|$; indeed,

$$\|f\| = 0 \Leftrightarrow f = 0 \, , \quad \|\alpha f\| = |\alpha| \, \|f\| \, , \quad \|f + g\| \leq \|f\| + \|g\| \, ,$$

for all $f, g \in C(\mathcal{X})$, $\alpha \in \mathbb{C}$. This norm defines the *uniform neighborhoods*

$$\mathcal{U}_\varepsilon(f) := \{g \in C(\mathcal{X}) \, : \, \|f - g\| \leq \varepsilon\} \, , \quad f \in C(\mathcal{X}) \, , \tag{2.43}$$

and equips $C(\mathcal{X})$ with a metric and a corresponding topology called *uniform topology, $\mathcal{T}_u$.*

2. A sequence $\{f_n\}_{n\in\mathbb{N}} \subset C(\mathcal{X})$ is a *Cauchy sequence* if, for any $\varepsilon > 0$ there exists $N \in \mathbb{N}$ such that $n, m \geq N \Longrightarrow \|f_n - f_m\| \leq \varepsilon$; since all Cauchy sequences in $C(\mathcal{X})$ converge uniformly to $f \in C(\mathcal{X})$, that is $\lim_n \|f - f_n\| = 0$ or $\lim_n f_n = f$, $C(\mathcal{X})$ is termed a *Banach algebra*. Also, $\|f^* f\| = \|f\|^2$, $\|f^*\| = \|f\|$ and $\|fg\| \leq \|f\| \, \|g\|$ for all $f, g \in C(\mathcal{X})$; this makes $C(\mathcal{X})$ a $C^*$ *algebra* (see Definition 5.2.1).
3. Because of assumed compactness of $\mathcal{X}$, the identity function $\mathbb{1}(x) = 1$ belongs to $C(\mathcal{X})$. When $\mathcal{X}$ is not compact, one considers the $^*$-algebra $C_0(\mathcal{X})$ consisting of the complex continuous functions on $\mathcal{X}$ vanishing at infinity. When equipped with the norm (2.42), $C_0(\mathcal{X})$ is a Banach algebra, but the identity function does not belong to it.

A description of dynamical systems by means of continuous functions is, however, too restrictive, in general. For instance, the corresponding $C^*$ algebras cannot contain observables related to yes/no questions like

is the state localized within a measurable subset (region) $A \in \mathcal{X}$ or not?

as these correspond to characteristic functions $\mathbf{1}_A$ of $A$ which are only measurable and not continuous. The Koopman-von Neumann formulation of Example 2.1.1 offers a natural way to enlarge the algebra of observables. In a quantum-like notation, we shall denote by $|\,\psi\,\rangle$ any function in $\mathbb{L}^2_\mu(\mathcal{X})$ and by $\langle\, x \,|\, \psi\,\rangle$ its value $\psi(x)$ at $x \in \mathcal{X}$. Functions $f \in C(\mathcal{X})$ can then be represented on $\mathbb{L}^2_\mu(\mathcal{X})$ as multiplication operators $M_f$:

$$\langle\, x \,|\, M_f \,|\psi\,\rangle = f(x)\psi(x) \;, \quad \forall \psi \in \mathbb{L}^2_\mu(\mathcal{X}) \;. \tag{2.44}$$

In the following, we shall identify, $C(\mathcal{X})$ and its representation by multiplication operators, that is we shall identify $M_f$ and $f$.

**Remarks 2.2.3.**

1. The maps $C(\mathcal{X}) \ni f \mapsto \mathcal{L}_\psi(f) := \|f|\,\psi\,\rangle\|$ are *semi-norms*. They define *strong-neighborhoods* on $C(\mathcal{X})$, that is neighborhoods in the so called *strong topology*, $\mathcal{T}_s$,
$$\mathcal{U}_\varepsilon^{\{\psi_j\}_{j=1}^n}(f) := \Big\{ g \in C(\mathcal{X}) \,:\, \|(f-g)|\,\psi_j\,\rangle\| \leq \varepsilon \;,\; 1 \leq j \leq n \Big\} \;. \tag{2.45}$$
Since $\|(f-g)|\,\psi\,\rangle\| \leq \|f-g\| \, \|\psi\|_2$, $g \in \mathcal{U}_{\varepsilon/\|\psi\|}(f) \Longrightarrow g \in \mathcal{U}_\varepsilon^\psi$; therefore, every strong-neighborhood contains a uniform neighborhood and is thus a uniform neighborhood itself; in general, however, there can be uniform neighborhoods which are not strong-neighborhoods, so that the uniform topology is finer than the strong topology, $\mathcal{T}_s \preceq \mathcal{T}_u$; namely, $\mathcal{T}_u$ has more neighborhoods. Practically speaking, a sequence $f_n \in C(\mathcal{X})$ converges strongly to $f \in C(\mathcal{X})$, $s - \lim_n f_n = f$, if

$\lim_{n\to\infty}\|(f-f_n)|\psi\rangle\| = 0 \quad \forall\psi \in \mathbb{L}^2_\mu(\mathcal{X})$, and, while all uniformly convergent sequences converge strongly, there can be strongly converging sequences which do not converge uniformly.

2. If $\{f_n\}_{n\in\mathbb{N}}$ converges with respect to $\mathcal{T}_u$, it converges also with respect to $\mathcal{T}_s$, but not vice versa; it follows that the *strong closure* of $C(\mathcal{X})$, that is $C(\mathcal{X})$ together with all its possible strong limit points, is strictly larger than $C(\mathcal{X})$. Indeed, it contains $C(\mathcal{X})$, simple functions and discontinuous functions $f$ that may jump arbitrarily but only on sets of zero measure [258]. Equip $\mathcal{X}$ with a $\sigma$-algebra and a measure $\mu$; then,

$$\|f\|_\infty := \inf\Big\{a \geq 0 : \mu\Big(\{x : |f(x)| \geq a\}\Big) = 0\Big\} ,$$

where $f$ is a measurable function on $\mathcal{X}$, defines a norm $\|\cdot\|_\infty$ called *essential norm.* If $\|f\|_\infty < \infty$, then $|f(x)| > \|f\|_\infty$ only on a set of zero measure; further, the following collection of measurable functions,

$$\mathbb{L}^\infty_\mu(\mathcal{X}) := \Big\{f : \|f\|_\infty < \infty\Big\} ,$$

is a $C^*$ algebra with respect to the essential norm known as the algebra of *essentially bounded functions.*

3. There is another topology on $C(\mathcal{X})$ which is inherited by its multiplicative action on $\mathbb{L}^2_\mu(\mathcal{X})$ and which is coarser than the strong topology, namely the *weak topology,* $\mathcal{T}_w \preceq \mathcal{T}_s \preceq \mathcal{T}_u$. It is generated by the semi-norms $\mathcal{L}_{\phi,\psi}(f) := |\langle\phi\,|\,M_f\,|\psi\rangle|$ which defines the *weak neighborhoods*

$$\mathcal{U}_\varepsilon^{\{(\phi_j,\psi_j)\}_{j=1}^n}(f) := \{g \in C(\mathcal{X}) \,:\, \mathcal{L}_{\phi_j,\psi_j}(f-g) \leq \varepsilon \,,\; 1 \leq j \leq n\} \,. \tag{2.46}$$

A sequence $f_n \in C(\mathcal{X})$ converges weakly to $f \in C(\mathcal{X})$, $w-\lim f_n = f$, if and only if $\lim_{n\to\infty} |\langle\phi|\,(f-f_n\,|\psi\rangle| = 0$ for all $\psi,\phi \in \mathbb{L}^2_\mu(\mathcal{X})$. As we shall see in the more general non-commutative context, the strong and the *weak closures* coincide. In the case of $C(\mathcal{X})$ they give rise to $\mathbb{L}^\infty_\mu(\mathcal{X})$ which has the structure of a so-called *von Neumann algebra.*

4. Actually, $\mathbb{L}^\infty_\mu(\mathcal{X})$ can be generated as the strong closure on $\mathbb{L}^2_\mu(\mathcal{X})$ of the algebra containing the characteristic functions of finer and finer partitions of $\mathcal{X}$. More precisely, one may consider a refining sequence $\{\mathcal{P}_n\}_{n\geq 0}$, $\mathcal{P}_n \preceq \mathcal{P}_{n+1}$, that generates the $\sigma$-algebra of $\mathcal{X}$ when $n \to +\infty$. Each $\mathcal{P}_n$ is a finite dimensional commutative algebra $\mathcal{A}_n$ whose elements are the step functions that are linear combinations of the characteristic functions of the finitely many atoms of $\mathcal{P}_n$; then,

$$\mathbb{L}^\infty_\mu(\mathcal{X}) = \overline{\bigcup_n \mathcal{A}_n}^{\,weak-closure} .$$

In order to complete the formulation of measure-theoretic triplets $(\mathcal{X}, T, \mu)$ into an algebraic framework, one has to endow $C(\mathcal{X})$ with a time-evolution corresponding to $T$ and a map $C(\mathcal{X}) \mapsto \mathbb{C}$ that play the role of $\mu$ by assigning mean values to continuous functions.

We shall consider invertible continuous dynamical maps $T$ on $\mathcal{X}$ and discrete-time dynamics. As in Example 2.1.2, any $f \in C(\mathcal{X})$ changes in time according to

$$f(x) \mapsto f(T^t x) = f \circ T^t(x) =: f_t(x) , \quad t \in \mathbb{Z} .$$

The map $\Theta_T : C(\mathcal{X}) \mapsto C(\mathcal{X})$, defined by $\Theta_T(f) = f \circ T$ is an *automorphism* of $C(\mathcal{X})$; namely, it is invertible and

$$\Theta_T(\alpha f + \beta g) = \alpha \Theta_T(f) + \beta \Theta_T(g) , \quad \Theta_T(fg) = \Theta_T(f)\, \Theta_T(g) . \quad (2.47)$$

Moreover, $\Theta_T$ preserves the uniform norm.

**Example 2.2.3.** In the Koopman-von Neumann formalism where functions $f \in C(\mathcal{X})$ are represented as multiplication operators, the Koopman operator $U_T$ implements unitarily the automorphism $\Theta_T$; indeed, using (2.4), for all $\psi \in \mathbb{L}^2_\mu(\mathcal{X})$ and $x \in \mathcal{X}$,

$$\begin{aligned} \langle x \,|\, U_T\, f\, U_T^\dagger \psi \rangle &= f(Tx) \langle Tx \,|\, U_T^\dagger \psi \rangle = f(Tx) \langle T^{-1} \circ Tx \,|\, \psi \rangle \\ &= \langle x \,|\, \Theta_T(f) \psi \rangle. \end{aligned}$$

Notice that $U_T$ cannot belong to $C(\mathcal{X})$, otherwise it would commute with all $f \in C(\mathcal{X})$ which would then be constant in time.

Concerning the possible states over $C(\mathcal{X})$ (see (2.2)), we shall consider the space $M(\mathcal{X})$ of regular Borel measures over $\mathcal{X}$ (see Remark 2.1.1.5). The simplest instances of elements of $M(\mathcal{X})$ are the evaluation functionals $\delta_x : C(\mathcal{X}) \mapsto \mathbb{C}$, defined by $\delta_x(f) := f(x)$, for all $x \in \mathcal{X}$ and $f \in C(\mathcal{X})$. These functionals can be seen as integration with respect to Dirac delta distributions and embody the fact that phase-space points are the simplest physical states: $\delta_x(f) = \int_{\mathcal{X}} \mathrm{d}y\; f(y)\, \delta(y - x)$. By making convex combinations of evaluation functionals one obtains more general *positive, normalized expectation functionals* over $C(\mathcal{X})$.

Actually, a theorem of Riesz [258] asserts that the action of any such functional is representable by integration with respect to a regular Borel measure in $M(\mathcal{X})$. In view of the physical interpretation of states as positive functionals that assign mean values to observables, it makes sense to identify measures $\mu \in M(\mathcal{X})$ and states $\omega_\mu : C(\mathcal{X}) \mapsto \mathbb{C}$ such that [3]

[3]For sake of notational convenience, we shall sometime employ the notation $\mu(f)$ for $\omega_\mu(f)$.

$$\mathcal{A} \ni f \mapsto \omega_\mu(f) = \int_{\mathcal{X}} \mathrm{d}\mu(x)\, f(x) , \quad \forall f \in C(\mathcal{X}) . \tag{2.48}$$

**Remarks 2.2.4.**

1. With two measures $\mu_{1,2}$ on a measure space $\mathcal{X}$ all *convex combinations* $p\mu_1+(1-p)\mu_2$ with $p \in [0,1]$ provide other measures; therefore, the space of states of classical systems is a convex set.
2. Given two measures $\mu_{1,2}$ on $\mathcal{X}$ equipped with a $\sigma$-algebra $\Sigma$, $\mu_1$ is said to be *absolutely continuous* with respect to $\mu_2$, $\mu_1 \preceq \mu_2$, if for any $B \in \Sigma$ $\mu_2(B) = 0 \Longrightarrow \mu_1(B) = 0$. Then, there exists a positive $f \in \mathbb{L}^1_\mu(\mathcal{X})$ such that $\mu_1(B) = \int_B \mathrm{d}\mu_2(x)\, f(x)$ for all $B \in \Sigma$. The density $f(x)$ is called *Radon-Nikodym derivative* and denoted by $\frac{\mathrm{d}\mu_1}{\mathrm{d}\mu_2}$. If also $\mu_2 \preceq \mu_1$ then $\mu_1$ and $\mu_2$ are said to be *equivalent*. Differently, $\mu_1$ and $\mu_2$ are called *mutually singular*, $\mu_1 \perp \mu_2$, if there exists $B \in \Sigma$ such that $\mu_1(B) = 0$ while $\mu_2(\mathcal{X} \backslash B) = 0$.
3. According to *Lebesgue decomposition theorem*, given two measures $\mu$ and $m$ on $\mathcal{X}$, there exists a unique choice of measures $\mu_{1,2}$ and of $p \in [0,1]$ such that $\mu = p\mu_1 + (1-p)\mu_2$ with $\mu_1 \preceq m$ and $\mu_2 \perp m$.
4. If $\mathcal{X} = \mathbb{T}^2$ as in Example 2.1.3, then any $\mathbb{L}^1_{\mathrm{d}\boldsymbol{r}}(\mathbb{T}^2) \ni \rho(\boldsymbol{r}) \geq 0$ with $\int_{\mathbb{T}^2} \mathrm{d}\boldsymbol{r}\, \rho(\boldsymbol{r}) = 1$ is the Radon-Nikodym derivative of a measure which is absolutely continuous with respect to $\mathrm{d}\boldsymbol{r}$. Vice versa, evaluation functionals $\delta_{\boldsymbol{r}}(f) = f(\boldsymbol{r})$ are singular measures with respect to $\mathrm{d}\boldsymbol{r}$.

Finally, a measure $\mu \in M(\mathcal{X})$ is $T$-invariant if the corresponding mean values are time-independent, $\omega_\mu(\Theta_T(f)) = \omega(f)$ or $\omega_\mu = \omega_\mu \circ \Theta_T$. Notice that (2.48) and (2.47) allows one to extend the state $\omega_\mu$ and the automorphism $\Theta_T$ to the von Neumann algebra of essentially bounded functions $\mathbb{L}^\infty_\mu(\mathcal{X})$.

**Definition 2.2.4.** *To any measure-theoretic triplet $(\mathcal{X}, T, \mu)$, where $\mathcal{X}$ is a compact metric space equipped with the Borel $\sigma$-algebra and $\mu \in M(\mathcal{X})$ is a $T$-invariant regular Borel measure, one can associate a $C^*$ algebraic triplet $(C(\mathcal{X}), \Theta_T, \omega_\mu)$ and a von Neumann triplet $(\mathbb{L}^\infty_\mu(\mathcal{X}), \Theta_T, \omega_\mu)$ where state $\omega_\mu$ and automorphism $\Theta_T$ are defined as in (2.48), respectively (2.47).*

### 2.2.2 Conditional Probabilities and Expectations

Given a measure space $(\mathcal{X}, \mu)$ with $\sigma$-algebra $\Sigma$, a finite partition $\mathcal{P} = \{P_i\}_{i=1}^p$ such that $\mu(P_i) > 0$ for all $i$, and $X \in \Sigma$, consider the following function

$$x \in \mathcal{X} \mapsto \mu(X|\mathcal{P})(x) := \frac{\mu(X \cap P_i)}{\mu(P_i)} \quad \text{if} \quad x \in P_i , \tag{2.49}$$

$$\text{such that} \quad \int_{P_i} \mathrm{d}\mu\,(x)\, \mu(X|\mathcal{P})(x) = \mu(X \cap P_i) . \tag{2.50}$$

It is the *conditional probability* of $X \in \Sigma$ *given the partition* $\mathcal{P}$ and represents the probability of the subset $X$ once it is known that $x$ belongs to one of the atoms of $\mathcal{P}$. This notion can be extended to the case of partitions with atoms $P$ such that $\mu(P) = 0$ by assigning a same fixed, arbitrary real value to $\mu(X|\mathcal{P})(x)$ when $x \in P$: in such a way one gets a family of *versions* of the conditional probability each of which satisfies (2.50) [61]. One can extend (2.49) and (2.50) and define probability distributions conditioned upon $\sigma$-subalgebras $\mathcal{T} \subseteq \Sigma$.

Consider an integrable function $f \in \mathbb{L}^1_\mu(\mathcal{X})$, the functional on $\mathcal{T}$ defined by $F(T) := \int_T \mathrm{d}\mu\,(x)\, f(x)$, $T \in \mathcal{T}$, is bounded, $\sigma$-additive and absolutely continuous with respect to $\mu$ (see Remarks 2.1.1.3 and 2.2.4.1); its Radon-Nikodym derivative $\frac{\mathrm{d}F}{\mathrm{d}\mu}(x) =: E(f|\mathcal{T})(x)$ such that

$$\int_T \mathrm{d}\mu\,(x)\, E(f|\mathcal{T})(x) = \int_T \mathrm{d}\mu\,(x)\, f(x) \qquad \forall T \in \mathcal{T}\;, \tag{2.51}$$

is $\mathcal{T}$-measurable and integrable and is called the *conditional expectation* of $f$ with respect to the $\sigma$-algebra $\mathcal{T}$.

By choosing as $f$ the characteristic function $1\!\!1_X$ of a subset $X \in \Sigma$ its conditional probability given $\mathcal{T}$ is thus defined by $\mu(X|\mathcal{T})(x) := E(1\!\!1_X|\mathcal{T})(x)$ and is such that

$$\int_T \mathrm{d}\mu\,(x)\, \mu(X|\mathcal{T})(x) = \mu(X \cap T) \qquad \forall X \in \Sigma\;,\; T \in \mathcal{T}\;. \tag{2.52}$$

Given a $\sigma$-subalgebra $\mathcal{T} \subseteq \Sigma$ consider the Abelian von Neumann algebra $\mathbb{L}^\infty_\mu(\mathcal{X}, \mathcal{T})$ consisting of the essentially bounded $\mathcal{T}$-measurable functions on $\mathcal{X}$ (see Remark 2.2.3.2). This is a subalgebra of the Abelian von Neumann algebra $\mathbb{L}^\infty_\mu(\mathcal{X})$ of the $\Sigma$-measurable essentially bounded functions on $\mathcal{X}$. Then, (2.51) makes the conditional expectation a linear map $E(\cdot|\mathcal{T}) : \mathbb{L}^\infty_\mu(\mathcal{X}) \mapsto \mathbb{L}^\infty_\mu(\mathcal{X}, \mathcal{T})$ which is linear, positive and a measure preserving projection, that is $\mu \circ E(\cdot|\mathcal{T}) = \mu$ and $E(E(f|\mathcal{T})|\mathcal{T}) = E(f|\mathcal{T})$. The first three assertions are evident, while idempotency is a corollary of the following more general property. Suppose $\mathcal{T}_1 \preceq \mathcal{T}_2$ are two $\sigma$-subalgebras of $\Sigma$ such that $T_1 \in \mathcal{T}_1 \Longrightarrow T_1 \in \mathcal{T}_2$ but not vice versa, in general. Then, if $f \in \mathbb{L}^\infty_\mu(\mathcal{X})$, (2.51) yields

$$\begin{aligned}\int_{T_1} \mathrm{d}\mu\,(x)\, E(E(f|\mathcal{T}_2)|\mathcal{T}_1) &= \int_{T_1} \mathrm{d}\mu\,(x)\, E(f|\mathcal{T}_2) = \int_{T_1} \mathrm{d}\mu\,(x)\, f(x)\\ &= \int_{T_1} \mathrm{d}\mu\,(x)\, E(f|\mathcal{T}_1)\;,\end{aligned}$$

for all $T_1 \in \mathcal{T}_1$, whence $\mathcal{T}_1 \preceq \mathcal{T}_2 \Longrightarrow E(E(f|\mathcal{T}_2)|\mathcal{T}_1) = E(f|\mathcal{T}_1)$.

**Proposition 2.2.1.** *If $f \in \mathbb{L}^\infty_\mu(\mathcal{X})$ and $g \in \mathbb{L}^\infty_\mu(\mathcal{X}, \mathcal{T})$, where $\mathcal{T} \subseteq \Sigma$, then*

$$E(g\, f|\mathcal{T}) = g\, E(f|\mathcal{T}) \; . \tag{2.53}$$

**Proof:** Suppose $g = \mathbf{1}_T$, the characteristic function of a subset $T \in \mathcal{T}$; then, for all $T_0 \in \mathcal{T}$, $T \cap T_0 \in \mathcal{T}$ and (2.51) yields

$$\int_{T_0} \mathrm{d}\mu\,(x)\, E(\mathbf{1}_T\, f|\mathcal{T})(x) = \int_{T \cap T_0} \mathrm{d}\mu\,(x)\, f(x) = \int_{T_0} \mathbf{1}_T(x)\, E(f|\mathcal{T})(x) \; .$$

Then, one concludes the proof by approximating $g \in \mathbb{L}^\infty_\mu(\mathcal{X}, \mathcal{T})$ with respect to the essential norm by means of simple functions. □

Given a refining sequence $\{\mathcal{T}_n\}_{n\in\mathbb{Z}}$, that is $n \leq m \Longrightarrow \mathcal{T}_n \subseteq \mathcal{T}_m \subseteq \Sigma$, we shall set $\mathcal{T}_+ := \bigvee_{n\in\mathbb{Z}}$ the smallest $\sigma$-subalgebra containing all the $\mathcal{T}_n$'s ($\mathcal{T}_n \uparrow \mathcal{T}_+$), respectively denote by $\mathcal{T}_- := \bigwedge_{n\in\mathbb{Z}} \mathcal{T}_n$ the largest $\sigma$-subalgebra contained in all the $\mathcal{T}_n$ ($\mathcal{T}_n \downarrow \mathcal{T}_-$). The proof of the following continuity properties can be found in [61] and [101].

**Theorem 2.2.1.** *Let $\mathcal{X}$ be a measure space equipped with a $\sigma$-algebra $\Sigma$ and a measure $\mu$; given a refining sequence of $\sigma$-subalgebras $\mathcal{T}_n$, then*

$$\lim_{n\to+\infty} E(f|\mathcal{T}_n) = E(f|\mathcal{T}_+) \; , \qquad \lim_{n\to-\infty} E(f|\mathcal{T}_n) = E(f|\mathcal{T}_-) \; ,$$

*for all $f \in \mathbb{L}^\infty_\mu(\mathcal{X})$.*

**Examples 2.2.4.**

1. If $\Sigma \supset \mathcal{T} := \mathcal{N}$, the trivial $\sigma$-algebra consisting of the empty set $\emptyset$ and the whole of $\mathcal{X}$; then, $E(f|\mathcal{N}) = \mu(f)\,\mathbb{1}$. On the other hand, if $\mathcal{T} = \Sigma$, then $E(f|\Sigma) = f$.
2. By inserting characteristic functions $f = \mathbf{1}_X$, $X \in \Sigma$, in the above theorem, one gets the following continuity properties of the conditional probabilities:
$$\lim_{n\to\pm\infty} \mu(X|\mathcal{T}_n)(x) = \mu(X|\mathcal{T}_\pm)(x) \; , \quad \mu - a.e.$$
for all$\Sigma$-measurable subsets of $\mathcal{X}$.
3. Consider the unit interval $[0,1)$ with the Borel $\sigma$-algebra $\Sigma$ and the Lebesgue measure $\mathrm{d}\mu\,(x) = \mathrm{d}x$; construct the measure algebra $\mathcal{T}_n$ generated by the partition $\mathcal{P}_n$ of $[0,1)$ into $2^n$ atoms $P_k = [k2^{-n}, (k+1)2^{-n})$. Then, $\mathcal{T}_n \uparrow \Sigma$ and [61]
$$E(f|\mathcal{T}_n)(x) = \sum_{k=0}^{2^n-1} \mathbf{1}_{P_k}(x)\, 2^n \int_{k2^{-n}}^{(k+1)2^{-k}} \mathrm{d}t\; f(t) \; ,$$
for all $f \in \mathbb{L}^\infty_{[0,1)}(\mathrm{d}t\,)$. For $n \to \infty$ the summand containing $x$ tends to the derivative of the integral at $x$ and thus to $f(x)$ $\mu$-a.e..

### 2.2.3 Dynamical Shifts and Classical Spin Chains

Dynamical shifts and symbolic models can be given an algebraic formulation in terms of classical spin chains. Consider a triplet $(\Omega_p^{\mathbb{Z}}, T_\sigma, \mu)$, that is a shift-dynamical system over doubly infinite sequences of symbols from an alphabet with $p$ elements that leaves invariant a measure $\mu$.

Let us associate to each symbol $j \in \{1, 2, \ldots, p\}$ a $p \times p$ matrix of the form

$$P_j = \begin{pmatrix} 0 & 0 & 0 & \cdots & 0 & 0 \\ 0 & 0 & 0 & \cdots & 0 & 0 \\ \cdot & \cdot & \cdot & \cdots & \cdot & \cdot \\ \cdot & \cdot & \cdot & \underbrace{1}_{\text{(j,j)}-\text{thentry}} & \cdot & \cdot \\ \cdot & \cdot & \cdot & \cdots & \cdot & \cdot \\ 0 & 0 & 0 & \cdots & 0 & 0 \end{pmatrix} .$$

Varying $1 \leq j \leq p$, we obtain an orthonormal family of orthogonal projections such that $P_i P_j = \delta_{ij} P_j$ and $\sum_{j=1}^p P_j = 1\!\!1_p$, where $1\!\!1_p$ denotes the $p \times p$ identity matrix; these projectors generate the *diagonal* $p \times p$ *matrix algebra* $D_p(\mathbb{C})$ [4] with elements

$$D = \begin{pmatrix} d_1 & 0 & 0 & \cdots & 0 & 0 \\ 0 & d_2 & 0 & \cdots & 0 & 0 \\ \cdot & \cdot & \cdot & \cdots & \cdot & \cdot \\ \cdot & \cdot & \cdot & \cdots & \cdot & \cdot \\ \cdot & \cdot & \cdot & \cdots & \cdot & \cdot \\ 0 & 0 & 0 & \cdots & 0 & d_p \end{pmatrix} = \sum_{j=1}^p d_j P_j . \tag{2.54}$$

To each label $j$ in a sequence $\boldsymbol{i} = \{i_j\}_{j\in\mathbb{Z}}$ one thus associates the diagonal matrix algebra $D_p(\mathbb{C})$: each of its minimal projectors thus corresponds to a simple cylinder. Extending the construction to generic cylinders $C^{([0,n-1]}_{i_0,i_1,\ldots,i_{n-1}}$ as in (2.24), these correspond to *tensor products* of projectors

$$P^{([0,n-1]}_{\boldsymbol{i}^{(n)}} := \bigotimes_{j=0}^{n-1} P_{i_j} , \qquad \boldsymbol{i}^{(n)} := i_0, i_1, \ldots, i_{n-1} . \tag{2.55}$$

Then, the natural matricial description that one associates to strings of length $n$ is the diagonal matrix algebra $\mathcal{D}^{(n)} = \mathcal{D}^{\otimes n} := \bigotimes_{j=0}^{n-1} (D_p(\mathbb{C}))_j$, namely the tensor product of $n$ copies of $D_p(\mathbb{C})$ whose elements are diagonal $p^n \times p^n$ matrices of the form

$$D^{(n)} := \sum_{\boldsymbol{i}^{(n)} \in \Omega_p^{(n)}} d(\boldsymbol{i}^{(n)})\, P^{[0,n-1]}_{\boldsymbol{i}^{(n)}} . \tag{2.56}$$

[4]These commutative matrix algebras are also called *Abelian* and projections as the $P_j$ are known as *minimal projectors*.

A suggestive physical picture is as follows: each matrix algebra $D_p(\mathbb{C})$ describes a classical spin, with $p$ possible states, in an infinite classical ferromagnet. Spins located at the lattice sites $-n \leq j \leq n$ are described by tensor products of the form $\mathcal{D}_{[-n,n]} := \bigotimes_{j=-n}^{n}(D_p(\mathbb{C}))_j$. These matrix algebras can be interpreted as algebras of observables for finite portions of the infinite ferromagnet by the embedding $\mathcal{D}_{[-n,n]} \mapsto \mathbb{1}_{-n-1]} \otimes \mathcal{D}_{[-n,n]} \otimes \mathbb{1}_{[n+1}$ into $\mathcal{D}_\infty := \bigcup_{n\geq 0} \mathcal{D}_{[-n,n]}$, where $\mathbb{1}_{-n-1]}$ and $\mathbb{1}_{[n+1}$ denote the tensor products of infinitely many identity matrices $\mathbb{1} \in D_p(\mathbb{C})$ located along the two-sided chain at sites from $-\infty$ up to $-n-1$, respectively from $n+1$ up to $+\infty$.

Each $\mathcal{D}_{[-n,n]}$ can be equipped with the standard *sup-norm* of matrix algebras (see (5.3)) [5]. The sup-norm inductively extends to $\mathcal{D}_\infty$ and allows to consider the uniform closure $\mathcal{D}_\mathbb{Z} := \overline{\bigcup_{n\in\mathbb{N}} \mathcal{D}_{[-n,n]}}^{\text{uniform}}$. This procedure is known as *$C^*$-inductive limit* [64] as it involves an increasing sequence of local algebras $\mathcal{D}_{[-n,n]}$; $\mathcal{D}_\mathbb{Z}$ provides a $C^*$ algebraic description of a *classical spin chain.*

Using (2.26), the left-shift along sequences gives rise to an algebraic shift map $\Theta_\sigma : \mathcal{D}_\mathbb{Z} \mapsto \mathcal{D}_\mathbb{Z}$ such that

$$\Theta_\sigma(\mathcal{D}_{[-n,n]}) = \mathcal{D}_{[-n+1,n+1]} \; . \tag{2.57}$$

Further, the local probability measures $\mu^{(n)} := \{p(\boldsymbol{i}^{(n)})\}_{\boldsymbol{i}^{(n)}\in\Omega_p^{(n)}}$ that yield the global $T_\sigma$-invariant state $\mu$ over $\Omega_p^\mathbb{Z}$ can be associated with diagonal matrices

$$\rho_\mu^{(n)} := \sum_{\boldsymbol{i}^{(n)}\in\Omega_p^{(n)}} p(\boldsymbol{i}^{(n)})\, P_{\boldsymbol{i}^{(n)}}^{[0,n-1]} \;, \tag{2.58}$$

by means of the *trace* operation (see (5.19)) which acting on any matrix returns the sum of its diagonal entries. In fact, multiplying $\rho_\mu^{(n)}$ with matrices as in (2.56), gives another matrix in $\mathcal{D}^{(n)}$ with diagonal elements $p(\boldsymbol{i}^{(n)})\, d(\boldsymbol{i}^{(n)})$, and one gets

$$\mathrm{Tr}\left(\rho_\mu^{(n)}\, D^{(n)}\right) = \sum_{\boldsymbol{i}^{(n)}\in\Omega_p^{(n)}} p(\boldsymbol{i}^{(n)})\, d(\boldsymbol{i}^{(n)}) \;,$$

whence $p(\boldsymbol{i}^{(n)}) = \mathrm{Tr}\left(\rho_\mu^{(n)}\, P_{\boldsymbol{i}^{(n)}}^{[0,n-1]}\right)$. Therefore, conditions (2.29)- (2.30) translate into the following algebraic relations to be satisfied by $\{\rho_\mu^{(n)}\}_{n\in\mathbb{N}}$:

$$\mathrm{Tr}_n(\rho_\mu^{(n)}) = \mathrm{Tr}_1(\rho_\mu^{(n)}) = \rho_\mu^{(n-1)} \;, \tag{2.59}$$

where $\mathrm{Tr}_j$ denotes the trace with respect to $j$-th factor. These conditions allows to consistently define a global state $\omega_\mu$ on the spin chain $\mathcal{D}_\mathbb{Z}$; this state

[5] The sup-norm of a diagonal matrix $D$ is the square root of the largest diagonal element of $D^\dagger D$.

is specified by its values as a positive expectation functional over local spin arrays where it coincides with the local states $\rho_\mu^{(n)}$ (which we shall encounter in the quantum setting as *density matrices*).

**Definition 2.2.5 (*Classical Spin Chains*).**
*The $C^*$ algebraic triplet $(\mathcal{D}_\mathbb{Z}, \Theta_\sigma, \omega_\mu)$ associated with a measure-theoretic triplet $(\Omega_p, T_\sigma, \mu)$ will be referred to as a classical spin chain.*

**Remark 2.2.5.** In Section 5.3.2, it will be proved that to all classical spin chains as defined above, there correspond algebraic triplets as in Definition 2.2.4. In particular, the von Neumann algebraic triplets arise when the $C^*$ triplets $(\mathcal{D}_\mathbb{Z}, \Theta_\sigma, \omega_\mu)$ are represented on a Hilbert space and enlarged by adding to them their weak-limit points.

**Example 2.2.5.** [113] Consider a Markov chain as in Example 2.1.5. Let

$$\rho = \sum_{i=1}^{d} p(i)\, P_i = \begin{pmatrix} p(1) & 0 & \cdots & 0 \\ 0 & p(2) & \cdots & 0 \\ \vdots & \vdots & \vdots & \vdots \\ 0 & 0 & \cdots & p(d) \end{pmatrix}$$

correspond to the probability measure $\mu = \{p(i)\}_{i=1}^d$. Define on the tensor product $D_d(\mathbb{C}) \otimes D_d(\mathbb{C})$ a linear map $\mathbb{E} : D_d(\mathbb{C}) \otimes D_d(\mathbb{C}) \mapsto D_d(\mathbb{C})$ by linear extension of the following action on tensor products of minimal projectors $P_i \in D_d(\mathbb{C})$,

$$P_i \otimes P_j \mapsto \mathbb{E}[P_i \otimes P_j] := p(j|i)\, P_i \ ,$$

where $P(j|i)$ are the transition probabilities of the Markov chain. From (2.34) and (2.35) and using that $\sum_{k=1}^d P_k = 1\!\!1$,

$$\mathbb{E}[1\!\!1 \otimes 1\!\!1] = \sum_{i,j=1}^{d} \mathbb{E}[P_i \otimes P_j] = \sum_{i,j=1}^{d} p(j|i)\, P_i = \sum_{i=1}^{d} P_i = 1\!\!1$$

$$\begin{aligned} \mathrm{Tr}\Big(\rho\,\mathbb{E}[1\!\!1 \otimes P_k]\Big) &= \sum_{i=1}^{d} \mathrm{Tr}\Big(\rho\,\mathbb{E}[P_i \otimes P_k]\Big) = \sum_{i=1}^{d} p(k|i)\,\mathrm{Tr}(\rho\, P_i) \\ &= \sum_{i=1}^{d} p(k|i)\, p(i) = p(k) = \mathrm{Tr}(\rho\, P_k) \ , \end{aligned}$$

for all $k = 1, 2, \ldots, d$. Furthermore, higher order probabilities as in (2.36) are iteratively obtained as:

$$p(i_0 i_1 \cdots i_{n-1}) = \mathrm{Tr}\Big(\rho\,\mathbb{E}[P_{i_0} \otimes \mathbb{E}[P_{i_1} \otimes \cdots \mathbb{E}[P_{i_{n-1}} \otimes 1\!\!1]]]\Big) \ .$$

For instance, to evaluate $\mathrm{Tr}\Big(\rho\,\mathbb{E}[P_{i_0}\otimes\mathbb{E}[P_{i_1}\otimes \mathbb{1}]]\Big)$ use $\sum_{k=1}^d P_k = \mathbb{1}$, then

$$\begin{aligned}\sum_{k=1}^d \mathrm{Tr}\Big(\rho\,\mathbb{E}[P_{i_0}\otimes\mathbb{E}[P_{i_1}\otimes P_k]]\Big) &= \sum_{k=1}^d P(k|i_1)\mathrm{Tr}\Big(\rho\,\mathbb{E}[P_{i_0}\otimes P_{i_1}]\Big)\\ &= \mathrm{Tr}\Big(\rho\,\mathbb{E}[P_{i_0}\otimes P_{i_1}]\Big) = p(i_1|i_0)\,\mathrm{Tr}(\rho\,P_{i_0})\\ &= p(i_1|i_0)\,p(i_0) = p(i_0 i_1)\ .\end{aligned}$$

Therefore, the local density matrices $\rho^{(n)}_\mu$ are the local restrictions $\omega_\mu\lceil\mathcal{D}^{(n)}$ of a global state $\omega_\mu$ on the classical spin chain $\mathcal{D}_\mathbb{Z}$ such that, for all $D_i\in D_d(\mathbb{C})$,

$$\omega_\mu\Big(D_1\otimes D_2\otimes\cdots D_n\Big) = \mathrm{Tr}\Big(\rho\,\mathbb{E}[D_1\otimes\mathbb{E}[D_2\otimes\cdots\mathbb{E}[D_n\otimes\mathbb{1}]\,]\,]\Big)\ .$$

## 2.3 Ergodicity and Mixing

The two uncoupled harmonic oscillators of Example 2.1.2 whose orbits fill the phase-space densely (see Remarks 2.1.2.2 and 2.1.2.3) are typical *ergodic systems.* Ergodic theory developed [167] from the attempt to explain why in thermodynamic systems time-averages of typical observables coincide with their mean values $\mu(f)$ with respect to equilibrium distributions. Intuitively, if an orbit fills the energy shell densely, evaluating the time-average of a function along such an orbit should indeed amount to integrating with respect to the Liouville measure restricted to the energy shell.

**Definition 2.3.1.** *Let $f:\mathcal{X}\to\mathbb{C}$ be a complex function associated to a dynamical system $(\mathcal{X},T,\mu)$;* time-averages *are defined by*

$$\overline{f}(x) := \lim_{t\to+\infty}\frac{1}{t}\sum_{s=0}^{t-1} f(T^s x)\ ,\quad \overline{f}(x) := \lim_{t\to+\infty}\frac{1}{t}\int_0^t \mathrm{d}s\, f(T_s x)$$

*in discrete, respectively continuous time.*

**Example 2.3.1.** Consider the uncoupled oscillators of Example 2.1.2 and a continuous function $f:\mathbb{T}^2\mapsto\mathbb{R}$. By means of (2.12), the discrete and continuous time averages yield

$$\overline{f}(\boldsymbol{\theta}) = \lim_{t\to+\infty}\sum_{\boldsymbol{n}\in\mathbb{Z}^2}\widehat{f}(\boldsymbol{n})\frac{\mathrm{e}^{it\sum_{i=1}^2\omega_i n_i}-1}{t(\mathrm{e}^{i\sum_{i=1}^2\omega_i n_i}-1)}\,e_{\boldsymbol{n}}(\boldsymbol{\theta}) = \sum_{\substack{\boldsymbol{n}\in\mathbb{Z}^2\\ \sum_{i=1}^2\omega_i n_i\in 2\pi\mathbb{Z}}}\widehat{f}(\boldsymbol{n})\,e_{\boldsymbol{n}}(\boldsymbol{\theta})\ ,$$

respectively

$$\overline{f}(\boldsymbol{\theta}) = \lim_{t \to +\infty} \sum_{\boldsymbol{n} \in \mathbb{Z}^2} \widehat{f}(\boldsymbol{n}) \frac{e^{it\sum_{i=1}^2 \omega_i n_i} - 1}{it\sum_{i=1}^2 \omega_i n_i} e_{\boldsymbol{n}}(\boldsymbol{\theta}) = \sum_{\substack{\boldsymbol{n} \in \mathbb{Z}^2 \\ \sum_{i=1}^2 \omega_i n_i = 0}} \widehat{f}(\boldsymbol{n})\, e_{\boldsymbol{n}}(\boldsymbol{\theta}) .$$

Then, besides ensuring that orbits fill $\mathbb{T}^2$ densely, the conditions in Remark 2.1.2.2 and 2.1.2.3 also imply that time-averages coincide with their mean values: $\overline{f}(\boldsymbol{\theta}) = \widehat{f}(\boldsymbol{0}) = \int_{\mathbb{T}^2} \mathrm{d}\boldsymbol{\theta}\, f(\boldsymbol{\theta}) = \mu(f)$.

A considerable break-through in ergodic theory was *Birkhoff's theorem* [6].

**Theorem 2.3.1.** *Given $\left(\mathcal{X}, T, \mu\right)$, let $f \in \mathbb{L}^1_\mu(\mathcal{X})$ be a complex $\mu$-summable function on $\mathcal{X}$. Then,*

1. *the time-average $\overline{f}(x)$ exists $\mu$-a.e. on $\mathcal{X}$;*
2. *the time-average $\overline{f}$ is $T$-invariant: $\overline{f} \circ T = f$ $\mu$-a.e.;*
3. *the time-average $\overline{f} \in \mathbb{L}^1_\mu(\mathcal{X})$ and $\mu(\overline{f}) = \mu(f)$.*

The proof [91, 61] of these important results hinges on the following lemma known as *maximal ergodic theorem.*

**Lemma 2.3.1.** *Given the dynamical triplet $\left(\mathcal{X}, T, \mu\right)$, for any $f \in \mathbb{L}^1_\mu(\mathcal{X})$, set $S^f_k(x) := \frac{1}{k}\sum_{j=0}^{k-1} f(T^j x)$ and $A^f := \left\{x \in \mathcal{X} \,:\, \sup_{k\geq 0} S^f_k(x) > 0\right\}$; then, $\int_{A^f} \mathrm{d}\mu\,(x)\, f(x) \geq 0$.*

**Proof:** Set $\Phi^{(1)}_n(x) := \max\left\{0, S^f_1(x), \ldots, S^f_n(x)\right\}$ and split $\mathcal{X}$ into the subset $A^f_n := \left\{x \,:\, \Phi^{(1)}_n(x) > 0\right\}$ and its complement where $\Phi^{(1)}_n(x) = 0$. Further, $\Phi^{(2)}_n(x) := \max\left\{S^f_1(x), \ldots, S^f_n(x)\right\} = \Phi^{(1)}_n(x)$ on $A^f_n$; also,

$$\begin{aligned}
\Phi^{(2)}_{n+1}(x) &= \max\left\{f(x), f(x) + f(Tx), \ldots, f(x) + f(Tx) + \cdots f(T^n x)\right\} \\
&= f(x) + \max\left\{0, f(Tx), \ldots, f(Tx) + \cdots f(T^n x)\right\} \\
&= f(x) + \Phi^{(1)}_n(Tx) .
\end{aligned}$$

Thus, since $\Phi^{(1)}_n(x)$ is non-negative, $\mu$ is $T$-invariant and $\Phi^{(2)}_{n+1}(x) \geq \Phi^{(2)}_n(x)$,

[6] Though the results presented below can be extended to dynamical systems in continuous time, we shall concentrate on discrete time dynamical systems.

$$\begin{aligned}\int_{A_n^f} d\mu(x)\, f(x) &= \int_{A_n^f} d\mu(x)\left(\Phi_{n+1}^{(2)}(x) - \Phi_n^{(1)}(Tx)\right)\\ &\geq \int_{A_n^f} d\mu(x)\,\Phi_n^{(2)}(x) - \int_{\mathcal{X}} d\mu(x)\Phi_n^{(1)}(Tx)\\ &= \int_{A_n^f} d\mu(x)\,\Phi_n^{(1)}(x) - \int_{\mathcal{X}} d\mu(x)\,\Phi_n^{(1)}(x) = 0\,.\end{aligned}$$

Then, the result follows for, when $n \to +\infty$, the points in $A_f$ that are not in $A_n^f$ form a set of vanishingly small measure $\mu$. □

**Proof of Theorem 2.3.1** Let $a < b \in \mathbb{R}$ and, using the notations of the previous lemma, set

$$E_{ab} := \left\{x \in \mathcal{X} : \liminf_{n\to+\infty} \frac{1}{n} S_n^f(x) < a < b < \limsup_{n\to+\infty} \frac{1}{n} S_n^f(x)\right\}.$$

Let $g_{ab}^{(1)}(x) := f(x) - b$ when $x \in E_{ab}$, otherwise $g_{ab}^{(1)}(x) = 0$; consider the set

$$\left\{x \in \mathcal{X} : \sup_n \frac{1}{n} S_n^{g_{ab}^{(1)}}(x) > 0\right\} = \left\{x \in \mathcal{X} : \sup_n \frac{1}{n} S_n^f(x) > b\right\}.$$

This set not only coincides with the set $A^{g_{ab}^{(1)}}$ as defined in Lemma 2.3.1, but it also equals $E_{ab}$; while the first property is contained in the definition of the set, the second one follows from the fact that, on one hand,

$$\limsup_{n\to+\infty} \frac{1}{n} S_n^f(x) > b \Longrightarrow \sup_{n\to+\infty} \frac{1}{n} S_n^f(x) > b \Longrightarrow E_{ab} \subseteq A^{g_{ab}^{(1)}}\,.$$

On the other hand, by definition of $E_{ab}$, if $x \notin E_{ab}$ also $T^n x \notin E_{ab}$ for all $n$, whence $g_{ab}^{(1)}(T^n x) = 0$, $S_n^{g_{ab}^{(1)}}(x) = 0$ and $x \notin A^{g_{ab}^{(1)}}$. Thus, Lemma 2.3.1 yields

$$\int_{A^{g_{ab}^{(1)}}} d\mu(x)\, g_{ab}^{(1)}(x) = \int_{E_{ab}} d\mu(x)\,(f(x) - b) \geq 0\,.$$

The same argument applied to the function $g_{ab}^{(2)}(x) := a - f(x)$ if $x \in E_{ab}$, otherwise $g_{ab}^{(2)}(x) = 0$, gives $\int_{E_{ab}} d\mu(x)\,(a - f(x)) \geq 0$, whence

$$b\,\mu(E_{ab}) \leq \int_{E_{ab}} d\mu(x)\, f(x) \leq a\,\mu(E_{ab})\,.$$

Since $a < b$ this can only be possible if $\mu(E_{ab}) = 0$; therefore, the limit $\hat{f}(x) := \lim_{n\to+\infty} \frac{1}{n} S_n^f(x)$ exists $\mu$-a.e. on $\mathcal{X}$. Namely, outside the union of all $E_{ab}$ with rational $a, b$, which is still a set of zero measure, the sequence $S_n^f(x)$ converges pointwise to $\hat{f}(x)$ which can however be $\pm\infty$.

The limit function is $T$-invariant by construction; moreover, $\hat{f} \in \mathbb{L}^1_\mu(\mathcal{X})$. Indeed,

$$\int_{\mathcal{X}} \mathrm{d}\mu\,(x) \left|\frac{1}{n} S^f_n(x)\right| \leq \int_{\mathcal{X}} \mathrm{d}\mu\,(x)\, |f(x)| \ ;$$

therefore, *Fatou's lemma* [7] yields

$$\int_X \mathrm{d}\mu\,(x)\, |\hat{f}(x)| \leq \liminf_{n\to+\infty} \int_{\mathcal{X}} \mathrm{d}\mu\,(x) \left|\frac{1}{n} S^f_n(x)\right| \leq \int_{\mathcal{X}} \mathrm{d}\mu\,(x)\, |f(x)| < +\infty \ .$$

Finally, choose $\lambda \geq 0$, let $g_\lambda(x) := \left|\frac{1}{n} S^f_n(x)\right| - \lambda$, and consider the set $A^{g_\lambda}$ as in Lemma 2.3.1 ; then,

$$\int_{\mathcal{X}} \mathrm{d}\mu\,(x) \left|\frac{1}{n} S^f_n(x) - \hat{f}(x)\right| \leq \int_{\mathcal{X}\setminus A^{g_\lambda}} \mathrm{d}\mu\,(x) \left|\frac{1}{n} S^f_n(x) - \hat{f}(x)\right| + \int_{A^{g_\lambda}} \mathrm{d}\mu\,(x) \left|\frac{1}{n} S^f_n(x)\right| + \int_{A^{g_\lambda}} \mathrm{d}\mu\,(x)\, |\hat{f}(x)| \ .$$

Consider the third integral, Lemma 2.3.1 yields $\mu(A^{g_\lambda}) \leq \frac{1}{\lambda} \|\hat{f}\|_1$; as to the second one, it can be estimated as follows

$$\int_{A^{g_\lambda}} \mathrm{d}\mu\,(x) \left|\frac{1}{n} S^f_n(x)\right| \leq \frac{1}{n} \sum_{k=0}^{n-1} \int_{|f(T^k x)|>\alpha} \mathrm{d}\mu\,(x)\, |f(T^k x)| + \alpha\, \mu(A^{g_\lambda})$$
$$= \int_{|f(x)|>\alpha} \mathrm{d}\mu\,(x)\, |f(x)| + \alpha\, \mu(A^{g_\lambda}) \ ,$$

for some fixed $\alpha \geq 0$. Now, $\mu(A^{g_\lambda})$ and thus the third integral can be made arbitrarily small by choosing an appropriate $\lambda$, as well as the second one by setting $\alpha$ large enough. Further, *Lebesgue dominated convergence theorem* [8], can be applied to $\int_{\mathcal{X}\setminus A^{g_\lambda}} \mathrm{d}\mu\,(x) \left|\frac{1}{n} S^f_n(x) - \hat{f}(x)\right|$ which becomes negligibly small when $n \longrightarrow +\infty$, whence

$$\int_{\mathcal{X}} \mathrm{d}\mu\,(x)\, \hat{f}(x) = \lim_{n\to+\infty} \int_{\mathcal{X}} \mathrm{d}\mu\,(x) \frac{1}{n} S^f_n(x) = \lim_{n\to+\infty} \frac{1}{n} \sum_{k=0}^{n-1} \int_{\mathcal{X}} \mathrm{d}\mu\,(x)\, f(T^k x)$$
$$= \int_{\mathcal{X}} \mathrm{d}\mu\,(x)\, f(x) \ .$$

---

[7] Fatou's lemma [258] asserts that if $f_n$ is a sequence of measurable functions on a measure space $\mathcal{X}$, then $\int_{\mathcal{X}} \mathrm{d}\mu \liminf_{n\to+\infty} f_n \leq \liminf_{n\to+\infty} \int_{\mathcal{X}} \mathrm{d}\mu\, f_n$.

[8] Lebesgue Dominated Convergence Theorem [258] asserts that if $\{f_n\}_{n\in\mathbb{N}}$ is a sequence of measurable functions on $\mathcal{X}$ such that the limit $f(x) = \lim_{n\to+\infty} f_n(x)$ exists for all $x \in \mathcal{X}$ and $|f_n(x)| \leq g(x)$ for all $x \in \mathcal{X}$ with $g \in \mathbb{L}^1_\mu(\mathcal{X})$, then $f \in \mathbb{L}^1_\mu(\mathcal{X})$ and $\int_{\mathcal{X}} \mathrm{d}\mu\,(x)\, f(x) = \lim_{n\to+\infty} \int_{\mathcal{X}} \mathrm{d}\mu\,(x)\, f_n(x)$ .

□

In the light of Birkhoff's ergodic theorem, we first give a general measure-theoretic definition of ergodicity and then consider its physical consequences.

**Definition 2.3.2.** *A dynamical system* $(\mathcal{X}, T, \mu)$ *is* ergodic *if for all measurable subsets* $T^{-1}(B) = B$ *implies* $\mu(B) = 0$ *or* $\mu(B) = 1$.

**Remarks 2.3.1.**

1. The first conclusion to be drawn from this definition is that ergodic systems cannot possess non-trivial $T$-invariant measurable functions (constants of the motion). Indeed, if $f : \mathcal{X} \to \mathbb{R}$ is such that $f \circ T = f$ then $N_a := \{x \in \mathcal{X} \,:\, f(x) = a\} \subseteq \mathcal{X}$ is measurable; moreover, as $x \in T^{-1}(N_a)$ implies $Tx \in N_a$, then $f(x) = f(Tx) = a$. Thus, $T^{-1}(N_a) \subseteq N_a$ and ergodicity forces $N_a$ to equal either $\mathcal{X}$ or $\emptyset$ $\mu$-a.e. for all $a \in \mathbb{R}$, whence $f(x) = c_f$ $\mu$-a.e. on $\mathcal{X}$.
2. If $f \in \mathbb{L}^1_\mu(\mathcal{X})$, its time-average $\overline{f}$ is $T$-invariant by point 2 in Birkhoff's theorem. If $(\mathcal{X}, T, \mu)$ is ergodic, from the previous remark and point 3 in Birkhoff's theorem, $\overline{f}(x) = c_{\overline{f}}$ $\mu - a.e$; thus, $\mu(f) = c_{\overline{f}} = \overline{f}(x)$ $\mu - a.e.$ on $\mathcal{X}$. Namely, ergodicity implies that time-averages and phase-averages (mean-values) of (summable) observables coincide. Vice versa, dynamical systems where time-averages and phase-averages coincide are ergodic because of Proposition 2.3.1 below.
3. If the only $T$-invariant measurable functions are constant almost everywhere on $\mathcal{X}$, then $(\mathcal{X}, T, \mu)$ is ergodic: in fact, the characteristic functions of $T$-invariant measurable subsets are $T$-invariant and must then be constant almost everywhere, namely equal either to 0 or to 1 $\mu$-a.e.
4. The average time spent within $B$ by almost all phase-points of an ergodic system equals the volume of $B$. Indeed, let $\overline{\mathbf{1}}^t_B(x) := \frac{1}{t}\sum_{s=0}^{t-1} \mathbf{1}_B(T^s_T x)$; count the mean number of times $B$ is crossed by the trajectory $\{T^n x\}_{n\in\mathbb{N}}$ during a span of time of length $t$. Then, ergodicity yields
$$\overline{\mathbf{1}_B}(x) = \lim_{t\to+\infty} \overline{\mathbf{1}}^t_B(x) = \mu(B) \qquad \mu - \text{a.e.}\ . \tag{2.60}$$

**Proposition 2.3.1.** *A dynamical system* $(\mathcal{X}, T, \mu)$ *is ergodic if and only if for all measurable* $A, B$ *it holds that*

$$\lim_{t\to\infty} \frac{1}{t}\sum_{s=0}^{t-1} \mu(A \cap T^{-s}(B)) = \mu(A)\mu(B)\ . \tag{2.61}$$

**Proof:** Consider $\mathbf{1}_A(x)\overline{\mathbf{1}}^t_B(x) = \frac{1}{t}\sum_{s=0}^{t-1} \mathbf{1}_{A\cap T^{-s}(B)}(x)$, by Birkhoff's theorem and Lebesgue dominated convergence theorem (see footnote 8) it follows that

$$\mu(\mathbf{1}_A\overline{\mathbf{1}_B}) = \lim_{t\to\infty}\frac{1}{t}\sum_{s=0}^{t-1}\mu(A\cap T^{-s}(B)) \ .$$

If the system is ergodic, then (2.61) follows from (2.60). If (2.61) holds, then $A = B = T^{-1}(B) \Longrightarrow \mu(B)^2 = \mu(B)$, whence $\mu(B)$ equals either 0 or 1. □

**Definition 2.3.3 (Mixing).** *A dynamical system $(\mathcal{X}, T, \mu)$ is* mixing *if and only if for all measurable subsets $A, B \subseteq \mathcal{X}$ it holds that*

$$\lim_{t\to+\infty}\mu(A\cap T^{-t}(B)) = \mu(A)\mu(B) \ . \tag{2.62}$$

The subsets $A\cap T^{-t}(B)$ consist of those points of $A$ that visit $B$ at time $t$; thus, (2.62) asserts that relative to the volume of any measurable subset $A$, the volume of points of $A$ that will eventually be in another measurable subset $B$ equals the volume of $B$. In other words, mixing dynamical systems are in the long run characterized by the uniform spreading of their measurable subsets; on the other hand (2.61) states that ergodicity amounts to a uniform spreading on average.

From a physical point of view, quantities like $\mu(A\cap T^{-t}(B))$ are *two-point correlation functions*; thus, mixing characterizes dynamical systems whose two-point correlation functions factorize asymptotically, whereas ergodicity corresponds to two-point correlation functions factorizing in the mean.

**Remarks 2.3.2.**

1. If $\lim_{n\to+\infty} a_n = a$, $a_n \in \mathbb{R}$, then, $\lim_{n\to+\infty}\frac{1}{n}\sum_{k=0}^{n-1} a_k = a$, whence (2.62) implies (2.61) and mixing implies ergodicity. The opposite is not in general true as the time-average can get rid of those $s$ for which $\mu(A\cap T^{-s}(B)) \neq \mu(A)\mu(B)$. There is a third asymptotic behavior, intermediate between ergodicity and mixing, known as *weak mixing* [313] and related to the fact that

$$\lim_{n\to\infty}|a_n - a| = 0 \Longrightarrow \lim_{n\to\infty}\frac{1}{n}\sum_{j=0}^{n-1}|a_n - a| = 0$$
$$\Longrightarrow \lim_{n\to\infty}\frac{1}{n}\left|\sum_{j=0}^{n-1}(a_n - a)\right| = 0 \ .$$

Weak mixing amounts to the request that

$$\lim_{t\to\infty}\frac{1}{t}\sum_{s=0}^{t-1}\left|\mu(A\cap T^{-s}(B)-\mu(A)\mu(B))\right|=0\ ;\tag{2.63}$$

it is implied by mixing and implies ergodicity.

2. Given an invertible map $T$, a stronger notion of mixing is formulated as follows [91]. Given any finite collection $\mathcal{S}_r := \{S_i\}_{i=1}^r$, $S_i \in \Sigma$, of measurable subsets of $\mathcal{X}$, denote by

$$\Sigma_n^\infty(\mathcal{S}_r) := \bigvee_{k\geq n}^{\infty} T^{-k}(\mathcal{S}_r)$$

the $\sigma$-algebra generated by all possible atoms of the form $T^{-k}(S_j)$ for $k \geq n$ and $S_j \in \mathcal{S}_r$. Then, $(\mathcal{X}, T, \mu)$ is said to be *K-mixing* if

$$\lim_{n\to\infty}\ \sup_{B\in\Sigma_n^\infty(\mathcal{S}_r)}\left|\mu(S_0\ \cap\ B)\ -\ \mu(S_0)\,\mu(B)\right|=0\ ,\tag{2.64}$$

for all $S_0, \mathcal{S}_r \in \Sigma$. Observe that $x \in \Sigma_n^\infty(\mathcal{S}_r)$ implies $T^{-k}x \in S_i$ at some time $k \geq n$ for some atom $S_i \in \mathcal{S}_r$; therefore, $K$-mixing amounts to the uniform statistical independence of any given measurable subset from the trajectories of any finite family of subsets if these are considered sufficiently far away in the past.

By using the density of the algebra of simple functions $\mathfrak{S}(\mathcal{X})$ in the Hilbert space $\mathbb{L}^2_\mu(\mathcal{X})$ as in Example 2.1.1, it is convenient to reformulate (2.61) and (2.62) in terms of square-summable functions. It is thus possible to study how those properties constrain the spectrum of the Koopman operator $U_T$.

**Proposition 2.3.2.** *A dynamical system* $(\mathcal{X}, T, \mu)$ *is*

1. *ergodic if and only if for all* $\psi, \phi \in \mathbb{L}^2_\mu(\mathcal{X})$

$$\lim_{t\to\infty}\frac{1}{t}\sum_{s=0}^{t-1}\mu(\psi\,\phi\circ T^s)=\mu(\psi)\mu(\phi)\ ;\tag{2.65}$$

2. *mixing if and only if for all* $\psi, \phi \in \mathbb{L}^2_\mu(\mathcal{X})$

$$\lim_{t\to+\infty}\mu(\psi\,\phi\circ T^t)=\mu(\psi)\mu(\phi)\ .\tag{2.66}$$

According to the Koopman-von Neumann formalism of Example 2.1.1, using (2.5), it turns out that $\mu(\psi\phi\circ T^t) = \langle\,\psi^*\,|\,U_T^t\,|\phi\,\rangle$. Therefore, ergodicity and mixing can conveniently be expressed as weak-limits, that is as limits with respect to the weak topology (see Remark 2.2.3.3). Then (2.65) and (2.66) are equivalent to

$$w - \lim_{t\to\infty} \frac{1}{t}\sum_{s=0}^{t-1} U_T^s = |\,\mathbb{1}\rangle\langle\,\mathbb{1}\,| \tag{2.67}$$

$$w - \lim_{t\to\infty} U_T^t = |\,\mathbb{1}\rangle\langle\,\mathbb{1}\,|\,. \tag{2.68}$$

The constant function $|\,\mathbb{1}\rangle$ is such that $U_T|\,\mathbb{1}\rangle = |\,\mathbb{1}\rangle$; if there exists $|\,\psi\rangle \neq |\,\mathbb{1}\rangle$ such that $U_T|\,\psi\rangle = |\,\psi\rangle$, then one can orthogonally decompose $|\,\psi\rangle = \alpha|\,\mathbb{1}\rangle + \beta|\,\phi\rangle$ with $\langle\,\phi\,|\,\mathbb{1}\rangle = 0$, $\|\phi\| = 1$ and $U_T|\,\phi\,\rangle = |\,\phi\,\rangle$. Thus,

$$1 = \lim_{t\to+\infty}\langle\,\phi\,|\,U_T^t\,|\phi\,\rangle \neq |\langle\,\phi\,|\,\mathbb{1}\rangle|^2 = 0\;,$$

whence (2.68) cannot hold. If (2.68) holds, a similar argument excludes the presence of eigenvectors $|\,\psi_\lambda\,\rangle$ such that $U_T|\,\psi_\lambda\,\rangle = e^{i\lambda}|\,\psi_\lambda\,\rangle$. Therefore,

**Proposition 2.3.3.** *A dynamical system $(\mathcal{X}, T, \mu)$ is mixing only if 1 is the only eigenvalue of its Koopman operator and it is not degenerate.*

In order to see the impact of ergodicity as expressed by (2.67) on the spectrum of $U_T$, we use [313]

**Proposition 2.3.4 (von Neumann Ergodic Theorem).** *Let $U_T$ be the unitary Koopman operator acting on the Hilbert space $\mathbb{H} := \mathbb{L}^2_\mu(\mathcal{X})$ of a dynamical triplet $(\mathcal{X}, T, \mu)$, with $T$ invertible. Let $A_t : \mathbb{H} \mapsto \mathbb{H}$ be defined by $A_t|\,\psi\,\rangle := \frac{1}{t}\sum_{s=0}^{t-1} U_T^s|\,\psi\,\rangle$, $\psi \in \mathbb{H}$, and let $P$ project onto the subspace $\mathbb{K}$ of vectors such that $U_T|\,\psi\,\rangle = |\,\psi\,\rangle$. Then, $\lim_{t\to+\infty}\|(A_t - P)\psi\| = 0$; in other words, $P$ is the strong limit (see Remark 2.2.3.1) of the sequence of operators $A_t$, $P = s - \lim_{t\to+\infty} A_t$.*

**Proof:** The subspace orthogonal to $\mathbb{K}$ is $(U_T - \mathbb{1})\mathbb{H}$; thus, for any $\psi \in \mathbb{H}$,

$$|\,\psi\,\rangle = P|\,\psi\,\rangle + (\mathbb{1} - P)|\,\psi\,\rangle = P|\,\psi\,\rangle + (U_T - \mathbb{1})|\,\phi\,\rangle\;,$$

for some $\phi \in \mathbb{H}$. Since $A_t(U_T - \mathbb{1}) = \dfrac{U_T^t - \mathbb{1}}{t}$, the result follows from

$$\|(A_t - P)|\,\psi\,\rangle\| \leq \frac{\|(U_T^t - \mathbb{1})|\,\phi\,\rangle\|}{t} \leq \frac{2\|\phi\|}{t}\;.$$

□

**Corollary 2.3.1.** *A dynamical system $(\mathcal{X}, T, \mu)$ is ergodic if and only if* 1 *is a non-degenerate eigenvalue of the Koopman operator $U_T$.*

**Proof:** Since strong convergence implies weak convergence, condition (2.67) means that ergodicity is equivalent to $P = |\,\mathbb{1}\rangle\langle\,\mathbb{1}\,|$. □

**Remarks 2.3.3.**

1. By substituting $\psi, \phi \in \mathbb{L}^2_\mu(\mathcal{X})$ with $\psi - \mu(\psi)$ and $\phi - \mu(\phi)$ (they also belong to $\mathbb{L}^2_\mu(\mathcal{X})$), ergodicity, respectively mixing amount to

$$\lim_{t\to\infty} \frac{1}{t} \sum_{s=0}^{t-1} \mu(\psi\, \phi \circ T^s) = 0 \ , \qquad \lim_{t\to+\infty} \mu(\psi\, \phi \circ T^t) = 0 \ , \tag{2.69}$$

   for all $\psi, \phi \in \mathbb{L}^2_\mu(\mathcal{X})$ with $\mu(\psi) = \mu(\phi) = 0$.
2. In case $T$ is not invertible, the Koopman operator is not unitary, but just an isometry, that is $U^\dagger U = \mathbb{1}$, while $UU^\dagger \neq \mathbb{1}$. If $T$ is invertible, time averages can be extended from $-\infty$ to $+\infty$ and, because of $T$-invariance of $\mu$, it does not matter which one of $\psi$ and $\phi$ is the time-evolving function in (2.65) and (2.66).

**Examples 2.3.2.**

1. Ergodic rotations as in Example 2.1.2 are never mixing for the Koopman operator has the exponential functions $e_{\boldsymbol{n}}(\boldsymbol{\theta})$ as eigenfunctions.
2. The system in Example 2.1.3 is mixing. Given $\psi, \phi \in \mathbb{L}^2_{\mathrm{d}\boldsymbol{r}}(\mathbb{T}^2)$ with $\mu(\psi) = \mu(\phi) = 0$, let $\varepsilon > 0$ and choose $L > 0$ such that $\|\psi - \psi_\varepsilon\| \le \varepsilon$ and $\|\phi - \phi_\varepsilon\| \le \varepsilon$, where $\psi_\varepsilon = \sum_{\|\boldsymbol{m}\|\le L} \widehat{\psi}(\boldsymbol{m}) e_{\boldsymbol{m}}$ and $\phi_\varepsilon = \sum_{\|\boldsymbol{n}\|\le L} \widehat{\phi}(\boldsymbol{n}) e_{\boldsymbol{n}}$. Then, using (2.22),

$$\begin{aligned} |\mu(\psi\, \phi \circ T^p_{\mathbb{A}})| &\le \varepsilon\,(\|\phi\| + \|\phi_\varepsilon\|) + |\mu(\psi_\varepsilon\, \phi_\varepsilon \circ T^p)| \\ &\le \varepsilon\,(\|\phi\| + \|\phi_\varepsilon\|) + \sum_{\substack{\|\boldsymbol{n}\|\le L \\ \|\mathbb{A}^{-p}\boldsymbol{n}\|\le L}} |\widehat{\psi}(\boldsymbol{n})|\, |\widehat{\phi}(\mathbb{A}^{-p}\boldsymbol{n})| \ . \end{aligned}$$

   When $p \to \infty$, hyperbolicity permits $\|\boldsymbol{n}\| \le L$ and $\|\mathbb{A}^{-p}\boldsymbol{n}\| \le L$ only if $\boldsymbol{n} = 0$; since $\mu(\phi) = \widehat{\phi}(0) = 0$, the Arnold Cat Map satisfies (2.66).
3. Conditions for the Markov chain in Example 2.1.5 to be mixing can be derived by considering two-point correlation functions involving cylinders, of the form

$$\mu\Big(C^{[0,p-1]}_{i_0\cdots i_{p-1}} \bigcap T_\sigma^{-t}(C^{[0,q-1]}_{j_0\cdots j_{q-1}})\Big) = \mu\Big(C^{[0,p-1]}_{i_0\cdots i_{p-1}} \bigcap C^{[t,t+q-1]}_{j_0 j_1\cdots j_{q-1}}\Big) \ .$$

   By means of the matrix $P = [p(i|j)]$ of transition probabilities and of (2.36), one writes

$$\mu\Big(C^{[0,p-1]}_{i_0\cdots i_{p-1}}\bigcap C^{[t,t+q-1]}_{j_0\cdots j_{q-1}}\Big) = \sum_{k_p,\cdots k_{t-1}} p(i_0\cdots i_{p-1}k_p\cdots k_{t-1}j_0\cdots j_{q-1})$$

$$= \sum_{k_p,,k_{t-1}=1}^{d} \Big(\prod_{a=0}^{q-2} P_{j_{a+1}j_a}\Big) P_{j_0 k_{t-1}} \left(\sum_{k_{p+1},\dots,k_{t-2}} \prod_{b=p}^{t-2} P_{k_{b+1}k_b}\right) \times$$

$$\times P_{k_p i_{p-1}} \Big(\prod_{c=0}^{p-2} P_{i_{c+1}i_c}\Big) p(i_0)$$

$$= \sum_{k_p,k_{t-1}=1}^{d} \Big(\prod_{a=0}^{q-2} P_{j_{a+1}j_a}\Big) P_{j_0 k_{t-1}} (P^{t+p-1})_{k_{t-1}k_p} P_{k_p i_{p-1}} \underbrace{p(i_0\cdots i_{p-1})}_{\mu(C_{i_0\cdots i_{p-1}})} .$$

Using (2.34) and (2.35), it follows that (see [61, 313])

$$\lim_{t\to+\infty} \mu\Big(C^{[0,p-1]}_{i_0\cdots i_{p-1}}\bigcap C^{[t,t+q-1]}_{j_0\cdots j_{q-1}}\Big) = \mu\Big(C^{[0,p-1]}_{i_0\cdots i_{p-1}}\Big)\,\mu\Big(C^{[0,q-1]}_{j_0\cdots j_{q-1}}\Big) ,$$

is achieved if and only if $\lim_{t\to+\infty}(P^t)_{ij} = p(i)$ for all $j = 1, 2, \dots, d$; while factorization in the mean (and thus ergodicity) holds if and only if, for all $j = 1, 2, \dots, d$,

$$Q_{ij} := \lim_{t\to+\infty} \frac{1}{t}\sum_{s=0}^{t-1}(P^s)_{ij} = p(i) .$$

4. The condition for mixing in the previous example is certainly satisfied by Bernoulli dynamical systems whose matrix of transition probabilities is $P = [p(i)]^d_{i,j=1}$ (see Example 2.1.5).

### 2.3.1 *K*-Systems

Consider an invertible Bernoulli system $(\Omega_p^{\mathbb{Z}}, T_\sigma, \mu)$, where the space of $p$-adic doubly infinite sequences $\boldsymbol{i} \in \Omega_p^{\mathbb{Z}}$ is equipped with the $\sigma$-algebra generated by cylinder sets and $\mu$ is a translation invariant product measure on $\Sigma$. Let $\mathcal{C}_{\{0\}} = \{C_j^{\{0\}}\}_{j=1}^d$ be the finite partition consisting of simple cylinders as in (2.25) and consider the $\sigma$-algebras

$$\mathcal{C}_{0]} := \bigvee_{j\geq 0} T_\sigma^{-j}(\mathcal{C}_{\{0\}}) = \bigvee_{j\geq 0} \mathcal{C}_{\{j\}} \tag{2.70}$$

$$\mathcal{C}_{n]} := T_\sigma^{n}(\mathcal{C}_{0]}) = \bigvee_{j\geq -n} \mathcal{C}_{\{j\}} \tag{2.71}$$

generated by union and intersections of cylinders of the form (see (2.26))

$$C^{[p,q]}_{\boldsymbol{i}^{(q-p+1)}} = \bigcap_{j=p}^{q} T_\sigma^{-j}(C^{\{j\}}_{i_j}) \;, \quad \boldsymbol{i}^{(q-p+1)} = i_p i_{p+1} \cdots i_q \in \Omega_p^{(q-p+1)} \;,$$

for any $q \geq p \geq -n$. From Section 2.1.1 and Examples 2.3.2.3-4, one deduces that:

$$i)\ \mathcal{C}_{n]} \subset \mathcal{C}_{n+1]} \;, \quad ii)\ \bigvee_{n\geq 0} \mathcal{C}_{n]} = \Sigma \;, \quad iii)\ \bigwedge_{n\geq 0} \mathcal{C}_{-n]} = \mathcal{N} \;, \tag{2.72}$$

where $\mathcal{N}$ is the trivial $\sigma$-algebra consisting only of the empty set $\emptyset$ and of $\Omega_d^{\mathbb{Z}}$, all equalities being understood up to sets of zero measure. Condition *ii)* expresses the fact that cylinder sets $\mathcal{C}_{[p,q]}$ with $p, q \in \mathbb{Z}$ generate $\Sigma$, while in condition *iii)*,

$$\bigwedge_{n\geq 0} \mathcal{C}_{-n]} = \bigwedge_{n\geq 0} \bigvee_{j\geq n} T_\sigma^{-j}(\mathcal{C}_{\{0\}})$$

denotes the largest $\sigma$-algebra, called *tail* of $\mathcal{C}_{\{0\}}$ (Tail $(\mathcal{C}_{\{0\}})$) contained in all $\mathcal{C}_{-n]}$ with $n \geq 0$.

Cylinders in $\mathcal{C}_{-n]}$ are of the form $C^{[p,p+q]}_{\boldsymbol{i}^{(q+1)}}$ with $p \geq n$, $q \geq 0$; they become subsets of Tail $(\mathcal{C}_{\{0\}})$ when $t \to +\infty$. Then, from the mixing relation in Remark 2.3.2.3, one deduces that the characteristic functions of these atoms go into the constant functions $\mu(C^{[0,q]}_{\boldsymbol{i}^{(q+1)}})\,\mathbb{1}$, asymptotically, whence condition *iii)*. Bernoulli shifts are particular instances of *Kolmogorov (K-)systems* [91] and $\mathcal{C}_{0]}$ a particular example of *K-partition.*

**Definition 2.3.4 (Classical $K$-systems).** *A discrete-time dynamical system $(\mathcal{X}, T, \mu)$ with $\sigma$-algebra $\Sigma$ is a $K$-system if there exists a $\sigma$-subalgebra (a so-called $K$-partition) $\Sigma_0 \subset \Sigma$ that gives rise to a nested $K$-sequence of $\sigma$-subalgebras $\Sigma_t := T^t(\Sigma_0)$ such that*

1. *$\Sigma_t := T^t(\Sigma_0) \subset \Sigma_{t+1}$ for all $t \in \mathbb{Z}$;*
2. *$\bigvee_{t\in\mathbb{Z}} \Sigma_t = \Sigma$;*
3. *$\bigwedge_{t\in\mathbb{Z}} \Sigma_t = \mathcal{N}$.*

For Bernoulli shifts, the partition $\mathcal{C}_{\{0\}}$ is such that $\bigvee_{n\in\mathbb{Z}} T_\sigma^n(\mathcal{C}_{\{0\}}) = \Sigma$ and Tail $(\mathcal{C}_{\{0\}}) = \mathcal{N}$: $\mathcal{C}_{\{0\}}$ is a *generating partition* with a *trivial tail.*

**Definition 2.3.5.** *Let $(\mathcal{X}, T, \mu)$ a measure theoretic dynamical triplet with $\Sigma$ as $\sigma$-algebra.*

1. *A finite, measurable partition $\mathcal{P}$ of $\mathcal{X}$ is called generating if (apart from sets of zero measure $\mu$)*

$$\bigvee_{j=-\infty}^{+\infty} T^j(\mathcal{P}) = \Sigma \quad (T \text{ invertible}) \text{ or } \quad \bigvee_{j=0}^{+\infty} T^j(\mathcal{P}) = \Sigma \quad (\text{otherwise}) \;.$$

2. *The **tail** of a finite measurable partition $\mathcal{P}$ is defined by*

$$\mathrm{Tail}\,(\mathcal{P}) := \bigwedge_{n\geq 0}\bigvee_{k\geq n} T_\sigma^{-k}(\mathcal{P}) \tag{2.73}$$

*and will be said to be **trivial** if* $\mathrm{Tail}\,(\mathcal{P}) = \mathcal{N}$*, that is if all its subsets equal* $\emptyset$ *or* $\mathcal{X}$ *up to sets of zero measure* $\mu$*.*

**Remark 2.3.4.** A generating partition $\mathcal{P} = \{P_i\}_{i\in I}$ consists of $\Sigma$-measurable atoms $P_i$ such that unions and intersections of their images $T^n(P_i)$, in the past, $n < 0$, and in the future, $n > 0$, generate $\Sigma$. Instead, the refinements $\mathcal{P}_{-n]} := \bigvee_{k=n}^{+\infty} T^{-k}(\mathcal{P})$ are the $\sigma$-subalgebras generated by the atoms in the past of $\mathcal{P}$ up to a discrete time $t = -n$. Since $\mathcal{P}_{-n-1]} \subseteq \mathcal{P}_{-n]}$, one can also loosely write $\mathrm{Tail}\,(\mathcal{P}) = \lim_{n\to+\infty} \mathcal{P}_{-n]}$ to indicate that the tail of $\mathcal{P}$ contains all measurable subsets generated by the remote past of $\mathcal{P}$. As such, tails are $T$-invariant.

From the preceding discussion concerning Bernoulli shifts, there clearly appears a relation between the triviality of the tails of partitions and the dynamical system mixing properties.

**Proposition 2.3.5.** *A dynamical system $(\mathcal{X}, T, \mu)$ is $K$-mixing (see Remark 2.3.2.2) if and only if all its finite partitions have trivial tails.*

**Proof:** Consider a finite partition $\mathcal{P}$ and its tail. By definition, $\mathrm{Tail}\,(\mathcal{P})$ is mapped into itself by $T$; thus, if in (2.64) $S_0 \in \mathrm{Tail}\,(\mathcal{P})$, then $S_0$ belongs to $\mathcal{P}_{-n]} := \bigvee_{j\geq n} T^{-j}(\mathcal{P})$ for all $n$ and thus to the $\sigma$-algebra $\Sigma_n^{\infty}(\mathcal{P})$, generated by $\mathcal{P}_{-n]}$. Therefore, one can choose $B = S_0$ in (2.64) which then yields $\mu(S_0 \cap S_0) = \mu(S_0)^2$ and, in turn, $\mathrm{Tail}\,(\mathcal{P}) = \mathcal{N}$.

Vice versa, let us choose as $\mathcal{S}_r$ in (2.64) a finite partition $\mathcal{P}$ [9] and consider the $\sigma$-subalgebras $\mathcal{P}_{-n]} \subseteq \Sigma$ generated by the infinite refinements $\bigvee_{k\geq n} T^{-k}(\mathcal{P})$. The corresponding conditional probabilities $\mu(S|\mathcal{P}_{-n]})(x)$, $S \in \Sigma$ (see (2.49) and (2.50)) are such that, for any $A_0 \in \Sigma$ and $B \in \mathcal{P}_{-n]}$,

$$\Big|\mu(A_0 \cap B)(x) - \mu(A_0)\mu(B)\Big| \leq \int_B \mathrm{d}\mu\,(x)\,\Big|\mu(A_0|\mathcal{P}_{-n]})(x) - \mu(A_0)\Big|\;.$$

Because of Theorem 2.2.1 and Examples 2.2.4.1,3, from $\mathcal{P}_{-n]} \downarrow \mathcal{N}$ it follows that $\mu(A_0|\mathcal{P}_{-n]})(x) \to \mu(A_0)$ $\mu$-a.e. when $n \to \infty$, whence $K$-mixing follows from Lebesgue dominated convergence theorem. □

In the next chapter, by using entropic tools, we shall show that all finite partitions of $K$-systems have trivial tails and are thus $K$-mixing; at this point it suffices to observe that

[9] Starting from the finite set $\mathcal{S}_r$ of measurable subsets $\{S_i\}_{i=1}^r$, one constructs the partition of $\mathcal{X}$ consisting of $S_0' := \bigcap_{i=1}^r S_i$, $S_i' := S_i \setminus S_0'$ and $S_{r+2}' := \mathcal{X}\setminus\bigcup_{i=0}^r S_i'$.

**Proposition 2.3.6.** *If a dynamical triplet* $(\mathcal{X}, T, \mu)$ *has a generating partition* $\mathcal{P}$ *with trivial tail,* $\mathrm{Tail}\,(\mathcal{P}) = \mathcal{N}$*, then it is a* $K$*-system.*

**Proof:** The $\sigma$-algebra $\mathcal{P}_{0]} := \bigvee_{n\geq 0} T^{-n}(\mathcal{P})$ is a $K$-partition. □

**Examples 2.3.3.**

1. As for Bernoulli shifts, also for the Markov shifts in Example 2.1.5, the partition $\mathcal{P} = \{P_{0,1}\}$ consisting of simple cylinders as in (2.25) satisfies condition *i)* and *ii)* in (2.72). However, the argument used when discussing the triviality of the tail for Bernoulli shifts shows that $\mathrm{Tail}\,(\mathcal{P}) = \mathcal{N}$ if and only if the Markov shifts are mixing in which case by Proposition 2.3.6 they are $K$-systems.
2. Consider Example 2.1.2 with the frequencies $\omega_{1,2}$ such that the system is ergodic (see Remarks 2.1.2.2 and 2.1.2.3). As a partition of $\mathbb{T}^2$, choose the Cartesian product $\mathcal{C}$ of the partitions of the 1-dimensional torus $\mathbb{T}$ into atoms $C_1 = \{0 \leq \theta < \pi\}$, $C_2 = \{\pi \leq \theta < 2\pi\}$. Because of ergodicity, the trajectories of the end points of the atoms $C_i \times C_j$ fill $\mathbb{T}^2$ densely and the intersections of their images $T^k(C_i \times C_j)$ under the dynamics become finer and finer and approximate better and better the Borel $\sigma$-algebras of $\mathbb{T}^2$. Actually, this already occurs if one restricts to $T^{-j}(\mathcal{C})$ with $j \geq 0$, namely $\bigvee_{j=0}^{+\infty} T^{-j}(\mathcal{C}) = \Sigma$. This also means that $\mathrm{Tail}\,(\mathcal{C}) = \Sigma$.
3. The partition in Example 2.2.2 of the two-torus into the vertical half-rectangles gives thinner and thinner vertical rectangles while moving into the past, and thinner and thinner horizontal rectangles into the future. Their intersections are squares of increasingly small side, by means of which one can approximate better and better every Borel subset of $\mathbb{T}^2$. The tail of such a partition is trivial due to the fact that the Baker map acts as a Bernoulli shift with respect to it.

In order to set the ground for a quantum extension of the notion of $K$-system (see Section 7.1.4), we operate a reformulation of the conditions in Definition (2.3.4) in terms of algebras of functions. Given a $K$-sequence $\{\Sigma_t := T^t(\Sigma_0)\}_{t\in\mathbb{Z}}$ of $\sigma$-subalgebras, consider the Abelian von Neumann subalgebras $\mathcal{M}_t := \mathbb{L}^\infty_\mu(\mathcal{X}, \Sigma_t) = \Theta^t_T[\mathcal{M}_0]$ consisting of the essentially bounded $\Sigma_t$-measurable functions on $\mathcal{X}$ (see Section 2.2.2). Then, one has

$$\mathcal{M}_t \subset \mathcal{M}_{t+1}\ ,\ \bigvee_{t\in\mathbb{Z}} \mathcal{M}_t = \mathcal{M}\ ,\ \bigwedge \mathcal{M}_n = \{\lambda \mathbb{1}\}\ ,$$

where the generation of $\mathcal{M}$ by $\bigvee$ is by strong-operator closure on the Hilbert space $\mathbb{L}^2_\mu(\mathcal{X})$, while $\bigwedge$ denotes set-theoretic intersection.

We shall see in Section 5.3.2 that unital Abelian von Neumann algebras $\mathcal{M}$ can always be identified with suitable $\mathbb{L}^\infty_\mu(\mathcal{X})$ and represented as multiplication operators on the Hilbert space $\mathbb{L}^2_\mu(\mathcal{X})$. It thus makes sense to provide an algebraic reformulation of Definition 2.3.4 (see Definition 2.2.4).

**Definition 2.3.6 (Classical Algebraic $K$-Systems).**
*A classical von Neumann algebraic triplet $(\mathcal{M}, \Theta_T, \omega)$ is an* algebraic $K$-system, *if there exists a von Neumann subalgebra $\mathcal{N}_0 \subseteq \mathcal{M}$ such that, setting $\mathcal{N}_t := \Theta_T^t(\mathcal{N}_0)$, $t \in \mathbb{Z}$,*

1. $\mathcal{N}_t \subset \mathcal{N}_{t+1}$ *for all* $t \in \mathbb{Z}$;
2. $\bigvee_{t\in\mathbb{Z}} \mathcal{N}_t = \mathcal{M}$;
3. $\bigwedge_{n\in\mathbb{Z}} \mathcal{N}_t = \{\lambda \mathbb{1}\}$.

*Any such sequence $\{\mathcal{N}_t\}_{t\in\mathbb{Z}}$ of von Neumann subalgebras of $\mathcal{M}$ will be called a* classical $K$-sequence.

**Remark 2.3.5.** The above definition can also be formulated in an Abelian $C^*$ algebraic context; there, $\mathcal{M}$ will be a $C^*$ algebra as well as the subalgebras of the $K$-sequence and $\bigvee_{t\in\mathbb{Z}} \mathcal{N}_t$ will denote the algebra generated by norm closure. The classical spin chains discussed in Section 2.2.3 are instances of classical $C^*$ $K$-systems: with $\mathcal{M} = \mathcal{D}_\mathbb{Z}$, $\mathcal{N}_0$ will be the left half-spin chain $\mathcal{D}_{0]}$ generated by the diagonal matrix algebras $\mathcal{D}_{[p,q]}$ with $p \leq q \leq 0$. Then, the algebraic $K$-sequence will consist of the subalgebras $\mathcal{D}_{t]} = \Theta_\sigma(\mathcal{D}_{0]})$ generated by the diagonal matrix algebras $\mathcal{D}_{[p,q]}$ with $p \leq q \leq t$.

Given a measure-theoretic $K$-system with a $K$-sequence of $\sigma$-algebras $\{\Sigma_n\}_{n\in\mathbb{Z}}$, instead of considering the von Neumann algebras $\mathcal{M}_n$, one may focus upon the Hilbert spaces $\mathbb{H}_n := \mathbb{L}^2_\mu(\mathcal{X}, \Sigma_n)$ of square-summable $\Sigma_n$-measurable functions on $\mathcal{X}$. From the conditions $i)$, $ii)$ and $iii)$ in Definition 2.3.4, it follows that

1. $\mathbb{H}_t \subset \mathbb{H}_{t+1}$ for all $t \in \mathbb{Z}$;
2. $\bigcup_{t\in\mathbb{Z}} \mathbb{H}_t = \mathbb{H}$;
3. $\bigcap_{t\in\mathbb{Z}} \mathbb{H}_t = \mathbb{C}\,\mathbb{1}$,

where $\mathbb{H} := \mathbb{L}^2_\mu(\mathcal{X})$ and $\mathbb{C}\,\mathbb{1}$ stands for the Hilbert space consisting of constant functions on $\mathcal{X}$ ($\mu$-a.e.). Since, according to the construction of the unitary Koopman operator in (2.4), $\mathbf{1}_{T(S_0)}(x) = \mathbf{1}_{S_0}(T^{-1}x) = (U_T^{-1}\mathbf{1}_{S_0})(x)$, it follows that $\mathbb{H}_t = U_T^{-t}\mathbb{H}_0$; whence, setting $\mathbb{K}_t := \mathbb{H}_{t+1} \ominus \mathbb{H}_n$,

$$t \neq s \Longrightarrow \mathbb{K}_t \perp \mathbb{H}_s \ , \quad \mathbb{H} = \bigoplus_{t\in\mathbb{Z}} \mathbb{K}_t \ . \tag{2.74}$$

By choosing an orthonormal basis $\{|\, f_j \,\rangle\}_{j\in J}$ in $\mathbb{K}_0$, one gets an orthonormal basis for $\mathbb{H}_t$ of the form $\{|\, e_{j,t} \,\rangle := U_T^{-t} |\, f_j \,\rangle\}_{j\in J}$ and thus one for $\mathbb{H}$ of the form $\{|\, e_{j,t} \,\rangle\}_{j\in J, t\in\mathbb{Z}}$. Any unitary operator $U$ on a separable Hilbert space $\mathbb{H}$ which generates an orthonormal basis of the previous form is said to have a *Lebesgue spectrum of multiplicity $J$.*

**Proposition 2.3.7.** *[91] For a $K$-system $(\mathcal{X}, T, \mu)$, $J$ is countably infinite.*

**Proof:** Since $\mathbb{H}_0 \subset \mathbb{H}_1$ there surely exists $f \in \mathbb{H}_1$ with $g := f - E(f|\Sigma_0)$, where $E(f|\Sigma_0)$ is the conditional expectation of $f$ with respect to $\Sigma_0$, such that $E(|g|^2|\Sigma_0) \neq 0$ on a $\Sigma_0$-measurable subset $S_0$ with $\mu(S_0) > 0$. Consider the function $G \in \mathbb{H}_1$ defined by $G(x) := \dfrac{g(x)}{\sqrt{E(|g|^2|\Sigma_0)(x)}} \mathbf{1}_{S_0}(x)$; from the properties of the conditional expectation it follows that

$$\begin{aligned} E(G|\Sigma_0)(x) &= \frac{E(g|\Sigma_0)(x)}{\sqrt{E(|g|^2|\Sigma_0)(x)}} \mathbf{1}_{S_0}(x) = 0 \quad (*) \\ E(|G|^2|\Sigma_0)(x) &= \frac{E(|g|^2|\Sigma_0)(x)}{E(|g|^2|\Sigma_0)(x)} \mathbf{1}_{S_0}(x) = \mathbf{1}_{S_0}(x) \quad (**) \ . \end{aligned}$$

Let $\{| e^0_k \rangle\}_{k\in\mathbb{N}}$ be an orthonormal basis in the Hilbert space $\mathbb{L}^2_\mu(S_0)$ of square summable functions supported within $S_0$ and set $| f^0_k \rangle := M_G | e^0_k \rangle$ where $M_G$ denotes the multiplication by $G$, namely $f^0_k(x) = G(x)\, e^0_k(x)$. Notice that $| f^0_k \rangle \in \mathbb{H}_1$; further, by using (2.51) and (2.53), it follows that $| f^0_k \rangle \perp \mathbb{H}_0$, whence $| f^0_k \rangle \in \mathbb{K}_0 = \mathbb{H}_1 \ominus \mathbb{H}_0$ as defined in (2.74). Indeed, let $| h_0 \rangle \in \mathbb{H}_0$, then $(*)$ yields

$$\begin{aligned} \langle h_0 | f^0_k \rangle &= \int_{S_0} \mathrm{d}\mu\,(x)\, h^*_0(x)\, G(x)\, e^0_k(x) = \int_{S_0} \mathrm{d}\mu\,(x)\, E(h^*_0\, G\, e^0_k|\Sigma_0)(x) \\ &= \int_{S_0} \mathrm{d}\mu\,(x)\, h^*_0(x)\, E(G|\Sigma_0)(x)\, e^0_k(x) = 0 \ . \end{aligned}$$

Also, the set $\{| f^0_k \rangle\}_{n\in\mathbb{N}}$ is an orthonormal basis for

$$\begin{aligned} \langle f^0_j | f^0_k \rangle &= \int_{S_0} \mathrm{d}\mu\,(x)\, (e^0_j)^*(x)\, |G(x)|^2\, e^0_k(x) \\ &= \int_{S_0} \mathrm{d}\mu\,(x)\, E\Big((e^0_j)^*\, |G|^2\, e^0_k|\Sigma_0\Big)(x) \\ &= \int_{S_0} \mathrm{d}\mu\,(x)\, (e^0_j)^*(x)\, E(|G|^2|\Sigma_0)(x)\, e^0_k(x) = \langle e^0_j | e^0_k \rangle = \delta_{jk} \ . \end{aligned}$$

Therefore, $\mathbb{K}_1$ must be an infinite dimensional separable Hilbert space. □

### 2.3.2 Ergodicity and Convexity

We conclude this section by considering some aspects of ergodicity and mixing in relation to continuous dynamics on compact, metric spaces and to the convex space $M(\mathcal{X}, T)$ of their regular, $T$-invariant Borel measures. The first result [313] states that ergodic measures are *extremal* in $M(\mathcal{X}, T)$, namely they cannot be decomposed into convex combinations of other measures in $M(\mathcal{X}, T)$. We shall make use of the algebraic setting of Definition 2.2.4.

**Proposition 2.3.8.** *An algebraic triplet $(C(\mathcal{X}),\Theta_T,\omega_\mu)$ is ergodic if and only if $M(\mathcal{X},T)\ni\omega_\mu=\lambda\omega_1+(1-\lambda)\omega_2$, $0<\lambda<1$, $\omega_{1,2}\in M(\mathcal{X},T)$ implies $\omega_\mu=\omega_{1,2}$.*

**Proof:** Suppose a $T$-invariant Borel measurable subset $E$ exists such that $0<\mu(E)<1$ and let $E^c:=\mathcal{X}\setminus E$. With $1\!\!1_E$ and $1\!\!1_{E^c}$ their characteristic functions and $\omega_\mu(1\!\!1_E)=\mu(E)$, the two states

$$C(\mathcal{X})\ni f\mapsto\omega_1(f)=\frac{\omega_\mu(1\!\!1_E f)}{\mu(E)}\ ,\quad C(\mathcal{X})\ni f\mapsto\omega_2(f)=\frac{\omega_\mu(1\!\!1_{E^c}f)}{1-\mu(E)}$$

are different and both in $M(\mathcal{X},T)$; furthermore, they decompose $\omega_\mu$. for $\omega_\mu=\mu(E)\,\omega_1+(1-\mu(E))\,\omega_2$.

Suppose $\omega_\mu$ can be decomposed as stated in the proposition, then the measure $\mu_1\in M(\mathcal{X},T)$ corresponding to $\omega_1$ is absolutely continuous with respect to $\mu$ (see Remark 2.2.4.2). Let $f_1(x)\geq 0$ be its Radon-Nikodym derivative. Consider the measurable subset $E=\{x\in\mathcal{X}:\,f(x)<1\}$. Observe that one can decompose $E=(E\cap T^{-1}(E))\cup(E\setminus T^{-1}(E))$ by means of disjoint subsets and, analogously, $T^{-1}(E)=(T^{-1}(E)\cap E)\cup(T^{-1}(E)\setminus E)$. As $\mu\circ T^{-1}=\mu$,

$$\begin{aligned}\omega_1(1\!\!1_E)=&\int_{E\cap T^{-1}(E)}\mathrm{d}\mu(x)\ f_1(x)+\int_{E\setminus T^{-1}(E)}\mathrm{d}\mu(x)\ f_1(x)=\omega_1(1\!\!1_{T^{-1}(E)})\\ -&\int_{T^{-1}(E)\cap E}\mathrm{d}\mu(x)\ f_1(x)+\int_{T^{-1}(E)\setminus E}\mathrm{d}\mu(x)\ f_1(x)\ .\end{aligned}$$

Therefore, as $f_1<1$ on $E$ while $f_1\geq 1$ outside it, it follows that

$$\begin{aligned}\mu(E\setminus T^{-1}(E))>&\int_{E\setminus T^{-1}(E)}\mathrm{d}\mu(x)\ f_1(x)=\int_{T^{-1}(E)\setminus E}\mathrm{d}\mu(x)\ f_1(x)\\ \geq&\ \mu(T^{-1}(E)\setminus E)\ .\end{aligned}$$

Then, $\mu(E\setminus T^{-1}(E))=\mu(T^{-1}(E)\setminus E)=0$ since

$$\begin{aligned}\mu(E)&=\mu(E\cap T^{-1}(E))+\mu(E\setminus T^{-1}(E))\\ &=\mu(T^{-1}(E))=\mu(T^{-1}(E)\cap E)+\mu(T^{-1}(E)\setminus E)\ .\end{aligned}$$

Thus, $E$ is $T$-invariant apart from sets of 0 measure $\mu$; if the system is ergodic, this implies either $\mu(E)=0$ or $\mu(E)=1$. The latter equality cannot hold, otherwise $1=\omega_1(1\!\!1)=\omega_1(1\!\!1_E)<\mu(E)=1$; thus, $\mu(E)=0$. The same argument applied to $F:=\{x\in\mathcal{X}:\,f_1(x)>1\}$, leads to $\mu(F)=0$ whence to $f_1(x)=1$ $\mu$-a.e. on $\mathcal{X}$ which implies $\omega_\mu=\omega_1$ and thus extremality. □

The second result is a refinement [313] of Proposition 2.3.2.

**Proposition 2.3.9.** *The triplet* $(C(\mathcal{X}), \Theta_T, \omega_\mu)$ *is ergodic, respectively mixing if and only if, for all* $f \in C(\mathcal{X})$ *and* $g \in \mathbb{L}^1_\mu(\mathcal{X})$,

$$\lim_{t\to\infty} \frac{1}{t} \sum_{s=0}^{t-1} \omega_\mu(f \circ \Theta_T^s\, g) = \omega_\mu(f)\omega_\mu(g) \tag{2.75}$$

$$\lim_{t\to\infty} \omega_\mu(f \circ \Theta_T^t\, g) = \omega_\mu(f)\omega_\mu(g) \; . \tag{2.76}$$

**Proof:** If (2.75) holds, it implies (2.65) by the fact that any $\psi \in \mathbb{L}^2_\mu(\mathcal{X})$ is also summable and can be approximated in $\mathbb{L}^2_\mu(\mathcal{X})$ by continuous functions. Vice versa (2.65) implies (2.75) as any $f \in C(\mathcal{X})$ also belongs to $\mathbb{L}^2_\mu(\mathcal{X})$ and any $f \in \mathbb{L}^1_\mu(\mathcal{X})$ can be approximated in $\mathbb{L}^1_\mu(\mathcal{X})$ by square-summable functions. The same considerations can be used to prove that (2.76) is equivalent to (2.66). □

**Example 2.3.4.** Given the algebraic triplet $(C(\mathcal{X}), \Theta_T, \omega_\mu)$, let $\nu$ be another state on $C(\mathcal{X})$ absolutely continuous with respect to $\omega_\mu$, but not $\Theta_T$-invariant, that is, for all $f \in C(\mathcal{X})$,

$$\nu(f) = \int_{\mathcal{X}} \mathrm{d}\mu\,(x)\, g_\nu(x)\, f(x) \; , \quad \nu(\mathbb{1}) = \int_{\mathcal{X}} \mathrm{d}\mu\,(x)\, g_\nu(x) = \omega_\mu(g_\nu) = 1 \; ,$$

with Radon-Nikodym derivative $g_\nu \neq g_\nu \circ \Theta_T \in \mathbb{L}^1_\mu(\mathcal{X})$. From a physical point of view, $\omega_\nu$ can be considered as a perturbation of the equilibrium state $\omega_\mu$. By duality (see (2.8)), for all $f \in C(\mathcal{X})$, $\nu_t(f) = \nu(f \circ \Theta_T^t)$ where $\nu_t := \nu \circ \Theta_T^{-t}$. If $(C(\mathcal{X}), \Theta_T, \omega_\mu)$ is mixing, then (2.76) implies

$$\lim_{t\to\infty} \nu_t(f) = \omega_\mu(f)\, \omega_\mu(g_\nu) = \omega_\mu(f) \; , \quad \forall f \in C(\mathcal{X}) \; .$$

Physical instances of measures that are absolutely continuous with respect to an invariant one are local perturbations of equilibrium states; then, being mixing guarantees that these perturbations fade away in time and provides a mathematical explanation of relaxation to equilibrium.

## 2.4 Information and Entropy

At its simplest, information theory is concerned with the description of two parties transmitting information to each other. Information is physical as it is encoded into physical carriers, *e.g.* electromagnetic waves, that undergo physical processes, *e.g.* interactions with an optical fiber. As long as the laws that describe these processes are those of classical physics, one talks of classical information theory.

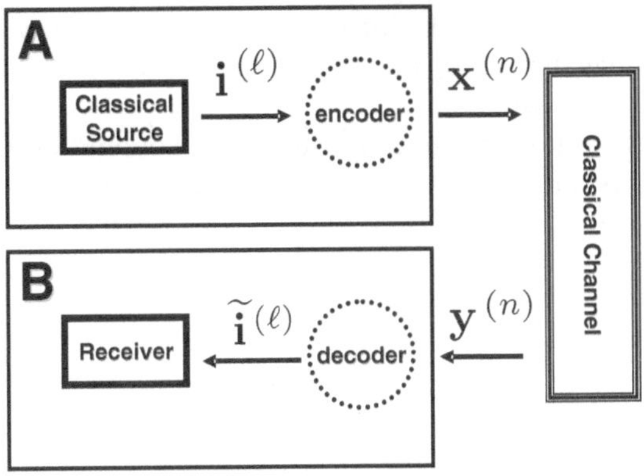


**Fig. 2.2.** Classical Transmission Channel

### 2.4.1 Transmission Channels

In the following, we shall consider two parties $A$ and $B$ exchanging signals according to the following typical scheme (see Figure 2.2):

1. At each use, a classical source emits symbols $i$ from an alphabet consisting, say, of integers $I_A = \{1, 2, \ldots, a\}$: symbols are emitted with probabilities $\pi_A = \{p_A(i)\}_{i=1}^a$.
2. After $\ell$ successive uses of the source, the source outputs are strings of length $\ell$, $\boldsymbol{i}^{(\ell)} := i_1 i_2 \cdots i_\ell \in \Omega_a^{(\ell)}$, emitted with probabilities $p_{A^{(\ell)}}(\boldsymbol{i}^{(\ell)})$. These strings can be interpreted as outcomes of a *random variable* $A^{(\ell)} := \bigvee_{i=1}^{\ell} A_i$ which is the *join* of $\ell$ successive random variables from the stochastic process $\{A_i\}_{i\in\mathbb{N}}$ ($A_1 := A$) associated with countably successive uses of the source. The random variable $A^{(\ell)}$ is distributed according to the probabilities $\pi_{A^{(\ell)}} = \left\{p_{A^{(\ell)}}(\boldsymbol{i}^{(\ell)})\right\}_{\boldsymbol{i}^{(\ell)}\in\Omega_a^{(\ell)}}$ that $\ell$ subsequent uses have actually emitted a given string of symbols.
3. The sender encodes the emitted strings $\boldsymbol{i}^{(\ell)}$ into strings of fixed length $n$, $\boldsymbol{x}^{(n)} = x_1 x_2 \cdots x_n \in \Omega_d^{(n)}$, consisting of symbols $x_i$ from another alphabet $I_X := \{1, 2, \ldots d\}$. The *encoding procedure* amounts to a map $\mathcal{E}^{(n)} : \Omega_a^{(\ell)} \mapsto \Omega_d^{(n)}$,
$$\mathcal{E}^{(n)} : \Omega_a^{(\ell)} \ni \boldsymbol{i}^{(\ell)} \mapsto \mathcal{E}^{(n)}(\boldsymbol{i}^{(\ell)}) = \boldsymbol{x}^{(n)} \in \Omega_d^{(n)} . \tag{2.77}$$
4. The code-words $\boldsymbol{x}^{(n)}$ are then sent to a receiver via a *transmission channel*, $\mathcal{C}^{(n)} = \mathcal{C} \times \mathcal{C} \times \cdots \mathcal{C}$, which transforms an input string $\boldsymbol{x}^{(n)}$ into an output string $\boldsymbol{y}^{(n)} = \mathcal{C}^{(n)}(\boldsymbol{x}^{(n)}) = y_1 y_2 \cdots y_n \in \Omega_\kappa^{(n)}$, consisting of symbols $y_i$ from, possibly, another alphabet $I_Y := \{1, 2, \ldots, \kappa\}$, according to a set of transition probabilities (compare Example 2.1.5)
$$p(\boldsymbol{y}^{(n)}|\boldsymbol{x}^{(n)}) \geq 0 , \qquad \sum_{\boldsymbol{y}^{(n)}\in\Omega_\kappa^{(n)}} p(\boldsymbol{y}^{(n)}|\boldsymbol{x}^{(n)}) = 1 . \tag{2.78}$$

These latter quantities take into account the possibility that the transmission channel be noisy and thus might randomly associate different outputs $\boldsymbol{y}^{(n)}$ to a same input $\boldsymbol{x}^{(n)}$.

5. The channel inputs and outputs are thus random variables $X^{(n)}$ and $Y^{(n)}$ with outcomes $\boldsymbol{x}^{(n)}$ and $\boldsymbol{y}^{(n)}$. If the code-words $\boldsymbol{x}^{(n)}$ occur with input probabilities $\pi_{X^{(n)}} = \{p_{X^{(n)}}(\boldsymbol{x}^{(n)})\}_{\boldsymbol{x}^{(n)} \in \Omega_d^{(n)}}$, the transition probabilities provide *joint probability distributions* for the joint random variables $X^{(n)} \vee Y^{(n)}$ given by

$$\pi_{X^{(n)} \vee Y^{(n)}} = \left\{ p_{X^{(n)} \vee Y^{(n)}}(\boldsymbol{x}^{(n)}, \boldsymbol{y}^{(n)}) \,:\, (\boldsymbol{x}^{(n)}, \boldsymbol{y}^{(n)}) \in \Omega_d^{(n)} \times \Omega_\kappa^{(n)} \right\}$$

$$p_{X^{(n)} \vee Y^{(n)}}(\boldsymbol{x}^{(n)}, \boldsymbol{y}^{(n)}) = p_{X^{(n)}}(\boldsymbol{x}^{(n)})\, p(\boldsymbol{y}^{(n)} | \boldsymbol{x}^{(n)}) \,. \tag{2.79}$$

Consequently the output random variables $Y^{(n)}$ are distributed according to the *marginal probability distributions* $\pi_{Y^{(n)}} = \{p_{Y^{(n)}}(\boldsymbol{y}^{(n)}\}_{\boldsymbol{y}^{(n)} \in \Omega_\kappa^{(n)}}$ where

$$p_{Y^{(n)}}(\boldsymbol{y}^{(n)}) := \sum_{\boldsymbol{x}^{(n)} \in \Omega_d^{(n)}} p_{X^{(n)} \vee Y^{(n)}}(\boldsymbol{x}^{(n)}, \boldsymbol{y}^{(n)}) \,. \tag{2.80}$$

6. At the receiving end of the transmission channel, the output string $\boldsymbol{y}^{(n)}$ goes through a *decoding procedure* whose aim is to retrieve the actual source output $\boldsymbol{i}^{(\ell)}$ that has been encoded into $\boldsymbol{x}^{(n)} = \mathcal{E}^{(n)}(\boldsymbol{i}^{(\ell)})$ from the received string $\boldsymbol{y}^{(n)} = \mathcal{C}^{(n)}(\boldsymbol{x}^{(n)})$. Decoding amounts to a map

$$\Omega_\kappa^{(n)} \ni \boldsymbol{y}^{(n)} \mapsto \mathcal{D}(\boldsymbol{y}^{(n)}) =: \tilde{\boldsymbol{i}}^{(\ell)} \in \Omega_a^{(\ell)} \,. \tag{2.81}$$

The whole procedure comprises the following steps

$$\begin{array}{ccccccc} \boldsymbol{i}^{(\ell)} & \overset{\mathcal{E}^{(n)}}{\longrightarrow} & \boldsymbol{x}^{(n)} & \overset{\mathcal{C}^{(n)}}{\longrightarrow} & \boldsymbol{y}^{(n)} & \overset{\mathcal{D}^{(n)}}{\longrightarrow} & \tilde{\boldsymbol{i}}^{(\ell)} \\ \in \Omega_a^{(\ell)} & & \in \Omega_d^{(n)} & & \in \Omega_\kappa^{(n)} & & \in \Omega_a^{(\ell)} \end{array}$$

The efficiency of the transmission is related to how much the decoded word $\tilde{\boldsymbol{i}}^{(\ell)}$ differs from the word $\boldsymbol{i}^{(\ell)}$ that, after being encoded into $\boldsymbol{x}^{(n)}$, has been sent through the noisy channel and received as $\boldsymbol{y}^{(n)}$. The task is thus to minimize decoding errors while keeping a non-vanishing number of bits transmitted per use of the channel.

In Section 3.2.2, we shall consider the class of *memoryless channels without feedback*; they act on input symbols in a way which is statistically independent from previous inputs and outputs. As such, they are completely specified by factorized transition probabilities:

$$p(\boldsymbol{y}^{(n)} | \boldsymbol{x}^{(n)}) = \prod_{i=1}^{n} p(y_i | x_i) \,. \tag{2.82}$$

**Examples 2.4.1 (Channels).**

1. **Noiseless Binary Channel**: two classical bits (*bits*) 0, 1 are emitted with probabilities $p_A(0)$, $p_A(1)$ and sent through a noiseless channel:
$$p(0|0) = p(1|1) = 1 \ , \quad p(0|1) = p(1|0) = 0 \ .$$
2. **Binary Symmetric Channel**: two *bits* are emitted with probabilities $p_A(0)$, $p_A(1)$ and sent through a channel which flips them according to
$$p(0|0) = p(1|1) = 1 - p > 0 \ , \quad p(1|0) = p(1|1) = p > 0 \ .$$
3. **Binary Erasure Channel**: two *bits* are emitted with probabilities $p_A(0)$, $p_A(1)$ and sent through a channel which does not flip them, but may erase anyone of them with a same probability $0 < \alpha < 1$. This action is described by a map $\mathcal{C}$ from the two-letter alphabet $\{0, 1\}$ onto the three-symbol alphabet $\{0, 1, 2\}$, where 2 stays for a junk symbol, and by transition probabilities
$$p(0|0) = p(1|1) = 1 - \alpha \ , \quad p(2|0) = p(2|1) = \alpha \ .$$

A noiseless channel is characterized by transition probabilities that equal 1 in correspondence to specific pairs of input and output strings otherwise they vanish. There are then no distortions in transmitting or storing information by means of these channels. In such cases, the question is whether the source information can be compressed and retrieved with negligible probability of error, the possibility of compression depending upon the presence of redundancies and regularities in the source.

More precisely, if the source emits binary strings of length $n$, one asks 1) whether for each *bit* from the source one can store $h < 1$ *bits* , still being able to reliably reconstruct the information emitted by the source from the $2^{h \times n}$ *bit* strings effectively retained and 2) which is the optimal compression rate $h$ achievable. This problem is addressed by Shannon's first theorem which asserts that, for stationary sources, the optimal rate is the their *entropy-rate*.

In the presence of noise in the transmission channel, the strategy is somehow the reverse with respect to noiseless transmission; in order to reduce the possibility of noise-induced errors, one introduces redundancies by multiple uses of the channel. The aim is to optimize the number of signals that can faithfully be transmitted by $n$ uses of the channel. Shannon's second theorem proves that this number can be made increase exponentially with $n$ at an optimal rate $R$, the *channel capacity*.

### 2.4.2 Stationary Information Sources

In most informational contexts, *stationary* sources are a reasonable description of the actual physical processes taking place. Stationarity means that the probability of a string $\boldsymbol{i}^{(\ell)}$ does not depend on when the source had emitted it, but only on the letters emitted. This is equivalent to (compare (2.30))

$$\sum_{i_1} p_{A^{(\ell)}}(i_1 i_2 \cdots i_\ell) = p_{A^{(\ell-1)}}(i_2 i_3 \cdots i_\ell) \ .$$

This condition goes together with the fact that the probability of a string of length $\ell - 1$ must be the sum of the probabilities of all words of length $\ell$ with the same first $\ell - 1$ symbols (compare (2.29)),

$$\sum_{i_\ell} p_{A^{(\ell)}}(i_1 i_2 \cdots i_\ell) = p_{A^{(\ell-1)}}(i_1 i_2 \cdots i_{\ell-1}) \ .$$

The similarities with shift dynamical systems now are apparent.

**Lemma 2.4.1.** *A classical, stationary information source corresponds to a stationary stochastic process* $\{A_i\}_{in\in\mathbb{N}}$*, where the random variables* $A_i$ *take on values in an alphabet* $I_A = \{1, 2, \ldots, a\}$ *and the joint random variables* $A^{(n)} := \bigvee_{i=1}^n A_i$ *are distributed with probability distributions* $\pi_{A^{(n)}}$ *satisfying appropriate compatibility and stationarity conditions.*

*Equivalently, a stationary classical source can be described by the measure-theoretic triplet* $(\Omega_a, T_\sigma, \mu_A)$*, where* $\mu_A = \mu_A \circ T_\sigma^{-1}$ *is a state on the set* $\Omega_a$ *of semi-infinite strings* $\boldsymbol{i} = \{i_j\}_{j\in\mathbb{N}}$*,* $i_j \in I_A$*, equipped with the left shift* $T_\sigma$*. The restrictions* $\mu_A^{(n)}$ *of* $\mu_A$ *to the sets of finite strings* $\Omega_a^{(n)}$ *are given by the probability distributions* $\pi_{A^{(n)}}$*.*

*Finally, a stationary source can be described as a* $C^*$ *triplet* $(\mathcal{D}_A, \Theta_\sigma, \Psi_A)$ *as in Definition 2.2.5, namely by a semi-infinite classical spin-chain consisting of a lattice of a-valued spins described, locally, by tensor products* $\mathcal{D}^{(n)} = \bigotimes_{j=0}^{n-1} D_a(\mathbb{C})$ *of diagonal* $a \times a$ *matrix algebras* $D_a(\mathbb{C})$*, equipped with an automorphism* $\Theta_\sigma$ *which amounts to the left shift along the chain and with a* $\Theta_\sigma$*-invariant state* $\Psi_A$ *such that (compare (2.56))*

$$\Psi_A \upharpoonright \mathcal{D}^{(n)} = \sum_{\boldsymbol{i}^{(n)} \in \Omega_a^{(n)}} p_{A^{(n)}}(\boldsymbol{i}^{(n)}) \, P_{\boldsymbol{i}^{(n)}}^{[0,n-1]} \ .$$

**Examples 2.4.2.**

1. **Bernoulli Sources** (see Example 2.1.4): the probabilities of strings are products of the probabilities of their symbols, $p_{A^{(n)}}(\boldsymbol{i}^{(n)}) = \prod_{j=1}^n p_A(i_j)$, that are statistically independent from each other.

2. **Markov Sources** (see Example 2.1.5): the probability of emission of the $n$-th symbol depends only on the $n-1$-th one, namely

$$\begin{aligned} p_{A^{(n)}}(\boldsymbol{i}^{(n)}) &= p(i_n|i_1, i_2, \cdots i_{n-1})\, p_{A^{(n-1)}}(i_1 i_2 \cdots i_{n-1}) \\ &= p(i_n|i_{n-1})\, p_{A^{(n-1)}}(i_1 i_2 \cdots i_{n-1}) \\ &= p(i_n|i_{n-1})\, p(i_{n-1}|i_{n-2}) \cdots p_A(i_2|i_1) p_A(i_1) \ , \end{aligned}$$

   where $p(i_n|i_1, i_2, \cdots i_{n-1})$ are the conditional probabilities for the occurrence of the $i_n$-th symbol if the symbols $i_1, i_2, \ldots, i_{n-1}$ have already occurred. Using (2.30), it follows that stationarity is equivalent to the probability vector $|\,\pi_A\,\rangle = \{p_A(i)\}_{i \in I_A}$ being eigenvector, relative to the eigenvalue 1, of the matrix of transition probabilities.

### 2.4.3 Shannon Entropy

Like an information source $A$ that emits symbols $j \in \{1, 2, \ldots, a\}$ with probabilities $p_A(j)$, also a partition $\mathcal{P} = \{P_i\}_{i=1}^p$ of the phase-space of a dynamical system $(\mathcal{X}, T, \mu)$ into atoms with volumes $\mu(P_i)$, can be interpreted as a classical random variable. In the latter case, randomness is related to the fact that the phase-point or state of the system is localized within the atom $P_i$ with probability $\mu(P_i)$.

The notion of entropy measures the amount of uncertainty about the outcomes of a random variable like $\mathcal{P}$ before the phase-point has been localized within a definite atom, for instance as a consequence of an observation or a measurement process of sort. Equivalently, entropy measures the amount of information, relative to the partition $\mathcal{P}$, that has been gained after the phase-point of the system has indeed been localized in one of its atoms.

**Definition 2.4.1 (*Shannon Entropy*).** *The Shannon entropy of a discrete random variable $A$ with probability distribution $\pi_A = \{p_A(j)\}_{j=1}^a$ is given by*

$$H(A) := -\sum_{j=1}^{a} p_A(j) \log p_A(j) = \sum_{j=1}^{a} \eta(p_A(j)) \ , \tag{2.83}$$

*where*

$$\eta(x) = \begin{cases} 0 & x = 0 \\ -x \log x & 0 < x \leq 1 \end{cases} \tag{2.84}$$

**Remark 2.4.1.** The Shannon entropy plays for discrete dynamical systems the role played by *Gibbs entropy* for continuous systems which is defined as [167, 300]

$$H_G(\rho) := -\int_{\mathcal{X}} \mathrm{d}x\; \rho(x)\, \log \rho(x) \ ,$$

for a state on the phase-space $\mathcal{X}$ with probability density $\rho(x)$.

The Shannon entropy is such that $H(A) = 0$ if and only if one outcome, say $j^*$, occurs with probability $p_A(j^*) = 1$, while $p_A(j) = 0$ for $j \neq j^*$; it reaches its maximum $H(A) = \log a$, when all outcomes are equiprobable, $p_A(j) = 1/a$. Indeed, the function $\eta(x)$ is concave, whence

$$x(\log x - \log y) \geq x - y \ , \quad \forall x, y \in [0,1] \ , \tag{2.85}$$

with equality holding if and only if $x = y$.

Let then $E$ be a random variable with $\pi_E = \{p_E(j) = 1/a\}_{j=1}^a$, then $H(E) = \log a$ and

$$H(A) - H(E) = -\sum_{j=1}^{a} p_A(j)\Big(\log p_A(j) + \log a\Big)\Big) \leq \sum_{j=1}^{a}(p_A(j) - 1/a) = 0 \ .$$

Given two random variables $A$ and $B$, we shall keep the notation used for the join of two partitions and denote by $A \vee B$ the random variable with *joint probability distribution*

$$\pi_{A\vee B} := \{p_{A\vee B}(i,j)\}_{i\in I_A, j\in I_B} \ , \ I_A = \{1,2,\ldots,a\} \ , \ I_B = \{1,2,\ldots,b\} \ . \tag{2.86}$$

By summing over the outcomes of $A$, respectively $B$, one obtains the *marginal probability distributions* $\pi_A := \{p_A(i)\}_{i\in I_A}$ and $\pi_B := \{p_B(j)\}_{j\in I_B}$, where

$$p_A(i) := \sum_{j=1}^{b} p_{A\vee B}(i,j) \ , \ p_B(j) := \sum_{i=1}^{a} p_{A\vee B}(i,j) \ . \tag{2.87}$$

**Lemma 2.4.2 (Subadditivity).** *Given two random variables $A$ and $B$,*

$$H(A \vee B) \leq H(A) + H(B) \ . \tag{2.88}$$

**Proof:** Given $\pi_{A\vee B}$ as in (2.86) and $\pi_A$ and $\pi_B$ as in (2.87), use (2.85) with $x = p_{A\vee B}(i,j)$ and $y = p_A(i)p_B(j)$,

$$\begin{aligned} H(A) + H(B) - H(A\vee B) &= \sum_{i\in I_A, j\in I_B} p_{A\vee B}(i,j) \log \frac{p_{A\vee B}(i,j)}{p_A(i)p_B(j)} \\ &\geq \sum_{i\in I_A, j\in I_B} (p_{A\vee B}(i,j) - p_A(i)p_B(j)) = 0 \ . \end{aligned}$$

□

**Remarks 2.4.2.**

1. As already observed, any finite (measurable) partition $\mathcal{P} = \{P_i\}_{I_\mathcal{P}}$ of $(\mathcal{X}, T, \mu)$ is a random variable $P$ whose outcomes correspond to the labels of the atoms to which the system phase-point happens to belong. The volumes of the atoms give the probabilities of such occurrences, so that a finite partition $\mathcal{P}$ also attributes to the random variable $P$ the natural probability distribution $\pi_P = \mu_\mathcal{P} = \{\mu(P_i\}_{i\in I_\mathcal{P}}$.
2. In analogy with Definition 2.2.1.2, a random variable $A$ is *finer* than a random variable $B$ ($B \preceq A$) if each outcome $j \in I_B$ of $B$ is determined by a subset $I_A^j \subseteq I_A$ of outcomes of $A$. It follows that, if $A$ has probability distribution $\pi_A = \{p_A(i)\}_{i\in I_A}$ and $B$ probability distribution $\pi_B = \{p_B(j)\}_{j\in I_B}$, $B \preceq A$ implies $p_B(j) = \sum_{i\in I_A^j} p_A(i)$.
3. According to Definition 2.2.1.3, the refinement $\mathcal{P} \vee \mathcal{Q}$ of two partitions $\mathcal{P} = \{P_i\}_{i\in I_P}$ and $\mathcal{Q} = \{Q_j\}_{j\in I_Q}$, is a random variable $P \vee Q$ with joint probability distribution $\mu_{\mathcal{P}\vee\mathcal{Q}} = \{\mu(P_i \cap Q_j)\}_{i\in I_P, j\in I_Q}$. $P \vee Q$ is finer than both random variables $P$ and $Q$; also, $\mathcal{Q} \preceq \mathcal{P} \Longrightarrow \mathcal{P} \vee \mathcal{Q} = \mathcal{P}$.

### 2.4.4 Conditional Entropy

Because of possible statistical correlations, the knowledge of a random variable $A$ may decrease the uncertainty about another random variable $B$; the less so, the more $A$ and $B$ are statistically independent. These intuitive arguments are formalized by introducing the notions of *conditional probability*, *conditional entropy* and *mutual information*.

Consider two random variables $A$ and $B$ with probability distributions $\pi_A = \{p_A(i)\}_{i\in I_A}$, respectively $\pi_B = \{p_B(j)\}_{j\in I_B}$, and joint probability $\pi_{A\vee B} = \{p_{A\vee B}(i,j)\}_{i\in I_A, j\in I_B}$. The quantity

$$p_{A|j}(i|j) := \frac{p_{A\vee B}(i,j)}{p_B(j)} \tag{2.89}$$

represents the probability of the outcome $A = i$ conditioned upon the outcome $B = j$. Altogether, $\pi_{A|B=j} = \{p_{A|j}(i|j)\}_{i=1}^a$ is the *conditional probability distribution* of $A$ conditioned upon the outcome $B = j$. The conditional probabilities are such that

$$p_{A|B=j}(i|j) \geq 0 \;, \quad \sum_{i=1}^{a} p_{A|B=j}(i|j) = 1 \quad \forall j = 1, 2, \ldots b \;.$$

The notion of conditional probability is naturally associated to that of *conditional entropy* which measures the amount of uncertainty about a random variable $A$ which is left once that relative to another one, $B$, has been removed.

**Definition 2.4.2 (*Conditional Entropy*).**

*Given two random variables A and B with probabilities* $\pi_A, \pi_B$ *as in (2.87) and joint probability* $\pi_{A\vee B}$ *as in (2.86), the conditional entropy of A with respect to B is*

$$H(A|B) = \sum_{j=1}^{b} p_B(j)\, H(A|B=j) \tag{2.90}$$

$$= -\sum_{j=1}^{b} p_B(j) \sum_{i=1}^{a} \frac{p_{A\vee B}(i,j)}{p_B(j)} \log \frac{p_{A\vee B}(i,j)}{p_B(j)}$$

$$= H(A\vee B) - H(B) , \tag{2.91}$$

*where* $H(A|B=j)$ *is the Shannon entropy corresponding to the conditional probability* $\pi_{A|B=j}$.

**Lemma 2.4.3.** *The conditional entropy fulfils*

$$\begin{aligned} &0 \leq H(A|B) \leq H(A) \\ &H(A\vee B|C) = H(A|C) + H(B|A\vee C) \leq H(A|C) + H(B|C) . \end{aligned}$$

**Proof:** The lower bound follows since the left hand side of (2.90) is positive, while the first upper bound is a consequence of (2.88) applied to (2.91). Further, using the latter relation one gets

$$\begin{aligned} H(A\vee B|C) &= H(A\vee B\vee C) - H(C) \\ &= H(A\vee C) - H(C) + H(A\vee B\vee C) - H(A\vee C) \\ &= H(A|C) + H(B|A\vee C) , \end{aligned}$$

while (2.88) applied to $H(A\vee B|C=k)$ gives the second upper bound. □

**Corollary 2.4.1.** $B \preceq A \Longrightarrow H(B) \leq H(A)$.

**Proof:** From Remark 2.4.2.3 it follows that $B \preceq A \Longrightarrow A\vee B = A$; thus,

$$H(A) = H(A\vee B) = H(A|B) + H(B) \geq H(B) .$$

□

**Example 2.4.3.** If $N$ denotes the (trivial) random variable with only one certain outcome, then $H(A|N) = H(A)$ for any other random variable $A$.

By the definition of conditional entropy, $H(A|B) = 0$ implies that in (2.90)

$$\sum_{i=1}^{a} \frac{p_{A\vee B}(i,j)}{p_B(j)} \log \frac{p_{A\vee B}(i,j)}{p_B(j)} = 0 \qquad \forall j .$$

Therefore, for fixed $j$, $p_{A \vee B}(i,j) = p_B(j)$ for only one $i \in I_A$; thus, for each fixed $i \in I_A$, the index set $I_B$ can be subdivided into disjoint subsets $I_B^i$ such that

$$p_A(i) = \sum_{j \in I_B^i} p_{A \vee B}(i,j) = \sum_{j \in I_B^i} p_B(j) .$$

That is (see Remark 2.4.2.2) $A \preceq B$; indeed, the outcomes of $A$ are determined by those of $B$. In other words, when knowing $B$ means knowing $A$, then $A \preceq B$. Viceversa, if $B$ is finer than $A$, then $H(A|B) = 0$; in fact,

$$A \vee B = B \Longrightarrow H(A|B) = H(A \vee B) - H(B) = 0 .$$

**Remarks 2.4.3.**

1. Conditioning can be extended to random variables $A_i$, $i = 1, 2, \ldots, n$. The probability of the events $A_i = a_i$, $i = p+1, \ldots, n$ conditioned on the events $A_i = a_i$, $i = 1, \ldots, p$ is given by

$$p(a_{p+1} \cdots a_n | a_1 \cdots a_p) := \frac{p(a_1 \cdots a_p a_{p+1} \cdots a_n)}{p(a_{p+1} \cdots a_n)} ,$$

where explicit reference to the random variables in $p(\cdots)$ has been omitted, for sake of simplicity. It follows that also the notion of conditional entropy can be extended to

$$H(A^{p+1} \vee A^{p+2} \ldots \vee A^n | A^1 \vee A^2 \ldots \vee A^p) := - \sum_{a_1, \ldots a_p} p(a_1, a_2, \ldots, a_p)$$

$$\times \sum_{a_{p+1}, \ldots, a_n} p(a_{p+1} \cdots a_n | a_1 \cdots a_p) \log p(a_{p+1} \cdots a_n | a_1 \cdots a_p) .$$

2. A sequence of random variables $\{A^j\}_{j \in \mathbb{N}}$ form a *Markov process* as in Example 2.4.2.2 if $p(a_n | a_1 \cdots a_{n-1}) = p(a_n | a_{n-1})$ for all $n \in \mathbb{N}$. In such a case $H(A^n | A^1 \vee \cdots A^{n-1}) = H(A^n | A^{n-1})$.
3. Since the conditional entropy is positive, it follows that

$$H(A \vee B) \geq \max\{H(A), H(B)\} ;$$

Both this observation and subadditivity (2.88) agree with the interpretation of the entropy as a measure of uncertainty. The latter is in fact greater about $A \vee B$ than about either $A$ or $B$, while, due to possible statistical correlations between $A$ and $B$, the uncertainty of $A \vee B$ is smaller than the sum of the uncertainties of $A$ and $B$ independently. Further, due to (2.85),

$$H(A \vee B) = H(A) + H(B)$$

if and only if $\pi_{A \vee B}$ factorizes into the product of the probabilities, namely if and only if $A$ and $B$ are statistically independent.

4. The second upper bound in Lemma 2.4.3 yields $H(B|A \vee C) \leq H(B|C)$; as $C \preceq A \vee C$, this inequality is a particular instance of the more general monotonicity property of the conditional entropy established in Corollary 2.4.2.

**Example 2.4.4.** Suppose a random variable is given by a finite partition $\mathcal{P} = \{P_i\}_{i=1}^d$ with atoms that are measurable with respect to a $\sigma$-algebra $\Sigma$ generated by a measure-algebra $\Sigma_0$ as in Example 2.2.1. Then, for any $\varepsilon > 0$, there exists a partition $\mathcal{Q} = \{Q_i\}_{i=1}^d$ with atoms $Q_i \in \Sigma_0$ such that $H(\mathcal{P}|\mathcal{Q}) < \varepsilon$. In fact, as showed in the example, one can always construct $\mathcal{Q}$ such that, for all $i = 1, 2, \ldots, d$, one has

$$\mu(P_i \,\Delta\, Q_i) \leq \delta \min_{1 \leq i \leq d} \frac{\mu(P_i)}{2} \,, \qquad 0 < \delta < 1 \,.$$

Now, $P_i \subseteq Q_i \cup (P_i \,\Delta\, Q_i)$ and $P_i \,\Delta\, Q_i = (P_i \cup Q_i) \setminus (P_i \cap Q_i)$ yield

$$\mu(P_i) \leq \mu(Q_i) + \delta \frac{\mu(P_i)}{2} \quad \text{and} \quad \mu(P_i \,\Delta\, Q_i) \geq \mu(Q_i) - \mu(P_i \cap Q_i) \,.$$

Thus, $\mu(Q_i) \geq \frac{\mu(P_i)}{2}$ and $\delta\, \mu(Q_i) \geq \mu(Q_i) - \mu(P_i \cap Q_i)$, whence

$$p_{\mathcal{P}|\mathcal{Q}=i}(i|i) := \frac{\mu(P_i \cap Q_i)}{\mu(Q_i)} \geq 1 - \delta \,.$$

Since $\pi_{\mathcal{P}|\mathcal{Q}=i}$ is a conditional probability, it follows that $p_{\mathcal{P}|\mathcal{Q}=i}(j|i) \leq \delta$ for $j \neq i$. Finally, choosing $\delta$ so that the continuous function $\eta(x)$ in (2.84) be such that $\eta(x) < \varepsilon/d$ when $0 \leq x \leq \delta$ and $1 - \delta \leq x \leq 1$, (2.91) yields

$$H(\mathcal{P}|\mathcal{Q}) = \sum_{i=1}^d \mu(Q_i) \sum_{j=1}^d \eta(p_{\mathcal{P}|\mathcal{Q}=i}(j|i)) \leq \varepsilon \,.$$

**Proposition 2.4.1 (*Strong Subadditivity*).**
*Given three discrete random variables $A$, $B$ and $C$, the following inequality holds,*

$$H(A \vee B \vee C) + H(B) \leq H(A \vee B) + H(B \vee C) \,, \qquad (2.92)$$

*together with those obtained by cyclic permutations of $A$, $B$ and $C$.*

**Proof:** Similarly to the proof of Lemma 2.88, the result follows by applying (2.85) as follows:

$$
\begin{aligned}
&H(A \vee B \vee C) + H(B) - H(A \vee B) - H(B \vee C) = \\
&= - \sum_{i \in I_A, j \in I_B, k \in I_C} p_{A \vee B \vee C}(i,j,k) \log \frac{p_{A \vee B \vee C}(i,j,k) p_B(j)}{p_{A \vee B}(i,j) p_{B \vee C}(j,k)} \\
&\leq - \sum_{i \in I_A, j \in I_B, k \in I_C} \left( p_{A \vee B \vee C}(i,j,k) - \frac{p_{A \vee B}(i,j) p_{B \vee C}(j,k)}{p_B(j)} \right) = 0 \, .
\end{aligned}
$$

□

As a consequence of strong subadditivity, the conditional entropy monotonically decreases upon refinement of its second argument.

**Corollary 2.4.2.** $B \preceq C \Longrightarrow H(A|C) \leq H(A|B)$.

**Proof:** From (2.92) and (2.91),

$$H(A \vee B \vee C) - H(B \vee C) = H(A|B \vee C) \leq H(A \vee B) - H(B) = H(A|B) \, .$$

The result follows since $B \preceq C \Longrightarrow B \vee C = C$. □

### 2.4.5 Mutual Information

A notion related to the conditional entropy is that of mutual information: it measures the amount of information about a random observable $A$ that can be achieved by knowing another random variable $B$.

**Definition 2.4.3 (*Mutual Information*).**
*Given two random variables $A$ and $B$, their mutual information is given by*

$$
\begin{aligned}
I(A;B) &:= H(A) + H(B) - H(A \vee B) \\
&= H(A) - H(A|B) = H(B) - H(B|A) \, .
\end{aligned}
\tag{2.93}
$$

The mutual information amounts to the *relative entropy* (also known as Kullback-Leibler distance or information divergence) of the joint probability distribution $\pi_{A \vee B}$ with respect to the product probability distribution $\tilde{\pi}_{A \vee B} = \{p_A(i) p_B(j)\}_{i \in I_A, j \in I_B}$ obtained from the marginal ones (see (2.87)):

$$S\left(\tilde{\pi}_{A \vee B}, \pi_{A \vee B}\right) := \sum_{ij} p_{A \vee B}(i,j) \log \frac{p_{A \vee B}(i,j)}{p_A(i) p_B(j)} \, . \tag{2.94}$$

Since $H(A)$ measures the unconditioned uncertainty about $A$ and $H(A|B)$ the uncertainty about $A$ if one knows $B$, their difference amounts to the knowledge of $A$ given by $B$. If $A$ and $B$ are statistically independent, knowing $B$ does not give any information about $A$, whence $H(A|B) = H(A)$, and $I(A;B) = 0$. On the other hand, if $A$ is finer than $B$, then knowing $A$ means knowing $B$, thus $B \preceq A \Longrightarrow H(B|A) = 0$ and $I(A;B) = H(B)$. Vice versa, $I(A;B) < H(B)$ means that $H(B|A) > 0$ or, in other words, that the knowledge of $B$ is unable to remove all the uncertainty about $A$.

An interesting inequality involves the mutual information in connection with three random variables $A$, $B$ and $C$ that form a so-called *Markov chain* $A \to B \to C$ [92]; namely (see Remark 2.4.3.2)

$$p_{C|A\vee B=(i,j)}(k|i,j) := \frac{p_{A\vee B\vee C}(i,j,k)}{p_{A\vee B}(i,j)} = p_{C|B=j}(k|j) = \frac{p_{B\vee C}(j,k)}{p_B(j)} .$$

Notice that $C$, $B$ and $A$ form a Markovian chain $C \to B \to A$, too; indeed, as $p_{C\vee B}(k,j) = p_{B\vee C}(j,k)$ it turns pout that

$$\begin{aligned} p_{A|C\vee B=(k,j)}(i|k,j) &:= \frac{p_{A\vee B\vee C}(i,j,k)}{p_{C\vee B}(k,j)} = p_{C|B=j}(k|j)\,\frac{p_{A\vee B}(i,j)}{p_{C\vee B}(k,j)} \\ &= \frac{p_{A\vee B}(i,j)}{p_B(j)} = p_{A|B=j}(i|j) . \end{aligned}$$

Using the latter property one can prove the so-called *data processing inequality* [92].

**Proposition 2.4.2.** $A \to B \to C \Longrightarrow I(A;C) \leq I(A;B)$.

**Proof:** From Definition 2.4.3, $I(A;B) - I(A;C) = H(A|C) - H(A|B)$ while the Markovianity assumption yields $H(A|B) = H(A|B \vee C)$ (see Remark 2.4.3.2), whence, from (2.4.1),

$$\begin{aligned} I(A;B) - I(A;C) &= H(A|C) - H(A|B \vee C) \\ &= H(A \vee C) + H(B \vee C) - H(A \vee B \vee C) - H(C) \geq 0 . \end{aligned}$$

□

The meaning of the data processing inequality is that the mutual information of two random variables $A$ and $B$ cannot be increased by any further processing of $B$ by a function $C = g(B)$, for this yields a Markov chain $A \to B \to C$.

**Example 2.4.5.** When dealing with noisy transmission channels, signals $a$ from a source described by a random variable $A$ are encoded into code-words $b(a)$ that give rise to another random variable $B = B(A)$. Then, the code-words are sent through the channel which outputs signals $c = c(b)$, providing a third random variable $C(B)$. Altogether, $A$, $B(A)$ and $C(B)$ form a Markovian chain $A \to B(A) \to C(B)$, as well as $C(B) \to B(A) \to A$; thus, we get the data-processing inequalities

$$I(A;C(B)) \leq I(A;B(A)) , \quad I(A;C(B)) \leq I(B(A);C(B)) . \tag{2.95}$$

### Bibliographical Notes

For a mathematical approach to classical dynamical systems and ergodic theory see [17, 61, 91, 199, 313]. For a more physical point of view on ergodic theory and related questions, one may consult [7, 106, 167, 300]. [16, 299] have been used as references for Hamiltonian mechanics and integrable systems, as well as [106, 129, 228, 271] for classical chaos. The review [62] discusses in detail the signatures of chaos in discrete classical systems.

For a modern overview of probability theory see [176].

For the notions of entropy and conditional entropy in a dynamical system context see [17, 61, 91, 164]; consult [92] for the same notions and that of mutual information from the point of view of information theory. As regards the relations and epistemological links between the entropy of Shannon and those of Gibbs and Boltzmann in a thermodynamical setting, see [167, 300, 314].

# 3 Dynamical Entropy and Information

Repeated uses of an information source or successive localizations of a trajectory with respect to a partition of phase-space, give rise to stochastic processes. Since equilibrium states $\mu$ give rise to shift-invariant probability distributions, the Shannon entropy is a constant of the motion: for instance, given the time-evolved partition $T^{-j}(\mathcal{P})$ in (2.38), one has $H_\mu(T^{-j}(\mathcal{P})) = H_\mu(\mathcal{P})$. Therefore, it is not the Shannon entropy, rather the *entropy rate* that is useful to quantify the degree of irregularity of the dynamics. Since its introduction as a mathematical tool, the notion of entropy rate or, more generally, of *dynamical entropy*, has been playing a major role in the theory of classical dynamical systems for it provides links among as different properties as *dynamical instability*, *informational compressibility* and *algorithmic complexity*.

## 3.1 Dynamical Entropy

As in Section 2.4.3, given a dynamical system corresponding to a measure-theoretic triplet $(\mathcal{X}, T, \mu)$, we will consider a coarse-graining of $\mathcal{X}$ by means of a finite, measurable partition $\mathcal{P} = \{P_i\}_{i=1}^p$ and identify $\mathcal{P}$ with the random variable (denoted by the same symbol) corresponding to the process of localization of the system phase-point (state) within one of the disjoint atoms $P_i$ that cover $\mathcal{X}$. The outcomes of $\mathcal{P}$ are the labels of the atoms and occur according to the discrete probability distribution $\mu_{\mathcal{P}} = \{p_\mu(i) := \mu(P_i)\}_{i=1}^p$. Further, the time-evolved partition $\mathcal{P}^j := T^{-j}(\mathcal{P})$ at time $j$ in (2.38) is identified with the $j$-th random variable of a stochastic process $\{\mathcal{P}^j\}_{j\in\mathbb{Z}}$. Thus, the refined partitions $\mathcal{P}^{(n)}$ with atoms $P^{(n)}_{\boldsymbol{i}^{(n)}}$ as in (2.40) correspond to joint random variables with discrete probability distributions as in (2.41),

$$\mu^{(n)}_{\mathcal{P}} = \left\{p^{(n)}_\mu(\boldsymbol{i}^{(n)}) := \mu(P^{(n)}_{\boldsymbol{i}^{(n)}})\right\}_{\boldsymbol{i}^{(n)}\in\Omega^{(n)}_p} ,$$

and Shannon entropies (comparing with (2.83), we explicitly indicate the dependence of the entropy from the measure and the partition)

$$H_\mu(\mathcal{P}^{(n)}) := -\sum_{\boldsymbol{i}^{(n)}\in\Omega^{(n)}_p} p^{(n)}_\mu(\boldsymbol{i}^{(n)}) \log p^{(n)}_\mu(\boldsymbol{i}^{(n)}) . \tag{3.1}$$

F. Benatti, *Dynamics, Information and Complexity in Quantum Systems*, Theoretical and Mathematical Physics, DOI 10.1007/978-1-4020-9306-7_3,

**Definition 3.1.1 (*Entropy Rate*).** *The entropy rate of $(\mathcal{X}, T, \mu)$ with respect to a finite, measurable partition $\mathcal{P}$ is given by*

$$\mathrm{h}^{\mathrm{KS}}_{\mu}(T, \mathcal{P}) := \lim_{n\to\infty} \frac{1}{n} H_\mu(\mathcal{P}^{(n)}) = \inf_n \frac{1}{n} H_\mu(\mathcal{P}^{(n)}) . \tag{3.2}$$

The above limit exists because of the stationarity of $\mu$,

$$H_\mu(\mathcal{P}^k) = H_\mu(\mathcal{P}) \qquad \forall\, k \geq 0 , \tag{3.3}$$

and because of the subadditivity of the Shannon entropy [313]. Together, they yield, for all $0 \leq p \leq n-1$,

$$\begin{aligned} H_\mu(\mathcal{P}^{(n)}) &\leq H_\mu(\mathcal{P}^{(p)}) + H_\mu\Big(\bigvee_{k=p}^{n-1} \mathcal{P}^k\Big) \\ &= H_\mu(\mathcal{P}^{(p)}) + H_\mu\Big(T^{-p}\Big(\bigvee_{k=0}^{n-p-1} \mathcal{P}^k\Big)\Big) = H_p + H_{n-p} , \end{aligned}$$

where $H_n := H_\mu(\mathcal{P}^{(n)})$. Fix $m \in \mathbb{N}$ and set $n = km + r$, $0 \leq r < m$; then, from (2.88)

$$\frac{H_n}{n} \leq \frac{H_m}{m} + \frac{H_r}{km + r} .$$

Since $m$ is fixed, when $n$ goes to infinity, $k$ goes to infinity as well, whence

$$\limsup_{n\to\infty} \frac{H_n}{n} \leq \frac{H_m}{m} .$$

Since $m$ is arbitrary, it follows that

$$\limsup_{n\to\infty} \frac{H_n}{n} \leq \inf_m \frac{H_m}{m} \leq \liminf_{n\to\infty} \frac{H_n}{n} .$$

The entropy rate can be expressed by means of the conditional entropy (2.91) of two partitions $H_\mu(\mathcal{P}|\mathcal{Q})$ in such a way that $h(\mu_{\mathcal{P}}, T_\sigma)$ measures to which extent the knowledge of the past of $\mathcal{P}$ may help to predict its future outcomes. Recursively using (2.91) and (3.3), one gets

$$H_\mu(\mathcal{P}^{(n)}) = H_\mu\Big(\mathcal{P}\Big|\bigvee_{j=1}^{n-1} \mathcal{P}^j\Big) + H_\mu(\mathcal{P}^{(n-1)}) = \sum_{i=1}^{n-1} H_\mu\Big(\mathcal{P}\Big|\bigvee_{j=1}^{n-i} \mathcal{P}^j\Big) + H_\mu(\mathcal{P})$$

$$\begin{aligned} H_\mu(\mathcal{P}^{(n)}) &= H_\mu\Big(\mathcal{P}^{n-1}\Big|\bigvee_{j=0}^{n-2} \mathcal{P}^j\Big) + H_\mu(\mathcal{P}^{(n-1)}) \\ &= \sum_{i=1}^{n-1} H_\mu\Big(\mathcal{P}^i\Big|\bigvee_{j=0}^{i-1} \mathcal{P}^j\Big) + H_\mu(\mathcal{P}) . \end{aligned}$$

Because of Corollary 2.4.2, the positive terms in the sums are monotonically decreasing with increasing $n$, thus, arguing as in Remark 2.3.2.1,

$$\mathrm{h}_\mu^{\mathrm{KS}}(T,\mathcal{P}) = \lim_{n\to\infty}\frac{1}{n}\sum_{i=1}^{n-1} H_\mu\Big(\mathcal{P}\Big|\bigvee_{j=1}^{n-i}\mathcal{P}^j\Big) = \lim_{n\to\infty} H_\mu\Big(\mathcal{P}\Big|\bigvee_{j=1}^{n}\mathcal{P}^j\Big) \tag{3.4}$$

$$\mathrm{h}_\mu^{\mathrm{KS}}(T,\mathcal{P}) = \lim_{n\to\infty}\frac{1}{n}\sum_{i=1}^{n-1} H_\mu\Big(\mathcal{P}^i\Big|\bigvee_{j=0}^{i-1}\mathcal{P}^j\Big) = \lim_{n\to\infty} H_\mu\Big(\mathcal{P}^n\Big|\bigvee_{j=0}^{n-1}\mathcal{P}^j\Big) . \tag{3.5}$$

Consider the first equality in (3.5): $\mathcal{P}^i$ is the random variable whose outcomes depend on which atom of $\mathcal{P}$ the system state is in at time $i$, while $\bigvee_{j=0}^{i-1}\mathcal{P}^j$ is the joint random variable relative to the atoms visited at previous times $0,1,\ldots,i-1$. Thus, the entropy rate corresponding to $\mathcal{P}$ is the average information about the next localization provided by the knowledge of all the previous ones.

**Remarks 3.1.1.**

1. Let $(\Omega_a, T_\sigma, \pi_A)$ describe a stationary information source. Then, the probabilities $\pi_{A^{(n)}}$, $A^{(n)} = \bigvee_{j=0}^{n-1} A_j$, together with the corresponding entropies $H(A^{(n)})$ refer to the statistical ensembles of strings of length $n$ emitted by the source $A$. As discussed in Section 2.4.3, repeated uses of the source $A$ can be described as a stochastic process $\{A_j\}_{j\in\mathbb{Z}}$ where $A_j$ is the random variable associated to the $j$ th use of the source. The entropy rate of the source $A$ is thus given by

$$h(A) := \lim_{n\to\infty}\frac{1}{n}H(A^{(n)}) . \tag{3.6}$$

   The entropy rate $h(A)$ of a stationary source is the entropy per symbol of the stationary stochastic process $\{A^j\}_{j\in\mathbb{Z}}$ generated by $A$.
2. Because of subadditivity (2.88), the entropy rate of a partition is always bounded by its Shannon entropy

$$\mathrm{h}_\mu^{\mathrm{KS}}(T,\mathcal{P}) \leq H_\mu(\mathcal{P}) . \tag{3.7}$$

   Furthermore, since $\sum_{\boldsymbol{i}^{(n)}\in\Omega_p^{(n)}} \mu(P^{(n)}_{\boldsymbol{i}^{(n)}}) = 1$, from (3.1), one gets the lower bound

$$H_\mu(\mathcal{P}^{(n)}) := -\sum_{\boldsymbol{i}^{(n)}\in\Omega_p^{(n)}} p_\mu^{(n)}(\boldsymbol{i}^{(n)}) \log p_\mu^{(n)}(\boldsymbol{i}^{(n)}) \geq -\log \sup_{P\in\mathcal{P}^{(n)}} \mu(P) ,$$

   whence [164]

$$\mathrm{h}_\mu^{\mathrm{KS}}(T,\mathcal{P}) \geq -\limsup_{n\to+\infty}\frac{1}{n}\log\sup_{P\in\mathcal{P}^{(n)}}\mu(P) . \tag{3.8}$$

3. If $\mathcal{Q} \preceq \mathcal{P}$, Corollary 2.4.1 implies $\mathrm{h}^{\mathrm{KS}}_{\mu}(T, \mathcal{Q}) \leq \mathrm{h}^{\mathrm{KS}}_{\mu}(T, \mathcal{P})$.
4. Since $\mu$ is $T$-invariant, the conditional entropy is stationary, namely

$$H_\mu(T^{-1}(\mathcal{P})|T^{-1}(\mathcal{Q})) = H_\mu(\mathcal{P}|\mathcal{Q}) \ .$$

Thus, if $T$ is invertible (3.5) can be rewritten as

$$\mathrm{h}^{\mathrm{KS}}_{\mu}(T, \mathcal{P}) = \lim_{n\to\infty} H_\mu\Big(\mathcal{P}\Big|\bigvee_{j=1}^{n-1} \mathcal{P}^{-j}\Big) \ .$$

5. Let $\mathcal{P}$ and $\mathcal{Q}$ be two partitions; then, using Corollary 2.4.1, the relation (2.91), Lemma 2.4.3, Corollary 2.4.2 and the previous remark, one derives

$$\begin{aligned} H_\mu(\mathcal{P}^{(n)}) &\leq H_\mu(\mathcal{P}^{(n)} \vee \mathcal{Q}^{(n)}) = H_\mu(\mathcal{Q}^{(n)}) + H_\mu(\mathcal{P}^{(n)}|\mathcal{Q}^{(n)}) \\ &\leq H_\mu(\mathcal{Q}^{(n)}) + \sum_{i=0}^{n-1} H_\mu(\mathcal{P}^i|\mathcal{Q}^{(n)}) \leq H_\mu(\mathcal{Q}^{(n)}) + \sum_{i=0}^{n-1} H_\mu(\mathcal{P}^i|\mathcal{Q}^i) \ , \end{aligned}$$

whence $H_\mu(\mathcal{P}^{(n)}) \leq H_\mu(\mathcal{Q}^{(n)}) + n\, H_\mu(\mathcal{P}|\mathcal{Q})$ implies

$$\mathrm{h}^{\mathrm{KS}}_{\mu}(T, \mathcal{P}) \leq \mathrm{h}^{\mathrm{KS}}_{\mu}(T, \mathcal{Q}) + H_\mu(\mathcal{P}|\mathcal{Q}) \ . \tag{3.9}$$

6. Given a partition $\mathcal{P}$, set $\mathcal{P}_{r,s} := \bigvee_{j=r}^{s} \mathcal{P}^j$, where $r \leq s$ and $r \geq 0$ if $T$ is not invertible. Notice that

$$\bigvee_{\ell=0}^{n-1} \mathcal{P}^\ell_{r,s} = \bigvee_{\ell=0}^{n-1} \bigvee_{j=r}^{s} \mathcal{P}^{j+\ell} = \bigvee_{\ell=r}^{s+n-1} \mathcal{P}^\ell = T^{-r}\left(\bigvee_{\ell=0}^{s+n-r-1} \mathcal{P}^\ell\right) \ ;$$

then, since $\mu$ is $T$-invariant, from

$$\frac{1}{n} H_\mu\Big(\bigvee_{\ell=0}^{n-1} \mathcal{P}^\ell_{r,s}\Big) = \frac{n+s-r}{n}\,\frac{1}{n+s-r} H_\mu\Big(\bigvee_{\ell=0}^{n+s-r-1} \mathcal{P}^\ell\Big)$$

it follows that $\mathrm{h}^{\mathrm{KS}}_{\mu}(T, \mathcal{P}_{r,s}) = \mathrm{h}^{\mathrm{KS}}_{\mu}(T, \mathcal{P})$. For instance, $s = -r = n$ gives

$$\mathrm{h}^{\mathrm{KS}}_{\mu}\left(T, \bigvee_{j=-n}^{n} \mathcal{P}^j\right) = \mathrm{h}^{\mathrm{KS}}_{\mu}(T, \mathcal{P}) \ , \quad \forall n \geq 0 \ . \tag{3.10}$$

7. As before, set $\mathcal{Q} = \bigvee_{j=0}^{k-1} \mathcal{P}^j$, $k \geq 1$; then, $\mathcal{P} \preceq \bigvee_{\ell=0}^{n-1} \mathcal{Q}^{k\ell} = \bigvee_{\ell=0}^{nk-1} \mathcal{P}^\ell$ and (3.9) yield

$$\frac{1}{k}\,\mathrm{h}^{\mathrm{KS}}_{\mu}(T^k, \mathcal{P}) \leq \frac{1}{k}\,\mathrm{h}^{\mathrm{KS}}_{\mu}(T^k, \mathcal{Q}) = \mathrm{h}^{\mathrm{KS}}_{\mu}(T, \mathcal{P}) \ . \tag{3.11}$$

8. After regrouping $\bigvee_{j=0}^{kn-1} T^{-j}(\mathcal{P}) = \bigvee_{i=0}^{n-1} \bigvee_{j=0}^{n-1} T^{-kj-i}(\mathcal{P})$, from subadditivity and $T$-invariance of $\mu$ it follows that
$$\frac{1}{kn} H_\mu\Big(\bigvee_{j=0}^{kn-1} \mathcal{P}^j\Big) \leq \frac{1}{kn} \sum_{i=0}^{n-1} H_\mu\Big(T^{-i} \circ \Big(\bigvee_{j=0}^{n-1} \mathcal{P}^{jk}\Big)\Big) = \frac{1}{n} H_\mu\big(\bigvee_{j=0}^{n-1} (\mathcal{P}^{kj}\big) ,$$
whence letting $n \to +\infty$ obtains $\mathrm{h}_\mu^{\mathrm{KS}}(T, \mathcal{P}) \leq \mathrm{h}_\mu^{\mathrm{KS}}(T^k, \mathcal{P})$.

The entropy rate relative to a given partition $\mathcal{P}$ of $(\mathcal{X}, T, \mu)$ strongly depends on the latter; for instance, if $\mathcal{N}$ is the trivial partition consisting only of $\mathcal{X}$ itself and the empty set, then $T^{-j}(\mathcal{N}) = \mathcal{N}$ for all $j \geq 0$, whence $\mathrm{h}_\mu^{\mathrm{KS}}(T, \mathcal{N}) = 0$. The obvious way of achieving an absolute entropy rate is to look for the greatest possible one; this leads to the notion of *dynamical entropy* also known as *Kolmogorov-Sinai entropy* (*KS* -entropy) or *metric entropy* [171, 172].

**Definition 3.1.2 (*KS* Entropy).** *The dynamical entropy of a classical dynamical system $(\mathcal{X}, T, \mu)$ is defined as*
$$\mathrm{h}_\mu^{\mathrm{KS}}(T) := \sup_{\mathcal{P}} \mathrm{h}_\mu^{\mathrm{KS}}(T, \mathcal{P}) ,$$
*where $\mathcal{P}$ is any finite, measurable partition.*

**Remark 3.1.2.** The dynamical entropy provides a quantity that remains invariant under isomorphisms between dynamical systems [61]. Two dynamical systems $(\mathcal{X}_{1,2}, T_{1,2}, \mu_{1,2})$ with $\sigma$-algebra $\Sigma_{1,2}$ are *isomorphic* if there exist subsets $\mathcal{X}_{1,2}^{(0)} \subseteq \mathcal{X}_{1,2}$ of measure $\mu_{1,2}(\mathcal{X}_{1,2}^{(0)}) = 1$ and a one-to-one map $\Phi : \mathcal{X}_1^{(0)} \mapsto \mathcal{X}_2^{(0)}$ such that

1. if $S_2 = \Phi(S_1)$ with $S_1 \in \mathcal{X}_1^{(0)}$, then $S_1 \in \Sigma_1$ if and only if $S_2 \in \Sigma_2$ and $\mu_1(S_1) = \mu_2(S_2)$, that is $\mu_2 \circ \Phi = \mu_1$ and $\mu_2 = \mu_1 \circ \Phi^{-1}$ relative to $\mathcal{X}_{1,2}^{(0)}$;
2. $\mathcal{X}_{1,2}^{(0)} \subseteq T_{1,2}^{-1}(\mathcal{X}_{1,2}^{(0)})$; namely, the specially selected subsets $\mathcal{X}_{1,2}^{(0)}$ must be mapped into themselves by the dynamics;
3. $\Phi(T_1 x_1) = T_2 \Phi(x_1)$, that is $T_2 \circ \Phi = \Phi \circ T_1$ and $\Phi^{-1} \circ T_2 = T_1 \circ \Phi^{-1}$ relative to $\mathcal{X}_{1,2}^{(0)}$.

Because of these properties, it turns out that, if $(\mathcal{X}_{1,2}, T_{1,2}, \mu_{1,2})$ are isomorphic, then $\mathrm{h}_{\mu_1}^{\mathrm{KS}}(T_1) = \mathrm{h}_{\mu_2}^{\mathrm{KS}}(T_2)$. The proof is as follows: if $\mathcal{X}_{1,2}^{(0)} = \mathcal{X}_{1,2}$, to any partition $\mathcal{P}_1$ of $\mathcal{X}_1$ there corresponds a partition $\mathcal{P}_2 := \Phi(\mathcal{P}_1)$ and vice versa, the same being true of the refined partitions $\mathcal{P}_1^{(n)}$ that are mapped into partitions $\bigvee_{j=0}^{n-1} \Phi \circ T_1^{-j}(\mathcal{P}_1) = \bigvee_{j=0}^{n-1} T_2^{-j}(\Phi(\mathcal{P}_1)) = \mathcal{P}_2^{(n)}$. The result thus follows since $\mu_1 \lceil \mathcal{P}_1^{(n)} = \mu_2 \lceil \mathcal{P}_2^{(n)} \Longrightarrow H_{\mu_1}(\mathcal{P}_1^{(n)}) = H_{\mu_2}(\mathcal{P}_2^{(n)})$.

If $\mathcal{X}_{1,2}^{(0)} \subset \mathcal{X}_{1,2}$, consider a finite, measurable partition $\mathcal{P}_1 = \{P_i^{(1)}\}_{i=1}^p$ of $\mathcal{X}_1$ and construct the partition $\mathcal{P}_2$ of $\mathcal{X}_2$ with atoms $P_i^{(2)} := \Phi(P_i^{(1)} \cap \mathcal{X}_1^{(0)})$, $i = 1, 2, \ldots, p$ and $P_{p+1}^{(2)} := \mathcal{X}_2 \setminus \mathcal{X}_2^{(0)}$. Since the latter atom has measure $\mu_2(P_{p+1}^{(2)}) = 0$, from the properties of $\mathcal{X}_{1,2}^{(0)}$ and the isomorphism $\Phi$, it turns out that $H_{\mu_2}(\mathcal{P}_2^{(n)}) = H_{\mu_1}(\mathcal{P}_1^{(n)})$. This gives $\mathrm{h}_{\mu_2}^{\mathrm{KS}}(T_2) \geq \mathrm{h}_{\mu_1}^{\mathrm{KS}}(T_1)$; indeed, $\mathcal{P}_1$ is a generic partition of $\mathcal{X}_1$, but $\mathcal{P}_2$ is not so for $\mathcal{X}_2$; the result thus follows by exchanging the roles of the two dynamical systems.

Concluding, two isomorphic dynamical systems must have the same dynamical entropy; since dynamical systems with the same dynamical entropy need not be isomorphic, the latter is not a *complete invariant* [61, 91].

**Example 3.1.1.** [61] Suppose the discrete-time dynamics of $(\mathcal{X}, T, \mu)$ is sampled by observing the time-evolving system not at each tick of the clock, rather every $k$ ticks; then

$$\mathrm{h}_{\mu}^{\mathrm{KS}}(T^k) = k\,\mathrm{h}_{\mu}^{\mathrm{KS}}(T) . \tag{3.12}$$

Indeed, consider Remark 3.1.1.7: since $\mathcal{Q}$ depends on $\mathcal{P}$ in a specific way, by varying $\mathcal{P}$, one does not in general exhaust the whole class of finite measurable partitions of $\mathcal{X}$. Then,

$$\mathrm{h}_{\mu}^{\mathrm{KS}}(T^k) \geq \sup_{\mathcal{P}} \mathrm{h}_{\mu}^{\mathrm{KS}}(T^k, \mathcal{Q}) = k\,\mathrm{h}_{\mu}^{\mathrm{KS}}(T) .$$

On the other hand, Remark 3.1.1.3 and $\mathcal{P} \preceq \mathcal{Q}$ yield

$$k\,\mathrm{h}_{\mu}^{\mathrm{KS}}(T, \mathcal{P}) = \mathrm{h}_{\mu}^{\mathrm{KS}}(T^k, \mathcal{Q}) \geq \mathrm{h}_{\mu}^{\mathrm{KS}}(T^k, \mathcal{P}) \Longrightarrow k\,\mathrm{h}_{\mu}^{\mathrm{KS}}(T) \geq \mathrm{h}_{\mu}^{\mathrm{KS}}(T^k) .$$

The technical difficulty of computing the sup in Definition 3.1.2 is overcome when there does exist a generating partition $\mathcal{P}$ (see Definition 2.3.5) such that, together with its images at different times $\mathcal{P}^j = T^{-j}(\mathcal{P})$, it provides refined partitions $\mathcal{P}^{(n)}$ that generate the $\sigma$-algebra $\Sigma$ of $\mathcal{X}$ when $n \to \infty$.

**Theorem 3.1.1 (Kolmogorov-Sinai Theorem).** *If the partition $\mathcal{P}$ is generating for $(\mathcal{X}, T, \mu)$, then $\mathrm{h}_{\mu}^{\mathrm{KS}}(T) = \mathrm{h}_{\mu}^{\mathrm{KS}}(T, \mathcal{P})$.*

**Proof:** Consider $T$ invertible (for $T$ not invertible the argument is the same) and a generic finite, measurable partition $\mathcal{Q}$; because of the assumption, using Example 2.4.4, for any $\varepsilon > 0$ one can find an $n \geq 0$ and a finite partition $\widetilde{\mathcal{P}} \preceq \mathcal{P}_{-n,n} := \bigvee_{j=-n}^{n} \mathcal{P}^j$, that is a partition generated by finite unions of atoms of $\mathcal{P}_{-n,n}$, such that the conditional entropy $H_\mu(\mathcal{Q}|\widetilde{\mathcal{P}}) \leq \varepsilon$. Therefore, from (3.9) and (3.10) together with Corollary 2.4.2, one derives

$$\mathrm{h}_\mu^{\mathrm{KS}}(T,\mathcal{Q}) \leq \mathrm{h}_\mu^{\mathrm{KS}}(T,\mathcal{P}_{-n,n}) + H_\mu\Big(\mathcal{Q}|\mathcal{P}_{-n,n}\Big) \leq \mathrm{h}_\mu^{\mathrm{KS}}(T,\mathcal{P}) + H_\mu(\mathcal{Q}|\widetilde{\mathcal{P}})$$
$$\leq \mathrm{h}_\mu^{\mathrm{KS}}(T,\mathcal{P}) + \varepsilon ,$$

whence, choosing $\mathcal{Q}$ such that $\mathrm{h}_\mu^{\mathrm{KS}}(T) \leq h_\mu(T,\mathcal{Q}) + \varepsilon$, it follows that

$$\mathrm{h}_\mu^{\mathrm{KS}}(T) - \varepsilon \leq \mathrm{h}_\mu^{\mathrm{KS}}(T,\mathcal{Q}) \leq \mathrm{h}_\mu^{\mathrm{KS}}(T,\mathcal{P}) + \varepsilon$$

with $\varepsilon > 0$ arbitrary. □

The following corollaries are often useful for concretely computing the dynamical entropy.

**Corollary 3.1.1.** *Suppose $\{\mathcal{P}_n\}_{n\in\mathbb{N}}$ is a sequence of finite partitions for $(\mathcal{X},T,\mu)$ of increasing finesse, $\mathcal{P}_n \preceq \mathcal{P}_{n+1}$, such that $\bigvee_n \mathcal{P}_n = \Sigma$. Then,*

$$\mathrm{h}_\mu^{\mathrm{KS}}(T) = \lim_{n\to\infty} \mathrm{h}_\mu^{\mathrm{KS}}(T,\mathcal{P}_n) \ .$$

**Proof:** Given $\varepsilon > 0$, let $\mathcal{Q}$ be a finite, measurable partition such that $\mathrm{h}_\mu^{\mathrm{KS}}(T) \leq \mathrm{h}_\mu^{\mathrm{KS}}(T,\mathcal{Q}) + \varepsilon$; from the assumption and Corollary 2.4.2 it follows that there exist $n \in \mathbb{N}$ and $\widetilde{\mathcal{Q}} \preceq \mathcal{P}_n$ such that

$$\mathrm{h}_\mu^{\mathrm{KS}}(T) - \varepsilon \leq \mathrm{h}_\mu^{\mathrm{KS}}(T,\mathcal{Q}) = \mathrm{h}_\mu^{\mathrm{KS}}(T,\mathcal{P}_n) + H_\mu(\mathcal{Q}|\mathcal{P}_n)$$
$$< \mathrm{h}_\mu^{\mathrm{KS}}(T,\mathcal{P}_n) + H_\mu(\mathcal{Q}|\widetilde{\mathcal{Q}})$$
$$\leq \mathrm{h}_\mu^{\mathrm{KS}}(T,\mathcal{P}_n) + \varepsilon \leq \mathrm{h}_\mu^{\mathrm{KS}}(T) + \varepsilon \ .$$

□

A similar argument as in the previous proof can be used to show

**Corollary 3.1.2.** *Given $(\mathcal{X},T,\mu)$, suppose $\Sigma_0$ is a measure algebra that generates the $\sigma$-algebra $\Sigma$ of $\mathcal{X}$. Then,*

$$\mathrm{h}_\mu^{\mathrm{KS}}(T) = \sup_{\mathcal{P}\subseteq\Sigma_0} \mathrm{h}_\mu^{\mathrm{KS}}(T,\mathcal{P}) \ .$$

**Examples 3.1.2.**

1. Given two dynamical systems $(\mathcal{X}_i,T_i,\mu_i)$, $i = 1,2$, their *direct product* $(\mathcal{X}_1\times\mathcal{X}_2, T_1\times T_2, \mu_1\times\mu_2)$ provides a new dynamical system $(\mathcal{X},T,\mu)$ consisting of two statistically and dynamically independent components. Concretely, $X := \mathcal{X}_1\times\mathcal{X}_2$ is the phase-space consisting of points $x = (x_1,x_2)$, $x_{1,2}\in\mathcal{X}_{1,2}$ and the dynamics $T$ is such that $Tx = (T_1x_1, T_2x_2)$. Furthermore, if $\Sigma_{1,2}$ are the $\sigma$-algebras of $\mathcal{X}_{1,2}$, then $\mathcal{X}$ remains equipped with the $\sigma$-algebra $\Sigma = \Sigma_1\times\Sigma_2$ of measurable sets of the form $S_1\times S_2$,

$S_{1,2} \in \Sigma_{1,2}$ and with the $T$-invariant measure on $\mathcal{X}$, $\mu = \mu_1 \times \mu_2$, defined by $\mu(S_1 \times S_2) = \mu_1(S_1)\mu_2(S_2)$. Then, [61]

$$\mathrm{h}^{\mathrm{KS}}_{\mu_1 \times \mu_2}(T_1 \times T_2) = \mathrm{h}^{\mathrm{KS}}_{\mu_1}(T_1) + \mathrm{h}^{\mathrm{KS}}_{\mu_2}(T_2) \ .$$

Indeed, $\Sigma$ is generated by the measure algebra $\bigcup_{\mathcal{P}_{1,2}} \mathcal{P}_1 \times \mathcal{P}_2$ where $\mathcal{P}_{1,2}$ are generic finite, measurable partitions in $\mathcal{X}_{1,2}$; thus, from Corollary 3.1.2 and statistical independence,

$$\begin{aligned}
\mathrm{h}^{\mathrm{KS}}_{\mu}(T) &= \sup_{\mathcal{P}_1 \times \mathcal{P}_2} \mathrm{h}^{\mathrm{KS}}_{\mu}(T, \mathcal{P}_1 \times \mathcal{P}_2) \\
&= \sup_{\mathcal{P}_1} \mathrm{h}^{\mathrm{KS}}_{\mu_1}(T_1, \mathcal{P}_1) + \sup_{\mathcal{P}_2} \mathrm{h}^{\mathrm{KS}}_{\mu_2}(T_2, \mathcal{P}_2) \\
&= \mathrm{h}^{\mathrm{KS}}_{\mu_1}(T_1) + \mathrm{h}^{\mathrm{KS}}_{\mu_2}(T_2) \ .
\end{aligned}$$

2. **Bernoulli Systems:** (see Example 2.1.4) let $\mu$ be a product measure such that $p^{(n)}(\boldsymbol{i}^{(n)}) = \prod_{j=0}^{n-1} p(i_j)$. As seen in Example 2.3.3.1, the partition $\mathcal{C}$ of $\Omega_p$ into $C^0_j := \Big\{ \boldsymbol{i} \in \Omega_p \ : \ i_j \in \{1, 2, \ldots, p\} \Big\}$ is generating for the $\sigma$-algebra of cylinders. Therefore,

$$\begin{aligned}
\mathrm{h}^{\mathrm{KS}}_{\mu}(T_\sigma) = \mathrm{h}^{\mathrm{KS}}_{\mu}(T_\sigma, \mathcal{C}) &= \lim_{n\to\infty} \frac{1}{n} H_\mu(\mathcal{C}^{(n)}) \\
&= - \lim_{n\to\infty} \frac{1}{n} \sum_{\boldsymbol{i}^{(n)} \in \Omega_p^{(n)}} \left( \prod_{j=0}^{n-1} p(i_j) \right) \sum_{j=1}^{p} \log p(i_j) \\
&= - \sum_{i=1}^{p} p(i) \log p(i) = H_\mu(\mathcal{C}) \ .
\end{aligned}$$

3. **Markov Processes:** Let the measure in $\Big(\Omega_p, T_\sigma, \mu\Big)$ be given, as in Example 2.4.2.2, by $p^{(n)}(\boldsymbol{i}^{(n)}) = p(i_0) \prod_{j=1}^{n} p(i_j | i_{j-1})$ on $\Omega_p^{(n)}$. Again, the partition $\mathcal{C}$ of the previous example is generating. Therefore, since $\sum_{i=1}^{p} p(i|j) = 1$,

$$\begin{aligned}
&\mathrm{h}^{\mathrm{KS}}_{\pi}(T_\sigma) = \mathrm{h}^{\mathrm{KS}}_{\pi}(T_\sigma, \mathcal{C}) = \lim_{n\to\infty} \frac{1}{n} H_\pi(\mathcal{C}^{(n)}) \\
&= - \lim_{n\to\infty} \frac{1}{n} \sum_{\boldsymbol{i}^{(n)} \in \Omega_p^{(n)}} p(i_0) \prod_{j=1}^{n-1} p(i_j | i_{j-1}) \Big( p(i_0) \prod_{k=1}^{n-1} \log_2 p(i_k | i_{k-1}) \Big) \\
&= - \sum_{i,j=1}^{p} p(i) p(j|i) \log_2 p(j|i) \ .
\end{aligned}$$

4. **Ergodic Rotations:** Consider the irrational rotations on the $\mathbb{T}^2$ described by the triplets $\left(\mathbb{T}^2, T, \mathrm{d}\boldsymbol{\theta}\right)$. As seen in Example 2.3.3.2, there is a generating partition $\mathcal{C}$ such that

$$\bigvee_{j=0}^{\infty} T^{-j}(\mathcal{C}) = \Sigma = T(\Sigma) = \bigvee_{j=1} T^{-j}(\mathcal{C}) \ ,$$

where the last two equalities follow from the invertibility of the dynamics $T$. Then, as in Example 2.4.4, for any $\varepsilon > 0$, one can find an $n \in \mathbb{N}$ and a partition $\widetilde{\mathcal{C}} \preceq \bigvee_{j=1}^{n} T^{-j}(\mathcal{C})$ such that $H_\mu(\mathcal{C}|\widetilde{\mathcal{C}}) \leq \varepsilon$. It thus follows from Corollary 2.4.2 that

$$H_\mu\left(\mathcal{C}\Big|\bigvee_{j=1}^{n} T^{-j}(\mathcal{C})\right) \leq H_\mu(\mathcal{C}|\widetilde{\mathcal{C}}) \leq \varepsilon \ ,$$

whence, from (3.4), $\mathrm{h}_\mu^{\mathrm{KS}}(T) = 0$. This very same argument holds for all reversible dynamical systems $\left(\mathcal{X}, T, \mu\right)$ that possess a partition $\mathcal{P}$ which generates the $\sigma$-algebra of $\mathcal{X}$ as $\Sigma = \bigvee_{j=0}^{\infty} \mathcal{P}^j$.

5. **Non-ergodic Rotations:** Unlike in the previous example, there exists $k \in \mathbb{N}$ such that $T^k = \mathbb{1}$, the trivial dynamics with $\mathrm{h}_\mu^{\mathrm{KS}}(\mathbb{1}) = 0$. Then, from Example 3.1.1, $0 = \mathrm{h}_\mu^{\mathrm{KS}}(\mathbb{1}) = \mathrm{h}_\mu^{\mathrm{KS}}(T^k) = k\, \mathrm{h}_\mu^{\mathrm{KS}}(T)$.

### *KS* entropy and Lyapounov Exponents

In Section 2.2, Lyapounov exponents (see Definition 2.1.2) have been introduced as indicators of hyperbolic behavior, that is of exponential separation of initially close trajectories. In Example 2.2.2 this has been calculated to be $\log 2$ for the Baker map, which is isomorphic to a Bernoulli shift $(\Omega_2, T_\sigma, \mu_B)$ with a balanced probability measure $\mu_B$; therefore, according to Example 3.1.2.2, the Lyapounov exponent equals the *KS* entropy for this system.

From an informational point of view this fact can be understood as being due to the loss of information along the direction where distances and thus errors increase exponentially fast [62, 271]. It is therefore plausible to expect that all possible expanding directions contribute with their Lyapounov exponents to the loss of information and thus to the *KS* entropy (see Remark 2.1.3.1). This is indeed the content of the following theorem [199]:

**Theorem 3.1.2 (Pesin Theorem).** *Let $(\mathcal{X}, T, \mu)$ be a smooth dynamical systems as in Remark 2.1.3.1; set* $\Lambda(x) := \sum_{j:\lambda^{(j)}(x)\geq 0} \lambda^{(j)}(x)\, \mathrm{dim} W_j(x)$*; then,*

$$\mathrm{h}_\mu^{\mathrm{KS}}(T) = \int_{\mathcal{X}} \mathrm{d}\mu(x)\, \Lambda(x) \ .$$

When the dynamical triplet $(\mathcal{X}, T, \mu)$ is ergodic, the Lyapounov exponents, which are constants of the motion, are constant almost everywhere on $\mathcal{X}$, whence *Pesin's equality* assumes the simpler expression

$$\mathrm{h}_\mu^{\mathrm{KS}}(T) = \sum_{j:\lambda^{(j)} \geq 0} \lambda^{(j)} .$$

A particular instance of Pesin's result applied to hyperbolic dynamical systems [313, 164] is provided by Example 8.2.4.

**Proposition 3.1.1.** *The KS entropy of the hyperbolic automorphisms of the torus with positive eigenvalues $\alpha^{\pm 1}$ of the matrix $\mathbb{A}$ is*

$$\mathrm{h}_\mu^{\mathrm{KS}}(T_{\mathbb{A}}) = \log \alpha .$$

Standard proofs of this result can be found in [164, 279]; here, we prefer to defer it to Chapter 8, where it will be obtained by means of a quantum dynamical entropy (see Proposition 8.2.7 and Remark 8.2.4).

### 3.1.1 Entropic $K$-systems

In Section 2.3.1, $K$-systems have been defined in terms of the existence of a $K$-sequence $\{\Sigma\}_{n\in\mathbb{Z}}$ of nested $\sigma$-subalgebras (see Definition 2.3.4) or of an algebraic $K$-sequence of nested Abelian von Neumann subalgebras (see Definition 2.3.6). We will now show that the algebraic characterization is equivalent to the following entropic properties, the link being the triviality of the tails of all finite partitions (see (2.73)).

**Theorem 3.1.3.** *[91, 216] Let $(\mathcal{X}, T, \mu)$ be a dynamical triplet, the following ones are equivalent properties:*

1. *there exists a $K$-sequence $\{\mathcal{P}_n\}_{n\in\mathbb{Z}}$ based upon a finite generating partition $\mathcal{P}$ (see Definition 2.3.5);*
2. $\mathrm{Tail}(\mathcal{Q}) = \mathcal{N}$ *for any finite measurable partition, where $\mathcal{N}$ is the trivial partition of $\mathcal{X}$;*
3. *for all finite measurable partitions $\mathcal{Q}$ of $\mathcal{X}$,*
$$\mathrm{h}_\mu^{\mathrm{KS}}(T, \mathcal{Q}) > 0 ; \tag{3.13}$$
4. *for all finite measurable partitions $\mathcal{Q}$ of $\mathcal{X}$*
$$\lim_{n\to+\infty} \mathrm{h}_\mu^{\mathrm{KS}}(T^n, \mathcal{Q}) = H_\mu(\mathcal{Q}) ; \tag{3.14}$$
5. *for any two finite measurable partitions $\mathcal{Q}_{1,2}$,*
$$\lim_{n\to+\infty} H_\mu\Big(\mathcal{Q}_1 | \bigvee_{k\geq n} T^{-k}(\mathcal{Q}_2)\Big) = H_\mu(\mathcal{Q}_1) ; \tag{3.15}$$

*6. for any two finite measurable partitions* $\mathcal{Q}_{1,2}$,

$$\lim_{n\to+\infty} H_\mu\Big(\mathcal{Q}_1 | \bigvee_{k\geq n} T^{-k}(\mathcal{Q}_2)\Big) = 0 \Longrightarrow \mathcal{Q}_1 = \mathcal{N} \ . \tag{3.16}$$

From the characterization of $K$-mixing by the triviality of the tails of all their finite partitions (condition (2) above), Proposition 2.3.5 gives

**Corollary 3.1.3.** *A dynamical triplet* $(\mathcal{X}, T\, mu)$ *with a finite generating partition is a* $K$*-system if and only if it is* $K$*-mixing.*

The key observation in the proof of Theorem 3.1.3 is the continuity of the conditional probabilities as stated in Theorem 2.2.1 and the continuity of entropies and conditional entropies with respect to their arguments. This fact allows us to recast (3.4) in the more suggestive form

$$\mathrm{h}^{\mathrm{KS}}_\mu(T, \mathcal{P}) = \lim_{n\to\infty} H_\mu\Big(\mathcal{P}\Big|\bigvee_{j=1}^{n} \mathcal{P}^j\Big) = H_\mu\Big(\mathcal{P}\Big|\bigvee_{j=1}^{+\infty} \mathcal{P}^j\Big) \ , \tag{3.17}$$

where $\mathcal{P}^j = T^{-j}(\mathcal{P})$. Also, by means of (2.73), in (3.15) and (3.16) one rewrites

$$\begin{aligned}\lim_{n\to+\infty} H_\mu\Big(\mathcal{Q}_1 | \bigvee_{k\geq n} T^{-k}(\mathcal{Q}_2)\Big) &= H_\mu\Big(\mathcal{Q}_1 | \lim_{n\to+\infty} \bigvee_{k\geq n} T^{-k}(\mathcal{Q}_2)\Big) \\ &= H_\mu\Big(\mathcal{Q}_1 | \mathrm{Tail}\,(\mathcal{Q}_2)\Big) \ . \end{aligned} \tag{3.18}$$

We shall also need the following two results [91].

**Lemma 3.1.1.** *Given two finite partitions* $\mathcal{Q}_{1,2}$,

$$H_\mu\Big(\mathcal{Q}_1 | \bigvee_{n=1}^{+\infty} T^{-n}(\mathcal{Q}_1) \vee \mathrm{Tail}\,(\mathcal{Q}_2)\Big) = \mathrm{h}^{\mathrm{KS}}_\mu(T, \mathcal{Q}_1) \ . \tag{3.19}$$

**Proof:** As a first step, observe that, given a finite partition $\mathcal{Q}$, repeatedly using (2.91) and (3.3) yield

$$\begin{aligned} H_\mu\Big(\bigvee_{k=1}^{n} T^k(\mathcal{Q})\Big|\bigvee_{j=1}^{+\infty} T^{-j}(\mathcal{Q})\Big) &= \sum_{k=0}^{n-1} H_\mu\Big(T^k(\mathcal{Q})\Big| \bigvee_{j=-k+1}^{+\infty} T^{-j}(\mathcal{Q})\Big) \\ &= n\, H_\mu\Big(\mathcal{Q}\Big|\bigvee_{j=1}^{+\infty} T^{-j}(\mathcal{Q})\Big) = n\, \mathrm{h}^{\mathrm{KS}}_\mu(T, \mathcal{Q}) \ . \end{aligned}$$

Then, for fixed $\varepsilon > 0$ and sufficiently large $n$, using Corollary 2.4.2 and Definition 3.1.1 one gets

$$\begin{aligned}
\mathrm{h}_\mu^{\mathrm{KS}}(T, \mathcal{Q}_1 \vee \mathcal{Q}_2) &= \frac{1}{n} \underbrace{H_\mu\Big(\bigvee_{k=0}^{n-1} T^k(\mathcal{Q}_1 \vee \mathcal{Q}_2)\Big| \bigvee_{j=1}^{+\infty} T^j(\mathcal{Q}_1 \vee \mathcal{Q}_2)\Big)}_{L_1(n)} \\
&\leq \frac{1}{n} \underbrace{H_\mu\Big(\bigvee_{k=0}^{n-1} T^k(\mathcal{Q}_1 \vee \mathcal{Q}_2)\Big| \bigvee_{j=1}^{+\infty} T^j(\mathcal{Q}_1)\Big)}_{L_2(n)} \\
&\leq \frac{1}{n} H_\mu\Big(\bigvee_{k=0}^{n-1} T^k(\mathcal{Q}_1 \vee \mathcal{Q}_2)\Big) \leq \mathrm{h}_\mu^{\mathrm{KS}}(T, \mathcal{Q}_1 \vee \mathcal{Q}_2) + \varepsilon\ .
\end{aligned}$$

Thus, $\lim_{n \to +\infty} \frac{L_1(n)}{n} = \lim_{n \to +\infty} \frac{L_2(n)}{n}$. Further, Lemma 2.91 yields

$$\begin{aligned}
L_1(n) = &\underbrace{H_\mu\Big(\bigvee_{k=0}^{n-1} T^k(\mathcal{Q}_1)\Big| \bigvee_{j=1}^{+\infty} T^j(\mathcal{Q}_1 \vee \mathcal{Q}_2)\Big)}_{L_{11}(n)} \\
&+ \underbrace{H_\mu\Big(\bigvee_{k=0}^{n-1} T^k(\mathcal{Q}_1)\Big| \bigvee_{j=1}^{+\infty} T^j(\mathcal{Q}_2) \vee \bigvee_{j=-n+1}^{+\infty} T^{-j}(\mathcal{Q}_1)\Big)}_{L_{12}(n)}
\end{aligned}$$

$$L_2(n) = \underbrace{H_\mu\Big(\bigvee_{k=0}^{n-1} T^k(\mathcal{Q}_1)\Big| \bigvee_{j=1}^{+\infty} T^j(\mathcal{Q}_1)\Big)}_{L_{21}(n)} + \underbrace{H_\mu\Big(\bigvee_{k=0}^{n-1} T^k(\mathcal{Q}_2)\Big| \bigvee_{j=-n+1}^{+\infty} T^j(\mathcal{Q}_1)\Big)}_{L_{22}(n)}\ .$$

Corollary 2.4.1 implies $L_{11}(n) \leq L_{21}(n)$ and $L_{11}(n) \leq L_{21}(n)$, then

$$\mathrm{h}_\mu^{\mathrm{KS}}(T, \mathcal{Q}_1) = \lim_{n \to +\infty} \frac{L_{21(n)}}{n} = \lim_{n \to +\infty} \frac{L_{11(n)}}{n}\ .$$

By applying (2.91) and then the argument that led to (3.4) one gets

$$\begin{aligned}
\mathrm{h}_\mu^{\mathrm{KS}}(T, \mathcal{Q}_1) &= \lim_{n \to +\infty} \frac{L_{11(n)}}{n} \\
&= \lim_{n \to +\infty} \frac{1}{n} \sum_{k=0}^{n-1} H_\mu\Big(T^k(\mathcal{Q}_1)\Big| \bigvee_{j=1}^{+\infty} T^{-j}(\mathcal{Q}_2) \vee \bigvee_{r=-k+1}^{+\infty} T^{-r}(\mathcal{Q}_1)\Big) \\
&= \lim_{n \to +\infty} \frac{1}{n} \sum_{k=0}^{n-1} H_\mu\Big(\mathcal{Q}_1\Big| \bigvee_{j=k+1}^{+\infty} T^{-j}(\mathcal{Q}_2) \vee \bigvee_{r=1}^{+\infty} T^{-r}(\mathcal{Q}_1)\Big)
\end{aligned}$$

$$= \lim_{n\to+\infty} H_\mu\Big(\mathcal{Q}_1\Big|\underbrace{\bigvee_{j=n}^{+\infty} T^{-j}(\mathcal{Q}_2)\vee\bigvee_{r=1}^{+\infty} T^{-r}(\mathcal{Q}_1)}_{\mathcal{C}_n}\Big) = H_\mu\Big(\mathcal{Q}_1\Big|\bigwedge_{n\geq 0}\mathcal{C}_n\Big)$$

$$\leq H_\mu\Big(\mathcal{Q}_1\Big|\mathrm{Tail}\,(\mathcal{Q}_2)\vee\bigvee_{r=1}^{+\infty} T^{-r}(\mathcal{Q}_1)\Big) \leq \mathrm{h}_\mu^{\mathrm{KS}}\,(T,\mathcal{Q}_1)\ .$$

The last equality follows from the fact that the partitions $\mathcal{C}_n$ (not finite in general) are such that $\mathcal{C}_n \preceq \mathcal{C}_{n-1}$, whereas for the last but one inequality Corollary 2.4.2 has been used and the fact that

$$\mathrm{Tail}\,(\mathcal{Q}_2)\vee\bigvee_{r=1}^{+\infty} T^{-r}(\mathcal{Q}_1) = \left(\bigwedge_{n\geq 0}\bigvee_{k\geq n} T^{-k}(\mathcal{Q}_2)\right)\vee\bigvee_{r=1}^{+\infty} T^{-r}(\mathcal{Q}_1) \preceq \bigwedge_{n\geq 0}\mathcal{C}_n\ .$$

□

**Lemma 3.1.2.** *Given two finite partitions* $\mathcal{Q}_{1,2}$,

$$\mathcal{Q}_2 \preceq \bigvee_{n\in\mathbb{Z}} T^n(\mathcal{Q}_1) \Longrightarrow \mathrm{Tail}\,(\mathcal{Q}_2) \preceq \mathrm{Tail}\,(\mathcal{Q}_1)\ . \tag{3.20}$$

**Proof:** We shall show that all partitions $\mathcal{Q} \preceq \mathrm{Tail}\,(\mathcal{Q}_2)$ are such that $\mathcal{Q} \preceq \mathrm{Tail}\,(\mathcal{Q}_1)$, too. Notice that $\mathcal{Q} \preceq \bigvee_{n\in\mathbb{Z}} T^n(\mathcal{Q}_1)$, by hypothesis. If

$$H_\mu\Big(\mathcal{P}\Big|\mathrm{Tail}\,(\mathcal{Q}_1)\vee\mathcal{Q}\Big) = H_\mu\Big(\mathcal{P}\Big|\mathrm{Tail}\,(\mathcal{Q}_1)\Big) \qquad (*)$$

for all finite partitions $\mathcal{P} \preceq \bigvee_{k=-n}^{n} T^k(\mathcal{Q}_1)$, then by approximating $\mathcal{Q}$ arbitrarily well by $\bigvee_{k=-n}^{n} T^k(\mathcal{Q}_1)$, continuity allows one to substitute $\mathcal{Q}$ for $\mathcal{P}$ in $(*)$. Then, (2.91) implies

$$H_\mu\Big(\mathcal{Q}\Big|\mathrm{Tail}\,(\mathcal{Q}_1)\vee\mathcal{Q}\Big) = H_\mu\Big(\mathcal{Q}\Big|\mathrm{Tail}\,(\mathcal{Q}_1)\Big) \Longrightarrow \mathcal{Q} \preceq \mathrm{Tail}\,(\mathcal{Q}_1)\ .$$

Equality $(*)$ is proved as follows: a repeated use of (2.91), together with the $T$-invariance of $\mathcal{Q}$ (see Remark 2.3.4) and Remark 3.1.1.4, yield

$$\underbrace{H_\mu\Big(\bigvee_{k=-n}^{n} T^k(\mathcal{Q}_1)\Big|\bigvee_{j=n+1}^{+\infty} T^{-j}(\mathcal{Q}_1)\vee\mathcal{Q}\Big)}_{L_1(n)} =$$

$$= \sum_{k=-n}^{n} H_\mu\Big(T^k(\mathcal{Q}_1)\Big|\bigvee_{j=-k+1}^{+\infty} T^{-j}(\mathcal{Q}_1)\vee\mathcal{Q}\Big) = 2n\, H_\mu\Big(\mathcal{Q}_1\Big|\bigvee_{k=1}^{+\infty} T^{-j}(\mathcal{Q}_1)\vee\mathcal{Q}\Big)$$

as well as

$$\underbrace{H_\mu\Big(\bigvee_{k=-n}^{n} T^k(\mathcal{Q}_1)\Big|\bigvee_{j=n+1}^{+\infty} T^{-j}(\mathcal{Q}_1)\Big)}_{L_2(n)} = 2n\, H_\mu\Big(\mathcal{Q}_1\Big|\bigvee_{k=1}^{+\infty} T^{-j}(\mathcal{Q}_1)\Big)$$

$$= 2n\, \mathrm{h}_\mu^{\mathrm{KS}}(T, \mathcal{Q}_1) \ .$$

Since $\mathcal{Q}$ is $T$-invariant, it coincides with Tail $(\mathcal{Q})$, whence Lemma 3.1.2 ensures that $L_1(n) = L_2(n)$. Furthermore, since

$$\mathcal{P} \preceq \bigvee_{k=-n}^{n} T^k(\mathcal{Q}_1) \Longrightarrow \mathcal{P} \vee \bigvee_{k=-n}^{n} T^k(\mathcal{Q}_1) = \bigvee_{k=-n}^{n} T^k(\mathcal{Q}_1) \ ,$$

by using (2.91) one gets

$$L_1(n) = \underbrace{H_\mu\Big(\mathcal{P}\Big|\bigvee_{j=n+1}^{+\infty} T^{-j}(\mathcal{Q}_1) \vee \mathcal{Q}\Big)}_{L_{11}(n)}$$

$$+ \underbrace{H_\mu\Big(\bigvee_{k=-n}^{n} T^k(\mathcal{Q}_1)\Big|\mathcal{P} \vee \bigvee_{j=n+1}^{+\infty} T^{-j}(\mathcal{Q}_1) \vee \mathcal{Q}\Big)}_{L_{12}(n)}$$

$$L_2(n) = \underbrace{H_\mu\Big(\mathcal{P}\Big|\bigvee_{j=n+1}^{+\infty} T^{-j}(\mathcal{Q}_1)\Big)}_{L_{21}(n)} + \underbrace{H_\mu\Big(\bigvee_{k=-n}^{n} T^k(\mathcal{Q}_1)\Big|\mathcal{P} \vee \bigvee_{j=n+1}^{+\infty} T^{-j}(\mathcal{Q}_1)\Big)}_{L_{22}(n)} \ .$$

Since $L_{11}(n) \leq L_{21}(n)$ and $L_{12}(n) \leq L_{22}(n)$, $L_1(n) = L_2(n)$ gives

$$H_\mu\Big(\mathcal{P}\Big|\bigvee_{j=n+1}^{+\infty} T^{-j}(\mathcal{Q}_1) \vee \mathcal{Q}\Big) = H_\mu\Big(\mathcal{P}\Big|\bigvee_{j=n+1}^{+\infty} T^{-j}(\mathcal{Q}_1)\Big)$$

for all $n \geq 0$. Therefore (see the proof of the previous lemma),

$$H_\mu\Big(\mathcal{P}\Big|\bigwedge_{n\geq 0}\Big(\bigvee_{j=n}^{+\infty} T^{-j}(\mathcal{Q}_1) \vee \mathcal{Q}\Big)\Big) = H_\mu\Big(\mathcal{P}\Big|\mathrm{Tail}\,(\mathcal{Q}_1)\Big) \ .$$

The equality (∗) thus follows from Corollary 2.4.2 and

$$\mathrm{Tail}\,(\mathcal{Q}_1) \vee \mathcal{Q} \preceq \bigwedge_{n\geq 0}\Big(\bigvee_{j=n}^{+\infty} T^{-j}(\mathcal{Q}_1) \vee \mathcal{Q}\Big) \qquad \text{so that}$$

$$H_\mu\Big(\mathcal{P}\Big|\mathrm{Tail}\,(\mathcal{Q}_1)\Big) \geq H_\mu\Big(\mathcal{P}\Big|\mathrm{Tail}\,(\mathcal{Q}_1) \vee \mathcal{Q}\Big)$$

$$\geq H_\mu\Big(\mathcal{P}\Big|\bigwedge_{n\geq 0}\Big(\bigvee_{j=n}^{+\infty} T^{-j}(\mathcal{Q}_1) \vee \mathcal{Q}\Big)\Big) = H_\mu\Big(\mathcal{P}\Big|\mathrm{Tail}\,(\mathcal{Q}_1)\Big) \ .$$

□

**Proof of Theorem 3.1.3** The equivalences will be proved according to the following scheme:

$$\begin{array}{ccccc} & & (5) & \Longrightarrow & (4) \\ & & \Uparrow & & \Downarrow \\ (1) & \Longleftrightarrow & (2) & \Longleftrightarrow & (3) \\ & & \Updownarrow & & \\ & & (6) & & \end{array}$$

(1) $\Longrightarrow$ (2): take $\mathcal{Q}_1$ has the $K$-partition $\mathcal{P}$, then Lemma 3.1.2 implies Tail $(Q) \preceq$ Tail $(\mathcal{P}) = \mathcal{N}$ for all finite partitions $\mathcal{Q}$.
(2) $\Longrightarrow$ (1): this is the content of Proposition 2.3.6.
(2) $\Longrightarrow$ (3): if $\mathrm{h}_\mu^{\mathrm{KS}}(T, \mathcal{Q}) = 0$ for a finite partition $\mathcal{Q}$, by means of (3.17) and the argument of Example 2.4.3 extended by continuity to non-finite contexts, one gets $\mathcal{Q} \preceq \bigvee_{n=1}^{+\infty} T^{-n}(\mathcal{Q})$. Then, $T^{-k}(\mathcal{Q}) \preceq \bigvee_{n=k+1}^{+\infty} T^{-n}(\mathcal{Q})$, for all $k \geq 0$, whence

$$\mathrm{Tail}\,(\mathcal{Q}) = \bigwedge_{k\geq 0} \bigvee_{n\geq k} T^{-n}(\mathcal{Q}) = \bigvee_{n=0}^{+\infty} T^{-n}(\mathcal{Q}) \succeq \mathcal{Q} \Longrightarrow \mathcal{Q} = \mathcal{N} \ .$$

(3) $\Longrightarrow$ (2): let $\mathcal{Q}_2$ be a finite partition with Tail $(Q_2) \neq \mathcal{N}$; Lemma 3.1.2 applied to $\mathcal{Q}_1 \preceq$ Tail $(Q_2)$ yields $\mathrm{h}_\mu^{\mathrm{KS}}(T, \mathcal{Q}_1) = 0$, whence $\mathcal{Q}_1 = \mathcal{N}$.
(2) $\Longrightarrow$ (5): using (3.18) one gets

$$H_\mu\Big(\mathcal{Q}_1 | \mathrm{Tail}\,(\mathcal{Q}_2)\Big) = H_\mu(\mathcal{Q}_1 | \mathcal{N}) = H_\mu(\mathcal{Q}_1) \ ,$$

for all finite partitions $\mathcal{Q}_{1,2}$ (see Example 2.4.3).
(2) $\Longrightarrow$ (6): follows from (2) $\Longrightarrow$ (5).
(6) $\Longrightarrow$ (2): suppose $\mathcal{Q}_2$ is a finite partition; if $\mathcal{Q}_1 \preceq$ Tail $(\mathcal{Q}_2)$, (3.18) yields $0 = H_\mu\Big(\mathcal{Q}_1 | \mathrm{Tail}\,(\mathcal{Q}_2)\Big)$. Thus, $\mathcal{Q}_1 = \mathcal{N}$ from (6), whence Tail $(\mathcal{Q}_1) = \mathcal{N}$.
(5) $\Longrightarrow$ (4): consider a finite partition $\mathcal{Q}$ and notice that

$$\bigvee_{k=n}^{+\infty} T^{-k}(\mathcal{Q}) \succeq \bigvee_{j=1}^{+\infty} T^{-jn}(\mathcal{Q}) \ .$$

Then, Corollary 2.4.2, (3.17) and (3.7) imply

$$\begin{aligned} H_\mu(\mathcal{Q}) &= \lim_{n\to+\infty} H_\mu\Big(\mathcal{Q} \Big| \bigvee_{k=n}^{+\infty} T^{-k}(\mathcal{Q})\Big) \\ &\leq \lim_{n\to+\infty} H_\mu\Big(\mathcal{Q} \Big| \bigvee_{j=1}^{+\infty} T^{-jn}(\mathcal{Q})\Big) = \mathrm{h}_\mu^{\mathrm{KS}}(T^n, \mathcal{Q}) \leq H_\mu(\mathcal{Q}) \ . \end{aligned}$$

(4) $\Longrightarrow$ (3): given a finite partition $\mathcal{Q} \neq \mathcal{N}$, choose $\varepsilon > 0$ in such a way that $H_\mu(\mathcal{Q}) - \varepsilon > 0$ and $n$ large enough to have $\mathrm{h}_\mu^{\mathrm{KS}}(T^n, \mathcal{Q}) \geq H_\mu(\mathcal{Q}) - \varepsilon$. Then, from (3.11) one derives

$$h_\mu^{KS}(T, \mathcal{Q}) \geq \frac{1}{n} h_\mu^{KS}(T^n, \mathcal{Q}) \geq \frac{H_\mu(\mathcal{Q}) - \varepsilon}{n} > 0 .$$

□

## 3.2 Codes and Shannon Theorems

As seen in Section 2.4 communication channels usually comprise a preliminary encoding of the source signals. In the following, we shall review some basic facts concerning the role of entropy in this context, with particular reference to compression of information and its transmission through noisy channels.

**Definitions 3.2.1 (Codes).**

1. *A code $\mathcal{E} : I_A \mapsto \Omega_d^*$ for a source $A$ with alphabet $I_A = \{1, 2, \ldots, a\}$ is any map which associates source symbols $i \in \mathrm{i}_A$ with strings of any lengths consisting of symbols $x \in I_X = \{1, 2, \ldots, d\}$:*

$$I_A \ni i \mapsto \mathcal{E}(i) = \boldsymbol{x}^{(n)} = x_1 x_2 \cdots x_n \in \Omega_d^* , \quad x_j \in \{1, 2, \ldots, d\} ,$$

   *where $\Omega_d^*$ denotes the set $\bigcup_{n \geq 1} \Omega_d^{(n)}$.*
2. *A code is* non-singular *if any two different source symbols $i, j \in I_A$ are mapped into different code-words $\mathcal{E}(i) \neq \mathcal{E}(j) \in \Omega_d^*$. In this way, any code-word corresponds to a unique source-symbol.*
3. *The extension of a code $\mathcal{E} : I_A \to \Omega_d^*$ to strings $\boldsymbol{i}^{(\ell)} = i_1 i_2 \cdots i_\ell \in \Omega_a^{(\ell)}$ of length $\ell$ is defined by* concatenation:

$$\Omega_a^{(\ell)} \ni \boldsymbol{i}^{(\ell)} \mapsto \mathcal{E}^{(\ell)}(\boldsymbol{i}^{(\ell)}) = \mathcal{E}(i_1)\mathcal{E}(i_2) \cdots \mathcal{E}(i_\ell) \in \Omega_d^* .$$

4. *A code $\mathcal{E}$ is* uniquely decodable *if its extensions $\mathcal{E}^{(\ell)}$ are non-singular.*
5. *A code $\mathcal{E}$ is a* prefix *or an* instantaneous code *if no code-word prefixes another code-word, that is if no code-word consists in code-symbols added to a code-word.*

**Examples 3.2.1.** [92]

1. Prefix-codes are uniquely decodable and uniquely decodable codes are non-singular.
2. Let $I_A = \{1, 2, 3\}$, $I_X = \{0, 1\}$; $\mathcal{E}(1) = 0$, $\mathcal{E}(2) = 00$, $\mathcal{E}(3) = 01$ is a non-singular code, but not an uniquely decodable one for $\mathcal{E}(11) = \mathcal{E}(2) = 00$.
3. The code $\mathcal{E}(1) = 0$, $\mathcal{E}(2) = 01$, $\mathcal{E}(3) = 11$, is not a prefix-code as $\mathcal{E}(1)$ prefixes $\mathcal{E}(2)$. However, it is uniquely decodable for the following reason. Suppose $\mathcal{E}(\boldsymbol{i}^{(\ell)}) = \mathcal{E}(\boldsymbol{j}^{(\ell)}) = \boldsymbol{x}^{(n)}$: if $x_1 = 1$ then $x_2 = 1$ and $i_1 = j_1 = 3$; if $x_1 = x_2 = 0$, then $i_1 = j_1 = 1$. Finally, if $x_1 = 0$ and $x_2 = 1$ then $i_1 = j_1 = 1$ when $x_3 = 1$, otherwise $i_1 = j_1 = 2$. In this way every string of code-words encodes a unique source-word.

4. The code $\mathcal{E}(1) = 0$, $\mathcal{E}(2) = 10$, $\mathcal{E}(3) = 11$ is such that no string can be prefix to another. Unlike in the previous one, in this case one need not check the next symbol in order to identify the corresponding source-symbol.

Prefix-codes are particularly important because the lengths of their code-words satisfy the following inequality.

**Proposition 3.2.1 (Kraft's Inequality).** *[92] If $\mathcal{E} : I_A \to \Omega_d^*$ is a prefix-code over the alphabet $I_X = \{1, \ldots, d\}$ for a source alphabet $I_A = \{1, 2, \ldots, a\}$ and $\ell_i$ denotes the length of the code-word $\mathcal{E}(i)$, then*

$$\sum_{i=1}^{a} d^{-\ell_i} \leq 1 . \tag{3.21}$$

*This inequality is known as Kraft inequality; vice versa, if a set of lengths $\ell_i$, $i = 1, 2, \ldots, a$ satisfies (3.21), then there exists a prefix-code $\mathcal{E} : I_A \to \Omega_d^*$.*

**Proof:** The lengths $\ell_i$ need not be all different; let them be ordered such that $\ell_1 < \ell_2 < \cdots < \ell_m$, $m \leq a$ and let $N_j$ be the number of source-symbols with code-words of length $\ell_j$. Necessarily, $N_1 \leq d^{\ell_1}$, otherwise there would be more source-symbols than words of length $\ell_1$ that encode them and the code would be singular. The prefix condition means that none of the $N_1$ code-words can prefix code-words of length $\ell_2$, whence $N_1\, d^{\ell_2 - \ell_1}$ code-words are no more available and non-singularity implies $N_2 \leq d^{\ell_2} - N_1\, d^{\ell_2 - \ell_1}$. Continuing, $N_2\, d^{\ell_3 - \ell_2}$ and $N_1\, d^{\ell_3 - \ell_1}$ cannot be used as code-words of length $\ell_3$, whence

$$N_3 \leq d^{\ell_3} - N_1\, d^{\ell_3 - \ell_1} - N_2\, d^{\ell_3 - \ell_2} .$$

Iterating the argument one gets a set of inequalities

$$N_j \leq d^{\ell_j} - \sum_{k=1}^{j-1} N_k\, d^{\ell_j - \ell_k} , \quad 1 \leq j \leq m ,$$

the last one ($j = m$) resulting in (3.21). Vice versa, if a set of $m$ different lengths $\ell_i$ satisfy the Kraft inequality, then they also satisfy the inequalities

$$\sum_{k=1}^{j} N_k\, d^{-\ell_k} \leq \sum_{i=1}^{a} d^{-\ell_i} \leq 1 \Longrightarrow N_j \leq d^{\ell_j} - \sum_{k=1}^{j-1} N_k\, d^{\ell_j - \ell_k} ,$$

for $1 \leq j \leq m$. Therefore, the source-symbols $i \in I_A$ can always be regrouped into subsets $I_A(j)$, each with $N_j$ elements, such that there are sufficiently many code-words to construct a prefix-code $I_A(j) \ni i \mapsto \mathcal{E}(i) \in \Omega_d^*$. □

**Example 3.2.2.** [92] Inequality (3.21) extends to countable prefix codes. Indeed, any $\boldsymbol{x}^{(n)} = (x_1, x_2, \ldots, x_n) \in \Omega_d^{(n)}$ can be associated with the interval $\Delta_{\boldsymbol{x}} := [0.x_1x_2\cdots x_n\,,\, 0.x_1x_2\cdots x_n + d^{-n}) \subset [0,1]$ by means of the $d$-nary expansion $x = \sum_{j=1}^{n} \frac{x_j}{d^j} =: 0.x_1x_2\cdots x_n$. Therefore, if a countable set $\{\boldsymbol{x}_i\}_{i\in\mathbb{N}}$ of code-words $\boldsymbol{x}_i^{(\ell_i)} \in \Omega_d^*$ with lengths $\ell_i$ have the prefix property, the corresponding intervals $\Delta_i$ of lengths $d^{-\ell_i}$ are all disjoint and the sum of their lengths cannot exceed 1. Viceversa, given a countable set of lengths satisfying the extended Kraft inequality

$$\sum_{i\in\mathbb{N}} d^{-\ell_i} \leq 1\,,$$

these can be assigned to disjoint dyadic intervals whose left ends can be used as code-words of a prefix-code.

Given a source $A$ some codes will prove more adapted to its statistical properties than others; for instance, it is convenient to assign shorter code-words to the symbols emitted with higher probability. In this context, a useful parameter is the following one.

**Definition 3.2.1 (Average Code Length).** *[92] Let $A$ be a source emitting symbols from the alphabet $I_A$ with probabilities $\pi = \{p(i)\}_{i\in I_A}$, the average length of a code $\mathcal{E} : I_A \to \Omega_d^*$ is defined by $L_\pi(\mathcal{E}) := \sum_{i=1}^a p(i)\ell_i$, where $\ell_i := \ell(\mathcal{E}(i))$ is the length of the code-word $\mathcal{E}(i)$ assigned to the i-th source-symbol.*

A way to optimize a code relative to a fixed source probability distribution is to try to achieve the shortest average length. If $\mathcal{E}$ is a prefix-code for which (3.21) becomes an equality, the optimal lengths are found by imposing that the quantity $L_\pi(\mathcal{E}) + \lambda\Big(\sum_{i=1}^a d^{-\ell_i} - 1\Big)$ be stationary upon variation of the lengths and of the Lagrange multiplier $\lambda$. Since $\sum_{i=1}^a p(i) = 1$, one gets $\lambda^* = -\log d$ and $\ell_i^* = -\log_d p(i)$, whence the corresponding average length equals the Shannon entropy in base $d$, $L^* = H_d(A)$. This is the smallest one achievable by a prefix code; indeed, with $D := \sum_{i=1}^a d^{-\ell_i} \leq 1$, by means of the relative entropy (2.94) and of (2.85), one estimates

$$\begin{aligned} L_\pi(\mathcal{E}) - L^* &= \sum_{i=1}^a p(i)\,(\ell_i + \log_d p(i)) = \sum_{i=1}^a p(i)\log_d\left(p(i)\frac{D}{d^{-\ell_i}}\right) - \log_d D \\ &= S(\widetilde{\pi}\,,\,\pi) - \log_d D \geq 0\,, \end{aligned} \tag{3.22}$$

where $\widetilde{\pi} = \{d^{-\ell_i}/D\}_{i=1}^a$. Since $\ell_i^*$ is not generally an integer, it cannot be directly used to construct an optimal code; however, set $\overline{\ell}_i := \lceil -\log_d p(i)\rceil$ [1], so that $\ell_i^* \leq \overline{\ell}_i < \ell_i^* + 1$ and

[1] $\lceil x\rceil$ denotes the smallest integer larger than $x \in \mathbb{R}_+$

$$\sum_{i=1}^{a} d^{-\overline{\ell}_i} \leq \sum_{i=1}^{a} d^{-\ell_i^*} = \sum_{i=1}^{a} p(i) = 1 \, .$$

According to Proposition 3.2.1, one can thus construct a prefix-code $\overline{\mathcal{E}}$ with average length $L_\pi(\overline{\mathcal{E}})$ such that

$$H_d(A) = L^* \leq L_\pi(\overline{\mathcal{E}}) = \sum_{i=1}^{a} p(i)\overline{\ell}_i < L^* + 1 = H_d(A) + 1 \, .$$

These upper and lower bounds also characterize the average code length $L_\pi(\mathcal{E}_{opt})$ of any optimal code for $L_\pi(\overline{\mathcal{E}}) \geq L_\pi(\mathcal{E}_{opt}) \geq L^*$.

**Example 3.2.3 (Shannon-Fano-Elias Code).** Let $A$ be a source that emits symbols $i \in I_A = \{1, 2, \ldots, a\}$ with probabilities $\pi = \{p(i)\}_{i=1}^a$ and assume, without loss of generality that $p(i) > 0$. Let $P(i) := \sum_{j=1}^{i} p(j)$; then, to each symbol $i$ there corresponds a jump from $P(i-1)$ to $P(i)$ and the value $Q(i) := P(i-1) + p(i)/2$ belonging to the corresponding step can be used to identify the $i$-th symbol. Since a code-word must contain a finite number of symbols, a suitable truncation of $Q(i)$ is necessary; for this the binary expansion of $Q(i)$ is used. Concretely, one assigns to the $i$-th symbol the code-word $\mathcal{E}(i) = x_1(i)x_2(i)\cdots x_{\ell_i}(i)$, where

$$-\log_2 p(i) + 1 \leq \ell_i := \lceil -\log_2 p(i) \rceil + 1 < -\log_2 p(i) + 2 \, , \tag{3.23}$$

and $x_j(i) \in \{0, 1\}$ are the binary coefficients of the expansion of $Q(i)$ truncated at the $\ell_i$-th digit:

$$Q(i) := \underbrace{\sum_{j=1}^{\ell_i} \frac{x_j(i)}{2^j}}_{\overline{Q}(i)} + \sum_{j=\ell_i+1} \frac{x_j(i)}{2^j} \leq \overline{Q}(i) + \frac{1}{2^{\ell_i}} < \overline{Q}(i) + \frac{p(i)}{2} < P(i) \, .$$

Since $P(i-1) < \overline{Q}(i) < P(i)$, $\overline{Q}(i)$ provides a code-word $\mathcal{E}(i)$ for the symbol $i$ of length $\ell_i < -\log_2 p(i) + 2$. Also, with the notation of Example 3.2.2, the binary intervals

$$\left[0.x_1(i)x_2(i)\cdots x_{\ell_i}(i) \, , \, 0.x_1(i)x_2(i)\cdots x_{\ell_i}(i) + 2^{-\ell_i}\right]$$

lie within the steps corresponding to different $i$'s and are thus disjoint. Then, $\mathcal{E}$ is a prefix-code with average length satisfying

$$H_2(A) \leq L_\pi(\mathcal{E}) = \sum_{i=1}^{a} p(i)\,\ell_i = \sum_{i=1}^{a} p(i)\Big(\lceil -\log_2 p(i) \rceil + 1\Big) < H_2(A) + 2 \, .$$

**Remark 3.2.1.** The difference between the average code-length and the entropy $H_d(A)$ in the case of the assignment $\ell_i = \lceil -\log_d p(i) \rceil$, can be eliminated asymptotically by coding not single source-symbols but whole blocks of them. In this case, given a stationary source $A$ and a prefix-code $\mathcal{E}$, one encodes strings of length $n$, $\boldsymbol{i}^{(n)} \in \Omega_a^{(n)}$, with code-words $\mathcal{E}^{(n)}(\boldsymbol{i}^{(n)}) \in \Omega_d^*$ of lengths $\ell_{\boldsymbol{i}^{(n)}}^{(n)}$ and *average code-length per symbol*

$$L_{\pi_{A^{(n)}}}(\mathcal{E}^{(n)}) := \frac{1}{n} \sum_{\boldsymbol{i}^{(n)} \in \Omega_a^{(n)}} p_{A^{(n)}}(\boldsymbol{i}^{(n)}) \, \ell_{\boldsymbol{i}^{(n)}}^{(n)} .$$

Then, the same argument developed for codings of single source-symbols yields the bounds

$$\frac{H_d(A^{(n)})}{n} \leq L_{\pi_{A^{(n)}},n}(\mathcal{E}^{(n)}) < \frac{H_d(A^{(n)})}{n} + \frac{1}{n} .$$

Taking the limit $n \to \infty$, one sees that the average code-length per symbol tends to the entropy rate (in base $d$) $h_d(A)$ of the source (see Remark 3.1.1.1). This simple result motivates the following interpretation:

The entropy rate of a stationary source represents the expected number of code-symbols needed to optimally describe the whole stochastic process corresponding to the source.

### 3.2.1 Source Compression

Storing or transmitting information consumes a certain amount of resources, like the number of uses of a channel or the allocation of memory. In order to minimize the costs, the strategy is to compress information as much as possible in such a way that it could be efficiently retrieved, that is with small probability of errors. We shall start with the case of binary Bernoulli sources $A$ emitting statistically independent signals (see (2.31)).

In such a case, the source amounts to a stochastic process $\{A^j\}_{j\in\mathbb{Z}}$ consisting of independent and identically distributed random variables, each with discrete probability distribution $\pi_A = \{p(i)\}_{i=1}^a$. Then, the mean value of the random variable

$$L_n(A) := -\frac{1}{n} \sum_{j=0}^{n-1} \log p(A^j) \tag{3.24}$$

is the Shannon entropy $H(A) = \sum_{\boldsymbol{i}^{(n)} \in \Omega_a^{(n)}} p(\boldsymbol{i}^{(n)}) \, L_n(\boldsymbol{i}^{(n)})$, while the variance equals $V_n(A) := \langle (L_n(A) - H(A))^2 \rangle = \frac{1}{n} \langle \log^2 p(A) \rangle - H^2(A)$.

**Lemma 3.2.1 (Tschebitcheff Inequality).** *Let $X$ be a random variable with outcomes $i = 1, 2, \ldots, d$, probability $\pi = \{p(i)\}_{i=1}^d$, mean value $M := \langle X \rangle$ and variance $V := \langle X^2 \rangle - M^2$, then Prob$\{|X - M| \geq \epsilon\} \leq \frac{V^2}{\epsilon^2}$ .*

**Proof:** The upper bound follows from

$$\text{Prob}\{|X - M| \geq \epsilon\} := \sum_{i:|i-M|\geq\epsilon} p(i) \leq \frac{1}{\epsilon^2} \sum_{i:|i-M|\geq\epsilon} p(i)(i-M)^2 .$$

□

With $X := L_n(A)$ as in (3.24), the previous Lemma yields

$$\text{Prob}\{|L_n(A) - H(A)| \geq \epsilon\} \leq \frac{1}{\epsilon^2 n}\langle \log^2 p(A) - H^2(A)\rangle .$$

Therefore, chosen $\epsilon > 0$ and $\delta > 0$, for $n$ sufficiently large, one can select *high probability subsets*

$$\mathcal{A}^{(n)}_{\epsilon,\delta} := \left\{\boldsymbol{i}^{(n)} \in \Omega^{(n)}_a : \left|-\frac{1}{n}\log p^{(n)}(\boldsymbol{i}^{(n)}) - H(A)\right| < \epsilon\right\} , \tag{3.25}$$

such that

$$\text{Prob}\Big(\mathcal{A}^{(n)}_{\epsilon,\delta}\Big) \geq 1-\delta , \qquad \text{Prob}\Big((\mathcal{A}^{(n)}_{\epsilon,\delta})^c\Big) \leq \delta , \tag{3.26}$$

where $(\mathcal{A}^{(n)}_{\epsilon,\delta})^c := \Omega^{(n)}_2 \setminus \mathcal{A}^{(n)}_{\epsilon,\delta}$ is the corresponding *low probability subset.*

**Proposition 3.2.2 (Asymptotic Equipartition Property (*AEP*)).** *For any $\epsilon > 0$ and $\delta > 0$, there exists $N_{\epsilon,\delta}$ such that, for all $n > N_{\epsilon,\delta}$, the high probability subsets $\mathcal{A}^{(n)}_{\epsilon,\delta} \subset \Omega^{(n)}_2$ are such that, for all $\boldsymbol{i}^{(n)} \in \mathcal{A}^{(n)}_{\epsilon,\delta}$,*

$$e^{-n(H(A)+\epsilon)} \leq p(\boldsymbol{i}^{(n)}) \leq e^{-n(H(A)-\epsilon)} , \tag{3.27}$$

*while, their cardinalities $\#(\mathcal{A}^{(n)}_{\epsilon,\delta})$ satisfy*

$$(1-\delta)e^{n(H(A)-\epsilon)} < \#(\mathcal{A}^{(n)}_{\epsilon,\delta}) \leq e^{n(H(A)+\epsilon)} . \tag{3.28}$$

**Proof:** The first statement follows from (3.25), while the second one is a consequence of (3.26) and of

$$1 = \sum_{\boldsymbol{i}^{(n)}\in\Omega_2} p(\boldsymbol{i}^{(n)}) \geq \sum_{\boldsymbol{i}^{(n)}\in\mathcal{A}^{(n)}_\epsilon} p(\boldsymbol{i}^{(n)}) \geq \#(\mathcal{A}^{(n)}_\epsilon)\, e^{-n(H(A)+\epsilon)}$$

$$1-\delta < \sum_{\boldsymbol{i}^{(n)}\in\mathcal{A}^{(n)}_\epsilon} p(\boldsymbol{i}^{(n)}) \leq \#(\mathcal{A}^{(n)}_\epsilon)\, e^{-n(H(A)-\epsilon)} .$$

□

Roughly speaking, the *AEP* states that, for large $n$, the binary strings of length $n$ can be subdivided into a high probability subspace $\mathcal{A}^{(n)}_{\epsilon,\delta}$ containing $\approx e^{nH(A)}$ strings each one of them occurring with probability $\approx e^{-nH(A)}$. Also, the closer the source entropy to 1, the closer $\mathcal{A}^{(n)}_{\epsilon,\delta}$ gets to $\Omega^{(n)}_2$.

For Bernoulli sources, the *AEP* amounts to $\frac{1}{n}\log p(\boldsymbol{i}^{(n)}) \to H(A)$ in probability. In fact, the *AEP* extends to ergodic sources and, more in general, to symbolic modeling of ergodic dynamical systems, with the Shannon entropy replaced by the entropy rate.

Let $\mathcal{P} = \{P_i\}_{i=1}^p$ denote a finite, measurable partition of a reversible ergodic triplet $(\mathcal{X}, T, \mu)$ and set $\mathcal{P}_r^s := \bigvee_{j=r}^s \mathcal{P}^j$, $\mathcal{P}^j := T^{-j}(\mathcal{P})$. Further, for any $x \in \mathcal{X}$ let $\mathcal{P}_r^s(x)$ denote the atom of the partition $\mathcal{P}_r^s$ that contains $x$: for $\mu$-almost all $x$ there is one and only one such atom. Notice that each $\mathcal{P}_r^s$ is a random variable on $\mathcal{X}$ such that

$$\mathcal{P}_r^s(x) = \bigcap_{j=r}^s T^{-j}(P_{i_j}) \Longleftrightarrow T^j x \in P_{i_j} \quad \forall j = r, r+1, \ldots, s\ .$$

Consider now the random variable

$$h_n(x) := -\frac{1}{n}\log \mu(\mathcal{P}_0^{n-1}(x))\ ; \tag{3.29}$$

with the notation of Section 3.1, its expectation is

$$\begin{aligned}\mu(h_n) &= \int_{\mathcal{X}} \mathrm{d}\mu\,(x)\, h_n(x) = -\frac{1}{n}\sum_{\boldsymbol{i}^{(n)}\in\Omega_p^{(n)}} \int_{P^{(n)}_{\boldsymbol{i}^{(n)}}} \mathrm{d}\mu\,(x)\log\mu(\mathcal{P}_0^{n-1}(x)) \\ &= -\frac{1}{n}\sum_{\boldsymbol{i}^{(n)}\in\Omega_p^{(n)}} \mu(P^{(n)}_{\boldsymbol{i}^{(n)}})\log\mu(P^{(n)}_{\boldsymbol{i}^{(n)}}) = \frac{1}{n}H_\mu(\mathcal{P}^{(n)})\ . \end{aligned} \tag{3.30}$$

Rewrite $h_n(x) = -\frac{1}{n}\sum_{k=1}^{n-1}\log\frac{\mu(\mathcal{P}_0^k(x))}{\mu(\mathcal{P}_0^{k-1}(x))} - \frac{1}{n}\log\mu(\mathcal{P}(x))$, $\mathcal{P}_0^0 = \mathcal{P}$, and observe that $\mathcal{P}_0^k(x) = \mathcal{P}_{-k}^0(T^k x)$ and $\mathcal{P}_0^{k-1}(x) = \mathcal{P}_{-k}^{-1}(T^k x)$, then

$$h_n(x) = \frac{1}{n}\sum_{k=0}^{n-1} g_k(T^k x) \quad \text{where} \tag{3.31}$$

$$g_0(x) := -\log\mu(\mathcal{P}(x))\ ,\ g_k(x) := \log\frac{\mu(\mathcal{P}_{-k}^0(x))}{\mu(\mathcal{P}_{-k}^{-1}(x))}\ . \tag{3.32}$$

All these functions are positive; furthermore, $0 \le g := \lim_k g_k$ exists almost everywhere and is integrable. In fact, let $f_k^i := g_k \lceil P_i$, that is

$$f_k^i(x) := -\log\frac{\mu(\mathcal{P}_{-k}^{-1}(x)\cap P_i)}{\mu(\mathcal{P}_{-k}^{-1}(x))}\ ;$$

then, the conditional probability (2.52) of the random variable $\mathcal{P}$ conditioned on the measure algebra generated by $\mathcal{P}_{-k}^{-1}$ reads

$$\mu\Big(\mathcal{P} = i \Big\| \mathcal{P}_{-k}^{-1}\Big)(x) = \mathrm{e}^{-f_k^i(x)} \ .$$

Since, from Theorem 2.2.1, $\lim_k f_k^i$ exists $\mu$-almost everywhere, the same is true of $g = \lim_k g_k$. Now, fix $a \in \mathbb{R}$ and define the following disjoint subsets of $\mathcal{X}$:

$$E_k := \Big\{x \quad \max_{1\leq j\leq k-1} g_j(x) \leq a < g_k(x)\Big\}$$

$$F_k^i := \Big\{x \quad \max_{1\leq j\leq k-1} f_j^i(x) \leq a < f_k^i(x)\Big\} \ .$$

Using the defining property (2.52) of conditional probabilities, one estimates

$$\mu(E_k) = \sum_{i=1}^p \mu(P_i \cap F_k^i) = \sum_{i=1}^p \int_{F_k^i} \mathrm{d}\mu\,(x)\,\mu\Big(\mathcal{P} = 1 \Big\| \mathcal{P}_{-k}^{-1}\Big)(x)$$

$$\leq \mathrm{e}^{-a}\mu(F_k^i) \quad \text{and}$$

$$\sum_{k=1}^\infty \mu(E_k) \leq \mathrm{e}^{-a} \sum_{i=1}^p \mu\left(\bigcup_{k=1}^\infty F_k^i\right) \leq p\,\mathrm{e}^{-a} \ .$$

Setting $\overline{g} := \sup_k g_k$ and $G_k := \{x \ : \ k < \overline{g}(x) \leq k+1\}$,

$$\mu(\overline{g}) = \sum_{k=0}^\infty \int_{G_k} \mathrm{d}\mu\,(x)\,\overline{g}(x) \leq \sum_{k=0}^\infty (k+1)\,\mathrm{e}^{-k} < +\infty \ ,$$

whence $\overline{g}$ and $g$ are both integrable.

**Example 3.2.4.** Consider the case of a bilateral Bernoulli shift as in Example 3.1.2.2. Then, $x = \boldsymbol{i} \in \Omega_a$, and, choosing as $\mathcal{P}$ the generating partition $\mathcal{C}$, one gets $g_k(\boldsymbol{i}) = g_0(\boldsymbol{i}) = -\log\mu(\mathcal{C}(\boldsymbol{i}))$. Therefore, the sum in (3.31) yields the time-average of $g_0$, whence one can apply Birkhoff's Theorem 2.3.1 and ergodicity to deduce that

$$\lim_{n\to\infty} h_n(\boldsymbol{i}) = H_\mu(\mathcal{C}) = \mathrm{h}_\mu^{\mathrm{KS}}\,(T_\sigma) \qquad \mu - a.e.$$

and that the asymptotic behavior $p(\boldsymbol{i}^{(n)}) \simeq \mathrm{e}^{-n\mathrm{h}_\mu^{\mathrm{KS}}(T_\sigma)}$ holds almost everywhere and not only in probability.

Despite the fact that, in general, the functions in (3.31) are different for different $k$'s and thus (3.31) is not a time-average as in Birkhoff's theorem, none the less the following result holds.

**Theorem 3.2.1 (Shannon-Mc Millan-Breiman Theorem).**
*Let $(\mathcal{X}, T, \mu)$ be a reversible, ergodic dynamical system, then, for all finite, measurable partitions $\mathcal{P} = \{P_i\}_{i=1}^p$,*

$$\lim_{n\to\infty} h_n(x) = \mathrm{h}_\mu^{\mathrm{KS}}\,(T, \mathcal{P}) \qquad \mu - a.e.$$

**Proof:** [61, 199] With the notation introduced in the preceding discussion, dominated convergence, $T$-invariance of $\mu$ and (3.30) together with (3.31) yield

$$\begin{aligned}\mu(g) &= \lim_{n\to\infty} \mu(g_n) = \lim_{n\to\infty} \frac{1}{n}\sum_{k=0}^{n-1} \mu(g_k) = \lim_{n\to\infty} \frac{1}{n}\sum_{k=0}^{n-1} \mu(g_k \circ T^k) \\ &= \lim_{n\to\infty} \mu(h_n) = \mathrm{h}^{\mathrm{KS}}_{\mu}(T, \mathcal{P}) \ .\end{aligned}$$

On the other hand, from ergodicity, $\mathrm{h}^{\mathrm{KS}}_{\mu}(T, \mathcal{P}) = \mu(g) = \lim_{n\to\infty} \frac{1}{n}\sum_{k=0}^{n-1} g(T^k x)$ $\mu - a.e.$, whence the theorem is proved by showing that

$$\lim_{n\to\infty} \frac{1}{n}\sum_{k=0}^{n-1} (g_k - g)(T^k x) = 0 \qquad \mu - a.e. \quad (*) \quad .$$

Consider $G_N(x) := \sup_{k\geq N} |g_k(x) - g(x)|$; these functions are integrable and $\lim_N G_N = 0$ $\mu$-a.e., thus, from ergodicity,

$$\begin{aligned}\limsup_{n\to\infty} \left|\frac{1}{n}\sum_{k=0}^{n-1} (g_k - g)(T^k x)\right| &\leq \limsup_{n\to\infty} \frac{1}{n}\sum_{k=0}^{n-1} \left|(g_k - g)(T^k x)\right| \\ \leq \limsup_{n\to\infty} \frac{1}{n}\sum_{k=0}^{n-1} G_N(T^k x) = \mu(G_N)\end{aligned}$$

$\mu$-a.e. and for all $N \in \mathbb{N}$ whence $(*)$. □

**Remark 3.2.2.** The Shannon-Mc Millan-Breiman theorem applied to an ergodic source allows a reformulation of the *AEP* in terms of the *KS* entropy. Indeed, choosing as $\mathcal{P}$ the standard generating partition as in Example 3.2.4, almost everywhere convergence of $p^{(n)}_A(\boldsymbol{i}^{(n)})$ to $\mathrm{e}^{-n\,h(A)}$, ensures that, given $\epsilon > 0$, and $\delta > 0$, for $n$ sufficiently large, the ensemble of strings of length $n$ can be subdivided into a high probability subspace $\mathcal{A}^{(n)}_\epsilon$ of probability $\approx 1$ containing $\approx \mathrm{e}^{n\,h(A)}$ strings.

The *AEP* allows the implementation of the following compression scheme of an ergodic binary source: one considers strings of length $n$, makes a list of those contained in a high probability subset $\mathcal{A}^{(n)}_\epsilon$ and assign them as a code their position in the list. Since $\mathcal{A}^{(n)}_\epsilon$ contains less than $2^{n(h(A)+\epsilon)}$ strings (entropies being conveniently computed with logarithms in base 2), the number of *bits* needed for the encoding is at the most

$$\lceil \log_2 2^{n(h(A)+\epsilon)} \rceil + 1 = \lceil h(A) + \epsilon \rceil + 1 \ ,$$

while the strings belonging to the complementary set $(\mathcal{A}_\epsilon^{(n)})^c$ may be encoded by a same integer, say $\#(\mathcal{A}_\epsilon^{(n)})+1$. Upon retrieval, the strings belonging to $\mathcal{A}_\epsilon^{(n)}$ are exactly identified by their code, but not those in $(\mathcal{A}_\epsilon^{(n)})^c$; however, since $\mathrm{Prob}((\mathcal{A}_\epsilon^{(n)})^c) \leq \delta$ and $\delta \to 0$ with $n \to \infty$, the larger $n$ gets, the lower is their probability of occurring. Therefore, the probability of error can be made vanishingly small by increasing $n$.

**Theorem 3.2.2 (Noiseless Coding Theorem).** *Let $A$ be an ergodic binary source with entropy rate $h(A)$: binary strings of length $n$ can be encoded by using $n\,R < n$ bits and vanishing probability of error if $R > h(A)$. If $R < h(A)$, then the probability of error goes to 1 with $n \to \infty$.*

**Proof:** The first part of the theorem follows from the previous discussion by applying the equipartition theorem with $R = h(A) + \epsilon$.

For the second part, let $R = h(A) - \epsilon$ and consider the high probability subset $\mathcal{A}_{\epsilon/2}^{(n)}$ together with its complement $(\mathcal{A}_{\epsilon/2}^{(n)})^c$. The probability of any subset $B$ of $\Omega_2^{(n)}$ containing $\lfloor 2^{nR} \rfloor$ [2] strings can be estimated as follows,

$$\begin{aligned}\mathrm{Prob}(B) &\leq \mathrm{Prob}\Big(B \cap (\mathcal{A}_{\epsilon/2}^{(n)})^c\Big) + \mathrm{Prob}\Big(B \cap \mathcal{A}_{\epsilon/2}^{(n)}\Big) \\ &\leq \delta + 2^{nR} 2^{-n(h(A)-\epsilon/2)} = \delta + 2^{-n\epsilon/2} ,\end{aligned}$$

where $\delta$ is a vanishingly small quantity given by the *AEP* . Thus, listing the strings belonging to a subset as $B$, one uses less than $h(A)$ *bit* per *bit* , but, when $n$ gets larger, the probability that an emitted string belong to $B$ gets vanishingly small and the probability of error close to 1. □

## Universal Source Codings

The compression protocols discussed in the previous section depends on the knowledge of the source statistics. Interestingly, encoding and decoding schemes exist which work equally well, namely with a same compression rate $R$, for all ergodic sources $A$ with an entropy rate $h(A) < R$, whatever their overall stationary probability distribution: these protocols provide *universal source codings.*

In the following, we shall consider Bernoulli sources [92], while the general case can be found discussed in [325, 168]. The method used is based on the concept of *type.*

Let $A$ be a stationary Bernoulli source emitting strings $\boldsymbol{i}^{(n)} = i_1 i_2 \cdots i_n \in \Omega_a^{(n)}$, $i_j \in I_A = \{1,2,\ldots,a\}$ according to compatible probability distributions $\pi_{A^{(n)}} = \Big\{p_A^{(n)}(\boldsymbol{i}^{(n)}) = \prod_{j=1}^n p_A(i_j)\Big\}$. We shall denote

[2] $\lfloor x \rfloor$ denotes the largest integer smaller than $x \in \mathbb{R}$.

1. by $N(j|\boldsymbol{i}^{(n)})$ the number of times $j \in I_A$ occurs in the string $\boldsymbol{i}^{(n)}$;
2. by $p(j|\boldsymbol{i}^{(n)}) := \frac{N(j|\boldsymbol{i}^{(n)})}{n}$ the so-called *empirical probability* generated by the string $\boldsymbol{i}^{(n)}$ and by $\Pi_{\boldsymbol{i}^{(n)}} := \{p(j|\boldsymbol{i}^{(n)})\}_{j=1}^{a}$ the corresponding *empirical distribution*. The latter is known as the *type* of $\boldsymbol{i}^{(n)}$: strings $\boldsymbol{i}^{(n)}$ whose symbols occur with same frequencies belong to a same type $\Pi^{(n)}$;
3. by $\mathcal{P}_n$ the set of all types $\Pi^{(n)}$;
4. by $T(\Pi^{(n)})$ the subset of all strings $\boldsymbol{i}^{(n)} \in \Omega_a^{(n)}$ with a same type $\Pi^{(n)}$.

The construction of universal codings is based on the following two bounds; of particular importance is the second one which states that the number of different types increases at most polynomially with $n$.

**Lemma 3.2.2.** *Let $\Pi^{(n)} \in \mathcal{P}_n$ be a type of $\Omega_a^{(n)}$ and let $H(\Pi^{(n)})$ be its Shannon entropy. The number of strings in $T(\Pi^{(n)})$ and the number of all possible types fulfil*

$$\#(T(\Pi^{(n)})) \leq 2^{nH(\Pi^{(n)})} \quad , \quad \#(\mathcal{P}_n) \leq (n+1)^a \ .$$

*Furthermore, the a-priori probability of $T(\Pi^{(n)})$ is such that*

$$\pi_A^{(n)}(T(\Pi^{(n)}) \leq 2^{-n\,S(\Pi^{(n)}\,,\,\pi_A)} \ ,$$

*where $S(\Pi^{(n)}\,,\,\pi_A)$ is the classical relative entropy (see (2.94)).*

**Proof:** Let $P^{(n)}$ be the following empirical probability distribution on $\Omega_a^{(n)}$,

$$P^{(n)}(\boldsymbol{i}^{(n)}) := \prod_{j=1}^{a} p(j|\boldsymbol{i}^{(n)})^{N(j|\boldsymbol{i}^{(n)})} = \prod_{j=1}^{a} 2^{np(j|\boldsymbol{i}^{(n)})\,\log_2 p(j|\boldsymbol{i}^{(n)})} = 2^{-nH(\Pi_{\boldsymbol{i}^{(n)}})} \ .$$

The probability of the type class $T(\Pi^{(n)})$ is certainly smaller than 1; thus,

$$1 \geq P^{(n)}\Big(T(\Pi^{(n)})\Big) = \sum_{\boldsymbol{i}^{(n)} \in T(\Pi^{(n)})} P^{(n)}(\boldsymbol{i}^{(n)}) = \#(T(\Pi^{(n)}))\,2^{-nH(\Pi^{(n)})}$$

yields the first estimate.

The second is a very loose upper bound: each type $\Pi_{\boldsymbol{i}^{(n)}}$ is entirely characterized by how many times each symbol $i \in I_A$ occurs in $\boldsymbol{i}^{(n)}$. Without constraints (that can only decrease the number of types) there are $n+1$ choices for each $i = 1, 2, \ldots, a$, namely 0, 1,..., $n$, whence the result.

Finally, the last bound is derived as follows: first, notice that

$$p_A^{(n)}(\boldsymbol{i}^{(n)}) = \prod_{j=1}^{n} p_A(i_j) = \prod_{\ell=1}^{a} p_A(\ell)^{N(\ell|\boldsymbol{i}^{(n)})} = \prod_{\ell=1}^{a} 2^{np(\ell|\boldsymbol{i}^{(n)}) \log_2 p_A(j)}$$

$$= 2^{n\sum_{\ell=1}^{a}\left(p(\ell|\boldsymbol{i}^{(n)}) \log_2 p(\ell|\boldsymbol{i}^{(n)}) - p(\ell|\boldsymbol{i}^{(n)}) \log_2 \frac{p(\ell|\boldsymbol{i}^{(n)})}{p_A(j)}\right)}$$

$$= 2^{-n\left(H(\Pi_{\boldsymbol{i}^{(n)}}) + S(\Pi_{\boldsymbol{i}^{(n)}}, \pi_A)\right)} .$$

Then, using the first upper bound,

$$\pi_A^{(n)}(T(\Pi^{(n)})) = \sum_{\boldsymbol{i}^{(n)} \in T(\Pi^{(n)})} p_A^{(n)}(\boldsymbol{i}^{(n)})$$

$$= \#(T(\Pi^{(n)}))\, 2^{-n\left(H(\Pi^{(n)}) + S(\Pi^{(n)}, \pi_A)\right)} \leq 2^{-n\, S(\Pi^{(n)}, \pi_A)} .$$

□

Because of the first bound in Lemma 3.2.2, at most $nR+1$ *bits* are needed to encode the label of a string $\boldsymbol{i}^{(n)}$ of type $\Pi^{(n)}$ with $H(\Pi^{(n)}) < R$, while at most $a \log_2(n+1) + 1$ *bits* ensures the encoding of the label specifying the type $P$ to which the string belongs (the $+1$ accounts for $R$ and $\log_2(n+1)$ not being integers). Therefore, in the limit $n \to \infty$, one expects a compression rate $R$ for all Bernoulli sources with $H(A) < R$.

**Definition 3.2.2 (Universal Codings).** *Let $R > 0$ and consider an encoding of a Bernoulli source A into binary strings of length $\lfloor nR \rfloor$, given by $\mathcal{E}^n : \Omega_a^{(n)} \to \Omega_2^{\lfloor nR \rfloor}$, followed by a decoding procedure $\mathcal{D} : \Omega_2^{\lfloor nR \rfloor} \to \Omega_a^{(n)}$. This gives a universal $(n, 2^{nR})$-code if the probability of error*

$$P_{err}^{(n)} := \pi_A^{(n)}\left(\left\{\boldsymbol{i}^{(n)} \ : \ \mathcal{D}^n \circ \mathcal{E}^n(\boldsymbol{i}^{(n)}) \neq \boldsymbol{i}^{(n)}\right\}\right)$$

*goes to 0 when $n \to \infty$ and $\mathcal{E}^n$, $\mathcal{D}^n$ do not depend on the Bernoulli source probability $\pi_A$.*

**Proposition 3.2.3.** *There exist universal source codings $(n, 2^{nR})$ for every Bernoulli source with $H(A) < R$.*

**Proof:** Given $R > 0$, let $R_n := R - a\dfrac{\log_2(n+1)}{n}$; using the first two bounds in Lemma 3.2.2, the subsets $\mathcal{A}^{(n)} := \left\{\boldsymbol{i}^{(n)} \in \Omega_a^{(n)} \ : \ H(\Pi_{\boldsymbol{i}^{(n)}}) \leq R_n\right\}$ have cardinalities such that

$$\#\left(\mathcal{A}^{(n)}\right) = \sum_{\substack{\Pi^{(n)} \in \mathcal{P}_n \\ H(\Pi^{(n)}) \leq R_n}} \#(T(\Pi^{(n)})) \leq \sum_{\substack{\Pi^{(n)} \in \mathcal{P}_n \\ H(\Pi^{(n)}) \leq R_n}} 2^{nH(\Pi^{(n)})}$$

$$\leq \sum_{\substack{\Pi^{(n)} \in \mathcal{P}_n \\ H(\Pi^{(n)}) \leq R_n}} 2^{n\, R_n} \leq (n+1)^a\, 2^{nR_n} = 2^{nR} .$$

Let $\mathcal{E}^n$ associate to strings in $\mathcal{A}^{(n)}$ their label in the list of such strings expressed in *bits* and let $\mathcal{D}^n$ be its inverse map. If $H(A) < R$, then, using the third bound in the previous lemma,

$$\begin{aligned} P_{err}^{(n)} &= 1 - \pi_A^{(n)}\left(\mathcal{A}^{(n)}\right) = \sum_{\substack{\Pi^{(n)} \in \mathcal{P}_n \\ H(\Pi^{(n)}) > R_n}} \pi^{(n)}\left(T(\Pi^{(n)})\right) \\ &\leq (n+1)^a \max\left\{\pi_A^{(n)}\left(T(\Pi^{(n)})\right) \ : \ H(\Pi^{(n)}) > R_n\right\} \\ &\leq (n+1)^a \, 2^{-n \min\left\{S(\Pi^{(n)}, \pi_A) \, : \, H(\Pi^{(n)}) > R_n\right\}} . \end{aligned}$$

Since $\lim_n R_n = R$ and $H(A) < R$, and the relative entropy $S(\pi_1, \pi_2) = 0$ iff $\pi_1 = \pi_2$, $P_{err}^n$ gets exponentially small for $n$ sufficiently large. □

### 3.2.2 Channel Capacity

Noiseless channels are an exception; usually, during transmission signals get distorted. It can thus happen that a channel outputs a same string $\boldsymbol{y}^{(n)}$ when presented with different input strings $\boldsymbol{x}_1^{(n)}$ and $\boldsymbol{x}_2^{(n)}$ which cannot then be decoded without errors. Like in compression, to counteract distortion one resorts to suitable encoding and decoding procedures of longer and longer strings; however, unlike in compression where redundancies are eliminated, in the presence of noise, the strategy is to introduce redundancies in order to lower the possibility that different input strings give rise to a same channel output.

**Example 3.2.5.** In Example 2.4.1, bits 0 and 1 can be converted into one another with probability $0 < p < 1/2$ by a binary symmetric channel $\mathcal{C}$. The probability of a wrong decoding can be lowered by encoding

$$\mathcal{E}(0) = \underbrace{00 \cdots 0}_{2n+1 \text{ times}} \quad , \quad \mathcal{E}(1) = \underbrace{11 \cdots 1}_{2n+1 \text{ times}} \quad .$$

Then, $2n+1$ uses of the channel output strings $\boldsymbol{i}^{(2n+1)} := \mathcal{C}^{(2n+1)} \circ \mathcal{E}(i)$ that can be decoded by a majority rule: let $N_{\boldsymbol{i}^{(2n+1)}}(0)$ denote the number of 0s in $\boldsymbol{i}^{(2n+1)}$, then

$$\mathcal{D}(\boldsymbol{i}^{(2n+1)}) = \begin{cases} 0 & \text{if} \quad N_{\boldsymbol{i}^{(2n+1)}}(0) > n \\ 1 & \text{if} \quad N_{\boldsymbol{i}^{(2n+1)}}(0) \leq n \end{cases} .$$

By such an encoding-decoding procedure one transmits one bit at the cost of $2n+1$ bits; an error occurs, that is $\mathcal{D} \circ \mathcal{C}^{(2n+1)} \circ \mathcal{E}(i) \neq i$, if $\geq n+1$ bits of $\mathcal{E}(i)$ are flipped by the channel $\mathcal{C}$. The probability of such an event, $\binom{2n+1}{n+1} p^{n+1}(1-p)^n$, vanishes with $n \to \infty$; unfortunately, the *transmission rate*, that is the number of bits transmitted per use of the channel, $\frac{1}{2n+1}$, vanishes, too.

In the following we shall consider channels $\mathcal{C}$ without memory and without feedback such that each of their uses is independent of the previous inputs and outputs. Further, $n$ uses of the channel $\mathcal{C}$ amount to a single use of a channel $\mathcal{C}^{(n)}$ which maps input strings $\boldsymbol{x}^{(n)} \in I_X^n$ consisting of $n$ symbols from an alphabet $I_X = \{1, 2, \ldots, n_X\}$ into output strings $\boldsymbol{y}^{(n)} \in I_Y^n$ consisting of symbols from an alphabet $I_Y = \{1, 2, \ldots, n_Y\}$. Input and output strings are conveniently described as realizations of stochastic processes $\{X_i\}_{i\in\mathbb{N}}$ and $\{Y_i\}_{i\in\mathbb{N}}$ with join random variables $X^{(n)} := \bigvee_{i=1}^n X_i$ and $Y^{(n)} := \bigvee_{j=1}^n Y_j$. The transitions $\boldsymbol{x}^{(n)} \mapsto \boldsymbol{y}^{(n)}$ occur with probabilities $p(\boldsymbol{y}^{(n)}|\boldsymbol{x}^{(n)})$ that factorize (see (2.82)) and are thus completely characterized by the single-use *transition probabilities* $p(y_j|x_i)$.

One of the great achievements of early information theory was obtained by Shannon who proved that codes exist such that the number $M(n)$ of distinguishable strings $\boldsymbol{x}^{(n)}$ increases with $n$ at a non-zero exponential rate $R$: $M(n) \approx 2^{Rn}$.

**Definition 3.2.3 (Channel Codes and Capacity).** *[92]*
*A code $(M, n)$, for a channel $\mathcal{C}$ consists of*

1. *a set $I_C := \{1, 2, \ldots M\}$;*
2. *an encoding $\mathcal{E} : I_C \mapsto I_X^n$ associating a code-word $\boldsymbol{x}^{(n)}(w) = \mathcal{E}(w)$ to any of the indices $w \in I_C$;*
3. *a decoding procedure $\mathcal{D} : I_Y^n \mapsto I_C$, $\mathcal{D}(\boldsymbol{y}^{(n)}(w)) =: \hat{w} \in I_C$, that returns $\hat{w} \in I_C$ given a channel output $\boldsymbol{y}^{(n)}(w) = \mathcal{C}^n(\boldsymbol{x}^{(n)}(w))$.*

*The rate of the code is defined by $R := \frac{\log_2 M}{2}$* [3]*. The probability of an error, $\hat{w} = \mathcal{D}(\boldsymbol{y}^{(n)}(w)) \neq w$, is*

$$\mathsf{e}_n(w) := \sum_{\boldsymbol{y}^{(n)} \in \Omega_{n_Y}^{(n)} : \mathcal{D}(\boldsymbol{y}^{(n)}) \neq w} p\Big(\boldsymbol{y}^{(n)}|\boldsymbol{x}^{(n)}(w)\Big) .$$

*The rate is said achievable if there exists a sequence of codes $(2^{nR}, n)$ with vanishing maximal probability of error $\mathsf{e}_n := \max_{w\in I_C} \mathsf{e}_n(w)$.*
*The capacity $C$ of the channel $\mathcal{C}$ is the largest of its achievable rates.*

**Remark 3.2.3.** For a memoryless channel, (2.82) holds; thus, if the probabilities of the input stochastic process $\{X_i\}_{i\in\mathbb{N}}$ factorize, so do the probabilities of the output stochastic process $\{Y_i\}_{i\in\mathbb{N}}$:

---

[3] For sake of simplicity, in the following $M = 2^{nR}$ will be identified with $\lceil 2^{nR} \rceil$, the smallest integer larger than $M$.

$$p_{Y^{(n)}}(\boldsymbol{y}^{(n)}) = \sum_{\boldsymbol{x}^{(n)} \in I_X^n} p(\boldsymbol{y}^{(n)}|\boldsymbol{x}^{(n)})\, p_{X^{(n)}}(\boldsymbol{x}^{(n)})$$
$$= \prod_{j=1}^n \sum_{x_j} p(y_j|x_j)\, p_X(x_j) = \prod_{j=1}^n p_Y(y_j) \ . \tag{3.33}$$

Shannon's result is that the mutual information (2.93) $I(X;Y)$ is an achievable rate and that the channel capacity is given by

$$C = \max_{\pi_X} I(X;Y) \ . \tag{3.34}$$

**Examples 3.2.6.**

1. Example 2.4.1.1: $p_B(i) = p_A(i)$, $i = 0,1$, implies $I(A;B) = H(A)$, whence capacity, $C = 1$, is attained at $\pi_A = \{1/2, 1/2\}$.
2. Example 2.4.1.2: with $H(p) := -p\log_2 p - (1-p)\log_2(1-p)$,

$$I(A;B) = H(B) + \sum_{i=0}^1 p_A(i) \sum_{j=0}^1 p(j|i)\log_2 p(j|i) = H(B) - H(p) \ ,$$

whence capacity $C = 1 - H(p)$ is attained at $\pi_A = \{1/2, 1/2\}$, since

$$p_B(0) = p_A(0)(1-p) + p_A(1)p = \frac{1}{2} \ , \quad p_B(1) = p_A(0)p + p_A(1)(1-p) = \frac{1}{2} \ .$$

3. Example 2.4.1.3: $p_B(1) = p_A(0)(1-\alpha)$, $p_B(2) = p_A(1)(1-\alpha)$ and $p_B(3) = \alpha(p_A(0) + p_A(1)) = \alpha$ yield

$$I(A;B) = H(B) - (p_A(0) + p_A(1))H(\alpha) = H(B) - H(\alpha)$$
$$= (1-\alpha)H(A) \ ,$$

whence capacity $C = (1-\alpha)$ is attained at $\pi_A = \{1/2, 1/2\}$.
4. The capacity in (3.2.3) refers to only one use of the channel $\mathcal{C}$; consider now the channel $\mathcal{C}^{(n)}$ acting on $\boldsymbol{x}^{(n)} \in I_X^n$ with outputs $\boldsymbol{y}^{(n)} \in I_Y^n$. The mutual information $I(X^{(n)}; Y^{(n)})$ of the corresponding random variables $X^{(n)}$ and $Y^{(n)}$ can be controlled by repeatedly using (2.93). From (2.82),

$$H(Y^{(n)}|X^{(n)}) = H(X^{(n)} \vee Y^{(n)}) - H(X^{(n)})$$
$$= H(Y_n|X^{(n)} \vee Y^{(n-1)}) + H(X^{(n)} \vee Y^{(n-1)}) - H(X^{(n)})$$
$$= \sum_{j=1}^n H(Y_j|X^{(n)} \vee Y^{(j-1)}) = \sum_{j=1}^n H(Y_j|X_j) \ .$$

Further, from (2.88),

$$I(X^{(n)};Y^{(n)}) = H(Y^{(n)}) - H(Y^{(n)}|X^{(n)}) \leq \sum_{j=1}^{n} \Big(H(Y_j) - H(Y_j|X_j)\Big) \leq nC \ . \tag{3.35}$$

Therefore, if $C^{(n)}$ denotes the capacity of the channel $\mathcal{C}^{(n)}$, the supremum over all input probability distributions gets $C^{(n)} \leq nC$. Actually, from Remark 3.2.3, equality is achieved by choosing a factorizing $\pi_{X^{(n)}}$ such that $p_{X^{(n)}}(\boldsymbol{x}^{(n)}) = \prod_{j=1}^{n} p_X(x_j)$, with $\pi_X$ the one achieving capacity $C$. Then, the output probabilities factorize too and thus $H(Y^{(n)}) = nH(Y)$.

The above relation between capacity and mutual information can be understood as follows. As showed in the last example, if $X^{(n)}$ consists of $n$ independent, identically distributed repetitions of $X$, then the same is true of $Y^{(n)}$ and $X^{(n)} \vee Y^{(n)}$ with respect to $Y$ and $X \vee Y$. With $H(X)$, $H(Y)$ and $H(X,Y)$ the corresponding entropies, based on the *AEP*, for large $n$ there are roughly $2^{nH(X)}$ $\pi_X$-typical inputs, $2^{nH(Y)}$ $\pi_Y$-typical outputs and $2^{nH(X,Y)}$ *jointly typical pairs* $(\boldsymbol{x}^{(n)}, \boldsymbol{y}^{(n)})$, that is typical with respect to $\pi_{X \vee Y}$. Of course, not all input-output pairs $(\boldsymbol{x}^{(n)}, \boldsymbol{y}^{(n)})$ with $\boldsymbol{x}^{(n)}$ $\pi_X$-typical and $\boldsymbol{y}^{(n)}$ $\pi_Y$-typical are jointly typical: this happens with probability roughly equal to

$$\frac{2^{nH(X,Y)}}{2^{nH(X)}\, 2^{nH(Y)}} = 2^{-nI(X;Y)} \ .$$

Therefore, in order to encounter a jointly typical pair with fixed output $\boldsymbol{y}^{(n)}$ one needs at least $2^{nI(X;Y)}$ inputs; in other words, one expects that encoding a number of input strings smaller than $2^{nI(X;Y)}$, none of them should be jointly typical with respect to a same $\boldsymbol{y}^{(n)}$. Vice versa, more than $2^{nI(X;Y)}$ inputs would start having a same jointly typical output and thus being not exactly identifiable. Memoryless channels with independent, identically distributed inputs are thus expected to have achievable rates $R \simeq I(X;Y)$.

In order to give a mathematical proof of the above intuitive argument, we start by extending the notion of typical strings.

**Definition 3.2.4 (Jointly-typical Strings).**
*Two strings $\boldsymbol{x}^{(n)} \in I_X^n$ and $\boldsymbol{y}^{(n)} \in I_Y^n$ are jointly typical if they belong to the subset $\mathcal{A}_\epsilon^{(n)} \subseteq I_X^n \times I_Y^n$ such that*

$$\Big|-\frac{1}{n}\log_2 p_{X^n}(\boldsymbol{x}^{(n)}) - H(X)\Big| < \epsilon$$

$$\Big|-\frac{1}{n}\log_2 p_{Y^n}(\boldsymbol{y}^{(n)}) - H(Y)\Big| < \epsilon$$

$$\Big|-\frac{1}{n}\log_2 p_{X^nY^n}(\boldsymbol{x}^{(n)}, \boldsymbol{y}^{(n)}) - H(X \vee Y)\Big| < \epsilon \ ,$$

*where* $0 < \epsilon \ll 1$.

Since (2.82) holds, the argument of the proof of Proposition 3.2.2 gives rise to a *jointly typical AEP*. Namely, let $\epsilon > 0$, for sufficiently large $n$'s the probability carried by subsets of strings violating any of the inequalities in the previous definition can be made smaller than $\epsilon/3$ so that $\mathrm{Prob}(\mathcal{A}_\epsilon^{(n)}) \geq 1-\epsilon$. Moreover, its cardinality fulfils

$$(1-\epsilon)\, 2^{n(H(X,Y)-\epsilon)} \leq \#(\mathcal{A}_\epsilon^{(n)}) \leq 2^{n(H(X\vee Y)+\epsilon)} , \qquad (3.36)$$

while the probabilities of strings $\boldsymbol{x}^{(n)}$, $\boldsymbol{y}^{(n)}$ and $(\boldsymbol{x}^{(n)}, \boldsymbol{y}^{(n)})$ satisfying the inequalities in Definition 3.2.4 fulfil

$$2^{-n(H(X)+\epsilon)} \leq p_{X^n}(\boldsymbol{x}^{(n)}) \leq 2^{-n(H(X)-\epsilon)} \qquad (3.37)$$

$$2^{-n(H(Y)+\epsilon)} \leq p_{Y^n}(\boldsymbol{y}^{(n)}) \leq 2^{-n(H(Y)-\epsilon)} \qquad (3.38)$$

$$2^{-n(H(X\vee Y)+\epsilon)} \leq p^{(n)}_{X^nY^n}(\boldsymbol{x}^{(n)}, \boldsymbol{y}^{(n)}) \leq 2^{-n(H(X\vee Y)-\epsilon)} , \qquad (3.39)$$

Then, $\mathrm{Prob}\Big(\Big\{(\boldsymbol{x}^{(n)}, \boldsymbol{y}^{(n)}) \in \mathcal{A}_\epsilon^{(n)}\Big\}\Big) = \sum_{(\boldsymbol{x}^{(n)}, \boldsymbol{y}^{(n)})\in\mathcal{A}_\epsilon^{(n)}} p_{X^n}(\boldsymbol{x}^{(n)})\, p_{Y^n}(\boldsymbol{y}^{(n)})$ can be bounded from below and above as follows:

$$(1-\epsilon)\, 2^{-n(I(X;Y)+3\epsilon)} \leq \mathrm{Prob}\Big(\Big\{(\boldsymbol{x}^{(n)}, \boldsymbol{y}^{(n)}) \in \mathcal{A}_\epsilon^{(n)}\Big\}\Big) \leq 2^{-n(I(X;Y)-3\epsilon)} . \qquad (3.40)$$

**Theorem 3.2.3 (Shannon Noisy-Channel Theorem).**
*All rates $R < C$, $C$ as in (3.34), are achievable and any sequence of codes $(nR, n)$ with the maximal error probability $\boldsymbol{e}_n \to 0$ must have $R \leq C$.*

**Proof that $\boldsymbol{e}_n \to 0 \Rightarrow R \leq C$ :** Suppose the signals $w \in \{1, 2, \ldots, M\}$, $M = \lceil 2^{nR} \rceil$, encoded by $(nR, n)$ into $\mathcal{E}(w) = \boldsymbol{x}^{(n)} \in I_X^n$, are equidistributed; let $W$ denote the random variable with outcomes $w$. Using (2.93), (2.95) with $C(B) = \mathcal{E}(W) = X^{(n)}$ and (3.35), it follows that

$$\begin{aligned} nR \leq \log_2 M = H(W) &= H\Big(W|Y^{(n)}\Big) + I\Big(W;Y^{(n)}\Big) \\ &\leq H\Big(W|Y^{(n)}\Big) + I\Big(X^{(n)};Y^{(n)}\Big) \leq H\Big(W|Y^{(n)}\Big) + nC\,. \end{aligned}$$

We need now connect $H\Big(W|Y^{(n)}\Big)$ to the error probability: this is done by means of the so-called *Fano's inequality.* By assumption the maximal error probability in Definition 3.2.3 goes to zero with $n$, so does the average error probability $\boldsymbol{e}_n^{av} := \frac{1}{M}\sum_{w\in I_C} \boldsymbol{e}_n(w)$. Let $E := \begin{cases} 0 & \hat{w} = w \\ 1 & \hat{w} \neq w \end{cases}$; $E$ is a random variable determined by $W$ and $Y^{(n)}$. Thus, using 2.91,

$$H\Big(W|Y^{(n)}\Big) = H\Big(E|W,Y^{(n)}\Big) + H\Big(W|Y^{(n)}\Big)$$
$$= H\Big(W|E,Y^{(n)}\Big) + H\Big(E|Y^{(n)}\Big) .$$

Now, from Remark 2.4.3.4, $H\Big(E|Y^{(n)}\Big) \leq H(E) \leq 1$. Further, $E = 0$ implies that $W$ is determined by $Y^{(n)}$ so that $H\Big(W|E = 0, Y^{(n)}\Big) = 0$, whereas if $E = 1$ then the cardinality of possible values of $W$ is $M - 1$. Therefore,

$$H\Big(W|E,Y^{(n)}\Big) = \sum_{i=0,1} \mathrm{Prob}(E = i) H\Big(W|E = i, Y^{(n)}\Big)$$
$$\leq e_n^{av} \log_2(M-1) \leq e_n^{av}\, nR \Longrightarrow H\Big(W|Y^{(n)}\Big) \leq 1 + e_n^{av}\, nR .$$

The result follows since $nR \leq 1 + e_{av}^{(n)}\, nR + nC$ implies $e_{av}^{(n)} \geq 1 - \frac{1}{nR} - \frac{C}{R}$ which in turn implies that $e_{av}^{(n)}$ cannot vanish with $n \to \infty$ if $R > C$. □

The proof of the first part of Theorem 3.2.3 relies on the following steps:

1. for $w \in \{1, 2, \ldots, M = 2^{nR}\}$, choose the code-word $\boldsymbol{x}^{(n)}(w)$ at random with probability $p_X^{(n)}(\boldsymbol{x}^{(n)}) = \prod_{i=1}^n p_{in}(x_i)$. This gives a *random code* of type $(nR, n)$ with overall probability $\mathrm{Prob}(\mathcal{E}) = \prod_{w=1}^{M} \prod_{i=1}^{n} p_{in}(x_i(w))$;
2. choose the symbols $w$ at random with the same probability $p(w) = M^{-1}$;
3. if $\mathcal{C}^{(n)}(\boldsymbol{x}^{(n)}) = \boldsymbol{y}^{(n)}$ and there is only one $\hat{w}$ such that $\mathcal{E}(\hat{w}) = \boldsymbol{x}^{(n)}(\hat{w})$ is jointly typical with $\boldsymbol{y}^{(n)}$, then associate with $\boldsymbol{y}^{(n)}$ the symbol $\hat{w}$, otherwise declare an error. This gives a decoding map $\boldsymbol{y}^{(n)} \mapsto \mathcal{D}(\boldsymbol{y}^{(n)}) = \widehat{w}$;
4. an error is also declared if $\mathcal{D}(\boldsymbol{y}^{(n)}) = \widehat{w} \neq w$ and $\mathcal{C}^n(\mathcal{E}(w)) = \boldsymbol{y}^{(n)}$.

**Proof that $(\mathbf{nR}, \mathbf{n})$ is achievable when $\mathbf{R} < \mathbf{C}$ :** Let $e_{\mathcal{E}}^{(n)}(w)$ be the probability of an error relative to a random code $\mathcal{E}$ and $e_{av}^{(n)}(\mathcal{E})$ the corresponding average error probability. Further, let

$$P(e) := \sum_{\mathcal{E}} \mathrm{Prob}(\mathcal{E})\, e_{av}^{(n)}(\mathcal{E}) = \frac{1}{M} \sum_{w=1}^{M} \sum_{\mathcal{E}} \mathrm{Prob}(\mathcal{E}) e_{\mathcal{E}}^{(n)}(w) \ :$$

this is the average error probability over all randomly generated codes. Then, every $w$ gives the same contribution to the error, so $P(e) = \sum_{\mathcal{E}} \mathrm{Prob}(\mathcal{E}) e_{\mathcal{E}}^{(n)}(1)$ with fixed $w = 1$. Let $F_w := \Big\{(\boldsymbol{x}^{(n)}(w), \boldsymbol{y}^{(n)}) \in \mathcal{A}_\epsilon^{(n)}\Big\}$, where $\mathcal{A}_\epsilon^{(n)}$ is a jointly-typical subspace. According to the rules of the game, if $\boldsymbol{y}^{(n)} = \mathcal{C}^n(\boldsymbol{x}^{(n)}(1))$, a decoding error occurs when

1. $(\boldsymbol{x}^{(n)}(1), \boldsymbol{y}^{(n)}) \notin \mathcal{A}_\epsilon^{(n)}$, that is when the input corresponding to $w = 1$ and the relative output are not jointly typical;

2. $(\boldsymbol{x}^{(n)}(i), \boldsymbol{y}^{(n)}) \in F_i$ for $i \neq 1$, that is when the output corresponding to $w = 1$ is jointly-typical with code-words associated to $w \neq 1$.

The overall average error probability can thus be estimated as follows:

$$P(e) = \mathrm{Prob}\Big((F_1)^c \cup \bigcup_{i=2}^{M} F_i\Big) \le \mathrm{Prob}((F_1)^c) + \sum_{i=1}^{M} \mathrm{Prob}(F_i) .$$

By the jointly-typical *AEP* , $F_1 \subseteq \mathcal{A}_\epsilon^{(n)} \implies \mathrm{Prob}((F_1)^c) \le \epsilon$ for $n$ large enough. Further, because of randomness of the code, the input $\boldsymbol{x}^{(n)}(i)$, $i \neq 1$, are statistically independent from $\boldsymbol{x}^{(n)}(1)$ and $\boldsymbol{y}^{(n)} = \mathcal{C}^n(\boldsymbol{x}^{(n)}(1))$. Then, the jointly-typical *AEP* also yields

$$\sum_{i=2}^{M} \mathrm{Prob}(F_i) \le (M-1)\, 2^{-n(I(X;Y)-3\epsilon)} \le 2^{-n(I(X;Y)-R-3\epsilon)} .$$

If $R < I(X;Y) - 3\epsilon$, the latter quantity gets $\le \epsilon$ for $n$ sufficiently large and thus $P(e) \le 2\epsilon$. This implies that there exists at least one code $\mathcal{E}^*$ with $e_{av}^{(n)} \le 2\epsilon$. By choosing for $X$ the distribution $\pi^*$ attaining capacity in (3.34), the condition for achieving the rate $R$ becomes $R < C$. Finally, at least half of the code-words $\boldsymbol{x}^{(n)}(w)$ of $\mathcal{E}^*$ must have $e^{(n)}(w) \le 4\epsilon$ otherwise $e_{av}^{(n)} > 2\epsilon$. Keeping only these ones, changes the rate from $R$ to $R(n) := R - 1/n$. The procedure thus yields a sequence of codes $(nR(n), n)$ such that $e^{(n)} \to 0$ and $R(n) \to R$ for all $R < C$. □

## Bibliographical Notes

Most of the results about the *KS* entropy have been drawn from [61]; other excellent books on the subject and its applications are [17, 91, 164]. In particular, the completeness of the *KS* entropy for Bernoulli systems is discussed in [91]. The book by [199] is a reference for Pesin's theory and the relations between the *KS* entropy and Lyapounov exponents (see also [106]).

In [62] one finds an extensive review of the role of *KS* entropy and Lyapounov exponents as regards the issue of predictability in continuous and discrete dynamical systems. In [18], the notion is discussed in relation to the broader notion of complexity.

The material on coding and compression has largely been drawn from [92, 254, 266].

# 4 Algorithmic Complexity

One of the intuitive notions which is most elusive from a mathematical point of view is that of randomness. Consider a string $\boldsymbol{i}^{(n)} \in \Omega_2^*$ emitted by a Bernoulli source with probabilities $p_{0,1}$; suppose that $n >> 1$ and that the number of 0s, $n(0)$, is nearly half the number of 1s, $n(1) \simeq 2n(0)$. One expects that, generically, the relative frequencies $n(i)/n$ tend to the probabilities $p_i$ with increasing $n$; indeed, only special, that is intuitively non-random, strings should fail such a *statistical test*. Therefore, one would call $\boldsymbol{i}^{(n)}$ random only if $p_0 = 1/3$ [305]. Of course, passing the frequency test is not enough; indeed, if $p_0 = 1/2$, both $\boldsymbol{i}^{(n)}$ consisting of $n/2$ subsequent pairs $0, 1$ and a string $\boldsymbol{j}^{(n)}$ of 0s and 1s distributed without any evident pattern occur with probability $2^{-n}$. However, because of its regularity, $\boldsymbol{i}^{(n)}$ would be called non-random and, vice versa, because of the absence of regular structures, $\boldsymbol{j}^{(n)}$ would be called random [92, 310].

Presence and absence of patterns seems to be a useful clue to defining which strings or sequences are random and which are not so; this property should somehow be related to the degree of compressibility so that one might wonder whether the entropy rate introduced in Section 3 could provide a natural measure of randomness. Also, by replacing the entropy rate with the dynamical *KS* entropy, one could define a classical dynamical system to be random or not on the basis of the compressibility of the best ones amongst its symbolic models. However, entropy rate and the *KS* entropy describe the average behavior of sources or of dynamical systems and say nothing about individual strings or individual trajectories.

Various attempts have been undertaken to tackle the problem of formalizing the intuitive notion of randomness of individual sequences $\boldsymbol{i} \in \Omega_2$. In [305], three relevant approaches are discussed: in the first one, randomness is identified with *stochasticness*, that is with the impossibility of devising a winning strategy when bets on the value of the next symbol $i_n$ of $\boldsymbol{i} \in \Omega_2^*$ are based on the knowledge of $i_i i_2 \cdots i_{n-1}$. In the second approach, randomness is identified with *chaoticness* that is with the absence of regular patterns in $\boldsymbol{i} \in \Omega_2$. In the third approach, randomness in a sequence $\boldsymbol{i} \in \Omega_2$ is identi-

F. Benatti, *Dynamics, Information and Complexity in Quantum Systems*, Theoretical and Mathematical Physics, DOI 10.1007/978-1-4020-9306-7_4,

fied with its *typicalness*, that is with the fact that it does not belong to any *effectively null* subset of $\Omega_2^*$ [1]

In the following we shall focus on the second approach which is also known as *algorithmic complexity theory*, and was developed independently and almost at the same time by Kolmogorov [173, 174], Chaitin [77] and Solomonoff [283, 284] in the early sixties. Algorithmic complexity theory involves as many subjects as mathematics, logics, computer science and physics [310, 73, 254]: we shall give a short overview of some of its aspects that are relevant for an extension of this notion to quantum dynamical systems.

## 4.1 Effective Descriptions

The main step towards a theory of randomness of individual strings was the observation that regular strings admit short *effective descriptions*, whereas irregular strings do not. By effective description of a (binary) target string it is meant any algorithm (binary program) that is computed by a suitable computer and makes it halt with the target string as output.

**Example 4.1.1.** Any string $\boldsymbol{i}^{(n)} = i_1 i_2 \cdots i_n$ consisting of $n$ bits can always be reproduced by processing the program

$$\textbf{PRINT} \quad i_1 i_2 \cdots i_n \ ,$$

which specifies the bits to print, one after the other.

This program amounts to the literal transcription of the target string. Clearly, one has to seek more clever ways to describe $\boldsymbol{i}^{(n)}$, that is shorter programs. In doing so, one is much helped by the presence of patterns; if $i_j = 0$ for all $1 \leq j \leq n$, the following simple program could be used:

---

[1] Let $\Omega_2^*$, the set of all binary sequences, be equipped with the $\sigma$-algebra generated by cylinder sets and with the uniform product probability distribution so that any cylinder $C_{\boldsymbol{i}}$ indexed by a string $\boldsymbol{i} \in \Omega_2^*$ of length length $\ell(\boldsymbol{i})$ has probability $\pi(C_{\boldsymbol{i}}) = 2^{-\ell(\boldsymbol{i})}$. A subset $A \subset \Omega_2^*$ is a null subset if for any $\varepsilon > 0$ there are cylinders $C_{\boldsymbol{i}_j}$, $\boldsymbol{i}_j \in \Omega_2^*$ such that $A \subset \bigcup_j C_{\boldsymbol{i}_j}$ and $\sum_j 2^{-\ell(\boldsymbol{i}_j)} \leq \varepsilon$. A subset $A \subset \Omega_2^*$ is an effectively null subset if the previous inequality is satisfied with the strings $\boldsymbol{i}_j$ that index the cylinders and $\varepsilon > 0$ (any rational number) both effectively computable by a suitable algorithm (for instance by a program processed by a computer) [305]. Intuitively, random sequences cannot be effectively reproducible and thus cannot belong to effectively null sets. Concretely, these latter sets consist of non-typical strings and correspond to effective statistical tests or *Martin-Löf tests* that, when failed, identify these non-random strings (an example is the frequency test mentioned in the discussion prior to this remark) [310]. In other words, a sequence is random according to the typicalness criterion if it passes all Martin-Löf tests. On the other hand, if typicalness were defined with reference to all possible null subsets, then there would be no typical sequences; indeed, any $\boldsymbol{i} \in \Omega_2$ belongs to the null subset of $\Omega_2$ consisting of the sequence itself.

**PRINT 0 $n$ TIMES .**

For large $n$, the length of such a program goes as $\log_2 n$, that is as the number of bits necessary to specify the length of the string $\ell(\boldsymbol{i}^{(n)}) = n$. This is also the case if, less trivially, the string $\boldsymbol{i}^{(n)}$ consists of a same pattern, $\boldsymbol{i}^{(q)}$ that repeats itself $\simeq n/q$ times. Indeed, what is to be specified is the length of the pattern at the cost of a fixed number, $\log_2 q$, of bits and the number of repetitions at the cost of $\simeq \log_2 n/q \simeq \log_2 n$ bits for $n \gg q$.

On the other hand, if $\boldsymbol{i}^{(n)}$ shows no pattern, there is no shorter effective description than literal transcription. In this case, the length of the effective description grows as $n$ and not as $\log_2 n$.

In the previous example, it is clear that one is interested in the shortest possible effective descriptions $s(\boldsymbol{i}^{(n)})$ of a given string $\boldsymbol{i}^{(n)}$: let $\mathrm{C}(\boldsymbol{i}^{(n)})$ denote the length of any of these shortest description, that is $\ell(s(\boldsymbol{i}^{(n)})) = \mathrm{C}(\boldsymbol{i}^{(n)})$.

The map $\boldsymbol{i}^{(n)} \mapsto s(\boldsymbol{i}^{(n)})$ is code for the ensemble of strings of length $n$. In Section 4.3, it will be showed that, by processing the effective descriptions by means of particular computing devices called *prefix machines* (in which case $\mathrm{C}(\boldsymbol{i}^{(n)})$ is denoted by $\mathrm{K}(\boldsymbol{i}^{(n)})$), the code becomes a prefix code (see Definition 3.2.1), so that the extended Kraft inequality (see Example 3.2.2) applies

$$\sum_{\boldsymbol{i}\in\Omega_2^*} 2^{-\mathrm{K}(\boldsymbol{i})} \leq 1 . \tag{4.1}$$

**Example 4.1.2 (Payoff Functions).** [120, 310] Suppose the government of a country claims that in the $j$-th one of $n$ successive elections it won with $0.99i_j$ percent of the votes, $i_j$ being any decimal digit for $j$ odd and the $j/2$ digit in the decimal expansion of $\pi$ for $j$ even. To defend itself from the accuse of fabricating the electoral results, the government replies that the probability $Q(\boldsymbol{i}^{(n)}) = 10^{-n}$ of such a string of decimal digits $\boldsymbol{i}^{(n)} = i_1 i_2 \cdots i_n$ is equal to that of any other string randomly obtained according to the uniform probability distribution over 10 symbols. This defense can be defeated by using the regularity of $\boldsymbol{i}^{(n)}$ to construct a suitable *payoff function* $t(\boldsymbol{i}^{(n)}|Q) \geq 0$, namely a non-negative function whose mean value is such that

$$\sum_{\boldsymbol{i}^{(n)}\in\Omega_{10}^{(n)}} 10^{-n} t(\boldsymbol{i}^{(n)}|Q) \leq 1 .$$

Its meaning is as follows: the accuser proposes the government to be payed $t(\boldsymbol{i}^{(n)}|Q)$ upon betting 1 on the outcome $\boldsymbol{i}^{(n)}$. This is a fair proposal for, if the outcomes $\boldsymbol{i}^{(n)}$ are distributed according to the uniform probability $Q$, the accuser average gain cannot be higher than 1.

However, if there is a pattern in $\boldsymbol{i}^{(n)}$, the accuser can construct a payoff function $t(\boldsymbol{i}^{(n)}|Q)$ that assumes high values on the strings with such a pattern. Concretely, for the half of the decimal digits of $\boldsymbol{i}^{(n)}$ that are randomly distributed according to $Q$, one needs $n/2 \log_2 10$ bits for its description; instead, for the remaining half that comes from an algorithm that computes successive approximations to $\pi$, a finite number, $C$, of bits [2] suffice. Then, one gets the following upper bound to the length of the shortest effective description computed by a prefix machine (see previous remark),

$$\mathrm{K}(\boldsymbol{i}^{(n)}) \leq \frac{n}{2} \log_2 10 + C \ .$$

Setting $t(\boldsymbol{i}^{(n)}|Q) := 2^{-\log_2 Q(\boldsymbol{i}^{(n)}) - \mathrm{K}(\boldsymbol{i}^{(n)})} = 10^n \, 2^{-\mathrm{K}(\boldsymbol{i}^{(n)})}$, one defines a payoff function; indeed, because of (4.1),

$$\sum_{\boldsymbol{i}^{(n)} \in \Omega_{10}^{(n)}} Q(\boldsymbol{i}^{(n)}) \, 2^{-\log_2 Q(\boldsymbol{i}^{(n)}) - \mathrm{K}(\boldsymbol{i}^{(n)})} = \sum_{\boldsymbol{i}^{(n)} \in \Omega_{10}^{(n)}} 2^{-\mathrm{K}(\boldsymbol{i}^{(n)})} \leq 1 \ .$$

While any fair Casino's owner should accept bets based on such a payoff function, the government cannot; indeed, by betting 1 on the digit of each one of $n$ successive elections, the accuser will pay $n$ to the government but receive $10^{n/2} 2^{-C}$ from it, quite an amount of money for large $n$. As the payoff function does depend only on the presence of a pattern, but not on its particular form, the accuser strategy does not require any a priori knowledge.

The aim of algorithmic complexity theory is an objective characterization of the randomness of individual strings in terms of the lengths of their shortest effective descriptions. It is thus necessary to eliminate the dependence of such lengths on the computers that process the corresponding programs. Indeed, given a same target string $\boldsymbol{i}^{(n)}$ two different computers $\mathfrak{V}_{1,2}$ will in general provide shortest descriptions $s_{1,2}(\boldsymbol{i}^{(n)})$ with different lengths $C_{1,2}(\boldsymbol{i}^{(n)})$. As explained in Proposition 4.1.1, this problem is overcome by resorting to effective descriptions processed by *universal computers*, namely by computers that are able to simulate the action of any other computing machine. The universal computers on which classical algorithmic complexity theory is based are the so-called *Universal Turing Machines* (*UTMs* ).

### 4.1.1 Classical Turing Machines

A *Turing Machine* (*TM* ) is a very basic (and abstract) model of computing device (see [310]) consisting of

1. a bi-infinite tape $\mathbf{T}$ subdivided into cells labeled by integers $i \in \mathbb{Z}$, each cell containing either a blank symbol $\#$ or a symbol $\sigma$ from a given alphabet $\widetilde{\Sigma}$. We shall set $\Sigma = \widetilde{\Sigma} \cup \#$;

[2] This number becomes negligible when $n$ increases.

2. a reading/write head **H** moving along the tape that, when positioned on the $i$-th cell, reads the symbol $\sigma_i \in \Sigma$, leaves it unchanged or changes it into $\sigma'_i \in \Sigma$ and then proceeds to either the cell $i+1$ to the right ($R$) or to the cell $i-1$ to the left ($L$);
3. a central processing unit **C** (*CPU*) capable of a finite number of control states $q_i \in Q := \{q_0, q_2 \ldots, q_{|Q|-1}\}$: at each computational step, the *CPU* state $q \in Q$ may remain the same or change into $q' \in Q$.

The list of possible moves defines a program for the *TM* ; formally, it amounts to a *transition function*

$$\delta : Q \times \Sigma \mapsto Q \times \Sigma \times \{L, R\} \ , \quad \delta(q, \sigma) = (q', \sigma', d) \ , \ d \in \{L, R\} \ . \tag{4.2}$$

As a consequence, any *TM* can be identified by the set of rules defining $\delta$. Each set of rules, that is any *TM* , corresponds to a certain task, a *computation*, to be performed on an input data string. Any computation can be assumed to start with the CPU control state in a chosen ready state $q_r$, the head positioned on a chosen 0-th cell and the input written on a finite number of cells extending from the 0-th one to its left, while all other cells to the left and to the right contain blank symbols. The computation then proceeds through a sequence of steps dictated by the transition function $\delta$, each one of them corresponding to a certain configuration of the *TM* that performs it.

**Definition 4.1.1 (*TM* configurations).** *At each step of a computation a classical configuration $c$ of a TM $\mathfrak{U}$ is a triplet*

$$C \ni c := \Big(q, \{\sigma_i\}_{i\in\mathbb{Z}}, k\Big) \in Q \times \Sigma^{\mathbb{Z}} \times \mathbb{Z} \ ,$$

*where in the infinite sequence $\{\sigma_i\}_{i\in\mathbb{Z}}$ of cell symbols only finitely many of them are such that $\sigma_i \neq \#$, while $q, k$ denote the state of the control unit and of the head position and $C$ the set of all configurations.*

In order to determine when a computation terminates, we assume that among the control states there is a special state, $q_f$, such that when the control unit is in the state $q_f$, then the output is read off from the position of the head to its right until the last $\sigma_i \neq \#$.

Because they consist of a finite set of rules involving finite sets of symbols, transition functions (and thus *TMs* ) can be encoded and numbered. Given a program $p$ (or the *TM* which computes it), its number $\gamma(p)$ in the enumeration of all programs (or *TMs* ) is known as *Gödel number of p* [93]. A universal Turing machine is any *TM* $\mathfrak{U}$ which, upon receiving the code of a *TM* $\mathfrak{V}$, is able to simulate $\mathfrak{V}$ on any input string.

**Example 4.1.3.** There are many possible ways to encode a transition function $\delta$; a simple one is as follows [128]: the control states $q_i$ and the symbols

$\sigma_j$ are identified by giving their positions $i$, $j$ in the respective lists $Q$ and $\Sigma$. These are then encoded as strings of as many 0's:

$$q_i \mapsto 0^i := \underbrace{00\cdots0}_{i \text{ times}} , \qquad \sigma_j \mapsto 0^j := \underbrace{00\cdots0}_{j \text{ times}} .$$

Thus, the rule $\delta(q_i, \sigma_j) = (q_k, \sigma_\ell, d)$ can be encoded as $0^i\,1\,0^j\,1\,0^k\,1\,0^\ell\,1\,0^{n(d)}$, where the 1s are used to separate the various entries (only sequences of 0s are entries corresponding to labels). These appear one after the other as they do in the given rule, while $n(d) = 1$ if $d = L$, $n(d) = 2$ if $d = R$. Then, the transition function (or, equivalently, the *TM* $\mathfrak{U}$ that performs the task specified by it) can be encoded as

$$\begin{aligned}1\,0^{|Q|}\;11\;0^{|\Sigma|}\;11\;&\underbrace{0^{i_1}\,1\,0^{j_1}\,1\,0^{k_1}\,1\,0^{\ell_1}\,1\,0^{n(d_1)}}_{\text{1st rule}}\;11\\ &\underbrace{0^{i_2}\,1\,0^{j_2}\,1\,0^{k_2}\,1\,0^{\ell_2}\,1\,0^{n(d_2)}}_{\text{2nd rule}}\;11\\ &\vdots\\ &\underbrace{0^{i_m}\,1\,0^{j_m}\,1\,0^{k_m}\,1\,0^{\ell_m}\,1\,0^{n(d_m)}}_{\text{last rule}}\;111\,,\end{aligned}$$

where the first two strings of 0s encode the total number of control states and of symbols, the pairs of 1s separate the rules, while the first and last 1 mark the beginning and the end of the list.

Suppose $f : \mathbb{N} \mapsto \mathbb{N}$ is a function from the integers to the integers; by passing to the binary representation of $n \in \mathbb{N}$, $f$ becomes a function from $\Omega_2^* \mapsto \Omega_2^*$. It is called *total* if its domain of definition is the whole of $\Omega_2^*$ (symbolically, $f(\boldsymbol{i}^{(n)}) \downarrow$ for all $\boldsymbol{i}^{(n)} \in \Omega_2^*$), *partial otherwise*, namely if there exist strings $\boldsymbol{i}^{(n)}$ on which $f$ is not defined (symbolically, $f(\boldsymbol{i}^{(n)}) \uparrow$ on these strings). The existence of an algorithm or an effective procedure which allows one to compute $f$ provides an intuitive and informal definition of *computable functions*; among others, a possible formalization of computability is as follows [93].

**Definition 4.1.2.** *A partial function $f : \Omega_2^* \mapsto \Omega_2^*$ is said to be* computable *if there is a Turing machine that on input $\boldsymbol{i} \in \Omega_2^*$ outputs $f(\boldsymbol{i})$.*

The so-called *Church-Turing thesis* asserts that the intuitively and informally defined set of computable functions coincides with those that are computable according to the previous definition [93, 128]. It is not a theorem, yet it could not be disproved as a conjecture; therefore, it is commonly accepted that the *TMs* provide a computational model which computes all what can be thought of being intuitively computable.

**Remark 4.1.1.** Given a computable partial function $f$, if $p_f$ is one of the (infinitely many) programs which compute it, one can assign $f$ the Gödel number $\gamma(p_f)$ of $p_f$ which is one of the (infinitely) many Gödel numbers of $f$ [93]. It follows that the computable functions form a countable set; this fact allows the use of Cantor's diagonal argument to construct a total function $f : \Omega_2^* \mapsto \Omega_2^*$ which is not computable. In order to show this, consider the enumeration as $\phi_j : \mathbb{N} \mapsto \mathbb{N}$ of all computable partial functions $f : \mathbb{N} \mapsto \mathbb{N}$ that can be constructed by choosing a definite Gödel number for each one of them. Then, the function defined by

$$\phi(n) = \begin{cases} \phi_n(n) + 1 & \text{if} \quad \phi_n(n) \downarrow \\ 0 & \text{if} \quad \phi_n(n) \uparrow \end{cases}$$

is total as $\phi \downarrow$ on all inputs. Furthermore, it cannot coincide with any $\phi_j$ for, if $\phi_j$ is defined on $j$, then $\phi(j) = \phi_j(j) + 1 \neq \phi_j(j)$.

**Example 4.1.4.** An important class of *TMs* are the *Probabilistic TMs* (*PTMs* ) which provide a more powerful classical model of computation than *TMs* [128]. They are defined by transition functions of the form

$$\delta : Q \times \Sigma \times Q \times \Sigma \times \{L, R\} \mapsto [0, 1] \tag{4.3}$$

$$(q, \sigma; q', \sigma', d) \mapsto \delta(q, \sigma; q', \sigma', d) \in [0, 1] \ , \qquad \sum_{q', \sigma', d} \delta(q, \sigma; q', \sigma', d) = 1 \ . \tag{4.4}$$

Namely, *PTMs* are defined by assigning the *probabilities* $\delta(q, \sigma; q', \sigma', d)$ with which the machine goes from a CPU control state $q \in Q$ and symbol read $\sigma \in \Sigma$ to a new control state $q'$, new symbol $\sigma'$ together with a subsequent head move $d \in \{L, R\}$. Therefore, given a starting configuration $c_i \in C$ the machine will move to a new configuration $c_j \in C$ with a certain *transition probability* $p_{ij} := p(c_i \to c_j)$, the successors of $c_i$ being all those $c_j$ with $p_{ij} \neq 0$. The transition probabilities satisfy $\sum_j p_{ij} = 1$; indeed, given a starting configuration $c_i$, the *PTM* will surely move to a subsequent one among those available to it. Each step performed by a *PTM* will then be described by a *transition matrix* $\pi = [p_{ij}]$.

Any computation performed by a *PTM* on an initial configuration $c_0$ can be seen as a tree whose nodes are the successor configurations and the branches connecting the leaves carry the relative non-zero transition probabilities. Each run of the machine defines a tree-level with its corresponding nodes; if a successor configuration at level $j$ appears more than once then the probability of its occurrence at that level is the sum of the probabilities leading to it through the various branches. As a simple instance of such a mechanism [128], consider an initial configuration $c_0$ branching into two different configurations $c_{11}$ and $c_{12}$ at level 1 with probabilities $p_{01} := p(c_0 \to c_{11})$

and $p_{02} := p(c_0 \to c_{12})$: $p_{01} + p_{02} = 1$. During the second step of the computation, the two configurations at level 1 branch into two configurations each: $c_{11}$ into $c_{21}$ and $c_{22}$ with probabilities $p_{11} := p(c_{11} \to c_{21})$, respectively $p_{12} := p(c_{11} \to c_{22})$, such that $p_{11} + p_{12} = 1$, while $c_{12}$ branches into $c_{23}$ and $c_{24}$ with probabilities $p_{23} := p(c_{12} \to c_{23})$, respectively $p_{24} := p(c_{12} \to c_{24})$, such that $p_{23} + p_{24} = 1$ (see Figure 4.1). Thus the probabilities of the four configurations are

$$p(c_{21}) = p_{01}\,p_{11}\ ,\ p(c_{22}) = p_{01}\,p_{12}\ ,\ p(c_{23}) = p_{02}\,p_{23}\ ,\ p(c_{24}) = p_{02}\,p_{24}\ .$$

If $c_{22} = c_{23} = c^*$ then the probability of $c^*$ is $p(c^*) = p(c_{22}) + p(c_{23})$.

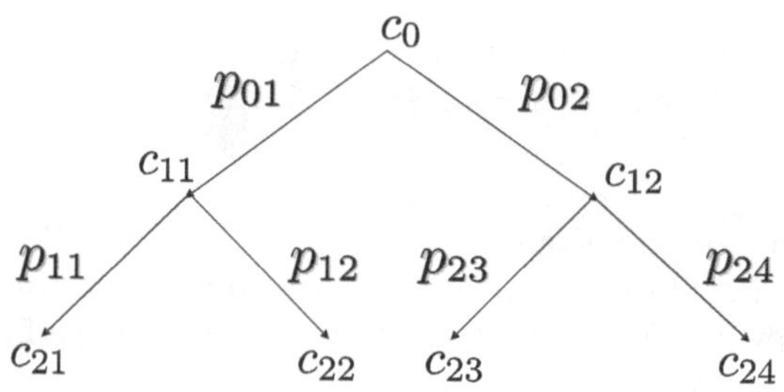


**Fig. 4.1.** Probabilistic Turing Machines: Level Tree

**Remark 4.1.2.** Within the class of *PTMs* , *TMs* are *deterministic* in the sense that the corresponding probabilities $\delta(q, \sigma; q', \sigma', d)$ equal 1 when the couples $(q, \sigma)$ and triplets $(q', \sigma', d)$ are connected by the rules (4.2), otherwise $\delta(q, \sigma; q', \sigma', d) = 0$. The computations performed by *TMs* correspond to deterministic classical processes, while those of *PTMs* correspond to stochastic classical processes (compare the ballistic and Brownian computers discussed in [51]); in other words, it is the laws of classical physics upon which the models of computations embodied by *TMs* and *PTMs* are based.

*PTMs* are important from the point of view of the so-called *computational complexity* [3] [128, 165]. All computational tasks need a certain amount of time to be performed and use a certain amount of memory (space); roughly speaking, computational complexity theory estimates how the amount of time and/or space required to perform a computation involving $n$ bits scales with $n$: if the time required to process $n$ bits goes as $n^\alpha$, $\alpha > 0$, one says that the computation has *polynomial computational complexity*, otherwise *superpolynomial* or *exponential.* When a computer $\mathfrak{U}$ simulates another computer $\mathfrak{V}$ that performs a certain task, there is an unavoidable overhead in space/time

[3]To be distinguished from the descriptional complexity.

resources due to the simulation. The latter is then called *efficient* if the overhead scales polynomially with respect to the space/time resources used by $\mathfrak{V}$. The *Classical Strong Church-Turing Thesis* [165] states that:

*Any realistic computational model can be efficiently simulated by a PTM .*

Namely, any computational model which is consistent with the laws of classical physics and which accounts for all necessary computational resources [4] only requires a polynomial space/time overhead to be simulated by a *PTM* . As the Church-Turing thesis (see Remark 4.1.1), also the strong Church-Turing thesis has survived all attempts to disprove it; however, as observed by Feynamn [116], this paradigm does not seem to be extendible to computational models based on quantum mechanics, for then classical physics appears unable to simulate their performances as efficiently.

### 4.1.2 Kolmogorov Complexity

In the following we shall restrict to the effective description of binary strings; using the notation of the previous section, we shall therefore consider *TMs* with the alphabet $\Sigma = \{0,1\} \cup \#$. Further, $\ell(p)$ will denote the length, that is the number of bits, of a program $p$ written as a binary string and $\mathfrak{U}(p)$ the result of $p$ being processed by a *TM* $\mathfrak{U}$.

**Definition 4.1.3 (Kolmogorov Complexity).** *The Kolmogorov complexity [92, 310] or plain algorithmic complexity of $\boldsymbol{i}^{(n)} \in \Omega_2^{(n)}$ is the length of the shortest binary program $p$ such that $\mathfrak{U}(p) = \boldsymbol{i}^{(n)}$:* [5]

$$C_{\mathfrak{U}}(\boldsymbol{i}^{(n)}) := \min\Big\{\ell(p) \,:\, \mathfrak{U}(p) = \boldsymbol{i}^{(n)}\Big\} \ .$$

Plain algorithmic complexity is thus seemingly related to the most efficient way individual strings can be compressed; indeed, by the previous definition, no effective description of a given string $\boldsymbol{i}^{(n)}$ can be shorter than programs with length equal to its algorithmic complexity $\mathrm{C}(\boldsymbol{i}^{(n)})$.

**Remark 4.1.3.** Unlike computational complexity (see Remark 4.1.2), algorithmic complexity is not concerned with the space/time resources needed to process certain programs, but only with their lengths, without restrictions on time and memory. ¿From the algorithmic point of view, only random strings are interesting, while those with simple effective descriptions are somewhat

[4]The adjective *realistic* refers to the fact that the time and space resources effectively needed should be explicitly declared [165].

[5]We shall conform to the notation of [310] which uses the letter $C$ for the Kolmogorov complexity and $K$ for the prefix complexity (see Section 4.3).

dull, despite the large amount of resources that may be needed to compute them. Indeed, there might be short effective descriptions that require a very long time to yield their targets, as for instance the DNA-encoding of human beings [310]. The attempts to fill this gap by considering algorithmic and computational complexity together has led to the notion of *logical depth* [310].

**Proposition 4.1.1.** *The following properties hold:*

1. *The plain algorithmic complexities of a same string* $\boldsymbol{i}^{(n)}$ *with respect to two different UTMs* $\mathfrak{U}_{1,2}$ *differ by a constant which* does not *depend on the string, but only on the UTMs .*
2. *The plain algorithmic complexity is upper bounded as follows*

$$C_{\mathfrak{U}}(\boldsymbol{i}^{(n)}) \leq A + \ell(\boldsymbol{i}^{(n)}) = A + n \ , \tag{4.5}$$

*where* $A$ *is a constant which does not depend on* $\boldsymbol{i}^{(n)}$*.*

3. *The number of strings* $\boldsymbol{i}^{(n)} \in \Omega^{(n)}$ *with plain algorithmic complexity strictly smaller than* $c > 0$ [6] *is bounded by*

$$\#\Big\{\boldsymbol{i}^{(n)} \in \Omega^{(n)} \ : \ C_{\mathfrak{U}}(\boldsymbol{i}^{(n)}) < c\Big\} \leq 2^c - 1 \ . \tag{4.6}$$

**Proof:** The proof of the first statement follows from the fact that $\mathfrak{U}_1$ can simulate $\mathfrak{U}_2$ and vice versa, for both are assumed to be universal. Given $\boldsymbol{i}^{(n)}$, let $p_1^*$ be such that $C_{\mathfrak{U}_1}(\boldsymbol{i}^{(n)}) = \ell(p_1^*)$ and let $P_{12}$ be the program, of length $\ell(P_{12}) = L_{12}$, which allows $\mathfrak{U}_2$ to simulate $\mathfrak{U}_1$. In order to make $\mathfrak{U}_2$ simulate $\mathfrak{U}_1$ on the input $p_1^*$, the programs $P_{12}$ and $p_1^*$ must be put together in way that $\mathfrak{U}_1$ knows when the simulation instructions end and the string to be processed starts. This is achieved by concatenating $P_{12}$ and $p_1^*$ as $q = p_1^* \beta(P_{12})$, where

$$\boldsymbol{i}^{(n)} = i_1 i_2 \cdots i_n \mapsto \beta(\boldsymbol{i}^{(n)}) = i_1 i_1 i_2 i_2 \cdots i_n i_n 01$$

is the encoding of a string obtained by repeating each of its bits twice and marking the end with a the pair of different bits 01: for this encoding one needs $\ell(\beta(P_{12})) = 2\,(L_{12} + 1)$ bits. In this way $\mathfrak{U}_2$ will first read $\beta(P_{12})$ being thus able to simulate $\mathfrak{U}_1$ on the subsequent portion $p_1^*$ of the program $q$. Therefore, from the definition of plain complexity, it follows that

$$C_{\mathfrak{U}_2}(\boldsymbol{i}^{(n)}) \leq \ell(q) \ + \ A \leq \ell(p^*) + 2\,(L_{12} + 1) \ + \ A \leq C_{\mathfrak{U}_1}(\boldsymbol{i}^{(n)}) + A_{12} \ .$$

Reversing the roles of $\mathfrak{U}_{1,2}$ one gets $C_{\mathfrak{U}_1}(\boldsymbol{i}^{(n)}) \leq C_{\mathfrak{U}_2}(\boldsymbol{i}^{(n)}) + A_{21}$; thus, $\left|C_{\mathfrak{U}_1}(\boldsymbol{i}^{(n)}) - C_{\mathfrak{U}_2}(\boldsymbol{i}^{(n)})\right| \leq A$, where $A$ is a suitable constant which does not depend on the input $\boldsymbol{i}^{(n)}$.

[6] If $c$ is not integer, $c$ is to be understood as $\lfloor c \rfloor$, the largest integer not larger than $c$: $\lfloor c \rfloor \leq c < \lfloor c \rfloor + 1$.

The first upper bound follows as in Example 4.1.1, from the effective description which tells $\mathfrak{U}$ to print the bits of $\boldsymbol{i}^{(n)}$ one after the other.

The second upper bound follows because the number of binary programs with length smaller than $c$ equals the number of binary strings with $\lfloor c \rfloor - 1$ digits at the most, whence

$$\#\{p : \ell(p) < c\} = \sum_{j=1}^{\lfloor c \rfloor - 1} 2^j = 2^{\lfloor c \rfloor} - 1 \leq 2^c - 1 \,.$$

□

**Example 4.1.5.** In order to improve the loose upper bound (4.5), given $\boldsymbol{i}^{(n)} \in \Omega_2^{(n)}$, let $k$ be the number of 1s among its bits; there are $\binom{n}{k}$ strings in $\Omega_2^{(n)}$ sharing this feature. They can be listed and each of them identified by its number $N_k(\boldsymbol{i}^{(n)})$ in the list; notice that no more than $\lceil \log_2 \binom{n}{k} \rceil$ bits are required to specify $N_k(\boldsymbol{i}^{(n)})$. One can thus construct an effective description of $\boldsymbol{i}^{(n)}$, by specifying $k$ and $N_k(\boldsymbol{i}^{(n)})$ in such a way that the *UTM* must be able to detach the specification of $k$, $p_k$, from that of $N_k(\boldsymbol{i}^{(n)})$. For this, one may do as in the proof of Proposition 4.1.1, by encoding $p_k$ as $\beta(p_k)$, the binary string obtained from $p_k$ by repeating each of its bits twice and marking the end by 01. Since, $\ell(\beta(p_k)) \leq 2\,(\log_2 k + 1)$, from Definition 4.1.3 it follows that

$$\mathrm{C}(\boldsymbol{i}^{(n)}) \leq \log_2 \binom{n}{k} + 2\,(\log_2 k + 1) \,.$$

The following upper bound holds [92],

$$\binom{n}{k} \leq 2^{n\, H_2(\frac{k}{n})} \,,$$

where $H_2(\frac{k}{n}) := -\frac{k}{n} \log_2 \frac{k}{n} \log_2 - (1 - \frac{k}{n} \log_2) \log_2 (1 - \frac{k}{n} \log_2)$, which can be derived by setting $p = k/n$ in

$$1 = \sum_{j=0}^{n} \binom{n}{j} \left(\frac{j}{n}\right)^j \left(1 - \frac{j}{n}\right)^{n-j} \geq \binom{n}{k} p^k (1-p)^{n-k} \,, \quad 0 \leq k \leq 1 \,.$$

Thus, $\frac{1}{n}\mathrm{C}(\boldsymbol{i}^{(n)}) \leq H_2(\frac{k}{n}) + 2\,\frac{\log_2 k + 1}{n}$. Consider now the strings $\boldsymbol{i}^{(n)}$ to be prefixes, that is the initial $n$ bits, of infinite binary sequences $\boldsymbol{i} \in \Omega_2$. Let $0 \leq p \leq 1$ be the probability of the bit 1; if $k/n \mapsto p$, then

$$\limsup_{n \to \infty} \frac{\mathrm{C}(\boldsymbol{i}^{(n)})}{n} \leq H_2(\pi) \,. \tag{4.7}$$

where $H_2(\pi)$ is the ($\log_2$) entropy rate of a Bernoulli binary source with probability $\pi = (p, 1-p)$. The upper bound in (4.5) is thus not a loose one for $p$ close to $1/2$.

**Example 4.1.6.** One would expect the algorithmic complexity of a pair $(\boldsymbol{i},\boldsymbol{j})$ of strings $\boldsymbol{i},\boldsymbol{j} \in \Omega_2^*$ to be smaller (apart from the usual additive constant independent of them) than the sum of the algorithmic complexities of $\boldsymbol{i}$ and $\boldsymbol{j}$, namely:

$$\mathrm{C}((\boldsymbol{i},\boldsymbol{j})) \leq \mathrm{C}(\boldsymbol{i}) + \mathrm{C}(\boldsymbol{j}) + C .$$

Intuitively, this should be so because one can always put together the shortest programs $p$, respectively $q$ for $\boldsymbol{i}$, respectively $\boldsymbol{j}$, in a program $pq$ which is an effective description of $(\boldsymbol{i},\boldsymbol{j})$. Unfortunately, the plain algorithmic complexity cannot enjoy the form of subadditivity expressed by the previous inequality.

Indeed, if $p, q$ are two programs such that $\mathrm{C}(\boldsymbol{i}) = \ell(p)$ and $\mathrm{C}(\boldsymbol{j}) = \ell(q)$, then any program using $p$ and $q$ to output the pair $(\boldsymbol{i},\boldsymbol{j})$ must separate $p$ from $p$, for instance by prefixing $p$ with its length $\ell(p)$ encoded by $\beta(\ell(p))$ (see the proof of Proposition 4.1.1) at the cost of $2(\log \ell(p)+1)$ extra bits. In this way, the reference *UTM* $\mathfrak{U}$ first computes $p$ generating $\boldsymbol{i}$, then computes $q$, generating $\boldsymbol{j}$ and finally outputs $(\boldsymbol{i},\boldsymbol{j})$. Thus, one estimates:

$$\begin{aligned}\mathrm{C}((\boldsymbol{i},\boldsymbol{j})) &\leq \ell(\beta(\ell(p))p) + \ell(q) + C_0 \\ &\leq \mathrm{C}(\boldsymbol{i}) + \mathrm{C}(\boldsymbol{j}) + 2\log_2 \ell(p) + C_1 ,\end{aligned}$$

where $C_{0,1}$ are additive constants independent of the strings considered.

The $\log_2 \ell(p)$ extra bits cannot in general be avoided by reducing it to an additive constant independent of the input string. Indeed [120, 310], let $\ell(\boldsymbol{i}) = n$, $\ell(\boldsymbol{j}) = m$ and set $k := n + m$; there are $(k+1)2^k$ pairs $(\boldsymbol{i},\boldsymbol{j})$ such that the concatenated string $\boldsymbol{ij} \in \Omega_2^{(k)}$. By setting $c = (k+1)2^k$ in (4.6), one gets that at least one pair $(\boldsymbol{i},\boldsymbol{j})$ of such strings satisfies

$$\mathrm{C}((\boldsymbol{i},\boldsymbol{j})) \geq k + \log_2(k+1) .$$

Then, using (4.5), from $k = n + m = \ell(\boldsymbol{i}) + \ell(\boldsymbol{j})$ it follows that

$$\mathrm{C}((\boldsymbol{i},\boldsymbol{j})) \geq \mathrm{C}(\boldsymbol{i}) + \mathrm{C}(\boldsymbol{j}) + \log_2(k+1) - C .$$

**Remarks 4.1.4.**

1. Since the algorithmic complexities of $\boldsymbol{i}^{(n)}$ with respect to two *UTMs* is a constant independent of the string, one can fix a *UTM* $\mathcal{U}$ once and for all and drop the reference to it in $C_{\mathcal{U}}(\boldsymbol{i}^{(n)})$.
2. The additive constant $A$ in (4.5) can in line of principle be very large; however, since it is the same for all target strings $\boldsymbol{i}^{(n)}$, it becomes less and less important with increasing $n$. The additive constant can even be got rid of if, as in Example 4.1.5, one considers infinite strings $\boldsymbol{i} \in \Omega_2$, their prefixes $\boldsymbol{i}^{(n)} \in \Omega_2^{(n)}$ and let $n \to \infty$ in the algorithmic complexity per symbol $\dfrac{\mathrm{C}(\boldsymbol{i}^{(n)})}{n}$ .

3. The bound (4.6) shows that the one in (4.5) is not too loose for large $n$. In fact, the fraction of strings $\boldsymbol{i}^{(n)}$ with complexity smaller than $n-c$, $0 \leq c \leq n$, can be estimated by

$$\frac{\#\left\{\boldsymbol{i}^{(n)} \in \Omega^{(n)} \ : \ \mathrm{C}(\boldsymbol{i}^{(n)}) \leq n-c\right\}}{2^n} < 2^{1-c} \, .$$

Therefore, when $n$ gets large, the number of strings with complexity significantly smaller than $n$ gets small.
4. In view of the previous remark, it is suggestive to define random those sequences $\boldsymbol{i} \in \Omega_2$ such their initial prefixes $\boldsymbol{i}^{(n)}$ fulfil $\mathrm{C}(\boldsymbol{i}^{(n)}) > n-c$ for all $n \in \mathbb{N}$, where $c$ is a constant independent of $n$. Unfortunately, the vary same reason why the plain complexity is not subadditive (see Example 4.1.6 makes this definition not very useful [310]. Fortunately, as we shall see in Section 4.3, by using prefix *TM*s to compute programs one replaces the algorithmic complexity $\mathrm{C}(\boldsymbol{i}^{(n)})$ with the so-called *prefix complexity* $\mathrm{K}(\boldsymbol{i}^{(n)})$ and, in so doing, restores subadditivity and makes $\mathrm{K}(\boldsymbol{i}^{(n)}) > n-c$ for all $n \in \mathbb{N}$ a good definition of random sequences $\boldsymbol{i} \in \Omega_2$ [310].

### Non-Computability of $\mathrm{C}(\boldsymbol{i}^{(n)})$

Algorithmic complexity is *not computable*; namely, there cannot exist an algorithm [7] able to compute the $\mathrm{C}(\boldsymbol{i}^{(n)})$ for all strings. Indeed [254], if such a program $q$ of length $\ell(q) < \infty$ existed, then, one could construct the following program $p$:

- Step 1: let $i_0$ equal the empty string;
- Step 2: generate the $k$-string $\boldsymbol{i}_k$ in the lexicographically ordered set of all binary strings, call for $q$ and compute $\mathrm{C}(\boldsymbol{i}_k)$;
- Step 3: if $\mathrm{C}(\boldsymbol{i}_k) > \ell(p)$ write $\boldsymbol{i}_k$ and halt else set $k = k+1$ and go to Step 2.

Since $q$, the program which computes the plain complexity of any input string, is assumed to exists, $p$ also exists. Moreover, it has finite length $\ell(p)$ and halts with the first binary string, say $\boldsymbol{i}_{k^*}$ in lexicographical order, as output. Since the its plain complexity exceeds $\ell(p)$, $p$ is an effective description of $\boldsymbol{i}_{k^*}$ that is strictly shorter than its shortest possible effective description, which is a contradiction.

**Remark 4.1.5.** [268] The non-computability of $\mathrm{C}(\boldsymbol{i}^{(n)})$ implies the *undecidability of the halting problem*, namely that there cannot exist an algorithm able to decide whether a *UTM* $\mathfrak{U}$ halts when processing a generic program

[7] A *TM* according to the Church-Turing thesis.

$p$. Indeed, if such an algorithm existed, then one could compute $\mathrm{C}(\boldsymbol{i}^{(n)})$ for all $\boldsymbol{i}^{(n)}$. Effectively, one would proceed by generating the binary strings in lexicographical order (each one of them is a program) and subsequently processing them in *dovetailed fashion* [310, 92]. That is, at stage 1, step 1 of program 1 is effected, at stage 2, step 2 of program 1 and step 1 of program 2, at stage $k$, step $k$ of program 1, step $k-j+1$ of program $j$, $1 \leq j \leq k$, and so on. At the $N$-th step, there will be three groups of programs,

those that have halted with $\mathfrak{U}(p) = \boldsymbol{i}^{(n)}$;
those that have halted with $\mathfrak{U}(p) \neq \boldsymbol{i}^{(n)}$;
those that are still being processed.

Notice that in the third group there might be shorter programs than those which have already halted. Let $p^*$ be one of the shortest in the first group. One cannot set $\mathrm{C}(\boldsymbol{i}^{(n)}) = \ell(p^*)$ because it cannot be excluded that a program $p$ in the third group, shorter than $p$, will halt later with $\mathfrak{U}(p) = \boldsymbol{i}^{(n)}$. However, if the halting problem could be decided, then one would exactly have this vital piece of information and, waiting long enough, would have a means to find the shortest one among those programs such that $\mathfrak{U}(p) = \boldsymbol{i}^{(n)}$.

In spite of the fact that the plain complexity is not computable, the previous remark provides a means to effectively approximate it from above; namely, one can construct a sequence of functions $C_t$ that can be computed by a *UTM* $\mathfrak{U}$ on any binary input string $\boldsymbol{i}^{(n)}$ and get closer to $\mathrm{C}(\boldsymbol{i}^{(n)})$ with increasing $n$ [120]. Let $\mathfrak{U}_t(p)$ denote the output of the computation by $\mathfrak{U}$ of a program $p$ that halts in $t$ steps. By processing in dovetailed fashion the programs of length $\ell(p) \leq t$, one can check whether during the first $t$ computational steps some of them has halted with output $\boldsymbol{i}^{(n)}$, in which case one sets

$$\widetilde{C}_t(\boldsymbol{i}^{n)}) := \min\{\ell(p) \leq t : \mathfrak{U}_t(p) = \boldsymbol{i}^{(n)}\}\ , \qquad \widetilde{C}_t(\boldsymbol{i}^{(n)}) = +\infty \quad \text{otherwise}\ .$$

Finally, with reference to the loose upperbound (4.5), let

$$C_t(\boldsymbol{i}^{(n)}) := \min\{\widetilde{C}_t(\boldsymbol{i}^{(n)})\, ,\, n + A\}\ .$$

The function $\widetilde{C}_t(\boldsymbol{i}^{(n)})$ can only decrease with increasing $t$; moreover, from Definition 4.1.3, $C_t(\boldsymbol{i}^{(n)}) \geq \mathrm{C}(\boldsymbol{i}^{(n)})$ so that it tends to the plain complexity of $\boldsymbol{i}^{(n)}$ monotonically from above. One says that the plain complexity is *semi-computable from above*. Notice that, although we know that the approximating values $C_t(\boldsymbol{i}^{(n)})$ tend to $\mathrm{C}(\boldsymbol{i}^{(n)})$ from above, yet we do not know how far from the actual value $\mathrm{C}(\boldsymbol{i}^{(n)})$ any given $C_t(\boldsymbol{i}^{(n)})$ might be.

**Definition 4.1.4.** *A real function $f$ on $\Omega_2^*$ is called semi-computable from above if there exists a non-increasing sequence of functions $\{f_k\}_{k\in\mathbb{N}}$ on $\Omega_2^*$*

*with rational values* [8] *such that they are computable in the sense of Definition 4.1.2 and* $\lim_{k\to\infty} f_k(\boldsymbol{i}^{(n)}) = f(\boldsymbol{i}^{(n)})$. *A real function* $f$ *on* $\Omega_2^*$ *is called* **semi-computable from below** *if* $-f$ *is semi-computable from above. A real function* $f$ *on* $\Omega_2^*$ *is* **computable** *if it is semi-computable both from above and below.*

**Remarks 4.1.6.**

1. The difference from semi-computable and computable functions can be understood as follows. If $f$ is computable then there exist two monotone sequences of rational-valued computable functions $\{f_k^{a,b}\}_{k\in\mathbb{N}}$, $f_k^a$ non-increasing and $f_k^b$ non-decreasing, such that

$$f(\boldsymbol{i}^{(n)}) = \lim_{k\to+\infty} f^{a,b}(\boldsymbol{i}^{(n)}) .$$

It follows that one can always estimate, for any $\boldsymbol{i} \in \Omega_2^*$, the distance between the computed values $f_k^{a,b}(\boldsymbol{i})$ and the actual value $f(\boldsymbol{i})$ by means of the computable difference $f_k^a(\boldsymbol{i}) - f_k^b(\boldsymbol{i})$.
2. The approximations $f_k(\boldsymbol{i})$ of a function $f(\boldsymbol{i})$ semi-computable from below can be seen as the result of a same program (binary string) $p_f$. When a reference *UTM* $\mathfrak{U}$ is presented with $p_f$, together with the binary representation $\boldsymbol{i}(k)$ of $k$ and an input string $\boldsymbol{i} \in \Omega_2^*$, it computes $f_k(\boldsymbol{i})$, that is $\mathfrak{U}(\langle p_f, \boldsymbol{i}(k), \boldsymbol{i}\rangle) = f_k(\boldsymbol{i})$, where $\langle p_f, \boldsymbol{i}(k), \boldsymbol{i}\rangle$ is the binary string which encodes and separates the various inputs. Consequently, as well as computable functions also semi-computable functions can be enumerated.

An interesting class of lower semi-computable functions consists of the so called *constructive semi-measures* [120, 310].

**Definition 4.1.5.** *A positive function* $\mu : \Omega_2^* \mapsto \mathbb{R}$ *is called a* **semi-measure** *if* $\sum_{\boldsymbol{i}\in\Omega_2^*} w(\boldsymbol{i}) \leq 1$ *and a* **constructive semi-measure** *if it is semi-computable from below. A constructive semi-measure* $\boldsymbol{m} : \Omega_2^* \mapsto \mathbb{R}$ *is called a* **universal semi-measure** *if for any constructive semi-measure* $\mu$ *there exists a constant* $C_\mu$ *such that*

$$C_\mu\, \mu(\boldsymbol{i}) \leq \boldsymbol{m}(\boldsymbol{i}) \qquad \forall \boldsymbol{i} \in \Omega_2^* .$$

Working with semi-measures $\mu$ instead of measures allows for more freedom; for instance constructive measures turn out to be automatically computable. Namely, if $f_k$ is a non-decreasing sequence of rational-valued computable functions that approximate $\mu$ from below and $\sum_{\boldsymbol{i}\in\Omega_2^*} \mu(\boldsymbol{i}^{(n)}) = 1$, one can construct a computable approximation $0 \leq \mu_k \leq \mu$ such that, given $\varepsilon > 0$, $\sum_{\boldsymbol{i}\in\Omega_2^*} \mu_k(\boldsymbol{i}^{(n)}) \geq 1 - \varepsilon$. Then, for all $\boldsymbol{i} \in \Omega_2^*$ it holds that

[8] Any $p/q$, $p, q \in \mathbb{N}$, can be written as a binary string $\langle p, q\rangle \in \Omega_2^*$.

$$|\mu(\boldsymbol{i}) - \mu_k(\boldsymbol{i})| \leq \sum_{\boldsymbol{i}\in\Omega_2^*} (\mu(\boldsymbol{i}) - \mu_k(\boldsymbol{i})) \leq \varepsilon \,.$$

**Example 4.1.7.** [120, 310] Constructive semi-measures can be enumerated (see Remark 4.1.6.2); let $\{\mu_n\}$ denote their list and let $\{\alpha(n)\}_{n\in\mathbb{N}}$ be lower semi-computable positive numbers such that $\sum_n \alpha(n) \leq 1$. Then

$$\boldsymbol{m} := \sum_n \alpha(n)\,\mu_n \geq \alpha(k)\,\mu_k \quad \forall\, \mu_k \,.$$

$\boldsymbol{m}$ is thus a dominating semi-measure, it is also constructive and thus universal in the sense of Definition 4.1.5; indeed, there exists a two-argument lower semi-computable function $\mu(\boldsymbol{i}^{(n)}, n)$ that reproduces all constructive semi-measure by varying $n \in \mathbb{N}$. The idea of the proof is as follows. Given a lower semi-computable function $f$ and a non-decreasing sequence of rational-valued approximations $f_k$, let $p_f$ the binary program that allows a reference *UTM* $\mathfrak{U}$ to compute them as outlined in Remark 4.1.6.2 and let $\{\boldsymbol{i}_j\}_{j\in\mathbb{N}}$ be the lexicographically ordered list of all binary strings. By computing them in dovetailed fashion, let then $U_p^t$ be the computable function defined by

$$U_{p_f}^k(\boldsymbol{i}_j) = \begin{cases} \mathfrak{U}(\langle p_f, \boldsymbol{i}(k), \boldsymbol{i}_j\rangle) & \text{if } \ell(\boldsymbol{i}_j) \leq k \\ 0 & \text{otherwise} \end{cases} .$$

Notice that $U_{p_f}^k \to f$ when $k \to +\infty$; then, consider the recursive effective procedure consisting of the following steps:

1. set $\mu_{p_f}^0(\boldsymbol{i}_j) = 0$;
2. set $k = k + 1$;
3. compute $\boldsymbol{i}_1, \boldsymbol{i}_2, \ldots, \boldsymbol{i}_k$ in dovetailed fashion; if some $U_{p_f}^k(\boldsymbol{i}_j)$ has not halted go to Step 5, else compute $\sum_{j=1}^k U_{p_f}^k(\boldsymbol{i}_j)$;
4. if $\sum_{j=1}^k U_{p_f}^k(\boldsymbol{i}_j) \leq 1$, set $\mu_{p_f}^k := U_{p_f}^k$, and go to Step 2, else
5. set $\mu_{p_f}^k := \mu_{p_f}^{k-1}$ and stop.

By construction, the function $\mu(\boldsymbol{i}^{(n)}, p_f) := \lim_{k\to+\infty} \mu_{p_f}^k(\boldsymbol{i}^{(n)})$ is lower semi-computable and a semi-measure; further, it coincides with $f$ if the latter is itself a constructive semi-measure.

### Algorithmic Complexity and Thermodynamics

Beside its many mathematical applications, algorithmic complexity has also been used to explore the relations between computation and thermodynamics [54, 51, 52, 268, 310]. As already remarked in this section, computing is a physical process and questions about its *thermodynamic cost* is surely of practical importance, but also of general interest as they amount to asking which computational steps are intrinsically irreversible and which ones can instead be performed reversibly [51]. As nicely illustrated in [268], trying to answer these questions brings together thermodynamics, computability theory and *Gödel incompleteness theorem*.

The starting step is the observation [187, 51, 116] that the only irreversible computer operations are *intrinsically logically irreversible*, namely those with outputs that do not uniquely identify the input. The most obvious instance of such operations is *erasure* and, as an oversimplified case, consider one molecule of gas contained in a cubic box of volume $V$ in which a freely moving piston can be used to confine the molecule on the left side of the box, a case which is read as a bit 1. The flip operation which turns 1 into 0 can be effected reversibly by slowly rotating the box around its vertical axis and thus exchanging its right and left sides.

In order to erase these two bits of information, the piston can be let loose so that free expansion (of one molecule) allows the molecule, which was confined in a volume $V/2$ before, to wander later within the whole volume $V$. If the process occurs isothermally at temperature $T$, the loss of information corresponding to the increase of the space at disposal corresponds to an increase in thermodynamical entropy and decrease of free energy (the internal energy does not change in isothermal processes):

$$\Delta S = \kappa \log 2 \ , \quad \Delta F = \Delta U - T\,\Delta S = -\kappa T \log 2 \ .$$

By extrapolating this simple observation, one is naturally led to the identification of free energy and free memory: one can consume free memory to store data instead of erasing them and in this wave saves free energy, or, vice versa, by consuming free energy in erasure processes one saves free memory [268].

Differently from erasure which can in no way be turned into a reversible operation, all other operations are only superficially irreversible and can be made reversible by adding enough supplementary information [51]. For instance binary addition maps the pairs $(0,0)$ and $(1,1)$ into 0 and pairs $(0,1)$ and $(1,0)$ into 1. Therefore, by reading off 0 (1) one cannot decide which couple of bits was the input; however, conserving the inputs and writing them together with their outputs turns the binary addition ($\oplus$) into a reversible operation:

$$\underbrace{\begin{array}{c} 0\oplus 0=0\\ 0\oplus 1=1\\ 1\oplus 0=1\\ 1\oplus 1=0\end{array}}_{\text{irreversible}} \ , \qquad \underbrace{\begin{array}{c} (0,0)\mapsto(0,0,0)\\ (0,1)\mapsto(0,1,1)\\ (1,0)\mapsto(1,0,1)\\ (1,1)\mapsto(1,1,0)\end{array}}_{\text{reversible}} \ .$$

Unfortunately, the redundant information that is used in order to make operations reversible has to be stored and this occupies free memory so that massive erasure operations are eventually needed, free energy consumed and heat waste generated. In order to minimize free energy consumption, one can first proceed to reversibly compress as much as possible the stored information to be erased. For instance, in the case of the binary addition, one can use only the first input bit since the second one can be recovered by binary subtraction ($\ominus$) from the output bit:

$$\underbrace{\begin{array}{ccc} (0,0) \mapsto (0,0) & , & 0 \ominus 0 = 0 \\ (0,1) \mapsto (0,1) & , & 1 \ominus 0 = 1 \\ (1,0) \mapsto (1,1) & , & 1 \ominus 1 = 0 \\ (1,1) \mapsto (1,0) & , & 0 \ominus 1 = 1 \end{array}}_{\text{still reversible}} \cdot$$

Suppose the occupied memory consists of a binary string $\boldsymbol{i}^{(n)}$, then the best compression achievable is given by the shortest binary program $p^*$ such that $\mathfrak{U}(p^*) = \boldsymbol{i}^{(n)}$ whose length is the Kolmogorov complexity $\mathrm{C}(\boldsymbol{i}^{(n)})$. Reversibly encoding $\boldsymbol{i}^{(n)}$ into $p^*$ and erasing the latter entails the optimal loss of free energy $\Delta_{opt}F = -\kappa T\,\mathrm{C}((\boldsymbol{i}^{(n)})\,\log 2$ to be compared with $\Delta F = -n\kappa T\,\log 2$.

These considerations suggest [326] that, when dealing with the thermodynamics of computation, the notion of entropy should be improved by the addition to the standard thermal contribution, $S_{th}$, of the one coming from the optimal erasure of the memory

$$S_{comp} = S_{th} + \kappa\,\mathrm{C}(M)\,\log 2\ ,$$

where $\mathrm{C}(M)$ is the algorithmic complexity of the computer memory. For instance, by using $S_{comp}$, the Maxwell's demon paradox [190] can be solved by observing [326] that $S_{therm}$ can indeed be diminished by the demon collecting together all fastest particles and transferring heath from lower to higher temperatures. However, storing all the information necessary to comparing particle velocities rapidly consumes free memory and asks for erasure thus restoring the second law of thermodynamics.

Unfortunately, the main problem with optimal compression is that it is based on the knowledge of the algorithmic complexity of the occupied memory which cannot always be computed. In few words, performing an optimal compression of the memory content before erasure is not always possible and there will always be an excess of free energy consumption. As this is ultimately due to the undecidability of the halting problem, this effect can be suggestively and not unduly called *Gödel friction* [268].

## 4.2 Algorithmic Complexity and Entropy Rate

Despite Remark 4.1.4.4, there is a sense in which the Kolmogorov complexity can be used to look at the individual trajectories of a classical dynamical system $(\mathcal{X}, T, \mu)$ and at their randomness, namely through their asymptotic complexity rate. As explained in Section 2.2, a partition $\mathcal{P}$ of $\mathcal{X}$ provides a symbolic model $(\widetilde{\Omega}_p, T_\sigma, \mu_{\mathcal{P}})$ whereby trajectories are reduced to sequences $\boldsymbol{i} \in \widetilde{\Omega}_p \subseteq \Omega_p$ of symbols from an alphabet with $p$ letters of which one can study the complexity of the prefixes $\boldsymbol{i}^{(n)} \in \Omega_p^{(n)}$ [9].

As for the Shannon entropy, when dealing with sequences, one may decide to focus not on the Kolmogorov complexity which generically diverges, rather upon its rate or *complexity per symbol* [7, 69].

**Definition 4.2.1.** *The complexity rate of a sequence $\boldsymbol{i} \in \widetilde{\Omega}_p$ is given by*

$$\mathrm{c}(\boldsymbol{i}) := \limsup_{n\to\infty} \frac{1}{n} \mathrm{C}(\boldsymbol{i}^{(n)}) ,$$

*where $\boldsymbol{i}^{(n)}$ is the initial prefix of $\boldsymbol{i}$ of length $n$.*

*Given a dynamical system $(\mathcal{X}, T, \mu)$ and a finite, measurable partition $\mathcal{P}$ of $\mathcal{X}$, let $\boldsymbol{i}(x) \in \Omega_p$ denote the symbolic trajectory that $\mathcal{P}$ associates to the trajectory $\{T^n x\}_{n\geq 0}$ issuing from $x \in \mathcal{X}$. Then, the complexity rate of $\{T^n x\}_{n\geq 0}$ with respect to $\mathcal{P}$ is $\mathrm{c}(x, \mathcal{P}) := \mathrm{c}(\boldsymbol{i}(x))$.*

To start with, we shall consider the case of a dynamical system which is itself already a symbolic model, namely a binary information source. An important result is that, typically, for sequences emitted by ergodic sources, the bound (4.7) becomes an equality in the limit.

**Theorem 4.2.1 (Brudno's Theorem).** *Let $(\Omega_2, T_\sigma, \pi)$ be a binary ergodic source with entropy rate $h(\pi)$. Then,*

$$\mathrm{c}(\boldsymbol{i}) = \lim_{n\to\infty} \frac{1}{n} \mathrm{C}(\boldsymbol{i}^{(n)}) = h(\pi) , \tag{4.8}$$

*for almost all $\boldsymbol{i} \in \Omega_2$ with respect to $\pi$.*

The proof [69, 318, 166] consists 1) in using the counting argument (4.6) and the *AEP* (Proposition 3.2.2) to show that

$$\liminf_{n\to\infty} \frac{1}{n} \mathrm{C}(\boldsymbol{i}) \geq h(\pi) \qquad \pi - \text{a.e} ; \tag{4.9}$$

and 2) in providing, for the initial prefixes $\boldsymbol{i}^{(n)}$ of $\pi$-almost all $\boldsymbol{i} \in \Omega_2^*$, an appropriate binary program $p_{\boldsymbol{i}}(n) \in \Omega_2^*$ such that $\displaystyle\lim_{n\to\infty} \frac{\ell(p_{\boldsymbol{i}}(n))}{n} \leq h(\pi)$ and $\mathfrak{U}(p_{\boldsymbol{i}}(n)) = \boldsymbol{i}^{(n)}$ whence

[9] In order to do this, one has to extend Definition 4.1.3 to the case of strings of symbols from generic finite alphabets. This is straightforward and will always be understood in the following.

$$\limsup_{n\to\infty}\frac{1}{n}\mathrm{C}(\boldsymbol{i}^{(n)})\leq h(\pi)\qquad \pi-\text{a.e}\ . \tag{4.10}$$

**Proof of the lower bound** : Because of the assumption of ergodicity, Theorem 3.2.1 allows us to use the *AEP* with the entropy rate $h(\pi)$ in place of the Shannon entropy $H(A)$. Let $A_\epsilon^{(n)}\subseteq\Omega_2^{(n)}$ be the set in (3.27) consisting of binary strings $\boldsymbol{i}^{(n)}$ such that

$$2^{-n(h(\pi)+\epsilon)}\leq\pi(\boldsymbol{i}^{(n)})\leq 2^{-n(h(\pi)-\epsilon)}\ ,$$

and $\hat{A}_\epsilon^{(n)}\subset\Omega_2$ the set of sequences whose initial prefixes of length $n$, $\boldsymbol{i}^{(n)}$ belong to $A_\epsilon^{(n)}$ and have complexity $\mathrm{C}(\boldsymbol{i}^{(n)})\leq n(h(\pi)-2\epsilon)$. From (4.6), it follows that

$$\begin{aligned}\pi\Big(\hat{A}_\epsilon^{(n)}\Big)&=\pi\Big(\Big\{\boldsymbol{i}^{(n)}\in A_\epsilon^{(n)}\ :\ \mathrm{C}(\boldsymbol{i}^{(n)})\leq n(h(\pi)-2\epsilon)\Big\}\Big)\\&\leq\#\Big(\hat{A}_\epsilon^{(n)}\Big)\cdot\max_{\boldsymbol{i}^{(n)}\in A_\epsilon^{(n)}}\pi(\boldsymbol{i}^{(n)})\\&\leq 2^{n(h(\pi)-2\epsilon)+1}\cdot 2^{-n(h(\pi)-\epsilon)}=2^{-n\varepsilon+1}\ .\end{aligned}$$

Since strings $\boldsymbol{i}^{(n)}\notin A_\epsilon^{(n)}$ may also have complexity $\mathrm{C}(\boldsymbol{i}^{(n)})\leq n(h(\pi)-2\epsilon)$, it is necessary to control their overall probability. Set $(\hat{A}_\epsilon^{(k)})^c:=\Omega_2\setminus\hat{A}_\epsilon^{(k)}$ and

$$\widetilde{A}_\epsilon^{(k)}:=\Big\{\boldsymbol{i}\in(\hat{A}_\epsilon^{(k)})^c:\ \mathrm{C}(\boldsymbol{i}^{(k)})\leq k(h(\pi)-2\epsilon)\Big\}\ ,\quad B_\epsilon^{(n)}:=\bigcup_{k\geq n}\widetilde{A}_\epsilon^{(k)}\ .$$

Since $\widetilde{A}_\epsilon^{(k)}\subset(\hat{A}_\epsilon^{(k)})^c$ implies $\pi\Big(B_\epsilon^{(n)}\Big)\leq\pi\Big(\bigcup_{k\geq n}(\hat{A}_\epsilon^{(k)})^c\Big)=1-\pi\Big(\bigcap_{k\geq n}\hat{A}_\epsilon^{(k)}\Big)$, it follows that the probability of the set of sequences whose initial prefixes have complexity $\mathrm{C}(\boldsymbol{i}^{(n)})\geq n(h(\pi)-2\varepsilon)$ is estimated from above by

$$\begin{aligned}\pi\Big(\bigcup_{k\geq n}\Big\{\hat{A}_\epsilon^{(k)}\cup\widetilde{A}_\epsilon^{(k)}\Big\}\Big)&\leq\pi\Big(\bigcup_{k\geq n}\hat{A}_\epsilon^{(k)}\Big)+\pi\Big(B_\epsilon^{(n)}\Big)\\&\leq\sum_{k\geq n}2^{-k\,\varepsilon+1}+\pi\Big(B_\epsilon^{(n)}\Big)\leq\frac{2^{-n\,\varepsilon+1}}{1-2^{-\varepsilon}}+1-\pi\Big(\bigcap_{k\geq n}\hat{A}_\epsilon^{(k)}\Big)\ .\end{aligned}$$

The set $\bigcap_{k\geq n}\hat{A}_\epsilon^{(k)}$ consists of sequences $\boldsymbol{i}\in\Omega_2$ whose initial prefixes are typical for all lengths $k\geq n$; therefore $\lim_{n\to\infty}\pi\Big(\bigcap_{k\geq n}\hat{A}_\epsilon^{(k)}\Big)=1$. It thus follows that $\inf_{n\geq n}\dfrac{\mathrm{C}(\boldsymbol{i}^{(n)})}{n}>h(\pi)-2\epsilon$ $\pi$-almost everywhere. Since $\varepsilon$ is arbitrary the lower bound follows. □

**Proof of the upper bound** : Given $\Omega_2^{(n)} \ni \boldsymbol{i}^{(n)} = i_1 i_2 \cdots i_n$, fix $0 < L < n$ and consider all strings of length $L$ made of consecutive bits of $\boldsymbol{i}^{(n)}$; there are $n - L + 1$ of them:

$$s_k := i_k i_2 \cdots i_{L+k-1} \ , \quad 1 \leq k \leq n - L + 1 \ . \qquad (*)$$

Let $\Omega_{\boldsymbol{i}^{(n)}}^{(L)}$ denote their set and let $N(s)$ be the number of occurrences of the string $s \in \Omega_{\boldsymbol{i}^{(n)}}^{(L)}$; $N(\boldsymbol{s})$ can be expressed as follows. Let $\boldsymbol{i} \in \Omega_2$ be any sequence with initial prefix of length $n$ equal to $\boldsymbol{i}^{(n)}$, then

$$N(s) = \sum_{j=0}^{n-L+1} \chi_s(T_\sigma^j(\boldsymbol{i})) \ , \qquad (**)$$

where $T_\sigma$ is the left shift and $\chi_s(T_\sigma^j(\boldsymbol{i}))$ is 1 if the initial prefix of length $L$ in $T_\sigma^j(\boldsymbol{i})$ equals $s$, 0 otherwise.

Given the $N(s)$, $s \in \Omega_{\boldsymbol{i}^{(n)}}^{(L)}$, one can thus construct a so-called *empirical probability distribution* $\pi_{\boldsymbol{i}^{(n)}}^{(L)}$ on $\Omega_{\boldsymbol{i}^{(n)}}^{(L)}$:

$$\pi_{\boldsymbol{i}^{(n)}}^{(L)} := \{p_n^{(L)}(s)\} \ , \quad p_n^{(L)}(s) := \frac{N(s)}{n - L + 1}$$

with corresponding Shannon entropy

$$H(\pi_{\boldsymbol{i}^{(n)}}^{(L)}) := - \sum_{s \in \Omega_{\boldsymbol{i}^{(n)}}^{(L)}} p_n^{(L)}(s) \log_2 p_n^{(L)}(s) \ .$$

Notice that the set of lengths $\ell(s) := \lceil -\log_2 p_n^{(L)}(s) \rceil$ is such that

$$-\log_2 p_n^{(L)}(s) \leq \ell(s) < -\log_2 p_n^{(L)}(s) + 1 \ ; \qquad (***)$$

therefore, they satisfy the Kraft inequality

$$\sum_{s \in \Omega_{\boldsymbol{i}^{(n)}}^{(L)}} 2^{-\ell(s)} \leq \sum_{s \in \Omega_{\boldsymbol{i}^{(n)}}^{(L)}} p_n^{(L)}(s) = 1 \ .$$

Because of Proposition 3.2.1, there thus exists a binary prefix code over the strings $s \in \Omega_{\boldsymbol{i}^{(n)}}^{(L)}$ consisting of codewords $w(s)$ of lengths $\ell(s) := \ell(w(s))$.

With $s_k$ as defined in $(*)$ above, for a given $1 \leq j \leq L - 1$, consider the adjacent strings of length $L$ of the form $s_{j+p_j L}$, $0 \leq p_j \leq p_j^{max}$. Since the first bit of $s_j$ is $i_j$ and the last bit of $s_{j+p_j^{max} L}$ is $i_{j+(p_j^{max}+1)L-1}$, then the bit not belonging to any $s_{j+p_j L}$ are $i_1 i_2 \cdots i_j$ and $i_{j+(p_j^{max}+1)L} i_{j+(p_j^{max}+1)L+1} \cdots i_n$, whence

$$j + (p_j^{max} + 1)L - 1 \leq n \Longrightarrow p_j^{max} \leq \frac{n - j - L + 1}{L} \ ,$$

for a total of no more than $2(L-1)$ bits. Also, since any $1 \leq k \leq n$ can be written as $k = j + pL$ with $1 \leq L-1$ and $0 \leq \left\lceil \frac{k}{L} \right\rceil$ uniquely determined, then, for different $1 \leq j \leq L-1$, the sets $S_j := \{s_{j+p_j L}\}_{p_1=0}^{p_j^{max}}$ do not overlap and $\bigcup_{j=1}^{L-1} S_j = \Omega^{(L)}_{\boldsymbol{i}^{(n)}}$.

Consider a program $Q_j$ that reconstructs $\boldsymbol{i}^{(n)}$ by specifying the codewords $w(s_{j+p_j L})$ plus the bits uncovered by them; its length can be bounded from above as follows:

$$\ell(Q_j) \leq C + 2(L-1) + \sum_{p_j=0}^{p_j^{max}} \ell(s_{j+p_j L}) ,$$

where $C$ is a constant independent of $j$ and of $L$. Further, $(***)$ entails the following bound for the plain algorithmic complexity of $\boldsymbol{i}^{(n)}$:

$$\begin{aligned}
\mathrm{C}(\boldsymbol{i}^{(n)}) &\leq \min_{1 \leq j \leq L-1} \ell(Q_j) \leq \frac{1}{L-1} \sum_{j=1}^{L-1} \ell(Q_j) \\
&\leq C + 2(L-1) + \frac{1}{L-1} \sum_{j=1}^{L-1} \sum_{p=0}^{p_j^{max}} \ell(s_{j+pL}) \\
&= C + 2(L-1) + \frac{1}{L-1} \sum_{s \in \Omega^{(L)}_{\boldsymbol{i}^{(n)}}} N(s)\ell(s) \\
&\leq C + 2(L-1) + \frac{n-L+1}{L-1} \underbrace{\sum_{s \in \Omega^{(L)}_{\boldsymbol{i}^{(n)}}} p_n^{(L)}(s)\Big(-\log_2 p_n^{(L)}(s) + 1\Big)}_{H(\pi^{(L)}_{\boldsymbol{i}^{(n)}})+1} .
\end{aligned}$$

From ergodicity and $(**)$, it follows that, when $n \to \infty$,

$$\frac{N(s)}{n-L+1} \longmapsto p(s) = \pi\Big(C_s^{[0,L-1]}\Big)$$

for $\pi$-almost all sequences $\boldsymbol{i} \in \Omega_2$, where $C_s^{[0,L-1]}$ is the cylinder set containing all $\boldsymbol{i} \in \Omega_2$ with $s \in \Omega^{(L)}_{\boldsymbol{i}^{(n)}}$ as initial prefix. Thus, when $n \to \infty$ $\pi^{(L)}_{\boldsymbol{i}^{(n)}}$ tends to the probability distribution $\pi^{(L)}$ over the partition $\mathcal{C}^{(L)} = \Big\{C^{[0,L-1]}_{s \in \Omega_2^{(L)}}\Big\}$ of $\Omega_2$ indexed by the strings $s \in \Omega_2^{(L)}$; then, by continuity,

$$\limsup_{n \to \infty} \frac{1}{n} \mathrm{C}(\boldsymbol{i}^{(n)}) \leq \frac{H(\mathcal{C}^{(L)}) + 1}{L-1} , \qquad \pi - \text{a.e} .$$

By taking $L \to \infty$, the upper bound follows (see Remark 3.1.1.1). □

The previous result that holds for ergodic binary information sources can easily be extended to generic ergodic sources and then to ergodic dynamical systems via Definition 4.2.1.

**Proposition 4.2.1.** *Let $(\mathcal{X}, T, \mu)$ be an ergodic dynamical system and $\mathcal{P}$ a finite, measurable partition of $\mathcal{X}$; then*

$$\mathrm{c}(x, \mathcal{P}) = \mathrm{h}_{\mu}^{\mathrm{KS}}(T, \mathcal{P}) \qquad \mu - a.e\ .$$

**Proof:** The partition $\mathcal{P}$ defines a symbolic model $(\widetilde{\Omega}_p, T_\sigma, \mu_{\mathcal{P}})$ which is an ergodic shift-dynamical system. The result follows since Brudno's theorem ensures that for $\mu_{\mathcal{P}}$-almost all $\boldsymbol{i} \in \widetilde{\Omega}_p$, hence for $\mu$-almost all $x \in \mathcal{X}$, it holds that $\mathrm{c}(\boldsymbol{i}) = h(\mu_{\mathcal{P}}) = \mathrm{h}_{\mu}^{\mathrm{KS}}(T, \mathcal{P})$. □

**Corollary 4.2.1.** *Let $(\mathcal{X}, T, \mu)$ be an ergodic dynamical system and $\mathcal{P}$ a finite, measurable generating partition of $\mathcal{X}$; then*

$$\mathrm{c}(x, \mathcal{P}) = \mathrm{h}_{\mu}^{\mathrm{KS}}(T) \qquad \mu - a.e\ .$$

## 4.3 Prefix Algorithmic Complexity

A way to eliminate the logarithmic correction that spoils the subadditivity of the plain algorithmic complexity (see Example 4.1.6) is to ask that the only acceptable programs for the *UTM* $\mathfrak{U}$ are the so-called *self-delimiting* ones, namely those containing the specification of their lengths, so that the *UTM* always knows when its input programs end. These programs have the *prefix property* that if $\mathfrak{U}$ halts on one of them, say $p$, then $p$ cannot be the prefix of any other halting program for $\mathfrak{U}$. Any *TM* that accepts only programs with the prefix property is called a *prefix TM* ; it can be showed [78] that there exist *prefix UTMs* capable of simulating the behavior of any other prefix *TM*. The consequences of the prefix constraint are far reaching. One first proceeds to define an adapted version of algorithmic complexity of binary strings (the extension to strings from different alphabets is straightforward).

**Definition 4.3.1 (Prefix Algorithmic Complexity).** *The prefix algorithmic complexity of $\boldsymbol{i}^{(n)} \in \Omega_2^{(n)}$ is the length of the shortest program p such that $\mathfrak{U}(p) = \boldsymbol{i}^{(n)}$, where $\mathfrak{U}$ is any chosen reference prefix UTM :*

$$\mathrm{K}(\boldsymbol{i}^{(n)}) = \min\left\{\ell(p)\ :\ \mathfrak{U}(p) = \boldsymbol{i}^{(n)}\ ,\ \mathfrak{U}\ \textit{a prefix UTM}\right\}\ .$$

**Remarks 4.3.1.**

1. A prefix *TM* can be figured out [78] as a *TM* with a control unit, two tapes and two reading-write heads. The first tape, the program tape, is entirely occupied by the program which is written as a binary string between two blank symbols marking its beginning and its end; the program is read by a head that can only read, halt and move right. The second tape, the work tape, is, as in the case of an ordinary *TM* , two-way infinite and the head on it can read, write 0,1, leave a blank #, halt or move both right and left. The computation starts with the head on the program tape scanning the first blank symbol, the other head on the 0-th cell of the work tape, only finitely many of its cells possibly carrying non-blank symbols, and with the control unit in its initial ready state $q_r$. Then, in agreement with the symbols read by the two heads and the control unit internal state, the head on the working tape erases and writes or does nothing and then moves left, right or stays, the head on the program tape either moves right or stays, while the control unit updates its internal state. The computation terminates if the reading head on the program tape reaches the end of the program, in which case, the output is what is written on the work tape to the right of the cell being scanned by the head until only cells with blank symbols are found. The program halts if and only if the head on the program tape reaches the end of the tape.
2. Since the set of programs with the prefix property is smaller than the set of all programs, then
$$\mathrm{C}(\boldsymbol{i}^{(n)}) \leq \mathrm{K}(\boldsymbol{i}^{(n)}) \ .$$
On the other hand, if $p$ is such that $\mathrm{C}(\boldsymbol{i}^{(n)}) = \ell(p)$, then, considering its self-delimiting encoding $p^* := \beta(\ell(p))p$, it follows that
$$\mathrm{K}(\boldsymbol{i}^{(n)}) \leq \ell(p^*) \leq \mathrm{C}(\boldsymbol{i}^{(n)}) + 2\log \ell(p) + C \ .$$
3. The prefix complexity is subadditive; in fact, if $p$ and $q$ are programs such that $\mathrm{K}(\boldsymbol{i}) = \ell(p)$ and $\mathrm{K}(\boldsymbol{j}) = \ell(q)$, with $\boldsymbol{i}, \boldsymbol{j} \in \Omega^*$, then, since $p$ and $q$ are now, by definition, self-delimiting, one has
$$\mathrm{K}(\boldsymbol{i}, \boldsymbol{j}) \leq \mathrm{K}(\boldsymbol{i}) + \mathrm{K}(\boldsymbol{j}) + C \ .$$
4. Unlike for the plain complexity (see Remark 4.1.4.4), one can rightly define *random* those sequences $\boldsymbol{i} \in \Omega_2$ for which
$$\mathrm{K}(\boldsymbol{i}^{(n)}) > n - c \ ,$$
for all their prefixes $\boldsymbol{i}^{(n)}$, that is all those sequences whose prefixes $\boldsymbol{i}^{(n)}$ have prefix complexity that increases at least as $n$. Indeed [310], it turns out that these sequences are those and only those passing all constructive statistical Martin-Löf tests checking whether they belong to effectively null sets (see footnote 1). In this sense, relative to the prefix definition

of algorithmic complexity, Levin's chaoticness and typicalness mentioned in the introduction to this section are equivalent characterization of randomness.

One of the most important consequences of working with prefix *UTMs* $\mathfrak{U}$ is that their halting programs $p$ form a set of prefix codes for the output strings $\mathfrak{U}(p) = \boldsymbol{i} \in \Omega_2^*$ and their lengths satisfy the extended Kraft inequality (3.2.2).

**Example 4.3.1.** Consider a prefix *UTM* $\mathfrak{U}$ and the so-called *Chaitin magic number* [80, 92, 50] defined by $\Omega = \sum_{p:\mathfrak{U}(p)\downarrow} 2^{-\ell(p)}$, where the sum runs over all halting programs $p$; because of the prefix property, $\Omega \leq 1$.

Let us consider the binary expansion of $\Omega$ which has infinitely many 0s if it is rational and suppose an algorithm exists that calculates the digits of $\Omega$. Then, the $n$-digit approximation $\Omega_n := \sum_{j=1}^{n} \frac{\omega_j}{2^j}$ is such that $\Omega_n > \Omega - 2^{-n}$. Then, one knows whether $\mathfrak{U}$ halts on programs of length $\leq n$.

Indeed, by listing them in lexicographical order and by processing them in dovetailed fashion, one can collect all programs $p_1, p_2, \ldots$ that halt until, after $T(n)$ computational steps,

$$S_n := \sum_{i=1}^{m(n)} 2^{-\ell(p_i)} \geq \Omega_n \ .$$

If $p$ is any program halting in more than $T(n)$ computational steps, one gets

$$\Omega \geq S_n + 2^{-\ell(p)} \geq \Omega_n + 2^{-\ell(p)} > \Omega + 2^{-\ell(p)} - 2^{-n}$$

Therefore, $\ell(p) > n$ so that if a program of length shorter than $n$ has not halted in $T(n)$ computational steps it will never halt.

Let $G(n)$ be the set of strings $\boldsymbol{i}_j := \mathfrak{U}(p_j)$, $j = 1, 2, \ldots, m(n)$, corresponding to the outputs of the programs that have halted in $T(n)$ computational steps and let $\boldsymbol{i}$ denote the first string (in a suitable order) not in $G(n)$. Such string must have prefix complexity $\mathrm{K}(\boldsymbol{i}) > n$: indeed, if $\mathrm{K}(\boldsymbol{i}) \leq n$, there would exist a program $p$ of length $\leq n$ such that $\mathfrak{U}(p) = \boldsymbol{i}$. However, from the previous discussion one deduces that also $p$ must have halted in $T(n)$ computational steps so that $\boldsymbol{i} \in G(n)$, too. Further, let $p^*$ be any shortest effective description of the string $\Omega^{(n)} := \omega_1\omega_2 \cdots \omega_n$ consisting of the first $n$ *bits* of $\Omega$, namely $\mathrm{K}(\Omega^{(n)}) = \ell(p^*)$. Then, by means of a fixed number $c$ of extra bits, one can use the knowledge of the $\Omega^{(n)}$ to recover $\boldsymbol{i}$, whence

$$n < \mathrm{K}(\boldsymbol{i}) \leq \ell(p^*) + c = \mathrm{K}(\Omega^{(n)}) + c \implies \mathrm{K}(\Omega^{(n)}) > n - c \quad \forall n \ .$$

Then $\Omega$ is a random sequence in the sense of Remark 4.3.1.4.

**Definition 4.3.2.** *Given a prefix UTM $\mathfrak{U}$, the map $\Omega_2^* \ni \boldsymbol{i} \mapsto P_{\mathfrak{U}}(\boldsymbol{i})$, where*

$$P_{\mathfrak{U}}(\boldsymbol{i}) := \sum_{p\,:\,\mathfrak{U}(p)=\boldsymbol{i}} 2^{-\ell(p)} \tag{4.11}$$

*defines a so-called the* universal probability *on $\Omega_2^*$.*

This definition makes sense, for, as a consequence of the prefix property, not only (4.1) holds, but it also turns out that

$$\sum_{\boldsymbol{i}\in\Omega_2^*} P_{\mathfrak{U}}(\boldsymbol{i}) = \sum_{\boldsymbol{i}\in\Omega_2^*} \sum_{p\,:\,\mathfrak{U}(p)=\boldsymbol{i}} 2^{-\ell(p)} \le 1 \ .$$

**Remarks 4.3.2.**

1. If a prefix *TM* $\mathfrak{A}$ halts on $p = 0$ and $q = 1$ with the strings $\boldsymbol{i}$ and $\boldsymbol{j}$ as outputs, then $P_{\mathfrak{A}}(\boldsymbol{i}) = P_{\mathfrak{A}}(\boldsymbol{j}) = 1/2$ since no other program can halt. Without the prefix restriction the sum in (4.11) would diverge simply because all programs prefixed by $p$ and $q$ would also output $\boldsymbol{i}$ and $\boldsymbol{j}$.
2. After division by $\sum_{\boldsymbol{i}\in\Omega^*} P_{\mathfrak{U}}(\boldsymbol{i})$, $P_{\mathfrak{U}}(\boldsymbol{i})$ represents the probability that $\boldsymbol{i}$ be the output of $\mathfrak{U}$ running a binary program $p$ of length $\ell(p)$ randomly chosen according to the Bernoulli uniform probability distribution that assigns probability $2^{-\ell(p)}$ to anyone of them. Since short programs have higher probabilities, random strings have smaller algorithmic probabilities than regular ones.
3. The probability $P_{\mathfrak{U}}$ is called universal (see Example 4.1.7) for the following reason. Let $\mathfrak{A}$ be any prefix *TM* and $q$ a program such that $\mathfrak{A}(q) = \boldsymbol{i} \in \Omega_2^*$; further, let $q'$ be a self-delimiting program of fixed length $L$ that makes $\mathfrak{U}$ simulate $\mathfrak{A}$ so that $\mathfrak{U}(q'q) = \mathfrak{A}(q) = \boldsymbol{i}$. Then,

$$P_{\mathfrak{U}}(\boldsymbol{i}) = \sum_{p\,:\,\mathfrak{U}(p)=\boldsymbol{i}} 2^{-\ell(p)} \ge \sum_{q\,:\,\mathfrak{U}(q'q)=\boldsymbol{i}} 2^{-\ell(q)-\ell(q')} = 2^{-L}\, P_{\mathfrak{A}}(\boldsymbol{i}) \ . \tag{4.12}$$

Suppose now $\pi = \{p(\boldsymbol{i})\}_{\boldsymbol{i}\in\Omega_2^*}$ to be a computable probability distribution over $\Omega_2^*$ (see Definition 4.1.2). Consider a prefix *TM* $\mathfrak{A}$ that does the following:

- it computes the probability distribution $\pi$;
- it encodes the strings $\boldsymbol{i} \in \Omega_2^*$ by means of the Shannon-Fano-Elias code corresponding to the computed $\pi$ (see Example 3.2.3);
- given a program $q \in \Omega_2^*$, it checks whether $q$ is the Shannon-Fano-Elias code for any $\boldsymbol{i} \in \Omega_2^*$; if so, it outputs $\boldsymbol{i}$.

Since the lengths of the code-words are as in (3.23), then, for all $\boldsymbol{i} \in \Omega_2^*$,

$$P_{\mathfrak{A}}(\boldsymbol{i}) = \sum_{\mathfrak{A}(q)=\boldsymbol{i}} 2^{-\ell(q)} \ge 4\,p(\boldsymbol{i}) \ .$$

For the prefix *UTM* $\mathfrak{U}$ to work as $\mathfrak{A}$, it is necessary to compute the probability distribution $\pi$, whence the program $q'$ in (4.12) is such that $L = \mathrm{K}(\pi) + L'$, where $\mathrm{K}(\pi)$ is the prefix complexity of $\pi$, where it is understood that the computable probability distribution $\pi$ is written as a binary string (denoted by the same symbol). Then, for all computable probability distributions $\pi$ on $\Omega_2^*$,

$$P_{\mathfrak{U}}(\boldsymbol{i}) \geq C\, 2^{-\mathrm{K}(\pi)}\, p(\boldsymbol{i}) \ , \tag{4.13}$$

with $C > 0$ a constant independent of $\boldsymbol{i}$ and $\pi$.

Universal probability, prefix complexity and Shannon entropy of computable probability distributions are intimately related. Given a prefix *UTM* $\mathfrak{U}$, the programs $p^*$ such that $\mathfrak{U}(p^*) = \boldsymbol{i} \in \Omega_2^*$ with $\ell(p^*) = \mathrm{K}(\boldsymbol{i})$ provide a prefix code such that

$$P_{\mathfrak{U}}(\boldsymbol{i}) = \sum_{\mathfrak{U}(p)=\boldsymbol{i}} 2^{-\ell(p)} \geq 2^{-\mathrm{K}(\boldsymbol{i})} \ . \tag{4.14}$$

Further, if the strings $\boldsymbol{i}$ are chosen at random with respect to a computable probability distribution $\pi$, then (3.22) implies that the corresponding average length, namely the average prefix complexity, satisfies

$$\sum_{\boldsymbol{i}\in\Omega_2^*} p(\boldsymbol{i})\,\mathrm{K}(\boldsymbol{i}) \ \geq \ H_2(\pi) \ = \ -\sum_{\boldsymbol{i}\in\Omega_2^*} p(\boldsymbol{i}) \log_2 p(\boldsymbol{i}) \ . \tag{4.15}$$

There might be infinitely many programs such that $\mathfrak{U}(p) = \boldsymbol{i}$, yet the lower bound in (4.14) is surprisingly good as the sum is actually dominated by the shortest programs for $\boldsymbol{i}$.

**Proposition 4.3.1.** *For all* $\boldsymbol{i} \in \Omega_2^*$, $P_{\mathfrak{U}}(\boldsymbol{i}) \leq C\, 2^{-\mathrm{K}(\boldsymbol{i})}$, *where* $C > 0$ *is a constant independent of* $\boldsymbol{i}$.

Together with (4.14), this result permits the identification (up to an additive constant) of the prefix complexity of a string with minus the logarithm of its universal probability.

**Corollary 4.3.1.** $\mathrm{K}(\boldsymbol{i}) = -\log_2 P_{\mathfrak{U}}(\boldsymbol{i}) + O(1)$.

There thus appears a similarity between the fact that the optimal codeword lengths with respect to a probability distribution $\pi = \{p(i)\}_{i\in I}$ are of the form $\ell_i^* = -\log_2 p(i)$ and the fact that the lengths of the shortest descriptions of binary strings practically amount to the logarithm of their universal probabilities. This similarity can be carried even further by examining the average complexity.

**Corollary 4.3.2.** *Given a computable probability distribution $\pi$ on $\Omega_2^*$, the corresponding average prefix complexity satisfies*

$$H_2(\pi) \leq \sum_{\boldsymbol{i}\in\Omega_2^*} p(\boldsymbol{i})\,\mathrm{K}(\boldsymbol{i}) \leq H_2(\pi) + \mathrm{K}(\pi) + C\ .$$

**Proof:** From Proposition 4.3.1 and (4.13)

$$\mathrm{K}(\boldsymbol{i}) \leq -\log_2 P_{\mathfrak{U}}(\boldsymbol{i}) + C' \leq -\log_2 p(\boldsymbol{i}) + \mathrm{K}(\pi) + C\ .$$

Multiplying by $p(\boldsymbol{i})$ and summing over $\boldsymbol{i} \in \Omega_2^*$ yields the upper bound, while (4.15) gives the lower bound. □

**Proof of Proposition 4.3.1** : The idea is to construct, for each $\boldsymbol{i} \in \Omega_2^*$, a set of programs $p$ of length $\ell(p) \leq -\log_2 P_{\mathfrak{U}}(\boldsymbol{i}) + C'$ with the prefix property such that $\mathfrak{U}(p) = \boldsymbol{i}$, so that $\mathrm{K}(\boldsymbol{i}) \leq \ell(p)$ would end the proof. Unfortunately, the argument of Remark 4.3.2.3 is not viable as the universal probability is not computable. However, as much as for the plain algorithmic complexity, the prefix complexity is semi-computable from above whence the universal probability results lower semi-computable because of Corollary 4.3.1; this turns out to be sufficient for constructing a prefix code with the desired property. Let all the programs (listed in lexicographical order) be run by $\mathfrak{U}$ in dovetail fashion and collect them in pairs $(p_k, \boldsymbol{x}_k)$ where $p_k$ is the program which halts at the $k$ step of the dovetailed computation with $\boldsymbol{x}_k \in \Omega_2^*$ as output. The quantity

$$P_{\mathfrak{U}}(k, \boldsymbol{x}_k = \boldsymbol{x}) := \sum_{\substack{(p_i, \boldsymbol{x}_i = \boldsymbol{x})\\ i\leq k}} 2^{-\ell(p_i)} \leq P_{\mathfrak{U}}(\boldsymbol{x})$$

is computable and tends to $P_{\mathfrak{U}}(\boldsymbol{x})$ along the subsequence $\{(p_k, \boldsymbol{x}_k = \boldsymbol{x})\}_k$; set $n_k := \lceil -\log_2 P_{\mathfrak{U}}(k, \boldsymbol{x}_k = \boldsymbol{x})\rceil$. Since

$$2^{-\ell_*(k)} \leq P_{\mathfrak{U}}(k, \boldsymbol{x}_k = \boldsymbol{x}) \leq 2^{-\ell_*(k)+1}\ ,$$

where $\ell_*(k)$ is the smallest length in the sum, it follows that $n_k = \ell_*(k)$. Given $(p_k, \boldsymbol{x}_k, n_k)$, this triplet is assigned to the first non-occupied node at the $(n_k + 1)$-th level of a binary tree; further, in order to enforce the prefix condition, all nodes stemming from it are made unavailable to further assignments. Since $n_k$ is not strictly monotonic, it may happen that different pairs $(p_i, \boldsymbol{x}_i = \boldsymbol{x}_k)$, $i \leq k$, have the same $n_k$; by eliminating all but the first pair with that value of $n_k$, no more than one node will be occupied by a triplet with the same $\boldsymbol{x}_k$ at level $n_k$. Therefore,

$$n_k \geq -\log_2 P_{\mathfrak{U}}(k, \boldsymbol{x}_k = \boldsymbol{x}) \geq -\log_2 P_{\mathfrak{U}}(\boldsymbol{x}) \Longrightarrow n_k = \lceil -\log_2 P_{\mathfrak{U}}(\boldsymbol{x})\rceil + r_k$$

with $r_k \geq 0$ and $r_k \neq r_j$ for $j \neq k$. To each $\boldsymbol{x} \in \Omega_2^*$ there correspond many assignments of triplets $(p_k, \boldsymbol{x}_k = \boldsymbol{x}, n_k)$ each one of them to one and only one node at level $n_k + 1$. The nodes thus provide binary code-words of length $n_k + 1$ for the triplets. In order to see that there are sufficiently many nodes to accommodate all triplets, we check that the lengths $n_k + 1$ satisfy the Kraft inequality (3.21). That this is indeed so follows from the fact that

$$\sum_{\boldsymbol{x}_k = \boldsymbol{x}} 2^{-n_k} = 2^{-\lceil -\log_2 P_{\mathfrak{U}}(\boldsymbol{x}) \rceil} \sum_{\boldsymbol{x}_k = \boldsymbol{x}} 2^{-r_k} \leq 2\, P_{\mathfrak{U}}(\boldsymbol{x}) \ ,$$

for all $\boldsymbol{x} \in \Omega_2^*$, whence

$$\sum_{\boldsymbol{x} \in \Omega_2^*} \sum_{\boldsymbol{x}_k = \boldsymbol{x}} 2^{-n_k - 1} \leq \sum_{\boldsymbol{x} \in \Omega_2^*} P_{\mathfrak{U}}(\boldsymbol{x}) \leq 1 \ .$$

The above algorithm allows the construction of a binary tree whereby any $\boldsymbol{x} \in \Omega_2^*$ can be identified with the binary string $\boldsymbol{i}(\boldsymbol{x}) \in \Omega_2^*$ corresponding to the lowest depth node assigned to its triplets $(p_k, \boldsymbol{x}_k = \boldsymbol{x}, n_k)$. The length of the code-word $\boldsymbol{i}(\boldsymbol{x}) \in \Omega_2^*$ is the smallest $n_k + 1$:

$$\ell(\boldsymbol{i}(\boldsymbol{x})) \leq \lceil -\log_2 P_{\mathfrak{U}}(\boldsymbol{x}) \rceil + 1 \leq -\log_2 P_{\mathfrak{U}}(\boldsymbol{x}) + 2 \ .$$

Finally, let $q$ be a program of fixed length $L$ that makes the prefix *UTM* $\mathfrak{U}$ generate the binary tree by dovetailed computation as specified above and let $q'$ be another program of fixed length $L'$ with the necessary instructions to $\mathfrak{U}$ such that, when presented with the code-word $q' q\, \boldsymbol{i}(\boldsymbol{x})$, $\mathfrak{U}$ computes the program in the triplet assigned to the node marked by $\boldsymbol{i}(\boldsymbol{x})$, writes the result and halts. By construction, the program $p$ in the triplet at the node $\boldsymbol{i}(\boldsymbol{x})$ is such that $\mathfrak{U}(p) = \boldsymbol{x}$; then

$$\mathrm{K}(\boldsymbol{x}) \leq \ell(q' q\, \boldsymbol{i}(\boldsymbol{x})) = -\log_2 P_{\mathfrak{U}}(\boldsymbol{x}) + L + L' + 2 \ .$$

□

**Remark 4.3.3.** Because of its construction the universal probability is a lower semi-computable semi-measure (see Definition 4.1.5), thus there exists a constant $C_P$ such that $C_P\, P_{\mathfrak{U}} \leq \boldsymbol{m}$, where $\boldsymbol{m}$ is the universal semi-measure constructed in Example 4.3.2. Furthermore, an argument similar to the one in the previous proof, extends the result in Corollary 4.3.1 to

$$\mathrm{K}(\boldsymbol{i}) = -\log_2 P_{\mathfrak{U}}(\boldsymbol{i}) + O(1) = -\log_2 \boldsymbol{m}(\boldsymbol{i}) + O(1) \ .$$

## Bibliographical Notes

The books [79]– [81] provide inspiring and motivating introductions to algorithmic complexity theory, as well as the reviews [327] and [305], the latter one being especially devoted to comparing several characterizations of randomness. A reference book is [310] which also provides a historical overview of the development of the theory, several applications to a variety of mathematical, logical and physical problems, as well as more recent advances. A more abstract presentation is offered by [73] whereas a handier introduction can be found in [120] and in [92]. In [7] algorithmic complexity is discussed in relation to stochastic processes, in [127] in relation to entropic tools in classical information theory and in [254] in relation with coding and statistical modeling. Finally, [93] presents an introduction to computable functions, recursion, the halting problem, Gödel's theorem and computational complexity. Possible uses of algorithmic complexity theory in relation to the predictability of discrete classical dynamical systems are discussed in [62]. Approaches to complexity issues in a broad sense can be found in [18, 294]

# Part II

# Quantum Dynamical Systems

In the second part of the book quantum dynamical systems with finite and infinite degrees of freedom are presented by using the algebraic approach to quantum statistical mechanics. The corresponding technical framework proves convenient for the extension of ergodic and information theory to non-commutative contexts.

# 5 Quantum Mechanics of Finite Degrees of Freedom

Quantum dynamical systems are described by means of non-commutative algebras of observables, by means of their time-evolution and by means of the expectation functionals that assign mean values to them. Classical dynamical systems can always be described in terms of phase-points and phase-trajectories; however, an algebraic formulation is always possible and has two advantages: on one hand, similarities and differences with respect to quantum dynamical systems become more evident and, on the other hand, one can infer from the algebraic reformulation of classical notions how to possibly extend them to the quantum setting.

With reference to information, the most important difference that one encounters passing from the commutative to the non-commutative setting is that the disturbances exerted on quantum systems by measurement processes cannot in general be made negligible, not even in line of principle.

## 5.1 Hilbert Space and Operator Algebras

In standard quantum mechanics, physical states are usually described by normalized vectors in separable Hilbert spaces, and the observables by self-adjoint linear operators acting on them. Here follows some notations and basic facts.

1. $|\psi\rangle$, $|\phi\rangle$, or $\psi$, $\phi$, and $|i\rangle$, with $i$ running on a suitable index set $I$, will denote (normalized) vectors in Hilbert spaces $\mathbb{H}$ and $P_\psi = |\psi\rangle\langle\psi|$ the associated *orthogonal projectors.*
2. The scalar product on $\mathbb{H}$, denoted by $\langle\psi|\phi\rangle$, linear in the second argument and anti-linear in the first one, satisfies the *Cauchy-Schwartz inequality*
$$|\langle\psi|\phi\rangle| \leq \|\psi\|\,\|\phi\| \ . \tag{5.1}$$
Any finite or countable set $\{\Psi_i\}_{i\in I} \subset \mathbb{H}$ such that $\langle\Psi_i|\Psi_j\rangle = \delta_{ij}$ and $|\psi\rangle = \sum_{i\in I}\langle\Psi_i|\psi\rangle|\Psi_i\rangle$ for all $\psi \in \mathbb{H}$, is an orthonormal basis (*ONB*) in $\mathbb{H}$. The corresponding projectors $P_i := |\Psi_i\rangle\langle\Psi_i|$ fulfil
$$\sum_{i\in I} P_i = \sum_{i\in I} |\Psi_i\rangle\langle\Psi_i| = \mathbb{1} \ , \tag{5.2}$$

F. Benatti, *Dynamics, Information and Complexity in Quantum Systems*, Theoretical and Mathematical Physics, DOI 10.1007/978-1-4020-9306-7_5,

where $\mathbb{1}$ denotes the identity operator on $\mathbb{H}$, $\mathbb{1}|\psi\rangle = |\psi\rangle$ for all $\psi \in \mathbb{H}$.

3. Given any linear operator $X$ on $\mathbb{H}$, its matrix elements with respect to $\psi, \phi \in \mathbb{H}$ will be denoted either as $\langle \psi | X\phi \rangle$ or as $\langle \psi | X | \phi \rangle$, depending on notational convenience.
4. $X^T$ and $X^*$ will denote transposition and complex conjugation with respect to a given *ONB* $\{\Psi_j\}_j$:

$$\langle \Psi_i | X^T \Psi_j \rangle = \langle \Psi_j | X\Psi_i \rangle , \qquad \langle \Psi_i | X^* \Psi_j \rangle = \langle \Psi_i | X\Psi_j \rangle^* .$$

Instead, $X^\dagger = (X^T)^* = (X^*)^T$ will represent the basis-independent adjoint of $X$:

$$\langle \psi | X^\dagger \phi \rangle = \langle X\psi | \phi \rangle = \langle \phi | X\psi \rangle^* \qquad \forall \psi, \phi \in \mathbb{H} .$$

Physical observables correspond to self-adjoint operators $X = X^\dagger$.

5. The *uniform norm*, $\|X\|$, of a linear operator $X$ on $\mathbb{H}$ is defined by

$$\|X\| := \sup_{\|\psi\|=1} \|X|\psi\rangle\| . \tag{5.3}$$

6. $X$ is *bounded* if $\| X \| < \infty$, in which case

$$\|X|\psi\rangle\| \leq \|X\| \, \|\psi\| , \qquad |\langle \phi | X\psi \rangle| \leq \|X\| \, \|\phi\| \, \|\psi\| . \tag{5.4}$$

Linear combinations of bounded operators are again bounded; their linear span will be denoted by $\mathbb{B}(\mathbb{H})$. The product of bounded operators is bounded, for $\|XY|\psi\rangle\| \leq \|X\| \, \|Y\| \, \|\psi\|$. Therefore, $\mathbb{B}(\mathbb{H})$ is a so-called $*$-algebra.

7. An operator $U \in \mathbb{B}(\mathbb{H})$ such that $U^\dagger U = \mathbb{1}$ is called an *isometry*; in general, $U\mathcal{U}^\dagger = (U U^\dagger)(U U^\dagger)$ is a projection, if also $U U^\dagger = \mathbb{1}$, then $U$ is a *unitary operator*. Isometries have $\|U\| = \sqrt{\|U^\dagger U\|} = \|\mathbb{1}\| = 1$.
8. The uniform norm defines on $\mathbb{B}(\mathbb{H})$ *uniform neighborhoods* of the form

$$\mathcal{U}_\varepsilon(X) = \{Y \in \mathbb{B}(\mathbb{H}) \; ; \; \|X - Y\| \leq \varepsilon\} , \qquad \varepsilon \geq 0 , \tag{5.5}$$

whence a sequence $X_n \in \mathbb{B}(\mathbb{H})$ *converges uniformly* to $X \in \mathbb{B}(\mathbb{H})$, $\lim_{n\to\infty} X_n = X$, if $\lim_{n\to\infty} \|X - X_n\| = 0$. The corresponding topology on $\mathbb{B}(\mathbb{H})$, $\tau_u$, is called *uniform topology.*

9. $\mathbb{B}(\mathbb{H})$ is complete with respect to the uniform topology, namely all sequences of operators which are of Cauchy type with respect to the uniform norm converge to an element of $\mathbb{B}(\mathbb{H})$. Therefore, $\mathbb{B}(\mathbb{H})$ is a so-called Banach $*$-algebra. Moreover, since the uniform norm fulfils

$$\|X^\dagger\| = \|X\| , \qquad \|X^\dagger X\| = \|X\|^2 , \tag{5.6}$$

$\mathbb{B}(\mathbb{H})$ is a *$C^*$-algebra* (see Section 5.2).

10. In the case of $n < \infty$ degrees of freedom, each one of them is described by a Hilbert space $\mathbb{H}_j$, $1 \leq j \leq n$. Altogether, their Hilbert space is the tensor product $\mathbb{H}^{(n)} = \bigotimes_{j=1}^{n} \mathbb{H}_j$, denoted by $\mathbb{H}^{\otimes n}$ when the Hilbert spaces $\mathbb{H}_j$ are copies of a same $\mathbb{H}$. Depending on notational convenience, its vectors will be denoted either by $|\,\psi\,\rangle = |\,\psi_1\,\rangle \otimes |\,\psi_2\,\rangle \otimes \cdots |\,\psi_n\,\rangle$ or by $|\,\psi\,\rangle = |\,\psi_1 \otimes \psi_2 \otimes \cdots \psi_n\,\rangle$, with scalar products $\langle\,\phi\,|\,\psi\,\rangle = \prod_{j=1}^{n} \langle\,\phi_j\,|\,\psi_j\,\rangle$. Bounded operators on $\mathbb{H}^{(n)}$ are linear combinations of tensor products of the form $X_1 \otimes X_2 \otimes \cdots X_n$, $X_j \in \mathbb{B}(\mathbb{H}_j)$; the associated $C^*$ algebra of bounded operators on $\mathbb{H}^{(n)}$ is $\mathbb{B}(\mathbb{H}^{(n)}) := \bigotimes_{j=1}^{n} \mathbb{B}(\mathbb{H}_j)$.
11. The *strong topology* on $\mathbb{B}(\mathbb{H})$, $\tau_s$, is the smallest topology with respect to which all semi-norms of the form $\mathcal{L}_\psi(X) := \|X|\,\psi\,\rangle\|$, $\psi \in \mathbb{H}$, are continuous; its *strong neighborhoods* are of the form

$$\mathcal{U}^s_\varepsilon(X) := \{Y \in \mathbb{B}(\mathbb{H}) \,:\, \mathcal{L}_{\psi_j}(Y - X) \leq \varepsilon \,,\; 1 \leq j \leq n\} \,, \tag{5.7}$$

for $\psi_j \in \mathbb{H}$, $n \in \mathbb{N}$ and $\varepsilon \geq 0$. A sequence $X_n \in \mathbb{B}(\mathbb{H})$ converges strongly to $X \in \mathbb{B}(\mathbb{H})$, $s - \lim_{n\to\infty} X_n = X$, if $\lim_{n\to\infty} \|(X_n - X)|\,\psi\,\rangle\| = 0$ for all $\psi \in \mathbb{H}$.
12. The *weak topology* on $\mathbb{B}(\mathbb{H})$, $\tau_w$, is the smallest topology with respect to which all semi-norms of the form $\mathcal{L}_{\phi,\psi}(X) := |\langle\,\phi\,|\,X\psi\,\rangle|$, $\phi\,,\,\psi \in \mathbb{H}$ are continuous; its *weak-neighborhoods* are of the form

$$\mathcal{U}^w_\varepsilon(X) := \{Y \in \mathbb{B}(\mathbb{H}) \,:\, \mathcal{L}_{\phi_j,\psi_j}(Y - X) \leq \varepsilon \,,\; 1 \leq j \leq n\} \,, \tag{5.8}$$

for $\psi_j, \phi_j \in \mathbb{H}$, $n \in \mathbb{N}$ and $\varepsilon \geq 0$. A sequence $X_n \subset \mathbb{B}(\mathbb{H})$ *converges weakly* to $X \in \mathbb{B}(\mathbb{H})$, $w - \lim_{n\to\infty} X_n = X$, if $\lim_{n\to\infty} |\langle\,\phi\,|\,(X_n - X)\psi\,\rangle| = 0$ for all $\phi, \psi \in \mathbb{H}$.
13. Since strong neighborhoods are uniform neighborhoods, but the reverse is not true when $\mathbb{H}$ is infinite dimensional, the uniform topology is in general finer than the strong one, that is $\tau_u$ has more neighborhoods than $\tau_s$: $\tau_s \preceq \tau_u$. The weak topology is in general coarser than the strong one; every weak neighborhood is also a strong neighborhood, but the reverse fails to be true in infinite dimensional $\mathbb{H}$. The norm, strong and weak topologies are equivalent in finite dimension.
14. Among other topologies on $\mathbb{B}(\mathbb{H})$ [64], one of some use in the following is the $\sigma$-*weak* topology, $\tau_{uw}$; it is finer than the weak topology for it is the smallest one that makes continuous the following semi-norms,

$$\mathcal{L}^{uw}_{\{\phi_n\},\{\psi_n\}}(X) = \sum_n |\langle\,\phi_n\,|\,X\,|\psi_n\,\rangle| \,, \tag{5.9}$$

where $\{\psi_n\}, \{\phi_n\} \subset \mathbb{H}$ are such that $\sum_n \|\psi_n\|^2 < \infty$ and $\sum_n \|\phi_n\|^2 < \infty$.

Most of the previous assertions are standard facts [64, 251, 300]; however, the various topologies on $\mathbb{B}(\mathbb{H})$ deserve a closer look.

**Remarks 5.1.1.**

1. That the uniform topology is finer than the strong topology can be seen as follows. Given any strong neighborhood $\mathcal{U}^s_\varepsilon(X)$, let $\alpha := \max_{1\leq i\leq n} \|\psi_i\|$, then $\mathcal{U}^u_{\varepsilon/\alpha}(X) \subseteq \mathcal{U}^s_\varepsilon(X)$; indeed,

$$Y \in \mathcal{U}^u_{\varepsilon/\alpha}(X) \Longrightarrow \|(X-Y)|\,\psi_i\,\rangle\| \leq \frac{\varepsilon}{\alpha}\|\psi_i\| \leq \varepsilon \Longrightarrow Y \in \mathcal{U}^s_\varepsilon(X) ,$$

whence $\mathcal{U}^s_\varepsilon(X)$ is a uniform neighborhood, too. In order to show that $\tau_u$ is in general strictly finer than $\tau_s$, it is sufficient to exhibit a sequence of operators in $\mathbb{B}(\mathbb{H})$ which converges strongly, but not uniformly. To this end, suppose $\mathbb{H}$ to be infinite dimensional, choose a *ONB* $\{\Psi_k\}_{k\in\mathbb{N}}$ with associated orthonormal projectors $P_k$ and construct $Q_N := \sum_{k=1}^N P_k$. Then, (5.2) reads $s-\lim_N Q_N = 1\!\!1$; namely, if $\psi \in \mathbb{H}$ and $c_\psi(i) = \langle\,\Psi_i\,|\,\psi\,\rangle$, then

$$\lim_{N\to\infty} \|(Q_N - 1\!\!1)|\,\psi\,\rangle\|^2 = \lim_{N\to\infty} \sum_{n\geq N+1}^{\infty} |c_\psi(n)|^2 = 0 .$$

On the other hand, $Q_N \to 1\!\!1$ cannot hold in the uniform sense, otherwise for any $\varepsilon \geq 0$ there would exist $N_0(\varepsilon)$ such that, if $N \geq N_0(\varepsilon)$, then $\|(Q_N - 1\!\!1)|\,\psi\,\rangle\| \leq \varepsilon$ uniformly in $\psi \in \mathbb{H}$, while $\|(Q_N - 1\!\!1)|\,\psi\,\rangle\| = \|\psi\|$ for all $\psi$ in the subspace orthogonal to that projected out by $Q_N$.
2. In like manner, the weak topology cannot have more neighborhoods than the strong topology. Given $\mathcal{U}^w_\varepsilon(X)$ as in (5.8), set $\beta := \max_{1\leq i\leq n} \|\phi_i\|$; then $\mathcal{U}^s_{\varepsilon/\beta}(X) \subseteq \mathcal{U}^w_\varepsilon(X)$; indeed, using (5.1),

$$Y \in \mathcal{U}^s_{\varepsilon/\alpha}(X) \Longrightarrow |\langle\,\phi_i\,|\,(X-Y)\,|\psi_i\,\rangle| \leq \frac{\varepsilon}{\beta}\|\phi_i\| \leq \varepsilon \Longrightarrow Y \in \mathcal{U}^w_\varepsilon(X) .$$

In general, $\tau_s$ is strictly finer than $\tau_w$. Let $\mathbb{H}$ be infinite dimensional and, given an *ONB* $\{\Psi_k\}_{k\in\mathbb{N}}$, consider the operator $X : \mathbb{H} \mapsto \mathbb{H}$ defined as the right shift along the *ONB* :

$$X|\,\Psi_k\,\rangle = |\,\Psi_{k+1}\,\rangle , \qquad X^\dagger|\,\Psi_k\,\rangle = \begin{cases} 0 & k=1 \\ |\,\Psi_{k-1}\rangle & k\geq 2 \end{cases} .$$

Note that $X^\dagger X|\,\Psi_k\,\rangle = |\,\Psi_k\,\rangle$ for all $k \in \mathbb{N}$ so that $X^\dagger X = 1\!\!1$; $X$ is an isometry with $XX^\dagger$ projecting onto the subspace orthogonal to $\Psi_1$. Furthermore, by expanding $\mathbb{H} \ni |\,\psi\,\rangle = \sum_{k=1}^\infty c_\psi(k)|\,\Psi_k\,\rangle$, $c_\psi(k) := \langle\,\Psi_k\,|\,\psi\,\rangle$, it turns out that

$$\|X^n|\,\psi\,\rangle\|^2 = \langle\,\psi\,|\,(X^\dagger)^n X^n\,|\psi\,\rangle = \|\psi\|^2 ,$$

whereas

$$w - \lim_{n\to\infty} X^n = 0 .$$

Indeed, given $\phi, \psi \in \mathbb{H}$, $\varepsilon > 0$ and $|\psi_K\rangle = \sum_{i=1}^{K} c_\psi(i) | \Psi_i \rangle$ such that $\| | \psi\rangle - |\psi_K\rangle \| \leq \varepsilon$, (5.1) yields

$$\left| \langle \phi | X^n | \psi \rangle \right| \leq \left| \langle \phi | X^n | \psi_K \rangle \right| + \varepsilon \|\psi\| = \left| \sum_{i=1}^{K} c_\phi^*(i) \, c_\psi(i+n) \right| + \varepsilon \|\psi\|$$
$$\leq \sqrt{\sum_{i=1}^{K} |c_\phi(i)|^2} \sqrt{\sum_{i=n+1}^{K+n} |c_\psi(i)|^2} + \varepsilon \|\psi\| \, ,$$

where the second square-root becomes negligibly small for sufficiently large $n$.

3. By adding to a subalgebra $\mathcal{A} \subseteq \mathbb{B}(\mathbb{H})$ its limit points with respect to a given topology $\tau$, one obtains its closure $\overline{\mathcal{A}}^{\tau}$. If of two topologies $\tau_{1,2}$ on $\mathcal{A} \subseteq \mathbb{B}(\mathbb{H})$, $\tau_1$ is coarser than $\tau_2$ ($\tau_1 \preceq \tau_2$), $\tau_1$ has less neighborhoods than $\tau_2$ and thus more convergent sequences; therefore, $\overline{\mathcal{A}}^{\tau_1} \supseteq \overline{\mathcal{A}}^{\tau_2}$. In particular, $\overline{\mathcal{A}}^{\tau_u}$ is a $C^*$-subalgebra of $\mathbb{B}(\mathbb{H})$; further, since $\tau_u \succeq \tau_s \succeq \tau_w$ it follows that $\overline{\mathcal{A}}^{\tau_u} \subseteq \overline{\mathcal{A}}^{\tau_s} \subseteq \overline{\mathcal{A}}^{\tau_w}$.
4. Given a $*$-subalgebra $\mathcal{A} \subset \mathbb{B}(\mathbb{H})$, consider the linear functionals $F : \mathcal{A}^{\tau_1} \mapsto \mathbb{C}$, respectively $F : \mathcal{A}^{\tau_1} \mapsto \mathbb{C}$, that are continuous with respect to two topologies $\tau_1 \preceq \tau_2$; more precisely, the preimages $F^{-1}(V)$ of open sets $V \subset \mathbb{C}$ are open sets in $\mathcal{A}^{\tau_1}$, respectively $\mathcal{A}^{\tau_2}$. Then, since not all open sets in $\mathcal{A}^{\tau_2}$ are open sets in $\mathcal{A}^{\tau_1}$, a $\tau_2$-continuous $F$, that is continuous with respect to the finer topology, may fail to be $\tau_1$-continuous, that is continuous with respect to the coarser topology. For instance, all weakly continuous linear functionals on $\mathcal{A} \subseteq \mathbb{B}(\mathbb{H})$ are strongly continuous but strong continuity does not in general ensure that a functional is also weakly continuous.
5. If $\mathcal{A}$ is a generic Banach algebra, its topological dual, $\mathcal{A}^*$ is the linear space $\mathcal{A}^*$ consisting of all linear functionals $F : \mathcal{A} \mapsto \mathbb{C}$ that are continuous on $\mathcal{A}$. Then, $\mathcal{A}^*$ can be equipped with the so-called *$w^*$-topology*, namely with the coarsest topology that makes continuous all semi-norms of the form

$$\mathcal{L}_X^{w^*}(F) = |F(X)| \qquad \forall X \in \mathcal{A} \, . \tag{5.10}$$

Its neighborhoods are of the form

$$\mathcal{U}_\varepsilon^{w^*}(F) := \{G \in \mathcal{A}^* \, : \, \mathcal{L}_{X_j}^{w^*}(G - F) \leq \varepsilon \, , \; 1 \leq j \leq n\} \, , \tag{5.11}$$

for any $X_j \in \mathcal{A}$, $n \in \mathbb{N}$ and $\varepsilon \geq 0$. A sequence $F_n \in \mathcal{A}^*$ *$w^*$-converges* to $F \in \mathcal{A}^*$, $w^* - \lim_{n\to\infty} F_n = F$, if $\lim_{n\to\infty} |F_n(X) - F(X)| = 0$ for all $X \in \mathcal{A}$.

## 5.2 $C^*$ Algebras

The bounded operators on a Hilbert space $\mathbb{H}$ form a Banach $*$-algebra with respect to the uniform norm (5.3); this norm fulfils the two equalities (5.6).

While the first one follows at once from (5.3) and the definition of adjoint, the second one is proved by using (5.1):

$$\|X\|^2 = \sup_{\|\psi\|=1} \langle \psi \,|\, X^\dagger X \psi \rangle \leq \|X^\dagger\| \, \|X\| \Longrightarrow \|X\| \leq \|X^\dagger\| \; ;$$

thus, exchanging $X$ and $X^\dagger$, yields $\|X\|^2 = \|X^\dagger|^2 = \|X^\dagger X\|$. From the fact that $\|XY\| \leq \|X\| \, \|Y\|$ (an inequality that follows at once from (5.3)), one gets $\|X^\dagger| = \|X\|$; in fact,

$$\|X\|^2 = \|X^\dagger X\| \leq \|X\| \, \|X^\dagger\| \Longrightarrow \|X\| \leq \|X|^\dagger \; ,$$

while $\|X^\dagger\|^2 = \|XX^\dagger\| \leq \|X\| \, \|X^\dagger| \Longrightarrow \|X^\dagger\| \leq \|X\|$.

More in general, let $\mathcal{A}$ be an algebra with an involution $\dagger : \mathcal{A} \mapsto \mathcal{A}$ such that

$$(\alpha\, A + \beta\, B)^\dagger = \alpha^* \, A^\dagger + \beta^* \, B^\dagger \; , \quad (AB)^\dagger = B^\dagger \, A^\dagger$$

for all $\alpha, \beta \in \mathbb{C}$ and $A, B \in \mathcal{A}$. Let $\mathcal{A}$ be complete with respect to a norm $\| \cdot \| : \mathcal{A} \mapsto \mathbb{R}_+$ such that

$$\|\alpha\, A\| = |\alpha| \, \|A\| \; , \quad \|A + B\| \leq \|A\| + \|B\| \; , \quad \|A\, B\| \leq \|A\| \, \|B\|$$

and $\|A\| = 0 \Longleftrightarrow A = 0$ for all $\alpha \in \mathbb{C}$ and $A, B \in \mathcal{A}$. If the norm further satisfes (5.6) it is called a $C^*$ *norm.*

**Definition 5.2.1.** *Any Banach $*$-algebra $\mathcal{A}$ with respect to a $C^*$ norm is called a $C^*$ algebra. $\mathcal{A}$ is called unital if it possesses an identity $\mathbb{1}$ such that $A\,\mathbb{1} = \mathbb{1}\, A = A$ for all $A \in \mathcal{A}$.*

**Examples 5.2.1.**

1. The commutative algebras $C(\mathcal{X})$, respectively $\mathbb{L}^\infty_\mu(\mathcal{X})$ of continuous, respectively essentially bounded functions over a compact phase-space $\mathcal{X}$ discussed in Section 2.2.1 are $C^*$ algebras with respect to the uniform, respectively essentially bounded norms.
2. Many instances of quantum systems are $N$-level systems; their Hilbert space is finite dimensional and thus can be taken as $\mathbb{H} = \mathbb{C}^N$, while their observables are Hermitian $N \times N$ matrices with complex entries. In such cases, the $C^*$ algebra of bounded operators $\mathbb{B}(\mathbb{H})$ is the full matrix algebra $M_N(\mathbb{C})$. Given an *ONB* $\{\Psi_j\}_{j=1}^N$ in $\mathbb{C}^N$, set $E_{ij} := |\,\Psi_i\,\rangle\langle\,\Psi_j\,|$, $i, j = 1, \ldots, N$; then,

$$E_{ij}|\,\psi\,\rangle = \langle\,\Psi_j\,|\,\psi\,\rangle\,|\,\Psi_i\,\rangle \; , \quad E_{ij}E_{k\ell} = \delta_{jk}E_{i\ell} \; , \quad \sum_{i=1}^N E_{ii} = \mathbb{1} \; . \tag{5.12}$$

Any set of matrices with these properties constitutes a set of *matrix units*, $E_{ii}$ being a complete set of orthogonal projections. Any $X \in M_N(\mathbb{C})$ can be thus expressed as a linear combination of the matrix units:

$$X = \sum_{i,j=1}^{N} E_{ii}\, X\, E_{jj} = \sum_{i,j=1}^{N} X_{ij}\, E_{ij} \ .$$

In the *standard representation* where the basis vectors have the form

$$|\, \Psi_j \,\rangle = (0 \quad 0 \quad \cdots \quad 1 \quad \cdots \quad 0 \quad 0)^T \ ,$$

with 1 in the $j$-th entry, the matrix units $E_{ij}$ are $N \times N$ matrices whose entries are all 0, but for the $ij$-th one which is equal to 1.

3. A typical scenario often encountered in quantum physics is as follows: a quantum system described by a generic (not necessarily finite-dimensional) Hilbert space $\mathbb{H}$ is coupled to an $N$-level system, the corresponding algebra being the tensor product $M_N(\mathbb{B}(\mathbb{H})) := M_N(\mathbb{C}) \otimes \mathbb{B}(\mathbb{H})$ consisting of operators of the form

$$\widetilde{X} = \sum_{i,j=1}^{N} E_{ij} \otimes X_{ij} = \begin{pmatrix} X_{11} & \cdots & X_{1N} \\ \cdot & \cdots & \cdot \\ \cdot & \cdots & \cdot \\ \cdot & \cdots & \cdot \\ X_{N1} & \cdots & X_{NN} \end{pmatrix} = [X_{ij}] \ , \tag{5.13}$$

where $X_{ij}$ are operators in $\mathbb{B}(\mathbb{H})$ and $E_{ij}$ are matrix units in standard form. The tensor product $M_N(\mathbb{B}(\mathbb{H}))$ is a $*$-algebra of operators acting on the Hilbert space $\widetilde{\mathbb{H}} := \mathbb{C}^N \otimes \mathbb{H}$ consisting of vectors

$$\widetilde{\mathbb{H}} \ni |\, \widetilde{\psi} \,\rangle = \sum_{i=1}^{N} |\, i \,\rangle \otimes |\, \psi_i \,\rangle = \begin{pmatrix} |\, \psi_1 \,\rangle \\ \vdots \\ |\, \psi_N \,\rangle \end{pmatrix} \ . \tag{5.14}$$

A uniform norm on $M_N(\mathbb{B}(\mathbb{H}))$ is defined by

$$\begin{aligned} \|\widetilde{X}\|^2 &= \sup_{\|\widetilde{\psi}\|=1} \langle\, \widetilde{\psi} \,|\, \widetilde{X}^\dagger \widetilde{X} \,|\widetilde{\psi}\,\rangle \\ &= \left\{ \sum_{i,j,k=1}^{N} \langle\, \psi_k \,|\, X_{ik}^\dagger X_{ij} \,|\psi_j\,\rangle \ : \ \sum_{i=1}^{N} \|\psi_i\|^2 = 1 \right\} \ . \end{aligned} \tag{5.15}$$

Indeed, it turns out that it satisfies (5.6); beside, $M_N(\mathbb{B}(\mathbb{H}))$ is a complete $*$-algebra with respect to it, thence a $C^*$ algebra.

4. Given $X, Y \in \mathbb{B}(\mathbb{H})$, $\mathbb{B}(\mathbb{H}) \ni [X, \, , Y] := XY - YX$ denotes their *commutator*. Let $\mathcal{V} \subseteq \mathbb{B}(\mathbb{H})$ be a linear self-adjoint subset, that is it contains the

adjoint of any of its elements. The *commutant* of $\mathcal{V}$, denoted by $\mathcal{V}'$, consists of all bounded operators that commute with all $X \in \mathcal{V}$. If $X' \in \mathcal{V}'$ also $(X')^\dagger \in \mathcal{V}'$; further, if $X', Y' \in \mathcal{V}'$, then

$$[X'Y', X] = X'[Y', X] + [X', X]Y' = 0 \quad \forall X \in \mathcal{V} .$$

Therefore, $X'Y' \in \mathcal{V}'$ and $\mathcal{V}'$ is a $*$-algebra; also, if a sequence of $X'_n \in \mathcal{V}'$ uniformly converges to $X \in \mathbb{B}(\mathbb{H})$, then,

$$\|[X', X]\| = \|[X' - X'_n, X]\| \leq 2\|X' - X'_n\| \, \|X\| ,$$

implies that $\mathcal{V}'$ is uniformly closed and thus a $C^*$ subalgebra of $\mathbb{B}(\mathbb{H})$.

5. If $\mathcal{A} \subseteq \mathbb{B}(\mathbb{H})$ is a $C^*$ subalgebra with commutant $\mathcal{A}'$, the *center* of $\mathcal{A}$, $\mathcal{Z}_\mathcal{A} := \mathcal{A} \cap \mathcal{A}'$, contains all those $A \in \mathcal{A}$ that commute among themselves and is thus an Abelian $C^*$ subalgebra of $\mathbb{B}(\mathbb{H})$.

Given a bounded operator $A$ in a unital $C^*$ algebra $\mathcal{A}$, $a - A$ is *invertible in* $\mathcal{A}$ ($a$ stands for $a\mathbb{1}$) if there exists $B := (a-B)^{-1} \in \mathcal{A}$ for which $(a-A)B = B(a-A) = \mathbb{1}$. The set of such $a \in \mathbb{C}$ is called the *resolvent set* of $A$ (Res$(A)$). Its complement is the *spectrum* of $A$ (Sp$(A)$).

**Examples 5.2.2.** [64]

1. Let $A \in \mathcal{A}$ and $a \in \mathbb{C}$ such that $\|A\| < |a|$, then by Taylor expansion

$$\frac{1}{a - A} = \frac{1}{a} \sum_{n=0}^{\infty} \left(\frac{A}{a}\right)^n$$

the series converges in norm and gives rise to a well-defined operator in $\mathcal{A}$. Therefore, Sp$(A)$ is contained in the subset of $a \in \mathbb{C}$ such that $|a| \leq \|A\|$. Furthermore, if $a_0 \in$ Res$(A)$ so that $(a - A)^{-1}$ exists in $\mathcal{A}$, choose $a \in \mathbb{C}$ such that $|a - a_0| < \|(a_0 - A)^{-1}\|$; then,

$$\frac{1}{a - A} = \frac{1}{(a - a_0) + a_0 - A} = \frac{1}{a_0 - A} \sum_{n=0}^{\infty} \left(\frac{a_0 - a}{a_0 - A}\right)^n$$

exists as well, whence Res$(A)$ is an open subset of $\mathbb{C}$ and Sp$(A)$ a closed subset of $\mathbb{C}$.

2. For $a, b \in \mathbb{C}$ and $A \in \mathcal{A}$, $a - (b - A)$ is invertible if and only if $(b - a) - A$ is invertible. Thus Sp$(b - A) = b -$ Sp$(A)$.
3. For $a \in \mathbb{C}$ and $A \in \mathcal{A}$; $a - A$ is invertible if and only if $a^* - A^\dagger$ is invertible, whence Sp$(A^\dagger) =$ Sp$(A)^*$, the conjugate set of Sp$(A)$.
4. If $A$ is invertible, using $A^{-1}$ one writes

$$a - A = aA(A^{-1} - a^{-1}) , \quad a^{-1} - A^{-1} = a^{-1}A^{-1}(A - a) .$$

Therefore, if $a - A$ is invertible, then $a^{-1} - A^{-1}$ turns out to be invertible too and vice versa; therefore, $\mathrm{Sp}(A^{-1}) = (\mathrm{Sp}(A))^{-1}$, the set consisting of the inverse of each element of $\mathrm{Sp}(A^{-1})$ (notice that $0 \notin \mathrm{Sp}(A)$ and $a \in \mathrm{Sp}(A) \Longrightarrow |a|^{-1} \leq \|A^{-1}\| < +\infty$).

5. The *spectral* radius $R(A)$ of $A \in \mathcal{A}$ is [64]

$$R(A) := \sup\{|\lambda| \,:\, \lambda \in \mathrm{Sp}(A)\} = \lim_{n \to +\infty} \|A^n\|^{1/n} .$$

An operator $A \in \mathcal{A}$ is *normal* if $A^\dagger A = A A^\dagger$; then, $R(A) = \|A\|$. Indeed, the $C^*$ properties of the norm yield

$$\begin{aligned} \|A^{2^n}\|^2 &= \|(A^\dagger)^{2^n} A^{2^n}\| = \|(A^\dagger A)^{2^n}\| = \|(A^\dagger A)^{2^{n-1}} (A^\dagger A)^{2^{n-1}}\| \\ &= \|(A^\dagger A)^{2^{n-1}}\|^2 = \|A^\dagger A\|^{2^n} = \|A\|^{2^{n+1}} , \quad \text{whence} \\ R(A) &= \lim_{n \to +\infty} \|A^{2^n}\|^{2^{-n}} = \|A\| . \end{aligned}$$

6. Self-adjoint $\mathcal{A} \ni A = A^\dagger$ are normal and hence $R(A) = \|A\|$.
7. If $U$ is unitary ($U^\dagger U = U U^\dagger = \mathbb{1}$) or isometric ($U^\dagger U = \mathbb{1}$), then

$$\begin{aligned} \|U^n\|^2 &= \|(U^\dagger)^n U^n\| = \|(U^\dagger)^{n-1} U^\dagger U U^{n-1}\| = \|(U^\dagger)^{n-1} U^{n-1}\| \\ &= \|U^\dagger U\| = \|\mathbb{1}\| = 1 . \end{aligned}$$

Therefore, $\mathrm{Sp}(U)$ is contained within the unit circle $\{x \in \mathbb{C} : |z|| \leq 1\}$. On the other hand, if $U$ is invertible, $U^{-1} = U^\dagger$ and the preceding point 3 implies that $\mathrm{Sp}(U) = \{z \in \mathbb{C} : |z| = 1\}$.

8. If $\mathcal{A} \ni A = A^\dagger$ and $|a|^{-1} > \|A\|$, then from point 1 above one deduces that $-i|a|^{-1} - A = -i|a|(1 - i|a|A)$ is invertible so that

$$\mathcal{A} \ni U := (1 + i|a|A)(1 - i|a|A)^{-1}$$

is a well defined unitary operator; moreover, the last point ensures that

$$\underbrace{\frac{1 - i|a|z}{1 + i|a|z}}_{w} - U = \frac{2i|a|}{1 + i|a|z}(A - z)(1 + i|a|A)^{-1}$$

is invertible whenever $|w| \neq 1$, namely whenever $\Im(z) \neq 0$. Therefore, $A - z$ is invertible and $\mathrm{Sp}(A) \subseteq [-\|A\|, \|A\|]$ because of points 3 and 6.

9. Let $P(z)$ be a polynomial of degree $n$ on $\mathbb{C}$, $A \in \mathcal{A}$ and for $a \in \mathbb{C}$ write

$$\begin{aligned} P(z) - a &= \alpha \prod_{i=1}^n (z - \alpha_i) , \qquad \alpha \, , \; \alpha_i \in \mathbb{C} \\ P(A) - a &= \alpha \prod_{i=1}^n (A - \alpha_i) . \end{aligned}$$

The operators $A - \alpha_i$ commute, thus $P(A) - a$ is not invertible if and only if at least one of them is not invertible, that is $a \in \mathrm{Sp}(P(A))$ if and only if at least one $\alpha_i \in \mathrm{Sp}(A)$. Since $P(\alpha_i) = a$, it follows that $\mathrm{Sp}(P(A)) = P(\mathrm{Sp}(A))$, the set of values attained by $P(z)$ on $\mathrm{Sp}(A)$.

10. Suppose $\mathcal{A} \ni A = A^\dagger$, from the previous result and point 8 it turns out that $\mathrm{Sp}(A^2) = (\mathrm{Sp}(A)^2) \subseteq [0, \|A\|^2]$.

**Remark 5.2.1.** If $A \in \mathcal{A}$ is self-adjoint, then by the density of the polynomials in the commutative $C^*$ algebra of continuous functions $f$ over $\mathbb{R}$, one can extend Example 5.2.2.8 to $f(\mathrm{Sp}(A)) = \mathrm{Sp}(f(A))$. This is the *spectral mapping theorem* [64, 324].

### 5.2.1 Positive Operators

Particularly important bounded self-adjoint operators are the positive ones, that is those whose spectrum consists of non-negative values; from a physical point of view, they represent observables that, when measured, always returns a positive outcome.

**Definition 5.2.2.** *An operator $A$ of a unital $C^*$ algebra $\mathcal{A}$ is* positive *($A \geq 0$) if $A = A^\dagger$ and $Sp(A) \subseteq \mathbb{R}_+$. Given $A, B \in \mathcal{A}$, one sets $A \geq B$ whenever $A - B \geq 0$.*

**Remark 5.2.2.** Positive operators $\mathcal{A} \ni A \geq 0$ are characterized by being of the form $A = B^\dagger B$, for some $B \in \mathcal{A}$ and by having a unique positive square-root $\sqrt{A}$ such that $A = \sqrt{A}\sqrt{A}$ [64] (see also Example 5.3.4.2).

When $\mathcal{A} = \mathbb{B}(\mathbb{H})$, the positivity of a self-adjoint operator $X \in \mathbb{B}(\mathbb{H})$ amounts to $\langle \psi | X\psi \rangle \geq 0$ for all $\psi \in \mathbb{H}$, which corresponds to the positivity of all its eigenvalues $x_i \in \mathbb{R}$ such that $(X - x_i)|\psi\rangle = 0$ for some $|\psi\rangle \in \mathbb{H}$. Denote by $|X| := \sqrt{X^\dagger X}$ the unique square-root of the positive operator $X^\dagger X$. The map $V : \mathrm{Ran}(|X|) \mapsto \mathrm{Ran}(X)$ defined by $V|X||\psi\rangle = X|\psi\rangle$, where $\mathrm{Ran}(X)$ denotes the *range* of $X \in \mathbb{B}(\mathbb{H})$, [1] is a *partial isometry,*

$$\|V|X||\psi\rangle\|^2 = \langle \psi | X^\dagger X | \psi \rangle\| = \||X||\psi\rangle\|^2 .$$

Let $U$ denote the *partial isometry* which equals the extension of $V$ on the closure of $\mathrm{Ran}(X)$ and 0 on $\mathrm{Ker}(|X|) = (\mathrm{Ran}(|X|))^\perp$, then $X = U|X|$ is the so-called *polar decomposition* of $X$ [64]. It is unique; namely, if $X = VB$ with $B \geq 0$ and $V$ is a partial isometry with $V = 0$ on $\mathrm{Ker}(B)$, then

[1]The range of $X \in \mathbb{B}(\mathbb{H})$ is the linear subset of vectors of the form $|\psi\rangle = X|\phi\rangle$ for some $\phi \in \mathbb{H}$. $\mathrm{Ran}(X^\dagger) \perp \mathrm{Ker}(X)$, where $\mathrm{Ker}(X)$ is the *kernel* of $X$ that is the closed subspace of vectors $\psi \in \mathbb{H}$ such that $X|\psi\rangle = 0$.

$$X^\dagger X = BV^\dagger VB = B^2 \Longrightarrow B = |X|$$

by the uniqueness of the square-root; further, $U = V$ for both annihilate $\text{Ran}(|X|)^\perp$. The projection $p := U^\dagger U$ is called the *initial projection* of $U$, while $q = UU^\dagger$ its *final projection.*

In the simplest cases $\mathbb{B}(\mathbb{H}) = M_N(\mathbb{C})$, then $|X| = \sqrt{X^\dagger X}$ can be spectralized, $|X| = \sum_{i=1} x_i |\Psi_i\rangle\langle\Psi_i|$. The eigenvalues $x_j \geq 0$ of $|X|$ are the so-called *singular values* of $X$, while its eigenvectors $\Psi_j$ form an *ONB* in $\mathbb{C}^N$. Using the polar decomposition, it turns out that any matrix can always be represented in terms of its singular values and of two, generally different, *ONBs* ,

$$X = U|X| = \sum_{j=1}^{N} x_j |\Phi_j\rangle\langle\Psi_j| \ , \quad |\Phi_j\rangle := U|\Psi_j\rangle \ . \tag{5.16}$$

Also, if $V$ is the unitary matrix that diagonalizes the Hermitian matrix $|X|$, $|X| = V\,D\,V^\dagger$, then $X = W\,D\,V^\dagger$ with $W := UV$ unitary.

**Examples 5.2.3.** [10, 296]

1. From Example 5.2.2.10 it turns out that the spectrum of the positive elements $A \geq 0$ of a $C^*$ algebra $\mathcal{A}$ is such that $\text{Sp}(A) \subseteq [0, \|A\|]$.
2. Suppose $A \geq B \geq 0$ for $A, B \in \mathcal{A}$; then, from the previous remark, $A - B = C^\dagger C$, whence

$$D^\dagger (A - B) D = (CD)^\dagger (CD) \geq 0 \Longrightarrow D^\dagger A D > D^\dagger B D \ ,$$

   for all $D \in \mathcal{A}$. A typical situation is when $A = P$, an orthogonal projection which is always $\leq \mathbb{1}$, then $D^\dagger P D \leq D^\dagger D$.
3. Let $\mathbb{B}(\mathbb{H}) \ni X := P_\psi - P_\phi = |\psi\rangle\langle\psi| - |\phi\rangle\langle\phi|$, $\psi \neq \phi \in \mathbb{H}$. One can always write

$$|\phi\rangle = \alpha|\psi\rangle + \beta|\psi^\perp\rangle \ , \langle\psi|\psi^\perp\rangle = 0 \ , \ \alpha := \langle\psi|\phi\rangle \ , \ \beta = \sqrt{1 - |\alpha|^2} \ .$$

   Then, on the subspace $\mathbb{K}$ spanned by $\psi$ and $\psi^\perp$, $X$ is represented by the $2 \times 2$ matrix

$$M_X = \begin{pmatrix} \beta^2 & -\beta\alpha \\ -\beta\alpha^* & -\beta^2 \end{pmatrix} \ .$$

   Thus $X$ has eigenvalues $\pm\beta$ and eigenprojectors

$$|\pm\rangle := \sqrt{\frac{1 \pm \beta}{2}}|\psi\rangle \mp \mathrm{e}^{-i\varphi}\sqrt{\frac{1 \mp \beta}{2}}|\psi^\perp\rangle,$$

   where $\mathrm{e}^{i\varphi}$ is the phase of $\alpha$. Therefore,

$$X = \beta(|+\rangle\langle+| - |-\rangle\langle-|) \ , \quad |X| = \beta\, Q_{\mathbb{K}} \ ,$$

   where $Q_{\mathbb{K}} = |+\rangle\langle+| + |-\rangle\langle-|$ projects onto $\mathbb{K}$. Further, $U = \beta^{-1}X$ is an isometry on $\mathbb{K}$ that vanishes on $\mathbb{K}^\perp$.

4. If $X = U\,|X|$ then $X^\dagger = |X|\,U^\dagger$ whence
$$X\,X^\dagger = U\,|X|^2\,U^\dagger = U\,X^\dagger\,X\,U^\dagger\ .$$
Therefore, $X^\dagger X$ and $X\,X^\dagger$ have the same eigenvalues with the same degeneracies apart, possibly, from the eigenvalue 0.
5. Suppose $0 < X \leq Y$, $X, Y \in \mathbb{B}(\mathbb{H})$, then $Y^{-1} \leq X^{-1}$. Indeed, $X^{-1}$ and $Y^{-1}$ exist; thus one can set $Z := Y^{-1/2}XY^{-1/2}$. Then, $Z \leq \mathbb{1}$. In fact, (see Definition 5.2.2) for all $\psi \in \mathbb{H}$
$$\begin{aligned}\langle\,\psi\,|\,Y^{-1/2}XY^{-1/2}\,|\psi\,\rangle &= \langle\,Y^{-1/2}\psi\,|\,X\,|Y^{-1/2}\psi\,\rangle\\ &\leq \langle\,Y^{-1/2}\psi\,|\,Y\,|Y^{-1/2}\psi\,\rangle = \langle\,\psi\,|\,\psi\,\rangle\ .\end{aligned}$$
By the same argument, multiplication of both sides of the inequality $Z \leq \mathbb{1}$ first by $Z^{-1} = Y^{1/2}X^{-1}Y^{1/2}$ yields $\mathbb{1} \leq Y^{1/2}X^{-1}Y^{1/2}$; one then multiplies both sides of this inequality by $Y^{-1/2}$.
6. Let $X \leq Y \in \mathbb{B}(\mathbb{H})$ be such that $\log X$ exists (see Remark 5.2.1 and Remark 5.3.4), then $\log X \leq \log Y$. This follows from the spectral calculus and the previous point, for $t + X \leq t + Y$ for all $t \geq 0$ and the fact that
$$\log\frac{x}{y} = \int_0^{+\infty} \mathrm{d}t\,\frac{1}{t+x} - \int_0^{+\infty} \mathrm{d}t\,\frac{1}{t+y}\ , \qquad \forall x, y > 0\ .$$
Then, $\log X - \log Y = \displaystyle\int_0^{+\infty} \mathrm{d}t\,\frac{1}{t+X} - \int_0^{+\infty} \mathrm{d}t\,\frac{1}{t+Y} \leq 0.$
7. Consider the setting of Example 5.2.1.3, that is the $C^*$ algebra $M_N(\mathbb{B}(\mathbb{H}))$; $\widetilde{X} \geq 0$ if and only if $\langle\,\widetilde{\psi}\,|\,\widetilde{X}\,|\widetilde{\psi}\,\rangle = \sum_{i,j=1}^N \langle\,\psi_i\,|\,X_{ij}\,|\psi_j\,\rangle \geq 0$ for all $\widetilde{\psi} \in \widetilde{\mathbb{H}}$. By arbitrarily choosing $Y_i \in \mathbb{B}(\mathbb{H})$, $\phi \in \mathbb{H}$ and setting $|\,\psi_i\,\rangle := Y_i|\,\phi\,\rangle$, it follows that $\widetilde{X} \geq 0$ if and only if
$$\sum_{i,j=1}^N Y_i^\dagger\,X_{ij}\,Y_j = \begin{pmatrix} Y_1^\dagger & \cdots & Y_N^\dagger\end{pmatrix}\begin{pmatrix} X_{11} & \cdots & X_{1N}\\ \vdots & \vdots & \vdots\\ X_{N1} & \cdots & X_{NN}\end{pmatrix}\begin{pmatrix} Y_1\\ \vdots\\ Y_N\end{pmatrix} \geq 0$$
for all $Y_i \in \mathbb{B}(\mathbb{H})$.
8. Also, $\widetilde{X} \geq 0$ if and only if $\widetilde{X} = \widetilde{Y}^\dagger\widetilde{Y}$, $\widetilde{Y} \in M_N(\mathbb{B}(\mathbb{H}))$. Then,
$$\widetilde{X} = \sum_{\substack{i,j=1\\k,\ell=1}}^N E_{ji}E_{k\ell} \otimes Y_{ij}^\dagger Y_{k\ell} = \sum_{k=1}^n \left(\sum_{j,\ell=1}^N E_{j\ell} \otimes Y_{kj}^\dagger Y_{k\ell}\right)\ .$$
Therefore, $\widetilde{X} \geq 0$ if and only if it is a sum of matrices of the form $[Y_i^\dagger Y_j]$, $Y_i \in \mathbb{B}(\mathbb{H})$.
9. Consider the $N \times N$ matrix $\widetilde{E} := [E_{ij}] \in M_N(M_N(\mathbb{C}))$ whose entries are matrix units $E_{ij} \in M_N(\mathbb{C})$. According to the previous point, $\widetilde{E} \geq 0$ ; indeed,

$$\widetilde{E} = \sum_{i,j=1}^{N} E_{ij} \otimes E_{ij} = \sum_{i,j=1}^{N} E_{ij} \otimes E^\dagger_{ki} E_{kj} \qquad \forall\, k = 1, 2, \ldots, N \ .$$

Let $\{|\,i\,\rangle\}_{i=1}^N$ be the *ONB* such that $E_{ij} = |\,i\,\rangle\langle\,j\,|$; then, $\widetilde{E}$ turns out to be proportional to the orthogonal projector $\widehat{P}^N_+$,

$$\widehat{P}^N_+ = |\widehat{\Psi}^N_+\rangle\langle\widehat{\Psi}^N_+| = \frac{1}{N}\sum_{i,j=1}^{N} |\,i\,\rangle\langle\,j\,| \otimes |\,i\,\rangle\langle\,j\,| = \frac{1}{N}\,\widetilde{E} \tag{5.17}$$

onto the *totally symmetric state*

$$\mathbb{C}^N \otimes \mathbb{C}^N \ni |\widehat{\Psi}^N_+\rangle := \frac{1}{\sqrt{N}}\sum_{i=1}^{N} |\,ii\,\rangle \ . \tag{5.18}$$

### Finite Dimensional algebras

A $C^*$ algebra $\mathcal{A}$ is finite dimensional if its dimension as a linear space is finite; as such it has an identity $\mathbb{1}$. In particular, its center $\mathcal{Z}_\mathcal{A}$ (see Example 5.2.1.5) is a finite dimensional algebra whose elements all commute: such an algebra is called *Abelian.* It is generated by minimal projections $\{P_i\}_{i=1}^n$ (see Example 5.3.4). Due to their orthogonality, $\mathcal{A} = \bigoplus_{i=1}^n \mathcal{A}_i$, where $\mathcal{A}_i := \mathcal{A}\,P_i$, with $P_i$ its identity operator, whence their centers are trivial and the $\mathcal{A}_i$ *simple* algebras. In fact, $\mathcal{A}_i$ cannot contain any non trivial ideal $\mathrm{i} \subseteq \mathcal{A}$, for, like $\mathcal{A}$, also $\mathrm{i}$ has an identity $E$ [296]; then, $XE \in \mathrm{i} \Longrightarrow EXE = XE$ so that $E$ commutes with all self-adjoint $X \in \mathcal{A}_i$ and thus belongs to its center: $E\,X = (X\,E)^\dagger = E\,X\,E = X\,E$.

We shall set $\mathcal{A} = \mathcal{A}_i$ and show that it is isomorphic to a matrix algebra by constructing an appropriate system of matrix units $\{E_{i,j}\}_{i,j=1}^d$.

Let $\mathcal{B} \subseteq \mathcal{A}$ be a maximally Abelian subalgebra with minimal projectors $\{Q_j\}_{j=1}^d$ such that $Q_jQ_k = \delta_{jk}Q_j$ and $\sum_{j=1}^d Q_j = \mathbb{1}$. For each $Q_j$, one can always choose $X_j \in \mathcal{A}$ such that $Y_j := Q_jX_jQ_1 \neq 0$; indeed, for all $X \in \mathcal{A}$, $\mathrm{i}_X := \{\sum_{i=1}^n X_i\,X\,Y_i \,:\, X_i, Y; i \in \mathcal{A}\}$ is an ideal of $\mathcal{A}$ which, as $\mathcal{A}$ is simple, must coincide with it.

Observe that $Y_j^\dagger Y_j = Q_1X_j^\dagger Q_jX_jQ_1$ and $Y_jY_j^\dagger = Q_jX_jQ_1X_j^\dagger Q_j$ commute with $\mathcal{B}$; since $\mathcal{B}$ is maximally Abelian, they belong to it. Thus, $Y_j^\dagger Y_j = \lambda_jQ_1$ and $Y_jY_j^\dagger = \mu_jQ_j$. Further, $\lambda_j = \mu_j > 0$, for $\|Y_j^\dagger Y_j\| = \|Y_jY_j^\dagger\|$. By setting $Z_j := Y_j/\sqrt{\lambda_j}$ and $E_{ij} := Z_iZ_j^\dagger$ it follows that $Z_j^\dagger Z_j = Q_1$, while $E_{jj} = Z_jZ_j^\dagger = Q_j$ for all $j$. Moreover,

$$E_{ij}E_{pq} = \frac{Q_iX_iQ_1X_j^\dagger Q_jQ_pX_pQ_1X_q^\dagger Q_q}{\sqrt{\lambda_i\lambda_j\lambda_p\lambda_q}}$$

$$= \delta_{jp} \frac{\overbrace{Q_i X_i Q_1}^{Y_i} \overbrace{Q_1 X_j^\dagger Q_j X_j Q_1}^{Y_j^\dagger Y_j} \overbrace{Q_1 X_q^\dagger Q_q}^{Y_q^\dagger}}{\sqrt{\lambda_i \lambda_q}\, \lambda_j} = \delta_{jp} E_{iq} \ .$$

Thus, the $E_{ij}$ are the required set of matrix units as they linearly span $\mathcal{A}$: for all $X \in \mathcal{A}$, $Z_i^\dagger X Z_j = (Q_1 X_i^\dagger Q_i X Q_j X_j Q_1)/\sqrt{\lambda_i \lambda_j} = \mu_{ij}(X) Q_1$, whence

$$X = \sum_{i,j=1}^d Q_i X Q_j = \sum_{i,j=1}^d Z_i Z_i^\dagger X Z_j Z_j^\dagger = \sum_{i,j=1}^d \mu_{ij}(X) Z_i Q_1 Z_j^\dagger$$
$$= \sum_{i,j=1}^d \mu_{ij}(X) E_{ij} \ .$$

Therefore, any finite dimensional $C^*$ algebra is isomorphic to the orthogonal sum of full matrix algebras: $\mathcal{A} \simeq \bigoplus_{i=1}^n M_{d_i}(\mathbb{C})$.

### Compact Operators

If $\mathbb{H}$ is infinite dimensional, the matricial structure of $M_N(\mathbb{C})$ carries over to the so-called *compact operators*, $\mathbb{B}_\infty(\mathbb{H})$. These are all $X \in \mathbb{B}(\mathbb{H})$ such that $|X|$ has a discrete spectrum of finitely degenerate eigenvalues that accumulate to 0, the only eigenvalue with possibly infinite degeneracy [2]. It turns out that these spectral properties are preserved by linear combinations and operator multiplication [251, 270].

Practically speaking, compact operators are obtained by closing with respect to the uniform norm the $*$-algebra of *finite rank operators*, that is of the linear span of all possible $X$ on $\mathbb{H}$ that are non-zero on finite dimensional subspaces, only, where they can be represented as usual matrices. As such the algebra of compact operators is a Banach $*$-algebra without identity operator for the only eigenvalue of $\mathbb{1}$ is infinitely degenerate.

### Trace-Class Operators

Consider a matrix algebra $M_N(\mathbb{C})$, the functional $\mathrm{Tr} : M_N(\mathbb{C}) \mapsto \mathbb{C}$,

$$M_N(\mathbb{C}) \ni X \mapsto \mathrm{Tr}(X) := \sum_{i=1}^N \langle \Psi_i \,|\, X \,|\Psi_i \rangle \ , \tag{5.19}$$

where $\{\Psi_j\}_{j=1}^N$ is any *ONB* in $\mathbb{C}^N$, defines a so-called *trace* on $M_N(\mathbb{C})$.

[2]The simplest example of compact operator is any projector $P = |\,\psi\,\rangle\langle\,\psi\,|$ which vanishes on the orthogonal complement of $\psi$, whence its zero eigenvalues is infinitely degenerate when $\mathbb{H}$ is infinite dimensional.

The trace is basis-independent; indeed, because of (5.2); given any two *ONB*s $\{\Phi_j\}_{j=1}^N$ and $\{\Psi_k\}_{k=1}^N$,

$$\sum_{j=1}^N \langle \Phi_j \,|\, X \,|\Phi_j \rangle = \sum_{j,k,\ell=1}^N \langle \Phi_j \,|\, \Psi_k \rangle \langle \Psi_k \,|\, X \,|\Psi_\ell \rangle \langle \Psi_\ell \,|\, \Phi_j \rangle$$
$$\sum_{k,\ell=1}^N \underbrace{\Big(\sum_{j=1}^N \langle \Psi_\ell \,|\, \Phi_j \rangle \langle \Phi_j \,|\, \Psi_k \rangle\Big)}_{\delta_{k\ell}} \langle \Psi_k \,|\, X \,|\Psi_\ell \rangle = \sum_{k=1}^N \langle \Psi_k \,|\, X \,|\Psi_k \rangle \ .$$

The trace of a matrix amounts to the sum of its diagonal entries, so $\mathrm{Tr}(X) \geq 0$ if $X \geq 0$. Further, it is *cyclic*; namely, for all $X, Y \in M_N(\mathbb{C})$, (5.2) yields

$$\mathrm{Tr}(XY) = \sum_{i=1}^N \langle \Psi_i \,|\, XY \,|\Psi_i \rangle = \sum_{i,j=1}^N \langle \Psi_i \,|\, X \,|\Psi_j \rangle \langle \Psi_j \,|\, Y \,|\Psi_i \rangle$$
$$= \sum_{i,j=1}^N \langle \Psi_j \,|\, Y \,|\Psi_i \rangle \langle \Psi_i \,|\, X \,|\Psi_j \rangle = \mathrm{Tr}(YX) \ . \tag{5.20}$$

Using the trace, one constructs the following map from $M_N(\mathbb{C})$ onto $\mathbb{R}^+$,

$$\|\cdot\|_1 \,:\, X \mapsto \|X\|_1 := \mathrm{Tr}|X| = \sum_{i=1}^N x_i \ , \tag{5.21}$$

where $x_j$ are the singular values of $X$ (see (5.16)). This map vanishes only if $X = 0$; also, from (5.4) and (5.16) it follows that

$$|\mathrm{Tr}(YX)| \leq \sum_{i=1}^N x_i |\langle \Psi_i \,|\, YU \,|\Psi_i \rangle| \leq \|Y\| \, \|X\|_1 \tag{5.22}$$
$$|\mathrm{Tr}X| = |\mathrm{Tr}(U|X|)| \leq \|X\|_1 \tag{5.23}$$
$$\|X + Z\|_1 = \mathrm{Tr}(U^\dagger(X+Z)) \leq \|X\|_1 + \|Z\|_1 \ . \tag{5.24}$$

Therefore, $\|\cdot\|_1$ is a norm on $M_N(\mathbb{C})$ called *trace-norm*.

If extended to $\mathbb{B}(\mathbb{H})$ with $\mathbb{H}$ infinite dimensional, the trace selects the linear subspace $\mathbb{B}_1(\mathbb{H}) \subset \mathbb{B}(\mathbb{H})$ of *trace-class operators*:

$$\mathbb{B}_1(\mathbb{H}) := \Big\{ X \in \mathbb{B}(\mathbb{H}) : \ \|X\|_1 < \infty \Big\} \ .$$

If $X \in \mathbb{B}_1(\mathbb{H})$ then, by the polar decomposition $X = \sum_n x_n | \phi_n \rangle \langle \psi_n |$, where $\{\phi_n\}$ and $\{\psi_n\}$ are two *ONB* in $\mathbb{H}$, $x_n$ are the eigenvalues of $|X| = \sqrt{X^\dagger X}$ and the sum converges in trace-norm; also, $\|X\|_1 = \sum_{i=1}^\infty x_i$.

Then, inequality (5.22) holds with $Y \in \mathbb{B}(\mathbb{H})$, $X \in \mathbb{B}_1(\mathbb{H})$, (5.23) and (5.24) with $X, Z \in \mathbb{B}_1(\mathbb{H})$. The trace-class operators thus form a $*$-algebra [3]; $\mathbb{B}_1(\mathbb{H})$ is also closed with respect to the trace-norm and thus a Banach $*$-algebra, without identity in infinite dimension for $\mathrm{Tr}(\mathbb{1})$ diverges [251, 270] .

**Example 5.2.4.** [64] Any $X \in \mathbb{B}(\mathbb{H})$ defines a linear functional

$$F_X : \mathbb{B}_1(\mathbb{H}) \mapsto \mathbb{C} , \qquad \rho \mapsto F_X(\rho) := \mathrm{Tr}(X\,\rho) \qquad \forall \rho \in \mathbb{B}_1(\mathbb{H}) ,$$

on $\mathbb{B}_1(\mathbb{H})$ which is bounded for $|F_X(\rho)| \leq \|X\|\,\|\rho\|_1$. Therefore, $\mathbb{B}(\mathbb{H})$ can be identified with a subspace of $\mathbb{B}_1(\mathbb{H})^*$, the topological dual of $\mathbb{B}_1(\mathbb{H})$, that is the (Banach) space consisting of all continuous linear functionals on $\mathbb{B}_1(\mathbb{H})$. Actually, $\mathbb{B}(\mathbb{H}) = \mathbb{B}_1(\mathbb{H})^*$. Indeed, let $F \in \mathbb{B}_1(\mathbb{H})^*$ and consider the bounded operator $|\,\phi\,\rangle\langle\,\psi\,|$ with $\phi,\, \psi \in \mathbb{H}$ not necessarily normalized. It is also trace-class; indeed, set $P_\psi := |\,\psi\,\rangle\langle\,\psi\,|/\|\psi\|^2$; then,

$$\|\,|\,\phi\,\rangle\langle\,\psi\,|\,\|_1 = \mathrm{Tr}\Big(\sqrt{\|\phi\|^2\,\|\psi\|^2\,P_\psi}\,\Big) = \|\phi\|\,\|\psi\| .$$

It thus follows that $|F(|\,\phi\,\rangle\langle\,\psi\,|)| \leq \|F\|\,\|\phi\|\,\|\psi\|$ for all $\phi, \psi \in \mathbb{H}$. Therefore, each $F \in \mathbb{B}_1(\mathbb{H})^*$ defines a so-called *sesquilinear form* on $\mathbb{H} \times \mathbb{H}$, linear in the first argument, antilinear in the second one and continuous with respect to both. Consequently, there exists an unique operator $X_F \in \mathbb{B}(\mathbb{H})$ such that $F(|\,\phi\,\rangle\langle\,\psi\,|) = \langle\,\psi\,|\,X_F\,|\phi\,\rangle$ for all $\phi, \psi \in \mathbb{H}$. Before proving this fact, we draw the conclusion; as already noticed, any $\rho \in \mathbb{B}_1(\mathbb{H})$ can be written as $\rho = \sum_n r_n |\,\phi_n\,\rangle\langle\,\psi_n\,|$ with the possibly infinite sum converging in trace-norm; thus, $\mathbb{B}_1(\mathbb{H})^* \subseteq \mathbb{B}(\mathbb{H})$ for

$$F(\rho) = \sum_n r_n\, F(|\,\phi_n\,\rangle\langle\,\psi_n\,|) = \sum_n r_n\,\langle\,\psi_n\,|\,X_F\,|\phi_n\,\rangle = \mathrm{Tr}(X_F\,\rho) .$$

The property of sesquilinear forms used above comes as follows: if $f : \mathbb{H} \mapsto \mathbb{C}$ is a continuous linear functional on $\mathbb{H}$, that is $|f(\psi)| \leq \|f\|\,\|\psi\|$, then $\mathrm{Ker}(f)$ is closed. Assume $\mathrm{Ker}(f) \neq \mathbb{H}$; if $\phi \in \mathrm{Ker}(f)^\perp$, $\|\phi\| = 1$, then $f(\phi) \neq 0$ and

$$\mathrm{Ker}(f) \ni |\,\chi\,\rangle := f(\phi)|\,\psi\,\rangle - f(\psi)|\,\phi\,\rangle \Longrightarrow f(\psi) = \langle\,\widetilde{\phi}\,|\,\psi\,\rangle ,$$

where $|\,\widetilde{\phi}\,\rangle := \overline{f(\phi)}|\,\phi\,\rangle$. It is easily seen that this vector is unique and that $\|f\| = |f(\phi)|$. Given a continuous sesquilinear form $f : \mathbb{H} \times \mathbb{H} \mapsto \mathbb{C}$, for each fixed $\psi \in \mathbb{H}$ it defines a continuous linear functional $f_\psi : \mathbb{H} \mapsto \mathbb{H}$; therefore, there exists a unique $|\,\chi_\psi\,\rangle \in \mathbb{H}$ such that $f(\phi, \psi) = f_\psi(\phi) = \langle\,\chi_\psi\,|\,\phi\,\rangle$. This allows to define a linear operator $X_f^\dagger \in \mathbb{B}(\mathbb{H})$ such that $X_f^\dagger|\,\psi\,\rangle = |\,\chi_\psi\,\rangle$ and, whence $f(\phi, \psi) = \langle\,\psi\,|\,X_f\,|\phi\,\rangle$.

[3] $\mathbb{B}_1(\mathbb{H})$ is a *two-sided ideal*, namely $YX, YX \in \mathbb{B}_1(\mathbb{H})$ whenever $X \in \mathbb{B}_1(\mathbb{H})$ and $Y \in \mathbb{B}(\mathbb{H})$.

**Remark 5.2.3.** As $\mathbb{B}(\mathbb{H})$ is the dual of $\mathbb{B}_1(\mathbb{H})$ it can be equipped with the corresponding $w^*$-topology (see Remark 5.1.1.5); namely, any $\rho \in \mathbb{B}_1(\mathbb{H})$ defines a linear functional $\mathbb{B}(\mathbb{H}) \ni X \mapsto \mathbb{E}_\rho(X) := \mathrm{Tr}(X\rho)$. The $w*$-topology on $\mathbb{B}(\mathbb{H})$ is the coarsest one with respect to which all semi-norms $\mathcal{L}_\rho(X) := |\mathrm{Tr}(\rho X)|$ are continuous. By comparing these semi-norms with those in (5.9), it turns out that the $w^*$ topology coincides with the $\sigma$-weak topology.

### Hilbert-Schmidt Operators

A second norm on $M_N(\mathbb{C})$, also based on the trace, is given by

$$\|\cdot\|_2 : X \mapsto \|X\|_2 := \sqrt{\mathrm{Tr}|X|^2} = \sqrt{\sum_{i=1}^N x_i^2} \ . \tag{5.25}$$

It is called *Hilbert-Schmidt norm*; unlike the trace-norm, it originates from a (Hilbert-Schmidt) scalar product

$$M_N(\mathbb{C}) \times M_N(\mathbb{C}) \ni (Y, X) \mapsto \mathrm{Tr}(Y^\dagger X) \ . \tag{5.26}$$

that satisfies $|\mathrm{Tr}(Y^\dagger X)| \leq \|Y\|_2 \|X\|_2$. In fact, using (5.16) and (5.1),

$$\begin{aligned} |\mathrm{Tr}(Y^\dagger X)| &\leq \sum_{i=1}^N x_i \Big| \langle \Psi_i \,|\, Y^\dagger U \,|\Psi_i \rangle \Big| \leq \sqrt{\sum_{i=1}^N x_i^2} \sqrt{\sum_{j=1}^N |\langle \Psi_j \,|\, Y^\dagger U \,|\Psi_j \rangle|^2} \\ &\leq \|X\|_2 \sqrt{\sum_{j=1}^N \langle \Psi_j \,|\, Y^\dagger U |\, \Psi_j \rangle \langle \Psi_j \,| U^\dagger Y \,|\Psi_j \rangle} \leq \|X\|_2 \, \|Y\|_2 \ , \end{aligned} \tag{5.27}$$

for $U|\Psi_j\rangle\langle\Psi_j|U^\dagger \leq \mathbb{1}$. When defined on $\mathbb{B}(\mathbb{H})$ with $\mathbb{H}$ infinite-dimensional, the Hilbert-Schmidt norm singles out the linear subspace $\mathbb{B}_2(\mathbb{H})$ of *Hilbert-Schmidt operators*

$$\mathbb{B}_2(\mathbb{H}) := \Big\{ X \in \mathbb{B}(\mathbb{H}) : \|X\|_2 < \infty \Big\} \ .$$

If $X \in \mathbb{B}_2(\mathbb{H})$, $\|X\|_2 = \sqrt{\sum_{i=1}^\infty x_i^2}$, with $x_i$ the singular values of $X$. Then, inequality (5.27) holds with $X, Y \in \mathbb{B}_2(\mathbb{H})$, respectively $Y \in \mathbb{B}(\mathbb{H})$, $X \in \mathbb{B}_2(\mathbb{H})$. $\mathbb{B}_2(\mathbb{H})$ is also close with respect to $\|\cdot\|_2$ and thus a Banach $*$-subalgebra of $\mathbb{B}(\mathbb{H})$ (actually also a two-sided ideal as $\mathbb{B}_1(\mathbb{H})$) without identity in the infinite dimensional case for $\|\mathbb{1}\|_2$ diverges [251, 270].

**Example 5.2.5.** Let $F_j$, $j = 1, 2, \ldots, N^2$, be a set of $N \times N$ matrices, orthogonal with respect to (5.26), $\mathrm{Tr}(F_j^\dagger F_k) = \delta_{jk}$: they form an *ONB* in $M_N(\mathbb{C})$. Indeed, as a Hilbert space equipped with (5.26), $M_N(\mathbb{C})$ has dimension $N^2$, therefore, for all $X \in M_N(\mathbb{C})$,

$$M_N(\mathbb{C}) \ni X = \sum_{j=1}^{N^2} \mathrm{Tr}(X\, F_j^\dagger)\, F_j \ . \tag{5.28}$$

Consider the linear map $\mathrm{Tr}_N : M_N(\mathbb{C}) \mapsto M_N(\mathbb{C})$ defined by

$$M_N(\mathbb{C}) \ni X \mapsto \mathrm{Tr}_N[X] := \mathrm{Tr}(X)\, \mathbb{1}_N \ . \tag{5.29}$$

We shall refer to it as *trace map*. Choose an *ONB* $\{|\,\alpha\,\rangle\}_{\alpha=1}^N$ in $\mathbb{C}^N$; the $N^2$ matrix units $E_{\alpha\beta} := |\,\alpha\,\rangle\langle\,\beta\,|$ also form an *ONB* in $M_N(\mathbb{C})$. Thus, there must exist a unitary matrix $U \in M_{N^2}(\mathbb{C})$ such that $E_{\alpha\beta} = \sum_{i=1}^{N^2} U_{\alpha\beta,i}\, F_i$. Therefore, the trace-map can be recast as

$$\begin{aligned}\mathrm{Tr}_N[X] &= \sum_{\alpha,\beta=1}^{N} E_{\beta\alpha}\, X\, E_{\alpha\beta} = \sum_{i,j=1}^{N^2} \sum_{\substack{\alpha,\beta=1\\ \gamma,\delta=1}}^{N} U^*_{\alpha\beta,i}\, U_{\alpha\beta,j}\, F_i^\dagger\, X\, F_j \\ &= \sum_{i=1}^{N^2} F_i^\dagger\, X\, F_i \ .\end{aligned} \tag{5.30}$$

**Remark 5.2.4.** The uniform, trace and Hilbert-Schmidt norms are all equivalent on finite-dimensional $\mathbb{H}$ and thus define equivalent topologies with the same converging sequences. Indeed, given $X \in M_N(\mathbb{C})$ its norm coincides with its largest singular values, $\|X\| = \max_{1\leq i\leq N} x_i$, then

$$\|X\| \leq \|X\|_1 \ , \ \|X\|_1 \leq N\, \|X\| \ , \ \|X\| \leq \|X\|_2 \ , \ \|X\|_2 \leq \sqrt{N}\, \|X\| \ .$$

Also, $\|X\|_2 \leq \|X\|_1$ for the sum of squares of positive numbers is smaller that the square of their sum; while from (5.1),

$$\|X\|_1 = \sum_{i=1}^N x_i \leq \sqrt{\sum_{i=1}^N 1^2}\sqrt{\sum_{j=1}^N x_j^2} = \sqrt{N}\, \|X\|_2 \ .$$

However, the trace and Hilbert-Schmidt norms are not $C^*$ norms; indeed, for any $N \in \mathbb{N}$,

$$\|X^\dagger X\|_1 = \sqrt{\sum_{i=1}^N x_i^2} \neq \sum_{i=1}^N x_i \ , \ \|X^\dagger X\|_2 = \sqrt{\sum_{i=1}^N x_i^4} \neq \sum_{i=1}^N x_i^2 \ .$$

Therefore, the trace-class and Hilbert-Schmidt operators form Banach $*$-algebras but not $C^*$ algebras.

### 5.2.2 Positive and Completely Positive Maps

Physical transformations of quantum systems are described by linear maps acting either on their observables or on their states. As already seen in the classical case, these two possibilities are *dual* to each other; in the first case states are not affected, in the second one, states change while the observables do not. The physically relevant request is that mean values do not depend on which of the two ways they are calculated.

In the classical setting, states must change while preserving their ultimate characteristic of being probability distributions; the maps which describe classical state transformations must thus be positivity preserving. In quantum mechanics things are more complicated and intriguing; it is indeed necessary to sharpen the notion of positive linear transformation. This latter is as follows.

**Definition 5.2.1 (Positive Maps).** *A linear map $\Lambda : \mathbb{B}(\mathbb{H}) \mapsto \mathbb{B}(\mathbb{H})$ is positive if and only if $\mathbb{B}(\mathbb{H}) \ni X \geq 0 \Longrightarrow \mathbb{B}(\mathbb{H}) \ni \Lambda[X] \geq 0$.*

Given a positive linear map $\Lambda$, one can always lift it to act on the operator-valued algebras $M_N(\mathbb{B}(\mathbb{H}))$ as

$$\mathrm{id}_N \otimes \Lambda : [X_{ij}] \mapsto [\Lambda[X_{ij}]] = \begin{pmatrix} \Lambda[X_{11}]\cdot & \cdots & \Lambda[X_{1N}] \\ \cdot & \cdots & \cdot \\ \cdot & \cdots & \cdot \\ \cdot & \cdots & \cdot \\ \Lambda[X_{N1}] & \cdots & \Lambda[X_{NN}] \end{pmatrix} . \tag{5.31}$$

One may then ask whether $\mathrm{id}_N \otimes \Lambda$ is positive, too.

**Definition 5.2.2 (Completely Positive Maps).**
*A linear map $\Lambda : \mathbb{B}(\mathbb{H}) \mapsto \mathbb{B}(\mathbb{H})$ is N-positive if and only if*

$$\mathrm{id}_N \otimes \Lambda : M_N(\mathbb{B}(\mathbb{H})) \mapsto M_N(\mathbb{B}(\mathbb{H}))$$

*is positive; $\Lambda$ is completely positive (CP) if and only if it is $N$-positive for all $N$. A linear map $\Lambda : \mathbb{B}(\mathbb{H}) \mapsto \mathbb{B}(\mathbb{H})$ is called a CPU map when it is CP and also unital, that is $\Lambda[\mathbb{1}] = \mathbb{1}$.*

**Examples 5.2.6.**

1. Positive maps are hermiticity-preserving, for $\Lambda[X^\dagger] = \Lambda[X]^\dagger$. Indeed, any $X \in \mathbb{B}(\mathbb{H})$ can be decomposed into Hermitian components, $X = \underbrace{\frac{X + X^\dagger}{2}}_{X_1} + i \underbrace{\frac{X - X^\dagger}{2i}}_{X_2}$. In turn, $X_{1,2}$ can be decomposed into positive components $X_{1,2} = X^+_{1,2} - X^-_{1,2}$, $X^\pm_{1,2} := \frac{|X_{1,2}| \pm X_{1,2}}{2}$. Thus,

$$\Lambda[X]^\dagger = \Lambda[X^+_1] - \Lambda[X^-_1] - i\,\Lambda[X^+_2] + i\,\Lambda[X^-_2] = \Lambda[X_1 - i\,X_2] = \Lambda[X^\dagger] .$$

2. Positivity corresponds to 1-positivity; complete positivity is stronger for there are positive maps which are not 2-positive, a renown example being transposition $\mathsf{T}_2 : M_2(\mathbb{C}) \mapsto M_2(\mathbb{C})$, $\mathsf{T}_2[X] = X^T$, with respect to a fixed *ONB*. We shall consider the $N$-dimensional case: $\mathsf{T}_N$ is positive for transposition does not affect the spectrum of a matrix, but the *partial transposition* $\mathrm{id}_N \otimes \mathsf{T}_N$ is not positive, whence $\mathsf{T}_N$ is not $N$-positive and thus not *CP* . This can be seen by considering the positive matrix $\widetilde{E} = [E_{ij}]$ of Example 5.2.3.9. Then,

$$V := \mathrm{id}_N \otimes \mathsf{T}_N[\widetilde{E}] = N\,\mathrm{id}_N \otimes \mathsf{T}_N[\widehat{P}^N_+] = \sum_{i,j=1}^N |\,i\,\rangle\langle\,j\,| \otimes |\,j\,\rangle\langle\,i\,| \tag{5.32}$$

acts as a *flip operator* on $\mathbb{C}^N \otimes \mathbb{C}^N$, that is $V\Big(|\,\psi\,\rangle \otimes |\,\phi\,\rangle\Big) = |\,\phi\,\rangle \otimes |\,\psi\,\rangle$. Since $V$ has eigenvalue 1 on the $N(N+1)/2$ dimensional subspace of symmetric states and $-1$ on the $N(N-1)/2$ dimensional subspace of anti-symmetric states, it is not positive and $\mathsf{T}_N$ not $N$-positive, hence not completely positive.

3. *CPU* maps satisfy the inequality

$$\Lambda[X^\dagger\, X] \geq \Lambda[X^\dagger]\Lambda[X] \geq 0\ . \tag{5.33}$$

In fact, from Example 5.2.3.8 it turns out that

$$\begin{pmatrix} \mathbb{1} & X \\ X^\dagger & X^\dagger X \end{pmatrix} = \begin{pmatrix} \mathbb{1} & 0 \\ X^\dagger & 0 \end{pmatrix} \begin{pmatrix} \mathbb{1} & X \\ 0 & 0 \end{pmatrix} \geq 0\ , \quad \forall X \in \mathbb{B}(\mathbb{H})\ .$$

Since $\Lambda$ is *CPU* , it follows that

$$\mathrm{id}_2 \otimes \Lambda \begin{pmatrix} \mathbb{1} & X \\ X^\dagger & X^\dagger X \end{pmatrix} = \begin{pmatrix} \mathbb{1} & \Lambda[X] \\ \Lambda[X^\dagger] & \Lambda[X^\dagger X] \end{pmatrix} \geq 0\ ,$$

whence, because of the previous point, and using Example 5.2.3.7,

$$\begin{aligned}(\,\Lambda[X]^\dagger \quad -\mathbb{1}\,) \begin{pmatrix} \mathbb{1} & \Lambda[X] \\ \Lambda[X^\dagger] & \Lambda[X^\dagger X] \end{pmatrix} \begin{pmatrix} \Lambda[X] \\ -\mathbb{1} \end{pmatrix} &= \Lambda[X^\dagger X] - \Lambda[X^\dagger]\Lambda[X] \\ &\geq 0\ .\end{aligned}$$

4. *CPU* maps are contractions: since $\mathbb{B}(\mathbb{H}) \ni X^\dagger X \leq \|X\|^2$, positivity, unitality and (5.33) yield $\Lambda[X]^\dagger\Lambda[X] \leq \Lambda[X^\dagger X] \leq \|X\|^2$, whence $\|\Lambda[X]\| \leq \|X\|$.
5. Let $\mathcal{B} \subseteq \mathcal{A}$ a $C^*$ subalgebra of a $C^*$ algebra $\mathcal{A}$, both assumed with identity, the map $\imath_{\mathcal{B}\mathcal{A}} : \mathcal{B} \mapsto \mathcal{A}$ denotes the *natural embedding* of $\mathcal{B}$ into $\mathcal{A}$; according to Examples 5.2.3.7 and 8, embeddings are *CPU* maps. Indeed, for all $A_i \in \mathcal{A}$, $B_j \in \mathcal{B}$ and $N \in \mathbb{N}$,

$$\sum_{i,j=1}^N A_i^\dagger\, \imath_{\mathcal{B}\mathcal{A}}[B_i^\dagger\, B_j]\, A_j = Z^\dagger\, Z \geq 0\ .$$

6. If $\mathcal{A}$ and $\mathcal{B}$ are $C^*$ algebras (with identity) and $\mathcal{A}$ is Abelian, then any positive map $\Lambda : \mathcal{A} \mapsto \mathcal{B}$ is *CP* . In order to check whether $\sum_{i,j=1}^n Y_i^\dagger \Lambda[X_i^\dagger X_j] Y_j \geq 0$ for all $X_{i,j} \in \mathcal{A}$, $Y_{i,j} \in \mathcal{B}$ and $n \in \mathbb{N}$ we use that $Y_{i,j}$ and $\Lambda[X_i^\dagger X_j]$ can be identified with functions $f(x) \in C(\mathcal{X})$ over a compact topological space $\mathcal{X}$. Then
$$\Big(\sum_{i,j=1}^n Y_i^\dagger \Lambda[X_i^\dagger X_j] Y_j\Big)(x) = \sum_{i,j=1}^n Y_i^*(x) \Lambda[X_i^\dagger X_j](x) Y_j(x)$$
$$= \Lambda[Z_x^\dagger Z_x](x) \geq 0 ,$$
where $Z_x := \sum_{i=1}^n Y_i(x)\, X_i \in \mathcal{A}$ for all $x \in \mathcal{X}$.
7. If $\mathcal{A}$ and $\mathcal{B}$ are $C^*$ algebras with identity and $\mathcal{B}$ is Abelian, then any positive map $\Lambda : \mathcal{A} \mapsto \mathcal{B}$ is *CP* . We take $\mathcal{B} = \mathbb{B}(\mathbb{H})$ and check whether $\sum_{i,j=1}^n \langle \psi_i \,|\, \Lambda[X_i^\dagger X_j] \,|\psi_j \rangle \geq 0$ for all $\psi_i \in \mathbb{H}$, $n \in \mathbb{N}$ and $X_i \in \mathcal{A}$ identified with functions in $C(\mathcal{X})$. By duality, $\Lambda^T[|\, \psi_j \rangle\langle \psi_i \,|]$ gives a complex measure on $\mathcal{X}$ such that
$$\sum_{i,j=1}^n \langle \psi_i \,|\, \Lambda[X_i^\dagger X_j] \,|\psi_j \rangle = \sum_{i,j=1}^n \int_{\mathcal{X}} \mathrm{d}\mu_{ij}(x)\, X_i^*(x) X_j(x) \geq 0 .$$
Indeed, for all $c_i \in \mathbb{C}$ and $|\, \Psi \rangle := \sum_{i=1}^n c_i |\, \psi_i \rangle$,
$$\sum_{i,j=1}^n c_i^* c_j \int_{\mathcal{X}} \mathrm{d}\mu_{i,j}(x) = \sum_{i,j=1}^n \langle \psi_i \,|\, \Lambda[\mathbb{1}] \,|\psi_j \rangle = \langle \Psi \,|\, \Lambda[\mathbb{1}] \,|\Psi \rangle \geq 0 .$$

In order to ascertain whether a linear map is completely positive, it seems necessary to check $N$-positivity for all $N \in \mathbb{N}$. Luckily, the following result [82, 83] shows that $N$-positive maps $\Lambda : M_N(\mathbb{C}) \to \mathbb{B}(\mathbb{H})$ are automatically *CP* .

**Theorem 5.2.1 (Choi).** *A linear map $\Lambda : M_N(\mathbb{C}) \to \mathbb{B}(\mathbb{H})$ is CP if and only if* $\mathrm{id}_N \otimes \Lambda[\widetilde{E}] \geq 0$*, where $\widetilde{E}$ is the matrix introduced in Example 5.2.3.9.*

**Proof:** As seen in Example 5.2.3.9, $\widetilde{E} \geq 0$; thus the "only" if part follows from the fact that if $\Lambda$ is *CP* it is $N$-positive (see Definition 5.2.2).

As regards the "if" part, in order to check that $\Lambda$ $N$-positive implies $\Lambda$ *CP* , we choose $M \in \mathbb{N}$ arbitrary and show that $\mathrm{id}_M \otimes \Lambda[\widetilde{X}] \geq 0$ for all $M_M(\mathbb{C}) \otimes \mathbb{B}(\mathbb{H}) \ni \widetilde{X} \geq 0$. Using Example 5.2.3.8, it is sufficient to show that $\sum_{i,j=1}^M Y_i^\dagger \Lambda[X_i^\dagger X_j] Y_j \geq 0$ for all choices of $Y_i \in \mathbb{B}(\mathbb{H})$ and $X_i \in M_N(\mathbb{C})$. Then, by writing $X_i = \sum_{k,\ell=1}^N x^{\,i}_{\,k\ell}\, E_{k\ell} \in M_N(\mathbb{C})$, from the assumed positivity of $[\Lambda[E_{ij}]]_{i,j=1}^N$ and Example 5.2.3.7 it turns out that

$$\sum_{i,j=1}^{M} Y_i^\dagger \Lambda[X_i^\dagger X_j] Y_j = \sum_{k,\ell,s=1}^{N} \left(\sum_{i=1}^{M} x_{k\ell}^i Y_i\right)^\dagger \Lambda[E_{\ell s}] \underbrace{\sum_{j=1}^{M} x_{ks}^j Y_j}_{Z_{ks}}$$

$$= \sum_{k=1}^{N} \sum_{\ell,s=1}^{N} Z_{k\ell}^\dagger \Lambda[E_{\ell s}] Z_{ks} \geq 0 ,$$

whence $\Lambda$ is $M$-positive for all $M \in \mathbb{N}$ and thus *CP* . □

**Remark 5.2.5.** The matrix $X_\Lambda := \mathrm{id}_N \otimes \Lambda[\widetilde{E}] \in M_N(M_M(\mathbb{C}))$ associated with any linear map $\Lambda : M_N(\mathbb{C}) \mapsto M_M(\mathbb{C})$ is known as *Choi matrix.* Theorem 5.2.1 can then be rephrased as: $\Lambda : M_N(\mathbb{C}) \mapsto M_M(\mathbb{C})$ is *CP* if and only if its Choi matrix is positive.

Vice versa, let $X = \sum_{i,j;s,r} X_{is,jr} E_{ij}^{(N)} \otimes E_{rs}^{(M)}$ be an $NM \times NM$ matrix, where $E_{ij}^{(N)}$ and $E_{rs}^{(M)}$ denote the matrix units in $M_N(\mathbb{C})$, respectively $M_M(\mathbb{C})$. The map $\Lambda_X : M_N(\mathbb{C}) \mapsto M_M(\mathbb{C})$ defined by linear extension of

$$E_{ij}^{(N)} \mapsto \Lambda_X[E_{ij}^{(N)}] := \sum_{r,s=1}^{M} E_{sr}^{(M)} X_{is,jr} , \tag{5.34}$$

is such that its Choi matrix is $X$. Denoting by $\mathcal{L}(N,M)$ the linear space of linear maps $\Lambda : M_N(\mathbb{C}) \mapsto M_M(\mathbb{C})$, the one-to-one relation

$$\mathcal{L}(N,M) \ni \Lambda \longleftrightarrow X \in M_{NM}(\mathbb{C})$$

is known as *Jamiołkowski isomorphism* [158].

If a map $\Lambda : M_N(\mathbb{C}) \mapsto \mathbb{B}(\mathbb{H})$ is only positive, its Choi matrix cannot be positive, but only *block positive*, namely only its mean values relative to product states in $\mathbb{C}^N \otimes \mathbb{H}$ are surely non-negative.

**Proposition 5.2.1.** *A linear map $\Lambda : M_N(\mathbb{C}) \mapsto \mathbb{B}(\mathbb{H})$ is positive if and only if $\langle \psi \otimes \phi \,|\, \mathrm{id}_N \otimes \Lambda[\widetilde{E}] \,|\psi \otimes \phi\rangle \geq 0$ for all $\psi \in \mathbb{C}^N$ and $\phi \in \mathbb{H}$.*

**Proof:** Positive matrices can be written as sums of projectors with positive coefficients, thus, according to Definition 5.2.1, $\Lambda : M_N(\mathbb{C}) \mapsto \mathbb{B}(\mathbb{H})$ is positive if and only if $\langle \phi \,|\, \Lambda[|\,\psi\rangle\langle\psi\,|] \,|\phi\rangle \geq 0$ for all $\phi \in \mathbb{H}$ and $\psi \in \mathbb{C}^N$. But,

$$\langle \phi \,|\, \Lambda[|\,\psi\rangle\langle\psi\,|] \,|\phi\rangle = \sum_{i,j=1}^{N_1} \psi^*(i)\psi(j) \langle \phi \,|\, \Lambda[E_{ij}] \,|\phi_2\rangle$$

$$= \langle \psi^* \otimes \phi \,|\, \sum_{i,j=1}^{N} E_{ij} \otimes \Lambda[E_{ij}] \,|\psi^* \otimes \phi\rangle$$

$$= \langle \psi^* \otimes \phi \,|\, \mathrm{id}_N \otimes \Lambda[\widetilde{E}] \,|\psi^* \otimes \phi\rangle ,$$

where $|\,\psi^*\,\rangle = \sum_{i=1}^N \psi^*(i)|\,i\,\rangle$ with respect to the fixed *ONB* in $\mathbb{C}^N$ such that $E_{ij} = |\,i\,\rangle\langle\,j\,|$. □

In finite dimension the structure of *CP* maps can be made explicit by means of the Hilbert-structure inherited by $M_N(\mathbb{C})$ from the Hilbert-Schmidt scalar product (5.26).

**Examples 5.2.7.**

1. Consider the linear space $\mathcal{L}(N,N)$ of linear maps $\Lambda : M_N(\mathbb{C}) \mapsto M_N(\mathbb{C})$; it can be given a Hilbert space structure by means of the Hilbert-Schmidt scalar product of their Choi matrices,

$$<< \Lambda_1|\Lambda_2 >>:= \mathrm{Tr}\Big(\mathrm{id}_N \otimes \Lambda_1[\widetilde{E}]^\dagger\, \mathrm{id}_N \otimes \Lambda_2[\widetilde{E}]\Big)\ .$$

Consider an *ONB* $\{F_i\}_{i=1}^{N^2}$ in $M_N(\mathbb{C})$ and the maps $\Phi_{ij} \in \mathcal{L}(N,N)$, defined by $\Phi_{ij}[X] := F_i^\dagger\, X\, F_j$. They satisfy $<< \Phi_{ij}\,|\,\Phi_{k\ell} >>= \delta_{ik}\delta_{j\ell}$ and therefore form an *ONB* in $\mathcal{L}(N,N)$, whence

$$\Lambda = \sum_{i,j=1}^{N^2} L_{ij}\, F_i^\dagger\, X\, F_j\ , \quad L_{ij} :=<< \Phi_{ij}\,|\,\Lambda >>= \mathrm{Tr}(F_i^\dagger\, \Lambda[F_j])\ , \tag{5.35}$$

for all $\Lambda \in \mathcal{L}(N,N)$. If $\Lambda$ preserves hermiticity, the $N^2 \times N^2$ matrix of coefficients $\Lambda_{ij}$ is Hermitian and can be diagonalized, $L_{ij} = \sum_{k=1}^{N^2} \ell_k V_{ki}^* V_{kj}$. Suppose the eigenvalues $\ell_k$ are positive, then

$$\Lambda[X] = \sum_{k=1}^{N^2} G_k^\dagger\, X\, G_k\ , \quad G_k := \sqrt{\ell_k}\sum_{j=1}^{N^2} V_{kj}\, F_j\ . \tag{5.36}$$

Using Example 5.2.3.8, maps $\Gamma \in \mathcal{L}(N,N)$ of the form $\Gamma[X] = G^\dagger\, X\, G$ are easily proved to be *CP*. Therefore, linear maps $\Lambda : M_N(\mathbb{C}) \mapsto M_N(\mathbb{C})$ of the form (5.36) are *CP* and *CPU* if $\sum_{k=1}^N G_k^\dagger G_k = \mathbb{1}_N$. Notice that the decomposition (5.36) is highly non-unique; another possible decomposition is indeed provided by (5.35) if the matrix $[L_{ij}]$ is positive.

2. Let $\mathrm{Tr}_N : M_N(\mathbb{C}) \mapsto M_N(\mathbb{C})$ be the trace map of Example 5.2.5 and consider the *reduction map* [151] $\Lambda : M_N(\mathbb{C}) \mapsto M_N(\mathbb{C})$,

$$\Lambda[X] = \mathrm{Tr}_N[X] - X\ . \tag{5.37}$$

$\Lambda$ is positive, but not *CP*; positivity follows since, if $X \geq 0$, $\mathrm{Tr}_N[X]$ is not smaller than any of the eigenvalues of $X$. On the other hand, using (5.17), the Choi matrix of $\Lambda$ turns out to be $\mathrm{id}_N \otimes \Lambda[\widetilde{E}] = \mathbb{1}_{N^2} - N\widehat{P}_+^N$; it has a negative eigenvalue $1 - N$, whence $\Lambda$ cannot be *CP*.

3. The transposition $\mathsf{T}_N : M_N(\mathbb{C}) \mapsto M_N(\mathbb{C})$ is the paradigm of a map which is positive but not *CP* (as it is not $N$-positive); by combining $\mathsf{T}_N$ with $\Lambda$ in (5.37), $\widehat{\Lambda} := \Lambda \circ \mathsf{T}_N : M_N(\mathbb{C}) \mapsto M_N(\mathbb{C})$ is *CP* . Indeed, using (5.32), $\mathrm{id}_N \otimes \widehat{\Lambda}[\widetilde{E}] = \mathrm{id}_N \otimes \Lambda[V] = \mathbb{1}_{N^2} - V \geq 0$, for the eigenvalues of the flip operator are $\pm 1$.

The next two results [287, 180] show that the *CP* maps are completely characterized by a structure as in (5.36).

**Theorem 5.2.2 (Stinespring Dilation).** *A unital map $\Lambda : \mathbb{B}(\mathbb{H}) \mapsto \mathbb{B}(\mathbb{K})$ is CPU if and only if there exists a triplet $(\mathbb{K}_\Lambda, \pi_\Lambda(\mathbb{B}(\mathbb{K})), V_\Lambda)$, where $\mathbb{K}_\Lambda$ is a Hilbert space, $V_\Lambda : \mathbb{K} \mapsto \mathbb{K}_\Lambda$ an isometry and $\pi_\Lambda : \mathbb{B}(\mathbb{H}) \mapsto \mathbb{B}(\mathbb{K}_\Lambda)$ a representation of $\mathbb{B}(\mathbb{H})$ on $\mathbb{K}_\Lambda$ such that*

$$\Lambda[X] = V_\Lambda^\dagger \, \pi_\Lambda(X) \, V_\Lambda \; . \tag{5.38}$$

*The triplet $(\mathbb{K}_\Lambda, \pi_\Lambda(\mathbb{B}(\mathbb{K})), V_\Lambda)$ is unique up to unitary equivalences.*

**Proof:** If $\Lambda$ has the form (5.38) with $V_\Lambda^\dagger V_\Lambda = \mathbb{1}_{\mathbb{K}}$, Example 5.2.3.8 shows that $\Lambda$ is *CPU* . To prove the converse, consider the linear span of all elements of the form $X \otimes \psi$, $X \in \mathbb{B}(\mathbb{H})$ and $\psi \in \mathbb{H}$,

$$\mathfrak{B} := \left\{ \sum_i X_i \otimes \psi_i \; : \; X_i \in \mathbb{B}(\mathbb{H}) \, , \, \psi_i \in \mathbb{H} \right\}$$

and the bilinear form $\langle \cdot | \cdot \rangle_\Lambda : \mathfrak{B} \times \mathfrak{B} \mapsto \mathbb{C}$ defined by

$$(X \otimes \psi, Y \otimes \phi) \mapsto \langle X \otimes \psi \, | \, Y \otimes \phi \rangle_\Lambda := \langle \psi \, | \, \Lambda[X^\dagger Y] \, | \phi \rangle \; . \tag{5.39}$$

If $\Lambda$ is *CP* , this bilinear form is positive on $\mathfrak{B} \times \mathfrak{B}$ (see Example 5.2.3.7),

$$\begin{aligned} \langle \sum_{i=1}^N X_i \otimes \psi_i \, | \, \sum_{j=1}^N X_j \otimes \psi_j \, \rangle_\Lambda &= \sum_{i,j=1}^N \langle \, \psi_i \, | \, \Lambda[X_i^\dagger X_j] \, | \psi_j \, \rangle \\ &= \langle \, \widetilde{\psi} \, | \sum_{i,j=1}^N E_{ij} \otimes \Lambda[E_{ij}] \, | \widetilde{\psi} \, \rangle \geq 0 \qquad \forall N \in \mathbb{N} \; . \end{aligned}$$

By considering the quotient of $\mathfrak{B}$ by the kernel of the bilinear form, (5.39) gives a scalar product on the linear span $\mathfrak{B}_\Lambda := \mathfrak{B}/\mathrm{Kern}(\langle \cdot | \cdot \rangle_\Lambda)$ of the equivalence classes $[X \otimes \psi]$ (see the discussion after Definition 5.3.5). Set $\mathbb{K}_\Lambda$ equal to the closure of $\mathfrak{B}_\Lambda$ with respect to the scalar product $\langle \cdot | \cdot \rangle_\Lambda$ and let $\pi_\Lambda$ represent $\mathbb{B}(\mathbb{H})$ on $\mathbb{K}_\Lambda$ by $\pi_\Lambda(X)[Y \otimes \phi] = [XY \otimes \phi]$. Then, the linear maps $V_\Lambda : \mathbb{K} \mapsto \mathbb{K}_\Lambda$ and $V_\Lambda^\dagger : \mathbb{K}_\Lambda \mapsto \mathbb{K}$,

$$V_\Lambda | \, \phi \, \rangle = [\mathbb{1} \otimes \phi] \; , \qquad V_\Lambda^\dagger [X \otimes \phi] = \Lambda[X] | \, \phi \, \rangle \; ,$$

define an isometry ($\Lambda$ is *CPU*) and $V_\Lambda^\dagger \pi(X) V_\Lambda | \, \phi \, \rangle = V_\Lambda^\dagger [X \otimes \phi] = \Lambda[X] | \, \phi \, \rangle$ for all $\phi \in \mathbb{H}$. □

**Example 5.2.8.** That *CPU* maps are *contractions* (see Example 5.2.6.4) comes easily from Stinespring dilation. In fact, $V_\Lambda$ is an isometry, thus $V_\Lambda V_\Lambda^\dagger \leq 1\!\!1$, whence

$$\Lambda[X^\dagger X] = V_\Lambda^\dagger \pi_\Lambda[X^\dagger]\pi_\Lambda[X]V_\Lambda \geq V_\Lambda^\dagger \pi_\Lambda[X^\dagger]V_\Lambda\, V_\Lambda^\dagger \pi_\Lambda[X]V_\Lambda = \Lambda[X]^\dagger \Lambda[X] \ .$$

**Proposition 5.2.1 (Kraus Representation).** $\Lambda : \mathbb{B}(\mathbb{H}) \mapsto \mathbb{B}(\mathbb{K}))$ *is CPU if and only if it admits a* Kraus representation *of the form*

$$\Lambda[X] = \sum_j G_j^\dagger X G_j \ , \tag{5.40}$$

*where the* Kraus operators $G_j : \mathbb{K} \mapsto \mathbb{H}$ *are such that, if infinite, the sum converges in the strong-operator topology.*

**Proof:** The Stinespring representation is of the form $\mathbb{B}(\mathbb{H}) \otimes 1\!\!1_{\widetilde{\mathbb{K}}}$ on $\mathbb{H} \otimes \widetilde{\mathbb{K}}$, for a finite dimensional or countably infinite Hilbert space $\widetilde{\mathbb{K}}$. Given an *ONB* $\{|\, j\,\rangle\}$ in $\widetilde{\mathbb{K}}$, the isometry $V_\Lambda : \mathbb{K} \mapsto \mathbb{H} \otimes \widetilde{\mathbb{K}}$ and its adjoint $V_\Lambda^\dagger : \mathbb{H} \otimes \widetilde{\mathbb{K}} \mapsto \mathbb{K}$ read

$$V_\Lambda |\,\phi\,\rangle = \sum_j G_j |\,\phi\,\rangle \otimes |\, j\,\rangle \ , \qquad V_\Lambda^\dagger |\,\psi \otimes \phi\,\rangle \sum_j \langle\, j\,|\,\phi\,\rangle\, G_j |\,\psi\,\rangle \ ,$$

with $G_j : \mathbb{K} \mapsto \mathbb{H}$ and $\sum_j G_j^\dagger G_j = 1\!\!1_{\mathbb{K}}$. □

**Remarks 5.2.6.**

1. Using (5.36), the composition of *CPU* maps results in a *CPU* map. Indeed, let $\Lambda_{12} : \mathbb{B}(\mathbb{H}_1) \mapsto \mathbb{B}(\mathbb{H}_2)$, $\Lambda_{12}[X] = \sum_j G_{12}^\dagger(j)\, X\, G_{12}(j)$, $X \in \mathbb{B}(\mathbb{H}_1)$, and $\Lambda_{23} : \mathbb{B}(\mathbb{H}_2) \mapsto \mathbb{B}(\mathbb{H}_3)$, $\Lambda_{23}[Y] = \sum_k G_{23}^\dagger(k)\, Y\, G_{23}(k)$, $Y \in \mathbb{B}(\mathbb{H}_2)$, then,
$$\Lambda_{23} \circ \Lambda_{12}[X] = \sum_{j,k} G_{13}^\dagger(jk)\, X\, G_{13}(jk) \ ,$$
with new Kraus operators $G_{13}(jk) := G_{12}(j) G_{23}(k)$.
2. Let $\Lambda : \mathbb{B}(\mathbb{H}) \mapsto \mathbb{B}(\mathbb{H})$ be *CPU* and $\mathsf{T}$ the transposition with respect to a fixed *ONB* in $\mathbb{H}$; while $\Lambda \circ \mathsf{T}$ need not be *CPU* , $\mathsf{T} \circ \Lambda \circ \mathsf{T}$ surely is; indeed, $\mathsf{T} \circ \Lambda \circ \mathsf{T}[X] = \sum_j \mathsf{T}\left[G_j^\dagger\, [X^T]\, G_j\right] = \sum_j G_j^T\, X\, G_j^*$, with $\mathsf{T}[X] =: X^T$, $X \in \mathbb{B}(\mathbb{H})$, the transposed of $X$ and $X^* := (X^\dagger)^T$ its conjugate.
3. In full generality, a *CPU* map $\Lambda : \mathbb{B}(\mathbb{H}) \mapsto \mathbb{B}(\mathbb{K})$ has the form
$$\Lambda[X] = \sum_{i,j} C_{ij}\, L_i^\dagger\, X\, L_j \ ,$$
with $\sum_{i,j} C_{ij} L_i^\dagger L_j = 1\!\!1_{\mathbb{K}}$ and $C = [C_{ij}]$ a positive matrix, from which the diagonal Kraus representation is achieved by diagonalization as in Example 5.2.7.

4. While the structure of completely positive maps is fully under control, it is not so for positive maps which are still somewhat elusive. For instance, if the coefficients matrix $C = [C_{ij}]$ is Hermitian but not positive, by grouping together its positive and negative eigenvalues, $c_k$, $\Lambda$ can always be written as the difference of two *CP* maps,

$$\Lambda[X] = \sum_{c_k \geq 0} c_k\, G_k^\dagger\, X\, G_k - \sum_{c_k < 0} |c_k|\, G_k^\dagger\, X\, G_k \ . \tag{5.41}$$

For instance, let $\{F_i\}_{i=1}^{N^2}$ be a Hilbert-Schmidt *ONB* in $M_N(\mathbb{C})$ with $F_1 = \mathbb{1}_N/N$; using (5.30), the reduction map in Example 5.2.7.2, which is positive, but not *CP* , reads $\Lambda[X] = \left(\frac{1}{N^2} - 1\right) X + \sum_{i=2}^{N^2} F_i^\dagger\, X\, F_i$.
If there are no negative $c_k$, then $\Lambda$ is completely positive; if not, no general rule exists to deduce from the $c_k$ whether $\Lambda$ is a positive map.

### Conditional Expectations

Particularly important *CPU* maps are the so-called *conditional expectations* which are the non-commutative counterparts of the Radon-Nikodym derivative in (2.51).

**Definition 5.2.3.** *[117] A positive, unital linear map $\mathbb{E} : \mathcal{A} \mapsto \mathcal{B} \subseteq \mathcal{A}$ where $\mathcal{A}$ and $\mathcal{B}$ are $C^*$ algebras with identity is a conditional expectation of $\mathcal{A}$ onto $\mathcal{B}$ if $\mathbb{E}[AB] = \mathbb{E}[A]B$ for all $A \in \mathcal{A}$ and $B \in \mathcal{B}$.*

**Proposition 5.2.2.** *Conditional expectations enjoy the following properties:*

$$\mathbb{E}[A]^\dagger = \mathbb{E}[A^\dagger] \qquad \forall\, A \in \mathcal{A} \tag{5.42}$$
$$\mathbb{E}[BA] = B\mathbb{E}[A] \qquad \forall\, A \in \mathcal{A},\ B \in \mathcal{B} \tag{5.43}$$
$$\mathbb{E} \circ \mathbb{E} = \mathbb{E} \tag{5.44}$$
$$\mathbb{E}[A^\dagger\, A] \geq \mathbb{E}[A]^\dagger\, \mathbb{E}[A] \qquad \forall\, A \in \mathcal{A} \tag{5.45}$$
$$\|\mathbb{E}\| = 1 \ . \tag{5.46}$$

*Further, $\mathbb{E}$ is a CPU map.*

**Proof:** Property (5.42) comes from positivity as in Example 5.2.6.1; property (5.43) is a consequence of (5.42):

$$\mathbb{E}[BA] = \mathbb{E}[(BA)^\dagger]^\dagger = \mathbb{E}[A^\dagger B^\dagger]^\dagger = (\mathbb{E}[A^\dagger]B^\dagger)^\dagger = B\mathbb{E}[A] \ .$$

Property (5.44) follows from $\mathbb{E}[\mathbb{1}\mathbb{E}[A]] = \mathbb{E}[A]$ for all $A \in \mathcal{A}$; in order to prove property (5.45) consider $A - \mathbb{E}[A]$ and use positivity and (5.43), then

$$0 \leq \mathbb{E}[(A - \mathbb{E}[A])^\dagger (A - \mathbb{E}[A])] = \mathbb{E}[A^\dagger\, A] \;-\; \mathbb{E}[A]^\dagger\, \mathbb{E}[A] \;.$$

Property (5.46) results from the previous property and positivity:

$$A^\dagger\, A \leq \mathbb{1}\|A\|^2 \Longrightarrow \mathbb{E}[A]^\dagger\, \mathbb{E}[A] \leq \mathbb{E}[A^\dagger\, A] \leq \|A\|^2 \;.$$

Complete positivity is a consequence of (5.43) which yields

$$\mathbb{E}[B_1 A B_2] = B_1 \mathbb{E}[A] B_2 \qquad \forall\, B_{1,2} \in \mathcal{B}\;,\; A \in \mathcal{A}\;.$$

Therefore, from Examples 5.2.3.7 and 8,

$$\sum_{i,j=1}^{N} B_i^\dagger\, \mathbb{E}[A_i^\dagger\, A_j]\, B_j = \sum_{i,j=1}^{N} \mathbb{E}[B_i^\dagger A_i^\dagger\, A_j B_j] = \mathbb{E}[Z^\dagger\, Z] \geq 0\;,$$

for all $B_i \in \mathcal{B}$, $A_j \in \mathcal{A}$ and $N \in \mathbb{N}$, where $Z := \sum_{j=1}^{N} A_j B_j$. □

Because of the properties 3 and 5, conditional expectations are also called *projections of norm one.*

**Remark 5.2.7.** In case $\mathcal{A}$ is a von Neumann algebra and $\mathcal{A}_0 \subseteq \mathcal{A}$ a von Neumann subalgebra, one call conditional expectations all projections of norm one which are also *normal.* This latter property of linear maps $\Lambda : \mathcal{A}_1 \mapsto \mathcal{A}_2$ between von Neumann algebras amounts to the following [64]. Let $\{A_\mu\}_\mu$ be an increasing net of operators in $\mathcal{A}$, that is a set of operators indexed by a set of indexes $\mu \in M$ equipped with a partial ordering $\preceq$ such that $\mu_1 \preceq \mu_2 \Longrightarrow A_{\mu_1} \leq A_{\mu_2}$. If the net $\{A_\mu\}_\mu$ has an upper bound, then it has a least upper bound $A \in \mathcal{A}$ to which the net converges strongly: $s - \lim_\mu A_\mu = A$. Then, $\Lambda$ is normal if for all nets $\{A_\mu\}_\mu$ with an upper bound $\lim_\mu \Lambda[A_\mu] = \Lambda[\lim_\mu A_\mu]$.

**Examples 5.2.9.**

1. Let $\{P_i\}_{i\in I} \in \mathbb{B}(\mathbb{H})$ be orthogonal projections $P_i P_j = \delta_{ij} P_i$ such that $\sum_{i\in I} P_i = \mathbb{1}$, then $\mathbb{E}[X] := \sum_{i\in I} P_i\, X\, P_i$ is a conditional expectation from $\mathbb{B}(\mathbb{H})$ onto the Abelian subalgebra $\mathcal{P}$ generated by the $P_i$. Indeed, it is positive, linear and writing $\mathcal{P} \ni P = \sum_{j\in I} p_j P_j$, it turns out that

$$\mathbb{E}[XP] = \sum_{i,j\in I} p_j P_i X P_j P_i = \sum_{i\in I} p_i P_i X P_i = \mathbb{E}[X] P \;.$$

2. Consider two finite-level systems $A$ and $B$ described by matrix algebras $M_{n_a}(\mathbb{C})$, respectively $M_{n_b}(\mathbb{C})$; let $\widehat{\mathrm{Tr}}_A$ denote the normalized trace map performed with respect to party $A$, namely

$$\widehat{\mathrm{Tr}}_A(X_A) := \frac{1}{n_a} \mathrm{Tr}_A(X_A)\, \mathbb{1}_A \;. \tag{5.47}$$

A linear map from the matrix algebra $M_{n_a}(\mathbb{C})\otimes M_{n_b}(\mathbb{C})$ of the compound system $A+B$ onto the subalgebra $1\!\!1_A \otimes M_{n_b}(\mathbb{C})$ is obtained by defining

$$\mathbb{E}[X_A \otimes X_B] := \widehat{\mathrm{Tr}}_A(X_A) \otimes X_B \qquad \forall\, A \in M_{n_a}(\mathbb{C})\;\; B \in M_{n_b}(\mathbb{C})$$

on tensor products and then by extending it by linearity and continuity to the whole of $M_{n_a}(\mathbb{C}) \otimes M_{n_b}(\mathbb{C})$. Any $0 \leq X \in M_{n_a}(\mathbb{C}) \otimes M_{n_b}(\mathbb{C})$ can be written as $\sum_i (X_A^i)^\dagger\, X_A^j \otimes E_{ij}^B$ by means of a system of matrix units $\{E_{ij}^B\}_{i,j=1}^{n_b}$ in $M_{n_b}(\mathbb{C})$ (see Examples 5.2.3.7 and 8). Thus, one verifies that $\mathbb{E}$ is a positive linear map; indeed,

$$\begin{aligned}\langle \Psi_B \,|\, \mathbb{E}[X]\, | \Psi_B \rangle &= \sum_{i,j=1}^{n_b} \widehat{\mathrm{Tr}}_A\Big( (X_A^i)^\dagger\; X_A^j \Big)\, \langle \Psi_B \,|\, E_{ij}^B \,| \Psi_B \rangle \\ &= \frac{1}{n_a} \mathrm{Tr}_A\Big( (Y_A)^\dagger\; Y_A \Big) \geq 0 \;,\end{aligned}$$

where $Y_A := \sum_{j=1}^{n_b} \Psi_B^j\, X_A^j$ with $\Psi_B^j$ the $j$-th component of $|\,\Psi_B\,\rangle \in \mathbb{C}^{n_b}$ along the *ONB* associated with the chosen matrix units. By writing the identity matrix $1\!\!1_{A+B} = \sum_{i=1}^{n_a} \sum_{j=1}^{n_b} E_{ii}^A \otimes E_{jj}^B$ where $\{E_{ij}^A\}_{i,j=1}^{n_A}$ is a system of matrix units in $M_{n_a}(\mathbb{C})$, one shows that $\mathbb{E}[1\!\!1_{A+B}] = 1\!\!1_{A+B}$, whence $\mathbb{E}$ is unital. Furthermore, as any $X \in M_{n_a}(\mathbb{C}) \otimes M_{n_b}(\mathbb{C})$ can be written in the form $X = \sum_\ell X_A^\ell \otimes X_B^\ell$, it turns out that

$$\mathbb{E}[X\, X_B] = \sum_\ell \widehat{\mathrm{Tr}}_A(X_A^\ell) \otimes X_B^\ell X_B = \mathbb{E}[X] X_B \;,$$

whence $\mathbb{E}$ is a conditional expectation from $M_{n_a}(\mathbb{C}) \otimes M_{n_b}(\mathbb{C})$ onto $1\!\!1_A \otimes M_{n_b}(\mathbb{C})$.

## 5.3 von Neumann Algebras

In this section, we consider in detail some techniques proper to von Neumann algebras which are $C^*$ subalgebras of $\mathbb{B}(\mathbb{H})$ that are also closed with respect to the strong and weak topologies [64, 293, 300].

**Definition 5.3.1.** *The* commutant *of a $C^*$ algebra $\mathcal{A} \subseteq \mathbb{B}(\mathbb{H})$ is the $C^*$ algebra $\mathbb{B}(\mathbb{H}) \supseteq \mathcal{A}' := \Big\{ X' \in \mathbb{B}(\mathbb{H}) \,:\, [X'\,,\,X] = 0\; \forall X \in \mathcal{A} \Big\}$, the* bicommutant *the $C^*$ algebra $\mathbb{B}(\mathbb{H}) \ni \mathcal{A}'' := \Big\{ X'' \in \mathbb{B}(\mathbb{H}) \,:\, [X''\,,\,X'] = 0\; \forall X' \in \mathcal{A}' \Big\}$.*

**Remark 5.3.1.** Beside being $C^*$ algebras, commutants and bicommutants are also closed with respect to both the strong and weak topology. Indeed, if

$X' = (s,w) - \lim_{n\to\infty} X'_n$ with $[X'_n, X] = 0$ for all $X \in \mathcal{A}$, then, because of the continuity of the scalar product,

$$\langle \psi \,|\, [X'\,,\, X]\,|\phi\,\rangle = \lim_{n\to\infty} \langle \psi \,|\, [X'\,,\, X_n]\,|\phi\,\rangle = 0$$

for all $\psi, \phi \in \mathbb{H}$, whence $X' \in \mathcal{A}'$. Therefore, $\mathcal{A}'$ and $\mathcal{A}''$ contain all those operators that can be constructed from operators in $\mathcal{A}$ via strong-limits and weak-limits. In particular, $\mathcal{A}'$ and $\mathcal{A}''$ contain the spectral projectors of any of their self-adjoint and unitary elements.

**Examples 5.3.1.**

1. If $\mathcal{A} \subseteq \mathcal{A}'$, all its elements commute with each other and $\mathcal{A}$ is an Abelian $C^*$ algebra, *maximally Abelian* if $\mathcal{A} = \mathcal{A}'$.
2. The center of $\mathcal{A}$ is the Abelian $C^*$ algebra $\mathcal{Z} := \mathcal{A} \cap \mathcal{A}'$.
3. Of the commutative algebras of section 2.2.1, the $C^*$ algebra of continuous functions, $C(\mathcal{X})$, is not maximally Abelian since it is properly contained within the $C^*$ algebra of essentially bounded functions, $\mathbb{L}^\infty_\mu(\mathcal{X})$; the latter is instead maximally Abelian [293].
4. If $\mathcal{A} = \mathbb{B}(\mathbb{H})$, only multiples of the identity operator can commute with all bounded operators on $\mathbb{H}$, that is $\mathbb{B}(\mathbb{H})' = \{\mathbb{1}\}$, the *trivial algebra.* On the contrary, $\{\mathbb{1}\}' = \mathbb{B}(\mathbb{H})'' = \mathbb{B}(\mathbb{H})$. The same is true for the $C^*$ algebra of compact operators $\mathcal{A} = \mathbb{B}_\infty(\mathbb{H})$, $\mathcal{A}' = \{\mathbb{1}\}$ for the identity is the only operator on $\mathbb{H}$ which commute with all finite-rank ones; however, unlike for $\mathbb{B}(\mathbb{H})$, $\mathbb{B}_\infty(\mathbb{H}) \subset \mathbb{B}(\mathbb{H}) = \mathbb{B}_\infty(\mathbb{H})''$ in infinite dimension.
5. Consider the operator-valued matrix algebra $M_N(\mathcal{A})$ consisting of $N \times N$ matrices with entries from a $C^*$ algebra $\mathcal{A} \subseteq \mathbb{B}(\mathbb{H})$. Let $\mathbb{1}_N \otimes \mathcal{A} \subset M_N(\mathcal{A})$ be the subalgebra whose elements have the form $\mathbb{1}_N \otimes X$, $X \in \mathcal{A}$. The request that $\Big[\mathbb{1}_N \otimes X\,,\, \sum_{ij=1}^N E_{ij} \otimes X_{ij}\Big] = \sum_{ij=1}^N E_{ij} \otimes [X\,,\, X_{ij}] = 0$ for all $X \in \mathcal{A}$ with $X_{ij} \in \mathbb{B}(\mathbb{H})$, implies $(\mathbb{1}_N \otimes \mathcal{A})' = M_N(\mathcal{A}')$. The bicommutant $(\mathbb{1}_N \otimes \mathcal{A})''$ can be identified by imposing that, for all $X' \in \mathcal{A}'$ and $1 \leq k \leq N$,
$$\Big[E_{kk} \otimes X'\,,\, Y\Big] = \sum_{j=1}^N \Big(E_{kj} \otimes X' X_{kj} \;-\; E_{jk} \otimes X_{jk} X'\Big) = 0\;,$$
where $Y = \sum_{ij=1}^N E_{ij} \otimes X_{ij} \in M_N(\mathbb{B}(\mathbb{H}))$. This forces $Y$ to be of the form $Y = \sum_{k=1}^N E_{kk} \otimes X_{kk}$ with $X_{kk} \in \mathcal{A}''$. Finally, $\Big[E_{ij} \otimes \mathbb{1}\,,\, Y\Big] = E_{ij} \otimes (X_{jj} - X_{ii}) = 0$ for all $i, j = 1, 2, \ldots, N$, yields $X_{ii} = X$ for all $i$, whence $(\mathbb{1} \otimes \mathcal{A})'' = \mathbb{1}_N \otimes \mathcal{A}''$.
6. Given $\psi \in \mathbb{H}$, let $\mathbb{H}^{\mathcal{A}}_\psi \subseteq \mathbb{H}$ be the closure of the linear span of vectors of the form $X|\,\psi\,\rangle$, $X \in \mathcal{A}$, with $\mathcal{A} \subseteq \mathbb{B}(\mathbb{H})$ a $C^*$ algebra, and by $P^{\mathcal{A}}_\psi : \mathbb{H} \mapsto \mathbb{H}^{\mathcal{A}}_\psi$ the corresponding orthogonal projector. Since $\mathcal{A}\mathbb{H}^{\mathcal{A}}_\psi \subseteq \mathbb{H}^{\mathcal{A}}_\psi$, it follows

that $P^{\mathcal{A}}_\psi X P^{\mathcal{A}}_\psi = X P^{\mathcal{A}}_\psi$ for all $X \in \mathcal{A}$, whence, by taking the adjoint $X P^{\mathcal{A}}_\psi = P^{\mathcal{A}}_\psi X P^{\mathcal{A}}_\psi = (P^{\mathcal{A}}_\psi X^\dagger P^{\mathcal{A}}_\psi)^\dagger = (X^\dagger P^{\mathcal{A}}_\psi)^\dagger = P^{\mathcal{A}}_\psi X$ for all $X \in \mathcal{A}$. Therefore, $P^{\mathcal{A}}_\psi \in \mathcal{A}'$ and $P^{\mathcal{A}'}_\psi \in \mathcal{A}''$ for all $\psi \in \mathbb{H}$.

**Definition 5.3.2.** *A vector $\psi \in \mathbb{H}$ is cyclic for a $C^*$ algebra $\mathcal{A} \subseteq \mathbb{B}(\mathbb{H})$ if $\mathbb{H}^{\mathcal{A}}_\psi = \mathbb{H}$, separating if $X|\,\psi\,\rangle = 0 \Longleftrightarrow X = 0$ for all $X \in \mathcal{A}$.*

Being cyclic and separating are related properties; indeed, suppose $\psi \in \mathbb{H}$ to be cyclic for a $C^*$ algebra with identity $\mathcal{A} \subseteq \mathbb{B}(\mathbb{H})$ and $X'|\,\psi\,\rangle = 0$ for $X' \in \mathcal{A}'$. Then, $0 = \mathcal{A}X'|\,\psi\,\rangle = X'\mathcal{A}|\,\psi\,\rangle$, whence $X' = 0$ for cyclicity of $\psi \in \mathbb{H}$ amounts to $\mathcal{A}|\,\psi\,\rangle$ being dense in $\mathbb{H}$. Vice versa, suppose $\psi$ to be separating for $\mathcal{A}'$, but not cyclic for $\mathcal{A}$; then, $\mathcal{A}' \ni \mathbb{1} - P^{\mathcal{A}}_\psi \neq 0$, but $P^{\mathcal{A}}_\psi|\,\psi\,\rangle = |\,\psi\,\rangle$ since we assumed $\mathbb{1} \in \mathcal{A}$, which is a contradiction.

**Lemma 5.3.1.** *Let $\mathcal{A} \subseteq \mathbb{B}(\mathbb{H})$ be a $C^*$ algebra with identity; then $\psi \in \mathbb{H}$ is cyclic for $\mathcal{A}$ if and only if it is separating for $\mathcal{A}'$.*

Cyclicity refers to the possibility that, by acting on some vectors with all the operators of a given algebra, one gets a dense subspace whose closure is the whole Hilbert space. This is the case with the vector $|\,\mathbb{1}\,\rangle \in \mathbb{L}^2_\mu(\mathcal{X})$ in the Koopman-von Neumann formulation of classical mechanics (see Example 2.1.1). By acting on $|\,\mathbb{1}\,\rangle$ with the simple functions, one gets a dense linear span, whose closure is the whole of $\mathbb{L}^2_\mu(\mathcal{X})$. The same is true using continuous functions $f \in C(\mathcal{X})$ or essentially bounded functions $g \in \mathbb{L}^\infty_\mu(\mathcal{X})$.

Differently, if a vector $\psi \in \mathbb{H}$ is not cyclic for $\mathcal{A} \subseteq \mathbb{B}(\mathbb{H})$, then $P^{\mathcal{A}}_\psi$ projects onto a proper *$\mathcal{A}$-invariant subspace* $\mathbb{H}^{\mathcal{A}}_\psi \subset \mathbb{H}$. The absence of proper invariant subspaces with respect to $\mathcal{A}$ is related to the triviality of the commutant $\mathcal{A}'$.

**Definition 5.3.3 (Irreducible Algebras).** *A $C^*$ algebra $\mathcal{A} \subseteq \mathbb{B}(\mathbb{H})$ is irreducible if only $\mathbb{H}$ if all $\psi \in \mathbb{H}$ are cyclic for it.*

**Lemma 5.3.2.** *A $C^*$ algebra $\mathcal{A} \subseteq \mathbb{B}(\mathbb{H})$ is irreducible if and only if $\mathcal{A}' = \{\mathbb{1}\}$.*

**Proof:** If $\mathcal{A}' = \{\mathbb{1}\}$ then $P^{\mathcal{A}}_\psi = \mathbb{1}$ for all $\psi \in \mathbb{H}$. If all $\psi \in \mathbb{H}$ are cyclic for $\mathcal{A}$ and $\mathcal{A}' \neq \{\mathbb{1}\}$, then, according to Remark 5.3.1, there exists a projection $\mathcal{A}' \ni Q \neq \mathbb{1}$. Therefore, if $Q|\,\psi\,\rangle = |\,\psi\,\rangle$, then $Q\mathcal{A}|\,\psi\,\rangle = \mathcal{A}|\,\psi\,\rangle$; thus, the closure of $\mathcal{A}|\,\psi\,\rangle$ cannot equal $\mathbb{H}$. □

Given the commutant and bicommutant of a $C^*$ algebra $\mathcal{A} \subseteq \mathbb{B}(\mathbb{H})$, we can continue and consider the commutant of the bicommutant and so on.

Notice that, if $\mathcal{A} \subseteq \mathcal{B}$ are two $C^*$ algebras acting on $\mathbb{H}$, then $\mathcal{B}' \subseteq \mathcal{A}'$. Thus, from $\mathcal{A} \subseteq \mathcal{A}''$ it follows that $\mathcal{A}''' \subseteq \mathcal{A}'$; however, $\mathcal{A}' \subseteq (\mathcal{A}')'' = \mathcal{A}'''$, whence

$$\mathcal{A} \subseteq \mathcal{A}'' = \mathcal{A}^{iv} = \mathcal{A}^{vi} = \cdots \,, \qquad \mathcal{A}' = \mathcal{A}''' = \mathcal{A}^{v} = \mathcal{A}^{vii} = \cdots \,. \tag{5.48}$$

The closure properties of commutants and bicommutants discussed in Remark 5.3.1 are typical of

**Definition 5.3.4 (von Neumann Algebras).** *A $C^*$ algebra $\mathcal{A} \subseteq \mathbb{B}(\mathbb{H})$ is called a von Neumann algebra if $\mathcal{A} = \mathcal{A}''$. In the following, we shall employ the symbol $\mathcal{M}$ to denote von Neumann algebras, while keeping $\mathcal{A}$ for $C^*$ algebras which are not von Neumann algebras. von Neumann algebras $\mathcal{M}$ with center (see Example 5.3.1.2) $\mathcal{Z} = \mathcal{M} \cap \mathcal{M}' = \{\lambda \mathbb{1}\}$ consisting of multiples of the identity are called factors.*

**Theorem 5.3.1 (von Neumann Bicommutant Theorem).** *A $C^*$ algebra $\mathcal{A} \subseteq \mathbb{B}(\mathbb{H})$ with identity $\mathbb{1}$, is a von Neumann algebra if and only if it is strongly and weakly closed.*

**Proof:** As the bicommutant $\mathcal{A}''$ is the commutant of the commutant it is strongly and weakly closed. Let $\mathcal{A}^w$, $\mathcal{A}^s$ denote the weak and strong closures of $\mathcal{A}$; the strong topology is finer than the weak one, thus $\mathcal{A}^s \subseteq \mathcal{A}^w \subseteq \mathcal{A}''$ (see Remark 5.1.1.3). Therefore, we need only show that if $\mathcal{A} = \mathcal{A}^s$ then $\mathcal{A} = \mathcal{A}''$; in other words, we have to prove that $\mathcal{A}$ is strongly dense in $\mathcal{A}''$, namely that in any strong neighborhood $U^s_\varepsilon(X'')$, $X'' \in \mathcal{A}''$, of the form (5.7), there is an $X \in \mathcal{A}$. In order to do so, as a first step, note that, according to Example 5.3.1.6, given $\psi \in \mathbb{H}$, $P^{\mathcal{A}}_\psi \in \mathcal{A}'$ and $P^{\mathcal{A}}_\psi | \psi \rangle = | \psi \rangle$ for $\mathbb{1} \in \mathcal{A}$; thus, $P^{\mathcal{A}}_\psi \mathcal{A}'' | \psi \rangle = \mathcal{A}'' P^{\mathcal{A}}_\psi | \psi \rangle = \mathcal{A}'' | \psi \rangle \subseteq \mathbb{H}^{\mathcal{A}}_\psi$; this implies that for any $\varepsilon \geq 0$ and $X'' \in \mathcal{A}''$ there exists $X \in \mathcal{A}$ such that $\|(X - X'')| \psi \rangle\| \leq \varepsilon$. The second step is to extend this result to generic strong neighborhoods; for this we use (5.15), Example 5.3.1.5 and the previous arguments with $\mathbb{1}_n \otimes \mathcal{A}$, $M_n(\mathcal{A}')$, $(\mathbb{1}_n \otimes \mathcal{A})'' = \mathbb{1}_n \otimes \mathcal{A}''$ and $\widetilde{\psi}$ replacing $\mathcal{A}$, $\mathcal{A}'$, $\mathcal{A}''$, respectively $\psi$. Then, for any $\varepsilon \geq 0$, $X'' \in \mathcal{A}''$ and $\widetilde{\psi} = \sum_{i=1}^n | i \rangle \otimes | \psi_i \rangle$, there exists $X \in \mathcal{A}$ such that $\sum_{i=1}^n \|(X'' - X)| \psi_i \rangle\| \leq \varepsilon$. □

**Examples 5.3.2.**

1. The previous proof shows that by considering the strong closure of a $C^*$ algebra $\mathcal{A} \subseteq \mathbb{B}(\mathbb{H})$ one obtains the bicommutant $\mathcal{A}'' \subseteq \mathbb{B}(\mathbb{H})$.
2. Since $\mathcal{M} \subseteq \mathcal{M}' \Longrightarrow \mathcal{Z} = \mathcal{M}$, Abelian von Neumann algebras can be factors only if trivial.
3. In the Koopman-von Neumann formalism, $C(\mathcal{X})$ is a $C^*$ but not a von Neumann algebra; actually $C(\mathcal{X})'' = \mathbb{L}^\infty_\mu(\mathcal{X})$ for the algebra of essentially bounded functions on $\mathbb{L}^2_\mu(\mathcal{X})$ is strongly closed by construction.

4. If $\mathcal{M}$ is an irreducible von Neumann Algebra (see Definition 5.3.3), then it is a factor, whereas the opposite is not true in general. However, if $\mathcal{M}$ contains a maximally Abelian von Neumann algebra $\mathcal{N} = \mathcal{N}' \subseteq \mathcal{M}$, then $\mathcal{M}' \subseteq \mathcal{N}$ and $\mathcal{Z} = \mathcal{M}'$, whence, if $\mathcal{M}$ is a factor it is also irreducible (see Lemma 5.3.2).
5. $\mathbb{B}(\mathbb{H})$ is a von Neumann algebra since any bounded operator can be constructed by closing the $*$ algebra of finite-rank operators on $\mathbb{H}$ in either the weak or strong topology. Their uniform closure instead yields the $C^*$ algebra $\mathbb{B}_\infty(\mathbb{H})$ of compact operators which is not a von Neumann algebra for $\mathbb{B}_\infty(\mathbb{H})'' = \mathbb{B}(\mathbb{H}) \supset \mathbb{B}_\infty(\mathbb{H})$, in infinite dimension.
6. Let $\mathcal{M} \subseteq \mathbb{B}(\mathbb{H})$ be a von Neumann algebra, $E \in \mathcal{M}$ an orthogonal projector and consider $E\,\mathcal{M}\,E$; this is a von Neumann algebra acting on $E\mathbb{H}$ with commutant $(E\,\mathcal{M}\,E)' = E\,\mathcal{M}'\,E (= E\mathcal{M}' = \mathcal{M}'\,E)$. Indeed, for all $X \in \mathcal{M}$ and $X' \in \mathcal{M}'$,

$$(EXE)(EX'E) = EXEX' = X'EXE = (EX'E)(EXE) ,$$

thus $E\,\mathcal{M}\,E \subseteq (E\,\mathcal{M}'\,E)'$. On the other hand, if $X = XE \in (E\,\mathcal{M}'\,E)'$ it commutes with $E$ and, for any $X' \in \mathcal{M}'$,

$$XX' = XEX' = XEX'E = EX'EX = X'X ,$$

whence $(E\,\mathcal{M}'\,E)' \subseteq E\,\mathcal{M}\,E$.
7. Suppose $\mathcal{M} \subseteq \mathbb{B}(\mathbb{H})$ is a factor von Neumann algebra and consider the algebra $\mathcal{M} \cup \mathcal{M}'$ consisting of operators of the form $\sum_j X_j X'_j$ with $X_j \in \mathcal{M}$ and $X'_j \in \mathcal{M}'$. Then, $(\mathcal{M} \cup \mathcal{M}')' = \mathcal{M}' \cap \mathcal{M}'' = \mathcal{M} \cap \mathcal{M}' = \{\lambda \mathbb{1}\}$, whence $(\mathcal{M} \cup \mathcal{M}')'' = \mathbb{B}(\mathbb{H})$.

### 5.3.1 States and *GNS* Representation

The $C^*$ algebras so far considered had concrete representations by means of bounded operators on given Hilbert spaces $\mathbb{H}$; more in general, the notion of $C^*$ algebra in Definition 5.2.1 can be formulated in purely algebraic terms. What one needs is the abstract setting at the beginning of Section 5.2 and an abstract definition of states, along the lines developed for classical systems in Section 2.2.1.

**Definition 5.3.5.** *A state on a $C^*$ algebra $\mathcal{A}$ is any positive, normalized linear map $\omega : \mathcal{A} \mapsto \mathbb{C}$; namely, $\omega(Y^\dagger Y) \geq 0$ for all $Y \in \mathcal{A}$ and $\omega(\mathbb{1}) = 1$. States are also known as* expectation functionals.

*A state $\omega$ is* pure *on $\mathcal{A}$ if the only positive, not necessarily normalized, functionals $\mu : \mathcal{A} \mapsto \mathbb{C}$ such that $\mu \leq \omega$ have the form $\mu = \lambda\omega$ for some $0 \leq \lambda \leq 1$.*

*States $\omega$ such that $\omega(X^\dagger X) = 0 \Longleftrightarrow X = 0$, $\mathcal{X} \in \mathcal{A}$, are called* faithful.

From positivity it follows that states are automatically continuous functionals; indeed, if $\omega$ is a state on $\mathcal{A}$, then

$$(\omega(X^\dagger Y))^* = \omega(Y^\dagger X) \ , \quad |\omega(X^\dagger Y)|^2 \leq \omega(X^\dagger X)\,\omega(Y^\dagger Y) \ . \tag{5.49}$$

In order to prove this, one chooses $\lambda \in \mathbb{C}$ and consider

$$0 \leq \omega((X+\lambda Y)^\dagger(X+\lambda Y)) = \omega(X^\dagger X)+\lambda\omega(X^\dagger Y)+\lambda^*\omega(Y^\dagger X)+|\lambda|^2\omega(Y^\dagger Y) \ .$$

Then, the equality comes from setting $\lambda = 1, i$, while the inequality from choosing $\lambda$ equal to the conjugate of the phase of $\omega(X^\dagger Y)$.

The bilinear map $\mathcal{A} \times \mathcal{A} \ni (X,Y) \mapsto \omega(X^\dagger Y)$ would be a scalar product on $\mathcal{A}$ as a linear space, were it not for the fact that, in general, $\omega(X^\dagger X) = 0$ even if $X \neq 0$. In order to circumvent this difficulty, one considers the set $\mathcal{I} := \Big\{X \in \mathcal{A} \ : \ \omega(X^\dagger X) = 0\Big\}$. Because of (5.49), $\mathcal{I}$ is a linear set and also close; therefore, one can consider the quotient $\mathcal{A}/\mathcal{I}$ consisting of the equivalence classes $[X]_\rho := \Big\{X + I \ : \ I \in \mathcal{I}\Big\}$, $X \in \mathcal{A}$. Since (5.49) gives $\omega((X+I_1)^\dagger(Y+I_2)) = \omega(X^\dagger Y)$ for all $I_{1,2} \in \mathcal{I}$, each class can be identified with a vector $|\,\Psi^\omega_X\,\rangle$, $\mathcal{I}$ corresponding to the null vector. It thus follows that (5.49) defines a true scalar product over the linear span of these vectors, $\langle\,\Psi^\omega_X\,|\,\Psi^\omega_Y\,\rangle := \omega(X^\dagger Y)$. Consequently, by closing the linear span with respect to the corresponding norm, one gets a Hilbert space $\mathbb{H}_\omega$.

Further, it is immediate to represent operators $X \in \mathcal{A}$ as linear operators $\pi_\omega(X)$ acting multiplicatively on $\mathbb{H}_\omega$,

$$X \mapsto \pi_\omega(X) \ , \quad \pi_\omega(X)|\,\Psi^\omega_Y\,\rangle = |\,\Psi^\omega_{XY}\,\rangle \ . \tag{5.50}$$

Since $\mathcal{I}$ is a two-sided ideal in $\mathcal{A}$, the null vector is mapped into the null vector and $\pi_\omega(X)$ is a well-defined linear operator on $\mathbb{H}_\omega$; it is also bounded, for (5.49) and $X^\dagger X \leq \|X\|^2 \mathbb{1}$ imply

$$\|\pi_\omega(X)|\,\Psi^\omega_Y\,\rangle\|^2 = \omega(Y^\dagger X^\dagger XY) \leq \|X\|^2\omega(Y^\dagger Y) = \|X\|^2\,\||\,\Psi^\omega_Y\,\rangle\|^2 \ .$$

Further, $\pi_\omega$ is a so-called $*$ *morphism*, that is

$$\pi_\omega(X^\dagger) = \pi_\omega(X)^\dagger \ , \quad \pi_\omega(XY) = \pi_\omega(X)\pi_\omega(Y) \qquad \forall X, Y \in \mathcal{A} \ .$$

Therefore, $\pi_\omega$ represents $\mathcal{A}$ as a subalgebra of the bounded operators on $\mathbb{H}_\omega$: $\mathcal{A} \mapsto \pi_\omega(\mathcal{A}) \subseteq \mathbb{B}(\mathbb{H}_\omega))$.

**Definition 5.3.6 (Representations).** *A* $*$homomorphism *(homomorphism for short) between two $C^*$ algebras $\mathcal{A}_{1,2}$ is a linear map $\pi : \mathcal{A}_1 \mapsto \mathcal{A}_2$ that preserves the algebraic relations and the adjoint operation:*

$$\pi(A_1^\dagger) = \pi(A_1)^\dagger \ , \ \pi(A_1 B_2) = \pi(A_1)\pi(B_1) \quad \forall\, A_1, B_1 \in \mathcal{A}_1 \ .$$

*When a homomorphism is invertible, it is called an* isomorphism, automorphism *if it maps $\mathcal{A}$ invertibly onto itself.*

*If $\mathcal{A}_1 = \mathcal{A}$ and $\mathcal{A}_2 = \mathbb{B}(\mathbb{H})$, then $\pi$ gives a representation $(\pi(\mathcal{A}), \mathbb{H})$ as a $C^*$ (sub)algebra of bounded operators on a Hilbert space $\mathbb{H}$. Two representations of $(\pi_{1,2}(\mathcal{A}), \mathbb{H}_{1,2})$ of $\mathcal{A}$ on two Hilbert spaces $\mathbb{H}_{1,2}$ are* equivalent *if there exists an isometry $U : \mathbb{H}_1 \mapsto \mathbb{H}_2$ such that $\pi_1(\mathcal{A}) = U^\dagger \pi_2(\mathcal{A})U$.*

According to Definition 5.3.2, the state $|\Psi^\omega_{\mathbb{1}}\rangle$ is *cyclic* for $\pi_\omega(\mathcal{A})$ on $\mathbb{H}_\omega$; in fact, $\pi_\omega(X)|\Psi^\omega_{\mathbb{1}}\rangle = |\Psi^\omega_X\rangle$ and the linear span of the vectors of the form $|\Psi^\omega_X\rangle$, $X \in \mathcal{A}$, is dense in $\mathbb{H}_\omega$, by construction. Also, the expectation associated with $\omega$ takes the form

$$\omega(X) = \langle \Psi^\omega_{\mathbb{1}} | \pi_\omega(X) | \Psi^\omega_{\mathbb{1}} \rangle , \quad X \in \mathcal{A} . \tag{5.51}$$

We shall set $|\Omega_\omega\rangle := |\Psi^\omega_{\mathbb{1}}\rangle$; from Definition 5.3.2 it also follows that $|\Omega_\omega\rangle$ is separating for the commutant $\pi_\omega(\mathcal{A})' \subseteq \mathbb{B}(\mathbb{H}_\omega)$

The previous approach is due to Gelfand, Naimark and Segal and is known as *GNS* construction. [64].

**Definition 5.3.7.** *Given the GNS triplet $(\mathbb{H}_\omega, \pi_\omega, \Omega_\omega)$, $\mathbb{H}_\omega$, $\pi_\omega$ and $\Omega$ will be called GNS Hilbert space, GNS representation and GNS vector, respectively.*

**Remarks 5.3.2.**

1. Any triplet $(\mathbb{H}_\nu, \pi_\nu, \Omega_\nu)$ with the *GNS* properties of $(\mathbb{H}_\omega, \pi_\omega, \Omega_\omega)$ is unitarily equivalent to it. Namely, there exists an isometry $U : \mathbb{H}_\nu \mapsto \mathbb{H}_\omega$ such that $U|\Omega_\nu\rangle = |\Omega_\omega\rangle$ and $\pi_\nu(X) = U^\dagger \pi_\omega(X) U$ for all $X \in \mathcal{A}$. Indeed, because of (5.51) that holds for both representations, the map $U : \mathbb{H}_\nu \mapsto \mathbb{H}_\omega$ defined by $U\pi_\nu(X)|\Omega\rangle = \pi_\omega(X)|\Omega_\omega\rangle$ is such that
$$\begin{aligned}\omega(X^\dagger Y) &= \langle \Omega_\omega | \pi_\omega(X)^\dagger \pi_\omega(Y) | \Omega_\omega \rangle = \langle \Omega_\nu | \pi_\nu(X)^\dagger U^\dagger U \pi_\nu(Y) | \Omega_\nu \rangle \\ &= \langle \Omega_\nu | \pi_\nu(X)^\dagger \pi_\nu(Y) | \Omega_\nu \rangle\end{aligned}$$
on the dense subsets $\pi_\omega(\mathcal{A})|\Omega\rangle \subseteq \mathbb{H}_\omega$ and $\pi_\nu(\mathcal{A})|\Omega_\nu\rangle \subseteq \mathbb{H}_\nu$. Then, $U$ extends to an isometry $U : \mathbb{H}_\omega \mapsto \mathbb{H}_\nu$; furthermore, on the dense subset of $\pi_\nu(Y)|\Omega_\nu\rangle$, $Y \in \mathcal{A}$,
$$\begin{aligned}U^\dagger \pi_\omega(X) U \pi_\nu(Y)|\Omega_\nu\rangle &= U^\dagger \pi_\omega(X)\pi_\omega(Y)|\Omega_\omega\rangle = U^\dagger \pi_\omega(XY)|\Omega_\omega\rangle \\ &= U^\dagger U \pi_\nu(XY)|\Omega_\nu\rangle = \pi_\nu(X)\pi_\nu(Y)|\Omega\rangle,\end{aligned}$$
that is $U^\dagger \pi_\omega(X) U = \pi_\nu(X)$ for all $X \in \mathcal{A}$.
2. As a $*$-homomorphism, the *GNS* representation $\pi_\omega$ preserves the $C^*$ properties of $\mathcal{A}$. Therefore $\pi_\omega(\mathcal{A})$ is a $C^*$ algebra, as well as its commutant $\pi_\omega(\mathcal{A})'$. The latter is also a von Neumann algebra, this need not be true of $\pi_\omega(\mathcal{A})$, but it is certainly so of the bicommutant $\pi_\omega(\mathcal{A})''$, that is of the strong closure of $\pi_\omega(\mathcal{A})$ on $\mathbb{H}_\omega$. If the center $\mathcal{Z}_\omega := \pi_\omega(\mathcal{A})'' \cap \pi_\omega(\mathcal{A})$ is trivial, that is it consists of the multiples of the identity only, then $\omega$ is called a *factor* or *primary state.*

3. If $\omega$ is a state on $\mathcal{A}$ and $\nu$ is a linear positive functional on it, majorized by $\omega$, $\nu \leq \omega$, then also $\nu$ satisfies a Cauchy-Schwartz inequality as $\omega$ in (5.49) [4], $|\nu(X^\dagger Y)|^2 \leq \nu(X^\dagger X)\nu(Y^\dagger Y) \leq \omega(X^\dagger X)\omega(Y^\dagger Y)$. Consequently, $\nu$ defines a continuous sesquilinear form on $\mathbb{H}_\omega \times \mathbb{H}_\omega$ so that, from Example 5.2.4, $\nu(X^\dagger Y) = \langle \Omega \,|\, \pi_\omega(X)^\dagger \, T' \, \pi_\omega(Y) \,|\Omega \rangle$, with $T' \in \mathbb{B}(\mathbb{H}_\omega)$. Further, from $0 \leq \nu(X^\dagger X) \leq \omega(X^\dagger X)$, for all $X \in \mathcal{A}$, one deduces that $0 \leq T' \leq \mathbb{1}$. Moreover, $T' \in \pi_\omega(\mathcal{A})'$; indeed,
$$\begin{aligned}\nu(X^\dagger Y Z) &= \langle \Omega \,|\, \pi_\omega(X)^\dagger T' \pi_\omega(Y)\pi_\omega(Z) \,|\Omega \rangle = \nu((Y^\dagger X)^\dagger Z) \\ &= \langle \Omega \,|\, \pi_\omega(X)^\dagger \pi_\omega(Y) T' \pi_\omega(Z) \,|\Omega \rangle \,,\end{aligned}$$
whence $[T' \,,\, \pi_\omega(Y)] = 0$ for all $Y \in \mathcal{A}$ since $\pi_\omega(\mathcal{A})|\,\Omega\,\rangle$ is dense in $\mathbb{H}_\omega$.
4. From the previous result, it turns out that $\pi_\omega(\mathcal{A})$ is an irreducible $C^*$ algebra (see Definition 5.3.3) if and only if $\omega$ is a pure state (see Definition 5.3.5). In fact, according to Lemma 5.3.2, $\pi_\omega(\mathcal{A})$ is irreducible if and only if $\pi_\omega(\mathcal{A})' = \{\lambda 1\}$. If $\pi_\omega(\mathcal{A})'$ is trivial then $\nu \leq \omega$ implies $T' = \lambda \mathbb{1}$, hence $\omega$ is pure. On the other hand, if $\pi_\omega(\mathcal{A})'$ is not trivial, then there exists some $\mathbb{1} \neq X' \in \pi_\omega(\mathcal{A})'$, so that also $X' + (X')^\dagger$ and its spectral projectors belong to the von Neumann algebra $\pi_\omega(\mathcal{A})'$. Therefore, there must be at least one non-trivial projector $P' \in \pi_\omega(\mathcal{A})'$; also, $\mathbb{1} - P' = Q' \in \pi_\omega(\mathcal{A})'$, so that one can decompose $\omega$ into a convex combination $\omega = \lambda\omega_{P'} + (1-\lambda)\omega_{Q'}$, where $\lambda := \langle \Omega \,|\, P' \,|\Omega \rangle$ while
$$\widetilde{\omega}_{P'} := \lambda\omega_{P'}(X) := \langle \Omega \,|\, P'\pi_\omega(X) \,|\Omega \rangle \tag{5.52}$$
$$\widetilde{\omega}_{Q'} := (1-\lambda)\omega_{Q'}(X) := \langle \Omega \,|\, Q'\pi_\omega(X) \,|\Omega \rangle \tag{5.53}$$
are positive, normalized linear functionals over $\mathcal{A}$ which are both majorized by $\omega$ but are not of the form $\lambda\omega$ (compare the analogous argument in the proof of Proposition 2.3.8).
5. The previous point is an example of the convex structure [20] of the space of states $\mathcal{S}(\mathcal{A})$ on a $C^*$ algebra $\mathcal{A}$. In more formal terms [64]: given a $C^*$ algebra $\mathcal{A}$ with identity, the set $\mathcal{S}(\mathcal{A})$ of its states is compact in the $w^*$ topology generated by the semi-norms $\mathcal{S}(\mathcal{A}) \ni \omega \mapsto \mathcal{L}_X(\omega) = |\omega(X)|$. Moreover, its extremal points are the pure states and $\mathcal{S}(\mathcal{A})$ is the $w^*$ closure of their convex hull.

### 5.3.2 $C^*$ and von Neumann Abelian algebras

Let $\mathcal{A}$ be an Abelian unital $C^*$ or von Neumann algebra. As discussed in Remarks 2.2.2.2,3, the algebra $C(\mathcal{X})$ of the continuous functions over a compact topological space $\mathcal{X}$ is a typical example of the first case, while the algebra $\mathbb{L}^\infty_\mu(\mathcal{X})$ of the essentially bounded functions is an instance of the second case. In this section we shall show that these two cases do in fact exhaust

[4] What matters in the derivation of (5.49) is positivity and not normalization.

all the possibilities: the main technique we shall use is the so-called *Gelfand transform*.

All multiplicative functionals $\chi : \mathcal{A} \mapsto \mathbb{C}$ such that

$$\chi(AB) = \chi(A)\chi(B) , \quad \chi(A^\dagger) = \chi(A)^* \qquad \forall\, A, B \in \mathcal{A}$$

are known as *characters* and their set will be denoted by $\mathcal{X}_\mathcal{A}$. It turns out that characters are states on $\mathcal{A}$; indeed,

$$\chi(A) = \chi(\mathbb{1}\,A) = \chi(\mathbb{1})\chi(A) \Longrightarrow \chi(\mathbb{1}) = 1 ,$$

while, if $\mathcal{A} \ni A \geq 0$ then $A = B^\dagger\, B$ (see Remark 5.2.2), whence

$$\chi(A) = \chi(B^\dagger\, B) = |\chi(B)|^2 \geq 0 .$$

Further, $A - a$, with $a \in \mathbb{C}$ and $A \in \mathcal{A}$, is invertible if there exists $B \in \mathcal{A}$ such that $B(A-a) = \mathbb{1}$; since, for any $\chi \in \mathcal{X}_\mathcal{A}$, $\chi(B)(\chi(A)-a) = 1$, it follows that if $A - a$ is invertible then $\chi(A) \neq a$ for all $\chi \in \mathcal{X}_\mathcal{A}$. Therefore, the spectrum of $A \in \mathcal{A}$ contains the values assumed on $A$ by the characters on $\mathcal{A}$:

$$\mathrm{Sp}(A) \supseteq \{\chi(A) \,:\, \chi \in \mathcal{X}_\mathcal{A}\} . \tag{5.54}$$

**Examples 5.3.3.**

1. Let $\mathcal{A} = C(\mathcal{X})$, then $\mathcal{X}_\mathcal{A}$ consists of the evaluation functionals (Dirac deltas) $\delta_x(f) = f(x)$ for all $x \in \mathcal{X}$, $f \in C(\mathcal{X})$.
2. Let $\mathcal{A} = D_N(\mathbb{C})$ the algebra of all diagonal matrices on $\mathbb{C}^N$ with respect to a given *ONB* $\{|\, i\,\rangle\}_{i=1}^N$, namely $\mathcal{A} \ni A = \sum_{i=1}^N A_i\, E_{ii}$, where $\{E_{ij}\}$ is the associated family of matrix units. Then, $\mathcal{X}_\mathcal{A}$ consists of the maps

$$\chi_j(A) := \mathrm{Tr}(AE_{jj}) = \langle\, j\,|\,A\,|j\,\rangle = A_j .$$

Indeed, $AB = \sum_{i,j=1}^N A_i B_j E_{ii} E_{jj} = \sum_{i=1}^N A_i B_i E_{ii}$ implies $\chi_j(AB) = A_j B_j = \chi_j(A)\,\chi_j(B)$ for all $A, B \in \mathcal{A}$. Notice that the multiplicative property cannot be true of any pure state $|\,\psi\,\rangle\langle\,\psi\,| \in M_N(\mathbb{C})$; in fact, for $|\,\psi\,\rangle = \alpha|\,i\,\rangle + \beta|\,j\,\rangle$

$$\langle\,\psi\,|\,AB\,|\psi\,\rangle = |\alpha|^2 A_i B_i + |\beta|^2 A_j B_j \qquad \text{while}$$
$$\langle\,\psi\,|\,A\,|\psi\,\rangle\,\langle\,\psi\,|\,B\,|\psi\,\rangle = |\alpha|^4 A_i B_i + |\beta|^4 A_j B_j + |\alpha|^2|\beta|^2(A_i B_j + A_j B_i) .$$

3. Characters behave as *tracial states* over $\mathcal{A}$, namely $\chi(AB) = \chi(BA)$. However, the only tracial state on $M_N(\mathbb{C})$ is given by

$$\tau(X) = \mathrm{Tr}(\frac{\mathbb{1}}{N}X) , \quad \text{so that} \quad \tau(XY) = \tau(YX) , \tag{5.55}$$

for all $X, Y \in M_N(\mathbb{C})$. It thus follows that there cannot be characters on $M_N(\mathbb{C})$. Indeed,

$$\tau(E_{ii}^2) = \tau(E_{ii}) = \frac{1}{N} \,, \quad \tau(E_{ii})\tau(E_{ii}) = \frac{1}{N^2} \;;$$

therefore the tracial state can be multiplicative only if $N = 1$. In order to show that the only tracial state on $M_N(\mathbb{C})$ is $\tau$, let us assume that there exists another state $\omega$ such that $\omega(XY) = \omega(YX)$ for all $X, Y \in M_N(\mathbb{C})$. Let $X = E_{ij}$ and $Y = E_{k\ell}$; because of (5.12), it turns out that

$$\omega(E_{ij}E_{k\ell}) = \delta_{jk}\omega(E_{i\ell}) = \omega(E_{k\ell}E_{ij}) = \delta_{i\ell}\omega(E_{kj}) \,.$$

Thus, $\omega(E_{ij}) = 0$ if $i \neq j$ and $\omega(E_{ii}) = \alpha$ for all $i = 1, 2, \ldots, N$ so that $\omega(\mathbb{1}) = 1 \Longrightarrow \alpha = 1/N$. In conclusion, $\omega$ acts as $\tau$ on a system of matrix units and must thus coincide with it.

The set of characters is a subset of the unit ball of the topological dual of $\mathcal{A}$ ($\mathcal{X}_{\mathcal{A}} \subset (\mathcal{A}^*)_1$); moreover, $\mathcal{X}_{\mathcal{A}}$ coincides with the set of pure states over $\mathcal{A}$. In fact, if $\omega$ is a pure state on $\mathcal{A}$, then in the *GNS* representation $\pi_\omega(A)$ is irreducible ($\pi_\omega(\mathcal{A})' = \{\lambda\mathbb{1}\}$), but then $\pi_\omega(A) = \lambda_A\mathbb{1}$ for all $A \in \mathcal{A}$ for Abelianness implies $\pi_\omega(A) \subseteq \pi_\omega(\mathcal{A})'$. Then, $\omega(A) = \langle\, \Omega_\omega \,|\, \pi_\omega(A) \,|\Omega_\omega\, \rangle = \lambda_A$ and

$$\omega(AB) = \langle\, \Omega_\omega \,|\, \pi_\omega(A)\pi_\omega(B) \,|\Omega_\omega\, \rangle = \lambda_A\lambda_B = \omega(A)\omega(B) \,,$$

whence $\omega \in \mathcal{X}_{\mathcal{A}}$.

Let $\mathcal{A}^*$ be endowed with the $w^*$-topology (see Remark 5.1.1.5), then $\mathcal{X}_{\mathcal{A}}$ is a $w^*$-closed subset of $(\mathcal{A}^*)_1$. In fact, if $\chi_n \in \mathcal{X}_{\mathcal{A}}$ $w^*$-converges to $\chi$, that is if $\chi_n(A) \to \chi(A)$ for all $A \in \mathcal{A}$, $\chi$ is linear and also multiplicative; indeed,

$$\begin{aligned}|\chi(AB) - \chi(A)\chi(B)| \leq {} & |\chi(AB) - \chi_n(AB)| \\ & + \|A\| \, |\chi_n(B) - \chi(B)| + \|B\| \, |\chi_n(A) - \chi(A)| \,.\end{aligned}$$

Therefore, by choosing $n$ large enough the left hand side of the inequality can be made arbitrarily small.

**Remark 5.3.3.** Once the topological dual $\mathcal{A}^*$ is equipped with the $w^*$-topology, its unit ball $(\mathcal{A}^*)_1$ is compact by the Banach-Alaoglu theorem [324]. As $\mathcal{X}_{\mathcal{A}}$ is a closed subset, it is also compact [324]; further, since the space of states is Hausdorff, so is $\mathcal{X}_{\mathcal{A}}$. One can thus consider the $C^*$ algebra $C(\mathcal{X}_{\mathcal{A}})$ of continuous functions over the set of characters and the corresponding properties. For instance, in the proof of Theorem 5.3.2, we shall profit from a theorem of Stone and Weierstrass [259] which states that the norm-closure of any algebra of complex functions on a compact Haussdorf space $\mathcal{X}$ that separates points and contains the identity coincides with $C(\mathcal{X})$.

**Definition 5.3.8 (Gelfand transform).** *The Gelfand transform is the map $\Gamma : \mathcal{A} \mapsto C(\mathcal{X}_\mathcal{A})$ from an Abelian $C^*$ algebra to the continuous functions over its characters, defined by*

$$\mathcal{A} \ni A \mapsto \Gamma[A](\chi) = \chi(A) \qquad \forall \chi \in \mathcal{X}_\mathcal{A} .$$

Notice that $\Gamma[A](\chi)$ is automatically continuous on $\mathcal{X}_A$ equipped with the $w^*$ topology inherited from $\mathcal{A}^*$. In full generality, the following property holds [64, 300, 324]

**Theorem 5.3.2.** *Any Abelian unital $C^*$ algebra $\mathcal{A}$ is isomorphic to $C(\mathcal{X}_\mathcal{A})$.*

**Proof:** The Gelfand transform is a $*$-homomorphism: linearity is evident, also $\Gamma[A^\dagger](\chi) = \chi(A^\dagger) = \chi(A)^* = (\Gamma[A](\chi))^*$. Moreover,

$$\Gamma[AB](\chi) = \chi(AB) = \chi(A)\chi(B) = (\Gamma[A]\Gamma[B])(\chi) .$$

It also preserves the norm; in fact,

$$\|\Gamma[A]\|^2 = \sup_{\chi \in \mathcal{X}_\mathcal{A}} |\Gamma[A](\chi)|^2 = \sup_{\chi \in \mathcal{X}_\mathcal{A}} |\chi(A)|^2 = \|A\|^2 ,$$

for all $A \in \mathcal{A}$. The latter equality is a consequence of the fact that $\mathcal{X}_\mathcal{A}$ coincides with the set of pure states over $\mathcal{A}$ and that [64, 300], for any $A \in \mathcal{A}$ one can always construct a pure state $\omega$ such that $\omega(A) = \|A\|$.

It thus follows that $\Gamma[A] = 0$ only if $A = 0$. Furthermore, $\Gamma[\mathbb{1}] = 1$ and, if $\chi_1 \neq \chi_2$, then $\chi_1(A) = \Gamma[A](\chi_1) \neq \chi_2(A) = \Gamma[A](\chi_2)$ for some $A \in \mathcal{A}$. One says that $\Gamma[\mathcal{A}]$ separates points of $\mathcal{X}_\mathcal{A}$; thus, the theorem of Stone and Weierstrass (see Remark 5.3.3) applies to $\Gamma[\mathcal{A}]$ so that $\Gamma[\mathcal{A}] = C(\mathcal{X}_\mathcal{A})$. □

**Remark 5.3.4.** [324] If $\mathcal{A}$ is a generic $C^*$ algebra and $X$ one of its normal elements ($X\,X^\dagger = X^\dagger\,X$), then one can consider the Abelian $C^*$ algebra $\mathcal{A}[X]$ generated by the norm closure of the $*$-algebra of polynomials in the commuting operators $\mathbb{1}$, $X$ and $X^\dagger$. Let us consider the Gelfand transform $\Gamma : \mathcal{A}[X] \mapsto C(\mathcal{X}_{\mathcal{A}[X]})$; to any function $f \in C(\mathcal{X}_{\mathcal{A}[X]})$ one associates a unique element $f(X) := \Gamma^{-1}[f] \in \mathcal{A}[X]$. This is known as *continuous functional calculus*. Consider, for instance, the function $f(z) := (\chi(X) - z)^{-1}$; then, by using a power series expansion and the isomorphic properties of $\Gamma$ and its inverse, one obtains

$$\Gamma\left[\frac{1}{X - z}\right] = f(z) , \quad f(X) = \Gamma^{-1}[f] = \frac{1}{X - z} ,$$

whenever $z > \|X\| = \sup_{\chi \in \mathcal{X}_{\mathcal{A}[X]}} |\chi(X)|$. Thus, from Example 5.2.2.5 it follows that the spectrum of a normal $X \in \mathcal{A}$ coincides with the values assumed on $X$ by the characters of $\mathcal{X}_{\mathcal{A}[X]}$: $\mathrm{Sp}(X) = \{\chi(X) \,:\, \chi \in \mathcal{X}_{\mathcal{A}[X]}\}$.

**Examples 5.3.4.**

1. If $\mathcal{A}$ is a finite-dimensional Abelian algebra, then $\mathcal{X}_\mathcal{A}$ contains finitely many points (characters) $\mathcal{X}_\mathcal{A} = \{\chi_i\}_{i=1}^a$. The maps $\widehat{\delta}_j : \mathcal{X}_\mathcal{A} \mapsto \mathbb{C}$ defined by $\widehat{\delta}_j(\chi_i) = \delta_{ij}$, are continuous with respect to the discrete topology on $\mathcal{X}_\mathcal{A}$. By inverting the Gelfand isomorphism, the corresponding elements $p_i := \Gamma^{-1}[\widehat{\delta}_i] \in \mathcal{A}$ are orthogonal projections:
$$p_i p_k = \Gamma^{-1}[\widehat{\delta}_i \widehat{\delta}_k] = \delta_{ik}\, \Gamma^{-1}[\widehat{\delta}_k] = \delta_{ik}\, p_k \ .$$
These are minimal projections of $\mathcal{A}$; namely, the only projections in $\mathcal{A}$ majorized by $p_i$ are the trivial one $p = 0$ and $p_i$ itself. Indeed, consider two projections $q, p$ in a generic unital $C^*$ algebra and suppose $q \leq p$; then, writing $p = q + (p - q)$, Example 5.2.3.2 yields
$$q \geq q\,p\,q = q + q(p-q)q \geq q \qquad \text{for} \quad p - q \geq 0 \ .$$
Therefore, $q(p-q)q = X^\dagger X = 0$ with $X = \sqrt{p-q}\,q$ whence $X = 0$ and $pq = q = qp$. If $p = p_i \in \mathcal{A}$ and $\mathcal{A} \ni q \leq p_i$, write $\Gamma[q] = \sum_{i=1}^n \pi_i(q)\,\widehat{\delta}_i$ with $\pi_i(q) \in \mathbb{R}_+$; it follows that
$$\Gamma[q] = \Gamma[qp_i] = \Gamma[q]\Gamma[p_i] = \pi_i(q)\widehat{\delta}_i \ .$$
Since $\Gamma[q] = \Gamma[q]^2$, one concludes that $\Gamma[q] = \widehat{\delta}_i$ whence $q = p_i$.
2. Consider $X \geq 0$, and the function $f(t) = \sqrt{t}$, $t \geq 0$. Since the spectrum of $X$ is contained in $[0, \|X\|]$ and thus also $\mathcal{X}_{\mathcal{A}[X]} \subseteq [0, \|X\|]$; from Remark 5.3.4 it thus follows that $Y := f[X] = \Gamma^{-1}[\sqrt{t}] \geq 0$ and $Y^2 = \Gamma^{-1}[t] = X$. We now show that the square-root of $X$ is unique; let $\mathcal{A} \ni Z > 0$ be such that $Z^2 = X$ and let $P_n(t)$ be a sequence of polynomials on $[0, \|X\|]$ converging uniformly to $\sqrt{t}$. Set $Q_n(t) := P_n(t^2)$: $\lim_{n\to+\infty} Q_n(t) = t$ uniformly on $[0, \|X\|]$. Furthermore, from Remark 5.3.4 and Remark 5.2.1,
$$\mathcal{X}_{\mathcal{A}[X]} = \mathrm{Sp}(X) = \mathrm{Sp}(Z^2) = (\mathrm{Sp}(Z))^2 = \{t^2 \,:\, t \in \mathrm{Sp}(Z)\} \ ,$$
whence
$$Z = \lim_{n\to+\infty} Q_n(Z) = \lim_{n\to+\infty} P_n(Z^2) = \lim_{n\to+\infty} f(X) = \sqrt{X} \ .$$
3. The Gelfand isomorphism maps the von Neumann algebra $\mathbb{L}^\infty_\mu(\mathcal{X})$ into $C(\mathcal{Y})$ where $\mathcal{Y}$ is a so-called *extremely disconnected* Hausdorff space whose open sets have open closures [324].

The preceding considerations can be extended to Abelian von Neumann algebras $\mathcal{M} \subset \mathbb{B}(\mathbb{H})$ on a separable Hilbert space $\mathbb{H}$ [324]. Since $\mathcal{M}$ is also a $C^*$

algebra, one considers the Gelfand isomorphism $\Gamma : \mathcal{M} \mapsto C(\mathcal{X}_{\mathcal{M}})$; suppose $\psi \in \mathbb{H}$ to be cyclic for $\mathcal{M}$ and define the linear functional $F : C(\mathcal{X}_{\mathcal{M}}) \mapsto \mathbb{C}$,

$$F(f) := \langle \psi \,|\, \Gamma^{-1}[f] \,|\psi\rangle \ .$$

This functional is positive since $\Gamma^{-1}$ is an isomorphism and preserves positivity; then, Riesz representation theorem [258] (see (2.48)) ensures that there exists a positive Borel measure $\mu$ on $\mathcal{X}_{\mathcal{M}}$ such that

$$\langle \psi \,|\, \Gamma^{-1}[f] \,|\psi\rangle = \int_{\mathcal{X}_{\mathcal{M}}} \mathrm{d}\mu\,(x)\, f(x) = \mu(f) \ .$$

The support of $\mu$ is the whole of $\mathcal{X}_{\mathcal{M}}$, otherwise there would exist $\mathcal{Y} \subset \mathcal{X}_{\mathcal{M}}$ and a positive continuous $f$, non-zero on $\mathcal{Y}$, such that

$$\mu(f) = 0 = \langle \sqrt{\Gamma^{-1}[f]}\psi \,|\, \sqrt{\Gamma^{-1}[f]}\psi\rangle \ .$$

Since $\psi$ is cyclic for $\mathcal{M}$, it is separating for $\mathcal{M}'$ and for $\mathcal{M} \subseteq \mathcal{M}'$; then $\sqrt{\Gamma^{-1}[f]}|\,\psi\rangle = 0$ implies $\Gamma^{-1}[f] = 0$ whence $f = 0$. For all $X \in \mathcal{M}$,

$$\int_{\mathcal{X}_{\mathcal{M}}} \mathrm{d}\mu\,(x)\, |\Gamma[X](x)|^2 = \langle \psi \,|\, \Gamma^{-1}[\Gamma[X^\dagger]\Gamma[X] \,|\psi\rangle = \| X|\,\psi\rangle\|^2 \ .$$

Then, one can construct a unitary operator $U : \mathbb{H} \mapsto \mathbb{K} := \mathbb{L}^2_\mu(\mathcal{X}_{\mathcal{M}})$ by extending to the $L^2$-closures of $\mathcal{M}|\,\psi\rangle$ and $C(\mathcal{X}_M)$ the linear operator defined on the latter spaces by $U : X|\,\psi\rangle \mapsto \Gamma[X]$. It turns out that

$$U\,X\,U^\dagger\,(\Gamma[Y]) = U\,X\,Y = \Gamma[X\,Y] = \Gamma[X]\,(\Gamma[Y]) \ ,$$

for all $X \in \mathcal{M}$, whence $U\,X\,U^\dagger$ is represented by a multiplication operator on $C(\mathcal{X}_{\mathcal{M}})$. This relation can be used to prove that $U\,\mathcal{M}\,U^\dagger$ is a von Neumann subalgebra of $\mathbb{B}(\mathbb{K})$ and since the algebra of multiplication by continuous functions is weakly dense in the von Neumann algebra of multiplication operators by functions in $\mathbb{L}^\infty_\mu(\mathcal{X}_{\mathcal{M}})$ it follows that this latter is isomorphic to $U\,\mathcal{M}\,U^\dagger$.

Even when a cyclic vector for the von Neumann algebra $\mathcal{M}$ does not exist, one can prove a similar result [324].

**Theorem 5.3.3.** *Every Abelian von Neumann algebra $\mathcal{M}$ acting on a separable Hilbert space $\mathbb{H}$ is isomorphic to some $\mathbb{L}^\infty_\mu(\mathcal{X})$, where $\mathcal{X}$ is a compact Hausdorff space and $\mu$ is a finite, positive Borel measure on $\mathcal{X}$ supported by the whole of $\mathcal{X}$.*

## 5.4 Quantum Systems with Finite Degrees of Freedom

The simplest quantum systems are 2-level systems (the *qubits* of quantum information): their states and observables are $2 \times 2$ matrices from $M_2(\mathbb{C})$

acting on the Hilbert space $\mathbb{C}^2$. Though simple, the 2 dimensional framework is sufficient to accommodate a variety of rather successful phenomenological descriptions as for spin 1/2 particles in magnetic contexts [248, 273, 280], for atoms whose ground and first excited states can be treated as isolated from the rest of the energy eigenvalues [87], for the polarization degree of freedom of photons [272, 312] and for the *strangeness degree of freedom* of neutral $K$ mesons [30, 31]. Recently, even macroscopic systems in particular ultracold atoms [191] have been started to be studied as spin 1/2 particles; this is the case for the low-lying energy states of Bose-Einstein condensates in double well potentials and superconducting boxes near resonance [197, 316]. The latest advances in the experimental manipulation of atomic systems have indeed provided concrete realizations of 2-level quantum systems and made them available for the actual verification of central issues of quantum information theory [6, 63].

The observables of 2-level systems are self-adjoint $2 \times 2$ matrices acting on the 2-dimensional Hilbert space $\mathbb{C}^2$; particularly important are the unitary and self-adjoint *Pauli matrices* $\sigma_{1,2,3}$ that satisfy the algebraic relations

$$\sigma_j \sigma_k = \delta_{jk} \mathbb{1}_2 + i\varepsilon_{jk\ell}\, \sigma_\ell \ , \tag{5.56}$$

where $\varepsilon_{jk\ell}$ is the antisymmetric 3-tensor, and $\mathbb{1}_2$ denotes the $2 \times 2$ identity matrix.

When normalized, $\widetilde{\sigma}_\mu := \sigma_\mu/\sqrt{2}$, they become an *ONB* with respect to the Hilbert-Schmidt scalar product (5.26), that is $\mathrm{Tr}(\widetilde{\sigma}_\mu \widetilde{\sigma}_\nu) = \delta_{\mu\nu}$. Thus, it turns out that any $X \in M_2(\mathbb{C})$ can be written as (see Example 5.2.5)

$$X = \sum_{\mu=0}^{3} (\mathrm{Tr}(\widetilde{\sigma}_\mu X))\, \widetilde{\sigma}_\mu \ . \tag{5.57}$$

It is customary to work within the representation of the eigenvectors of $\sigma_3$, $|\,0\,\rangle = \begin{pmatrix} 1 \\ 0 \end{pmatrix}$ and $|\,1\,\rangle = \begin{pmatrix} 0 \\ 1 \end{pmatrix}$ (the states of a particle with spin 1/2 pointing down, respectively up along the $z$ direction in space). Then, the Pauli matrices have the standard form

$$\sigma_1 = \begin{pmatrix} 0 & 1 \\ 1 & 0 \end{pmatrix} \ , \quad \sigma_2 = \begin{pmatrix} 0 & -i \\ i & 0 \end{pmatrix} \ , \quad \sigma_3 = \begin{pmatrix} 1 & 0 \\ 0 & -1 \end{pmatrix}$$

so that $\sigma_1|\,0\,\rangle = |\,1\,\rangle$, $\sigma_1|\,1\,\rangle = |\,0\,\rangle$ and $\sigma_2|\,0\,\rangle = -i|\,1\,\rangle$, $\sigma_2|\,1\,\rangle = i|\,0\,\rangle$.

The action of $\sigma_1$ on the standard *ONB* amounts to a spin flip; if 0 and 1 were classical spin states encoding bits, then $\sigma_1$ would implement the *NOT* logical operation: $0 \mapsto 1$, $1 \mapsto 0$ or $i \mapsto i \oplus 1$, where $\oplus$ denotes the binary addition (addition mod 2). The *ONB* associated with $\sigma_1$ consists of

$$|\,\pm\,\rangle := \frac{|\,0\,\rangle \pm |\,1\,\rangle}{\sqrt{2}} = \frac{1}{\sqrt{2}} \begin{pmatrix} 1 \\ \pm 1 \end{pmatrix} \ , \quad \sigma_1|\,\pm\,\rangle = \pm|\,\pm\,\rangle \ .$$

Their *ONB* is unitarily related to the standard one by the *Hadamard rotation*

$$U_H := \frac{1}{\sqrt{2}}\begin{pmatrix} 1 & 1 \\ 1 & -1 \end{pmatrix} = U_H^\dagger = U_H^{-1} \ , \ U_H | \, i \, \rangle = \frac{1}{\sqrt{2}} \sum_{j=0}^{1} (-1)^{ij} \, | \, j \, \rangle \ . \quad (5.58)$$

A system consisting of $n$ spins $1/2$ is described by the matrix algebra $M_{2^n}(\mathbb{C}) = (M_2(\mathbb{C}))^{\otimes n}$; denoting by $\sigma^i_\mu$, $\mu = 0, 1, 2, 3$, the Pauli matrices of the $i$-th spin, the elements of $M_{2^n}(\mathbb{C})$ are linear combinations of operators of the form $\bigotimes_{i=1}^n \sigma^i_{\mu_i}$ acting on $(\mathbb{C}^2)^{\otimes n}$. The so-called *computational basis* of quantum information consists of tensor products of eigenvectors of $\sigma_3$,

$$| \, \boldsymbol{i}^{(n)} \, \rangle = | \, i_1 i_2 \cdots i_n \, \rangle = | \, i_1 \, \rangle \otimes | \, i_2 \, \rangle \otimes \cdots | \, i_n \, \rangle \ , \quad i_j \in \{0, 1\} \ ,$$

that are in one-to-one correspondence with bit-strings $\boldsymbol{i}^{(n)} \in \Omega_2^{(n)}$.

One may interpret them as orthogonal configurations of quantum spins located at the integer sites $0 \le \ell \le n$ of an infinite 1-dimensional lattice. Of course, unlike for classical spins, in this case linear combinations of these configurations are also possible physical states. Thus, a one dimensional array of $n$ spins $1/2$ provide a non-commutative counterpart to classical spin-chains of finite length $n$. Interestingly, their algebra $M_{2^n}(\mathbb{C})$ also describes $n$ degrees of freedom satisfying *Canonical Anticommutation Relations* (*CAR*).

**Example 5.4.1 (Finite Spin Systems: *CAR*).** From (5.56) it follows that different Pauli matrices anticommute,

$$\Big\{\sigma_j \, , \, \sigma_k\Big\} := \sigma_j \sigma_k + \sigma_k \sigma_j = 2\delta_{jk} \, \mathbb{1} \ .$$

Set $\sigma_+ := \dfrac{\sigma_1 + i\sigma_2}{2} = \begin{pmatrix} 0 & 1 \\ 0 & 0 \end{pmatrix}$ and $\sigma_- := \dfrac{\sigma_1 - i\sigma_2}{2} = \begin{pmatrix} 0 & 0 \\ 1 & 0 \end{pmatrix}$. These matrices fulfil the following algebraic relations

$$\Big\{\sigma_+ \, , \, \sigma_-\Big\} = 1 \ , \quad \Big[\sigma_+ \, , \, \sigma_-\Big] = \sigma_3 \quad (5.59)$$

$$\sigma_\pm^2 = 0 \ , \quad \Big[\sigma_3 \, , \, \sigma_+\Big] = 2\sigma_+ \ , \quad \Big[\sigma_3 \, , \, \sigma_-\Big] = -2\sigma_- \ . \quad (5.60)$$

With $| \, 0 \, \rangle$, $| \, 1 \, \rangle$ the eigenvectors of $\sigma_3$, $\sigma_- | \, 1 \, \rangle = 0$, while $\sigma_+ | \, 1 \, \rangle = | \, 0 \, \rangle$.

In the case of $n$-spin $1/2$ systems, let $\sigma^i_\# = \sigma^i_3 \, , \, \sigma^i_\pm \, , \, \mathbb{1}_i$ denote the spin operators relative to the $i$-th spin. They are identified as elements of $M_{2^n}(\mathbb{C})$ by embedding them as

$$M_2(\mathbb{C}) \ni \sigma^i_\# \mapsto \mathbb{1}_{[1,i-1]} \otimes \sigma^i_\# \otimes \mathbb{1}_{[i+1,n]} \ , \quad (5.61)$$

where $\mathbb{1}_{[j,k]} := \mathbb{1}_j \otimes \mathbb{1}_{j+1} \otimes \cdots \mathbb{1}_k$ is the tensor product of as many identity matrices $\mathbb{1} \in M_2(\mathbb{C})$ as the sites in the subset $[j, k]$. Then, setting

$$a_i := \left(\bigotimes_{j=1}^{i-1} \sigma_3^j\right) \otimes \sigma_-^i \otimes \mathbb{1}_{[i+1,N]} \ , \ a_i^\dagger := \left(\bigotimes_{j=1}^{i-1} \sigma_3^j\right) \otimes \sigma_+^i \otimes \mathbb{1}_{[i+1,N]} \ ,$$

one obtains operators that obey the *CAR* of $n$ *Fermionic* degrees of freedom:

$$\left\{a_i \, , \, a_j^\dagger\right\} := a_i a_j^\dagger + a_j^\dagger a_i = \delta_{ij} \ , \quad \left\{a_i \, , \, a_j\right\} = \left\{a_i^\dagger \, , \, a_j^\dagger\right\} = 0 \ . \qquad (5.62)$$

Further, the vector state $|\,1\,\rangle^{\otimes n} := \underbrace{|\,1\,\rangle \otimes |\,1\,\rangle \otimes \cdots |\,1\,\rangle}_{n \ times}$ consisting of $n$ spins all pointing down, behaves as the *vacuum state* for it is annihilated by all $a_i$, while $a_i^\dagger$, acting on it, creates the vector state with the $i$-th spin pointing up,

$$a_i |\,1\,\rangle^{\otimes n} = 0 \ , \quad a_i^\dagger |\,1\,\rangle^{\otimes n} = |\,1 \cdots 1 \underbrace{\ 0\ }_{ithsite} 1 \cdots 1\,\rangle \ .$$

There cannot be more than one Fermion for each $i$ as from (5.62) $(a_i^\dagger)^2 = 0$. Thus, products of the form $\prod_{j=1}^n (a_j^\dagger)^{i_j}$, where $i_j = 0, 1$ and $(a_j^\dagger)^0 = \mathbb{1}$, create the computational basis,

$$\prod_{j=1}^{n} (a_j^\dagger)^{i_j \oplus 1} |\,1\,\rangle^{\otimes n} = |\,i_1 i_2 \cdots i_n\,\rangle \ ,$$

where $\oplus$ denotes summation mod 2. Since

$$\begin{aligned} [AB, C] &= ABC - CAB = A(BC + CB) - (AC + CA)B \\ &= A\{C \, , \, B\} + \{A \, , \, C\}B \ , \end{aligned} \qquad (5.63)$$

the *number operator*

$$\begin{aligned} \widehat{N} &:= \sum_{i=1}^{n} a_i^\dagger a_i = \sum_{i=1}^{n} \mathbb{1}_{[1,i-1]} \otimes \sigma_+^i \sigma_-^i \otimes \mathbb{1}_{[i+1,N]} \\ &= \sum_{i=1}^{n} \mathbb{1}_{[1,i-1]} \otimes \frac{\mathbb{1}_i + \sigma_3^i}{2} \otimes \mathbb{1}_{[i+1,n]} \end{aligned} \qquad (5.64)$$

satisfies $\widehat{N} |\,1\,\rangle^{\otimes n} = 0$ and

$$[\widehat{N} \, , \, a_i] = -a_i \ , \qquad [\widehat{N} \, , \, a_i^\dagger] = a_i^\dagger \ , \qquad (5.65)$$

whence $\widehat{N} |\,\boldsymbol{i}^{(n)}\,\rangle = (\sum_{j=1}^n i_j) |\,\boldsymbol{i}^{(n)}\,\rangle$. Therefore, the basis vectors $|\,\boldsymbol{i}^{(n)}\,\rangle$ are *occupation number states* that is eigenstates of $\widehat{N}$; they span the so-called *Fock space* $\mathbb{H}_F^{(n)} = |\,vac\,\rangle \oplus \bigoplus_{k=1}^{n} \mathbb{H}_k$, where $|\,vac\,\rangle = |\,1\,\rangle^{\otimes n}$ is the vacuum state and $\mathbb{H}_k$ is the Hilbert space corresponding to $k$ Fermi degrees of freedom.

As much as one can construct $n$ Fermi creation and annihilation operators satisfying the *CAR* out of $n$ spin 1/2 operators, so one can obtain the $n$ spin algebra $M_{2^n}(\mathbb{C})$ out of the creation and annihilation operators of $n$ Fermi degrees of freedom. Indeed, from (5.64) one derives $\sigma_3^i = 2a_i^\dagger a_i - \mathbb{1}$ and

$$\sigma_-^i = \Big(\prod_{j=1}^{i-1}(2a_i^\dagger a_i - \mathbb{1})\Big)\, a_i \ , \quad \sigma_+^i = \Big(\prod_{j=1}^{i-1}(2a_i^\dagger a_i - \mathbb{1})\Big)\, a_i^\dagger \ .$$

These relations are known as *Jordan-Wigner transformations* [295]. They show that the algebra of $n$ Fermi degrees of freedom is isomorphic to that of $n$ spins 1/2, $M_{2^n}(\mathbb{C})$. It is important to notice that one Fermi degree of freedom correspond to a totally delocalized spin operator.

**Remark 5.4.1.** [290, 291] The kinematical description of $n$ Fermionic degrees of freedom is abstractly provided by a set of $n$ operators $a_j$, $a_j^\dagger$ satisfying the *CAR* (5.62) together with the algebra $\mathcal{A}_F$ comprising all polynomials $P(a_j, a_j^\dagger)$ constructed with them. The Fock one is a concrete representation of the $a_j$, $a_j^\dagger$ as annihilation and creation operators on a Hilbert space with a distinguished vector $|\, vac\,\rangle$, the vacuum state, such that $a_j|\, vac\,\rangle = 0$ for all $j = 1, 2, \ldots, n$.

Because of the *CAR* relations, in any representation $\pi$ on a Hilbert space $\mathbb{H}$, $\pi(a_j)$ and $\pi(a_j^\dagger)$ are bounded operators with respect to uniform norm:

$$\pi(a_j^\dagger a_j) + \pi(a_j a_j^\dagger) = \mathbb{1} \geq \pi(a_j^\dagger a_j) \Longrightarrow \|\pi(a_j)\| \leq 1 \ . \tag{5.66}$$

Also, any two irreducible representations $(\pi_{1,2}(\mathcal{A}_F), \mathbb{H}_{1,2})$ of the *CAR* of finitely many Fermions are unitarily equivalent to the Fock one (see Definition 5.3.6). Indeed, from the previous example, we know that both representations are isomorphic to $M_{2^n}(\mathbb{C})$ for $n$ Fermions; so, the positive operators $\pi_i(\widehat{N}) := \sum_{j=1}^n \pi_i(a_j)^\dagger \pi(a_j)$ have discrete integer spectrum with an eigenvalue 0. In fact, if $\pi_i(\widehat{N})|\,\psi_i\,\rangle = \lambda|\,\psi_i^\lambda\,\rangle$, then (5.65) implies

$$\begin{aligned}[\widehat{N}\,,\, a_i^n] &= a_i\,[\widehat{N}\,,\, a_i^{n-1}] + [\widehat{N}\,,\, a_i]\, a_i^{n-1} \\ &= a_i\,[\widehat{N}\,,\, a_i^{n-1}] - a_i^n = -n a_i^n \ ,\end{aligned}$$

so that

$$\begin{aligned}\pi_i(\widehat{N})\pi(a_i)^n|\,\psi_i^\lambda\,\rangle &= \lambda\pi(a_i)^n|\,\psi_i^\lambda\,\rangle + \Big[\pi_i(\widehat{N})\,,\, \pi(a_i)^n\Big]|\,\psi_i^\lambda\,\rangle \\ &= (\lambda - n)\pi(a_i)^n|\,\psi_i^\lambda\,\rangle \ .\end{aligned}$$

Thus the spectrum is discrete with 0 as its smallest eigenvalue; the corresponding eigenvector $|\,\psi_i^0\,\rangle$ is annihilated by all $\pi_i(a_j)$ and is unique as implied by the assumed irreducibility of the representation $\pi_i$. In fact, the linear span $\pi_i(\mathcal{A}_F)|\,\psi_i^0\,\rangle$ is dense in $\mathbb{H}_i$ (see Definition 5.3.3); therefore, if $\pi_i(a_j)|\,\phi_i\,\rangle = 0$ for all $j = 1, 2, \ldots, n$, then the same should hold for its component $|\,\phi_i^\perp\,\rangle$ orthogonal to $|\,\psi_i^0\,\rangle$; therefore, $\langle\,\phi_i^\perp\,|\,\pi_i(P(a_j, a_j^\dagger)\,|\psi_i^0\,\rangle = 0$ for all polynomials in annihilation and creator operators, whence $|\,\phi_i^\perp\,\rangle = 0$ as it would be orthogonal to a dense subset in $\mathbb{H}_i$. The eigenvector $|\,\psi_i^0\,\rangle$ is the vacuum for the Fock representation $\pi_i$; let $U : \mathbb{H}_1 \mapsto \mathbb{H}_2$ be such that

$$U|\,\psi_1^0\,\rangle = |\,\psi_2^0\,\rangle\ , \quad U\pi_1(P(a_j, a_j^\dagger))|\,\psi_1^0\,\rangle = \pi_2(P(a_j, a_j^\dagger))|\,\psi_2^0\,\rangle\ .$$

Since the scalar products $\langle\,\psi_i^0\,|\,\pi_i(P')^\dagger\pi_i(P'')\,|\psi_i^0\,\rangle$, with $P'$ and $P''$ arbitrary polynomials, are completely determined by the *CAR* relations, their values do not depend on the representation chosen; that is

$$\begin{aligned}\langle\,\psi_1^0\,|\,\pi_1(P')^\dagger\pi_1(P'')\,|\psi_1^0\,\rangle &= \langle\,\psi_2^0\,|\,\pi_2(P')^\dagger\pi_2(P'')\,|\psi_2^0\,\rangle\\ &= \langle\,\psi_1^0\,|\,\pi_1(P')^\dagger U^\dagger\,U\pi_1(P'')\,|\psi_1^0\,\rangle\ .\end{aligned}$$

on a dense set, whence $U$ extends to an isometry from $\mathbb{H}_1$ to $\mathbb{H}_2$.

Of course, not all quantum systems with finite degrees of freedom are finite level systems. A free quantum particle in one dimension or a one-dimensional quantum harmonic oscillator are systems with one degree of freedom, but they are described by means of the infinite dimensional Hilbert space of square-summable complex functions over $\mathbb{R}$. In quantum information, these systems are sometimes referred to as *continuous variable systems* in contrast to spin-like systems whose variables (observables) are instead discrete ($N \times N$ matrices).

For continuous variable systems the standard kinematics is more appropriately given in terms of unitary groups of translations in position and momentum. The algebraic relations between them are known as *Canonical Commutation Relations* (*CCR*).

Consider a Hamiltonian classical system with $f$ degrees of freedom and canonical coordinates $\mathbb{R}^{2f} \ni (\boldsymbol{q}, \boldsymbol{p})$, $\boldsymbol{q} = (q_1, q_2, \ldots, q_f)$, $\boldsymbol{p} = (p_1, p_2, \ldots, p_f)$. In standard quantization, one introduces unbounded, densely defined, self-adjoint *position* and *momentum* operators $(\widehat{q}_i, \widehat{p}_i)$ on $\mathbb{H} = \mathbb{L}^2_{\mathrm{d}\boldsymbol{q}}(\mathbb{R}^f)$ defined, in the so-called *position representation*, by

$$(\widehat{q}_i\psi)(\boldsymbol{q}) = q_i\psi(\boldsymbol{q})\ , \quad (\widehat{p}_i\psi)(\boldsymbol{q}) = -i\,\partial_{q_i}\psi(\boldsymbol{q})\ , \quad \psi \in \mathbb{H}\ . \tag{5.67}$$

On a common dense domain, they satisfy the *standard commutation relations* (with $\hbar = 1$)

$$\Big[\widehat{q}_i\,,\,\widehat{p}_j\Big] = i\delta_{ij}\ , \quad \Big[\widehat{q}_i\,,\,\widehat{q}_j\Big] = \Big[\widehat{p}_i\,,\,\widehat{p}_j\Big] = 0\ . \tag{5.68}$$

Unlike, Fermionic annihilation and creation operators, the operators $\widehat{q}_i$ and $\widehat{p}_i$ have continuous spectrum and cannot be bounded. Indeed, from (5.68),

$$\left[\widehat{q}_i\,,\,\widehat{p}_i^n\right] = i\,n\,\widehat{p}_i^{n-1} \Longrightarrow 2\|\widehat{q}_i\|\,\|\widehat{p}_i\| \geq n\;, \tag{5.69}$$

for all integer $n$. Because of unboudedness, one introduces the one-parameter groups of unitary operators $\{U_i(q)\}_{q\in\mathbb{R}}$ and $\{V_i(p)\}_{p\in\mathbb{R}}$,

$$\Big(U_i(q)\psi\Big)(\boldsymbol{q}) = \psi(\boldsymbol{q}+\boldsymbol{q}_i)\;,\quad \Big(V_i(p)\psi\Big)(\boldsymbol{q}) = \mathrm{e}^{ipq_i}\psi(\boldsymbol{q})\;, \tag{5.70}$$

where $(\boldsymbol{q}_i)_j = \delta_{ij}\,q$, in position representation, while

$$\Big(U_i(q)\psi\Big)(\boldsymbol{p}) = \mathrm{e}^{iqp_i}\psi(\boldsymbol{p})\;,\quad \Big(V_i(p)\psi\Big)(\boldsymbol{p}) = \psi(\boldsymbol{p}-\boldsymbol{p}_i)\;, \tag{5.71}$$

in momentum representation, with $(\boldsymbol{p}_i)_j = \delta_{ij}\,p$.

These semi-groups are continuous with respect to the strong-operator topology, whence, by Stone theorem [300], they are generated by self-adjoint operators $\widehat{q}_i$ and $\widehat{p}_i$

$$U_i(q) := \exp\left(i\,q\,\widehat{p}_i\right)\;,\quad V_i(p) := \exp\left(i\,p\,\widehat{q}_i\right)\;. \tag{5.72}$$

By writing them as formal series, it can be checked that they implement translations in position, respectively momentum:

$$U_i(q)\,\widehat{q}_i\,U_i^\dagger(q) \;=\; \widehat{q}_i \,+\, q\;,\quad V_i(p)\,\widehat{p}_i\,U_i^\dagger(p) \;=\; \widehat{p}_i \,-\, p\;. \tag{5.73}$$

The *CCR* can thus be recast as

$$\begin{cases} U_i(q)\,V_j(p) &= V_j(p)\,U_i(q) \quad i\neq j \\ U_i(q)\,V_i(p) &= \mathrm{e}^{i\,qp}\,V_i(p)\,U_i(q) \quad . \end{cases} \tag{5.74}$$

Set $\widehat{\boldsymbol{q}} := (\widehat{q}_1, \widehat{q}_2, \ldots, \widehat{q}_f)$, $\widehat{\boldsymbol{p}} := (\widehat{p}_1, \widehat{p}_2, \ldots, \widehat{p}_f)$, $\widehat{\boldsymbol{r}} := (\widehat{\boldsymbol{q}}, \widehat{\boldsymbol{p}})$, and

$$W(\boldsymbol{r}) := \mathrm{e}^{i\,(\boldsymbol{q}\cdot\widehat{\boldsymbol{p}}+\boldsymbol{p}\cdot\widehat{\boldsymbol{q}})} = \mathrm{e}^{i\,\boldsymbol{r}\cdot(\Sigma_1\widehat{\boldsymbol{r}})}\;, \tag{5.75}$$

where $\cdot$ denotes the usual scalar product and $\Sigma_1 := \begin{pmatrix} 0 & 1\!\!1_f \\ 1\!\!1_f & 0 \end{pmatrix}$ with $1\!\!1_f$ is the $f\times f$ identity matrix. These operators are known as *Weyl operators*; by the Campbell-Hausdorff formula,

$$\exp\left(A+B\right) = \exp\left(-\frac{1}{2}[A,B]\right)\,\exp\left(A\right)\,\exp\left(B\right)\;, \tag{5.76}$$

that holds when $[A\,,\,B]$ is a multiple of the identity, they can be recast in the form

$$W(\boldsymbol{r}) = \mathrm{e}^{-\frac{i}{2}\boldsymbol{q}\cdot\boldsymbol{p}}\;\mathrm{e}^{i\,\boldsymbol{q}\cdot\widehat{\boldsymbol{p}}}\,\mathrm{e}^{i\,\boldsymbol{p}\cdot\widehat{\boldsymbol{q}}}\;. \tag{5.77}$$

The Weyl operators satisfy $W(\boldsymbol{r})^\dagger = W(-\boldsymbol{r})$ and the composition law

$$W(\boldsymbol{r}_1)\,W(\boldsymbol{r}_2) \;=\; \mathrm{e}^{\frac{i}{2}\sigma(\boldsymbol{r}_1,\boldsymbol{r}_2)}\,W(\boldsymbol{r}_1+\boldsymbol{r}_2)\;, \tag{5.78}$$

where

$$\sigma(\boldsymbol{r}_1,\boldsymbol{r}_2) := \boldsymbol{q}_1\cdot\boldsymbol{p}_2 - \boldsymbol{p}_1\cdot\boldsymbol{q}_2 = \boldsymbol{r}\cdot(\mathbb{J}_f\boldsymbol{r})\;,\quad \mathbb{J}_f = \begin{pmatrix} 0 & \mathbb{1}_f \\ -\mathbb{1}_f & 0\end{pmatrix}\;, \tag{5.79}$$

is the *symplectic form* characteristic of the *Weyl relations.* It thus follows that the $*$ algebra $\mathcal{W}$ generated by linear combinations and products of Weyl operators coincides with their linear span.

**Remark 5.4.2.** Given the $*$ algebra $\mathcal{W}$, one looks for its closure with respect to a suitable topology; it turns out that the $C^*$ algebra that arises from the uniform norm is too small for physical purposes [300]. For instance, one would like that two Weyl operators $W(\boldsymbol{r}_{1,2})$ be close to each other when $\|\boldsymbol{r}_1-\boldsymbol{r}_2\|\to 0$; however, whenever $\boldsymbol{r}_1\neq\boldsymbol{r}_2$,

$$\|W(\boldsymbol{r}_1)-W(\boldsymbol{r}_2)\| = \|\mathbb{1}-W^\dagger(\boldsymbol{r}_1)W(\boldsymbol{r}_2)\| = 2\;,$$

since unitary operators have norm 1.

The fact that the translation groups $\{U_i(q)\}_{q\in\mathbb{R}}$, $\{V_i(p)\}_{p\in\mathbb{R}}$ are strongly continuous, makes it a natural choice to consider the closure $\overline{\mathcal{W}}$ of $\mathcal{W}$ with respect to the strong-operator topology. Similarly to the *CAR* , also for the *CCR* of finitely many degrees of freedom all irreducible (strongly continuous) representations are unitarily equivalent. In order to show this, one uses the so-called *Weyl-transform* of a function $f\in\mathbb{L}^1_{\mathrm{d}\boldsymbol{r}}(\mathbb{R}^{2f})$ [300, 142],

$$f\mapsto\overline{\mathcal{W}}\ni W(f) := \int \mathrm{d}\boldsymbol{r}\; f(\boldsymbol{r})\,W(-\boldsymbol{r})\;. \tag{5.80}$$

The Weyl transform is such that $W^\dagger(f) = W(f^+)$, where $f^+(\boldsymbol{r}) := f^*(-\boldsymbol{r})$ and, by using (5.78), $W(f_1)W(f_2) = W(f_1\times f_2)$ where

$$(f_1\times f_2)(\boldsymbol{r}) := \int \mathrm{d}\boldsymbol{w}\; f_1(\boldsymbol{w})f_2(\boldsymbol{r}-\boldsymbol{w})\,\mathrm{e}^{\frac{i}{2}\sigma(\boldsymbol{w},\boldsymbol{r})}\;.$$

Let $P := W(g)$ where $g(\boldsymbol{r}) := (\sqrt{2\pi})^{-f}\exp(-\frac{1}{4}\|\boldsymbol{r}\|^2)$, then,

$$P = P^\dagger\;,\qquad P\,W(\boldsymbol{r})\,P \;=\; \mathrm{e}^{-\frac{1}{4}\|\boldsymbol{r}\|^2}\,P\;,$$

whence $P$ is a projection for $P\neq 0$. Indeed, if $W(f) = 0$, choose $\psi,\phi\in\mathbb{H}$ such that $I_{\psi,\phi} := \langle\,\psi\,|\,P\,|\phi\,\rangle\neq 0$; then, for all $\boldsymbol{r}\in\mathbb{R}^{2f}$,

$$\begin{aligned}0 &= \langle\,\psi\,|\,PW^\dagger(\boldsymbol{r})W(f)W(\boldsymbol{r})P\,|\phi\,\rangle = \int \mathrm{d}\boldsymbol{w}\; f(\boldsymbol{w})\,\mathrm{e}^{i\sigma(\boldsymbol{r},\boldsymbol{w})}\langle\,\psi\,|\,PW(\boldsymbol{w})P\,|\phi\,\rangle \\ &= I_{\psi,\phi}\int \mathrm{d}\boldsymbol{w}\; f(\boldsymbol{w})\,\mathrm{e}^{-\frac{1}{4}\|\boldsymbol{w}\|^2}\,\mathrm{e}^{i\sigma(\boldsymbol{r},\boldsymbol{w})}\;.\end{aligned}$$

Since the integral is the Fourier transform of $f(\boldsymbol{w}) \exp(-\frac{1}{4}\|\boldsymbol{w}\|^2)$, it follows that $f(\boldsymbol{r}) = 0$ almost everywhere, which is not true of $g(\boldsymbol{r})$.

Let $\mathbb{K} \subseteq \mathbb{H}$ denote the subspace projected out by $P$ and consider an *ONB* $\{\phi_a\}$ in $\mathbb{K}$; since

$$\begin{aligned}\langle W(\boldsymbol{r}_1)\phi_a \,|\, W(\boldsymbol{r}_2)\phi_b \rangle &= \langle P\phi_a \,|\, W^\dagger(\boldsymbol{r}_1)W(\boldsymbol{r}_2) \,|P\phi_b \rangle \\ &= \langle \phi_a \,|\, \phi_b \rangle \, \mathrm{e}^{\frac{i}{2}\sigma(\boldsymbol{r}_2,\boldsymbol{r}_1)} \, \mathrm{e}^{-\frac{1}{4}\|\boldsymbol{r}_1-\boldsymbol{r}_2\|^2} \,, \qquad (5.81)\end{aligned}$$

the closures $\mathbb{K}_a$ of the linear spans of vectors of the form $W(\boldsymbol{r})|\,\phi_a\rangle$, $\boldsymbol{r} \in \mathbb{R}^{2f}$, are mutually orthogonal. Each vector in $\mathbb{K}_a$ is cyclic for $\overline{\mathcal{W}}$ which is then irreducibly represented on it. Further, $\mathbb{K} = \mathbb{H}$; in fact, the orthogonal complement $\mathbb{K}_\perp$ is also invariant under $\overline{\mathcal{W}}$. Restricting $\overline{\mathcal{W}}$ onto $\mathbb{K}_\perp$ yields another representation, $\overline{\mathcal{W}}_\perp$, such that the maps $f \mapsto W_\perp(f) := W(f) \lceil \mathbb{K}_\perp$ are injective. But this contradicts the fact that $W_\perp(g) = P \lceil \mathbb{K}_\perp = 0$.

Therefore, every strongly continuous representation of the *CCR* decomposes into an orthogonal sum of irreducible representations. Further, the relation

$$PW(\boldsymbol{r})|\,\phi_a\rangle = PW(\boldsymbol{r})P|\,\phi_a\rangle = \mathrm{e}^{-\frac{1}{4}\|\boldsymbol{r}\|^2}|\,\phi_a\rangle$$

extends linearly to the whole of $\mathbb{K}_a$, whence $P$ acts as a multiple of the identity on each of the invariant subspaces $\mathbb{K}_a$. Consider any two irreducible representations $\overline{\mathcal{W}}_{a,b}$ with their orthogonal projections $P_{a,b} = W_{a,b}(g)$ onto the cyclic vectors $P_{a,b}|\,\phi_{a,b}\rangle = |\,\phi_{a,b}\rangle$ and their representation Hilbert spaces $\mathbb{K}_{a,b}$. By linear extension, define the operator $U_{ab} : \mathbb{K}_a \mapsto \mathbb{K}_b$ such that $U_{ab}W_a(\boldsymbol{r})|\,\phi_a\rangle := W_b(\boldsymbol{r})|\,\phi_b\rangle$, for all $\boldsymbol{r} \in \mathbb{R}^{2f}$; from (5.81)

$$\begin{aligned}\langle W_a(\boldsymbol{r}_1)\phi_a \,|\, W_a(\boldsymbol{r}_2)\phi_a \rangle &= \langle W_b(\boldsymbol{r}_1)\phi_b \,|\, W_b(\boldsymbol{r}_2)\phi_b \rangle \\ &= \langle W(\boldsymbol{r}_1)\phi_a \,|\, U_{ab}^\dagger U_{ab} \,|W(\boldsymbol{r}_2)\phi_a \rangle .\end{aligned}$$

This relation extends to $\mathbb{K}_{a,b}$, whence $U_{ab}$ is an isometry such that

$$U_{ab}^\dagger \, W_a(\boldsymbol{r}) \, U_{ab} = W_b(\boldsymbol{r}) \,, \qquad \forall \boldsymbol{r} \in \mathbb{R}^{2f} \,,$$

so that the two irreducible representations $\overline{\mathcal{W}}_{a,b}$ are unitarily equivalent.

As we shall see in Chapter 7, for infinitely many degrees of freedom there are inequivalent irreducible representations of the *CCR* ; this is true also for finitely many degrees of freedom when the symplectic manifold is not $\mathbb{R}^{2f}$, rather a torus [300, 291] or when strong continuity is relaxed. This is the case for a discrete formulation of the Weyl relations that holds for discrete variable quantum systems and turns out to be extremely useful for quantizing hyperbolic dynamics of the kind studied in Example 2.1.3.

**Example 5.4.2 (Weyl Relations: Finite Dimension).**

Actions similar to translations in position and momentum can also be defined for finite level systems, that is on Hilbert space $\mathbb{H} = \mathbb{C}^N$. Let

$\{|k\rangle\}_{k=0}^{N-1}$ be an *ONB* , fix $\alpha_{u,v} \in [0,1]$ and consider the following matrices $U_N, V_N \in M_N(\mathbb{C})$

$$U_N := e^{\frac{2\pi}{N} i\alpha_u} \sum_{k=0}^{N-1} e^{\frac{2\pi}{N} ik} |k\rangle\langle k| \,, \quad V_N := e^{\frac{2\pi}{N} i\alpha_v} \sum_{k=0}^{N-1} |k\rangle\langle k-1| \,,$$

together with the identification $|j\rangle = |j \bmod N\rangle$. These operators are unitary and

$$U_N|\ell\rangle = e^{\frac{2\pi}{N} i(\alpha_u+\ell)}|\ell\rangle \,, \quad V_N|\ell\rangle = e^{\frac{2\pi}{N} i\alpha_v}|\ell+1\rangle \,. \tag{5.82}$$

Thus, setting $\boldsymbol{n} := (n_1, n_2) \in \mathbb{Z}^2$, $U_N$ and $V_N$ satisfy the *discrete Weyl relations*

$$U_N^{n_1} V_N^{n_2} = e^{\frac{2\pi}{N} i n_1 n_2} V_N^{n_1} U_N^{n_2} \,. \tag{5.83}$$

Further, like in the continuous case, it is convenient to introduce the *discrete Weyl operators*

$$W_N(\boldsymbol{n}) := e^{-i\frac{\pi}{N} n_1 n_2} U_N^{n_1} V_N^{n_2} \,. \tag{5.84}$$

They satisfy $W_N^\dagger(\boldsymbol{n}) = W(-\boldsymbol{n})$ and the composition law

$$W_N(\boldsymbol{n})W_N(\boldsymbol{m}) = e^{i\frac{\pi}{N}\sigma(\boldsymbol{n},\boldsymbol{m})} W_N(\boldsymbol{n}+\boldsymbol{m}) \,, \tag{5.85}$$

with $\sigma(\boldsymbol{n},\boldsymbol{m}) := n_1 m_2 - n_2 m_1$. Since $\|\,[U_N\,,\,V_N]\,\| = 2|\sin\frac{\pi}{N}|$, letting $N \to \infty$ one expects to recover the commutative structure of Example 2.1.3.

In order to be compatible with a finite dimensional Hilbert space $\mathbb{C}^N$, powers as $U_N^N$ and $V_N^N$ must be proportional to the $N \times N$ identity matrix $\mathbb{1}_N$; in particular,

$$U_N^N = e^{2\pi i\alpha_u}\,\mathbb{1}_N \,, \quad V_N^N = e^{2\pi i\alpha_v}\,\mathbb{1}_N \,. \tag{5.86}$$

Different choices of $\alpha_{u,v}$ label different irreducible representations $\mathcal{W}_N^{\alpha_{u,v}}$; they play a role in the quantization of classical discrete maps as in Example 2.1.3 [98] (see Example 5.6.1). These representations cannot be equivalent, otherwise, for $\alpha_u \neq \alpha'_u$, there would exist an isometry $T$ such that

$$T^\dagger\, U_{N,\alpha_u}^N\, T = e^{2\pi i\alpha_u} = U_{N,\alpha'_u}^N = e^{2\pi i\alpha_u} \,,$$

where $U_{N,\alpha}$ denotes the operator $U_N$ fulfilling a specific rule (5.86).

When normalized, the discrete Weyl operators form an *ONB* in $M_N(\mathbb{C})$. Indeed, using (5.84) and (5.82), it turns out that

$$\begin{aligned}
\mathrm{Tr}\Big(W_N(\boldsymbol{n})\Big) &= \sum_{\ell=0}^{N-1} e^{-i\frac{\pi}{N} n_1 n_2} \langle \ell | U_N^{n_1} V_N^{n_2} | \ell\rangle \\
&= \sum_{\ell=0}^{N-1} e^{-i\frac{\pi}{N}(n_1 n_2 + 2n_1(\alpha_u+\ell) - 2n_2\alpha_v)} \langle \ell | \ell + n_2\rangle \\
&= \delta_{n_2 0} \sum_{\ell=0}^{N-1} e^{-\frac{2\pi i}{N}(\alpha_u+\ell)} = N\,\delta_{\boldsymbol{n},\boldsymbol{0}} \,.
\end{aligned} \tag{5.87}$$

This in turn yields

$$\mathrm{Tr}\Big(W_N^\dagger(\boldsymbol{n})W_N(\boldsymbol{m})\Big) = N\,\delta_{\boldsymbol{n},\boldsymbol{m}}\ , \tag{5.88}$$

whence (see Example 5.2.5)

$$X = \frac{1}{N}\sum_{\boldsymbol{n}\in\mathbb{Z}_N^2}\Big(\mathrm{Tr}\Big(W_N^\dagger(\boldsymbol{n})\,X\Big)\Big)\,W_N(\boldsymbol{n}) \qquad \forall X\in M_N(\mathbb{C})\ , \tag{5.89}$$

where $\mathbb{Z}_N^2 := \{\boldsymbol{n} = (n_1, n_2) : 0 \le n_i \le N-1\}$.

Returning to continuous variable systems, a particular vector state in $\mathbb{H} = \mathbb{L}^2(\mathbb{R}^f)$ is given by the Gaussian function

$$g(\boldsymbol{q}) = (\pi)^{-f/4}\exp(-\frac{\boldsymbol{q}^2}{2})\ , \qquad (\widehat{q}_i + i\widehat{p}_i)|\,g\,\rangle = 0\ , \tag{5.90}$$

for all $i = 1, 2, \ldots, f$. Notice that in momentum representation $\hat{g}(\boldsymbol{p})$, obtained from $g(\boldsymbol{q})$ by Fourier transform, has the same Gaussian form as the latter. For reasons which will become immediately clear we shall refer to the Gaussian state as to the *vacuum* and set $|\,vac\,\rangle := |\,g\,\rangle$.

**Example 5.4.3 (*CCR* : Annihilation and Creation Operators).**
Given $f$ canonical pairs $(\widehat{q}_i, \widehat{p}_i)$, using (5.68) one shows that the operators

$$a_i = \frac{\widehat{q}_i + i\widehat{p}_i}{\sqrt{2}}\ , \quad a_i^\dagger = \frac{\widehat{q}_i - i\widehat{p}_i}{\sqrt{2}}\ , \tag{5.91}$$

satisfy the *CCR* that describe *Bosonic* degrees of freedom,

$$\Big[a_i, a_j^\dagger\Big] = \delta_{ij}\ , \quad \Big[a_i\,, a_j\Big] = \Big[a_i^\dagger\,, a_j^\dagger\Big] = 0\ . \tag{5.92}$$

The Gaussian function $|\,vac\,\rangle$ plays thus the role of the vacuum for the *CCR* as it is annihilated by all $a_i$, $a_i|\,vac\,\rangle = 0$. Since

$$a_i^\dagger a_i\,(a_i^\dagger)^{n_i}|\,vac\,\rangle = a_i^\dagger[a_i,\,(a_i^\dagger)^{n_i}]|\,vac\,\rangle = n_i\,(a_i^\dagger)^{n_i}|\,vac\,\rangle\ ,$$

the vectors $|\,\boldsymbol{k}\,\rangle := |\,k_1, k_2, \ldots, k_f\,\rangle = \prod_{i=1}^f \frac{(a_i^\dagger)^{k_i}}{\sqrt{k_i!}}|\,vac\,\rangle$ are such that

$$a_i|\,\boldsymbol{k}\,\rangle = \sqrt{k_i}|\,\boldsymbol{k} - 1_i\,\rangle\ , \quad a_i^\dagger|\,\boldsymbol{k}\,\rangle = \sqrt{k_i + 1_i}|\,\boldsymbol{k} + 1_i\,\rangle\ , \tag{5.93}$$

where $\boldsymbol{k} \pm 1_i := (k_1, \ldots, k_{i-1}, k_i \pm 1, k_{i+1}, \ldots, k_f)$. They are the orthonormal eigenvectors of the *number operator*,

$$\widehat{N} := \sum_{i=1}^{f} a_i^\dagger a_i \ , \qquad \widehat{N}|\,\boldsymbol{k}\,\rangle = (\sum_{i=1}^{f} k_i)|\,\boldsymbol{k}\,\rangle \ . \tag{5.94}$$

The *occupation number states* $|\,\boldsymbol{k}\,\rangle$ span the *Fock* space for the $f$ Bosonic modes (degrees of freedom), $\mathbb{H}_F^{(f)} = |\,vac\,\rangle \oplus \bigoplus_{n=1}^{f} \mathbb{H}_n$, where $\mathbb{H}_n$ is the Hilbert space of $n$ modes. Unlike for Fermions, the number operator is unbounded and the Bosonic Fock space is infinite dimensional for such is the Hilbert space of each mode.

By introducing the $2f$-dimensional operator valued vectors

$$\boldsymbol{A} := (a_1, \ldots, a_f; a_1^\dagger, \ldots, a_f^\dagger) \ , \quad \boldsymbol{A}^\dagger := (a_1^\dagger, \ldots, a_f^\dagger; a_1, \ldots, a_f) \ , \tag{5.95}$$

the Weyl operators (5.75) can be rewritten as

$$W(\boldsymbol{r}) = \mathrm{e}^{\boldsymbol{Z}^* \cdot \boldsymbol{A}} = \prod_{j=1}^{f} \mathrm{e}^{a_j \frac{q_j + ip_j}{\sqrt{2}} - a_j^\dagger \frac{q_j - ip_j}{\sqrt{2}}} \quad \text{where} \tag{5.96}$$

$$\boldsymbol{Z} = (\boldsymbol{z}^*, -\boldsymbol{z}) \ , \quad \boldsymbol{z} = \frac{\boldsymbol{q} + i\boldsymbol{p}}{\sqrt{2}} \in \mathbb{C}^f \ , \tag{5.97}$$

while the *CCR* relations (5.78) become

$$\mathrm{e}^{\boldsymbol{Z}_1^* \cdot \boldsymbol{A}} \, \mathrm{e}^{\boldsymbol{Z}_2^* \cdot \boldsymbol{A}} = \mathrm{e}^{-\frac{1}{2}\boldsymbol{Z}_i^* \cdot (\Sigma_3 \boldsymbol{Z}_j)} \, \mathrm{e}^{(\boldsymbol{Z}_1^* + \boldsymbol{Z}_2^*) \cdot \boldsymbol{A}} \ ,$$

where $\Sigma_3 = \begin{pmatrix} \mathbb{1}_f & 0 \\ 0 & -\mathbb{1}_f \end{pmatrix}$ and (5.97) yields

$$\boldsymbol{Z}_1^* \cdot (\Sigma_3 \boldsymbol{Z}_2) = -2i\Im(\boldsymbol{z}_1^* \cdot \boldsymbol{z}_2) = -2i\,\sigma(\boldsymbol{r}_1, \boldsymbol{r}_2) \ . \tag{5.98}$$

Of particular interest are the so-called *displacement operators*

$$D(\boldsymbol{z}) = \mathrm{e}^{\boldsymbol{z} \cdot \boldsymbol{a}^\dagger \, - \, \boldsymbol{z}^* \cdot \boldsymbol{a}} = \mathrm{e}^{-\frac{|\boldsymbol{z}|^2}{2}} \, \mathrm{e}^{\boldsymbol{z} \cdot \boldsymbol{a}^\dagger} \, \mathrm{e}^{-\boldsymbol{z}^* \cdot \boldsymbol{a}} \ , \tag{5.99}$$

where $\boldsymbol{z} := \{z_i\}_{i=1}^f \in \mathbb{C}^f$ and (5.76) has been used. Their action is as follows,

$$D(\boldsymbol{z})^\dagger \, a_j \, D(\boldsymbol{z}) = \sum_{k_j=0}^{\infty} \frac{1}{k_j!} d_{z_j}^{k_j}[a_j] = a_j \, + \, z_j \ , \quad j = 1, 2, \ldots, n \ , \tag{5.100}$$

where $d_{z_j}^{k_j}$ denotes the map $d_{z_j}[\cdot] = [-z_j \, a^\dagger \, + \, z_j^* \, a_j \, , \, \cdot\,]$ applied $k_j$ times.

Given $\boldsymbol{z} \in \mathbb{C}^f$ and the corresponding displacement operator $D(\boldsymbol{z})$, using (5.96) and (5.97), one finds that it corresponds to the Weyl operator $W(\boldsymbol{r}(\boldsymbol{z}))$ with $\boldsymbol{r}(\boldsymbol{z}) = \sqrt{2}(-\Re(\boldsymbol{z}), \Im(\boldsymbol{z}))$, whence, via (5.78), one computes

$$D(\boldsymbol{z}_1) \, D(\boldsymbol{z}_2) = \mathrm{e}^{-\frac{i}{2}\Im(\boldsymbol{Z}_1^* \cdot (\Sigma_3 \boldsymbol{Z}_2))} \, D(\boldsymbol{z}_1 + \boldsymbol{z}_2) \ . \tag{5.101}$$

## 5.5 Quantum States

Hilbert space vectors as those encountered in the previous sections are the simplest possible instances of quantum states: once it is known that a quantum system is in a physical state described by $\psi \in \mathbb{H}$, then the system observables $X = X^\dagger \in \mathbb{B}(\mathbb{H})$ have mean-values, or *expectations*, $\langle X \rangle_\psi := \langle \psi | X | \psi \rangle$. With $\mathbb{B}(\mathbb{H}) \ni P_\psi := |\psi\rangle\langle\psi|$ the orthogonal projector onto $|\psi\rangle$, using the trace (5.19), one writes $\langle X \rangle_\psi = \mathrm{Tr}(P_\psi X)$.

One-dimensional projections $P_\psi$ are known as *pure states* and are the most informative about the system they describe; they are quantum counterparts to the classical evaluation functionals $\delta_x(f) = f(x)$ of section 2.2.1. Also in quantum mechanics, however, what is often practically achievable is not the specification of a precise vector state, but only that the system physical state corresponds to a projector $P_j$ occurring with a certain weight $0 \le \lambda_j \le 1$ within a statistical ensemble $J$ of projectors such that $\sum_{j\in J} \lambda_j = 1$. In such a case, the state of the system is a *mixed state*, namely a mixture of pure states; relatively to them, observables have mean-values that are linear convex combinations of pure state mean-values:

$$\langle X \rangle_\rho := \sum_{j\in J} \lambda_j \langle \psi_j | X | \psi_j \rangle = \mathrm{Tr}(\rho X) \ , \quad \rho := \sum_{j\in J} \lambda_j |\psi_j\rangle\langle\psi_j| \ . \tag{5.102}$$

As a linear convex combination (weighted sum) of projectors, $\rho$ is a positive operator of trace 1, known as *density matrix*.

**Definition 5.5.1 (Density Matrices).** *Any positive trace-class operator* $\rho \in \mathbb{B}_1(\mathbb{H})$ *with* $\mathrm{Tr}\rho = 1$ *describes a mixed state; let* $\rho = \sum_{j=1} r_j |r_j\rangle\langle r_j|$ *be its spectral representation with* $1 \ge r_j \ge 0$, $\sum_j r_j = 1$. *Then,* $\rho$ *defines a positive, linear and normalized functional on* $\mathbb{B}(\mathbb{H})$*:*

$$\mathbb{B}(\mathbb{H}) \ni X \mapsto \omega_\rho(X) := \mathrm{Tr}(\rho X) = \sum_j r_j \langle \phi_j | X | \phi_j \rangle \ . \tag{5.103}$$

*The set of all density matrices over the Hilbert space* $\mathbb{H}$ *of a quantum system* $S$ *will be denoted by* $\mathcal{S}(S)$ *or by* $\mathbb{B}_1^+(\mathbb{H})$ *and called* **state-space**.

*Its extremal points, those which cannot be decomposed into convex combinations of other states, are called pure states.*

**Remark 5.5.1.** The eigenvalue $r_j$ of $\rho$ in (5.103) represents the probability to find the system in the state $|r_j\rangle$ once it is known that it is described by the density matrix $\rho$. However, a mixture as in (5.102) can correspond to a convex combination of non-orthogonal projectors $P_j = |\psi_j\rangle\langle\psi_j|$; this fact points to two crucial aspects that mark a substantial difference with respect to classical phase-space probability distributions: 1) a same $\rho$ corresponds to different mixtures and 2) the weights $\lambda_j$ of the mixture are interpretable as probabilities if and only if

$$\lambda_k = \langle \psi_k \,|\, \rho \,|\psi_k \rangle = \sum_{j\in J} \lambda_j \,|\langle \psi_j \,|\, \psi_k \rangle|^2 \Longleftrightarrow \langle \psi_j \,|\, \psi_k \rangle = \delta_{jk} \;,$$

that is if and only if the vector states corresponding to a given physical mixture are the eigenvectors of the associated density matrix.

**Examples 5.5.1.**

1. The geometry of the state-space of two level systems can be simply visualized. Using (5.57), the density matrices $\rho \in M_2(\mathbb{C})$ read

$$\rho = \begin{pmatrix} r & s \\ s^* & 1-r \end{pmatrix} = \frac{1}{2}(\mathbb{1} + \boldsymbol{\rho}\cdot\boldsymbol{\sigma}) = \frac{1}{2}\begin{pmatrix} 1+\rho_3 & \rho_1 - i\rho_2 \\ \rho_1 + i\rho_2 & 1-\rho_3 \end{pmatrix} \;, \tag{5.104}$$

with $0 \leq r \leq 1$ and $r(1-r) \geq |s|^2$ for $\rho \geq 0$. Thus, $\boldsymbol{\rho} \in \mathbb{R}^3$ has length

$$0 \leq \|\boldsymbol{\rho}\|^2 = 1 - 4\mathrm{Det}(\rho) \leq 1 \;.$$

Thus, the density matrices of two-level systems are identified by the vector $\boldsymbol{\rho}$, the so-called *Bloch vector*, inside the 3-dimensional sphere. The pure states are uniquely associated with points on its surface, while orthogonal states are connected by a diameter.

2. By expanding the exponential operators in (5.99) as power series and acting on the vacuum state, the resulting pure state,

$$|\, \boldsymbol{z} \,\rangle := D(\boldsymbol{z})|\, vac \,\rangle = \mathrm{e}^{-\frac{|\boldsymbol{z}|^2}{2}}\, \mathrm{e}^{\boldsymbol{z}\cdot\boldsymbol{a}^\dagger}\,|\, vac \,\rangle = \mathrm{e}^{-\frac{|\boldsymbol{z}|^2}{2}} \prod_{j=1}^{n} \sum_{k_j=0}^{\infty} \frac{z_j^{k_j}}{\sqrt{k_j!}} |\, \boldsymbol{k} \,\rangle \;, \tag{5.105}$$

is a so-called *coherent state* [312], that is an eigenstate of the annihilation operators

$$a_i |\, \boldsymbol{z} \,\rangle \;=\; z_i |\, \boldsymbol{z} \,\rangle \qquad \forall i = 1, 2, \ldots, n \;. \tag{5.106}$$

It thus follows that the squared-moduli of the components of $\boldsymbol{z}$ are mean-occupation numbers: $\langle\, \boldsymbol{z} \,|\, \widehat{N} \,|\boldsymbol{z}\,\rangle = \sum_{j=1}^{f} |z_j|^2$. Coherent states cannot be orthogonal to each other for there are uncountably many of them, indeed, from (5.101), one derives

$$\langle\, \boldsymbol{z}^1 \,|\, \boldsymbol{z}^2 \,\rangle = \langle\, 0 \,|\, D^\dagger(-\boldsymbol{z}^1) D(\boldsymbol{z}^2) \,|0\,\rangle = \mathrm{e}^{i\Im((\boldsymbol{z}^1)^*\cdot\boldsymbol{z}^2)}\mathrm{e}^{-\frac{1}{2}|\boldsymbol{z}^1 - \boldsymbol{z}^2|^2} \;. \tag{5.107}$$

Nevertheless, with $z_j = r_j \exp{(i\vartheta_j)}$ and $\mathrm{d}\boldsymbol{z} \;= \prod_{j=1}^{n} r_j \mathrm{d}r_j \,\mathrm{d}\vartheta_j$ ,

$$\frac{1}{\pi^f}\int_{\mathbb{R}^f} \mathrm{d}\boldsymbol{z}\,|\,\boldsymbol{z}\,\rangle\langle\,\boldsymbol{z}\,| = \prod_{j=1}^{f}\sum_{p_j,q_j=0}^{\infty} \frac{|\,p_j\,\rangle\langle\,q_j\,|}{\sqrt{p_j!q_j!}}\int_0^{\infty} r_j \mathrm{d}r_j\,\mathrm{e}^{-r_j^2} r_j^{p_j+q_j}\,\times$$

$$\times\frac{1}{\pi}\int_0^{2\pi}\mathrm{d}\vartheta_j\,\mathrm{e}^{i\vartheta_j(q_j-p_j)} = \prod_{j=1}^{f}\sum_{p_j=0}^{\infty}|\,p_j\,\rangle\langle\,p_j\,| = \mathbb{1}\,. \qquad (5.108)$$

Namely, coherent states form an *overcomplete* set in the Fock space $\mathbb{H}_F^{(f)}$.

Let $a^\dagger$ represent the creation operator of a photon in a single mode; a coherent state $|\,z\,\rangle = \mathrm{e}^{-|z|^2/2}\sum_{n=0}^{\infty}\frac{z^n}{\sqrt{n!}}|\,n\,\rangle$ corresponds to a Poisson distribution over the mode number states, $|\langle\,n\,|\,z\,\rangle|^2 = \frac{|z|^{2n}}{n!}\mathrm{e}^{-|z|^2}$.

3. Passing from annihilation and creation operators to position and momentum ones, by inverting (5.97) the complex parameters $\boldsymbol{z}\in\mathbb{C}^f$ correspond to points $\boldsymbol{r} = (\boldsymbol{q},\boldsymbol{p}) \in \mathbb{R}^{2f}$ in phase-space, where $\boldsymbol{q} := \sqrt{2}\Re(\boldsymbol{z})$ and $\boldsymbol{p} := \sqrt{2}\Im(\boldsymbol{z})$. Coherent states are then characterized by gaussian localization both in $\boldsymbol{q}$ and $\boldsymbol{p}$; indeed, if $\sqrt{2}\boldsymbol{z}_0 = \boldsymbol{q}_0 + i\boldsymbol{p}_0$, then from the discussion preceding equation (5.101), using (5.77), (5.70) and (5.90) one gets, in position representation,

$$\langle\,\boldsymbol{q}\,|\,\boldsymbol{z}_0\,\rangle = \langle\,\boldsymbol{q}\,|\,W((-\boldsymbol{q}_0,\boldsymbol{p}_0))\,|vac\,\rangle = \mathrm{e}^{i\boldsymbol{p}_0\cdot(\boldsymbol{q}-\boldsymbol{q}_0/2)}\,\frac{\mathrm{e}^{-\|\boldsymbol{q}-\boldsymbol{q}_0\|^2/2}}{\pi^{f/4}}\,, \qquad (5.109)$$

while, in momentum representation (see (5.71)),

$$\langle\,\boldsymbol{p}\,|\,\boldsymbol{z}_0\,\rangle = \mathrm{e}^{-i\boldsymbol{q}_0\cdot(\boldsymbol{p}-\boldsymbol{p}_0/2)}\,\frac{\mathrm{e}^{-\|\boldsymbol{p}-\boldsymbol{p}_0\|^2/2}}{\pi^{f/4}}\,. \qquad (5.110)$$

The phase-space localization properties of coherent states make them useful tools for studying the quasi-classical behavior of quantum states [138]. In particular, given a density matrix $\rho$ for a continuous variable system with $f$ degrees of freedom, one can compare its statistical properties with those of the function $R_\rho(\boldsymbol{q},\boldsymbol{p}) := \langle\,\boldsymbol{z}\,|\,\rho\,|\boldsymbol{z}\,\rangle$ [300] which is positive, since $\rho\geq 0$, and normalized because of (5.108), thence a well-defined phase-space probability density. Viceversa, given a phase-space density $R(\boldsymbol{q},\boldsymbol{p})$ one can naturally associate to it a density matrix, $\rho_R$, which is diagonal with respect to the overcomplete set of coherent states:

$$\rho_R = \int \mathrm{d}\boldsymbol{z}\;R(\boldsymbol{z})\,|\,\boldsymbol{z}\,\rangle\langle\,\boldsymbol{z}\,| = \int_{\mathbb{R}^{2f}}\frac{\mathrm{d}\boldsymbol{x}\,\mathrm{d}\boldsymbol{y}}{(2\pi)^f}\,R(\boldsymbol{x},\boldsymbol{y})\,|\,\boldsymbol{x}+i\boldsymbol{y}\,\rangle\langle\,\boldsymbol{x}+i\boldsymbol{y}\,|\,. \qquad (5.111)$$

Most density matrices $\rho$ admit a so-called $P$-representation [121] as above in terms of a function $R(\boldsymbol{x},\boldsymbol{y})$ that is summable and normalized, but, in general, not positive and thus not a phase-space density.

The classical character of coherent states stands in sharp contrast with that of number eigenstates: this behavior shows up most evidently when photons interact with *beam-splitters* [123].

### Beamsplitters

A beam splitter is an optical device that is used to divide an incoming classical light beam of intensity $I$ along a spatial direction 1 into a reflected beam of intensity $R \times I$, with reflection coefficient $R$ along an orthogonal direction 2, and a transmitted beam of intensity $T \times I$, with transmission coefficient $T$ along the incoming direction 1. In absence of absorption and dissipation, $R + T = 1$.

Quantum mechanically, one associates photon modes to the two spatial directions; in an effective two-dimensional description, a generic single photon state $|\,\psi\,\rangle$ incident upon the beam splitter is a superposition of single-photon basis states $|\,1\,\rangle$, $|\,2\,\rangle$ describing photons impinging on the beam-splitter along the directions 1 and 2.

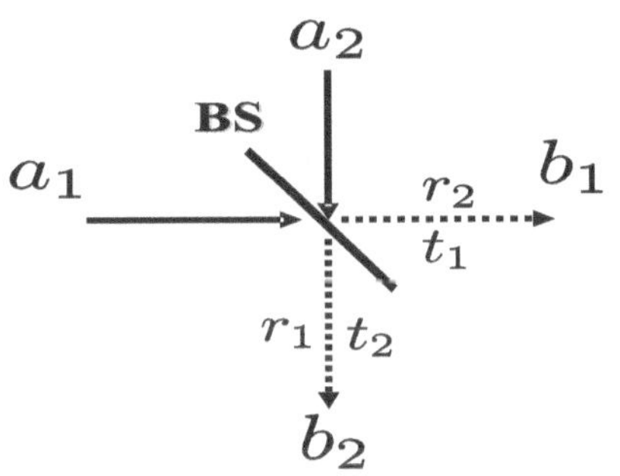


**Fig. 5.1.** Beam Splitter

In absence of dissipation, the interaction with the beam splitter produces outgoing photon states according to the rules

$$|\,1'\,\rangle := U|\,1\,\rangle = t_1|\,1\,\rangle + r_2|\,2\,\rangle\ , \qquad |\,2'\,\rangle := U|\,2\,\rangle = r_1|\,1\,\rangle + t_2|\,2\,\rangle\ ,$$

where $r_{1,2}, t_{1,2} \in \mathbb{C}$ are the reflection and transmission amplitudes along the directions $1, 2$. Therefore, a natural matrix $U = \begin{pmatrix} t_1 & r_2 \\ r_1 & t_2 \end{pmatrix}$ appears which is unitary. In fact, by using creation operators $a_i^\dagger$ for the two modes, one writes $|\,i\,\rangle = a_i^\dagger|\,vac\rangle$, where the vacuum $|\,vac\rangle$ is the state with no photons. Setting $|\,1'\,\rangle := b_1^\dagger|\,vac\rangle$ and $|\,2'\,\rangle = b_2^\dagger|\,vac\rangle$, yields $b_1^\dagger = t_1 a_1^\dagger + r_2 a_2^\dagger$, $b_2^\dagger = r_1 a_1^\dagger + t_2 a_2^\dagger$ and the Hermitian conjugate linear relations. As the $b_i^\#$ create and annihilate new photon states, they must comply with the *CCR* (5.92), whence

$$[b_1\,,\,b_1^\dagger]=|t_1|^2+|r_2|^2=1=[b_1\,,\,b_1^\dagger]=|t_1|^2+|r_2|^2\;,\;[b_1\,,\,b_2^\dagger]=t_1^*r_1+r_2^*t_2=0\,.$$

The whole physical process can thus be characterized by the matrix

$$U:=\begin{pmatrix} t_1 & r_2\\ r_1 & t_2\end{pmatrix}=\begin{pmatrix} te^{i\psi_1} & re^{i\phi_2}\\ re^{i\phi_1} & te^{i\psi_2}\end{pmatrix}\,,$$

where $r^2+t^2=1$ and $(\phi_1+\phi_2)-(\psi_1+\psi_2)=\pi$. For sake of simplicity, we shall set $t=r=1/\sqrt{2}$, $\psi_1=\psi_2=0$ and $\phi_1=\phi$, so that $\phi_2=\pi-\phi$; in this case, the matrix $U$ reads

$$U=\begin{pmatrix} t_1 & r_2\\ r_1 & t_2\end{pmatrix}=\frac{1}{\sqrt{2}}\begin{pmatrix} 1 & e^{i\phi}\\ -e^{-i\phi} & 1\end{pmatrix}\,. \qquad (5.112)$$

It describes a so-called $50:50$ beam-splitter that rotates by $\phi$ and $\pi-\phi$ the reflected and transmitted beams. The transformation $a_{1,2}\mapsto b_{1,2}$ can be unitarily implemented, namely we can explicitly construct the operator $\widehat{U}$ that sends $|\,1\,\rangle$, $|\,2\,\rangle$ into $|\,1'\,\rangle$, $|\,2'\,\rangle$. Since the beam-splitter does nothing to the vacuum, namely $\widehat{U}|\,vac\,\rangle=\widehat{U}^\dagger|\,vac\,\rangle=|\,vac\,\rangle$, the action of $\widehat{U}$ must be such that $b_{1,2}=\widehat{U}\,a_{1,2}\,\widehat{U}^\dagger$. Let us consider the operator

$$\widehat{U}(z):=e^{z\,a_1a_2^\dagger-z^*\,a_1^\dagger a_2}\;,\quad z=|z|e^{i\alpha}\;;$$

it is unitary and its action can be computed as for the displacement operators in (5.100). Namely

$$\widehat{U}(z)\,a_i^\dagger\,\widehat{U}(z)^\dagger=\sum_{k=0}^{\infty}\frac{1}{k!}d_z^k[a_j^\dagger]\;,\quad d_z[\cdot]=[z\,a_1a_2^\dagger-z^*\,a_1^\dagger a_2\,,\,\cdot\,]\,.$$

Since $d_z[a_1^\dagger]=z\,a_2^\dagger$ and $d_z[a_2^\dagger]=-z^*\,a_1^\dagger$, the infinite sums can be explicitly computed, the result being

$$\widehat{U}(z)\,a_1^\dagger\,\widehat{U}(z)^\dagger=(\cos|z|)\,a_1^\dagger\;-\;e^{i\alpha}\,(\sin|z|)\,a_2^\dagger$$
$$\widehat{U}(z)\,a_2^\dagger\,\widehat{U}(z)^\dagger=e^{-i\alpha}\,(\sin|z|)\,a_1^\dagger\;+\;(\cos|z|)\,a_2^\dagger\,.$$

Therefore, the matrix (5.112) corresponds to $|z|=\pi/4$ and $\alpha=\pi+\phi$; set $\widehat{U}:=\widehat{U}(-\pi/4\exp{(i\phi)})$. In terms of $\widehat{U}$ it is now easy to check that an incoming photon along direction 1 emerges in a superposition of states,

$$|\,1'\,\rangle=b_1^\dagger|\,vac\,\rangle=\widehat{U}^\dagger a_1^\dagger\widehat{U}|\,vac\,\rangle=\frac{a_1^\dagger+e^{i\phi}a_2^\dagger}{\sqrt{2}}|\,vac\,\rangle=\frac{|\,1\,\rangle+e^{i\phi}|\,2\,\rangle}{\sqrt{2}}\,,$$

Instead, for a coherent state $|\,\alpha\,\rangle=D(\alpha)|\,vac\,\rangle$, $\alpha\in\mathbb{C}$, one has

$$\widehat{U}|\,\alpha\,\rangle=\widehat{U}\,D(\alpha)\,\widehat{U}^\dagger|\,vac\,\rangle=e^{\alpha\,\widehat{U}\,a_1^\dagger\,\widehat{U}^\dagger-\alpha^*\,\widehat{U}\,a_1\,\widehat{U}^\dagger}|\,vac\,\rangle$$
$$=e^{\frac{\alpha}{\sqrt{2}}\,a_1^\dagger-\frac{\alpha^*}{\sqrt{2}}a_1}\,e^{\frac{\alpha}{\sqrt{2}}e^{i\phi}a_2^\dagger-\frac{\alpha^*}{\sqrt{2}}e^{-i\phi}a_2}|\,vac\,\rangle=|\,\frac{\alpha}{\sqrt{2}}\,\rangle_1\otimes|\,e^{i\phi}\frac{\alpha}{\sqrt{2}}\,\rangle_2\,.$$

This means that an incoming coherent state gets split into a transmitted coherent state $|\,\frac{\alpha}{\sqrt{2}}\,\rangle_1$ of intensity $|\alpha|^2/2$ and a reflected/phase-shifted coherent state $|\,e^{i\phi}\frac{\alpha}{\sqrt{2}}\,\rangle_2$ of intensity $|\alpha|^2/2$, exactly as with classical light.

On the other hand, a purely quantum effect results from a beam splitter acting on a state $|\,1_1 1_2\,\rangle$ consisting of two photons coming from the orthogonal directions 1, 2. It is transformed into $|\,1'_1 1'_2\,\rangle = b_1^\dagger b_2^\dagger |\,vac\,\rangle$; explicitly,

$$|\,1'_1 1'_2\,\rangle = \widehat{U}\,a_1^\dagger \widehat{U}\,\widehat{U}^\dagger\,a_2^\dagger\,\widehat{U}^{\,\dagger}|\,vac\,\rangle = \frac{1}{2}(a_1^\dagger + e^{i\phi}a_2^\dagger)(-e^{-i\phi}a_1^\dagger + a_2^\dagger)|\,vac\,\rangle$$
$$= \frac{1}{2}(-e^{i\phi}(a_1^\dagger)^2 + a_1^\dagger a_2^\dagger - a_2^\dagger a_1^\dagger + e^{i\phi}(a_2^\dagger)^2)|\,vac\,\rangle = \frac{|\,2_1\,\rangle + |\,2_2\,\rangle}{\sqrt{2}}\,.$$

The outgoing state thus consists in a superposition of states with both photons moving along a same direction; then, photons will always be found together either along direction 1, $|\,2_1\,\rangle$, or along direction 2, $|\,2_2\,\rangle$, with the same probability. On the contrary, no experiment can reveal one photon along direction 1 and the other photon along direction 2. This is because the amplitude for $|\,1_1 1_2\,\rangle$ is the sum of the amplitudes of all processes leading to this state; in the present case these are reflections along either directions 1 and 2 with amplitudes $r_1 r_2 = -1/2$ and transmissions along either directions 1 and 2 with amplitudes $t_1 t_2 = 1/2$. These processes interfere destructively.

The same kind of effect appears in single photon experiments with Mach-Zender interferometers.

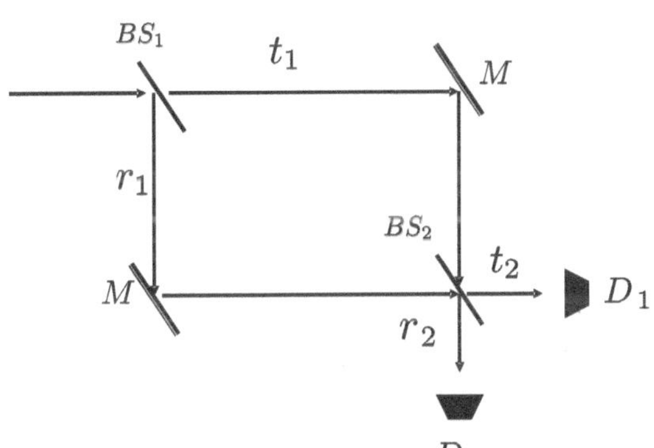


**Fig. 5.2.** Mach-Zender Interferometer

In a configuration as in Figure 5.2, an incoming photon is either reflected or transmitted at a beam-splitter $BS_1$ with amplitudes $r_1$ and $t_1$, reflected by perfectly reflecting mirrors $M$ and then either reflected or transmitted with amplitudes $r_2$ and $t_2$ at a second beam-splitter $BS_2$. The outgoing photons are then counted by detectors $D_{1,2}$. The probability $P_1$ of a photon being detected at $D_1$ is determined by the amplitude at $D_1$ which is the sum of those of the processes "reflection at $BS_1$ + transmission at $BS_2$" and "transmission at $BS_1$ + reflection at $BS_2$", that is $P_1 = |r_1 t_2 + t_1 r_2|^2$. Analogously, the processes "transmission at $BS_1$ + transmission at $BS_2$" and "reflection at $BS_1$ + reflection at $BS_2$" contribute to the detection probability at $D_2$, $P_2 = |t_1 t_2 + r_1 r_2|^2$. One can visualize the entire process by means of a binary tree with one level for each beam-splitter.

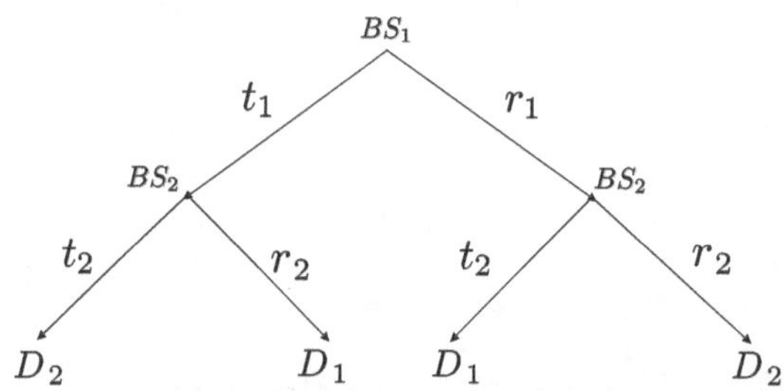


**Fig. 5.3.** Mach-Zender Interferometer: Binary Tree

If both beam-splitters act through a same operator $U$ as before, then, taking into account the impinging directions,

$$P_1 = \left|\frac{1}{2}\left(-e^{-i\phi} + e^{i\phi}\right)\right|^2 = \sin^2\phi\ , \quad P_2 = \left|\frac{1}{2}\left(1 + e^{2i\phi}\right)\right|^2 = \cos^2\phi\ .$$

An interference pattern thus emerges which depends on the phase-shift $\phi$ and can then be experimentally controlled.

### Uncertainty Relations

Beside being necessary for the consistency of the statistical interpretation of quantum mechanics, the positivity of quantum states is, together with non-commutativity, at the origin of the Heisenberg uncertainty relations.

Consider the *CCR* for $f$ degrees of freedom; with the notation of (5.75), let $\rho \in \mathbb{B}_1^+(\mathbb{H})$ be a density matrix such that all first moments $r_i := \mathrm{Tr}(\rho\,\widehat{r}_i)$ and all second moments $\mathrm{Tr}(\rho\,\widehat{r}_i\widehat{r}_j)$ are finite. Then, the $2f \times 2f$ real matrices

$$\widetilde{C}^\rho := \Big[\mathrm{Tr}\Big(\rho\,(\widehat{r}_i - r_i)(\widehat{r}_j - r_j)\Big)\Big]_{i,j=1}^{2f}\ , \quad (\widetilde{C}^\rho)^T := \Big[\mathrm{Tr}\Big(\rho\,(\widehat{r}_j - r_j)(\widehat{r}_i - r_i)\Big)\Big]_{i,j=1}^{2f}$$

are both positive. In fact,

$$\langle\, u \,|\, \widetilde{C}^\rho \,|\, u\,\rangle = \sum_{i,j=1}^{2f} u_i^* u_j\, \mathrm{Tr}\Big(\rho\,(\widehat{r}_i - r_i)(\widehat{r}_j - r_j)\Big) = \mathrm{Tr}(\rho\, X^\dagger X) \geq 0\ ,$$

for all $u \in \mathbb{C}^{2f}$, where $X := \sum_{i=1}^{2f} u_i(\widehat{r}_i - r_i)$. Using commutators $([\cdot\,,\,\cdot])$ and anti-commutators $(\{\cdot\,,\,\cdot\})$, one finds

$$(\widehat{r}_i - r_i)(\widehat{r}_j - r_j) = \frac{1}{2}\Big\{(\widehat{r}_i - r_i)\,,\,(\widehat{r}_j - r_j)\Big\} + \frac{1}{2}\Big[\widehat{r}_i\,,\,\widehat{r}_j\Big]$$
$$(\widehat{r}_j - r_j)(\widehat{r}_i - r_i) = \frac{1}{2}\Big\{(\widehat{r}_i - r_i)\,,\,(\widehat{r}_j - r_j)\Big\} - \frac{1}{2}\Big[\widehat{r}_i\,,\,\widehat{r}_j\Big]\,.$$

With $\mathbb{J}_f$ as in (5.79), the *CCR* (5.68) read $\Big[\widehat{r}_i\,,\,\widehat{r}_j\Big] = i\,(\mathbb{J}_f)_{ij}$; it thus turns out that the *correlation matrix*

$$C^\rho := [C^\rho_{ij}]_{i,j=1}^{2f} = \frac{\widetilde{C}^\rho + (\widetilde{C}^\rho)^T}{2}\ ,\quad C^\rho_{ij} := \frac{1}{2}\mathrm{Tr}\Big(\rho\Big\{(\widehat{r}_i - r_i)\,,\,(\widehat{r}_j - r_j)\Big\}\ ,$$

beside being positive, must also satisfy

$$C^\rho \pm \frac{1}{2}\Big[\mathrm{Tr}\Big(\rho\Big[\widehat{r}_i\,,\,\widehat{r}_j\Big]\Big)\Big] = C^\rho \pm i\frac{\mathbb{J}_f}{2} \geq 0\,. \tag{5.113}$$

Let $f = 1$ and choose $\rho = |\,\psi\,\rangle\langle\,\psi\,|$; then,

$$C^\rho \pm \frac{i}{2}\begin{pmatrix} 0 & 1 \\ -1 & 0 \end{pmatrix} =$$
$$= \begin{pmatrix} \langle\,\psi\,|\,(\Delta\widehat{q})^2\,|\psi\,\rangle & \langle\,\psi\,|\,\{\widehat{q}\,,\,\widehat{p}\}\,|\psi\,\rangle/2 \pm i/2 \\ \langle\,\psi\,|\,\{\widehat{q}\,,\,\widehat{p}\}\,|\psi\,\rangle/2 \mp i/2 & \langle\,\psi\,|\,(\Delta\widehat{p})^2\,|\psi\,\rangle \end{pmatrix}\,,$$

where $\Delta\widehat{q} := \widehat{q} - \langle\,\psi\,|\,\widehat{q}\,|\psi\,\rangle$ and $\Delta\widehat{p} := \widehat{p} - \langle\,\psi\,|\,\widehat{p}\,|\psi\,\rangle$. Therefore, (5.113) implies

$$\langle\,\psi\,|\,\Delta^2\widehat{q}\,|\psi\,\rangle\,\langle\,\psi\,|\,\Delta^2\widehat{p}\,|\psi\,\rangle \geq \frac{\langle\,\psi\,|\,\{\widehat{q}\,,\widehat{p}\}\,|\psi\,\rangle^2}{4} + \frac{1}{4} \geq \frac{1}{4}\,.$$

These are the uncertainty relations for conjugate position and momentum.

In terms of Bosonic annihilation and creation operators, using (5.91) and (5.95), the correlation matrix reads

$$V^\rho := U_1^\dagger\, C^\rho\, U_1 = \frac{1}{2}\Big[\mathrm{Tr}\Big(\rho\Big\{A_i - \langle A_i\rangle_\rho\,,\,A_j^\dagger - \langle A_j^\dagger\rangle_\rho\Big\}\Big)\Big]_{i,j=1}^{2f}\ , \tag{5.114}$$

where $U_1 = \frac{1}{\sqrt{2}}\begin{pmatrix} 1 & 1 \\ -i & i \end{pmatrix} \otimes \mathbb{1}_f$ and $\langle A_i^\#\rangle_\rho := \mathrm{Tr}\Big(\rho\,A_i^\#\Big)$. Notice that the matrices $V^\rho$,

$$\widetilde{V}^\rho = \frac{1}{2}\Big[\mathrm{Tr}\Big(\rho\Big(A_i - \langle A_i\rangle_\rho\Big)\Big(A_j^\dagger - \langle A_j^\dagger\rangle_\rho\Big)\Big]_{i,j=1}^{2f}$$

and the transposed

$$(\widetilde{V}^\rho)^T = \frac{1}{2}\Big[\mathrm{Tr}\Big(\rho\Big(A_j^\dagger - \langle A_j^\dagger\rangle_\rho\Big)\Big(A_i - \langle A_i\rangle_\rho\Big)\Big]_{i,j=1}^{2f}$$

are also positive. Since $\left[A_i\,,\,A_j^\dagger\right] = \delta_{ij}$ if $1 \leq i \leq f$, while $\left[A_i\,,\,A_j^\dagger\right] = -\delta_{ij}$ if $1+f \leq i \leq 2f$, and

$$\frac{1}{2}\left\{A_i\,,\,A_j^\dagger\right\} = A_i A_j^\dagger - \frac{1}{2}\left[A_i\,,\,A_j^\dagger\right] = A_j^\dagger A_i + \frac{1}{2}\left[A_i\,,\,A_j^\dagger\right]\ ,$$

it follows that

$$V^\rho \geq 0\ , \quad V^\rho \pm \frac{1}{2}\Sigma_3 \geq 0\ , \qquad \Sigma_3 := \begin{pmatrix} \mathbb{1}_f & 0 \\ 0 & -\mathbb{1}_f \end{pmatrix}\ . \tag{5.115}$$

The correlation matrix of a coherent state $\rho = |\,\boldsymbol{z}\,\rangle\langle\,\boldsymbol{z}\,|$ as in (5.105) is particularly simple; indeed, by virtue of $a_i|\,\boldsymbol{z}\,\rangle = z_i|\,\boldsymbol{z}\,\rangle$, it turns out that

$$\langle\,\boldsymbol{z}\,|\,a_i^\dagger a_j\,|\boldsymbol{z}\,\rangle = z_i^* z_j\ ,\ \langle\,\boldsymbol{z}\,|\,a_i a_j\,|\boldsymbol{z}\,\rangle = z_i z_j\ ,\ \langle\,\boldsymbol{z}\,|\,a_i a_j^\dagger\,|\boldsymbol{z}\,\rangle = \delta_{ij} + z_i z_j^*\ ,$$

whence $V^\rho = \dfrac{1}{2}\begin{pmatrix} \mathbb{1}_f & 0 \\ 0 & \mathbb{1}_f \end{pmatrix}$.

### Gaussian States

In classical probability theory, continuous probability distributions $\mu$ on $\mathbb{R}^n$ can be described in terms of their Fourier transforms or *characteristic functions* [157]

$$F_\mu(\boldsymbol{\xi}) := \int \mathrm{d}\mu(\boldsymbol{x})\ \mathrm{e}^{i\boldsymbol{\xi}\cdot\boldsymbol{x}}\ .$$

In this way, the moments of the probability distribution can be obtained by differentiating $F_\mu(\boldsymbol{\xi})$ at $\boldsymbol{\xi} = 0$. Similarly, let $\rho$ be a state of a continuous variable system with $f$ degrees of freedom equipped with a strongly continuous representation of the *CCR* in the form (5.96), then the characteristic function of $\rho$ is given by

$$F_\rho^C(\boldsymbol{r}) := \mathrm{Tr}\Big(\rho\, W(\boldsymbol{r})\Big) = \mathrm{Tr}\Big(\rho\,\mathrm{e}^{\boldsymbol{Z}^*\cdot\boldsymbol{A}}\Big) =: F_\rho^V(\boldsymbol{z})\ , \tag{5.116}$$

where (5.96) and (5.97) have been used.

**Remark 5.5.2.** The characteristic function $F_\rho^C(\boldsymbol{r})$ is the inverse of the Weyl-transform (5.80); indeed,

$$\rho = \int \frac{\mathrm{d}\boldsymbol{r}}{(2\pi)^f}\, F_\rho^C(\boldsymbol{r})\, W(-\boldsymbol{r})\ ,$$

where the convergence of the integral is understood with respect to the weak-operator topology on the representation Hilbert space. The easiest way to see this is to call $X$ the right hand side of the previous equality and calculate its matrix elements in the position representation (5.67) using (5.70):

$$\begin{aligned}
\langle \boldsymbol{q}_1 | X | \boldsymbol{q}_2 \rangle &= \int \frac{\mathrm{d}\boldsymbol{r}}{(2\pi)^f} F_\rho^C(\boldsymbol{r}) \langle \boldsymbol{q}_1 | W(-\boldsymbol{r}) | \boldsymbol{q}_2 \rangle \\
&= \int \frac{\mathrm{d}\boldsymbol{q}\, \mathrm{d}\boldsymbol{p}}{(2\pi)^f} F_\rho^C(\boldsymbol{r})\, \delta(\boldsymbol{q} - \boldsymbol{q}_1 + \boldsymbol{q}_2) \mathrm{e}^{-i\boldsymbol{p}\cdot(\frac{1}{2}\boldsymbol{q}+\boldsymbol{q}_2)} \\
&= \int \frac{\mathrm{d}\boldsymbol{p}}{(2\pi)^f} F_\rho^C(\boldsymbol{q}_1 - \boldsymbol{q}_2, \boldsymbol{p})\, \mathrm{e}^{-\frac{i}{2}\boldsymbol{p}\cdot(\boldsymbol{q}_1+\boldsymbol{q}_2)} \ .
\end{aligned}$$

By computing the trace in position representation, one gets

$$\begin{aligned}
F_\rho^C(\boldsymbol{q}_1 - \boldsymbol{q}_2, \boldsymbol{p}) &= \mathrm{Tr}\Big(\rho\, W(\boldsymbol{q}_1 - \boldsymbol{q}_2, \boldsymbol{p})\Big) \\
&= \int \mathrm{d}\boldsymbol{x} \, \langle \boldsymbol{x} | \rho | \boldsymbol{x} - \boldsymbol{q}_1 + \boldsymbol{q}_2 \rangle \, \mathrm{e}^{-i\boldsymbol{p}\cdot(\frac{\boldsymbol{q}_1-\boldsymbol{q}_2}{2} - \boldsymbol{x})} \ ,
\end{aligned}$$

whence, from the representation $\displaystyle\int \frac{\mathrm{d}\boldsymbol{p}}{(2\pi)^f} \mathrm{e}^{i\boldsymbol{p}\cdot\boldsymbol{q}} = \delta(\boldsymbol{q})$ of the Dirac delta,

$$\langle \boldsymbol{q}_1 | X | \boldsymbol{q}_2 \rangle = \int \mathrm{d}\boldsymbol{x} \int \frac{\mathrm{d}\boldsymbol{p}}{(2\pi)^f}\, \mathrm{e}^{i\boldsymbol{p}\cdot(\boldsymbol{q}-\boldsymbol{x})} \langle \boldsymbol{x} | \rho | \boldsymbol{x} - \boldsymbol{q}_1 + \boldsymbol{q}_2 \rangle = \langle \boldsymbol{q}_1 | \rho | \boldsymbol{q}_2 \rangle \ .$$

Taking derivatives of $F_\rho^C(\boldsymbol{r})$ at $\boldsymbol{r} = 0$ with respect to the real variables $q_i, p_i$, respectively of $F_\rho^V(\boldsymbol{z})$ at $\boldsymbol{z} = 0$ with respect to the complex variables $z_i$, $z_i^*$, one gets the expectation values of all products of position and momentum coordinates, respectively of annihilation and creation operators.

For instance, the first moments arise as follows,

$$\partial_{z_i} F_\rho^V(\boldsymbol{z})\Big|_{\boldsymbol{z}=0} = \mathrm{Tr}(\rho\, a_i) \ , \quad \partial_{z_i^*} F_\rho^V(\boldsymbol{z})\Big|_{\boldsymbol{z}=0} = -\mathrm{Tr}(\rho\, a_i^\dagger) \ , \tag{5.117}$$

while second moments can be extracted from

$$\partial^2_{z_i z_j} F_\rho^V(\boldsymbol{z})\Big|_{\boldsymbol{z}=0} = \mathrm{Tr}\Big(\rho\, a_j a_i\Big) = \frac{1}{2}\mathrm{Tr}\Big(\rho\Big\{A_i\,,\, A_j^\dagger\Big\}\Big) \tag{5.118}$$

for $1 \leq i \leq f$, $1 + f \leq j \leq 2f$,

$$\partial^2_{z_i^* z_j^*} F_\rho^V(\boldsymbol{z})\Big|_{\boldsymbol{z}=0} = \mathrm{Tr}\Big(\rho\, a_j^\dagger a_i^\dagger\Big) = \frac{1}{2}\mathrm{Tr}\Big(\rho\Big\{A_i\,,\, A_j^\dagger\Big\}\Big) \tag{5.119}$$

for $1 + f \leq i \leq 2f$, $1 \leq j \leq f$,

$$\partial^2_{z_i^* z_j} F_\rho^V(\boldsymbol{z})\Big|_{\boldsymbol{z}=0} = \frac{\delta_{ij}}{2} - \mathrm{Tr}\Big(\rho\, a_j a_i^\dagger\Big) = -\frac{1}{2}\mathrm{Tr}\Big(\rho\Big\{A_i\,,\, A_j^\dagger\Big\}\Big) \tag{5.120}$$

for $1 + f \leq i \leq 2f$, $1 + f \leq j \leq 2f$,

$$\partial^2_{z_i z_j^*} F_\rho^V(\boldsymbol{z})\Big|_{\boldsymbol{z}=0} = \frac{\delta_{ij}}{2} - \mathrm{Tr}\Big(\rho\, a_i a_j^\dagger\Big) = -\frac{1}{2}\mathrm{Tr}\Big(\rho\Big\{A_i\,,\, A_j^\dagger\Big\}\Big) \tag{5.121}$$

for $1 \leq i \leq f$, $1 \leq j \leq f$.

The link with the correlation matrix (5.114) is apparent; indeed, the previous moments arise from a gaussian characteristic function of the form

$$F^V_\rho(\boldsymbol{z}) = \mathrm{e}^{\boldsymbol{Z}^*\cdot\langle \boldsymbol{A}\rangle_\rho - \frac{1}{2}\boldsymbol{Z}^*\cdot(V^\rho \boldsymbol{Z})} , \tag{5.122}$$

where $\langle \boldsymbol{A}\rangle_\rho := \{\mathrm{Tr}(\rho A_i)\}_{i=1}^{2f}$ and the vectors $\boldsymbol{Z}$, $\boldsymbol{A}$ are as in (5.97) and (5.95). Notice that (5.114) implies that the sesquilinear form

$$(\boldsymbol{Z}_1, \boldsymbol{Z}_2) \mapsto \boldsymbol{Z}_1^* \cdot (V^\rho \boldsymbol{Z}_2)$$

is symmetric and positive.

Taking into account (5.91) and (5.97), one passes from the complex vector $\boldsymbol{Z} = (\boldsymbol{z}^*, -\boldsymbol{z}) \in \mathbb{C}^{2f}$ to the vector $\boldsymbol{r} = (\boldsymbol{q}, \boldsymbol{p})$ of canonical position and momentum coordinates by means of

$$\boldsymbol{Z} = U_2 \boldsymbol{r} , \quad U_2 := \frac{1}{\sqrt{2}} \begin{pmatrix} 1 & -i \\ -1 & -i \end{pmatrix} \otimes \mathbb{1}_f .$$

Since $U_2^\dagger U_1 = -i\Sigma_1 = -i \begin{pmatrix} 0 & \mathbb{1}_f \\ \mathbb{1}_f & 0 \end{pmatrix}$, where $U_1$ is the matrix in (5.114), $F^V_\rho(\boldsymbol{z})$ becomes the following Gaussian function of $\boldsymbol{r} \in \mathbb{R}^{2f}$,

$$F^C_\rho(\boldsymbol{r}) = \mathrm{e}^{i\boldsymbol{r}\cdot(\Sigma_1\langle \widehat{\boldsymbol{r}}\rangle_\rho) - \frac{1}{2}\boldsymbol{r}\cdot(\Sigma_1 C^\rho \Sigma_1 \boldsymbol{r})} , \tag{5.123}$$

where $\langle \widehat{\boldsymbol{r}}\rangle_\rho := \{\mathrm{Tr}(\rho \widehat{r}_i)\}_{i=1}^{2f}$.

As in classical probability, the Gaussian form of the characteristic function is such that higher moments are determined by first and second moments. Obviously, not all quantum states have this property, if they do have it, they are called *Gaussian states.*

**Example 5.5.2.** Coherent states $\rho = |\,\boldsymbol{u}\,\rangle\langle\,\boldsymbol{u}\,| = D(\boldsymbol{u})|\,0\,\rangle\langle\,0\,|D^\dagger(\boldsymbol{u})$, $\boldsymbol{u} \in \mathbb{C}^f$ are Gaussian; indeed, using (5.100),

$$\begin{aligned} F^V_\rho(\boldsymbol{z}) &= \langle\,0\,|\,D^\dagger(\boldsymbol{u})\mathrm{e}^{\boldsymbol{Z}^*\cdot\boldsymbol{A}} D(\boldsymbol{u})\,|0\,\rangle = \langle\,0\,|\,\mathrm{e}^{\boldsymbol{Z}^*\cdot(\boldsymbol{A}+\boldsymbol{U})}\,|0\,\rangle \\ &= \mathrm{e}^{\boldsymbol{Z}^*\cdot\boldsymbol{U}}\,\langle\,0\,|\,\mathrm{e}^{\boldsymbol{Z}^*\cdot\boldsymbol{A}}\,|0\,\rangle = \mathrm{e}^{\boldsymbol{Z}^*\cdot\boldsymbol{U} - \frac{1}{2}\|\boldsymbol{z}\|^2} , \quad \boldsymbol{U} = (\boldsymbol{u}, \boldsymbol{u}^*) \in \mathbb{C}^{2f} . \end{aligned}$$

The first moments are $\boldsymbol{u} = \langle\,\boldsymbol{u}\,|\,\boldsymbol{a}\,|\boldsymbol{u}\,\rangle$, $\boldsymbol{u}^* = \langle\,\boldsymbol{u}\,|\,\boldsymbol{a}^\dagger\,|\boldsymbol{u}\,\rangle$, the correlation matrix $V^\rho = \frac{1}{2}\begin{pmatrix} \mathbb{1}_f & 0 \\ 0 & \mathbb{1}_f \end{pmatrix}$.

From its characteristic function, one can easily determine whether a given Gaussian state $\rho$ is pure; indeed, by using Remark 5.5.2, one computes

$$\begin{aligned} \mathrm{Tr}(\rho^2) &= \int \frac{\mathrm{d}\boldsymbol{r}_1\,\mathrm{d}\boldsymbol{r}_2}{(2\pi)^{2f}}\, F^C_\rho(\boldsymbol{r}_1)\, F^C_\rho(\boldsymbol{r}_2)\, \mathrm{Tr}\Big(W(-\boldsymbol{r}_1)W(-\boldsymbol{r}_2)\Big) \\ &= \int \frac{\mathrm{d}\boldsymbol{r}}{(2\pi)^f}\, F^C_\rho(\boldsymbol{r})\, F^C_\rho(-\boldsymbol{r}) = \int \frac{\mathrm{d}\boldsymbol{r}}{(2\pi)^f}\, \mathrm{e}^{-\boldsymbol{r}\cdot(C^\rho \boldsymbol{r})} = \frac{1}{\sqrt{4^f \mathrm{Det}(C^\rho)}} . \end{aligned}$$

Therefore, $\rho$ is pure if and only if $\mathrm{Det}(C^\rho) = 4^{-f}$.

A part from their first moments which can always be set equal to 0 by a suitable shift operated by a displacement operator $D(\boldsymbol{z})$ in (5.99), Gaussian states are completely determined by their correlation matrix. An interesting question is the following one: given a Gaussian function

$$F(\boldsymbol{z}) = \mathrm{e}^{\boldsymbol{Z}^* \cdot \boldsymbol{M}} \, \mathrm{e}^{-\frac{1}{2}\boldsymbol{Z}^* \cdot (V\boldsymbol{Z})} \ , \tag{5.124}$$

with $\boldsymbol{M} \in \mathbb{C}^{2f}$ an assigned complex vector and $V$ an assigned $(2f) \times (2f)$ positive matrix such that the associated sesquilinear form is symmetric,

$$\boldsymbol{Z}_1^* \cdot (V\boldsymbol{Z}_2) = \boldsymbol{Z}_2^* \cdot (V\boldsymbol{Z}_1) \ , \tag{5.125}$$

is $F(\boldsymbol{z})$ the characteristic function of Gaussian state $\rho$ with correlation matrix $V$ and first moments given by the components of $\boldsymbol{M}$?

The answer is that it is so if and only if $V$ satisfies the conditions (5.115). While necessity descends from the uncertainty relations, sufficiency comes instead from the following general result [143].

**Proposition 5.5.1.** *A function $\mathbb{R}^{2f} \ni \boldsymbol{r} \mapsto F^C(\boldsymbol{r})$ ($\mathbb{C}^{2f} \ni \boldsymbol{z} \mapsto F^V(\boldsymbol{z})$) is the characteristic function of a quantum state $\rho$ of $f$ degrees of freedom satisfying the CCR if and only if 1) $F^C(0) = 1$ ($F^V(0) = 1$), 2) $F^C(\boldsymbol{r})$ ($F^V(\boldsymbol{z})$) is continuous at $\boldsymbol{r} = 0$ ($\boldsymbol{z} = 0$) and 3) for any $n$-tuple $\{\boldsymbol{r}_i\}_{i=1}^n$, $\boldsymbol{r}_i \in \mathbb{R}^{2f}$, ($\{\boldsymbol{z}_i\}_{i=1}^n$) the $n \times n$ matrix $\mathcal{F}^C$ ($\mathcal{F}^V$) with entries (see (5.98))*

$$\mathcal{F}_{ij}^C = \mathrm{e}^{-\frac{i}{2}\sigma(\boldsymbol{r}_i,\boldsymbol{r}_j)} \, F^C(\boldsymbol{r}_j - \boldsymbol{r}_i) \quad \left(\mathcal{F}_{ij}^V = \mathrm{e}^{\frac{1}{2}\boldsymbol{Z}_i^* \cdot (\Sigma_3 \boldsymbol{Z}_j)} \, F^C(\boldsymbol{z}_j - \boldsymbol{z}_i)\right)$$

*is positive definite.*

We postpone the proof of the proposition and instead show that if the positive $(2f) \times (2f)$ matrix $V$ in (5.124) satisfies $V \pm \frac{1}{2}\Sigma_3 \geq 0$, then there is a (Gaussian) state $\rho$ with $F(\boldsymbol{z})$ as characteristic function.

We just need to consider condition 3) and prove that

$$\sum_{i,j=1}^n u_i^* u_j \, \mathrm{e}^{(\boldsymbol{Z}_j^* - \boldsymbol{Z}_i^*) \cdot \boldsymbol{M}} \, \mathrm{e}^{-\frac{1}{2}(\boldsymbol{Z}_j^* - \boldsymbol{Z}_i^*) \cdot (V(\boldsymbol{Z}_j - \boldsymbol{Z}_i))} \, \mathrm{e}^{\frac{1}{2}\boldsymbol{Z}_i^* \cdot (\Sigma_3 \boldsymbol{Z}_j)} \geq 0 \ ,$$

for all choices of $n$ complex vectors $\boldsymbol{z}_i \in \mathbb{C}^f$ in $\boldsymbol{Z}_i = (\boldsymbol{z}_i^*, -\boldsymbol{z}_i)$. Because of (5.125), we shall thus show that

$$\sum_{i,j=1}^n w_i^* w_j \, \mathrm{e}^{\boldsymbol{Z}_i^* \cdot ((V + \frac{1}{2}\Sigma_3)\boldsymbol{Z}_j)} \geq 0 \ , \quad \text{where} \quad w_i := \mathrm{e}^{\boldsymbol{Z}_i^* \cdot \boldsymbol{M} - \frac{1}{2}\boldsymbol{Z}_i^* \cdot (V\boldsymbol{Z}_i)} \ .$$

Since $V + \frac{1}{2}\Sigma_3$ is positive, the same is true of the $n \times n$ Hermitian, positive definite matrix $A := [A_{ij}]$, with entries $A_{ij} := \boldsymbol{Z}_i^* \cdot ((V + \frac{1}{2}\Sigma_3)\boldsymbol{Z}_j)$. Then, consider the spectral decomposition of $A$ with eigenvalues $a_\ell \geq 0$, $\ell = 1, 2, \ldots, n$, so that $A_{ij} = \sum_{\ell=1}^n a_\ell \psi_{\ell i} \psi_{\ell j}^*$, where $\psi_{\ell i}$ is the $i$-th component of the $\ell$-th eigenvector of $A$; then,

$$\sum_{i,j=1}^n w_i^* w_j \,\mathrm{e}^{A_{ij}} = \sum_{k=0}^\infty \frac{1}{k!} \sum_{\ell_1,\ell_2,\ldots \ell_k=1}^n \prod_{j=1}^k a_{\ell_j} \left| \sum_{i=1}^n w_i \prod_{r=1}^k \psi_{\ell_r i}^* \right|^2 \geq 0 .$$

**Proof of Proposition 5.5.1** If $F^V(\boldsymbol{z}) = F_\rho^V(\boldsymbol{z})$ in (5.116), then condition 1) in the statement of the proposition is satisfied because $\mathrm{Tr}(\rho) = 1$, while condition 2) is fulfilled since we assumed a strongly-continuous representation of the *CCR* and $\rho$ is a trace-class operator, so that

$$\left|F^V(\boldsymbol{z}) - 1\right| \leq \sum_j r_j \left| \langle r_j \,|\, \mathrm{e}^{\boldsymbol{Z}^* \cdot \boldsymbol{A}} - \mathbb{1} \,|\, r_j \rangle \right| ,$$

where $r_j$ and $|\, r_j \rangle$ are eigenvalues and eigenvectors of $\rho$. As regards condition 3), observe that using (5.116) and (5.78), it turns out that

$$\sum_{i,j=1}^n u_i^* u_j \mathcal{F}_{ij}^V = \mathrm{Tr}\Big(\rho X^\dagger X\Big) \geq 0 ,$$

where $X := \sum_{i=1}^n u_i \exp(\boldsymbol{Z}_i^* \cdot \boldsymbol{A})$.

The sufficiency of conditions 1), 2) and 3) is shown by using them to construct a strongly continuous representation of the *CCR* by Weyl operators $W(\boldsymbol{r})$ on a Hilbert space $\mathbb{H}$ and a density matrix $\rho$ on $\mathbb{H}$ such that $F^C(\boldsymbol{r})$ is of the form (5.116) [143]. One starts by defining the operators

$$\Big(W_0(\boldsymbol{r})\psi\Big)(\boldsymbol{w}) = \mathrm{e}^{-\frac{i}{2}\sigma(\boldsymbol{r},\boldsymbol{w})}\psi(\boldsymbol{w} + \boldsymbol{r}) , \quad \boldsymbol{r} \in \mathbb{R}^{2f} , \tag{5.126}$$

on the functions on $\mathbb{R}^{2f}$; these operators satisfy the *CCR* (5.78):

$$\Big(W_0(\boldsymbol{r}_1)\, W_0(\boldsymbol{r}_2)\psi\Big)(\boldsymbol{w}) = \mathrm{e}^{-\frac{i}{2}\sigma(\boldsymbol{r}_1,\boldsymbol{r}_2)} \Big(W_0(\boldsymbol{r}_1 + \boldsymbol{r}_2)\psi\Big)(\boldsymbol{w}) .$$

Then, one considers the linear span $\mathbb{K}_0$ of all functions on $\mathbb{R}^{2f}$ of the form

$$\Psi_C(\boldsymbol{w}) = \sum_{k=1}^n c_k \exp(-\frac{i}{2}\sigma(\boldsymbol{r}_k, \boldsymbol{w})) ;$$

for all finite $n \in \mathbb{N}$, and defines on it the sesquilinear form

$$\langle \Psi_{C_1} \,|\, \Psi_{C_2} \rangle_F := \sum_{i=1}^{n_1} \sum_{j=1}^{n_2} (c_i^1)^* c_j^2 \, F^C(\boldsymbol{r}_j - \boldsymbol{r}_i)\, \mathrm{e}^{-\frac{i}{2}\sigma(\boldsymbol{r}_i,\boldsymbol{r}_j)} ,$$

where $\boldsymbol{z}_{i,j}$ are related to $\boldsymbol{r}_{i,j}$ via (5.97). Because of the positive semi-definiteness of $\mathcal{F}^C$, $\langle \Psi_C \,|\, \Psi_C \rangle_F \geq 0$; thus, in analogy to the *GNS* construction, one takes the quotient of $\mathbb{K}_0$ by the kernel consisting of those $\Psi_C$ such that $\langle \Psi_C | \Psi_C \rangle_F = 0$ and then its completion with respect to the scalar product defined on the quotient by $\langle \cdot | \cdot \rangle_F$. This gives a Hilbert space $\mathbb{K}$ containing the constant function $\mathbb{1}(\boldsymbol{r}) = 1$ on $\mathbb{R}^{2f}$ for $\langle \mathbb{1} | \mathbb{1} \rangle_F = 1$; similarly to the *GNS* vector, $| \mathbb{1} \rangle$ is cyclic for the family of operators $W_0(\boldsymbol{r})$ since

$$\Psi_C(\boldsymbol{w}) = \sum_{k=1}^{n} c_k \exp(-\frac{i}{2}\sigma(\boldsymbol{r}_k, \boldsymbol{w})) = \sum_{k=1}^{n} c_k \Big(W_0(\boldsymbol{r}_k)\, \mathbb{1}\Big)(\boldsymbol{w}) \ ,$$

and

$$\langle \mathbb{1} | W_0(\boldsymbol{r}) | \mathbb{1} \rangle_F = \langle \mathbb{1} | \mathrm{e}^{-\frac{i}{2}\sigma(\boldsymbol{r},\cdot)} \rangle_F = F(\boldsymbol{z}) \ . \tag{5.127}$$

Further, (5.126) yields

$$\Big(W_0(\boldsymbol{r})\Psi_{C_1}\Big)(\boldsymbol{w}) = \sum_{k=1}^{n} c_k \mathrm{e}^{-\frac{i}{2}\sigma(\boldsymbol{r}_k, \boldsymbol{r})}\, \mathrm{e}^{-\frac{i}{2}\sigma(\boldsymbol{r}_k + \boldsymbol{r}, \boldsymbol{w})} \ ,$$

whence

$$\begin{aligned} \langle W_0(\boldsymbol{r})\Psi_{C_1} | W_0(\boldsymbol{r})\Psi_{C_2} \rangle_F &= \sum_{i,j} (c_i^1)^* c_j^2\, \mathrm{e}^{\frac{i}{2}\sigma(\boldsymbol{r}_i^1, \boldsymbol{r})}\, \mathrm{e}^{-\frac{i}{2}\sigma(\boldsymbol{r}_j^2, \boldsymbol{r})} \\ &\quad \times F^C(\boldsymbol{r}_j^2 - \boldsymbol{r}_i^1)\, \mathrm{e}^{-\frac{i}{2}\sigma(\boldsymbol{r}_i^1 + \boldsymbol{r}, \boldsymbol{r}_j^2 + \boldsymbol{r})} \\ &= \sum_{i,j} (c_i^1)^* c_j^2\, F^C(\boldsymbol{r}_j^2 - \boldsymbol{r}_i^1)\, \mathrm{e}^{-\frac{i}{2}\sigma(\boldsymbol{r}_i^1, \boldsymbol{r}_j^2)} \\ &= \langle \Psi_{C_1} | \Psi_{C_2} \rangle_F \ . \end{aligned}$$

The operators $W_0(\boldsymbol{r})$ can thus be extended to unitary operators on $\mathbb{K}$ where they provide a representation of the *CCR* . If the latter is strongly continuous, then, from Remark 5.4.2, it reduces to an orthogonal sum of unitarily equivalent irreducible representations. Namely, there exists an irreducible representation of the *CCR* on a Hilbert space $\mathbb{H}$ by Weyl operators $W(\boldsymbol{r})$ and an isometry $U : \mathbb{K} \mapsto \widetilde{\mathbb{H}} := \bigoplus_n \mathbb{H}$ such that

$$W_0(\boldsymbol{r}) = U^\dagger\, \widetilde{W}(\boldsymbol{r})\, U \ , \quad \widetilde{W}(\boldsymbol{r}) := \bigoplus_n W(\boldsymbol{r}) \ .$$

Also, $U | \mathbb{1} \rangle = \bigoplus_n | \psi_n \rangle$ is a normalized vector in $\widetilde{\mathbb{H}}$, namely $\sum_n \|\psi_n\|^2 = 1$, so that $\rho := \sum_n \lambda_n P_n$ is a density matrix on $\mathbb{H}$, where $\lambda_n := \|\psi_n\|^2$ and $P_n := | \widehat{\psi}_n \rangle\langle \widehat{\psi}_n |$ with $\widehat{\psi}_n := \psi_n / \|\psi_n\|$. Now, from (5.127) it follows that

$$\begin{aligned} \mathrm{Tr}\Big(\rho\, W(\boldsymbol{r})\Big) &= \sum_n \langle \psi_n | W(\boldsymbol{r}) | \psi_n \rangle = \langle U\mathbb{1} | \widetilde{W}(\boldsymbol{r}) | U\mathbb{1} \rangle \\ &= \langle \mathbb{1} | W_0(\boldsymbol{r}) | \mathbb{1} \rangle = F(\boldsymbol{z}) \ , \end{aligned}$$

which concludes the proof. In order to show that the representation of the *CCR* by the Weyl operators $W_0(\boldsymbol{r})$ on $\mathbb{K}$ is strongly continuous, we shall first show that the condition 1), 2) and 3) ensure uniform continuity of $F^C(\boldsymbol{r})$ at any $\boldsymbol{r}$. Indeed, choosing vectors $\boldsymbol{r}_1 = 0$, $\boldsymbol{r}_2 = \boldsymbol{u}$ and $\boldsymbol{r}_3 = \boldsymbol{w}$, it turns out that

$$\mathcal{F}^C = \begin{pmatrix} 1 & F^C(\boldsymbol{u}) & F^C(\boldsymbol{w}) \\ F^C(-\boldsymbol{u}) & 1 & F^C(\boldsymbol{w}-\boldsymbol{u})\mathrm{e}^{-\frac{i}{2}\sigma(\boldsymbol{u},\boldsymbol{w})} \\ F^C(-\boldsymbol{w}) & F^C(\boldsymbol{u}-\boldsymbol{w})\mathrm{e}^{\frac{i}{2}\sigma(\boldsymbol{u},\boldsymbol{w})} & 1 \end{pmatrix} ;$$

its positivity implies $F^C(-\boldsymbol{u}) = F^C(\boldsymbol{u})^*$, $|F^C(\boldsymbol{u})| \leq 1$ and

$$\begin{aligned} \Big|F^C(\boldsymbol{u}) - F^C(\boldsymbol{w})\Big| &\leq 1 - \Big|F^C(\boldsymbol{u}-\boldsymbol{w})\Big|^2 \\ &\quad - 2\Re\Big\{(F^C)^*(\boldsymbol{u})F^C(\boldsymbol{w})\Big[1 - F^C(\boldsymbol{u}-\boldsymbol{w})\,\mathrm{e}^{\frac{i}{2}\sigma(\boldsymbol{u},\boldsymbol{w})}\Big]\Big\} \\ &\leq 4\Big|1 - F^C(\boldsymbol{u}-\boldsymbol{w})\,\mathrm{e}^{\frac{i}{2}\sigma(\boldsymbol{u},\boldsymbol{w})}\Big| \ . \end{aligned}$$

Then, the strong continuity of the representation is a consequence of the fact that, for all $\Psi_C \in \mathbb{K}_0$, the contribution

$$\langle \Psi_C \,|\, W_0(\boldsymbol{r})\,|\Psi_C \rangle_F = \sum_{i,j} c_i^* c_j \,\mathrm{e}^{-\frac{i}{2}(\sigma(\boldsymbol{r}_j,\boldsymbol{r})+\sigma(\boldsymbol{r}_i,\boldsymbol{r}_j+\boldsymbol{r}))}\, F^C(\boldsymbol{r}_j + \boldsymbol{r} - \boldsymbol{r}_i)$$

goes to $\langle \Psi_C \,|\, \Psi_C \rangle_F$ when $\boldsymbol{r} \to 0$ in the equality

$$\|(W_0(\boldsymbol{r}) - \mathbb{1})|\,\Psi_C\rangle\|_F^2 = 2\Big(\langle \Psi_C \,|\, \Psi_C \rangle_F - \Re\langle \Psi_C \,|\, W_0(\boldsymbol{r})\,|\Psi_C \rangle_F\Big)$$

and that this result can be extended to the whole of $\mathbb{K}$. □

**Examples 5.5.3 (Two-Mode Gaussian States).**

1. Consider two bosonic modes ($\widehat{\boldsymbol{r}} = (\widehat{q}_1, \widehat{q}_2, \widehat{p}_1, \widehat{p}_2)$) in a state $\rho$ with Gaussian characteristic function,

$$F^C_\rho(\boldsymbol{r}) = \mathrm{e}^{-\frac{1}{2}\boldsymbol{r}\cdot(\Sigma_1 C^\rho \Sigma_1 \boldsymbol{r})} \ , \ \mathbb{R}^4 \ni \boldsymbol{r} = (\boldsymbol{q}_1, \boldsymbol{q}_2, \boldsymbol{p}_1, \boldsymbol{p}_2) \ . \qquad (5.128)$$

With respect to (5.123) $\langle\widehat{\boldsymbol{r}}\rangle_\rho = 0$, a case which can always be attained by suitably translating $\rho$. This is more easily ascertained in terms of creation and annihilation operators as in (5.122); indeed, if $\rho$ is such that

$$\langle \boldsymbol{A}\rangle_\rho := \{\mathrm{Tr}(\rho\, A_i)\}_{i=1}^4 =: \boldsymbol{U} = (\boldsymbol{u}, \boldsymbol{u}^*) \neq 0 \ ,$$

consider $\widetilde{\rho} := D^\dagger(\boldsymbol{u})\,\rho\, D(\boldsymbol{u})$, with $D(\boldsymbol{u})$ as in (5.99) (see also Example 5.5.2). Using (5.100) and (5.116),

$$\begin{aligned} F^V_{\widetilde{\rho}}(\boldsymbol{z}) &= \mathrm{Tr}\Big(\rho\, D(\boldsymbol{u})\,\mathrm{e}^{\boldsymbol{Z}^*\cdot\boldsymbol{A}}\, D^\dagger(\boldsymbol{u})\Big) = \mathrm{e}^{-\boldsymbol{Z}^*\cdot\boldsymbol{U}}\, F^V_{\widetilde{\rho}}(\boldsymbol{z}) \\ &= \mathrm{e}^{-\boldsymbol{Z}^*\cdot\boldsymbol{U}}\,\mathrm{e}^{\boldsymbol{Z}^*\cdot\boldsymbol{U} - \frac{1}{2}\boldsymbol{Z}^*\cdot(V^\rho\boldsymbol{Z})} = \mathrm{e}^{-\frac{1}{2}\boldsymbol{Z}^*\cdot(V^\rho\boldsymbol{Z})} \ . \end{aligned}$$

It is convenient to rearrange $\widehat{\boldsymbol{r}}$ as $\widehat{\boldsymbol{R}} = M\widehat{\boldsymbol{r}}$, $M = M^{-1} = M^T$,

$$\widehat{\boldsymbol{R}} = \begin{pmatrix} \widehat{q}_1 \\ \widehat{p}_1 \\ \widehat{q}_2 \\ \widehat{p}_2 \end{pmatrix} = \underbrace{\begin{pmatrix} 1 & 0 & 0 & 0 \\ 0 & 0 & 1 & 0 \\ 0 & 1 & 0 & 0 \\ 0 & 0 & 0 & 1 \end{pmatrix}}_{M} \begin{pmatrix} \widehat{q}_1 \\ \widehat{q}_2 \\ \widehat{p}_1 \\ \widehat{p}_2 \end{pmatrix} . \tag{5.129}$$

As a consequence, the *CCR* (5.68) now read

$$[\widehat{R}_i\,,\,\widehat{R}_j] = i\Omega_{ij}\ , \qquad \Omega := M\,\mathbb{J}_2\,M = \begin{pmatrix} 0 & 1 & 0 & 0 \\ -1 & 0 & 0 & 0 \\ 0 & 0 & 1 & 0 \\ 0 & 0 & -1 & 0 \end{pmatrix} , \tag{5.130}$$

and the characteristic function (5.128) becomes

$$F^C_\rho(\boldsymbol{r}) =: G(\boldsymbol{R}) = \mathrm{e}^{-\frac{1}{2}\boldsymbol{R}\cdot(\widehat{\Sigma}_1\,\mathcal{V}\widehat{\Sigma}_1\boldsymbol{R})} , \tag{5.131}$$

where $\widehat{\Sigma}_1 := \begin{pmatrix} 0 & 0 & 0 & 1 \\ 0 & 0 & 1 & 0 \\ 0 & 1 & 0 & 0 \\ 1 & 0 & 0 & 0 \end{pmatrix}$ and $\mathcal{V} = \left[\frac{1}{2}\mathrm{Tr}\Big(\rho\Big\{\widehat{R}_i\,,\,\widehat{R}_j\Big\}\Big)\right]_{i,1=1}^4$. More explicitly, a same argument as the one that led to (5.113) shows that $C$ is a positive real $4\times 4$ of the form

$$\mathcal{V} = \begin{pmatrix} A & C \\ C^T & B \end{pmatrix} \tag{5.132}$$

$$0 \le A := \begin{pmatrix} \mathrm{Tr}(\rho\,\widehat{q}_1^2) & \frac{1}{2}\mathrm{Tr}(\rho\,\{\widehat{q}_1\,,\,\widehat{p}_1\}) \\ \frac{1}{2}\mathrm{Tr}(\rho\,\{\widehat{q}_1\,,\,\widehat{p}_1^2\}) & \mathrm{Tr}(\rho\,\widehat{q}_1^2) \end{pmatrix} \tag{5.133}$$

$$0 \le B := \begin{pmatrix} \mathrm{Tr}(\rho\,\widehat{q}_2^2) & \frac{1}{2}\mathrm{Tr}(\rho\,\{\widehat{q}_2\,,\,\widehat{p}_2\}) \\ \frac{1}{2}\mathrm{Tr}(\rho\,\{\widehat{q}_2\,,\,\widehat{p}_2\}) & \mathrm{Tr}(\rho\,\widehat{q}_2^2) \end{pmatrix} \tag{5.134}$$

$$C := \begin{pmatrix} \mathrm{Tr}(\rho\,\widehat{q}_1\widehat{q}_2) & \mathrm{Tr}(\rho\,\widehat{q}_1\widehat{p}_2) \\ \mathrm{Tr}(\rho\,\widehat{p}_1\widehat{q}_2) & \mathrm{Tr}(\rho\,\widehat{p}_1\widehat{p}_2) \end{pmatrix} . \tag{5.135}$$

By multiplying (5.113) on both sides by $M$, the necessary and sufficient conditions for $\mathcal{V}$ to be the correlation matrix of a gaussian state are

$$\mathcal{V} \pm \frac{i}{2}\Omega \ge 0 \,. \tag{5.136}$$

2. Every $2\times 2$ real matrix $S$ of determinant 1 such that $S\,\mathbb{J}_2\,S^T = \mathbb{J}_2$, where $\mathbb{J}_2 = \begin{pmatrix} 0 & 1 \\ -1 & 0 \end{pmatrix}$ is the symplectic matrix (2.6) for one degree of freedom, can be used to define new one-mode canonical operators, $\widehat{\boldsymbol{r}}' = S\widehat{\boldsymbol{r}}$, $\widehat{\boldsymbol{r}} = \begin{pmatrix} \widehat{q} \\ \widehat{p} \end{pmatrix}$, $\widehat{\boldsymbol{r}}' := \begin{pmatrix} \widehat{q}' \\ \widehat{p}' \end{pmatrix}$, $[\widehat{r}'_i\,,\,\widehat{R}'_j] = i(\mathbb{J}_2)_{ij}$. This fact allows for

a greatly simplification of the basic structure of the correlation matrix $\mathcal{V}$ [278, 115]. As a first step, consider a positive, real symmetric $2 \times 2$ matrix $X$ with $\alpha := \sqrt{\det(X)}$ and define $S := \sqrt{X/\alpha}$; it turns out that $X = S^T \begin{pmatrix} \alpha & 0 \\ 0 & \alpha \end{pmatrix} S$. Furthermore, since $\sqrt{X}$ is symmetric and $\mathbb{J}_2$ antisymmetric, $\sqrt{X}\,\mathbb{J}_2\,\sqrt{X}$ is antisymmetric and thus proportional to $\mathbb{J}_2$. Its determinant is $\alpha$ whence $\sqrt{X}\,\mathbb{J}_2\,\sqrt{X} = \alpha\,\mathbb{J}_2$ and $S\,\mathbb{J}_2\,S^T = \mathbb{J}_2$ [5]. As a second step, use this result and let $S_{A,B}$ effect the symplectic diagonalization of the positive, real matrices $A, B$ in $\mathcal{V}$, then

$$\mathcal{V} = \begin{pmatrix} S_A & 0 \\ 0 & S_B \end{pmatrix} \begin{pmatrix} \alpha\,\mathbb{1}_2 & \widetilde{C} \\ \widetilde{C}^T & \beta\,\mathbb{1}_2 \end{pmatrix} \begin{pmatrix} S_A^T & 0 \\ 0 & S_B^T \end{pmatrix} .$$

As a third and last step, notice that the real matrix $\widetilde{C}$ can be written as $\widetilde{C} = U \begin{pmatrix} c_1 & 0 \\ 0 & c_2 \end{pmatrix} V^T$, where $c_{1,2}$ are its singular values (see (5.16)) and $U, V$ are two orthogonal matrices. If their determinant is 1 then they also preserves $\mathbb{J}_2$, otherwise set $\widetilde{U} := U\sigma_3$, $\widetilde{V} := V\sigma_3$ and

$$\begin{pmatrix} \gamma_1 & 0 \\ 0 & \gamma_2 \end{pmatrix} := \sigma_3 \begin{pmatrix} c_1 & 0 \\ 0 & c_2 \end{pmatrix} \sigma_3 ,$$

where now $\gamma_{1,2}$ need not be both positive. Therefore, any two-mode correlation matrix can be written as [278, 103]

$$\mathcal{V} = \begin{pmatrix} U_A & 0 \\ 0 & U_B \end{pmatrix} \underbrace{\begin{pmatrix} \alpha & 0 & \gamma_1 & 0 \\ 0 & \alpha & 0 & \gamma_2 \\ \gamma_1 & 0 & \beta & 0 \\ 0 & \gamma_2 & 0 & \beta \end{pmatrix}}_{\mathcal{V}_0} \begin{pmatrix} U_A^T & 0 \\ 0 & U_B^T \end{pmatrix} , \tag{5.137}$$

by means of matrices $U_{A,B}$ such that $U_{A,B}\,\mathbb{J}_2\,U_{A,B}^T = \mathbb{J}_2$. In conclusion, by suitably changing canonical coordinates, one can always reduce $\mathcal{V}$ to the standard form $\mathcal{V}_0 = \begin{pmatrix} A_0 & C_0 \\ C_0^T & B_0 \end{pmatrix}$, where $A_0 =: \alpha\,\mathbb{1}_2$, $B_0 := \beta\,\mathbb{1}_2$ and $C_0 := \begin{pmatrix} \gamma_1 & 0 \\ 0 & \gamma_2 \end{pmatrix}$. Then, for $\mathcal{V}_0$, by imposing the positivity of the principal minors of $\mathcal{V}_0 \pm \frac{i}{2}\Omega$, the condition (5.136) amounts to

$$\frac{1}{4} + I_4 \geq I_1 + I_2 + 2\,I_3 , \tag{5.138}$$

where $I_1 = \alpha^2 = \mathrm{Det}(A_0)$, $I_2 = \beta^2 = \mathrm{Det}(B_0)$, $I_3 = \gamma_1\gamma_2 = \mathrm{Det}(C_0)$ and

[5] The above argument is the simplest formulation of a more general theorem of Williamson on the symplectic diagonalization of positive, real $2f \times 2f$ matrices [115]

$$I_4 = (\alpha\beta - \gamma_1^2)\,(\alpha\beta - \gamma_2^2) = \det(\mathcal{V}_0)\ .$$

As the determinants $I_j$ are invariant under transformations as those leading from $\mathcal{V}$ to $\mathcal{V}_0$, that is $I_1 = \mathrm{Det}(A)$, $I_2 = \mathrm{Det}(B)$, $I_3 = \mathrm{Det}(C)$ and $I_4 = \mathrm{Det}(\mathcal{V})$, inequality (5.138) is necessary and sufficient to ensure that a positive, real $4 \times 4$ matrix $\mathcal{V}$ as in (5.132)– (5.135) be the correlation matrix of a two-mode Gaussian state.

3. Consider a two-mode Gaussian state $\rho$ as in (5.111); a necessary and sufficient condition such that $R_\rho(q_1, q_2; p_1, p_2) \geq 0$ is that the correlation matrix $\mathcal{V}$ satisfy

$$\mathcal{V} \pm \frac{\mathbb{1}_4}{2} \geq 0\ .$$

Indeed, using the argument at the end of Example 5.4.3, from the *CCR* relations (5.78) and (5.107) the characteristic function of $\rho_R$ results

$$\begin{aligned} E_{\rho_R}(\boldsymbol{r}) &= \int_{\mathbb{R}^4} \frac{\mathrm{d}\boldsymbol{x}\,\mathrm{d}\boldsymbol{y}}{(2\pi)^2}\, R(\boldsymbol{x},\boldsymbol{y})\ \times \\ &\qquad \times \langle\, vac\,|\, W(\sqrt{2}(\boldsymbol{x},-\boldsymbol{y}))\, W(\boldsymbol{q},\boldsymbol{p})\, W(\sqrt{2}(-\boldsymbol{x},\boldsymbol{y}))\,|vac\,\rangle \\ &= \int_{\mathbb{R}^4} \frac{\mathrm{d}\boldsymbol{x}\,\mathrm{d}\boldsymbol{y}}{(2\pi)^2}\, R(\boldsymbol{x},\boldsymbol{y})\, \mathrm{e}^{i\sqrt{2}\,(\boldsymbol{q}\cdot\boldsymbol{y}+\boldsymbol{p}\cdot\boldsymbol{x})}\,\langle\, vac\,|\, W(\boldsymbol{q},\boldsymbol{p})\,|vac\,\rangle \\ &= \mathrm{e}^{-\frac{1}{4}(\|\boldsymbol{q}\|^2+\|\boldsymbol{p}\|^2)} \int_{\mathbb{R}^4} \frac{\mathrm{d}\boldsymbol{x}\,\mathrm{d}\boldsymbol{y}}{(2\pi)^2}\, R(\boldsymbol{x},\boldsymbol{y})\, \mathrm{e}^{i\sqrt{2}\,(\boldsymbol{q}\cdot\boldsymbol{y}+\boldsymbol{p}\cdot\boldsymbol{x})}\ . \end{aligned} \tag{5.139}$$

Because of the argument developed in the first one of the above examples, we can assume $\rho_R$ to be a Gaussian function with $\langle\hat{\boldsymbol{r}}\rangle_{\rho_R} = 0$; thus, by Fourier transform and using (5.131) the result follows from $\|\boldsymbol{R}\|^2 = \|\boldsymbol{r}\|^2 = \|\boldsymbol{q}\|^2 + \|\boldsymbol{p}\|^2$ and

$$\begin{aligned} R(\boldsymbol{u}) &= \int_{\mathbb{R}^4} \frac{\mathrm{d}\boldsymbol{R}}{\pi^2}\, \mathrm{e}^{-i\sqrt{2}\,\boldsymbol{u}\cdot(\Sigma_1 M\boldsymbol{R})}\, G(\boldsymbol{R})\, \mathrm{e}^{\frac{\|\boldsymbol{R}\|^2}{4}} \\ &= \int_{\mathbb{R}^4} \frac{\mathrm{d}\boldsymbol{R}}{\pi^2}\, \mathrm{e}^{-i\sqrt{2}\,\boldsymbol{u}\cdot(\Sigma_1 M\boldsymbol{R})}\, \mathrm{e}^{-\frac{1}{2}\boldsymbol{R}\cdot\left(\widehat{\Sigma}_1(\mathcal{V}-\frac{1}{2}\mathbb{1}_4)\widehat{\Sigma}_1\boldsymbol{R}\right)}\ , \end{aligned}$$

where $\boldsymbol{R} := (q_1, p_1, q_2, p_2) \in \mathbb{R}^4$, $M_4(\mathbb{C}) \ni \Sigma_1 = \begin{pmatrix} 0 & \mathbb{1}_2 \\ \mathbb{1}_2 & 0 \end{pmatrix}$ and $M_4(\mathbb{C}) \ni M$ is the matrix in (5.129).

**Remark 5.5.3.** Matrices as those in the second example above form the so-called symplectic group for $f = 1$ degrees of freedom [265]; the symplectic group for $f > 1$ degrees of freedom consists of real matrices $S \in M_f(\mathbb{R})$ of determinant 1 such that preserve the symplectic matrix $\mathbb{J}_f$, that is $S\,\mathbb{J}_f\,S^T = \mathbb{J}_f$. Setting as before $\widehat{\boldsymbol{r}}' := S\widehat{\boldsymbol{r}}$, the *CCR* are respected, namely $[\widehat{r}'_i\,,\,\widehat{r}'_j] = i(\mathbb{J}_f)_{ij}$. Therefore, because of the unitary equivalence of the *CCR* representations (see Remark 5.4.2), there exists a unitary operator $U(S)$ on the representation Hilbert space $\mathbb{H} = \mathbb{L}^2_{d\boldsymbol{q}}(\mathbb{R}^f)$ such that $\widehat{\boldsymbol{r}}' = U^\dagger(S)\,\widehat{\boldsymbol{r}}\,U(S)$. From (5.75), the Weyl operators transform as

$$U^\dagger(S)\,W(\boldsymbol{r})\,U(S) = \mathrm{e}^{i\,\boldsymbol{r}\cdot(\Sigma_1 S\widehat{\boldsymbol{r}})} = \mathrm{e}^{i\,(\widetilde{S}^T\boldsymbol{r})\cdot(\Sigma_1\widehat{\boldsymbol{r}})} = W(\widetilde{S}^T\boldsymbol{r})\ , \tag{5.140}$$

where, if $S = \begin{pmatrix} A & C \\ C^T & B \end{pmatrix}$ with $A, B, C \in M_f(\mathbb{R})$, then $\widetilde{S} = \begin{pmatrix} D & C \\ B & A \end{pmatrix}$ while the transposed $\widetilde{S}^T$ equals $\begin{pmatrix} D^T & C^T \\ C & A^T \end{pmatrix}$.

Let $\rho \in \mathbb{B}_1^+(\mathbb{H})$ be a density matrix for the $f$ Bosonic degrees of freedom and consider the state $\rho_S := U(S)\,\rho\,U^\dagger(S)$ obtained by operating the symplectic transformation of the canonical operators; because of (5.140), their characteristic functions (5.116) of $\rho$ and $\rho_S$ are related by $E_{\rho_S}(\boldsymbol{r}) = E_\rho(\widetilde{S}^T\boldsymbol{r})$. With $\boldsymbol{r} = (\boldsymbol{q}, \boldsymbol{p}), \boldsymbol{u} = (\boldsymbol{x}, \boldsymbol{y}) \in \mathbb{R}^{2f}$, (5.139) generalizes to

$$E_\rho(\boldsymbol{r}) = \mathrm{e}^{-\|\boldsymbol{r}\|^2/4} \int_{\mathbb{R}^{2f}} \frac{\mathrm{d}\boldsymbol{u}}{(2\pi)^f}\, R_\rho(\boldsymbol{u})\, \mathrm{e}^{\sqrt{2}\boldsymbol{u}\cdot(\Sigma_1\boldsymbol{r})}\ ,$$

where $\Sigma_1 := \begin{pmatrix} \mathbb{O}_f & \mathbb{1}_f \\ \mathbb{1}_f & \mathbb{O}_f \end{pmatrix}$, whence the corresponding functions $R_{\rho_S}(\boldsymbol{u})$ and $R_\rho(\boldsymbol{u})$ in the $P$-representation (5.111) are related by

$$\begin{aligned} &\mathrm{e}^{-\|\boldsymbol{r}\|^2/4} \int_{\mathbb{R}^{2f}} \frac{\mathrm{d}\boldsymbol{u}}{(2\pi)^f}\, R_{\rho_S}(\boldsymbol{u})\, \mathrm{e}^{i\sqrt{2}\boldsymbol{u}\cdot(\Sigma_1\boldsymbol{r})} = \\ &= \mathrm{e}^{-\|\widetilde{S}^T\boldsymbol{r}\|^2/4} \int_{\mathbb{R}^{2f}} \frac{\mathrm{d}\boldsymbol{u}}{(2\pi)^f}\, R_\rho(S^{-1}\boldsymbol{u})\, \mathrm{e}^{i\sqrt{2}\boldsymbol{u}\cdot(\Sigma_1\boldsymbol{r})}\ . \end{aligned} \tag{5.141}$$

### 5.5.1 States in the Algebraic Approach

As we have seen in Example 5.2.4 and Remark 5.2.3, expectations as in Definition 5.5.1 provide semi-norms that equip $\mathbb{B}(\mathbb{H})$ with a $w^*$ topology which is equivalent to the $\sigma$-weak topology. On the other hand, in Section 5.3.1, states have been defined as positive, normalized linear functionals on $C^*$ algebras which are continuous with respect to the uniform topology of $\mathbb{B}(\mathbb{H})$. Since this topology is finer and thus has more open subsets than the $\sigma$-weak one, in general expectations need not be also $\sigma$-weak continuous. The following result [64] characterizes the space of density matrices within the more general space of states on $\mathbb{B}(\mathbb{H})$.

**Proposition 5.5.2.** *If $\mathcal{M} \subseteq \mathbb{B}(\mathbb{H})$ is a von Neumann algebra with identity, all $\sigma$-weakly continuous expectations on $\mathcal{M}$ have the form $\mathcal{M} \ni X \mapsto \mathrm{Tr}(\rho\,X)$, with $\rho \in \mathbb{B}_1(\mathbb{H})$ a density matrix.*

**Proof:** From Example 5.2.4, any $\sigma$-weak functional $F : \mathcal{M} \mapsto \mathbb{C}$ takes the form $F(X) = \sum_n \langle \psi_n \,|\, X \,|\phi_n \rangle$ where $\{\psi_n\}$ and $\{\phi_n\}$ are sequences of vectors in $\mathbb{H}$ such that $\sum_n \|\psi\|^2 \leq \infty$ and $\sum_n \|\phi\|^2 \leq \infty$.

Set $|\,\widetilde{\psi}\,\rangle = \bigoplus_n |\,\psi_n\,\rangle$, $|\,\widetilde{\phi}\,\rangle = \bigoplus_n |\,\phi_n\,\rangle$ and consider the following representation of $\mathbb{B}(\mathbb{H})$ on their Hilbert space $\widetilde{\mathbb{H}}$, $\pi(X)|\,\widetilde{\psi}\,\rangle = \bigoplus_n (X|\,\psi_n\,\rangle)$, $X \in \mathbb{B}(\mathbb{H})$. Then, $F(X) = \langle\,\widetilde{\psi}\,|\,\pi(X)\,|\widetilde{\phi}\,\rangle$. If $F$ is positive and $\mathcal{M} \ni X \geq 0$, then

$$\begin{aligned} F(X) &= \frac{1}{4}\Big(\langle\,\widetilde{\psi}+\widetilde{\phi}\,|\,\pi(X)\,|\widetilde{\psi}+\widetilde{\phi}\,\rangle - \langle\,\widetilde{\psi}-\widetilde{\phi}\,|\,\pi(X)\,|\widetilde{\psi}-\widetilde{\phi}\,\rangle\Big) \\ &\leq \frac{1}{4}\langle\,\widetilde{\psi}+\widetilde{\phi}\,|\,\pi(X)\,|\widetilde{\psi}+\widetilde{\phi}\,\rangle\ . \end{aligned}$$

By considering the *GNS* construction based on the vector state $|\,\widetilde{\psi}+\widetilde{\phi}\,\rangle$ (once normalized), Remark 5.3.2.3 implies the existence of $0 \leq T' = (S')^\dagger S' \leq \mathbb{1}/4$ in the commutant of $\pi(\mathcal{M})$. Since $S'$ maps $\widetilde{\mathbb{H}}$ into itself,

$$\begin{aligned} F(X) &= \langle\,\widetilde{\psi}+\widetilde{\phi}\,|\,T'\pi(X)\,|\widetilde{\psi}+\widetilde{\phi}\,\rangle = \langle\,S'(\widetilde{\psi}+\widetilde{\phi})\,|\,\pi(X)\,|S'(\widetilde{\psi}+\widetilde{\phi})\,\rangle \\ &= \sum_n \langle\,\chi_n\,|\,X\,|\chi_n\,\rangle\ , \qquad \forall X \in \mathcal{M}\ . \end{aligned}$$

Set $\rho := \sum_n |\,\chi_n\,\rangle\langle\,\chi_n\,|$; this operator is positive. If $F(\mathbb{1}) = 1$, then $\mathrm{Tr}(\rho) = 1$, whence $\rho \in \mathbb{B}_1^+(\mathbb{H})$ and $F(X) = \mathrm{Tr}(\rho X)$ for all $X \in \mathcal{M}$. □

**Remark 5.5.4.** [64] As functionals, density matrices are *normal* as their $\sigma$-weak continuity is equivalent to the property of normal linear maps outlined in Remark 5.2.7.

Because of the convexity of the space of states (see Remark 5.3.2.5), one has that

**Proposition 5.5.1.** *The space of states $\mathcal{S}(S)$ of a quantum system $S$ is convex and a same density matrix can in general be decomposed into infinitely many different convex combinations of other density matrices, unless it is a pure state which is thus extremal in $\mathcal{S}(S)$.*

**Proof:** Take any set of $0 \leq X_j \in \mathbb{B}(\mathbb{H})$, $j \in J$, such that $\sum_{j\in J} X_j = \mathbb{1}$; as $\rho \geq 0$, in terms of its spectral decomposition $\rho = \sum_k r_k|\,\psi_k\,\rangle\langle\,\psi_k\,|$, its unique (positive) square-root is given by $\sqrt{\rho} = \sum_k \sqrt{r_k}|\,\phi_k\,\rangle\langle\,\phi_k\,|$. Then,

$$\rho = \sum_{j\in J} \lambda_j\,\rho_j\ , \quad \rho_j := \frac{\sqrt{\rho}\,X_j\sqrt{\rho}}{\lambda_j}\ , \quad \lambda_j = \mathrm{Tr}(\rho\,X_j)\ . \tag{5.142}$$

Thus the same density matrix $\rho$ describes mixtures whose components are described by density matrices $\rho_j$; in turn, these can also be decomposed unless they are projectors, as, in this case, $\rho = |\,\psi\,\rangle\langle\,\psi\,|$ implies

$$\sqrt{\rho}X_j\sqrt{\rho} = \langle\,\psi\,|\,X_j\,|\psi\,\rangle\,|\,\psi\,\rangle\langle\,\psi\,|$$

so that $\rho_j = \rho$ for all choices of $X_j \geq 0$. □

**Example 5.5.4.** [156] Consider two generic decompositions of a same density matrix $\rho \in \mathcal{S}(S)$ into non-orthogonal one-dimensional projectors,

$$\rho = \sum_{p=1}^{P} \underbrace{\sqrt{\rho}\,|\,\psi_p\,\rangle\langle\,\psi_p\,|\,\sqrt{\rho}}_{|\,w_p\,\rangle\langle\,w_p\,|}\,, \quad \rho = \sum_{q=1}^{Q} \underbrace{\sqrt{\rho}\,|\,\phi_q\,\rangle\langle\,\phi_q\,|\,\sqrt{\rho}}_{|\,z_q\,\rangle\langle\,z_q\,|}\,,$$

with $\sum_{p=1}^{P}|\,\psi_p\,\rangle\langle\,\psi_p\,| = 1$ and $\sum_{q=1}^{Q}|\,\phi_q\,\rangle\langle\,\phi_q\,| = 1$. By means of the spectral representation $\rho = \sum_{j=1}^{N} r_j|\,r_j\,\rangle\langle\,r_j\,|$, setting $|\,v_j\,\rangle := \sqrt{r_j}\,|\,r_j\,\rangle$, one gets

$$|\,w_p\,\rangle = \sum_{j=1}^{N} \underbrace{\langle\,r_j\,|\,\psi_p\,\rangle}_{(W^\dagger)_{jp}}\,|\,v_j\,\rangle\,, \qquad |\,z_q\,\rangle = \sum_{j=1}^{N} \underbrace{\langle\,r_j\,|\,\phi_q\,\rangle}_{(Z^\dagger)_{jq}}\,|\,v_j\,\rangle\,.$$

The $P\times N$ matrix $W : \mathbb{C}^N \mapsto \mathbb{C}^P$ with entries $W_{pj} := \langle\,\psi_p\,|\,r_j\,\rangle$ and the $Q\times N$ matrix $Z : \mathbb{C}^N \mapsto \mathbb{C}^Q$ with entries $Z_{qj} := \langle\,\phi_q\,|\,r_j\,\rangle$ are such that $W^\dagger W = 1_N$ and $Z^\dagger Z = 1_N$. Then $|\,v_\ell\,\rangle = \sum_{p=1}^{P} W_{p\ell}|\,w_p\,\rangle$ and $|\,z_q\,\rangle = \sum_{p=1}^{P} V_{pq}|\,w_p\,\rangle$, where $V = WZ^\dagger : \mathbb{C}^Q \mapsto \mathbb{C}^P$. Thus, any two decompositions of $\rho$ into projections are related by a $P \times Q$ matrix $V$ such that $VV^\dagger = WW^\dagger$ and $V^\dagger V = ZZ^\dagger$; therefore, if $P = Q = \mathrm{rank}(\rho)$, then $W$, $Z$ and hence $V$ are unitary matrices on the support of $\rho$. Also, if $P = \mathrm{rank}(\rho)$, but $Q$ is arbitrary, then $V$ is an isometry such that $V^\dagger V = 1_N$.

In Section 5.3.1, states on $C^*$ algebras have been used to construct Hilbert space representations; in the present setting, a representation on a concrete Hilbert space $\mathbb{H}$ is a priori given, it is nevertheless instructive to consider pure and mixed states of a finite dimensional quantum system $S$. In such cases, the *GNS* representation amounts to what in quantum information is known as *mixed state purification.*

Let $\rho \in M_N(\mathbb{C})$ be a density matrix and consider its spectral representation $\rho = \sum_{j=1}^{N} r_j|\,r_j\,\rangle\langle\,r_j\,|$, some eigenvalues possibly being equal to zero. To $\rho$ one associates the state vector $|\,\sqrt{\rho}\rangle \in \mathbb{C}^N \otimes \mathbb{C}^N$ given by

$$|\,\sqrt{\rho}\rangle := \sum_{j=1}^{N} \sqrt{r_j}\,|\,r_j\,\rangle \otimes |\,r_j\,\rangle\,. \tag{5.143}$$

Given $X \in M_N(\mathbb{C})$, let it be represented by $\pi(X) = X \otimes 1\!\!1_N$ on $\mathbb{C}^N \otimes \mathbb{C}^N$,

$$\begin{aligned} X \otimes 1\!\!1_N|\,\sqrt{\rho}\rangle &= \sum_{j=1}^{N} \sqrt{r_j}\,(X|\,r_j\,\rangle) \otimes |\,r_j\,\rangle = \sum_{j,k=1}^{N} \sqrt{r_j}\,\langle\,r_k\,|\,X\,|r_j\,\rangle\,|\,r_k\,\rangle \otimes |\,r_j\,\rangle \\ &= |\,X\sqrt{\rho}\rangle\,, \end{aligned} \tag{5.144}$$

whence $\langle\,\sqrt{\rho}\,|\,X \otimes 1\!\!1_N\,|\sqrt{\rho}\rangle = \langle\,\sqrt{\rho}\,|\,X\sqrt{\rho}\rangle = \mathrm{Tr}(\rho X)$.

### Pure States

If $\rho = |\psi\rangle\langle\psi|$, $\psi \in \mathbb{C}^N$, then $|\sqrt{\rho}\rangle := |\psi\rangle \otimes |\psi\rangle \in \mathbb{C}^N \otimes \mathbb{C}^N$ and $\pi(X)|\sqrt{\rho}\rangle = X|\psi\rangle \otimes |\psi\rangle$, for all $X \in M_N(\mathbb{C})$, whence the *GNS* Hilbert space is (isomorphic to) $\mathbb{C}^N$. Indeed, by taking the quotient of $M_N(\mathbb{C})$ with respect to the set

$$\mathcal{I} := \left\{X \in M_N(\mathbb{C}) \,:\, \langle\psi|X^\dagger X|\psi\rangle = 0\right\} ,$$

the equivalence classes $|\Psi_X^\rho\rangle$ are identified with operators of the form $X|\psi\rangle\langle\psi|$. By varying $X$, the *GNS* Hilbert space $\mathbb{H}_\rho$ is generated by vectors $|\phi\rangle\langle\psi|$ for all $\phi \in \mathbb{C}^N$ and is thus isomorphic to $\mathbb{C}^N$. It follows that the *GNS* representation $\pi_\rho(M_N(\mathbb{C}))$ is unitarily equivalent to $M_N(\mathbb{C})$ and thus irreducible in agreement with Remark 5.3.2.

### Faithful Density Matrices

At the opposite end with respect to pure states, let us consider a density matrix $\rho \in M_N(\mathbb{C})$ with eigenvalues $r_j$ all different from zero, so that $\mathrm{Tr}(\rho X^\dagger X) = 0 \Longleftrightarrow X = 0$. According to Definition 5.3.5, $\rho$ is faithful.

Matrices $X \in M_N(\mathbb{C})$ becomes $N^2$-dimensional vectors whose components are their matrix elements with respect to the *ONB* consisting of the eigenvectors of $\rho$, $|X\rangle = \sum_{i,j=1}^N \langle r_i|X|r_j\rangle\,|r_i\rangle \otimes |r_j\rangle$. Also, by varying $X \in M_N(\mathbb{C})$, the linear span of vectors of the form $|X\sqrt{\rho}\rangle$ is dense in $\mathbb{C}^N \otimes \mathbb{C}^N$. Let $\psi = \sum_{i,j=1}^N \psi_{ij}|r_i\rangle \otimes |r_j\rangle \in \mathbb{C}^N \otimes \mathbb{C}^N$ and $X = |r_p\rangle\langle r_q|$, then $\langle\psi|X \otimes \mathbb{1}_N|\sqrt{\rho}\rangle = \psi^*_{pq}\sqrt{r_q}$. Therefore, if $\psi$ is orthogonal to the linear span of $|X\sqrt{\rho}\rangle$ then, $\psi^*_{pq} = 0$ for all $p, q$ as $\rho$ is faithful.

Because of Remark 5.3.2.1, the triplet $(\mathbb{C}^N \otimes \mathbb{C}^N, \pi, |\sqrt{\rho}\rangle)$ is unitarily equivalent to the *GNS* triplet $(\mathbb{H}_\rho, \pi_\rho, \Omega_\rho)$ corresponding to the expectation functional $\omega_\rho : M_N(\mathbb{C}) \ni X \mapsto \omega_\rho(X) := \mathrm{Tr}(\rho X)$.

The matrix algebra $M_N(\mathbb{C})$ is represented by $M_N(\mathbb{C}) \otimes \mathbb{1}_N$ on $\mathbb{H}_\rho$; so, its commutant is $\pi_\rho(M_N(\mathbb{C}))' = \mathbb{1}_N \otimes M_N(\mathbb{C})$, $\pi_\rho(M_N(\mathbb{C}))$ has trivial center and is thus a factor. The action of the commutant is given by

$$\mathbb{1}_N \otimes X|\sqrt{\rho}\rangle = \sum_{j=1}^N \sqrt{r_j}\,|r_j\rangle \otimes (X|r_j\rangle) = |\sqrt{\rho}X^T\rangle , \tag{5.145}$$

where $X^T$ denotes the transposition of $X$ with respect to the eigenbasis of $\rho$.

We can now look at the decomposers in (5.142) from the point of view of Remark 5.3.2.3. Given a convex decomposition $\rho = \sum_{j\in J}\lambda_j\sigma_j$, every $\sigma_j$ corresponds to a unique $0 \le X'_j$ in the commutant $\pi(M_N(\mathbb{C}))'$, thence to a unique $0 \le X_j \in M_N(\mathbb{C})$, such that $X'_j = \mathbb{1}_N \otimes X^*_j$ and

$$\begin{aligned}\lambda_j\sigma_j(X) &= \langle\sqrt{\rho}|\pi(X)X'_j|\sqrt{\rho}\rangle = \langle\sqrt{\rho}\,|X \otimes X^*_j\,|\sqrt{\rho}\rangle\\ &= \langle\sqrt{\rho}|X\sqrt{\rho}X_j\rangle = \mathrm{Tr}(\sqrt{\rho}X_j\sqrt{\rho}X) ,\end{aligned} \tag{5.146}$$

whence $\lambda_j = \mathrm{Tr}(\rho X_j)$ and $\sigma_j = (\sqrt{\rho}\, X_j\, \sqrt{\rho})/\lambda_j$.

**Example 5.5.5.** Consider a two-level system equipped with the density matrix $\rho = \frac{1}{2}\begin{pmatrix} 1-s & 0 \\ 0 & 1+s \end{pmatrix}$, $0 \leq s \leq 1$; as *GNS* vector we can take its purification (5.143)

$$|\sqrt{\rho}\,\rangle = \sqrt{\frac{1-s}{2}}\,|\,0\,\rangle \otimes |\,0\,\rangle + \sqrt{\frac{1+s}{2}}\,|\,1\,\rangle \otimes |\,1\,\rangle\ ,$$

where $|\,0\,\rangle$, $|\,1\,\rangle$ are the eigenstates of $\rho$. The corresponding *GNS* representation is $\pi_\rho(M_2(\mathbb{C})) = M_2(\mathbb{C}) \otimes \mathbb{1}_2$ with *GNS* Hilbert space $\mathbb{C}^4$,

$$\langle\,\sqrt{\rho}\,|\,X \otimes \mathbb{1}\,|\sqrt{\rho}\,\rangle = \frac{1-s}{2}\langle\,0\,|\,X\,|0\,\rangle + \frac{1+s}{2}\langle\,1\,|\,X\,|1\,\rangle = \mathrm{Tr}(\rho\, X)\ ,$$

for all $X \in M_2(\mathbb{C})$. The commutant is $\pi_\rho(M_2(\mathbb{C}))' = \mathbb{1}_2 \otimes M_2(\mathbb{C})$ so that $\pi_\rho$ is reducible and a factor since $\pi_\rho(M_2(\mathbb{C}))'' = \pi_\rho(M_2(\mathbb{C}))$ whence its center (see Definition 5.3.4) is trivial, $\mathcal{Z}_\rho = \pi_\rho(M_2(\mathbb{C}))'' \cap \pi_\rho(M_2(\mathbb{C})) = \{\lambda\mathbb{1}\}$.

### Modular Theory

The *GNS* state of any faithful density matrix is separating for $\pi_\rho(M_N(\mathbb{C}))$, namely $\pi_\rho(X)|\,\sqrt{\rho}\,\rangle = 0 \iff X = 0$, and thus cyclic for the commutant $\pi_\rho(M_N(\mathbb{C}))'$ (see Lemma 5.3.1).

We shall now give the fundamentals of the so-called modular theory that looks particularly simple for finite-level systems.

Let $\rho$ be a faithful states and identify its *GNS* triplet $(\mathbb{H}_\rho, \pi_\rho, \Omega_\rho)$ with $(\mathbb{C}^N \otimes \mathbb{C}^N, M_N(\mathbb{C}) \otimes \mathbb{1}_N, |\,\sqrt{\rho}\,\rangle)$. The so-called *modular conjugation* is the antilinear map $J_\rho : \mathbb{C}^N \otimes \mathbb{C}^N \mapsto \mathbb{C}^N \otimes \mathbb{C}^N$ such that

$$|\,\psi\,\rangle := \sum_{i,j=1}^{N} \psi_{ij}\,|\,r_i\,\rangle \otimes |\,r_j\,\rangle \mapsto J_\rho|\,\psi\,\rangle = \sum_{i,j=1}^{N} \psi^*_{ij}\,|\,r_j\,\rangle \otimes |\,r_i\,\rangle\ . \tag{5.147}$$

It satisfies $J_\rho^2 = \mathbb{1}$ and $J_\rho|\,\sqrt{\rho}\,\rangle = |\,\sqrt{\rho}\,\rangle$; furthermore,

$$\begin{aligned} J_\rho|\,X\sqrt{\rho}\,\rangle &= J_\rho\,(X \otimes \mathbb{1}_N)\,J_\rho|\,\sqrt{\rho}\,\rangle = \sum_{i,k=1}^{N} \sqrt{r_i}\,(\langle\,r_k\,|\,X\,|r_i\,\rangle)^*\,|\,r_k\,\rangle \otimes |\,r_i\,\rangle \\ &= \mathbb{1}_N \otimes X^*|\,\sqrt{\rho}\,\rangle = |\,\sqrt{\rho}X^\dagger\,\rangle\ , \end{aligned} \tag{5.148}$$

where $X^*$ is the conjugate of $X$ with respect to the *ONB* $\{|\,r_j\,\rangle\}_{j=1}^N$, that is $\langle\,r_k\,|\,X^*\,|r_j\,\rangle = (\langle\,r_k\,|\,X\,|r_j\,\rangle)^*$. Given $X, Y, Z \in M_N(\mathbb{C})$, one explicitly computes

$$\begin{aligned}
\Big[J_\rho(X \otimes 1\!\!1_N)J_\rho\,,\, Y \otimes 1\!\!1_N\Big]|\, Z\sqrt{\rho}\,\rangle &= \\
&= J_\rho(X \otimes 1\!\!1_N)J_\rho|\, YZ\sqrt{\rho}\,\rangle - (Y \otimes 1\!\!1_N)J_\rho(X \otimes 1\!\!1_N)|\,\sqrt{\rho}Z^\dagger\,\rangle \\
&= J_\rho|\, X\sqrt{\rho}(YZ)^\dagger\,\rangle - (Y \otimes 1\!\!1_N)|\, Z\sqrt{\rho}X^\dagger\,\rangle \\
&= |\, YZ\sqrt{\rho}X^\dagger\,\rangle - |\, YZ\sqrt{\rho}X^\dagger\,\rangle = 0\ .
\end{aligned}$$

Thus, $J_\rho(X \otimes 1\!\!1_N)J_\rho$ belongs to the commutant $1\!\!1_N \otimes M_N(\mathbb{C}) = \pi_\rho(M_N(\mathbb{C}))'$ of $M_N(\mathbb{C}) \otimes 1\!\!1_N = \pi_\rho(M_N(\mathbb{C}))$; since the *GNS* vector $|\sqrt{\rho}\rangle$ is cyclic for $\pi_\rho(M_N(\mathbb{C}))$ it is separating for $\pi_\rho(M_N(\mathbb{C}))'$. Therefore, from (5.148),

$$J_\rho\, X \otimes 1\!\!1_N\, J_\rho = 1\!\!1_N \otimes X^*\ , \tag{5.149}$$

whence $J_\rho$ antilinearly embeds $\pi_\rho(M_N(\mathbb{C}))$ into its commutant $\pi_\rho(M_N(\mathbb{C}))'$. Actually, the embedding is an anti-isomorphism,

$$J_\rho \pi_\rho(M_N(\mathbb{C}))J_\rho = \pi_\rho(M_N(\mathbb{C}))'\ . \tag{5.150}$$

Indeed, using (5.145), for any $S' \in \pi_\rho(M_N(\mathbb{C}))'$ and $Z \in M_N(\mathbb{C})$, it holds

$$S'|\sqrt{\rho}\rangle = |\, Y_{S'}\sqrt{\rho}\rangle = |\sqrt{\rho}(\sqrt{\rho}Y^\dagger_{S'}\frac{1}{\sqrt{\rho}})^\dagger\,\rangle = J_\rho\left(\frac{1}{\sqrt{\rho}}Y^T_{S'}\sqrt{\rho}\right) \otimes 1\!\!1_N\, J_\rho|\sqrt{\rho}\rangle\ ,$$

where the first inequality is due to the fact that $S'|\sqrt{\rho}\rangle$ is a vector in $\mathbb{H}_\rho$ which can be obtained by acting on the *GNS* vector with some $\pi_\rho(Y_{S'})$. Then,

$$S' = J_\rho\left(\frac{1}{\sqrt{\rho}}Y^T_{S'}\sqrt{\rho}\right) \otimes 1\!\!1_N\, J_\rho\ .$$

A related notion is that of *modular operator*

$$\Delta_\rho := \rho \otimes \rho^{-1}\ , \tag{5.151}$$

which, according to (5.148), is such that,

$$J_\rho \Delta^{1/2}_\rho|\, X\sqrt{\rho}\rangle = J_\rho\sqrt{\rho} \otimes \frac{1}{\sqrt{\rho}}|\, X\sqrt{\rho}\,\rangle = J_\rho|\sqrt{\rho}X\,\rangle = |\, X^\dagger\sqrt{\rho}\rangle\ . \tag{5.152}$$

### 5.5.2 Density Matrices and von Neumann Entropy

From the point of view of the *GNS* construction, pure states can be distinguished from mixed states because their *GNS* representations are non-irreducible factors for the latter, while they are irreducible factors for the former. There are however handier ways to sort these states out; perhaps the easiest is to consider $\rho^2$: if more than one eigenvalue of $\rho$ is non zero, then $\rho$ is mixed since then $\rho^2 \neq \rho$. Indeed, the spectrum of a mixed state $\rho$ has a richer structure than that of any one-dimensional projection.

The possibility of comparing, in some cases, the eigenvalues of two density matrices $\rho_{1,2} \in \mathbb{B}_1^+(\mathbb{H})$ comes form the so-called *minmax principle* of which we give a short sketch (for a general formulation and proof see [252]). We shall consider the eigenvalues listed in decreasing order; namely, in the spectral decompositions $\mathbb{B}_1^+(\mathbb{H}) \ni \rho = \sum_i r_i \,|\, r_i \,\rangle\langle\, r_i \,|$, with eigenvalues repeated according to their multiplicities, we shall take $r_i \geq r_{i+1}$, for all $i$. Because of the ordering, it turns out that

$$r_i = \sup\Big\{ \langle\, \psi \,|\, \rho \,|\psi \,\rangle \,:\, \|\psi\| = 1\,,\, |\, \psi \,\rangle \perp \{|\, r_1 \,\rangle, |\, r_2 \,\rangle, \ldots, |\, r_{i-1} \,\rangle\} \Big\} \,. \tag{5.153}$$

Indeed, for a $\psi$ specified as above, $\langle\, \psi \,|\, \rho \,|\psi \,\rangle = \sum_{k \geq i} r_k |\langle\, \psi \,|\, r_k \,\rangle|^2 \leq r_i$ and $r_i$ is achieved by choosing $|\, \psi \,\rangle = |\, r_i \,\rangle$. The *minmax principle* asserts that

$$r_i = \inf_{\{\phi_j\}_{j=1}^{i-1}} \sup\Big\{ \langle\, \psi \,|\, \rho \,|\psi \,\rangle \,:\, \|\psi\| = 1\,,\, |\, \psi \,\rangle \perp \{|\, \phi_1 \,\rangle, |\, \phi_2 \,\rangle, \ldots, |\, \phi_{i-1} \,\rangle\} \Big\} \,, \tag{5.154}$$

where $\{\phi_j\}_{j=1}^{i-1}$ is any set of vectors in $\mathbb{H}$.

In order to prove this relation, we denote by $U_\rho(\{\phi_j\}_{j=1}^{i-1})$ the argument of the inf and show that $U_\rho(\{\phi_j\}_{j=1}^{i-1}) \geq r_i$, whence the result follows for $r_i$ is achieved by choosing $|\, \phi_j \,\rangle = |\, r_j \,\rangle$, $j = 1, 2, \ldots, i-1$. Now, there surely exists a normalized vector $|\, \Psi \,\rangle = \sum_{k=1}^{i} c_k |\, r_k \,\rangle \perp \{\phi_j\}_{j=1}^{i-1}$; indeed, if $P$ projects onto the linear span of $\{|\, r_j \,\rangle\}_{j=1}^{i}$, the vectors $\{P |\, \phi_\ell \,\rangle\}_{\ell=1}^{i-1}$ span at most an $(i-1)$-dimensional subspace. But then,

$$U_\rho(\{\phi_j\}_{j=1}^{i-1}) \geq \langle\, \Psi \,|\, \rho \,|\Psi \,\rangle = \sum_{k=1}^{i} r_k \,|c_k|^2 \geq r_i \sum_{k=1}^{i} |c_k|^2 = r_i \,.$$

From the minmax principle, it follows that $\rho_1 \geq \rho_2 \Longrightarrow e_i(\rho_1) \geq e_i(\rho_2)$, where $e_i(\rho)$ is the $i$-th one in the ordered list of eigenvalues of $\rho$. Further, for generic $\rho_{1,2} \in \mathbb{B}_1^+(\mathbb{H})$, the minmax principle provides an upper bound to the differences $|e_i(\rho_1) - e_i(\rho_2)|$ in terms of the trace-norm (5.21).

**Example 5.5.6.** Given $\rho_{1,2} \in \mathbb{B}_1^+(\mathbb{H})$, decompose $\rho_1 - \rho_2 = R_+ - R_-$, where $R_\pm$ are positive orthogonal operators, so that

$$\|\rho_1 - \rho_2\|_1 = \mathrm{Tr}(R_+ + R_-) = \mathrm{Tr}(2R - \rho_1 - \rho_2) \,,$$

where $R := \rho_1 + R_- = \rho_2 + R_+ \geq \rho_{1,2}$. Let $r_i$, respectively $r_i^{1,2}$ be the eigenvalues of $R$, respectively $\rho_{1,2}$ listed in decreasing order; then, by the minmax principle, $r_i \geq r_i^{1,2}$ for all $i$. Thus, $2r_i - r_i^1 - r_i^2 \geq |r_i^1 - r_i^2| \Longrightarrow \sum_i |r_i^1 - r_i^2| \leq \|\rho_1 - \rho_2\|_1$.

The spectrum of a density matrix is a classical probability distribution: this hints at the possibility of quantifying its information content of by means of the Shannon entropy of such a distribution. This leads to the notion of von Neumann entropy of a state $\rho \in \mathbb{B}_1^+(\mathbb{H})$ [226].

**Definition 5.5.2 (von Neumann Entropy).**

*Given $\rho \in \mathcal{S}(S)$ with spectral decomposition (degenerate eigenvalues are repeated according with their multiplicity and with chosen orthogonal one-dimensional eigenprojectors) $\rho = \sum_j r_j \,|\, r_j \,\rangle\langle\, r_j \,|$, the von Neumann entropy of $\rho$ is the Shannon entropy of the probability distribution corresponding to its spectrum: $S(\rho) = -\mathrm{Tr}(\rho \log \rho) = -\sum_j r_j \log r_j$.*

The following are some of the properties of the von Neumann entropy: they show that it plays a role similar to that of the Shannon entropy in a classical context. Other properties more related to composite systems will be discussed in Section 5.5.3.

**Proposition 5.5.3.** *Let $\rho \in \mathbb{B}_1^+(\mathbb{H})$ be a density matrix. If $\mathbb{H} = \mathbb{C}^N$, then the von Neumann entropy is bounded by the entropy of the state $\mathbb{1}_N/N$:*

$$0 \leq S(\rho) \leq \log N \ . \tag{5.155}$$

*The von Neumann entropy is concave; that is, given weights $\lambda_i \geq 0$, $i \in I$, $\sum_{i\in I} \lambda_i = 1$ and density matrices $\rho_i \in \mathbb{B}_1^+(\mathbb{H})$,*

$$\sum_{i\in I} \lambda_i S(\rho_i) \leq S\left(\sum_{i\in I} \lambda_i \rho_i\right) \leq \sum_{i\in I} \lambda_i S(\rho_i) + \sum_{i\in I} \eta(\lambda_i) \ , \tag{5.156}$$

*where $\eta$ is the concave function (2.84). If $\mathbb{H} = \mathbb{C}^N$, the von Neumann entropy is continuous on $\mathbb{B}_1^+(\mathbb{H})$ with respect to the trace-norm; namely, if $\rho_{1,2} \in \mathbb{B}_1^+(\mathbb{H})$ are such that $\|\rho_1 - \rho_2\|_1 \leq 1/e$; then they satisfy the so-called Fannes inequality*

$$|S(\rho_1) - S(\rho_2)| \leq \|\rho_1 - \rho\|_1 \log N + \eta(\|\rho_1 - \rho_2\|_1) \ . \tag{5.157}$$

**Proof:** Since $S(\rho) - \log N = -\sum_{i=1}^N r_i(\log r_i - \log 1/N)$, boundedness follows from (2.85). The lower bound in (5.156) comes from the concavity of $\eta(x)$; indeed, let $r_j$ and $|\, r_j \,\rangle$, respectively $r_j^i$ and $|\, r_j^i \,\rangle$ be eigenvalues and eigenvectors of $\rho := \sum_{i\in I} \lambda_i \rho_i$, respectively $\rho_i$. Then,

$$r_j = \sum_{i\in I} \lambda_i \langle\, r_j \,|\, \rho_i \,|r_j \,\rangle = \sum_{i\in I} \lambda_i \sum_k r_k^i \,|\langle\, r_k^i \,|\, r_j \,\rangle|^2 \ ,$$

with $\sum_{i\in I,k} \lambda_i \,|\langle\, r_k^i \,|\, r_j \,\rangle|^2 = 1$. Therefore,

$$S(\rho) = \sum_j \eta(r_j) \geq \sum_{i\in I,k} \lambda_i \,|\langle\, r_k^i \,|\, r_j \,\rangle|^2 \,\eta(r_k^i) = \sum_{i\in I} \lambda_i \sum_k \eta(r_k^i) = \sum_{i\in i} \lambda_i S(\rho_i) \ .$$

On the other hand, from Example (5.2.3).9,

$$\lambda_i\rho_i \leq \rho \Longrightarrow \lambda_i\rho_i \log(\lambda_i\rho_i) = \lambda_i\rho_i \log\rho_i - \eta(\lambda_i) \leq \lambda_i\rho_i \log\rho \ .$$

Fannes inequality follows from the fact that $|\eta(u) - \eta(v)| \leq \eta(|u-v|)$ if $|u-v| \leq 1/e$ and from Example 5.5.6 which implies that

$$s_k := |r_k^1 - r_k^2| \leq \sum_{k=1}^N s_k =: S \leq \|\rho_1 - \rho_2\|_1 \leq 1/e \ ,$$

where $r_k^{1,2}$ are the ordered eigenvalues of $\rho_{1,2}$. Then,

$$\begin{aligned}|S(\rho_1) - S(\rho_2)| &\leq \sum_{k=1}^N |\eta(r_k^1) - \eta(r_k^2)| \leq \sum_{k=1}^N \eta(s_k) = S\sum_{k=1}^N \eta(\frac{s_k}{S}) + \eta(S) \\ &\leq \|\rho_1 - \rho_2\|_1 \log N + \eta(\|\rho_1 - \rho_2\|_1) \ ,\end{aligned}$$

for $\eta(x)$ increases when $x \in [0, 1/e]$. We complete the proof by showing that $|\eta(u) - \eta(v)| \leq \eta(|u-v|)$ indeed holds when $|u-v| \leq 1/e$ (see [222]). The function $f(x) := \eta(x + (u-v)) - \eta(x)$ decreases for $u - v \geq 0$, thus $f(0) = \eta(u-v) \geq f(v) = \eta(u) - \eta(v)$. If $u - v \leq 1/e$, $\eta(u-v) \geq u - v$, while the increasing function $g(t) := t + \eta(t)$ gives $g(u) = u + \eta(u) \geq v + \eta(v)$ and thus $\eta(u) - \eta(v) \geq v - u$ which implies $\eta(u-v) \geq \eta(v) - \eta(u)$. □

**Remarks 5.5.5.** 1. The second inequality in (5.156) becomes an equality if and only if the ranges of the matrices $\rho_i$ are orthogonal to each other; indeed, in such a case the eigenvectors of different $\rho_i$'s are orthogonal so that their spectral decompositions give the spectral decomposition of $\rho$,

$$\rho = \sum_{ik} \lambda_i r_i^k |\, r_i^k \rangle\langle r_i^k \,| \Longrightarrow S(\rho) = \sum_{ik} \lambda_i r_i^k \log(\lambda_i r_i^k) \ .$$

2. In the case of an infinite dimensional Hilbert space, the von Neumann entropy is only *lower semicontinuous*: if a sequence of density matrices $\sigma_n$ tends to a density matrix $\sigma$ in trace norm, then $S(\sigma) \leq \lim_n S(\sigma_n)$, in general. As an example [300], take $\sigma_n := (1 - \frac{1}{n})\rho + \frac{1}{n}\rho_n$, where $S(\rho) = 0$ and $S(\rho_n)$ increases like $n$. Then, $\|\sigma_n - \rho\|_1 \leq 2/n \to 0$ when $n \to +\infty$; also, by (5.156),

$$S(\sigma_n) \geq (1 - \frac{1}{n})\, S(\rho) + \frac{1}{n} S(\rho_n) = \frac{1}{n} S(\rho_n) \to c > 0 = S(\rho) \ .$$

The least mixed states, the pure states, are 1-dimensional projectors for which $r_1 = 1$ while all other states have $r_1 < 1$. This suggests the following

**Definition 5.5.1.** *[300] A density matrix $\rho_1 \in \mathcal{S}(S)$ is said to be more mixed than another density matrix $\rho_2 \in \mathcal{S}(S)$, $\rho_1 \succeq \rho_2$, if their decreasingly ordered eigenvalues $e_i(\rho_{1,2})$, $j = 1, 2, \ldots, d$, satisfy*

$$\sum_{i=1}^{k} e_i(\rho_1) \leq \sum_{i=1}^{k} e_i(\rho_2) \ , \quad k = 1, 2, \ldots, d \ .$$

The relation $\succeq$ is a total ordering among density matrices of two level systems; this is because $e_1(\rho_{1,2}) + e_2(\rho_{1,2}) = 1$ for any pair of density matrices $\rho_{1,2} \in M_2(\mathbb{C})$. Therefore, $\rho_1 \succeq \rho_2 \Longleftrightarrow e_1(\rho_1) \leq e_1(\rho_2)$.

Unfortunately, for higher dimensional systems $\succeq$ is only a partial ordering; for instance, consider the following density matrices $\rho_{1,2} \in M_3(\mathbb{C})$,

$$\rho_1 = \begin{pmatrix} 1/2 & 0 & 0 \\ 0 & 1/2 & 0 \\ 0 & 0 & 0 \end{pmatrix} \ , \quad \rho_2 = \begin{pmatrix} 2/3 & 0 & 0 \\ 0 & 1/6 & 0 \\ 0 & 0 & 1/6 \end{pmatrix} \ .$$

Then, $e_1(\rho_1) = 1/2 < e_1(\rho_2) = 2/3$, but $e_1(\rho_1) + e_2(\rho_1) > e_1(\rho_2) + e_2(\rho_2)$.

The following proposition provides a helpful tool that allows, in some cases, to establish whether two density matrices are in the $\succeq$ order.

**Proposition 5.5.4 (Ky Fan Inequality).** *Given $\rho \in \mathbb{B}_1^+(\mathbb{H})$, it holds that*

$$\sum_{j=1}^{k} e_j(\rho) = \max\Big\{ \mathrm{Tr}(P\rho) \ : \ P^2 = P = P^\dagger \ , \ dim(P\mathbb{H}) = k \Big\} \ . \tag{5.158}$$

**Proof:** Let $\{|\,\phi_j\,\rangle\}_{j=1}^k$ be an orthonormal set in $\mathbb{H}$, $\mathbb{K}$ the subspace they generate and $P := \sum_{j=1}^k |\,\phi_j\,\rangle\langle\,\phi_j\,|$ the corresponding orthogonal projector. If $\{|\,\psi_i\,\rangle\}_{i=1}^k$ is any other *ONB* in $\mathbb{K}$, then $\displaystyle\sum_{j=1}^{k} \langle\,\phi_j\,|\,\rho\,|\phi_j\,\rangle = \sum_{j=1}^{k} \langle\,\psi_j\,|\,\rho\,|\psi_j\,\rangle$. Let $|\,r_i\,\rangle$ be the eigenprojectors of $\rho$ corresponding to the eigenvalues $e_i(\rho) =: r_i$ listed in decreasing order and consider the subspace spanned by $\{|\,r_i\,\rangle\}_{i=1}^{k-1}$. A same argument as in the proof of the minmax principle (5.154), ensures the existence of some $|\,\psi_k\,\rangle \in \mathbb{K}$ orthogonal to it; analogously, there must exist

$$|\,\psi_{k-1}\,\rangle \in \mathbb{K} \perp \{|\,r_i\,\rangle\}_{i=1}^{k-2} \cup |\,\psi_k\,\rangle\}$$

and so on. Thus, one collects $\{|\,\psi_j\,\rangle\}_{j=1}^k \in \mathbb{K}$ such that

$$|\,\psi_j\,\rangle \perp \{|\,r_1\,\rangle, \ldots, |\,r_{j-1}\,\rangle, |\,\psi_{j+1}\,\rangle \ldots, |\,\psi_k\,\rangle\}$$

and (5.154) yields $\mathrm{Tr}(P\,\rho) = \sum_{i=1}^k \langle\,\psi_i\,|\,\rho\,|\psi_i\,\rangle \leq \sum_{i=1}^k e_i(\rho)$. □

**Examples 5.5.7.**

1. Given any $\rho \in \mathcal{S}(S)$ and unitary matrices $U_j \in M_N(\mathbb{C})$, $j \in J$, together with weights $0 < \lambda_j < 1$, $\sum_{j\in J} \lambda_j = 1$, set $\widetilde{\rho} := \sum_{j\in J} \lambda_j\, U_j\, \rho\, U_j^\dagger$. If $\rho = \sum_{i=1}^N r_i |\, r_i\,\rangle\langle\, r_i\,|$, the unitarily rotated matrices $U_j\,\rho\, U_j^\dagger$ have the same spectrum for $U_j\,\rho\, U_j^\dagger = \sum_{i=1}^{N} r_i |\,\tilde{r}_i\,\rangle\langle\,\tilde{r}_i\,|$ and $|\,\tilde{r}_i\,\rangle := U|\,r_i\,\rangle$ form an *ONB* . Further, their convex combination $\widetilde{\rho} \succeq \rho$; indeed, let $P$ be the projector achieving the maximum in (5.158), $\sum_{i=1}^k e_i(\widetilde{\rho}) = \mathrm{Tr}(P\widetilde{\rho})$; then
$$\sum_{i=1}^k e_i(\widetilde{\rho}) = \sum_{j\in J} \lambda_j\, \mathrm{Tr}(U_j^\dagger\, P\, U_j\, \rho) \le (\sum_{j\in J} \lambda_j) \sum_{i=1}^k e_i(\rho) = \sum_{i=1}^k e_i(\rho)\ ,$$
for the projectors $U_j^\dagger\, P\, U_j$ need not achieve the maximum in (5.158).
2. Consider the convex set $\mathbb{B}_1^+(\mathbb{H})$ of all density matrices of a system $S$ described by a Hilbert space $\mathbb{H}$, not necessarily finite dimensional, and let $\mathcal{S}_{ord}(S)$ be totally ordered: the most mixed $\rho \in \mathcal{S}_{ord}(S)$ have the largest entropy [300, 314]. Indeed, $\rho_1 \succeq \rho_2 \Longrightarrow S(\rho_1) \ge S(\rho_2)$. In order to show this, set $\alpha_i := e_i(\rho_1)$; then, for all $N \ge 1$ and $\alpha_N > 0$,
$$\sum_{k=1}^{N-1} \Big(\sum_{i=1}^{k} \alpha_i\Big) (\log \alpha_k - \log \alpha_{k+1}) + \log \alpha_N \ = \sum_{k=1}^{N} \alpha_k \log \alpha_k\ .$$
Set $\beta_i := e_i(\rho_2)$; since $\rho_1 \succeq \rho_2 \Longrightarrow \sum_{i=1}^k \alpha_i \le \sum_{i=1}^k \beta_i$ for all $k \ge 1$, using (2.85), one finds
$$\begin{aligned}\sum_{k=1}^{N} -\alpha_k \log \alpha_k &\ge -\sum_{k=1}^{N-1} \Big(\sum_{i=1}^{k} \beta_i\Big) (\log \alpha_k - \log \alpha_{k+1}) \ + \ \log \alpha_N \\ &= -\sum_{k=1}^{N} \beta_k \log \alpha_k \ge -\sum_{k=1}^{N} \beta_k \log \beta_k + \sum_{k=1}^{N} (\beta_k - \alpha_k)\ ,\end{aligned}$$
for all $N \ge 1$, whence $S\,(\rho_1) \ge S\,(\rho_2)$, for $\sum_k \beta_k = \sum_k \alpha_k = 1$.

### 5.5.3 Composite Systems

In quantum information, physical systems $S$ consisting of several subsystems, $S = S_i + S_2 + \cdots S_n$, are called *multi-partite*.

If each of the constituent subsystems is described by a Hilbert space $\mathbb{H}_i$, the Hilbert space of $S$ is $\mathbb{H}^{(n)} = \bigotimes_{i=1}^n \mathbb{H}_i$ and its observables are Hermitian elements of the $C^*$ algebra $\mathbb{B}(\mathbb{H}^{(n)}) = \bigotimes_{i=1}^n \mathbb{B}(\mathbb{H}_i)$.

Given a multi-partite state $\rho \in \mathcal{S}(S)$, *marginal states* $\rho_{i_1 i_2 \cdots i_k}$ for all possible choices of subsystems $S_{i_1} + S_{i_2} + \ldots + S_{i_k}$ are obtained by *partial tracing* over the Hilbert spaces $\mathbb{H}_\ell$ whose indices are different from the selected ones $i_1, i_2, \ldots, i_k$, namely

$$\rho_{i_1 i_2 \cdots i_k} = \mathrm{Tr}_{j_1} \mathrm{Tr}_{j_2} \cdots \mathrm{Tr}_{j_{n-k}}(\rho) \ , \quad j_\ell \neq i_1, i_2, \ldots, i_k \ , \ \ell = 1, 2, \ldots, n-k \ ,$$

where $\mathrm{Tr}_j(\rho) = \sum_k \langle \psi_k^{(j)} \,|\, \rho \,|\psi_k^{(j)} \rangle$ denotes the trace computed with respect to any *ONB* $\{|\, \psi_k^{(j)} \rangle\} \in \mathbb{H}_j$ and yields a density matrix acting on the Hilbert $\mathbb{H}^{(n-1)} = \bigotimes_{1=i\neq j}^{n} \mathbb{H}_i$.

In particular, the states of bipartite systems $S = S_1 + S_2$, are described by density matrices $\rho_{12} \in \mathbb{B}_1^+(\mathbb{H}^{(2)})$ with marginal states

$$\rho_1 = \mathrm{Tr}_2 \rho_{12} := \sum_j \langle \psi_j^{(2)} \,|\, \rho_{12} \,|\psi_j^{(2)} \rangle \ , \ \rho_2 = \mathrm{Tr}_1 \rho_{12} := \sum_j \langle \psi_j^{(1)} \,|\, \rho_{12} \,|\psi_j^{(1)} \rangle \ .$$

**Proposition 5.5.5.** *The marginal states of any pure state $\rho_{12} \in \mathcal{S}(S_1 + S_2)$ have the same eigenvalues with the same multiplicity, apart from the zero eigenvalue, and thus the same von Neumann entropy.*

**Proof:** Let $|\, \psi_{12} \rangle \in \mathbb{H}$ be the vector onto which the pure state $\rho_{12}$ projects and $r_j^{(1)}$, $|\, r_j^{(1)} \rangle$, the non-zero eigenvalues (repeated according to their multiplicities) and eigenvectors of the marginal density matrix $\rho_1 = \mathrm{Tr}_2 \rho$. Using the corresponding *ONB* $\{|\, r_j^{(1)} \rangle\}$ in $\mathbb{H}_1$ and any other *ONB* $\{|\, \phi_k^{(2)} \rangle\}$ in $\mathbb{H}_2$, one can expand

$$|\, \psi_{12} \rangle = \sum_{j,k} C_{jk} |\, r_k^{(1)} \rangle \otimes |\, \psi_k^{(2)} \rangle = \sum_j |\, r_j^{(1)} \rangle \otimes |\, \phi_j^{(2)} \rangle \ ,$$

where $|\, \phi_j^{(2)} \rangle := \sum_k C_{jk} |\, \phi_k^{(2)} \rangle$ need not be either orthogonal or normalized. Then,

$$\rho_1 = \sum_j r_j^{(1)} |\, r_j^{(1)} \rangle \langle r_j^{(1)} \,| = \sum_{j,k} \langle \phi_j^{(2)} \,|\, \phi_k^{(2)} \rangle \,|\, r_k^{(1)} \rangle \langle r_j^{(1)} \,| \ ,$$

whence $\langle \phi_j^{(2)} \,|\, \phi_k^{(2)} \rangle = \delta_{jk} r_j^{(1)}$. Setting $|\, r_j^{(2)} \rangle := |\, \phi_j^{(2)} \rangle / \sqrt{r_j^{(1)}}$ yields

$$|\, \psi_{12} \rangle = \sum_j \sqrt{r_j^{(1)}} \,|\, r_j^{(1)} \rangle \otimes |\, r_j^{(2)} \rangle \ . \tag{5.159}$$

It thus follows that $\rho_2 = \sum_j r_j^{(1)} |\, r_j^{(2)} \rangle \langle r_j^{(2)} \,|$, whence $S(\rho_1) = S(\rho_2)$. □

**Remark 5.5.6.** The expression (5.159) yields the so-called *Schmidt decomposition* of bipartite pure states $|\Psi_{12}\rangle \in \mathbb{H}_1 \otimes \mathbb{H}_2$ into a linear combination with positive coefficients of tensor products of equally indexed states from two *ONB*s of the two subsystems. The degeneracy of the 0 eigenvalue accounts for the possibly different dimensions of the Hilbert spaces $\mathbb{H}_{1,2}$.

**Example 5.5.8.** Let a bipartite system $S = S_1 + S_2$ consist of a 2-level system $S_1$ and an $N$-levl system $S_2$. Let $\{X_i\}_{i=1}^n \in M_2(\mathbb{C})$ be a set of matrices such that $\sum_{i=1}^n X_i^\dagger X_i = \mathbb{1}_2$, consider a fixed vector $\psi \in \mathbb{C}^2$ and the vector state $\mathbb{C}^2 \otimes \mathbb{C}^N \ni |\Psi\rangle := \sum_{i=1}^n X_i|\psi\rangle \otimes |i\rangle$, where $\{|i\rangle\}_{i=1}^N$ is an *ONB* for $S_2$. The normalized vector $|\Psi\rangle$ yields marginal states

$$M_2(\mathbb{C}) \ni \rho_1 = \mathrm{Tr}_2(|\Psi\rangle\langle\Psi|) = \sum_{i=1}^N X_i|\psi\rangle\langle\psi|X_i^\dagger$$

$$M_N(\mathbb{C}) \ni \rho_2 = \mathrm{Tr}_1(|\Psi\rangle\langle\Psi|) = \sum_{i,j=1}^N \langle\psi|X_j^\dagger X_i|\psi\rangle\,|i\rangle\langle j|\,.$$

Let $r_a$, $|a\rangle$, $a = 1, 2$, be the eigenvalues, respectively eigenvectors of $\rho_1$; by expanding $X_i|\psi\rangle = \sum_{a=1}^2 c_{ia}|a\rangle$, it turns out that $|\Psi\rangle = \sum_{a=1}^2 |a\rangle \otimes |\widetilde{\phi}_a\rangle$, where $|\widetilde{\phi}_a\rangle := \sum_{i=1}^N c_{ia}|i\rangle$. The vectors $|\phi_a\rangle := |\widetilde{\phi}_a\rangle/\|\widetilde{\phi}_a\|$ are orthonormal and the 0 eigenvalue of $\rho_2 = r_1|\phi_1\rangle\langle\phi_1| + r_2|\phi_2\rangle\langle\phi_2|$ is $(N-2)$-degenerate.

We end this section with a list of properties of the von Neumann entropy which pertain to composite systems.

**Proposition 5.5.6.** *Let $S = S_1 + S_2$ be a composite system with Hilbert space $\mathbb{H} = \mathbb{H}_1 \otimes \mathbb{H}_2$, $\mathbb{H}_{1,2}$ of dimension $d_{1,2}$. The von Neumann entropy is* **additive** *on product states $\mathbb{B}_1^+(\mathbb{H}) \ni \rho_{12} = \rho_1 \otimes \rho_2$,*

$$S(\rho_{12}) = S(\rho_1) + S(\rho_2)\ . \tag{5.160}$$

*Given $\rho_{12} \in \mathbb{B}_1^+(\mathbb{H})$, let $\mathbb{B}_1^+(\mathbb{H}_1) \ni \rho_1 := \mathrm{Tr}_2\rho_{12}$ and $\mathbb{B}_1^+(\mathbb{H}_2) \ni \rho_2 := \mathrm{Tr}_1\rho_{12}$ be the marginal states. Then*

$$|S(\rho_1) - S(\rho_2)| \leq S(\rho_{12}) \leq S(\rho_1) + S(\rho_2)\ ; \tag{5.161}$$

*The second inequality expresses that von Neumann entropy is* **subadditive***; more in general, the von Neumann entropy is* **strongly subadditive***. Namely, let $S = S_1 + S_2 + S_3$ be a tripartite system described by a state $\rho_{123} \in \mathbb{B}_1^+(\mathbb{H})$ with $\mathbb{H} = \mathbb{H}_1 \otimes \mathbb{H}_2 \otimes \mathbb{H}_3$. For any cyclic permutation $(i, j, k)$ of $(1, 2, 3)$ let $\mathbb{B}_1^+(\mathbb{H}_i \otimes \mathbb{H}_j) \ni \rho_{ij} := \mathrm{Tr}_k\rho_{123}$, $\mathbb{B}_1^+(\mathbb{H}_i) \ni \rho_i := \mathrm{Tr}_{jk}\rho_{123}$ be marginal states. Then [192, 314]*

$$S(\rho_{123}) + S(\rho_j) \leq S(\rho_{ij}) + S(\rho_{jk})\ . \tag{5.162}$$

*Furthermore, the differences between the von Neumann entropies of* $\rho_{12}$ *and of the marginal states* $\rho_{1,2}$, $S(\rho_{12}) - S(\rho_{1,2})$, *are concave:*

$$S\left(\sum_j \lambda_j \rho_{12}^{(j)}\right) - S\left(\sum_j \lambda_j \rho_{1,2}^{(j)}\right) \geq \sum_j \lambda_j \left(S\left(\rho_{12}^{(j)}\right) - S\left(\rho_{1,2}^{(j)}\right)\right. , \quad (5.163)$$

*where* $\lambda_j > 0$ *and* $\sum_j \lambda_j = 1$.

**Proof:** Additivity comes from the fact that the spectrum of $\rho_{12} = \rho_1 \otimes \rho_2$ consists of the products of the eigenvalues of $\rho_1$ and $\rho_2$.

Assume strong subadditivity holds and let $\rho_{AB} = \sum_i r_i^{AB} |\, r_i^{AB}\,\rangle\langle\, r_i^{AB}\,|$ be a density matrix on $\mathbb{H}_A \otimes \mathbb{H}_B$ and

$$|\,\sqrt{\rho_{AB}}\,\rangle = \sum_i \sqrt{r_i^{AB}}\; |\, r_i^{AB}\,\rangle \otimes |\, r_i^{AB}\,\rangle \in (\mathbb{H}_A \otimes \mathbb{H}_B) \otimes (\mathbb{H}_A \otimes \mathbb{H}_B)$$

the corresponding *GNS* state. Set $\mathbb{H}_1 := \mathbb{H}_A$, $\mathbb{H}_2 := \mathbb{H}_B$, in the first factor, $\mathbb{H}_3 := \mathbb{H}_A \otimes \mathbb{H}_B$ in the second one and

$$\rho_{123} := |\,\sqrt{\rho_{AB}}\,\rangle\langle\,\sqrt{\rho_{AB}}\,| = \sum_{i,j} \sqrt{r_i^{AB} r_j^{AB}}\; |\, r_i^{AB}\,\rangle\langle\, r_j^{AB}\,| \otimes |\, r_i^{AB}\,\rangle\langle\, r_j^{AB}\,| \,.$$

Then, $\rho_3 = \mathrm{Tr}_{12}\rho_{123} = \rho_{AB} = \mathrm{Tr}_3\rho_{123} = \rho_{12}$, therefore, $\rho_1 = \mathrm{Tr}_{23}\rho_{123} = \rho_A$, $\rho_2 = \mathrm{Tr}_{13}\rho_{123} = \rho_B$. Also, because of purity, $S(\rho_{123}) = 0$, thus (5.162) and Proposition 5.5.5 yield

$$S(\rho_{AB}) = S(\rho_3) \leq S(\rho_{13}) + S(\rho_{23}) = S(\rho_2) + S(\rho_1) = S(\rho_A) + S(\rho_B)$$

which implies subadditivity. The lower bound in (5.161) follows instead from

$$S(\rho_A) = S(\rho_1) \leq S(\rho_{13}) + S(\rho_{12}) = S(\rho_2) + S(\rho_3) - S(\rho_B) + S(\rho_{AB})$$
$$S(\rho_B) = S(\rho_2) \leq S(\rho_{12}) + S(\rho_{23}) = S(\rho_3) + S(\rho_1) = S(\rho_{AB}) + S(\rho_A) \,.$$

In order to prove strong subadditivity, we introduce the *quantum relative entropy* of two density matrices $\rho$ and $\sigma$ (see Definition 6.3.1)

$$S(\rho\,,\sigma) := \mathrm{Tr}\Big(\rho\,\log\rho\; - \;\rho\,\log\sigma\Big)$$

which is well defined when $\sigma|\,\psi\,\rangle = 0 \Longrightarrow \rho|\,\psi\,\rangle = 0$.

We shall prove in Section 6.3 that $S(\rho\,,\sigma)$ does not increase under maps like the partial traces, thence

$$S\left(\rho_{12}\,, \frac{\mathbb{1}_1}{d_1}\otimes\rho_2\right) = S\left(\mathrm{Tr}_3\rho_{123}\,, \mathrm{Tr}_3\Big(\frac{\mathbb{1}_1}{d_1}\otimes\rho_{23}\Big)\right) \leq S\left(\rho_{123}\,, \frac{\mathbb{1}_1}{d_1}\otimes\rho_{23}\right) \,.$$

On the other hand,

$$S\left(\rho_{12}\,,\,\frac{\mathbb{1}_1}{d_1}\otimes\rho_2\right)=-S\left(\rho_{12}\right)+S\left(\rho_2\right)+\log d_1 \qquad \text{and}$$

$$S\left(\rho_{123}\,,\,\frac{\mathbb{1}_1}{d_1}\otimes\rho_{23}\right)=-S\left(\rho_{123}\right)+S\left(\rho_{23}\right)+\log d_1\ ,$$

whence $\Big(S\left(\rho_{123}\right)-S\left(\rho_{23}\right)\Big)-\Big(S\left(\rho_{12}\right)-S\left(\rho_{2}\right)\Big)\leq 0$.

The concavity (5.163) follows from another property of the relative entropy which shall be discussed in Section 6.3, namely its *joint convexity*:

$$S\left(\sum_j\lambda_j\rho^{(j)}\,,\,\sum_j\lambda_j\sigma^{(j)}\right)\leq\sum_j\lambda_j S\left(\rho^{(j)}\,,\,\sigma^{(j)}\right)\ ,$$

where $\lambda_j>0$ and $\sum_j\lambda_j=1$. Let $\rho^{(j)}:=\rho^{(j)}_{12}$ and $\sigma^{(j)}:=\rho^{(j)}_1\otimes\frac{\mathbb{1}_2}{d_2}$, with $\rho^{(j)}_1:=\mathrm{Tr}_2\rho^{(j)}_{12}$; then,

$$\begin{aligned}
&S\left(\sum_j\lambda_j\rho^{(j)}_{12}\,,\,\sum_j\lambda_j\rho^{(j)}_1\otimes\frac{\mathbb{1}_2}{d_2}\right)=\\
&\qquad\qquad -S\left(\sum_j\lambda_j\rho^{(j)}_{12}\right)+S\left(\sum_j\lambda_j\rho^{(j)}_1\right)+\log d_2\\
&\leq\sum_j\lambda_j S\left(\rho^{j}_{12}\,,\,\rho^{(j)}_1\otimes\frac{\mathbb{1}_2}{d_2}\right)=\sum_j\lambda_j\Big(S\left(\rho^{(j)}_1\right)-S\left(\rho^{(j)}_{12}\right)\Big)+\log d_2\ .
\end{aligned}$$

□

### 5.5.4 Entangled States

One of the most puzzling and fascinating aspects of quantum mechanics is its *non-locality* embodied by the concept of *quantum entanglement* [152]. This is a property of certain quantum states of composite systems, called *entangled*, which are such that their constituting subsystems cannot be attributed properties of their own, not even with a certain probability. As a paradigm of such states, consider the following vector state of two spin 1/2 particles (the simplest instance of the vector states (5.18) in Example 5.2.3.9),

$$|\,\Psi_{00}\,\rangle:=\frac{|\,00\,\rangle+|\,11\,\rangle}{\sqrt{2}}\ , \tag{5.164}$$

where $|\,0\,\rangle$ and $|\,1\,\rangle$ are eigenstates of the Pauli matrix $\sigma_3$. By looking at the corresponding projector,

$$
\begin{aligned}
|\,\Psi_{00}\,\rangle\langle\,\Psi_{00}\,| &= \frac{1}{2}\Big(|\,0\,\rangle\langle\,0\,|\otimes|\,0\,\rangle\langle\,0\,| + |\,1\,\rangle\langle\,1\,|\otimes|\,1\,\rangle\langle\,1\,|\Big) \\
&+ \frac{1}{2}\Big(|\,0\,\rangle\langle\,1\,|\otimes|\,0\,\rangle\langle\,1\,| + |\,1\,\rangle\langle\,0\,|\otimes|\,1\,\rangle\langle\,0\,|\Big) ,
\end{aligned}
$$

one sees that, while the first line is the density matrix of an equally distributed mixture of both spins pointing up and down along the $z$ direction, the interference term in the second line forbids to attribute these two properties with probability 1/2 to the component spins. By rotating the orthonormal projectors $(1 \pm \sigma_3)/2$ into any two other orthonormal pairs $(\mathbb{1} \pm \boldsymbol{n}_1 \cdot \boldsymbol{\sigma})/2$ and $(\mathbb{1} \pm \boldsymbol{n}_2 \cdot \boldsymbol{\sigma})/2$, the same obstruction occurs along any two directions $\boldsymbol{n}_{1,2}$.

**Example 5.5.9 (Bell States).** [224] The symmetric vector (5.164) is the first one in the so-called *Bell basis* of $\mathbb{C}^2 \otimes \mathbb{C}^2$ of which the others read

$$
|\,\hat{\Psi}_{01}\,\rangle = \frac{|\,01\,\rangle + |\,10\,\rangle}{\sqrt{2}} \ , \ |\,\hat{\Psi}_{10}\,\rangle = \frac{|\,00\,\rangle - |\,11\,\rangle}{\sqrt{2}} \ , \ |\,\hat{\Psi}_{11}\,\rangle = \frac{|\,00\,\rangle - |\,11\,\rangle}{\sqrt{2}} \ .
$$

In quantum information 2-level systems are called *qubits*, unitary actions on them *quantum gates* and nets of unitary gates *quantum circuits*. The Bell states can be created out of a separable pure state of two *qubits* by means of *local* and *non-local operations*, according to the quantum circuit in Figure 5.4.

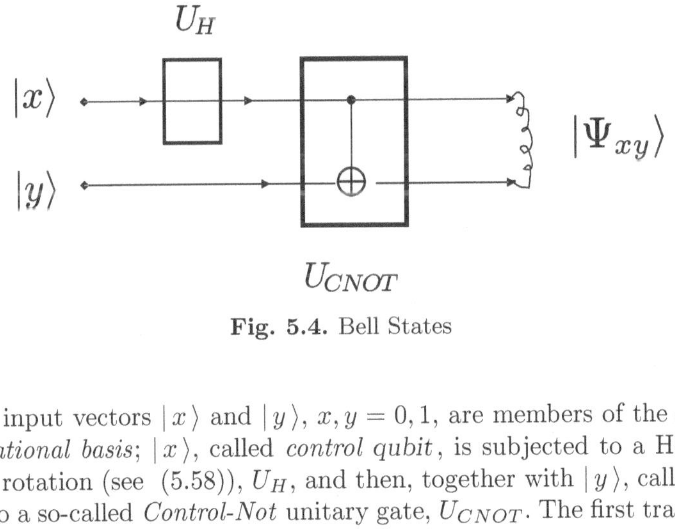


**Fig. 5.4.** Bell States

The input vectors $|\,x\,\rangle$ and $|\,y\,\rangle$, $x, y = 0, 1$, are members of the so-called *computational basis*; $|\,x\,\rangle$, called *control qubit*, is subjected to a Hadamard unitary rotation (see (5.58)), $U_H$, and then, together with $|\,y\,\rangle$, called *target qubit*, to a so-called *Control-Not* unitary gate, $U_{CNOT}$. The first transformation affects one of the two *qubits* only via the matrix $M_4(\mathbb{C}) \ni U_H \otimes \mathbb{1}_2$ and is thus local; the second one involves both *qubits* in a *non-local way*. Indeed, the unitary matrix $U_{CNOT} \in M_4(\mathbb{C})$ implements the classical *CNOT* gate, $CNOT(x, y) = (x, y \oplus x)$, that acting on pairs $(x, y)$ of *bits* leaves the control *bit* unchanged and adds it to the target *bit*:

$$CNOT(00) = (00) \quad CNOT(01) = (01)$$
$$CNOT(10) = (11) \quad CNOT(11) = (10) \; .$$

If we substitute pairs of *bits* with tensor products of computational basis vectors, the Pauli matrix $\sigma_1$ flips the $|\,x\,\rangle$ so that the same relations are implemented on $\mathbb{C}^2 \otimes \mathbb{C}^2$ by

$$U_{CNOT} := |\,0\,\rangle\langle\,0\,| \otimes \mathbb{1} + |\,1\,\rangle\langle\,1\,| \otimes \sigma_1 = \begin{pmatrix} \mathbb{1} & 0 \\ 0 & \sigma_1 \end{pmatrix} \; . \tag{5.165}$$

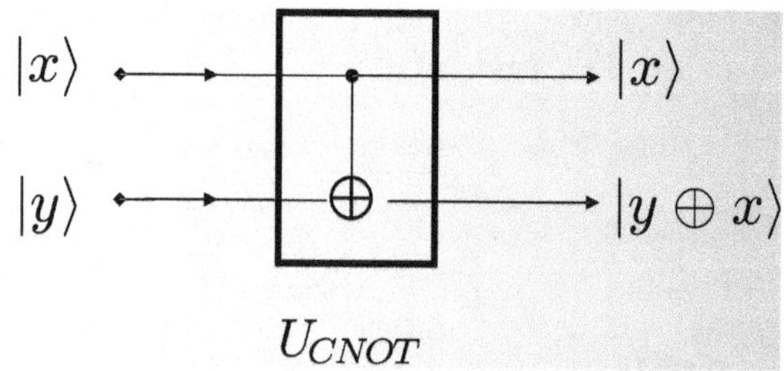


**Fig. 5.5.** *CNOT* Gate

Let $|\,\hat{\Psi}_{xy}\,\rangle := U_{CNOT} H \otimes \mathbb{1}|\,xy\,\rangle$; by a Hadamard rotation (5.58), one gets

$$|\,\hat{\Psi}_{xy}\,\rangle = \frac{1}{\sqrt{2}} \sum_{i=0}^{1} (-1)^{ix} |\,i, y \oplus i\,\rangle = \frac{|\,0, y\,\rangle + (-1)^x |\,1, y \oplus 1\,\rangle}{\sqrt{2}} \; ,$$

and, by varying $x, y \in \{0, 1\}$, one obtains the Bell basis.

Entanglement is a purely quantum phenomenon, with no classical counterpart; it has from the start attracted a lot of scientific and, unfortunately, also pseudo-scientific interest; one of the great merits of quantum information is to have promoted entanglement to the status of a physical resource for performing informational and computational tasks otherwise impossible in a purely commutative context.

In the following, we shall mainly focus upon bipartite discrete quantum systems consisting of two parties described by means of finite dimensional Hilbert spaces $\mathbb{C}^{d_1}$ and $\mathbb{C}^{d_2}$, respectively. Within the state-space $\mathcal{S}(S_1 + S_2)$, one distinguishes *separable* from *entangled* states.

**Definition 5.5.3 (Separable and Entangled States).** *A density matrix $\rho \in \mathcal{S}(S_1 + S_2)$ is* separable *if and only if it can be approximated in trace norm by a linear convex combination of tensor products of density matrices:*

$$\rho = \sum_{(i_1,i_2)\in I_1\times I_2} \lambda_{i_1 i_2}\, \rho^1_{i_1} \otimes \rho^2_{i_2} \;, \quad \lambda_{i_1 i_2} \geq 0 \;, \quad \sum_{(i_1,i_2)\in I_1\times I_2} \lambda_{i_1 i_2} = 1 \;. \quad (5.166)$$

*Those $\rho \in \mathcal{S}(S_1 + S_2)$ which cannot be written in a factorized form as in (5.166) are called* **entangled** *or* **non-separable** *states.*

**Remark 5.5.7.** Pure separable states are of the form $|\,\psi\,\rangle = |\,\psi_1\,\rangle \otimes |\,\psi_2\,\rangle$ for some $\psi_{1,2} \in \mathbb{C}^{d_{1,2}}$, otherwise they are entangled. The set $\mathcal{S}_{sep}(S)$ of separable states of the bipartite system $S$ is the closure of the convex hull of its separable pure states (see Remark 5.3.2.5).

In order to judge whether a pure bipartite state is entangled or separable is sufficient to look at its marginal density matrices.

**Proposition 5.5.7.** *A vector state $|\,\Psi_{12}\,\rangle \in \mathbb{C}^{d_1} \otimes \mathbb{C}^{d_2}$ of a bipartite system $S_1 + S_2$ is separable if and only if its the marginal states $\rho_{1,2}$ are pure.*

**Proof:** If $|\,\Psi_{12}\,\rangle$ is separable, its projector is the tensor product of two projectors and partial tracing yields one of them. Vice versa, if the marginal density matrices are not projectors, then the Hilbert-Schmidt decomposition (5.159) contains more than one pair and $|\,\Psi_{12}\,\rangle$ is entangled. □

The structure of separable states is apparent from (5.166): they can be obtained by mixing with weights $\lambda_{i_1 i_2}$ otherwise independent states of $S_1$ and $S_2$, the only possible correlations between them being those relative to the probability distribution $\{\lambda_{i_1 i_2}\}$ associated with the weights and thus of purely classical nature. Instead, pure entangled states carry correlations that are purely quantum mechanical.

**Examples 5.5.10.**

1. Consider a bipartite system consisting of two $d$-level systems in the state (5.18) which generalizes the two *qubit* symmetric state (5.164). From partial tracing the projector $\widehat{P}^d_+ = |\widehat{\Psi}^d_+\rangle\langle\widehat{\Psi}^d_+|$, one gets

$$\rho_1 = \rho_2 = \frac{1}{d}\sum_{i=1}^{d} |\,i\,\rangle\langle\,i\,| = \frac{\mathbb{1}}{d} \;,$$

namely the totally mixed state for both parties. Thus, the von Neumann entropy of the bipartite state $\widehat{P}^d_+$ is smaller than that of either its components, $0 = S\left(\widehat{P}^d_+\right) \leq S\left(\rho_{1,2}\right) = \log d$, which is maximal, instead. In other terms, the information content of the entangled pure state $\widehat{P}^d_+$ is smaller than that of its constituent parties. This holds for all entangled pure states. In order to see this, one uses the Schmidt-decomposition (5.159) and Proposition 5.5.5: if $\rho_{12} = |\,\Psi_{12}\,\rangle\langle\,\Psi_{12}\,|$, then $S(\rho_{12}) = 0$, and $S(\rho_1) = S(\rho_2) = -\sum_{i=1}^{N_1} r_i^1 \log_2 r_i > 0$ unless $\rho_{1,2}$ are pure states and thus $\rho_{12}$ separable.

2. The previous observation makes the statistical properties of pure entangled states incompatible with classical ones; indeed, by (2.88) one knows that the Shannon entropy of a bipartite classical system (described by two random variables) cannot be less than that of any of its marginal distributions. This classical behavior is characteristic of all separable states; namely, the von Neumann entropy of all separable bipartite states cannot be smaller than the von Neumann entropy of their marginal states. This fact follows from (5.163); in fact, consider a separable state

$$\rho_{12} = \sum_{ij} \lambda_{ij} \rho_i^{(1)} \otimes \rho_j^{(2)} \in \mathbb{B}_1^+(\mathbb{H}_1 \otimes \mathbb{H}_2)$$

so that $\rho_1 = \sum_i \lambda_i^{(1)} \rho_i^{(1)}$, $\lambda_i^{(1)} := \sum_j \lambda_{ij}$; then, by applying (5.163), (5.160) and the positivity of the von Neumann entropy (see (5.155)), one gets

$$\begin{aligned} S\left(\rho_{12}\right) - S\left(\rho_1\right) &\geq \sum_{ij} \lambda_{ij} \left( S\left(\rho_i^{(1)} \otimes \rho_j^{(2)}\right) - S\left(\rho_i^{(1)}\right) \right) \\ &= \sum_{ij} \lambda_{ij} S\left(\rho_j^{(2)}\right) \geq 0 \,. \end{aligned}$$

3. The so-called *GHZ* states are entangled pure states of tripartite systems consisting of 3 *qubits* : $|\,\Phi_\pm\,\rangle := \dfrac{|\,000\,\rangle \pm |\,111\,\rangle}{\sqrt{2}}$ in the computational basis. Though entangled as tripartite states, all their two *qubit* marginal states are separable, for instance

$$\rho_{13}^{\pm} := \frac{1}{2} \mathrm{Tr}_2 \Big( \sum_{i,j=0}^{1} (-1)^{i+j} \,|\, iii \,\rangle\langle\, jjj \,|\Big) = \frac{1}{2} \sum_{i=0}^{1} |\, i \,\rangle\langle\, i \,| \otimes |\, i \,\rangle\langle\, i \,| \,.$$

From $|\,\Phi_+\,\rangle$ one obtains an *ONB* in $\mathbb{H}^{(3)} = (\mathbb{C}^2)^{\otimes 3}$ by acting locally with the Pauli matrices,

$$\begin{aligned} |\,\Psi_{abc}\,\rangle &:= \sigma_1^a \otimes \sigma_1^b \otimes \sigma_3^c |\,\Phi_+\,\rangle \,, \quad a,b,c = 0,1 \\ \langle\,\Psi_{def}\,|\,\Psi_{abc}\,\rangle &= \frac{1}{2} \sum_{i,j=0}^{1} \langle\, i \,|\, \sigma_1^{d+a} \,|j\,\rangle \langle\, i \,|\, \sigma_1^{e+b} \,|j\,\rangle \underbrace{\langle\, i \,|\, \sigma_3^{f+c} \,|j\,\rangle}_{\propto \delta_{ij}} \\ &= \frac{1}{2} \sum_{i=0}^{1} (-1)^{f+c} \langle\, i \,|\, \sigma_1^{d+a} \,|i\,\rangle \langle\, i \,|\, \sigma_1^{e+b} \,|i\,\rangle = \delta_{ad}\delta_{be}\delta_{cf} \,. \end{aligned}$$

## 5.6 Dynamics and State-Transformations

The standard time-evolution of quantum systems is typically described by a strongly continuous one-parameter family of unitary operators $\{U_t\}_{t\in\mathbb{R}}$ on a Hilbert space $\mathbb{H}$, fulfilling the group composition law $U_tU_s = U_{t+s}$, for all $t, s \in \mathbb{R}$. By Stone's theorem [300], the group is generated by a self-adjoint operator $H$ on $\mathbb{H}$, the Hamiltonian, such that, for all $\psi \in \mathbb{H}$, $(\hbar = 1)$

$$\partial_t|\,\psi_t\,\rangle = -i\,H|\,\psi_t\,\rangle\ , \quad |\,\psi_t\,\rangle = U_t|\,\psi\,\rangle\ , \quad U_t = \mathrm{e}^{-itH}\ . \tag{5.167}$$

This type of time-evolution equation is proper to the so-called *closed quantum systems*. As any other physical system, also closed systems $S$ are in contact with the environment $E$ which contains them; however their mutual interactions are negligible and the dynamics of $S$ is independent of $E$ and is reversible. When the interactions between $S$ and $E$ cannot be neglected, it may nevertheless be possible to derive a closed dynamics for the system $S$ alone which nevertheless accounts for the presence of the environment. In such cases, $S$ is known as an *open quantum system* and its so-called *reduced dynamics* is irreversible and incorporates noisy and dissipative effects due to the presence of $E$.

The Schrödinger time-evolution easily extends from vector states to mixtures. Since pure states $|\,\psi\,\rangle\langle\,\psi\,|$ evolve into pure states $U_t|\,\psi\,\rangle\langle\,\psi\,|U_t^\dagger$, extension to convex combinations of pure states yields the *Liouville equation*

$$\partial_t\rho_t = -i\Big[H\,,\,\rho_t\Big]\ , \tag{5.168}$$

with formal solution $\rho_t = U_t\,\rho\,U_t^*$ for any initial state $\rho \in \mathbb{B}_1^+(\mathbb{H})$. By duality (compare (2.9) and (2.7)), if $X \in \mathbb{B}(\mathbb{H})$ then $\mathrm{Tr}(\rho_t\,X) = \mathrm{Tr}(\rho\,X_t)$ and one gets the Heisenberg time-evolution equation for the operators

$$\partial_t X_t = i\Big[H\,,\,X_t\Big]\ . \tag{5.169}$$

This gives rise to a one parameter family $\{\mathcal{U}_t\}_{t\in\mathbb{R}}$ of automorphisms of $\mathbb{B}(\mathbb{H})$,

$$X \mapsto \mathcal{U}_t[X] := X_t = U_t^\dagger\,X\,U_t\ , \tag{5.170}$$

that preserve hermiticity and products of operators,

$$\mathcal{U}_t[X^\dagger] = \mathcal{U}_t[X]^\dagger\ , \quad \mathcal{U}_t[XY] = \mathcal{U}_t[X]\mathcal{U}_t[Y] \qquad \forall\ X, Y \in \mathbb{B}(\mathbb{H})\ .$$

As automorphisms, these linear maps are positive, and also completely positive as their action is of the Kraus-type discussed in Proposition 5.2.1. By duality the action of $\mathcal{U}_t$ is transferred to the action of the dual $\mathcal{U}_t^+$ on $\mathbb{B}_1^+(\mathbb{H})$: $\mathcal{U}_t^+[\rho] = U_t\rho U_t^\dagger$; $\mathcal{U}_t$ preserves the trace, $\mathrm{Tr}(\mathcal{U}_t[\rho]) = 1$, and sends projectors into projectors,

$$P^2 = P = P^\dagger \Longrightarrow (\mathcal{U}_t^+[P])^2 = \mathcal{U}_t^+[P] = \mathcal{U}_t^+[P]^\dagger \; .$$

A state $\rho$ is an equilibrium state if and only if it commutes with the Hamiltonian that generates the dynamics, $\mathcal{U}_t^+(\rho) = \rho \Leftrightarrow [H\,,\,\rho] = 0$. However, if a state $\rho$ changes in the course of time under a time-evolution implemented by unitary operators, its spectrum does not. As a consequence, as much as the Gibbs entropy of classical probability distributions evolving under a Hamiltonian flux on phase-space is a constant of the motion, the von Neumann entropy is always preserved by the Schrödinger-Liouville time-evolution, $S(\mathcal{U}_t^+[\rho]) = S(\rho)$.

**Examples 5.6.1.**

1. We have seen that density matrices $\rho$ of 2-level systems are identified by their Bloch vectors $\boldsymbol{\rho} \in \mathbb{R}^3$. By denoting them as kets $|\,\boldsymbol{\rho}\,\rangle$, the linear action of the commutator on the right hand side of (5.168) corresponds to a $3\times 3$ matrix acting on $|\,\boldsymbol{\rho}\,\rangle$, whence the Liouville equation can be recast in the form $\partial_t|\,\boldsymbol{\rho}\,\rangle = -2\mathcal{H}\,|\,\boldsymbol{\rho}\,\rangle$. Since $[\mathbb{1}\,,\,\rho] = 0$, it is no restriction to take the Hamiltonian of the form $H = \boldsymbol{\omega}\cdot\boldsymbol{\sigma}$ with $\boldsymbol{\omega} = (\omega_1, \omega_2, \omega_3) \in \mathbb{R}^3$, $\boldsymbol{\sigma} = (\sigma_1, \sigma_2, \sigma_3)$. Then, the algebraic relations (5.56) yield

$$\frac{\mathrm{d}\rho_i}{\mathrm{d}t} = 2\sum_{j,k=1}^{3} \varepsilon_{ijk}\omega_j\rho_k \Longrightarrow \mathcal{H} = \begin{pmatrix} 0 & \omega_3 & -\omega_2 \\ -\omega_3 & 0 & \omega_1 \\ 0 & \omega_2 & -\omega_1 \end{pmatrix} .$$

   Thus, Bloch vectors rotate with angular velocity $\omega = \|\boldsymbol{\omega}\|$ around the direction of $\boldsymbol{\omega}$; their lengths are then constant, pure states remain pure and the surface of the Bloch sphere is mapped into itself.
   Suppose $\boldsymbol{\omega} = (0, 0, \omega)$, then, by series expansion,

$$U_t = \mathrm{e}^{-it\omega\sigma_3} = \cos\omega t \,+\, i\,\sigma_3 \sin\omega t \;;$$

   thus, $\sigma_3(t) := U_t^\dagger \sigma_3\, U_t = \sigma_3$, while

$$\sigma_+(t) := U_t^\dagger\, \sigma_+\, U_t = \sigma_+\,(\cos 2\omega t \,+\, i\,\sin 2\omega t) = \sigma_+\,\mathrm{e}^{2it\omega} \; .$$

   The same result more directly follows from the fact that

$$\sigma_3^k\sigma_+ = \sigma_3^{k-1}\,\sigma_+\,\sigma_3 \,+\, 2\sigma_+ = \sigma_+\,(\sigma_3 + 2)^k$$

   and that this relation extends to functions $f(\sigma_3)$ that can be expanded as power series, namely $f(\sigma_3)\sigma_+ = \sigma_+ f(\sigma_3 + 2)$.
2. Consider an array of $N$ spins 1/2 equipped with the Hamiltonian [300]

$$H_N = \sum_{j=1}^{N} B\mu_j\sigma^j \,+\, \sum_{j=1}^{N}\sum_{i=1}^{N-1} \varepsilon(i)\,\sigma^j\,\sigma^{j+i} = \sum_{j=1}^{N}(H_j \,+\, H_j^{int}) \; ,$$

where, as in Example 5.4.1, $\sigma^j$ denotes the Pauli matrix $\sigma_3$ at site $j$ and products of Pauli matrices at different sites denote tensor products of commuting operators. The single sum corresponds to the spins being coupled to a vertical constant magnetic field $B$, while the double sum describes spin-spin interactions whose range is regulated by the coupling constants $\varepsilon(i)$ which only depend on the distance between spins.
Let the array be provided with periodic boundary conditions $\sigma^i_\# = \sigma^{i+N}_\#$ ($\sigma_\# = \sigma_{3,\pm}$); then, the $j$-th spin interacts with the same strength with those symmetrically placed on its left and right hand side. Suppose $N$ odd, it follows that $H^{int}_j$ can be recast as

$$H^{int}_j = \sum_{i=1}^{(N-1)/2} \varepsilon(i)\Big(\sigma^{j-i}\sigma_j + \sigma^j\sigma^{j+i}\Big) .$$

Using the previous example and the fact that $\sigma^k_\#$ commutes with all spin operators from sites different from $k$, one obtains

$$\sigma^j(t) := \mathrm{e}^{itH_N}\,\sigma^j\,\mathrm{e}^{-itH_N} = \sigma^j \tag{5.171}$$

$$\begin{aligned}\sigma^j_+(t) &:= \mathrm{e}^{itH_N}\,\sigma^j_\pm\,\mathrm{e}^{-itH_N} = \mathrm{e}^{it(H_j+H^{int}_j)}\,\sigma^j_+\,\mathrm{e}^{-it(H_j+H^{int}_j)} \\ &= \sigma^j_+\,\mathrm{e}^{2it(B\mu_j+\sum_{i=1}^{(N-1)/2}\varepsilon(i)(\sigma^{j-i}+\sigma^{j+i}))} \\ &= \sigma^j_+\,\mathrm{e}^{2itB\mu_j}\prod_{i=1}^{(N-1)/2}\Big(\Big(\cos 2t\varepsilon(i)+\sigma^{j-i}\sin 2\varepsilon(i)\Big)\times \\ &\qquad \times\Big(\cos 2t\varepsilon(i)+\sigma^{j+i}\sin 2\varepsilon(i)\Big)\Big)\end{aligned} \tag{5.172}$$

while $\sigma^j_-(t)$ is obtained by taking the adjoint of $\sigma^j_+(t)$. Let the spin system be endowed with a state which is the tensor product of equal pure states as in (5.104) each of them for each one of the sites

$$\rho^{\otimes N} = \bigotimes_{n=1}^{N}\rho\,,\quad \rho := \frac{1}{2}\begin{pmatrix} 1+s & \sqrt{1-s^2}\mathrm{e}^{i\phi} \\ \sqrt{1-s^2}\mathrm{e}^{-i\phi} & 1-s\end{pmatrix} .$$

This state is not invariant under the time automorphism in (5.171) and (5.172): indeed, $\rho^{\otimes N}(\sigma^j(t)) = s$ for all $j$, but

$$\begin{aligned}\rho^{\otimes N}(\sigma_+(t)) &= \mathrm{e}^{2itB\mu_j}\,\rho(\sigma^j_+)\prod_{i=1}^{(N-1)/2}\Big(\Big(\cos 2t\varepsilon(i)+\rho(\sigma^{j-i})\sin 2\varepsilon(i)\Big)\times \\ &\qquad\times\Big(\cos 2t\varepsilon(i)+\rho(\sigma^{j+i})\sin 2\varepsilon(i)\Big)\Big) \\ &= \mathrm{e}^{2itB\mu_j}\,\frac{\sqrt{1-s^2}}{2}\,f^2_N(s,t)\,, \\ f_N(s,t) &:= \prod_{i=1}^{(N-1)/2}\Big(\cos 2\varepsilon(i)t + i\,s\,\sin 2\varepsilon(i)t\Big) .\end{aligned}$$

If the coupling constants decrease exponentially with the spin distance, $\varepsilon(i) = 2^{-(i+1)}$, and $s = 0$, then

$$f_N(0,t) = \prod_{i=1}^{(N-1)/2} \cos\frac{t}{2^i}$$

and the observables $\sigma_\pm^j$ show a recurrence time that increases as $2^{(N-1)/2}$,

$$\rho^{\otimes N}(\sigma_+^j(t)) = \mathrm{e}^{2itB\mu_j}\,\frac{1}{2}\,f_N^2(0,t)\ . \tag{5.173}$$

3. Consider $f$ uncoupled harmonic oscillators of masses $m = 1$ and frequencies $\omega_i$ described by the algebra of Weyl operators (5.75) $W(\boldsymbol{r})$, $\boldsymbol{r} = (\boldsymbol{q}, \boldsymbol{p}) \in \mathbb{R}^{2f}$ and by the Hamiltonian operator

$$H_f = \sum_{j=1}^{f}\Big(\frac{\widehat{p}_i^2}{2} + \frac{\omega_i^2}{2}\widehat{q}_i^2\Big)\ .$$

Using (5.68), the Heisenberg equations of motion (5.169) for position and momentum operators $\widehat{\boldsymbol{r}} = (\widehat{\boldsymbol{q}}, \widehat{\boldsymbol{p}})$ read

$$\frac{\mathrm{d}\widehat{\boldsymbol{q}}}{\mathrm{d}t} = \widehat{\boldsymbol{p}}\ ,\quad \frac{\mathrm{d}\widehat{\boldsymbol{p}}}{\mathrm{d}t} = -\Omega^2\widehat{\boldsymbol{q}}\ ,$$

where $\Omega^2$ is the diagonal $f \times f$ matrix $\Omega^2 = \mathrm{diag}(\omega_1^2, \omega_2^2, \ldots, \omega_f^2)$. They are solved by

$$\widehat{\boldsymbol{r}}_t := \mathcal{U}_t[\widehat{\boldsymbol{r}}] = \mathrm{e}^{itH}\,\widehat{\boldsymbol{r}}\,\mathrm{e}^{-itH} = A_t\widehat{\boldsymbol{r}}\ ,\quad A_t = \begin{pmatrix} \cos\Omega t & \Omega^{-1}\sin\Omega t \\ -\Omega\sin\Omega t & \cos\Omega t \end{pmatrix}\ .$$

Because of linearity, it turns out that the time-evolution maps Weyl operators into Weyl operators,

$$\begin{aligned} W(\boldsymbol{r}) = \mathrm{e}^{i\boldsymbol{r}\cdot(\Sigma_1\widehat{\boldsymbol{r}})} &\mapsto \mathcal{U}_t[W(\boldsymbol{r})] =: W_t(\boldsymbol{r}) = \mathrm{e}^{i\boldsymbol{r}\cdot(\Sigma_1\widehat{\boldsymbol{r}}_t)} = \mathrm{e}^{i(A_t\boldsymbol{r})\cdot(\Sigma_1\widehat{\boldsymbol{r}})} \\ &= W(\boldsymbol{r}_t), \end{aligned} \tag{5.174}$$

where $\boldsymbol{r}_t$ solves the Hamilton equations for $f$ classical harmonic oscillators. Using the notation (5.95), one passes to annihilation and creation operators via the relations

$$\boldsymbol{A} = \frac{1}{\sqrt{2}}\begin{pmatrix} \Omega^{1/2} & i\,\Omega^{-1/2} \\ \Omega^{1/2} & -i\,\Omega^{-1/2} \end{pmatrix}\widehat{\boldsymbol{r}}\ .$$

The Hamiltonian then reads $H = \dfrac{1}{2}\sum_{i=1}^{f}\omega_i\,a_i^\dagger a_i$, so that

$$a_i(t) = \mathcal{U}_t[a_i] = a_i\,\mathrm{e}^{-i\omega_i t}\ ,\quad a_i^\dagger(t) = \mathcal{U}_t[a_i^\dagger] = a_i^\dagger\,\mathrm{e}^{i\omega_i t}\ .$$

4. Example 5.4.2 provides the right algebraic setting for quantizing classical dynamical systems as those studied in Example 2.1.3. As much as in that case, the quantum time-evolution will be described by a one-parameter group $\{\Theta_N^t\}_{t\in\mathbb{Z}}$ consisting of integer powers of an automorphism $\Theta_N : M_N(\mathbb{C}) \mapsto M_N(\mathbb{C})$.
Let $\mathbb{A} = \begin{pmatrix} a & b \\ c & d \end{pmatrix}$ of Example 2.1.3 be an evolution matrix with integer components and determinant equal to one. According to (2.22), the time-evolution of the exponential functions reads

$$(U_A e_{\boldsymbol{n}})(\boldsymbol{r}) = \mathrm{e}^{2\pi\, i\, \boldsymbol{n}\cdot(\mathbb{A}\boldsymbol{r})} = \mathrm{e}^{2\pi\, i\,(\mathbb{A}^T\boldsymbol{n})\cdot\boldsymbol{r}} = e_{\mathbb{A}^T\boldsymbol{n}}(\boldsymbol{r})\ , \quad \mathbb{A}^T = \begin{pmatrix} a & c \\ b & d \end{pmatrix} .$$

If one identifies $e_{\boldsymbol{m}}$ with $W_0(\boldsymbol{m})$, then the discrete Weyl relations (5.85) can be read as a *non-commutative deformation* of the fact that the exponential functions commute:

$$W_0(\boldsymbol{n})\, W_0(\boldsymbol{m}) = W_0(\boldsymbol{n}+\boldsymbol{m}) = W_0(\boldsymbol{m})\, W_0(\boldsymbol{n})\ . \tag{5.175}$$

It is thus natural to define the automorphism as

$$\Theta_N[W_N(\boldsymbol{n})] = W_N(\mathbb{A}^T\boldsymbol{n})\ , \tag{5.176}$$

and extend it linearly to the whole of $M_N(\mathbb{C})$.
In order to be an automorphism, $\Theta$ has to fulfil $\Theta_N[\mathbb{1}_N]$; then, from (5.84) and (5.86),

$$\begin{aligned}
\Theta_N[U_N^N] &= \mathrm{e}^{2\pi\, i\,\alpha_u} = W_N(N\mathbb{A}(1,0)) = W_N(N(a,b)) \\
&= \mathrm{e}^{2\pi\, i(a\alpha_u + b\alpha_v - \frac{N}{2}ab)} \\
\Theta_N[V_N^N] &= \mathrm{e}^{2\pi\, i\,\alpha_v} = W_N(N\mathbb{A}(0,1)) = W_N(N(c,d)) \\
&= \mathrm{e}^{2\pi\, i(c\alpha_u + d\alpha_v - \frac{N}{2}cd)}\ .
\end{aligned}$$

It thus follows that a discrete representation $\mathcal{W}_N^{\alpha_{u,v}}$ of the *CCR* has to be chosen such that

$$\begin{pmatrix} a & b \\ c & d \end{pmatrix}\begin{pmatrix} \alpha_u \\ \alpha_v \end{pmatrix} = \begin{pmatrix} \alpha_u \\ \alpha_v \end{pmatrix} + \frac{N}{2}\begin{pmatrix} ab \\ cd \end{pmatrix} \mod 1\ .$$

For instance, when in Example 2.1.3 $\alpha = -1$, then $\mathbb{A} = \begin{pmatrix} 1 & 1 \\ 1 & 2 \end{pmatrix} = \mathbb{A}^T$ and the choice $\alpha_{u,v} = N/2 \mod 1$ yields a finite-dimensional quantization of the Arnold Cat Map [98, 135, 135].
Furthermore, the automorphism $\Theta_N$ is implemented by a unitary operator $S_N : \mathbb{C}^N \mapsto \mathbb{C}^N$ which can be determined by means of the equation (5.89) once it is represented with respect to the chosen basis $\{|\, j\,\rangle\}_{j=0}^{N-1}$:

$$
\begin{aligned}
\langle k \,|\, S_N \,|\ell\rangle &= \sum_{p,q=0}^{N-1} \mathcal{S}_{k\ell,pq} \,\langle q \,|\, S_N \,|p\rangle \\
\mathcal{S}_{k\ell,pq} &:= \frac{1}{N} \sum_{\boldsymbol{n}\in\mathbb{Z}_N^2} \langle p|\, W_N(-\boldsymbol{n})\,|q\rangle\,\langle k \,|\, W_N((-\boldsymbol{n})\,|\ell\rangle \ .
\end{aligned}
$$

### Thermal States

In the following, we shall focus upon states of quantum systems that are left invariant by the dynamics, namely we shall be interested in equilibrium states. A well-known class of such states is represented by the *thermal* or *Gibbs states*:

$$
\rho_\beta := \frac{\mathrm{e}^{-\beta H}}{\mathcal{Z}_\beta} \ , \qquad \mathcal{Z}_\beta := \mathrm{Tr}\Big(\mathrm{e}^{-\beta H}\Big) \ , \tag{5.177}
$$

at inverse temperature $T^{-1} = \beta$ relative to a Hamiltonian $H$. Let us consider a finite level system and the two-point time-correlation functions

$$
F_{XY}(t) := \mathrm{Tr}\Big(\rho_\beta \,\mathcal{U}_t[X]\,Y\Big) \ , \quad G_{XY}(t) := \mathrm{Tr}\Big(\rho_\beta\, Y\, \mathcal{U}_t[X]\Big) \tag{5.178}
$$

for all $X, Y \in M_N(\mathbb{C})$, where the dynamical maps $\mathcal{U}_t$, $t \in \mathbb{R}$, are generated by (5.169). Simple manipulations based on the cyclicity of the trace show that

$$
\begin{aligned}
F_{XY}(t) &= \mathrm{Tr}\Big(Y\,\rho_\beta\,\mathcal{U}_t[X]\Big) = \mathrm{Tr}\Big(\rho_\beta\, Y\,\rho_\beta\,\mathcal{U}_t[X]\,\rho_\beta^{-1}\Big) \\
&= \mathrm{Tr}\Big(\rho_\beta\,\mathrm{e}^{i\,H\,(t+i\beta)}\,X\,\mathrm{e}^{-i\,H\,(t+i\beta)}\Big) = G_{XY}(t+i\beta) \ .
\end{aligned} \tag{5.179}
$$

The above equality expresses the Kubo-Marting-Schwinger (*KMS* ) conditions in their simplest form [183, 203] and Gibbs states as in (5.177) are the simplest instances of *KMS states.*

**Remarks 5.6.1.**

1. Only the Gibbs state $\rho_\beta \in M_N(\mathbb{C})$ can have two-point correlation functions satisfying (5.179); in fact,

   $$
   \mathrm{Tr}\Big(\rho\,X\,Y\Big) = \mathrm{Tr}\Big(\rho\,Y\,\mathcal{U}_{i\beta}[X]\Big) = \mathrm{Tr}\Big(\mathcal{U}_{i\beta}[X]\,\rho\,Y\Big)
   $$

   for all $Y \in M_N(\mathbb{C})$ yields

   $$
   \rho\,X = \mathrm{e}^{-\beta H}\,X\,\mathrm{e}^{\beta H}\,\rho \Longrightarrow \big[\mathrm{e}^{\beta H}\,\rho\,,\,X\big] = 0
   $$

   for all $X \in M_N(\mathbb{C})$ whence $\rho \propto \mathrm{e}^{-\beta H}$. If the *KMS* condition are taken as a signature of thermal equilibrium, the conclusion to be drawn from this example is that for finite degrees of freedom there can be only one equilibrium state at a given temperature. Therefore, in order to mathematically describe phase-transitions one needs infinitely many degrees of freedom [300].

2. At infinite temperature $\beta = 0$ and the Gibbs states reduce to a *tracial state* (see Example 5.3.3.3): $\tau(X) = \mathrm{Tr}(\frac{\mathbb{1}}{N}X)$.
3. We have seen that faithful density matrices $\rho > 0$ are naturally associated with the modular operator (5.151). The modular operator defines the *modular automorphisms* $\sigma_t^\rho : \pi_\rho(M_N(\mathbb{C})) \mapsto \pi_\rho(M_N(\mathbb{C}))$, $t \in \mathbb{R}$, given by

$$\sigma_t^\rho[\pi_\rho(X)] = \Delta_\rho^{it}\,\pi_\rho(X)\,\Delta^{-it} = \rho^{it}\otimes\rho^{-it}\,\pi_\rho(X)\,\rho^{it}\otimes\rho^{-it} . \quad (5.180)$$

They from a group, the *modular group*, and preserve the *GNS* state, $\sigma_t^\rho|\,\sqrt{\rho}\,\rangle = |\,\sqrt{\rho}\,\rangle$, so that (5.152) reads

$$\sigma_{-i/2}^\rho(\pi_\rho(X))|\,\sqrt{\rho}\,\rangle = J_\rho\pi_\rho(X)^\dagger|\,\sqrt{\rho}\,\rangle = |\,\sqrt{\rho}X\,\rangle .$$

Further, it turns out that $\rho$ is a *KMS* state at inverse temperature $\beta = 1$ with respect to $\sigma_{-t}^\rho$,

$$\begin{aligned}\langle\,\sqrt{\rho}\,|\,\sigma_{-t}^\rho[\pi_\rho(X)]\pi_\rho(Y)\,|\sqrt{\rho}\,\rangle &= \mathrm{Tr}\Big(\rho^{1-it}\,X\,\rho^{it}\,Y\Big)\\ &= \mathrm{Tr}\Big(\rho Y\,\rho^{-i(t+i)}\,X\,\rho^{i(t+i)}\Big) = \langle\,\sqrt{\rho}\,|\,\pi_\rho(Y)\sigma_{-(t+i)}^\rho[\pi_\rho(X)]\,|\sqrt{\rho}\,\rangle .\end{aligned}$$

By means of the modular group, when a faithful $\rho$ is decomposed into a linear convex combination of other density matrices $\sigma_j$, $\rho = \sum_j \lambda_j\sigma_j$, its decomposers in (5.146) can be recast as follows,

$$\begin{aligned}\lambda_j\sigma_j(X) &= \langle\,\sqrt{\rho}\,|\,\pi_\rho(X)\,|\sqrt{\rho}X_j\,\rangle = \langle\,\sqrt{\rho}\,|\,\pi_\rho(X)\,\sigma_{-i/2}^\rho[\pi_\rho(X_j)]\,|\sqrt{\rho}\,\rangle\\ &= \langle\,\sqrt{\rho}\,|\,\sigma_{i/2}^\rho[\pi_\rho(X_j)]\,\pi_\rho(X)\,|\sqrt{\rho}\,\rangle . \end{aligned} \quad (5.181)$$

The two-point correlation functions $F_{XY}(t)$, respectively $G_{XY}(t)$ can be analytically extended to the strip $\{t + iy : -\beta < y < 0\}$, respectively to the strip $\{t + iy : 0 < y < \beta\}$ where they are continuous and bounded, including the boundaries where they satisfy the *KMS* conditions (5.179). When the Gibbs state (5.177) is a density matrix, $\rho_\beta \in \mathbb{B}_1^+(\mathbb{H})$, these properties which are almost obvious in the case of finite-level systems, can be extended to systems with an infinite dimensional Hilbert space $\mathbb{H}$ [107, 300].

**Examples 5.6.2.**

1. **Spin** $1/2$: The density matrix in Example 5.5.5 corresponds to Gibbs states with Hamiltonian $H = \omega\,\sigma_3$ and temperature $\beta^{-1}$ such that

$$\rho = \frac{1}{2}\begin{pmatrix}1-s & 0\\ 0 & 1+s\end{pmatrix} = \frac{1}{2\cosh\beta\omega}\,\mathrm{e}^{-\beta\omega\sigma_3} ,$$

with $s = \tanh \beta\omega$. Therefore, the modular group consists of

$$\Delta_\rho^t = \rho^{it} \otimes \rho^{-it} = \mathrm{e}^{-it\beta\omega\sigma_3} \otimes \mathrm{e}^{it\beta\omega\sigma_3} = \mathrm{e}^{-it\beta\omega(\sigma_3\otimes\mathbb{1}_2 - \mathbb{1}_2\otimes\sigma_3)} .$$

2. **Fermions:** Let $N$ Fermionic modes, described by creation and annihilation operators $a_i^\#$, $i = 1, 2, \dots, N$, satisfying the *CAR* $\{a_i\,, a_j^\dagger\} = \delta_{ij}$, be equipped with a Hamiltonian operator

$$H = \sum_{i=1}^N \varepsilon_i\, a_i^\dagger\, a_i .$$

This can be regarded as the *second quantization* of an $N$-level, *one-particle Hamiltonian* $h = \sum_{i=1}^N \varepsilon_i |\, i\,\rangle\langle\, i\,| \in M_N(\mathbb{C})$ and $a_i^\dagger$ as the creation operator of a Fermion in the eigenstate $|\, i\,\rangle$ with energy $\varepsilon_i$.
The *partition function* $Z_\beta$ is easily calculated since for each mode the occupation number states are $|\,0\,\rangle$ and $|\,1\,\rangle$ (see Example 5.4.1):

$$Z_\beta = \mathrm{Tr}\Big(\mathrm{e}^{-\beta H}\Big) = \prod_{i=1}^N \sum_{n_i=0}^1 \mathrm{e}^{-\beta\varepsilon_i\, n_i} = \prod_{i=1}^N \Big(1 + \mathrm{e}^{-\beta\varepsilon_i}\Big) ,$$

whence the Gibbs state of $N$ non-interacting Fermions read

$$\rho_\beta = \prod_{i=1}^N \Big(1 + \mathrm{e}^{-\beta\varepsilon_i}\Big)^{-1} \mathrm{e}^{-\beta H} .$$

In thermodynamics, Gibbs states are *canonical equilibrium states*, $\rho_\beta^C$, while *gran-canonical states* have the form

$$\rho_\beta^{GC} = \prod_{i=1}^N \Big(1 + \mathrm{e}^{-\beta(\varepsilon_i - \mu)}\Big)^{-1} \mathrm{e}^{-\beta\,(H - \mu\, N)} ,$$

where $\mu$ is the *chemical potential* and $\widehat{N} = \sum_{i=1} a_i^\dagger\, a_i$ is the number operator. Two-point expectations read

$$\mathrm{Tr}(\rho_\beta^{GC}\, a_i^\dagger\, a_j) = \delta_{ij}\, \frac{z}{\mathrm{e}^{\beta\varepsilon_i} + z} , \tag{5.182}$$

where $z = \mathrm{e}^{\mu\beta}$ is the so called fugacity. As regards higher order correlation functions, by means of the *CCR* anyone of them can be reduced to sums of expectations with equal numbers of annihilation and creation operators matching in pairs:

$$\mathrm{Tr}(\rho_\beta\, a_{i_p}^\dagger \cdots a_{i_1}^\dagger\, a_{j_1} \cdots a_{j_q}) = \delta_{pq} \mathrm{Det}\Big(\Big[\mathrm{Tr}(\rho\, a_{i_k}^\dagger\, a_{j_\ell})\Big]_{k,\ell=1}^N\Big) . \tag{5.183}$$

By suitably shifting the one-particle Hamiltonian, one may always assume the lowest eigenvalue (ground state energy) to be 0, whence

$$0 \le \mathrm{Tr}\Big(\rho^{GC}_\beta\, a^\dagger_1\, a_1\Big) = \frac{z}{1+z} \le 1 \ ,$$

for the *CAR* imply $\|a^\dagger a\| \le 1$, so that $z \ge 0$.

3. **Bosons:** Let $a^\#_i$, $i = 1, 2, \dots, N$, satisfy the *CCR* $[a_i\,, a^\dagger_j] = \delta_{ij}$ of a system of $N$ Bosonic modes with a second-quantized Hamiltonian

$$H = \sum_{i=1}^N \varepsilon_i\, a^\dagger_i\, a_i \ , \quad \varepsilon_i \ge 0 \ .$$

The *partition function* reads

$$Z_\beta = \mathrm{Tr}\Big(\mathrm{e}^{-\beta H}\Big) = \prod_{i=1}^N \sum_{n_i=0}^\infty \mathrm{e}^{-\beta \varepsilon_i\, n_i} = \prod_{i=1}^N \Big(1 - \mathrm{e}^{-\beta\varepsilon_i}\Big)^{-1} \ ,$$

and the canonical and gran-canonical equilibrium states have the form

$$\rho^C_\beta = \prod_{i=1}^N \Big(1 - \mathrm{e}^{-\beta\varepsilon_i}\Big)\, \mathrm{e}^{-\beta H} \ , \quad \rho^{GC}_\beta = \prod_{i=1}^N \Big(1 - \mathrm{e}^{-\beta(\varepsilon_i - \mu)}\Big)\, \mathrm{e}^{-\beta\,(H - \mu \widehat{N})} \ .$$

The Bose two-point correlation functions read

$$\mathrm{Tr}(\rho^{GC}_\beta\, a^\dagger_i\, a_j) = \delta_{ij}\, \frac{z}{\mathrm{e}^{\beta\varepsilon_i} - z} \ , \tag{5.184}$$

while $2N$-point ones are of the form

$$\mathrm{Tr}(\rho_\beta\, a^\dagger_{i_p} \cdots a^\dagger_{i_1}\, a_{j_1} \cdots a_{j_q}) = \delta_{pq} \mathrm{Per}\Big(\Big[\mathrm{Tr}(\rho\, a^\dagger_{i_k}\, a_{j_\ell})\Big]_{k,\ell=1}^N\Big) \ , \tag{5.185}$$

where, unlike for Fermions, $2N$ point correlation functions do not assign different signs to different permutations whence a *permanent* appears instead of a determinant. Furthermore, with the ground state energy set equal to 0,

$$\mathrm{Tr}\Big(\rho^{GC}_\beta\, a^\dagger_1\, a_1\Big) = \frac{z}{1-z} \Longrightarrow 0 \le z < 1 \ .$$

The fact that when $z \to 1$, the ground level can be infinitely populated is the source of the phenomenon of *Bose-Einstein condensation.* Canonical and gran-canonical $N$ Bose states as $\rho^{GC}$ are Gaussian (see Example 5.5.2); indeed, the characteristic functions (5.116) of $\rho^C_\beta$ equals

$$\mathrm{Tr}\Big(\rho_\beta\, W(\boldsymbol{z})\Big) = \prod_{i=1}^N \Big(1 - \mathrm{e}^{-\beta\varepsilon_i}\Big)\, \mathrm{Tr}\Big(\mathrm{e}^{-\beta\varepsilon_i\, a^\dagger_i\, a_i} \mathrm{e}^{z_i a_i - z^*_i a^\dagger_i}\Big) \ .$$

In order to compute the trace, it is convenient to split $W(\boldsymbol{z})$ as in (5.77), and to use the overcomplete basis of coherent states (5.108) expressed as in (5.105); this yields

$$\mathrm{Tr}\Big(\mathrm{e}^{-\beta\varepsilon_i\, a_i^\dagger\, a_i}\mathrm{e}^{z_i a_i - z_i^* a_i^\dagger}\Big) = \mathrm{e}^{-|z_i|^2/2}\mathrm{Tr}\Big(\mathrm{e}^{z_i a_i}\mathrm{e}^{-\beta\varepsilon_i a_i^\dagger a_i}\mathrm{e}^{-z_i^* a_i^\dagger}\Big)$$

$$= \mathrm{e}^{-|z_i|^2/2}\int \frac{\mathrm{d}w_i}{\pi}\langle\, w_i\,|\,\mathrm{e}^{z_i a_i}\mathrm{e}^{-\beta\varepsilon_i a_i^\dagger a_i}\mathrm{e}^{-z_i^* a_i^\dagger}\,|w_i\,\rangle$$

$$= \mathrm{e}^{-|z_i|^2/2}\int \frac{\mathrm{d}w_i}{\pi}\,\mathrm{e}^{-|w_i|^2}\,\langle\, vac\,|\,\mathrm{e}^{(z_i+w_i^*)a_i}\mathrm{e}^{-\beta\varepsilon_i a_i^\dagger a_i}\mathrm{e}^{-(z_i^*-w_i)a_i^\dagger}\,|vac\rangle\ .$$

Taking into account that (see (5.100)), for any $\alpha\in\mathbb{C}$,

$$\mathrm{e}^{\alpha\,\varepsilon\, a^\dagger\, a}\, a\,\mathrm{e}^{-\alpha\,\varepsilon\, a^\dagger\, a} = \sum_{k=0}^{\infty}\frac{\alpha^k}{k!}\underbrace{[H\,,[H\,,\cdots[\,H\,,a]]\cdots]}_{k\,\text{times}} = \mathrm{e}^{-\alpha\varepsilon}\,a$$

$$\mathrm{e}^{\alpha\,\varepsilon\, a^\dagger\, a}\, a^\dagger\,\mathrm{e}^{-\alpha\,\varepsilon\, a^\dagger\, a} = \mathrm{e}^{\alpha\varepsilon}\,a^\dagger$$

and that (5.76) yields

$$\mathrm{e}^{\alpha\, a}\,\mathrm{e}^{\gamma\, a^\dagger} = \mathrm{e}^{\alpha\gamma/2}\,\mathrm{e}^{\alpha\, a+\gamma\, a^\dagger} = \mathrm{e}^{\alpha\gamma}\,\mathrm{e}^{\gamma\, a^\dagger}\,\mathrm{e}^{\alpha\, a}\ ,\qquad \alpha\in\mathbb{C}\ ,$$

one obtains, by Gaussian integration,

$$\int\frac{\mathrm{d}w_i}{\pi}\,\mathrm{e}^{-|w_i|^2}\,\langle\, vac\,|\,\mathrm{e}^{(z_i+w_i^*)a_i}\mathrm{e}^{-\beta\varepsilon_i a_i^\dagger a_i}\mathrm{e}^{-(z_i^*-w_i)a_i^\dagger}\,|vac\rangle =$$

$$= \int\frac{\mathrm{d}w_i}{\pi}\,\mathrm{e}^{-|w_i|^2-(z_i+w_i^*)(z_i^*-w_i)\mathrm{e}^{-\beta\varepsilon_i}} = \frac{\mathrm{e}^{-|z_i|^2\mathrm{e}^{-\beta\varepsilon_i}(1-\mathrm{e}^{-\beta\varepsilon_i})^{-1}}}{1-\mathrm{e}^{-\beta\varepsilon_i}}\ .$$

Therefore, one derives the form of the correlation matrix (5.122) from

$$\mathrm{Tr}\Big(\rho_\beta\, W(\boldsymbol{z})\Big) = \exp\Big(-\frac{1}{2}\sum_{i=1}^{N}|z_i|^2\coth\frac{\beta\varepsilon_i}{2}\Big)$$

$$= \exp\left(-\frac{1}{4}(\boldsymbol{z},-\boldsymbol{z}^*)\begin{pmatrix}\coth\frac{\beta\, h}{2} & 0\\ 0 & \coth\frac{\beta\, h}{2}\end{pmatrix}\begin{pmatrix}\boldsymbol{z}^*\\ -\boldsymbol{z}\end{pmatrix}\right)\ ,\qquad (5.186)$$

where $h=\sum_{i=2}^{N}\varepsilon_i|\,i\,\rangle\langle\, i\,|$ $(\varepsilon_1=0)$ has been used.

4. The relation (5.87) in Example 5.4.2, allows to equip the quantized hyperbolic automorphisms of the torus $\mathbb{T}^2$ with the $\Theta_N$-invariant state $\omega_N$ defined by

$$\omega_N(W_N(\boldsymbol{n})) := \frac{1}{N}\mathrm{Tr}(W_N(\boldsymbol{n})) = \delta_{\boldsymbol{n}\boldsymbol{0}}\qquad (5.187)$$

on the Weyl operators and extended by linearity to their linear span, where it amounts to the normalized trace.

### 5.6.1 Quantum Operations

A major departure from classical mechanics is represented by the role played in quantum mechanics by the measurement processes where a microscopic system, $S$, on which the measurement is performed, interacts with a (usually) macroscopic system, $E$, the measuring apparatus.

In a classical, commutative context it is always possible, at least in line of principle, to make negligible the effects on $S$ due to its interaction with $E$; instead, in quantum mechanics states are generically unavoidably perturbed when undergoing a measurement process. The standard way the quantum mechanical perturbations are taken into account is via the so-called *wave-packet reduction postulate*; in its simplest formulation it goes as follows. Let $X = X^\dagger$ be an observable with discrete, finite and non-degenerate spectrum, say $X = \sum_{j=1}^d x_j P_j$, $P_j := |\psi_j\rangle\langle\psi_j|$. Upon measuring $X$ on a system $S$, the outcomes are the eigenvalues $x_j$; the measurement process can be schematized as follows: a beam of copies of a same system $S$, all prepared so as to be described by a same state $\rho$, are sent through an apparatus that measures the eigenvalues $x_j$ leaving the system state in the corresponding eigenprojections $P_j$ and direct them towards a screen with $d$ slits. By opening the $j$th slit, the others being kept closed, only those systems on which the eigenvalue $x_j$ has been measured are collected. Suppose $N_j$ of the $N$ systems that interacted with the apparatus reach the screen through the $j$-th slit; then, the ratio $N_j/N$ approximates the quantity

$$p_j^\rho := \mathrm{Tr}(\rho\, P_j) = \langle\psi_j|\rho|\psi_j\rangle$$

when $N$ becomes sufficiently large. If no selection is operated, that is if all the $d$ slits are left open, after sufficiently many repetitions of the experiment with the same state preparation, the collected mixture of systems is described by the projections $P_j$ weighted with the corresponding mean values $p_j^\rho$. Thus, a typical non-selective measurement process changes the state as follows:

$$\rho \mapsto \sum_{j=1}^d p_j^\rho\, P_j = \sum_{j=1}^d |\psi_j\rangle\langle\psi_j|\rho|\psi_j\rangle\langle\psi_j| = \underbrace{\sum_{j=1}^d P_j\,\rho\, P_j}_{\mathbb{F}_{\mathcal{P}}[\rho]} . \tag{5.188}$$

The map $\mathbb{F}_{\mathcal{P}}$ is linear on the state-space $\mathcal{S}(S)$ and transforms states into states: indeed, $\mathbb{F}_{\mathcal{P}}[\rho] \geq 0$ and $\mathrm{Tr}(\mathbb{F}_{\mathcal{P}}[\rho] = \mathrm{Tr}(\rho) = 1$, as one can check by using the cyclicity of the trace and the fact that the $P_j$'s constitute a resolution of the identity, $\sum_j P_j = \mathbb{1}$.

In general, the instantaneous change from $\rho$ into $\mathbb{F}_{\mathcal{P}}[\rho]$ transforms pure states into mixtures and may intuitively be associated with the loss of information due to the interaction with the many degrees of freedom of the macroscopic measuring apparatus. As a consequence, contrary to classical mechanics, quantum mechanics distinguishes between two state-changes, a reversible one due to the Liouville time-evolution and an irreversible one, the wave-packet reduction, describing the action of measurement processes.

**Remark 5.6.2.** Effectively, while being subjected to a measurement, any quantum micro-system is to be considered as an open quantum system dynamically and statistically correlated with the (infinitely) many, degrees of freedom of the measure instrument. This many-body interaction is usually not controllable and the phenomenological description of its overall effects is via maps as in (5.188). In particular, a measurement process of an observable on a system whose state $|\,\psi\,\rangle$ is a coherent superposition of the observable eigenstates, $|\,\psi\,\rangle = \sum_{j=1}^{d} c_j|\,\psi_j\,\rangle$, transforms it into a mixture $\rho = \sum_{j=1}^{d} |c_j|^2|\,\psi_j\,\rangle\langle\,\psi_j\,|$, with consequent loss of coherence.

The existence of two basic quantum time-evolutions, one reversible typical of closed quantum systems, the other one irreversible and related to measurement processes, is unsatisfactory from an epistemological point of view. All the more so, since the irreversible macroscopic behavior of system plus apparatus should be deducible from the reversible dynamics of their constituent microsystems. Alongside with the problem of reconciling thermodynamical irreversibility with microscopic reversibility, quantum mechanics raises the question of how to reconcile a reversible microscopic dynamics which preserves the purity of states with an irreversible macroscopic one which transforms pure states into mixtures. A number of approaches have been developed to attack this problem, for a thorough review of one of them which is based on a modification of microscopic dynamics by the insertion of a decoherent mechanism with negligible effects on microsystems, but substantial ones on macrosystems see [21].

It is convenient to extend the notion of wave-packet reduction to that of *positive operator-valued measures* (*POVM*).

The key property of a map as in (5.188) is its structure and the use on the right and left of $\rho$ of operators such that $\sum_j P_j^2\Big(= \sum_j P_j\Big) = 1$. The generalization is quite natural.

**Definition 5.6.1 (Partitions of Unity).** *Let $E_j \in \mathbb{B}(\mathbb{H})$, $j \in J$, be a selection of operators such that $\sum_{j\in J} E_j^\dagger E_j = \mathbb{1}$: it is usually referred to as a POVM or a partition of unity. One associates to it the linear map $\mathbb{F}_{\mathcal{E}} : \mathcal{S}(S) \mapsto \mathcal{S}(S)$,*

$$\mathbb{F}_{\mathcal{E}}[\rho] = \sum_{j\in J} E_j\, \rho\, E_j^\dagger \ . \tag{5.189}$$

In Example 5.6.4, we shall discuss the interpretation of generic *POVMs* in relation to measurement processes; for the moment, it suffices to stress that the operators forming *POVMs* need neither be self-adjoint nor orthogonal projections. While the von Neumann entropy is constant under the Liouville time-evolution; on the contrary, under a generic *POVM*, it can increase or decrease.

**Example 5.6.3.** If $\rho = Q$ is a pure state (one-dimensional projection), then $S(Q) = 0$, while under the action of a wave-packet reduction, $\mathbb{F}_{\mathcal{E}}[Q]$ gets mixed and $S(\mathbb{F}_{\mathcal{E}}[Q]) \geq 0$. However, if one starts with a mixed $\rho$, $S(\mathbb{F}_{\mathcal{E}}[\rho])$ can be smaller than $S(\rho)$: take for instance $E_1 = |1\rangle\langle 0|$, $E_2 = |1\rangle\langle 1|$, where $|0\rangle$, $|1\rangle$ are a basis in $\mathbb{C}^2$, then, $E_1^\dagger E_1 + E_2^\dagger E_2 = \mathbb{1}$ and, for all $\rho \in \mathcal{S}(S)$,

$$\mathbb{F}_{\mathcal{E}}[\rho] = |1\rangle\langle 0|\rho|0\rangle\langle 1| + |1\rangle\langle 1|\rho|1\rangle\langle 1| = |1\rangle\langle 1|\,\mathrm{Tr}(\rho) = |1\rangle\langle 1|\ .$$

Thus, for any given mixed $\rho$, $S(\rho) > S(\mathbb{F}_{\mathcal{E}}[\rho]) = 0$.

As we shall see in Section 6.1, the use of generic *POVMs* , that is not made of orthogonal projections, is practically useful when one wants to distinguish between non-orthogonal quantum states. However, consider the statement

measuring the (orthonormal) *POVM* $\mathcal{P} := \{P_j\}_{j\in J} \in \mathbb{B}(\mathbb{H})$ on the system $S$ in the state $\rho$ corresponds to the irreversible map $\rho \mapsto \mathbb{F}_{\mathcal{P}}[\rho] = \sum_{j\in J} P_j \rho P_j$.

This has an acceptable interpretation in physical terms for the orthogonality of the $P_j$'s reduces the experimental measure of $\mathcal{P}$ to an experiment with $\#(J)$ slits. The same argument does not directly work when projective *POVMs* $\mathcal{P}$ are substituted with generic $\mathcal{E} := \{E_j\}_{j\in J}$, $\sum_{j\in J} E_j^\dagger E_j = \mathbb{1}$. Consider the statement

measuring a generic *POVM* $\mathcal{E} := \{E_j\}_{j\in J} \in \mathbb{B}(\mathbb{H})$ on the system $S$ in the state $\rho$ corresponds to the irreversible map $\rho \mapsto \mathbb{F}_{\mathcal{E}}[\rho] = \sum_{j\in J} E_j \rho E_j^\dagger$.

In order to give it a meaning, one has to specify what is measured and on which system; indeed, the non-orthogonality of the $E_j$'s makes untenable the straightforward interpretation accorded to projective measurements. An answer to the above question is given in terms of *couplings to ancillas* and partial tracing.

**Example 5.6.4.** [224] Let $\mathcal{E} := \{E_j\}_{j\in J} \subseteq \mathbb{B}(\mathbb{H})$ be a *POVM* for a system $S$. Let $R$ be an auxiliary system described by a Hilbert space $\mathbb{K}$ which provides an abstract quantum description of an instrument to which $S$ is coupled during a measurement. A schematic description of a measurement process associated with $\mathcal{E}$ is as follows:

1. there exist orthonormal bases, $\{|\psi_j\rangle\} \in \mathbb{H}$ and $\{|k\rangle\}_{k\geq 0} \in \mathbb{K}$, with $|0\rangle$ corresponding to the ready-state of the measurement apparatus;
2. there is a unitary time-evolution operator $U_t$ on $\mathbb{H}\otimes\mathbb{K}$ such that, at the end of the process, at time $t = T$ say, for any initial $\psi \in \mathbb{H}$, one has

$$U_T|\psi\rangle \otimes |0\rangle = \sum_{j\in J} E_j|\psi\rangle \otimes |j\rangle =: |\Psi\rangle\ .$$

The unitary operator $U_T$ is well-defined: indeed, the right hand side of the above equality can be taken as a definition of $U_T$ as a linear operator from $\mathbb{H}\otimes|\,0\,\rangle$ into $\mathbb{H}\otimes\mathbb{K}$. Since $\sum_{j\in J} E_j^\dagger E_j = \mathbb{1}_S$, where $\mathbb{1}_S$ denotes the identity operator on $\mathbb{H}$, scalar products of vectors in the subspace $\mathbb{H}\otimes|\,0\,\rangle$ are preserved and the isometry $U_T$ can be extended to a unitary operator on $\mathbb{H}\otimes\mathbb{K}$. Let the compound system $S+R$ be in the state $\Psi$, according to the postulate of wave-packet reduction, by measuring the eigenprojectors $\mathbb{1}_S\otimes P_k$, $P_k := |\,k\,\rangle\langle\,k\,|$, $k\in J$, the outcoming (not-normalized) states

$$\mathbb{1}_S\otimes P_k\,|\,\Psi\,\rangle\langle\,\Psi\,|\mathbb{1}_S\otimes P_k = \Big(E_k|\,\psi\,\rangle\langle\,\psi\,|E_k^\dagger\Big)\otimes P_k\ ,$$

are obtained with probabilities

$$\pi_k(\psi) := \langle\,\Psi\,|\,\mathbb{1}_S\otimes P_k\,|\Psi\,\rangle = \langle\,\psi\,|\,E_k^\dagger E_k\,|\psi\,\rangle\ .$$

By disregarding the "instrument" $R$, the overall effect of the entire process on the system $S$ alone is as follows:

- by measuring the projections $\mathbb{1}_S\otimes P_k$ on $S+R$ after the action of $U_T$ on $|\,\psi\,\rangle\otimes|\,0\,\rangle$, the state $|\,\psi\,\rangle$ changes into the normalized states

$$|\,\widetilde{\psi}_k\,\rangle := \frac{E_k|\,\psi\,\rangle\langle\,\psi\,|E_k^\dagger}{\sqrt{\langle\,\psi\,|\,E_k^\dagger E_k^\dagger\,|\psi\,\rangle}}\ ,$$

  with probabilities $\pi_k(\psi)$;
- without selection, the overall effect is

$$|\,\psi\,\rangle\langle\,\psi\,|\mapsto\sum_{j\in J}\pi_j(\psi)\,|\,\widetilde{\psi}_j\,\rangle\langle\,\widetilde{\psi}_j\,| = \sum_{j\in J} E_j\,|\,\psi\,\rangle\langle\,\psi\,|\,E_j^\dagger = \mathbb{F}_{\mathcal{E}}[|\,\psi\,\rangle\langle\,\psi\,|]\ .$$

- The process described by (5.189) is obtained by linear extension of the action of $\mathbb{F}_{\mathcal{E}}$ from projectors to mixtures of projectors.

**Remark 5.6.3.** The previous one is an example of *dilation* of a *CP* map to a unitary evolution on a larger system from which the former is obtained by partial tracing [109]. In general, any *POVM* $\mathcal{E} := \{E_j\}_{j\in J}\subseteq\mathbb{B}(\mathbb{H})$ can be dilated to a projective *POVM* $\mathcal{P} = \{P_j\}_{j\in J}$ consisting of orthogonal projectors $P_j$ on a larger Hilbert space $\mathbb{K}$ [143]. For *POVMs* such that $\mathrm{card}(J)=d$, the proof goes as follows: consider the Hilbert space $\mathbb{K}_{\mathcal{E}} = \mathbb{H}\otimes\mathbb{C}^d$ linearly spanned by vectors of the form

$$|\,\Psi\,\rangle_{\mathcal{E}} := \sum_{j=1}^d E_j|\,\psi_j\,\rangle\otimes|\,j\,\rangle\ ,$$

where $|\,\psi_j\,\rangle\in\mathbb{H}$ and $\{|\,j\,\rangle\}_{j=1}^d$ is an *ONB* in the auxiliary Hilbert space $\mathbb{C}^d$. Let $|\,\phi\,\rangle\in\mathbb{H}$ and set $|\,\Psi_\phi\,\rangle_{\mathcal{E}} := \sum_{j=1}^d E_j|\,\phi\,\rangle\otimes|\,j\,\rangle$. The operators $P_j$ on $\mathbb{K}_{\mathcal{E}}$ defined by

$$P_j|\Psi\rangle = E_j|\psi_j\rangle \otimes |j\rangle$$

are orthogonal projections such that $\sum_{j=1}^d P_j = \mathbb{1}$ on $\mathbb{K}_\mathcal{E}$ and the projective *POVM* $\mathcal{P} = \{P_j\}_{j=1}^d \subseteq \mathbb{B}(\mathbb{K}_\mathcal{E})$ is such that

$$\sum_{j=1}^d P_j\,|\Psi_\phi\rangle_{\mathcal{E}\mathcal{E}}\langle\Psi_\phi\,|\,P_j = \sum_{j=1}^d E_j\,|\phi\rangle\langle\phi\,|\,E_j^\dagger \otimes |j\rangle\langle j\,|$$

whence (5.189) results by tracing over the auxiliary Hilbert pace $\mathbb{C}^d$.

### 5.6.2 Open Quantum Dynamics

Despite the practical impossibility of describing the interaction between a micro-system and a macro-system during a measurement process, it is not without hope to try a dynamical derivation of the wave-packet reduction (5.189). The idea is that the latter is a time asymptotic effect of a many-body interaction whose time-scale is much shorter than the duration of the process. The phenomenological description of the process cannot be given in terms of an automorphism $\mathcal{U}_t$: on one hand, $\mathcal{U}_t$ is reversible, while the wave-packet reduction is not, on the other hand $\mathcal{U}_t$ cannot transform pure into mixed states.

A straightforward way to extend the quantum time-evolution beyond the reversible one generated by the unitary Liouville equation is to add some extra structure to (5.168). One observes that the commutator corresponds to a linear action on the state-space $\mathcal{S}(S)$, and that the generated dynamical maps $\mathcal{U}_t$ satisfy the composition law $\mathcal{U}_t \circ \mathcal{U}_s = \mathcal{U}_{t+s}$ for all $s, t \in \mathbb{R}$.

A sensible step is to modify (5.168) by adding to the commutator a linear term that breaks time-reversibility and generates a semi-group $\Gamma_t$, $t \geq 0$, of linear maps obeying a forward-in-time composition law $\Gamma_t \circ \Gamma_s = \Gamma_{t+s}$, where now $s, t \geq 0$. Namely, one tries to substitute (5.168) with a time-evolution equation of the form

$$\partial_t \rho(t) = L_H[\rho(t)] + D[\rho(t)] \,. \tag{5.190}$$

Formally, the semi-group of linear maps $\{\Gamma_t\}_{t\geq 0}$, solutions of (5.190), is obtained by exponentiating the generator:

$$\rho \mapsto \rho(t) = \Gamma_t[\rho] \,, \quad \Gamma_t := \mathrm{e}^{t\,L} \,, \quad L = L_H + D \,. \tag{5.191}$$

Not all linear maps $D$ lead to physically consistent irreversible time-evolutions $\Gamma_t$; the following conditions result necessary:

1. $\mathrm{Tr}(D[\rho]) = 0$: since $\mathrm{Tr}([H\,,\,\rho_t]) = 0$, this implies trace-conservation $\partial_t \mathrm{Tr}(\rho_t) = 0$;
2. $D[\rho]^\dagger = D[\rho]$: this guarantees preservation of hermiticity;

3. the positivity $\Gamma_t[\rho]$ must be preserved at all times $t \geq 0$.

While the first condition can be relaxed, for instance in the case of decaying systems [11], which we shall not consider, the other two conditions are instead necessary to ensure that $\Gamma_t$ map density matrices into density matrices. However, positivity-preservation alone does not suffice for the full physical consistency of $\Gamma_t$, the stronger property of complete positivity discussed in Section 5.2.2 turns out to be necessary. Despite its mathematical origin, this notion is deeply rooted in quantum physics. Its importance was firstly appreciated in the theory of open quantum systems [11, 96, 181].

Equations of the form (5.190) that lead to semi-groups of dynamical maps that break time-reversibility are usually derived when one thinks of $S$ as a subsystem immersed in a large (infinite) reservoir, or heat bath, $R$. Practically, one deals with a situation similar to the one in Example 5.6.4 and uses a partial tracing technique: the system $S$ is not closed, but coupled to a large system $R$. The system $S+R$ is described by the tensor product Hilbert space $\mathbb{H} \otimes \mathbb{K}$, its states $\rho_{S+R}$ are density matrices on such a space and evolve in time according to the unitary time-evolution

$$\rho_{S+R} \mapsto \rho_{S+R}(t) := \mathcal{U}_t^{S+R}[\rho_{S+R}] = U_{S+R}(t)\, \rho_{S+R}\, U_{S+R}^\dagger(t),$$

generated through (5.168) by a Hamiltonian of the form

$$H = H_S + H_R + \lambda H_I \ , \tag{5.192}$$

where $H_{S,R}$ are the Hamiltonian operators describing the reversible time-evolutions of system and reservoir alone, while $H_I$ takes into account their interactions with $\lambda$ an adimensional coupling constant.

The interaction Hamiltonian is such that there are practically no hopes to arrive at an explicit unitary time-evolution $U_{S+R}(t)$. On the other hand, one is interested in the dynamics of the open system $S$ alone. Furthermore, in many situations of physical interest, one may reasonably assume that there are no statistical correlations between system and reservoir at $t = 0$; namely, the initial state of the compound system can be taken of the *factorized* form $\rho_{S+R} = \rho_S \otimes \rho_R$ of the initial condition. Then, by tracing over the environment degrees of freedom, one obtains a one-parameter family of dynamical maps

$$\rho_S \mapsto \rho_S(t) := \mathrm{Tr}_{\mathbb{K}}(\rho_{S+R}(t)) \tag{5.193}$$

on the state-space of $S$ which is called *reduced dynamics.*

Together with the fixed form of the initial condition, the elimination of the environment degrees of freedom by means of the partial trace $\mathrm{Tr}_{\mathbb{K}}$ makes the evolution irreversible. The factorized initial condition does get entangled in the course of time, so that, in general, the family $\{\rho_S(t)\}_{t \geq 0}$ of states satisfies a highly complicated integro-differential evolution equation of the form

$$\partial_t \rho_S(t) = \int_0^t \mathrm{d}s\, L_s^\lambda[\rho_S(t-s)] \ , \tag{5.194}$$

in which the linear operator $L_s^\lambda$ on the state space $\mathcal{S}(S)$ of the system exhibits memory effects that account for the entanglement of the system with the reservoir from time $t = 0$ to time $t > 0$. Before dealing with how one can eliminate the memory effects and get an evolution equation that generate a semi-group of maps on $\mathcal{S}(S)$, it is convenient to examine (5.193) in some more detail.

Let for sake of simplicity assume that the initial state of the reservoir is described by a density matrix be $\rho_R = \sum_j p_j | \psi_j^R \rangle\langle \psi_j^R |$, where $p_j \geq 0$, $\sum_j p_j = 1$, and the $\psi_j^R$ form an orthonormal basis in $\mathbb{K}$. We use them to calculate the partial trace $\mathrm{Tr}_{\mathbb{K}}$:

$$\rho_S(t) = \sum_{ij} p_j \langle \psi_i^R | U_{S+R}(t) | \psi_j^R \rangle \, \rho_S \, \langle \psi_j^R | U_{S+R}^\dagger(t) | \psi_i^R \rangle \, .$$

Notice that the matrix elements provide operators

$$V_{ij}(t) := \sqrt{p_j} \langle \psi_i^R | U_{S+R}(t) | \psi_j^R \rangle \, : \, \mathbb{H} \mapsto \mathbb{H} \, ,$$

so that the reduced dynamics corresponds to maps

$$\rho_S \mapsto \Lambda_t[\rho_S] := \sum_{ij} V_{ij}(t) \, \rho_S \, V_{ij}^\dagger(t) \, . \tag{5.195}$$

According to Proposition 5.2.1, the resulting dynamical maps are *CPU* with the $V_{ij}(t)$ as Kraus operators. However, they do not form a semi-group because of the memory effects built in the integro-differential equation they satisfy. Under the hypothesis of a very weak coupling between $S$ and $R$ ($\lambda << 1$), a semi-group reduced dynamics is obtained by performing suitable *Markov approximations*, the most straightforward being the substitution

$$\int_0^t \mathrm{d}s \, L_s^\lambda[\rho_S(t-s)] \mapsto L[\rho_S(t)] := \left( \int_0^{+\infty} \mathrm{d}s \, L_s^\lambda \right) [\rho_S(t)] \, .$$

Since the memory effects have been eliminated, $L$ is a generator corresponding to a Liouville equation or *master equation* as in (5.191).

In the so-called *weak-coupling limit* [105], the Markov approximation sketched above can be understood as follows: an expansion to second order in the small coupling constant $\lambda$ shows that (5.194) becomes

$$\partial_t \rho_S(t) = -i[H_S + \lambda^2 H_1 \, , \, \rho_S(t)] + \lambda^2 \int_0^t \mathrm{d}s \, D(s)[\rho_S(t-s)] \, ,$$

where $D[\cdot]$ acts linearly on the state space $\mathcal{S}(S)$. The effects due to the presence of the reservoir are thus visible on a time-scale $\tau = t\lambda^2$ which is slow as $\lambda \ll 1$; by rescaling the evolution equation reads

$$\partial_\tau \rho_S(\tau\lambda^{-2}) = -i[\lambda^{-2} H_S + H_1 \, , \, \rho_S(\tau\lambda^{-2})] + \int_0^{\tau\lambda^{-2}} \mathrm{d}s \, D(s)[\rho_S(\tau\lambda^{-2} - s)] \, .$$

Then, by letting $\lambda \to 0$ one replaces the upper integration limit by $+\infty$ and neglects $s$ in comparison with $\tau\lambda^{-2}$ in the argument of the state appearing in the integral. The problem with too naive Markovian approximations as this one is that very rarely they lead to irreversible evolutions that are positivity preserving [105]: most derivations provide time-evolutions that are not positive and generate physically inconsistent negative probabilities. For instance, the wild oscillations due to the system Hamiltonian term $\lambda^{-2}H_S$ when $\lambda \to 0$ makes intuitively plausible using an ergodic average to smooth away too fast effects [95].

### Irreversible Dynamics within the Bloch Sphere

With respect to Example 5.6.1.1, it proves convenient to represent density matrices $\rho \in M_2(\mathbb{C})$ by Bloch vectors with one more component corresponding to the coefficient of $\sigma_0$ in the expansion (5.104).

We shall identify $\rho$ as a 4-dimensional ket $\mathbb{R}^4 \ni |\,\rho\,\rangle := (1, \rho_1, \rho_2, \rho_3)$. As a consequence, the linear action of the generator $L : \rho \mapsto L[\rho]$ in (5.191) corresponds to a $4 \times 4$ matrix $\mathcal{L} = [\mathcal{L}_{\mu\nu}]$ acting on $|\,\rho\,\rangle$. The Liouville equation (5.168) thus becomes

$$\partial_t|\,\rho\,\rangle = -2(\mathcal{H} + \mathcal{D}) \;,$$

with $-2\mathcal{H}$ and $-2\mathcal{D}$ $4 \times 4$ matrices corresponding to the commutator $L_H$ and the added term $D$ in (5.190) ($-2$ has been inserted for convenience). Concerning the matrix $\mathcal{D}$, the request of trace and hermiticity preservation imposes $\mathcal{D}_{0j} = 0$, $j = 1, 2, 3$, and $\mathcal{D}_{\mu\nu} \in \mathbb{R}$. By splitting $\mathcal{D}$ into the sum of a symmetric and antisymmetric matrix, the latter corresponds to a Hamiltonian contribution that can be incorporated into $\mathcal{H}$. Thus, one remains with a purely dissipative matrix

$$\mathcal{D} = \begin{pmatrix} 0 & 0 & 0 & 0 \\ u & a & b & c \\ v & b & \alpha & \beta \\ w & c & \beta & \gamma \end{pmatrix} \;, \tag{5.196}$$

with 9 real parameters which depends on the phenomenology of the system-environment interaction and can, in line of principle, be tested in dedicated experiments [32].

By exponentiating $\mathcal{L}$, one gets a one-parameter semi-group of $4 \times 4$ matrices, $\{\mathcal{G}_t\}_{t\geq 0}$, such that $\mathcal{G}_t = \mathrm{e}^{-2t(\mathcal{H}+\mathcal{D})}$ which corresponds to the semi-group $\{\Gamma_t\}_{t\geq 0}$ on the state-space $\mathcal{S}(S)$ given by

$$\rho \mapsto \rho(t) = \Gamma_t[\rho] = \sum_{\mu=0}^{3} \rho_\mu(t)\,\sigma_\mu \;, \quad \rho_\mu(t) = (\mathcal{G}_t\rho)_\mu \;.$$

Since the trace is preserved at all times, checking positivity preservation amounts to checking whether $\mathrm{Det}[\rho(t)] \geq 0$ for all $t \geq 0$ and for all initial $\rho$.

Since the contributions of the anti-symmetric $\mathcal{H}$ cancel out, the time-derivative of the determinant reads

$$\dot{D}[\rho] := \frac{\mathrm{dDet}[\rho(t)]}{\mathrm{d}t}\Big|_{t=0} = 2\Big(\sum_{i,j=1}^{3} \mathcal{D}_{ij}\rho_i\rho_j + \sum_{j=1}^{3}\mathcal{D}_{j0}\rho_j\Big) .$$

Let $\rho$ be a pure state $P(\boldsymbol{n}) := \frac{\mathbb{1}_2 + \boldsymbol{n}\cdot\boldsymbol{\sigma}}{2}$, then $\mathrm{Det}[P(\boldsymbol{n})] = 0$. Therefore, $\Gamma_t[P(\boldsymbol{n})] \geq 0$ asks for $\dot{D}[P(\boldsymbol{n})] \geq 0$ and the same must also be true for the orthogonal projector $P(-\boldsymbol{n})$. By summing $\dot{D}[P(\boldsymbol{n})] \geq 0$ and $\dot{D}[P(-\boldsymbol{n})] \geq 0$ and varying $\boldsymbol{n}$ in the unit sphere, positivity is preserved only if

$$\mathcal{D}^{(3)} = \begin{pmatrix} a & b & c \\ b & \alpha & \beta \\ c & \beta & \gamma \end{pmatrix} \geq 0 . \tag{5.197}$$

The positivity of $\mathcal{D}^{(3)}$ is necessary for positivity preservation, but not sufficient, the reason being that $\dot{D}[P(\boldsymbol{n})] < 0$ can follow because of the extra term $\sum_{j=1}^{3}\mathcal{D}_{j0}\rho_j$. However, it becomes also sufficient when we ask that $\Gamma_t$ increase the von Neumann entropy of any initial state, as this is equivalent to $u = v = w = 0$ in $\mathcal{D}$. Indeed, given any initial $\rho$, let it be spectralized as

$$\rho = r_1\frac{\mathbb{1} + \boldsymbol{n}\cdot\boldsymbol{\sigma}}{2} + r_2\frac{\mathbb{1} - \boldsymbol{n}\cdot\boldsymbol{\sigma}}{2} ,$$

with $0 \leq r_{1,2} \leq 1$, $\boldsymbol{n} \in \mathbb{R}^3$ and $\sum_{j=1}^{3} n_j^2 = 1$. Then, one explicitly computes

$$\begin{aligned}\dot{S}(\rho) := \frac{\mathrm{d}S(\rho(t))}{\mathrm{d}t}\Big|_{t=0} &= -\mathrm{Tr}\Big(\frac{\mathrm{d}\rho(t)}{\mathrm{d}t}\Big|_{t=0}\ln\rho\Big) \\ &\quad - 2\Big\{(r_1 - r_2)\langle\boldsymbol{n}\,|\,\mathcal{D}^{(3)}\,|\boldsymbol{n}\rangle + \sum_{j=1}\mathcal{D}_{j0}n_j\Big\}\ln\frac{r_1}{r_2} .\end{aligned}$$

If $\mathcal{D}_{j0} = 0$ for $j = 1, 2, 3$, then the time-derivative is positive because of the positivity of $\mathcal{D}$ and due to the fact that $(r_1 - r_2)\ln r_1/r_2 \geq 0$; if not, one can always find a $\rho$ for which $\dot{S}(\rho) < 0$: it suffices to choose $r_1 - r_2$ sufficiently small and adjust $\boldsymbol{n}$ to make negative the second term in the previous expression.

**Example 5.6.5.** Let us consider the following simple master equation for a 2-level system $S$,

$$\frac{\partial\rho(t)}{\partial t} = \frac{1}{2}(\sigma_1\,\rho\,\sigma_1 - \sigma_2\,\rho\,\sigma_2 + \sigma_3\,\rho\,\sigma_3 - \rho) .$$

Using (5.56), $L[\mathbb{1}] = L[\sigma_1] = L[\sigma_3] = 0$, while $L[\sigma_2] = -2\,\sigma_2$; therefore, the generated semi-group $\gamma_t = \exp(t\,L)$ is such that

$$\gamma_t[\rho] = \frac{1}{2}\Big(\mathbb{1} + \rho_1\sigma_1 + \mathrm{e}^{-2t}\rho_2\sigma_2 + \rho_3\sigma_3\Big) .$$

Since the matrix in (5.197) is now $\mathcal{D}^{(3)} = \begin{pmatrix} 0 & 0 & 0 \\ 0 & 1 & 0 \\ 0 & 0 & 0 \end{pmatrix}$, the necessary condition for positivity preservation is satisfied. Further, the Bloch vector at time $t$, $\boldsymbol{\rho}(t)$, is such that $\|\boldsymbol{\rho}(t)\| \leq \|\boldsymbol{\rho}\|$; therefore, any initial density matrix remains a density matrix.

Suppose the 2-level $S$ system evolving under $\gamma_t$ is statistically coupled to another 2-level system $S'$ that has no evolution of its own. Then, one has to consider states of the composite system $S' + S$ that evolve in time under the semi-group of maps of the form $\Gamma_t = \mathrm{id}_2 \otimes \gamma_t$, that is $\Gamma_t$ lifts the action of $\gamma_t$ from $M_2(\mathbb{C})$ to $M_2(M_2(\mathbb{C})) = M_2(\mathbb{C}) \otimes M_2(\mathbb{C})$ as in Section 5.2.2.

Among the possible initial conditions for $\Gamma_t$ there is the Bell state $|\hat{\Psi}_{00}\rangle$ in (5.164); we know from (5.17) that the corresponding projector $\widehat{P}^2_+$ is proportional to the matrix $\widetilde{E} = [E_{ij}] \in M_2(M_2(\mathbb{C}))$ whose entries are matrix units in $M_2(\mathbb{C})$. By writing $E_{11} = (\mathbb{1} + \sigma_3)/2$, $E_{12} = (\sigma_1 + i\sigma_2)/2$ and $E_{22} = (\mathbb{1} - \sigma_3)/2$ in terms of the Pauli matrices, it turns out that

$$\widehat{P}^2_+ = \frac{1}{4}\Big(\mathbb{1} \otimes \mathbb{1} + \sigma_1 \otimes \sigma_1 - \sigma_2 \otimes \sigma_2 + \sigma_3 \otimes \sigma_3\Big) .$$

Then, setting $\lambda_t := \exp(-2t)$, under the time-evolution $\Gamma_t$, $\widehat{P}^2_+$ evolves into

$$\Gamma_t[\widehat{P}^2_+] = \frac{1}{4}\Big(\mathbb{1} \otimes \mathbb{1} + \sigma_1 \otimes \sigma_1 - \lambda_t \sigma_2 \otimes \sigma_2 + \sigma_3 \otimes \sigma_3\Big)$$

$$= \frac{1}{4}\begin{pmatrix} \mathbb{1} + \sigma_3 & \sigma_1 + i\lambda_t\sigma_2 \\ \sigma_1 - i\lambda_t\sigma_2 & \mathbb{1} - \sigma_3 \end{pmatrix} = \frac{1}{4}\begin{pmatrix} 2 & 0 & 0 & 1+\lambda_t \\ 0 & 0 & 1-\lambda_t & 0 \\ 0 & 1-\lambda_t & 0 & 0 \\ 1+\lambda_t & 0 & 0 & 2 \end{pmatrix} ,$$

which is not positive definite for any $t > 0$, for it always shows a negative eigenvalue $(\lambda_t - 1)/4$.

The physical meaning of the previous example is that, though $\gamma_t$ is a meaningful time-evolution for one 2-level system, $\Gamma_t$ is not so for two 2-level system as there exists a state of the two together which does not remain positive definite. Notice that the state which exposes the problem is entangled; indeed, any separable state, as in Definition 5.5.3 would remain positive under $\Gamma_t$: as $\gamma_t[\rho] \geq 0$ for all $\rho \in \mathcal{S}(S)$,

$$\Gamma_t\left[\sum_{ij} \lambda_{ij} \rho'_i \otimes \rho_j\right] = \sum_{ij} \lambda_{ij} \rho'_i \otimes \gamma_t[\rho_j] \geq 0 .$$

The importance of Theorem 5.2.1 is now apparent: physical transformations of an $N$-level system $S$ cannot be described by linear maps $\Lambda$ that are only positivity preserving, they must also be completely positive. Otherwise, by coupling $S$ with another $N$-level system $S'$, one would obtain a map $\mathrm{id}_N \otimes \Lambda$ which would map the initial entangled state $\widehat{P}_+^N$ into a non-positive definite matrix.

The standard quantum time-evolution $\mathcal{U}_t$ is automatically in Kraus form, thus completely positive and free from inconsistencies with respect to statistical couplings to ancillas. It is only when performing a Markovian approximation that one must check that complete positivity be guaranteed by the procedure [95, 96, 105, 285].

### 5.6.3 Quantum Dynamical Semigroups

Positivity and complete positivity depend on the dissipative term $D[\rho]$ added to the commutator in (5.190): it turns out that when one asks that the generated semi-group consist of completely positive maps, then the form of the generator is completely fixed.

**Theorem 5.6.1.** *[126] Let $\{\gamma_t\}_{t\geq 0}$ be a one-parameter semi-group of hermiticity preserving, unital linear maps $\gamma_t : M_d(\mathbb{C}) \mapsto M_d(\mathbb{C})$ such that $\lim_{t\to 0}\gamma_t = \mathrm{id}_d$ with respect to the norm-topology. Then,*

1. *the semi-group has the form $\gamma_t = \exp(tL)$ with generator*

$$L[X] = i\Big[H\,,\,X\Big] + \sum_{a,b=1}^{d^2-1} C_{ab}\Big(F_a^\dagger\,X\,F_b - \frac{1}{2}\Big\{F_a^\dagger F_b\,,\,X\Big\}\Big)\ ,$$

*where the matrices $F_a$ form an ONB in $M_d(\mathbb{C})$ with respect to the Hilbert-Schmidt scalar product with $F_{d^2} = \mathbb{1}_d/\sqrt{d}$ ($\mathrm{Tr}(F_a) = 0$ if $a \neq d^2-1$)) and the $(d^2-1)\times(d^2-1)$ matrix $C := [C_{ab}]$, called Kossakowski matrix, is Hermitian.*

2. *The maps $\gamma_t$ are completely positive if and only if $[C_{ab}]$ is a positive matrix.*

**Proof:** From Example 5.2.7.1, the linear maps $\mathcal{F}_{ab} : M_d(\mathbb{C}) \mapsto M_d(\mathbb{C})$ defined by $\mathcal{F}_{ab}[X] := F_a^\dagger\,X\,F_b$, $a,b = 1,2,\ldots,d^2$, form an orthonormal basis in the $d^2$ dimensional linear space of all linear operators on $M_d(\mathbb{C})$ equipped with the Hilbert-Schmidt scalar product of the associated Choi matrices. It follows that the generator $L$ can be expanded as $L = \sum_{a,b=1}^{d^2} L_{ab}\,\mathcal{F}_{ab}$. Then, the request that the generated semi-group preserve hermiticity implies that $L[X]^\dagger = L[X^\dagger]$ for all $X \in M_d(\mathbb{C})$ which in turn yields $L_{ab}^* = L_{ba}$. Now, after rewriting

$$L[X] = F\,X + X\,F^\dagger + \sum_{a,b=1}^{d^2-1} L_{ab}\,F^d ag_a\,X\,F_b\ ,\quad F := \frac{1}{\sqrt{d}}\sum_{a=1}^{d^2} L_{ad^2}\,F_a^\dagger\ ,$$

and separating $F$ into its Hermitian components, $F = K + iH$, where

$$K := \frac{1}{2}\sum_{a=1}^{d^2}(L_{ad^2}\,F_a^\dagger \,+\, L_{d^2a}\,F_a)\ ,\ H := \frac{1}{2i}\sum_{a=1}^{d^2}(L_{ad^2}\,F_a^\dagger \,-\, L_{d^2a}\,F_a)\ ,$$

one concludes that

$$L[X] = i[H\,,\,X] + (K\,X + X\,K) + \sum_{a,b=1}^{d^2-1} L_{ab}\,F_a^\dagger\,X\,F_b\ .$$

The first statement of the theorem follows by further imposing unitality, that is that $\gamma_t[\mathbb{1}] = \mathbb{1}$ for all $t \geq 0$. One thus gets that $L[\mathbb{1}] = 0$, which further imposes $K = -\frac{1}{2}\sum_{a,b=1}^{d^2-1} L_{ab}\,F_a^\dagger\,F_b$.

According to Theorem 5.2.1, $\gamma_t$ is a *CPU* map on $M_d(\mathbb{C})$ if and only if $\Gamma_t := \mathrm{id}_d \otimes \gamma_t$ is a positive, unital map on $M_d(\mathbb{C}) \otimes M_d(\mathbb{C})$. Notice that the maps $\Gamma_t$ form a norm-continuous semi-group with generator $L_{12} := \mathrm{id}_d \otimes L$; then, according to [177, 178] (see also [64]), the maps $\mathrm{id}_d \otimes \gamma_t$ are positive if and only if

$$I(\psi,\phi) := \langle\,\psi\,|\,L_{12}[|\,\phi\,\rangle\langle\,\phi\,|]\,|\psi\,\rangle \geq 0$$

for all orthogonal $\psi, \phi \in \mathbb{C}^d \otimes \mathbb{C}^d$. Since $\langle\,\psi\,|\,\phi\,\rangle = 0$, it follows that

$$I(\psi,\phi) = \sum_{a,b=1}^{d^2-1} C_{ab}\left(\langle\,\psi\,|\,\mathbb{1}_d \otimes F_a^\dagger\,|\phi\,\rangle\langle\,\phi\,|\,\mathbb{1}_d \otimes F_b\,|\psi\,\rangle\right)\ .$$

Then, it proves convenient to define the $d^2 \times d^2$ matrices $\Psi = [\psi_{ij}]$ and $\Phi = [\phi_{ij}]$ where $\psi_{ij}$ and $\phi_{ij}$ are the components of the vectors $\psi$ and $\phi$ with respect to a fixed *ONB* $\{|\,i,j\,\rangle\}_{i,j=1}^d$ in $\mathbb{C}^d \otimes \mathbb{C}^2$. Notice that $\langle\,\psi\,|\,\phi\,\rangle = \mathrm{Tr}(\Psi^\dagger\Phi)$. By introducing the vectors $|\,v\,\rangle \in \mathbb{C}^{d^2-1}$ with components given by

$$v_a := \langle\,\phi\,|\,\mathbb{1}_d \otimes F_a\,|\psi\,\rangle = \mathrm{Tr}(F_a(\Phi^\dagger\Psi)^T)\ ,$$

one then rewrites $I(\psi,\phi) = \sum_{a,b=1}^{d^2-1} C_{ab}\,v_a^* v_b = \langle\,v\,|\,C\,|v\,\rangle$. If $C = [C_{ab}] \geq 0$, then $I(\psi,\phi) \geq 0$ for all orthogonal $\psi, \phi \in \mathbb{C}^d \otimes \mathbb{C}^d$, whence $\Gamma_t$ is positive.

Vice versa, given a generic vector $|\,v\,\rangle \in \mathbb{C}^{d^2-1}$, the traceless matrix $M_d(\mathbb{C}) \ni \Psi := \sum_{a=1}^{d^2-1} v_a\,F_a$ corresponds to a vector $\psi \in \mathbb{C}^d \otimes \mathbb{C}^d$ that is orthogonal to the non-normalized totally symmetric vector $|\Psi_+^d\rangle = \sum_{i=1}^{d^2-1}|\,ii\,\rangle$. If $\Gamma_t$ is positive, then $I(\psi,\Psi_+^d) = \langle\,v\,|\,C\,|v\,\rangle \geq 0$ for all $|\,v\,\rangle \in \mathbb{C}^{d^2-1}$ whence $C \geq 0$. □

**Remarks 5.6.4.**

1. If $\{\gamma_t\}_{t\geq 0}$ is a norm-continuous one-parameter semi-group of positive maps $\gamma_t : M_d(\mathbb{C}) \mapsto M_d(\mathbb{C})$ with generator $L$ and $\psi, \phi \in \mathbb{C}^d$ are orthogonal vectors, then
$$0 \leq \langle \psi \,|\, \gamma_t[|\,\phi\rangle\langle\,\phi\,|]\,|\psi\rangle = t\,\langle\,\psi\,|\,L[|\,\phi\rangle\langle\,\phi\,|]\,|\psi\rangle$$
to first order in $t$. This yields the only if part of the theorem used in the previous proof; it turns out that this condition is also sufficient for the maps $\gamma_t$ to be positive [177, 178, 64].
2. The extension of Theorem 5.6.1 from $M_d(\mathbb{C})$ to $\mathbb{B}(\mathbb{H})$ with $\mathbb{H}$ an infinite dimensional Hilbert space, has been provided by [193] under the assumption that $\|L[X]\| \leq \|L\|\,\|X\|$ for all $X \in \mathbb{B}(\mathbb{H})$, namely that the generator $L$ be bounded on $\mathbb{B}(\mathbb{H})$.
3. By duality, one gets the following time-evolution equation for the states of the open quantum system $S$:
$$L^+[\rho] = -i\Big[H\,,\,\rho\Big] + \sum_{a,b=1}^{d^2-1} C_{ab}\Big(F_b\,\rho\,F_a^\dagger - \frac{1}{2}\Big\{F_a^\dagger F_b\,,\,\rho\Big\}\Big)\,.$$
Since the $\gamma_t$ are unital, their dual maps $\gamma_t^+$ preserve the trace of $\rho$.
4. If $\gamma_t$ is *CP* , the expression $\sum_{a,b+1}^{d^2-1} C_{ab}\,F_b\,\rho\,F_a^\dagger$ can be put in Kraus form as in (5.36). Such a term corresponds to what in the classical Brownian motion is the diffusive effect due to the presence of a white-noise. It is indeed sometimes called *quantum noise* which is also in agreement with the effects of generic *POVMs* on quantum states [121].
5. Beside the noise contribution, the remaining part of the generator has the form
$$-i(H - \frac{i}{2}K)\,\rho + i\,\rho\,(H + \frac{i}{2}K)\,, \qquad K = \frac{1}{2}\sum_{a,b=1}^{d^2-1} L_{ab}\,F_a^\dagger\,F_b\,.$$
This expression corresponds to the typical phenomenological description of the time-evolution of decaying systems; in particular, $K$ is a damping term due to probability that goes irreversibly from the system $S$ to its decay products.
6. Regarding the generated maps $\gamma_t = \exp(tL)$, there are no general results on the form of the Kossakowski matrix $C = [C_{ij}]$ able to ensure that the $\gamma_t$ be positive; the only available general expression is for $d = 2$ [178].

Semigroups consisting of completely positive maps are called *quantum dynamical semi-groups*. Their derivation as Markovian approximations of an underlying reversible many-body dynamics mainly follows three schemes, the already mentioned weak-coupling limit, the *singular-coupling limit* [126, 125, 230] and the *low density limit* [104]. All of them work when the time scales of the system $S$ and of the reservoir $R$ are clearly distinguishable. The weak-coupling limit is the one most frequently encountered in the literature since the beginning of the theory of open quantum systems and also the one which, if not performed with due accuracy [96], leads to semi-groups of maps which are not completely positive and thus to physical inconsistencies in relation to entanglement.

**Example 5.6.6.** [27] Let $d = 2$; in such a case, by choosing the orthonormal basis of Pauli matrices $F_j = \sigma_j/\sqrt{2}$'s, $j = 1, 2, 3$, the dissipative contribution to the semi-group generator reads

$$L_D[\rho] = \sum_{i,j=0}^{3} C_{ij}\Big[\sigma_i\rho\sigma_j - \frac{1}{2}\Big\{\sigma_j\sigma_i\,,\,\rho\Big\}\Big]\,.$$

For sake of simplicity, we shall restrict to entropy-increasing semi-groups. Then, we can consider the matrix $\mathcal{D}$ in (5.196) whose entries read

$$\begin{aligned} a &= C_{22} + C_{33}\ ,\ \alpha = C_{11} + C_{33}\ ,\ \gamma = C_{11} + C_{22} \\ b &= -C_{12}\ ,\ c = -C_{13}\ ,\ \beta = -C_{23}\ . \end{aligned}$$

Thus, the positivity of $[C_{ij}]$, which, according to the previous theorem, is necessary and sufficient for the complete positivity of $\Gamma_t$, results in the necessary and sufficient inequalities for $a, b, c, \alpha, \beta, \gamma$:

$$\begin{aligned} &2R \equiv \alpha + \gamma - a \geq 0\ , \quad RS \geq b^2 \\ &2S \equiv a + \gamma - \alpha \geq 0\ , \quad RT \geq c^2 \\ &2T \equiv a + \alpha - \gamma \geq 0\ , \quad ST \geq \beta^2 \\ &RST \geq 2\,bc\beta + R\beta^2 + Sc^2 + Tb^2\ . \end{aligned}$$

These constraints are much stronger than those coming from positivity alone, that is from $\mathcal{D}^{(3)} \geq 0$ in (5.197) which yields

$$a \geq 0\ ,\ \alpha \geq 0\ ,\ \gamma \geq 0\ ,\ a\alpha \geq b^2\ ,\ a\gamma \geq c^2\ ,\ \alpha\gamma \geq \beta^2$$

and $\mathrm{Det}\mathcal{D}^{(3)} \geq 0$. As a concrete example, take $a = \alpha$ and $\beta = b = c = 0$; so that the Kossakowski-Lindblad generator reads

$$L[\rho] = \frac{\gamma}{2}(\sigma_1\rho\sigma_1 - \rho) + \frac{\gamma}{2}(\sigma_2\rho\sigma_2 - \rho) + \frac{2\alpha - \gamma}{2}(\sigma_3\rho\sigma_3 - \rho)\ ,$$

whence $L[\sigma_{1,2}] = -\alpha\,\sigma_{1,2}$ and $L[\sigma_3] = -\gamma\,\sigma_3$. It follows that, when $\alpha, \gamma > 0$, the generated semi-group $\gamma_t$ describes a decay process towards $\rho_\infty = \mathbb{1}/2$ with different rates for the diagonal and off-diagonal elements of $\gamma_t[\rho]$.

Indeed, setting $\mu_t := \exp(-\gamma t)$ and $\lambda_t := \exp(-\alpha t)$, it turns out that

$$\gamma_t[\rho] = \frac{1}{2}\Big(\mathbb{1} + \lambda_t(\rho_1\sigma_1 + \rho_2\sigma_2) + \mu_t\sigma_3\Big) = \frac{1}{2}\begin{pmatrix} 1+\mu_t\rho_3 & \lambda_t(\rho_1 - i\rho_2) \\ \lambda_t(\rho_1 + i\rho_2) & 1-\mu_t\rho_3 \end{pmatrix} .$$

On the other hand, consider $\Gamma_t = \mathrm{id}_2 \otimes \gamma_t$ and $\widehat{P}^2_+$ as in Example 5.6.5, it turns out that

$$\begin{aligned} \Gamma_t[\widehat{P}^2_+] &= \frac{1}{4}\Big(\mathbb{1}\otimes\mathbb{1} + \lambda_t\,(\sigma_1\otimes\sigma_1 - \sigma_2\otimes\sigma_2) + \mu_t\,\sigma_3\otimes\sigma_3\Big) \\ &= \frac{1}{4}\begin{pmatrix} 1+\mu_t & 0 & 0 & 2\lambda_t \\ 0 & 1-\mu_t & 0 & 0 \\ 0 & 0 & 1-\mu_t & 0 \\ 2\lambda_t & 0 & 0 & 1+\lambda_t \end{pmatrix} . \end{aligned}$$

This matrix is positive definite if and only if $1+\mu_t \geq 2\lambda_t$. This is implied by the complete positivity condition $\gamma \leq 2\alpha$, whereas if $\gamma > 2\alpha$, when $t \to 0$ one gets

$$1 + \mu_t - 2\lambda_t \simeq t(2\alpha - \gamma) < 0 .$$

In conclusion, only complete positivity guarantees full physical consistency with respect to statistical couplings with other systems. However, this imposes a hierarchy, $\gamma \leq 2\alpha$, upon the decay rates of the entries of the dissipatively evolving state $\gamma_t[\rho]$, which should otherwise only be positive.

### 5.6.4 Physical Operations and Positive Maps

The argument behind the request of complete positivity on state transformations is that one can never exclude that the system $S$ undergoing the transformation is indeed entangled with an *ancilla system* $S'$, even without any effective sign of statistical correlations. Though plausible, this point of view is not always accepted [235]; after all, the mere possibility of entanglement with an uncontrollable ancilla would then, via complete positivity, constrain the decay properties. Consider, for instance, an actual experiments where optically active molecules interact weakly with a heat bath; they can effectively be described as 2-level systems. The relaxation to equilibrium of their optical activity can accordingly be predicted by an appropriate master equation. Clearly, the fact that the optical activity may depend on whether the molecules are entangled with some other system out of any experimental control sounds admittedly weird [72, 185, 186, 292].

However, most of the objections to complete positivity do not consider the entanglement issue for they all focus upon single open quantum systems in heat baths. If, however, two optically active molecules in a same environment are considered, the entanglement issue comes to the fore. If the two molecules do not interact between themselves, but are weakly coupled to their environment, it is sensible to describe their open dynamics by a semigroup of dynamical maps of the form $\Gamma_t = \gamma_t \otimes \gamma_t$, where $\gamma_t$ is the reduced dynamics of a single molecule. These dynamical maps differ from $\mathrm{id}_2 \otimes \gamma_t$ in Examples 5.6.5 and 5.6.6.

Notice that in going from $\mathrm{id}_d \otimes \gamma_t$ to $\Gamma_t = \gamma_t \otimes \gamma_t$ one effectively goes from the possible existence of statistical correlations between the system $S_d$ and another system of the same type which is somewhat uncontrollable, to a concrete scenario when one has two statistically coupled systems in a same environment. The following result on one hand extends Theorem 5.6.1 and on the other stresses once more the fact that complete positivity is not just a mathematical option without physical meaning, rather an unavoidable constraint on all sensible Markovian approximations.

**Proposition 5.6.1.** *[37] Let $\{\gamma_t\}_{t\geq}$ be a norm-continuous semi-group of dynamical maps $\gamma_t : M_d(\mathbb{C}) \mapsto M_d(\mathbb{C})$ with generator as in Theorem 5.6.1. Then, the linear maps $\Gamma_t = \gamma_t \otimes \gamma_t$ form a norm-continuous semi-group on $M_d(\mathbb{C}) \otimes M_d(\mathbb{C})$ and preserve positivity if and only if $\gamma_t$ is a CPU map for all $t \geq 0$.*

**Proof:** One implication is straightforward: if $\gamma_t$ is a *CPU* map, then $\gamma_t \otimes \mathrm{id}_2$ and $\mathrm{id}_1 \otimes \gamma_t$ are positive and such is $\Gamma_t$.

For the other implication, notice that, in view of the assumptions, the one-parameter family $\{\Gamma_t\}_{t\geq 0}$ is a norm-continuous semi-group with generator $L_{12} = L \otimes \mathrm{id}_d + \mathrm{id}_d \otimes L$. Then we argue as in the proof of the second part of Theorem 5.6.1 and show that

$$I(\psi, \phi) := \langle \psi \,|\, L_{12}[|\, \phi \rangle\langle \phi \,|]\, |\psi\rangle \geq 0$$

for all orthogonal $\psi, \phi \in \mathbb{C}^d \otimes \mathbb{C}^d$. Since $\langle \psi \,|\, \phi \rangle = 0$, it follows that

$$I(\psi, \phi) = \sum_{a,b=1}^{d^2-1} C_{ab} \Big( \langle \psi \,|\, F_a^\dagger \otimes \mathbb{1}_d \,|\phi\rangle\langle \phi \,|\, F_b \otimes \mathbb{1}_d \,|\psi\rangle$$

$$+ \langle \psi \,|\, \mathbb{1}_d \otimes F_a^\dagger \,|\phi\rangle\langle \phi \,|\, \mathbb{1}_1 \otimes F_b \,|\psi\rangle \Big) .$$

By means of the matrices $\Psi = [\psi_{ij}]$ and $\Phi = [\phi_{ij}]$ associated to the vectors $\psi, \phi \in \mathbb{C}^d \otimes \mathbb{C}^d$ as explained in the proof of Theorem 5.6.1, one introduces the vectors $|\, w\rangle, |\, v\rangle \in \mathbb{C}^{d^2-1}$ with components given by

$$w_b := \langle \phi \,|\, F_b \otimes \mathbb{1}_d \,|\psi\rangle = \mathrm{Tr}(F_b(\Psi\Phi^\dagger)) \, , \; v_b := \langle \phi \,|\, \mathbb{1}_d \otimes F_b \,|\psi\rangle = \mathrm{Tr}(F_b(\Phi^\dagger\Psi)^T) \, ,$$

and rewrites

$$I(\psi, \phi) = \sum_{a,b=1}^{d^2-1} C_{ab} \, (w_a^* w_b + v_a^* v_b) = \langle w \,|\, C \,|w\rangle + \langle v \,|\, C \,|v\rangle \qquad (*) \, .$$

Given $|\,w\,\rangle \in \mathbb{C}^{d^2-1}$, construct $M_d(\mathbb{C}) \ni W := \sum_{a=1}^{d^2-1} w_a\, F_a$. Since a matrix and its transposed are always similar [134], let $Y \in M_d(\mathbb{C})$ be such that $W^T = Y\,W\,Y^{-1}$ and define $\Phi := Y^{-1}$, $\Psi^\dagger := Y\,W$ so that

$$\Phi\Psi^\dagger = W\ , \quad (\Psi^\dagger\Phi)^T = (Y\,W\,Y^{-1})^T = W \quad \text{and} \quad |\,w\,\rangle = |\,v\,\rangle\ ,$$

whence $(*)$ becomes $I(\psi,\phi) = 2\langle\, w\,|\,C\,|w\,\rangle \geq 0$. Observe that $|\,w\,\rangle \in \mathbb{C}^{d^2-1}$ is generic and that to any such vector one can associate orthogonal vectors $\psi, \phi \in \mathbb{C}^d \otimes \mathbb{C}^d$ through the matrices $\Psi, \Phi$ as described above [6]. Thus, $I(\psi,\phi) \geq 0$ for any such pair implies $C := [C_{ab}] \geq 0$. □

**Remarks 5.6.5.**

1. If positive, $\Gamma_t = \gamma_t \otimes \gamma_t$ is also *CP*; indeed, using (5.36), it turns out that $\gamma_t[X] = \sum_j V_j^\dagger(t)\,X, V_j(t)$, $X \in M_d(\mathbb{C})$. As a consequence,
$$\Gamma_t[X] = \sum_{j,k} V_j^\dagger(t) \otimes V_k^\dagger(t)\,X\,V_j(t) \otimes V_k(T)\ .$$
2. The equivalence between the complete positivity of $\gamma_t$ and the positivity of $\Gamma_t = \gamma_t \otimes \gamma_t$ does not extend to the tensor products of generic $\gamma_t^{(1,2)}$; indeed, in Proposition 6.2.2 it will be shown that $\Gamma_t = \gamma_t^{(1)} \otimes \gamma_t^{(2)}$ can be positive without $\gamma_t^{(1,2)}$ being both *CPU* maps.

Once a semi-group reduced dynamics is accepted as a phenomenological time-evolution under certain physical conditions as those compatible with, for instance, the weak-coupling limit scenario, there is only one possible way to get rid of the complete positivity constraint. One has to rely upon the existence of physical mechanisms that eliminate those initial entangled states that, like the symmetric projector $\widehat{P}^d_+$, would otherwise be cast out of the state of space by $\Gamma_t$ when $\gamma_t$ is not completely positive [122, 124, 292, 319].

In quantum information the situation is physically clearer and complete positivity compulsory. In fact, the state transformations that are commonly considered do not from dynamical semi-groups arising from suitable Markovian approximations, rather they are maps as in Definition 5.6.1. Indeed, the simplest operations are local state transformations that two parties operate on shared entangled states as $P^d_+$. In order to be physically consistent, these local operations must correspond to completely positive maps.

What then of positive maps? If, on one hand, the existence of entanglement in nature forbids their use as dynamical maps that describe actually occurring physical processes, on the other hand, as we shall see in the following chapter, they are extremely useful as *entanglement witnesses*.

---

[6]Notice that, as $\langle\,\psi\,|\,\phi\,\rangle = 0$, the matrices $\Phi\Psi^\dagger$ and $\Psi^\dagger\Phi$ are traceless.

## Bibliographical Notes

The books [253, 270, 251] provide introductions to most of the aspects of Hilbert space techniques, bounded, compact, Hilbert-Schmidt and trace-class operators, while [20] is an excellent introduction to convexity.

The book [117] is a handy overview of many salient facts concerning $C^*$ and von Neumann algebras, while [64] presents the theory of $C^*$ and von Neumann algebras in detail with an eye to their applications to quantum statistical mechanics in [65]. The book [300] does the same, though in a somewhat more condensed way: here one can find a handy presentation of the modular theory which has been reference for this chapter, whereas a more thorough one can be found in [293]. The books [90, 97] provide mathematical introductions of $C^*$ and von Neumann algebras, of which [161, 162] offer an exhaustive overview.

For physically oriented reviews of entropy related concepts with their mathematical properties see [10, 314] and [300]. An exhaustive mathematical presentation of both von Neumann and relative entropy and of their applications to physics can be found in [226].

The books [11, 132, 96, 285] are historical references for the theory of open quantum systems; more recent developments and applications can be found in [27, 29, 14, 68, 315].

# 6 Quantum Information Theory

In the last years, a considerable amount of theoretical and experimental studies have been focussing on the impact that quantum mechanics may have on computer science, information theory and cryptography. We shall loosely refer to this vast and variegated field as quantum information [71, 48, 100, 128, 224, 242, 152, 239, 307]. In the following, we shall briefly touch upon a small fraction of its many achievements.

## 6.1 Quantum Information Theory

Why quantum information? Is it not classical information sufficiently powerful a theory to satisfy our needs? The answer is that it will indeed be so until computational models and information transmission protocols are based on classical physics. Indeed, information is physical [187, 55] for it is carried by physical entities, transmitted and manipulated by physical means; as a consequence, any actual information processing protocol will rely upon a model describing the physical processes involved. Since Nature is considered to be ultimately quantal, one is inevitably led to consider a scenario in which quantum mechanics will set the rules of the game also in dealing with information and its manifold aspects.

Roughly speaking, the issue at stake is the use of *qubits* instead of *bits* as fundamental informational resources so that one has the whole Bloch sphere of two-level system states at disposal instead of just the two states (up and down along the $z$-direction) that are available to classical spins.

When the information that is manipulated regards computational processes, the question is whether *Quantum Turing Machines* (*QTMs*), that is computing devices based on the laws of quantum mechanics, might perform better than classical Turing machines. A breakthrough was indeed the discovery that relevant speedups can be gained by quantum algorithms because of the huge parallel computation made available by the possibility of linearly superposing *qubits* states.

Truly, from an abstract perspective, as much as classical mechanics is contained in quantum mechanics, also classical information, computation and cryptography may be thought of as commutative versions of more general theories, still in their infancy, that are to be soundly formulated within a

F. Benatti, *Dynamics, Information and Complexity in Quantum Systems*, Theoretical and Mathematical Physics, DOI 10.1007/978-1-4020-9306-7_6,

quantum, non-commutative, framework. However, the need to elaborate these more general theories is not only justified in line of principle, but comes from concrete facts. The pace at which every two years electronic devices double their efficiency (the so-called Moore's law) and decrease in size is such that non-classical effects will soon appear and quantum mechanics will become necessary to cope with them.

If information is carried by *qubits* , then the possible reversible operations to which they can be subjected are all those corresponding to unitary matrices in $M_2(\mathbb{C})$: these are called *quantum gates.* In the classical case, the only non-trivial gate on *bits* 0, 1 is that which flips them, $0 \mapsto 1$, $1 \mapsto 0$. Consider, for instance, the Hadamard transformation in (5.58); its $n$-fold tensor product $U_H^{\otimes n}$ acting on $|\,0\,\rangle^{\otimes n}$ produces a uniform linear combination of kets labeled by the $2^n$ binary strings $\boldsymbol{i}^{(n)} \in \Omega_2^{(n)}$, in one stroke:

$$U_H^{\otimes n}|\,0^{\otimes n}\,\rangle = \frac{1}{2^{n/2}} \sum_{\boldsymbol{i}^{(n)} \in \Omega_2^{(n)}} |\,\boldsymbol{i}^{(n)}\,\rangle\ . \tag{6.1}$$

Linearity is at the basis of *quantum parallelism*: suppose that the computation of a binary function $\mathbf{f} : \Omega_2^{(n)} \mapsto \Omega_2^{(n)}$ on $n$ *bits* with $n$-*bit* strings as images, $\boldsymbol{i}^{(n)} \mapsto (\mathbf{f}(\boldsymbol{i}^{(n)}))^{(n)}$, can be operated by means of a unitary transformation

$$|\,\boldsymbol{i}^{(n)}\,\rangle \mapsto U_f|\,\boldsymbol{i}^{(n)}\,\rangle = |\,(\mathbf{f}(\boldsymbol{i}^{(n)}))^{(n)}\,\rangle$$

on $n$ *qubits*. By $U_f$ $\mathbf{f}$ is computed on all strings at once, as follows:

$$U_f U_H^{\otimes n}|\,0\,\rangle^{\otimes n} = \frac{1}{2^{n/2}} \sum_{\boldsymbol{i}^{(n)} \in \Omega_2^{(n)}} |\,(\mathbf{f}(\boldsymbol{i}^{(n)}))^{(n)}\,\rangle\ .$$

The linear structure of quantum mechanics seems to provide a more powerful setting than the classical scenario; however, the extraction of information out of quantum states is a much more delicate problem than with binary strings.

Any computation performed by a *QTM* on $n$ *qubits* must correspond to a unitary operator on $(\mathbb{C}^2)^{\otimes n}$; then, a *quantum algorithm* acting on an initial state of the $n$ *qubits* would halt in a linear combination of all possible computational basis vectors in $(\mathbb{C}^2)^{\otimes n}$, each one of them corresponding to a classical $n$-bit string $\boldsymbol{i}^{(n)}$ occurring with a certain amplitude $C(\boldsymbol{i}^{(n)})$. If the solution of a problem is a specific binary string $\boldsymbol{i}^{(n)}$, an efficient quantum computation must associate to that string a very high probability, $|C(\boldsymbol{i}^{(n)})|^2 \approx 1$. Only in this case the solution would show up with almost certainty from a measurement in the computational basis.

**Example 6.1.1 (Deutsch-Josza Algorithm).** Let $f : \Omega_2^{(n)} \mapsto \{0, 1\}$ be a binary function that is known to be either constant or balanced, that is $f(\boldsymbol{i}^{(n)}) = 0$ on half of the $n$-digit strings and $f(\boldsymbol{i}^{(n)}) = 1$ on the other half.

The task is to decide between the two possibilities. Classically, the only way to ascertain whether $f$ is constant or not is to compute it on $2^{n-1}+1$ strings, that is on half plus one of them; this is because one can compute always 0 or always 1 on $2^n/2$ strings in a row without the function being constant, so that only one more computation can settle the question. On the other hand, if the *bit* strings $\boldsymbol{i}^{(n)}$ could indeed be treated as computational basis vectors $|\,\boldsymbol{i}^{(n)}\,\rangle$ in the Hilbert space $\mathbb{H}^{(n)} = (\mathbb{C}^2)^{\otimes n}$ of $n$ *qubits*, then the following quantum algorithm would answer the question in just one trial. It is based on a generalization of the *CNOT* gate of Example 5.5.9; instead of one control *qubit*, there are $n$ of them all prepared in the same state $|\,0\,\rangle$ together with one target *qubit* in the state $|\,1\,\rangle$. As seen in (6.1),

$$|\,\Psi\,\rangle := U_H^{\otimes(n+1)}|\,0\,\rangle^{\otimes n}\otimes|\,1\,\rangle = \left(\frac{1}{\sqrt{2}}\right)^n \sum_{\boldsymbol{i}^{(n)}\in\Omega_2^{(n)}} |\,\boldsymbol{i}^{(n)}\,\rangle \otimes \frac{|\,0\,\rangle - |\,1\,\rangle}{\sqrt{2}}\ .$$

The matrix $M_{2^n}(\mathbb{C})\otimes M_2(\mathbb{C}) \ni U_f = \sum_{\boldsymbol{j}^{(n)}\in\Omega_2^{(n)}} |\,\boldsymbol{j}^{(n)}\,\rangle\langle\,\boldsymbol{j}^{(n)}\,| \otimes \sigma_1^{f(\boldsymbol{j}^{(n)})}$ is unitary and flips the last *qubit* only if $f(\boldsymbol{j}^{(n)}) = 1$. This yields [1]

$$\begin{aligned}|\,\Psi_f\,\rangle := U_f|\,\Psi\,\rangle &= \left(\frac{1}{\sqrt{2}}\right)^n \sum_{\boldsymbol{i}^{(n)}\in\Omega_2^{(n)}} |\,\boldsymbol{i}^{(n)}\,\rangle \otimes \frac{|\,0\oplus f(\boldsymbol{i}^{(n)})\,\rangle - |\,1\oplus f(\boldsymbol{i}^{(n)})\,\rangle}{\sqrt{2}} \\ &= \left(\left(\frac{1}{\sqrt{2}}\right)^n \sum_{\boldsymbol{i}^{(n)}\in\Omega_2^{(n)}} (-1)^{f(\boldsymbol{i}^{(n)})}\,|\,\boldsymbol{i}^{(n)}\,\rangle\right) \otimes \frac{|\,0\,\rangle - |\,1\,\rangle}{\sqrt{2}}\ .\end{aligned}$$

Applying the Hadamard rotation on the first $n$ *qubits* (see (5.58)), one gets

$$\begin{aligned}|\,\widetilde{\Psi}\,\rangle := U_H^{\otimes n}\otimes \mathbb{1}|\,\Psi_f\,\rangle = \left(\frac{1}{\sqrt{2}}\right)^n \Bigg( &\sum_{\substack{\boldsymbol{j}^{(n)}\\ \boldsymbol{i}^{(n)}\in\Omega_2^{(n)}}} (-1)^{f(\boldsymbol{i}^{(n)})+\boldsymbol{i}^{(n)}\cdot\boldsymbol{j}^{(n)}}\,|\,\boldsymbol{j}^{(n)}\,\rangle\Bigg) \otimes \\ &\otimes \frac{|\,0\,\rangle - |\,1\,\rangle}{\sqrt{2}}\ ,\end{aligned}$$

where $\boldsymbol{i}^{(n)}\cdot\boldsymbol{j}^{(n)} := \sum_{k=1}^n i_k j_k$. Since projecting onto $|\,0\,\rangle^{\otimes n}$ yields

$$\left(\left(\frac{1}{\sqrt{2}}\right)^n \sum_{\boldsymbol{i}^{(n)}\in\Omega_2^{(n)}} (-1)^{f(\boldsymbol{i}^{(n)})}\right) |\,0\,\rangle^{\otimes n} \otimes \frac{|\,0\,\rangle - |\,1\,\rangle}{\sqrt{2}}\ ,$$

the amplitude of $|\,0\,\rangle^{\otimes n}$ in $|\,\widetilde{\Psi}\rangle$ is 0 if $f$ is balanced, $\pm 1$ if $f$ is constant. Therefore, after operating the circuit one has just to perform a measurement in the computational basis $\{|\,\boldsymbol{i}^{(n)}\,\rangle\}_{\boldsymbol{i}^{(n)}\in\Omega_2^{(n)}}$ of the first $n$ *qubits* : if $|\,0\,\rangle^{\otimes n}$ occurs the binary function is constant, otherwise it is balanced.

---

[1]The *CNOT* gate (5.165) corresponds to choosing $f(i) = i$, $i = 0, 1$.

The preceding discussion regards instances of classical information being encoded into quantum states and manipulated by quantum gates. Similarly, classical information might be stored or transmitted by quantum means and the question is then how to retrieve it with high reliability or *fidelity*. Sending classical information encoded into non-orthogonal quantum states has indeed the advantage of being protected against undetected eavesdropping.

**Example 6.1.2 (No-Cloning).** Suppose a sender $A$ encodes the *bits* 0 and 1 into the *qubits* $|\,\psi_0\,\rangle, |\,\psi_1\,\rangle \in \mathbb{C}^2$ with $|\,\psi_0\,\rangle \neq |\,\psi_1\,\rangle$ and sends them to a receiver $B$. If a spy $E$ wants to access this amount of information without being spotted, he/she has to read the transmitted state without changing it, otherwise sender and receiver might get alerted. A way to do this is for the spy to intercept the message during transmission and to copy it by means of a unitary operator $U_E$ acting as follows $U_E(|\,\psi\,\rangle \otimes |\,e\,\rangle) = |\,\psi\,\rangle \otimes |\,\psi\,\rangle$. But unitarity implies

$$\langle\,\psi_0\,|\,\psi_1\,\rangle = \langle\,\psi_0 \otimes e\,|\,\psi_1 \otimes e\,\rangle = \langle\,\psi_0 \otimes e\,|\,U_E^\dagger U_E\,|\psi_1 \otimes e\,\rangle = \Big(\langle\,\psi_0\,|\,\psi_1\,\rangle\Big)^2 ,$$

whence $\psi_{0,1}$, not being equal, must be orthogonal. Therefore, if the code states $\psi_{0,1}$ are chosen not to be orthogonal, the spy cannot copy them without alterations. This argument goes under the name of *no-cloning theorem* and asserts that there cannot exist a unitary operator $U$ that implements the operation of copying two generic quantum states, unless they are orthogonal. Indeed, if such an unitary operator $U$ existed, then, on the linear combinations of two orthogonal states $|\,\psi\,\rangle, |\,\phi\,\rangle$,

$$\begin{aligned} U\Big((\alpha\,|\,\psi\,\rangle + \beta\,|\,\phi\,\rangle) \otimes |\,e\,\rangle\Big) &= \alpha\,|\,\psi\,\rangle \otimes |\,\psi\,\rangle + \beta\,|\,\phi\,\rangle \otimes |\,\phi\,\rangle \\ &= \Big(\alpha\,|\,\psi\,\rangle + \beta\,|\,\phi\,\rangle\Big) \otimes \Big(\alpha\,|\,\psi\,\rangle + \beta,|\,\phi\,\rangle\Big) \\ &= |\alpha|^2\,|\,\psi\,\rangle \otimes |\,\psi\,\rangle + |\beta|^2\,|\,\phi\,\rangle \otimes |\,\phi\,\rangle \\ &+ \alpha\beta\,\Big(|\,\psi\,\rangle \otimes |\,\phi\,\rangle + |\,\phi\,\rangle \otimes |\,\psi\,\rangle\Big) . \end{aligned}$$

This can only be true if either $\alpha = 0$ or $\beta = 0$ as one can see by scalar multiplication by $|\,\psi\,\rangle \otimes |\,\phi\,\rangle$.

In Section 5.5.4, it has already been emphasized the central role of entanglement as a resource for quantum informational tasks. Among the applications of entangled states to information transmission are the protocols for the so-called *dense coding* and *teleportation*. In the first case, 2 *bits* can be sent with one use of an entangled quantum channel, which points to the possibility of achieving higher channel capacities if channel behave quantum mechanically. In the second case, quantum states can be transferred between distant parties sharing an entangled state by means of *local quantum operations* and *classical communication* (known as *LOCC* operations).

**Example 6.1.3 (Dense Coding).** If sender $A$ and receiver $B$ share the entangled state (5.164), $A$ can encode the pairs of *bits* $(xy)$ into the Bell states of Example 5.5.9 by local operations performed on his *qubit*, only. Indeed, the states $|\,\hat{\Psi}_{xy}\,\rangle$ result from acting with the Pauli matrices on the first *qubit* of $|\hat{\Psi}_{00}\rangle$; explicitly

$$|\,\hat{\Psi}_{xy}\,\rangle = (\sigma_3^x \sigma_1^y) \otimes \mathbb{1}|\hat{\Psi}_{00}\rangle \ , \quad x, y = 0, 1 \ .$$

Then, if $A$ and $B$ share $|\hat{\Psi}_{00}\rangle$, in order to send $B$ two *bits* $(x, y)$ of classical information, $A$ acts on his *qubit* with $\sigma_3^x \sigma_1^y$ and sends it to $B$. When both *qubits* are with him, $B$ has them in the state $|\,\hat{\Psi}_{xy}\,\rangle$; by performing a measurement in the Bell basis, he can thus recover the pair $(xy)$. Roughly speaking, one can transmit two *bits* at the price of 1 *qubit*, that is by just one use of the entangled quantum channel represented by $|\hat{\Psi}_{00}\rangle$ and its local modifications.

**Example 6.1.4 (Teleportation).** Suppose $A$ has two *qubits*, denoted by $1, 2$, the first one in the state $|\,\psi\,\rangle_1 = \alpha|\,0\,\rangle_1 + \beta|\,1\,\rangle_1$ and the second one being one party in the symmetric Bell state $|\,\hat{\Psi}_{00}\,\rangle_{23} = \frac{1}{\sqrt{2}}\sum_{i=0}^{1}|\,i\,\rangle_2 \otimes |\,i\,\rangle_3$ together with a third *qubit* (3) of $B$. Let $B$ perform a Hadamard rotation on his *qubit* in $|\hat{\Psi}_{00}\rangle$, changing the entangled state into (see Figure 6.1)

$$|\,\Phi\,\rangle_{23} := (\mathbb{1} \otimes U_H)|\hat{\Psi}_{00}\rangle_{23} = \frac{1}{2}\sum_{i=0}^{1}|\,i\,\rangle_2 \otimes U_H|\,i\,\rangle_3 \ .$$

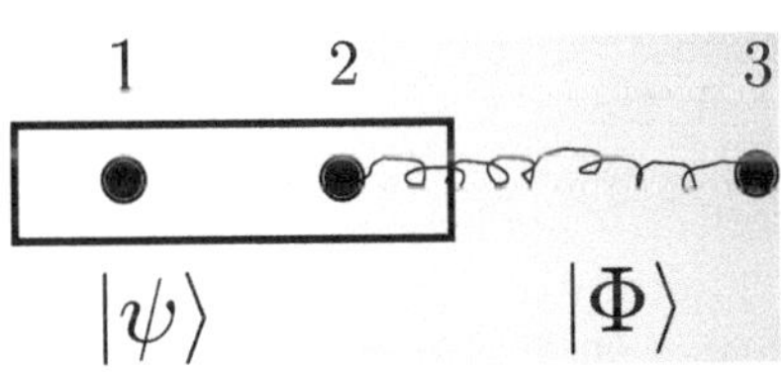


**Fig. 6.1.** Teleportation

The state $|\,\psi\,\rangle_1$ can now be teleported from $A$ to $B$ becoming $|\,\psi\,\rangle_3$. The protocol is as follows; $A$ performs on his two *qubits* $1, 2$ a measurement in the *ONB* $\{|\,\Psi^\mu\,\rangle_{12}\}_{\mu=0}^{3}$ of $\mathbb{C}^2 \otimes \mathbb{C}^2$, where $|\,\Psi_\mu\,\rangle_{12} := \sigma_\mu \otimes U_H|\hat{\Psi}_{00}\rangle_{12}$ with $\langle\,\Psi_\mu\,|\,\Psi_\nu\,\rangle = \frac{1}{2}\mathrm{Tr}(\sigma_\mu\sigma_\nu) = \delta_{\mu\nu}$. Notice that the amplitude of $|\,\Psi_\mu\,\rangle_{12}$ in the state $|\,\psi\,\rangle_1 \otimes |\,\Phi\,\rangle_{23}$ is

$${}_{12}\langle\,\Psi_\mu\,|(|\,\psi\,\rangle_1 \otimes |\,\Phi\,\rangle_{23}) = \frac{1}{2}\sum_{i,j=0}^{1}\langle\,i\,|\,\sigma_\mu\,|\psi\,\rangle\,\langle\,i\,|\,U_H\,|j\,\rangle\,U_H|\,j\,\rangle_3$$

$$= \frac{1}{2} \sum_{i=0}^{1} \langle i \,|\, \sigma_\mu \,|\psi\rangle \, U_H^2 |\, i \,\rangle_3 = \sigma_\mu |\, \psi \,\rangle_3 ,$$

where it has been used that $U_H = U_H^\dagger$ and $U_H^2 = \mathbb{1}$. Thus, after $A$ has classically (that is by means of a classical channel) communicated to $B$ the result of his local measurement, $B$ knows his *qubit* to be in the state $\sigma_\mu |\, \psi \,\rangle_3$, whence by a local rotation by $\sigma_\mu$ he gets his third *qubit* in the state $|\, \psi \,\rangle_3$.

The procedure does not violate no-cloning for the state that appears at $B$'s end, disappears from $A$'s end. Neither does it violate Einstein's locality; indeed, before classical communication of the actually measured index $\mu$, $B$'s state is the equidistributed mixture of the four possibilities corresponding to the four different measurement outcomes of $A$; explicitly, using Example 5.2.5 with the normalized Pauli matrices $\sigma_\mu/\sqrt{2}$ as *ONB*,

$$\rho = \frac{1}{4} \sum_{\mu=0}^{3} \sigma_\mu |\, \psi \,\rangle\langle\, \psi \,| \sigma_\mu = \frac{\mathbb{1}}{2} \, .$$

On the other hand, before $A$'s measurement, the marginal state of $B$ is

$$\begin{aligned} \rho_3 &= \mathrm{Tr}_{1,2}(|\, \psi \,\rangle_{11}\langle\, \psi \,| \otimes |\, \Phi \,\rangle_{2323}\langle\, \Phi \,|) \\ &= \mathrm{Tr}\Big(|\, \psi \,\rangle_{11}\langle\, \psi \,|\Big) \, \frac{1}{2} \sum_{i,j=0}^{1} \mathrm{Tr}\Big(|\, i \,\rangle_{22}\langle\, j \,|\Big) \, U_H |\, i \,\rangle_{33}\langle\, j \,| U_H = \frac{\mathbb{1}}{2} \, . \end{aligned}$$

Notice that the net effect of quantum teleportation is to get the third *qubit* in the rotated state $\sigma_\mu |\, \psi \,\rangle$ by means of a measurement in the *ONB* $\{|\, \Psi_\mu \,\rangle_{12}\}_{\mu=0}^3$ peformed on *qubits* 1 and 2, when the state of $1, 2, 3$ is $|\, \psi \,\rangle_1 \otimes |\, \Phi \,\rangle_{23}$.

**Example 6.1.5.** In order to implement a two-*qubit* gate like the unitary $U_{CNOT}$ on two target *qubit* states $\psi_{1,2}$, one adds to them three pairs of *qubits* each of which in the same entangled state $|\, \Phi \,\rangle$ introduced in the previous example. Thus, one deals with a multipartite entangled state [249, 323, 71]

$$|\, \Psi \,\rangle := |\, \psi_1 \,\rangle_1 \otimes |\, \psi_2 \,\rangle_2 \otimes |\, \Phi \,\rangle_{34} \otimes |\, \Phi \,\rangle_{57} \otimes |\, \Phi \,\rangle_{68}$$

corresponding to the scheme in Figure 6.2.

Then, measurements are performed on *qubits* $(1, 3, 5)$ and $(2, 4, 6)$ in the *ONB*s obtained from the *GHZ* vectors as in Example 5.5.10.3. By projecting $|\, \Psi \,\rangle$ onto the 6 *qubit* state $|\, \Psi_{abc} \,\rangle_{135} \otimes |\, \Psi_{def} \,\rangle_{468}$, one computes

$$\Big({}_{135}\langle\, \Psi_{abc} \,| \otimes {}_{246}\langle\, \Psi_{def} \,|\Big) |\, \Psi \,\rangle = \left(\frac{1}{\sqrt{2}}\right)^5 \sum_{\substack{r,s=0 \\ i,j,k=0}}^{1} \langle\, r \,|\, \sigma_1^a \,|\psi_1\,\rangle \langle\, r \,|\, \sigma_1^b \,| i \,\rangle \langle\, r \,|\, \sigma_3^c \,| j \,\rangle$$

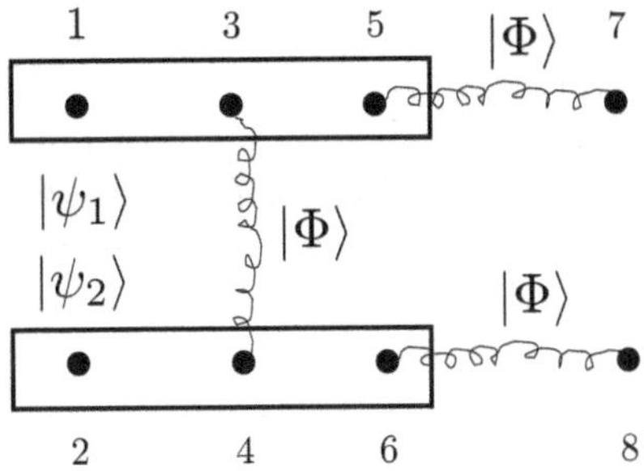


**Fig. 6.2.** One way Quantum Computation

$$\times \langle\, s \,|\, \sigma_1^d \,|\psi_2\,\rangle\, \langle\, s \,|\, \sigma_1^e U_H \,|i\,\rangle\, \langle\, s \,|\, \sigma_3^f \,|k\,\rangle\; U_H |\, j\,\rangle_7 \otimes U_H |\, k\,\rangle_8$$

$$= \left(\frac{1}{\sqrt{2}}\right)^5 \sum_{r,s=0}^{1} \langle\, r \,|\, \sigma_1^a \,|\psi_1\,\rangle\, \langle\, s \,|\, \sigma_1^d \,|\psi_2\,\rangle \langle\, s \,|\, \sigma_1^e U_H \sigma_1^b \,|r\,\rangle\; U_H \sigma_3^c |\, r\,\rangle_7 \otimes U_H \sigma_3^f |\, s\,\rangle_8$$

$$= \left(\frac{1}{\sqrt{2}}\right)^5 \left(U_H \sigma_3^c \otimes U_H \sigma_3^f\right) U_Z^{eb} \left(\sigma_1^a \otimes \sigma_1^d\right) |\, \psi_1\,\rangle_7 \otimes |\, \psi_2\,\rangle_8 \;,$$

where in summing over $i, j, k$ it has been used that, under transposition, $\sigma_{1,3}^T = \sigma_{1,3}$. Thus, a part from local unitary rotations, by measuring in the chosen *ONB* one implements the unitary transformation

$$U_Z^{eb} := \sum_{r,s=0}^{1} \langle\, s \,|\, \sigma_1^e U_H \sigma_1^b \,|r\,\rangle\, |\, r\,\rangle\langle\, r\,| \otimes |\, s\,\rangle\langle\, s\,|$$

$$= \sum_{r,s=0}^{1} \langle\, s \oplus e \,|\, U_H \,|r \oplus b\,\rangle\, |\, r\,\rangle\langle\, r\,| \otimes |\, s\,\rangle\langle\, s\,|$$

on the state $|\, \psi_1\,\rangle_7 \otimes |\, \psi_2\,\rangle_8$ of the pair of *qubits* that remain unaffected by the measurement. In particular, choosing $a = b = c = d = e = f = 0$, it turns out that $\sqrt{2}\, U_Z^{00} = |\, 0\,\rangle\langle\, 0\,| \otimes \mathbb{1} + |\, 1\,\rangle\langle\, 1\,| \otimes \sigma_3 = \mathbb{1} \otimes U_H\, U_{CNOT}\, \mathbb{1} \otimes U_H$, namely $\sqrt{2}\, U_Z^{00}$ amounts to the *CNOT* gate unitary matrix apart from unitary, local rotations.

## 6.2 Bipartite Entanglement

We have seen in Section 5.5.4 that, by looking at its marginal states, one knows whether a pure bipartite state is entangled or not. For density matrices entanglement detection is by far more difficult; only in low dimension the problem has been completely solved by the so-called Peres-Horodecki criterion [236, 148, 152].

**Proposition 6.2.1.** *Let a bipartite system $S_1+S_2$ be described by the algebra $M_{d_1}(\mathbb{C}) \otimes M_{d_2}(\mathbb{C})$, a state $\rho_{12} \in \mathcal{S}(S_1+S_2)$ is entangled if and only there exists a positive map $\Lambda : M_{d_2}(\mathbb{C}) \mapsto M_{d_1}(\mathbb{C})$ such that $\mathrm{id}_{d_1} \otimes \Lambda^+[\rho_{12}]$ is not positive definite, where $\Lambda^+ : \mathbb{B}_1^+(\mathbb{C}^{d_2}) \mapsto \mathbb{B}_1^+(\mathbb{C}^{d_1})$ is the dual map of $\Lambda$ from the space of states $\mathcal{S}(S_2) = \mathbb{B}_1^+(\mathbb{C}^{d_2})$ to the space of states $\mathcal{S}(S_1) = \mathbb{B}_1^+(\mathbb{C}^{d_1})$.*

**Proof:** The set $\mathcal{S}_{sep}(S_1+S_2)$ of separable states over $M_{d_1}(\mathbb{C}) \otimes M_{d_2}(\mathbb{C})$ is the closure in trace-norm of the convex hull of pure separable states (see Remark 5.5.7). By the Hahn-Banach theorem [258], $\mathcal{S}_{sep}(S_1+S_2)$ can be strictly separated from any entangled state $\rho_{ent}$ by a hyperplane, that is by a continuous linear functional $\mathcal{R} : \mathcal{S}(S_1+S_2) \mapsto \mathbb{R}$ and a real constant $a$ such that $\mathcal{R}(\rho_{ent}) < a \leq \mathcal{R}(\rho_{sep})$. As the trace-norm and the Hilbert-Schmidt topology are equivalent in finite dimension, using the argument of Example 5.2.4, the action of $\mathcal{R}$ can be represented by means of $R = R^\dagger \in M_{d_1}(\mathbb{C}) \otimes M_{d_2}(\mathbb{C})$ such that $\mathcal{R}(\rho) = \mathrm{Tr}(R\,\rho)$. Setting $S := R' - a\mathbb{1}$, it thus follows that $\rho \in M_{d_1}(\mathbb{C}) \otimes M_{d_2}(\mathbb{C})$ is entangled if and only if there exists $S \in M_{d_1}(\mathbb{C}) \otimes M_{d_2}(\mathbb{C})$ such that $\mathrm{Tr}(S\,\rho) < 0$ while $\mathrm{Tr}(S\,\rho_{sep}) \geq 0$ for all $\rho_{sep} \in \mathcal{S}_{sep}(S_1+S_2)$.

Furthermore, to any such matrix, the Jamiołkowski isomorphism (see Remark 5.2.5) associates a positive map $\Lambda_S : M_{d_1}(\mathbb{C}) \mapsto M_{d_2}(\mathbb{C})$ with $S$ as Choi matrix. Let $\Lambda_S^+ : \mathbb{B}_1^+(\mathbb{C}^{d_2}) \mapsto \mathbb{B}_1^+(\mathbb{C}^{d_1})$ be its dual such that

$$\mathrm{Tr}(S\,\rho) = \mathrm{Tr}\left(\mathrm{id}_{d_1} \otimes \Lambda_S[P_+^{d_1}]\,\rho\right) = \mathrm{Tr}\left(P_+^{d_1}\mathrm{id}_{d_1} \otimes \Lambda_S^+[\rho]\right) ,$$

for all $\rho \in \mathcal{S}(S_1+S_2)$. If $\rho$ is an entangled state such that $\mathrm{Tr}(S\,\rho) < 0$, then $\mathrm{id}_{d_1} \otimes \Lambda_S^+[\rho]$ cannot be positive definite. Vice versa, if $\mathrm{id}_{d_1} \otimes \Lambda^+[\rho] \geq 0$ for all positive $\Lambda : M_{d_2}(\mathbb{C}) \mapsto M_{d_2}(\mathbb{C})$, then $\rho \in \mathcal{S}_{sep}(S_1+S_2)$. □

As a consequence of the previous argument, a map $\Lambda : M_{d_2}(\mathbb{C}) \mapsto M_{d_1}(\mathbb{C})$ is a witness of the entanglement of the state $\rho \in \mathcal{S}(S_1+S_2)$ if $\mathrm{id}_{d_1} \otimes \Lambda^+$ turns $\rho$ into a non-positive matrix. Therefore, $\Lambda$ cannot be a *CP* map; however it preserves positivity. Indeed, the Choi matrix $L \in M_{d_1}(\mathbb{C}) \otimes M_{d_2}(\mathbb{C})$ associated to the dual map $\Lambda^+$ is block positive for $\mathrm{Tr}(L\,\rho) \geq 0$ whenever $\rho$ is separable, that is $\langle\,\psi\otimes\phi\,|\,L\,|\psi\otimes\phi\,\rangle$ for all $\psi \in \mathbb{C}^{d_1}$ and $\phi \in \mathbb{C}^{d_2}$, whence $\Lambda^+$ is a positive map.

Unfortunately, as already noticed (see Remark 5.2.6.3), unlike *CP* maps for which Proposition 5.2.1 holds, positive linear maps still lack a complete characterization. Consequently, given an entangled state $\rho \in \mathcal{S}(S_1+S_2)$ it is usually rather difficult to find a corresponding entanglement witness $\Lambda$. A relatively understood sub-class of positive maps is the following one.

**Definition 6.2.1 (Decomposable Maps).** *A map $\Lambda : \mathbb{B}(\mathbb{H}) \mapsto \mathbb{B}(\mathbb{K})$ is* decomposable *if it is positive and $\Lambda = \Lambda_1 + \Lambda_2 \circ T_{\mathbb{H}}$, with $\Lambda_{1,2}$ CP maps and $T_{\mathbb{H}}$ the transposition on $\mathbb{B}(\mathbb{H})$ with respect to a fixed orthonormal basis in $\mathbb{H}$.*

Let $(d_1, d_2) = (2,2), (2,3), (3,2)$, then a theorem of Woronowicz [321] asserts that all positive maps $\Lambda : M_{d_1}(\mathbb{C}) \mapsto M_{d_2}(\mathbb{C})$ are decomposable. This fact makes transposition an exhaustive entanglement witness in low dimension; in other words, for the stated dimensions, those states that remain *positive under partial transposition*, are separable and viceversa.

**Corollary 6.2.1.** *If in the previous proposition* $(d_1, d_2) = (2,2), (2,3), (3,2)$, *then,* $\rho_{12} \in \mathcal{S}(S_1 + S_2)$ *is entangled if and only if* $T^{(2)}[\rho_{12}]$ *is not positive-definite, where* $T^{(2)} := \mathrm{id}_{d_1} \otimes T_{d_2}$ *denotes partial transposition on the second factor.*

**Proof:** If $\rho \in \mathcal{S}(S_1 + S_2)$ is separable then $\mathsf{T}^{(2)}[\rho] \geq 0$ for transposition is a positive map. Vice versa, because of the assumption, Woronowicz theorem ensures that any positive map is decomposable. Therefore, if $\mathsf{T}^{(2)}[\rho] \geq 0$, it turns out that, for all positive $\Lambda : M_{d_2}(\mathbb{C}) \mapsto M_{d_2}(\mathbb{C})$,

$$\mathrm{id}_{d_1} \otimes \Lambda[\rho] = \mathrm{id}_{d_1} \otimes \Lambda_1[\rho] + \mathrm{id}_{d_1} \otimes \Lambda_2[\mathsf{T}^{(2)}[\rho]] \geq 0 ,$$

as $\Lambda_{1,2}$ are *CP* maps. □

**Remarks 6.2.1.**

1. Though partial transposition as transposition are to be defined with respect to a chosen *ONB*, the spectrum of an operator is basis-independent; therefore, the non-positivity of $\mathsf{T}^{(2)}[\rho_{12}]$, thence the entanglement of $\rho_{12}$, does not depend on the *ONB* with respect to which the partial transposition is performed.
2. Those states which remain positive under partial transposition are called *PPT* states, otherwise *NPT* states, namely *negative under partial transposition.* Woronowicz theorem does not extend to higher dimension; there are instances of non-decomposable positive maps already for $d_1 = d_2 = 3$ [83, 130, 288, 321]; as a consequence partial transposition is not an exhaustive entanglement witness in higher dimension. In other words, all *NPT* states are entangled, but there can exist *PPT* entangled states [149, 297].
3. No pure bipartite state can be *PPT* entangled; indeed, by Proposition 5.5.7, entangled vector states $|\Psi_{12}\rangle \in \mathbb{C}^{d_1} \otimes \mathbb{C}^{d_2}$ have a Schmidt decomposition (5.159) of the form $|\Psi_{12}\rangle = \sum_{j=1}^{d} \sqrt{\lambda_j} |\psi_j^{(1)}\rangle \otimes |\psi_j^{(2)}\rangle$, where $d := \max\{d_1, d_2\}$, $\{|\psi_j^{(1,2)}\rangle\}_{j=1}^d$ are orthonormal sets in the Hilbert spaces $\mathbb{C}^{d_{1,2}}$ and the Schmidt coefficients $\lambda_j > 0$ for at least two indices. Set $P_{12} := |\Psi_{12}\rangle\langle\Psi_{12}|$; the partial transposition with respect to the *ONB* having $\{|\psi_j^{(2)}\rangle\}_{j=1}^d$ among its elements yields

$$R_{12} := \mathsf{T}^{(2)}[P_{12}] = \sum_{i,j=1}^{d} \sqrt{\lambda_j \lambda_j}\, |\,\psi_i^{(1)}\,\rangle\langle\,\psi_j^{(1)}\,| \otimes |\,\psi_j^{(2)}\,\rangle\langle\,\psi_i^{(2)}\,| \,.$$

Let $\lambda_{12} > 0$, then

$$R_{12}\,\frac{|\,\psi_1^{(1)}\psi_2^{(2)}\,\rangle - |\,\psi_2^{(1)}\psi_1^{(2)}\,\rangle}{\sqrt{2}} = -\sqrt{\lambda_1\lambda_2}\,\frac{|\,\psi_1^{(1)}\psi_2^{(2)}\,\rangle - |\,\psi_2^{(1)}\psi_1^{(2)}\,\rangle}{\sqrt{2}} \,.$$

Thus $\mathsf{T}^{(2)}[P_{12}]$ cannot be positive if $P_{12}$ is entangled.

4. The entanglement of *PPT* entangled density matrices cannot be detected by decomposable positive maps as one can see from an argument similar to the one used in the proof of Corollary 6.2.1. An instance of such states will be discussed in Example 6.2.4.

The following ones are families of bipartite states over $\mathbb{C}^d \otimes \mathbb{C}^d$, $d \geq 2$, where *PPT* states are always separable.

**Examples 6.2.1.**

1. **Werner States** [317] This is a class of $d^2 \times d^2$ density matrices on $\mathbb{C}^d \otimes \mathbb{C}^d$ of the form $\rho_W = \alpha \mathbb{1}_{d^2} + \beta V$ where $V$ is the flip operator (see (5.32)) and $W := \mathrm{Tr}(\rho_W V)$ (*).
As the eigenvalues of $V$ are $\pm 1$, those of $\rho_W$ are $\alpha \pm \beta$ and must be positive. Also, $V^2 = \mathbb{1}_{d^2}$ and $\mathrm{Tr}V = \sum_{i,j=1}^N \langle\, ij \,|\, V \,|ij\,\rangle = \sum_{i,j=1}^N |\langle\, i \,|\, j\,\rangle|^2 = d$; thus, normalization and (*) yield $\alpha\, d^2 + \beta\, d = 1$ and $W = \alpha\, d + \beta\, d^2$, whence

$$\alpha = \frac{d-W}{d(d^2-1)} \,,\ \beta = \frac{dW-1}{d(d^2-1)} \;;\ \alpha+\beta = \frac{1+W}{d(d+1)} \,,\ \alpha-\beta = \frac{1-W}{d(d-1}$$

$$\rho_W = \frac{d(d-W)}{d^2-1}\,\frac{\mathbb{1}_{d^2}}{d^2} + \frac{dW-1}{d(d^2-1)} V \,,\quad -1 \leq W \leq 1 \,. \tag{6.2}$$

If $\rho_W$ is separable as in (5.166), by spectralizing the contributing density matrices, it can always be recast as $\rho_W = \sum_{ij} \mu_{ij} |\,\psi_i^1\,\rangle\langle\,\psi_i^1\,| \otimes |\,\psi_j^2\,\rangle\langle\,\psi_j^2\,|$, $\mu_{ij} \geq 0$, $\sum_{ij}\mu_{ij} = 1$. Then,

$$W = \mathrm{Tr}(\rho_W V) = \sum_{ij} \mu_{ij} \langle\,\psi_i^1 \otimes \psi_j^2 \,|\, V \,|\psi_i^1 \otimes \psi_j^2\,\rangle = \sum_{ij} \mu_{ij} |\langle\,\psi_i^1 \,|\, \psi_j^2\,\rangle| \geq 0 \,.$$

This is a necessary condition for the separability of Werner states in dimension $d$; it is also sufficient; the clue is that Werner states are exactly those states on $\mathbb{C}^{d^2}$ that commute with all unitaries of the form $U \otimes U$ with $U$ a unitary matrix in $M_d(\mathbb{C})$. Practically, since $V(A \otimes B)V = B \otimes A$ for all $A, B \in M_d(\mathbb{C})$, all Werner states have the form

$$\rho_W = \int_{\mathcal{U}} \mathrm{d}U \, (U \otimes U) \, \rho \, (U^\dagger \otimes U^\dagger) \, ,$$

where $\mathrm{d}U$ is the normalized, invariant Haar measure over the unitary group $\mathcal{U}$ on $\mathbb{C}^{d\,2}$; furthermore, $W = \mathrm{Tr}(\rho V)$.
If $1 \geq W \geq 0$, let $\psi \in \mathbb{C}^d$, $|\phi\rangle = \sqrt{W}\,|\psi\rangle + \sqrt{1-W}\,|\psi^\perp\rangle \in \mathbb{C}^d$, with $\langle\psi|\psi^\perp\rangle = 0$ and set $\rho := |\phi\rangle\langle\phi| \otimes |\psi\rangle\langle\psi|$. Thus, $\rho_W$ arises by twirling a separable state with tensor products of local unitaries, therefore it is itself separable, as local actions cannot create entanglement.
The necessary and sufficient condition for separability, $W \geq 0$, turns out to be equivalent to $\rho_W$ being positive under partial transposition. Indeed, by applying $\mathsf{T}^{(2)} = \mathrm{id}_d \otimes \mathsf{T}_d$ to $\rho_W$ one gets

$$\mathsf{T}^{(2)}[\rho_W] = \frac{d-W}{d(d^2-1)} \, \mathbb{1}_{d^2} + \frac{dW-1}{d^2-1} \widehat{P}^d_+ \, .$$

Its eigenvalues $(d-W)/(d^3-d) \geq 0$ and $W/d$ are positive if and only if $W \geq 0$.

2. **Isotropic States** [151] This is a class of $d^2 \times d^2$ density matrices on $\mathbb{C}^d \otimes \mathbb{C}^d$ which are related to Werner states by partial transposition. They have the form $\rho_F = \alpha \mathbb{1}_{d^2} + \beta\,\widehat{P}^d_+$ and are uniquely identified by the parameter $0 \leq F := \mathrm{Tr}(\rho_F \widehat{P}^d_+)$ (*).
Like for Werner states, positivity, normalization and (*) yield $\alpha \geq 0$, $\alpha\, d^2 + \beta = 1$ and $1 \geq F = \alpha + \beta \geq 0$, whence isotropic states are mixtures of the totally depolarized state on $\mathbb{C}^{d^2}$ and of the totally symmetric state,

$$\rho_F = \frac{d^2(1-F)}{d^2-1} \frac{\mathbb{1}_{d^2}}{d^2} + \frac{d^2F-1}{d^2-1} \widehat{P}^d_+ \, . \tag{6.3}$$

Since $\langle\psi\otimes\phi|\,\widehat{P}^d_+\,|\psi\otimes\phi\rangle = |\langle\psi|\phi^*\rangle|^2$, where $\psi^*$ is the vector in $\mathbb{C}^d$ with complex conjugate components with respect to $\psi$, if $\rho_F$ is separable, then (see the previous example)

$$F = \mathrm{Tr}(\rho_F \widehat{P}^d_+) = \frac{1}{d} \sum_{ij} \mu_{ij} |\langle\psi^1_i\,|\,(\phi^2_j)^*\rangle|^2 \leq \frac{1}{d} \, .$$

As for Werner states, $0 \leq F \leq 1/d$ is necessary and also sufficient for separability. The reason is that isotropic states are all and only those

---

[2] The particular convex combination of states $(U \otimes U)\,\rho\,(U^\dagger \otimes U^\dagger)$ appearing in the integral is known as *twirling*. Twirled $\rho$ are such that, for all unitary $V$,

$$V \otimes V \left( \int_{\mathcal{U}} \mathrm{d}U \, U \otimes U \, \rho \, U^\dagger \otimes U^\dagger \right) V^\dagger \otimes V^\dagger = \int_{\mathcal{U}} \mathrm{d}U \, \underbrace{VU}_{W} \otimes VU \, \rho \, (UV)^\dagger \otimes (UV)^\dagger$$

$$= \int_{\mathcal{U}} \mathrm{d}(V^\dagger W) \, W \otimes W \, \rho \, W^\dagger \otimes W^\dagger = \int_{\mathcal{U}} \mathrm{d}U \, U \otimes U \, \rho \, U^\dagger \otimes U^\dagger \, ,$$

for the Haar measure satisfies $\mathrm{d}(VU) = \mathrm{d}U$ for all unitary $V$.

$d^2 \times d^2$ density matrices which commute with local unitaries of the form $U \otimes U^*$, where $U$ is any unitary matrix in $M_d(\mathbb{C})$ and $U^*$ denotes its complex conjugate (not its adjoint). Moreover, one can show that, since $(U \otimes U^*)\widehat{P}^d_+(U^\dagger \otimes U^T) = \widehat{P}^d_+$, any isotropic $\rho_F$ arises from a twirling of the form

$$\rho_F = \int_{\mathcal{U}} \mathrm{d}U \, (U \otimes U^*) \, \rho \, (U^\dagger \otimes U^T) \, ,$$

where $U^T$ denotes the transposition of $U$ and $\rho$ is such that $\mathrm{Tr}(\rho \widehat{P}^d_+) = F$. If $Fd \leq 1$, set $|\,\phi\,\rangle = \sqrt{dF}\,|\,\psi\,\rangle + \sqrt{1-dF}\,|\,\psi^\perp\,\rangle$ and choose $\rho = |\,\psi\,\rangle\langle\,\psi\,| \otimes |\,\phi\,\rangle\langle\,\phi\,|$; then, $\rho_F$ can be obtained by twirling a separable state and is thus itself separable.
The above necessary and sufficient conditions for separability coincides with the isotropic states being positive under partial transposition. Indeed,

$$\mathsf{T}^{(2)}[\rho_F] = \frac{1-F}{d^2-1}\,\mathbb{1}_{d^2} + \frac{d^2F-1}{d(d^2-1)}\,V$$

has positive eigenvalues $(dF+1)/(d^2+d) \geq 0$ and $(1-dF)/(d^2-d)$ if and only if $0 \leq F \leq 1/d$.

### Distillability and Bound Entanglement

Entangled states of two *qubits* as the Bell states (see Example 5.5.9) are called maximally entangled. Consider a pure state $|\,\Psi_{12}\,\rangle \in \mathbb{C}^d \otimes \mathbb{C}^d$ of a bipartite system consisting of two copies of a same system; as we shall see, there are good reasons to measure the amount of entanglement of $\rho_{12}$ by means of the von Neumann entropy of any of its two marginal density matrices $\rho^{(1,2)}_{\Psi_{12}} := \mathrm{Tr}_{2,1}\Big(|\,\Psi_{12}\,\rangle\langle\,\Psi_{12}\,|\Big)$ (see Proposition 5.5.5).

**Definition 6.2.1 (Pure State Entanglement).** *Let $|\,\Psi_{12}\,\rangle \in \mathbb{C}^d \otimes \mathbb{C}^d$ be a pure state of the bipartite system $S_1 + S_2$; the entanglement of $|\,\Psi_{12}\,\rangle$ is*

$$\mathrm{E_F}[|\,\Psi_{12}\,\rangle\langle\,\Psi_{12}\,|] := S\left(\rho^{(1,2)}_{\Psi_{12}}\right) \, . \tag{6.4}$$

According to Example 5.5.10.1, all Bell states have marginal states that are the tracial state with maximal von Neumann entropy: $E[\hat{\Psi}_{xy}] = \log 2$.

The presence of uncontrollable interactions with the environment in which a bipartite system may be immersed usually spoils its maximally entangled states. For instance, the so-called *singlet state* $|\,\Psi\,\rangle_- := \dfrac{|\,01\,\rangle - |\,10\,\rangle}{\sqrt{2}}$ might be rotated into $|\,\Psi\,\rangle = \alpha|\,01\,\rangle + \beta|\,10\,\rangle$, with $|\alpha|^2 + |\beta|^2 = 1$ and $0 < |\alpha| < 1$, so that

$$\rho_\Psi^{(1)} = \begin{pmatrix} |\alpha|^2 & 0 \\ 0 & |\beta|^2 \end{pmatrix} , \quad E[\Psi] = -|\alpha|^2 \log |\alpha|^2 - |\beta|^2 \log |\beta|^2 < \log 2 .$$

It may even be turned into a mixed state because the environment usually acts as a source of noise and dissipation. A measure of the entanglement of $\rho_{12} \in \mathcal{S}(S_1 + S_2)$ is as follows [3].

**Definition 6.2.2 (Entanglement of Formation).** *[53] The entanglement of formation of a state $\rho_{12}$ of a bipartite system $S_1 + S_2$ is the least average pure state entanglement over all convex decompositions of $\rho_{12}$,*

$$\mathrm{E_F}[\rho_{12}] := \min \left\{ \sum_j \lambda_j S\left(\rho^{(1,2)}_{\psi^j_{12}}\right) , \rho_{12} = \sum_j \lambda_j | \psi^j_{12} \rangle\langle \psi^j_{12} | \right\} , \qquad (6.5)$$

*where $\lambda_j > 0$ and $\sum_j \lambda_j = 1$.*

Maximal entanglement is an important resource in quantum information, but also a highly degradable one; of particular importance are then those quantum protocols that enable to distil maximally entangled states out of non-maximally entangled ones by means of *LOCC* [4]. The basic scheme of a distillation protocol is as follows: given $m$ copies of $\rho_{12} \in \mathbb{B}_1^+(\mathbb{C}^d \otimes \mathbb{C}^d)$, one tries to maximize the number $n$ of copies of the singlet state projection $P_- := | \Psi_- \rangle\langle \Psi_- |$ that can be obtained by means of local operations and classical communication:

$$\underbrace{\rho_{12} \otimes \rho_{12} \otimes \cdots \rho_{12}}_{m\, times} \overset{LOCC}{\longmapsto} \underbrace{P_- \otimes P_- \otimes \cdots P_-}_{n\, times} .$$

The *LOCC* defining the distillation protocols amount to maps of the form

$$\rho_{12}^{\otimes m} \mapsto \rho_{12}^{(n)} = \frac{1}{N_I} \sum_{i \in I} A_{1i} \otimes A_{2i}\, \rho_{12}^{\otimes m}\, A_{1i}^\dagger \otimes A_{2i}^\dagger , \qquad (6.6)$$

where

$$N_I := \sum_{i \in I} \mathrm{Tr}\left(A_{1i}^\dagger A_{1i} \otimes A_{2i}^\dagger A_{2i}\, \rho_{12}^{\otimes m}\right) ,$$

while $A_{ji} : (\mathbb{C}^d)^{\otimes m} \mapsto (\mathbb{C}^2)^{\otimes n}$, $j = 1, 2$.

---

[3] Various entanglement measures have appeared while quantum entanglement theory has been developing, for a review and the related literature see the contribution by M.B. Plenio and S.S. Virmani in [71].

[4] For a review of entanglement distillation and the other topics of this Section see [70] and the contributions by A. Sen, U. Sen, M.Lewenstein et al., W. Dür and H.-J. Briegel, and P. Horodecki in [71]

Practically, one seeks distillation protocols that output states $\rho^{(n)}$ whose distance from $P_-^{\otimes n}$ (for instance, with respect to the trace-norm (5.21)) vanishes when $m \to +\infty$, while the ratio $n/m$ is the highest possible. The optimal ratio, denoted by $\mathrm{E_D}[\rho_{12}]$, is called *entanglement of distillation* and represents the maximal asymptotic fraction of singlets per $\rho_{12}$ that one can achieve by *LOCC*. In other words, one can hope to distil at most $n \simeq m\,\mathrm{E_D}[\rho_{12}]]$ singlets $P_-$ out of $m$ $\rho_{12}$ when $m$ gets sufficiently large.

It turns out that *PPT* states $\rho_{12}$ cannot be distilled [150]; when $\rho_{12}$ is separable this is obvious since one cannot create non-local quantum correlations by means of local operations and classical communication. The interesting point is that one has to distinguish between *free entanglement*, the entanglement which can be distilled, and *bound entanglement*, that which cannot be distilled. The result just quoted can be rephrased by saying that the entanglement of *PPT* entangled states is bound. This can be seen as follows: if the entanglement in $\rho_{12}$ is distillable, then for some $m$, the state $\rho_{12}^{(n)}$ in (6.6) must be an entangled state of 2 *qubits*. whence an *NPT* state according to Corollary 6.2.1. This implies that, for at least one index $i_0 \in I$, the (non-normalized) state

$$\widetilde{\rho}_{12} := A_{1i_0} \otimes A_{2i_0}\, \rho_{12}^{\otimes m}\, A_{1i_0}^\dagger \otimes A_{2i_0}^\dagger \tag{6.7}$$

is *NPT*. Observe that, for such $m$ and $i_0$, $A_{ji_0} : (\mathbb{C}^d)^{\otimes m} \mapsto \mathbb{C}^2$; therefore, one can always write $A_{ji_0} = \sum_{k=0}^{1} |\,k\,\rangle\langle\,\psi_{jk}\,|$, where $|\,\psi_{jk}\,\rangle \in (\mathbb{C}^d)^{\otimes m}$ and $|\,k\,\rangle$, $k = 0, 1$, is any chosen basis in $\mathbb{C}^2$. Let $Q_j$ be the projections onto the subspaces of $(\mathbb{C}^d)^{\otimes m}$ spanned by $|\,\psi_{j0}\,\rangle$ and $|\,\psi_{j1}\,\rangle$; then,

$$\widetilde{\rho}_{12} := A_{1i_0} \otimes A_{2i_0}\, Q_1 \otimes Q_2\, \rho_{12}^{\otimes m}\, Q_1 \otimes Q_2\, A_{1i_0}^\dagger \otimes A_{2i_0}^\dagger$$

implies that $\rho'_{12} := Q_1 \otimes Q_2\, \rho_{12}^{\otimes m}\, Q_1 \otimes Q_2$ must be entangled, otherwise its separability would be preserved when passing to $\widetilde{\rho}_{12}$.

Consider now the orthonormal bases $\{|\,b_{jn}\,\rangle\}_{n=1}^{d^m}$ in $(\mathbb{C}^d)^{\otimes m}$, $j = 1, 2$, such that $Q_j = |\,b_{j1}\,\rangle\langle\,b_{j1}\,| + |\,b_{j2}\,\rangle\langle\,b_{j2}\,|$; in the corresponding representation $\rho'_{12}$ is a $4 \times 4$ matrix acting on the subspace $\mathbb{K}$ spanned by the product states $|\,b_{1i}b_{2j}\,\rangle$, $i, j = 1, 2$. Since it corresponds to an entangled state, by partial transposition with respect to the *ONB* $\{|\,b_{2j}\,\rangle\}_{j=1}^2$, $\mathsf{T}_2[\rho'_{12}]$ cannot be positive semi-definite. Therefore, there must exist $|\,\Phi\,\rangle \in \mathbb{K}$ such that

$$\langle\,\Phi\,|\,\mathsf{T}_2[\rho'_{12}]\,|\Phi\,\rangle = \sum_{i,j;k,\ell=1}^{2} \Phi_{ij}^*\Phi_{k\ell}\,\langle\,b_{1i}b_{2j}\,|\,\mathsf{T}_2[\rho'_{12}]\,|b_{1k}b_{2\ell}\,\rangle$$

$$= \sum_{i,j;k,\ell=1}^{2} \Phi_{ij}^*\Phi_{k\ell}\,\langle\,b_{1i}b_{2\ell}\,|\,\rho'_{12}\,|b_{1k}b_{2j}\,\rangle = \sum_{i,j;k,\ell=1}^{2} \Phi_{ij}^*\Phi_{k\ell}\,\langle\,b_{1i}b_{2\ell}\,|\,\rho_{12}^{\otimes m}\,|b_{1k}b_{2j}\,\rangle$$

$$= \langle\,\Phi\,|\,\mathsf{T}[\rho_{12}^{\otimes m}]\,|\Phi\,\rangle < 0\,,$$

where $\mathsf{T}[\rho_{12}^{\otimes m}]$ is now the partial transposition with respect to the whole *ONB* $\{|\,b_{2j}\,\rangle\}_{j=1}^{d^m}$. Also, the last equality follows because $|\,\Phi\,\rangle$ is supported by the subspace $\mathbb{K}$ corresponding to the orthogonal projection $Q_1 \otimes Q_2$ and thus has vanishing projections onto all $|\, b_{1i} b_{2j} \,\rangle$ unless $i, j = 1, 2$.

Since *NPT* is a property which does not depend on the basis chosen to compute the partial transposition (see Remark 6.2.1.1), fix the bases $\{|\, e_{jk} \,\rangle\}_{k=1}^d$ in $\mathbb{C}^d$ and choose in $(\mathbb{C}^d \otimes \mathbb{C}^d)^{\otimes m}$ the product basis consisting of vectors $|\, e_{1k_1} e_{1k_2} \cdots e_{1k_m} \,\rangle \otimes |\, e_{2\ell_1} e_{2\ell_2} \cdots e_{2\ell_m} \,\rangle$. Then, $\mathsf{T}[\rho_{12}^{\otimes m}] = (\mathsf{T}_2[\rho_{12}])^{\otimes m}$; one thus concludes that $\rho_{12}$ is distillable only if $\rho_{12}$ is *NPT* .

**Remark 6.2.2.** From Remark 6.2.1.3 we know that no pure *PPT* entangled state can exist; it turns out that their entanglement is always distillable and thus free. Whether the entanglement of generic *NPT* states is also free, that is whether all *NPT* states are distillable, is one of the open problems in quantum information theory [71, 152].

### Entanglement Cost

One of the first questions in quantum information has been whether, by means of *LOCC* one can turn a pure state $|\, \Psi_{12} \,\rangle$ of a bipartite system into another pure state $|\, \Phi_{12} \,\rangle$. The answer is that this is possible if and only if the marginal states $\rho_{\Psi_{12}}^{(1,2)}$ are more mixed than those of $|\, \Phi_{12} \,\rangle$ in the sense of Definition 5.5.1 [224].

If one considers asymptotic *LOCC* protocols where $m$ copies of a state $|\, \Psi_{12} \,\rangle$ are turned into $n$ copies of a state $|\, \Phi_{12} \,\rangle$ with vanishing error when $m \to +\infty$, then the transformation of $|\, \Psi_{12} \,\rangle$ into $|\, \Phi_{12} \,\rangle$ is possible if and only if [71]

$$\frac{n}{m} \leq \frac{\mathrm{E_F}[\Psi_{12}]}{\mathrm{E_F}[\Phi_{12}]} \ .$$

Since $\mathrm{E_F}[\Psi_-] = 1$ (we shall use $\log_2$ in the following), one can always asymptotically distil $n \leq m\, \mathrm{E_F}[\Psi_{12}]$ copies of $P_-$ out of $m$ copies of any pure bipartite entangled state $\Psi_{12}$.

Furthermore, the reverse operation is also possible; namely, protocols have been devised which invert distillation and, by using $m$ copies of the singlet state $P_-$, form, by means of *LOCC*, $n$ copies of a bipartite pure state $\Psi_{12}$. Actually, like in the case of entanglement distillation, one considers the asymptotic minimal ratio $m/n$ when $n \to +\infty$ and $\rho_{12}^{\otimes n}$ is better and better approximated (within a suitable distance) by a suitable *LOCC* operation acting on $P_-^{\otimes m}$ [137]. The optimal asymptotic ratio, denoted by $\mathrm{E_C}[\rho_{12}]$ is called the *entanglement cost* of $\rho_{12}$; it represents the minimal fraction of singlet that is needed to create one bipartite system in the state $\rho_{12}$. In other words, for large $n$, one can create $n$ copies of $\rho_{12}$ only acting with *LOCC* on no less than $n\, \mathrm{E_C}[\rho_{12}]$ singlets.

In [224] a distillation protocol $\Lambda_D$ is constructed which asymptotically yields $\mathrm{E_F}[\Psi_{12}] = S\left(\rho_{\Psi_{12}}^{(1)}\right)$ singlets per bipartite entangled state $\rho_{12}$ (see (6.4)) and a formation protocol $\Lambda_F$ that asymptotically yields one copy of $\rho_{12}$ at the cost of $\mathrm{E_F}[\Psi_{12}]$ singlets. It turns out that the entanglement cost and the entanglement of distillation equal the pure state entanglement of formation.

By definition, $\mathrm{E_C}[\Psi_{12}] \leq \mathrm{E_F}[\Psi_{12}] \leq \mathrm{E_D}[\Psi_{12}]$; if $\mathrm{E_F}[\Psi_{12}] < \mathrm{E_D}[\Psi_{12}]$; then, by means of the protocol $\Lambda_F$ one could asymptotically obtain $\frac{m}{\mathrm{E_F}[\Psi_{12}]}$ copies of $\Psi_{12}$ out of $m$ copies of $P_-$ and then, using an optimal distillation protocol, extract from them $m\frac{\mathrm{E_D}[\Psi_{12}]}{\mathrm{E_F}[\Psi_{12}]} > m$ copies of $P_-$. This is impossible as one cannot increase the amount of entanglement by deterministic *LOCC* [5]. Analogously, if $\mathrm{E_C}[\Psi_{12}] < \mathrm{E_F}[\Psi_{12}]$, then one could use an optimal creation protocol to obtain $\frac{m}{\mathrm{E_C}[\Psi_{12}]}$ copies of $\rho_{12}$ out of $m$ copies of $P_-$ (for $m$ large) and then use the distillation protocol $\Lambda_D$ to extract from them $m\frac{\mathrm{E_F}[\Psi_{12}]}{\mathrm{E_C}[\Psi_{12}]} > m$ copies of $P_-$.

For pure states, forming entangled states from singlets and distilling singlets from entangled states are reversible operations; it is not so for mixed states and the reason for this peculiar kind of irreversibility is bound entanglement [152, 150, 71].

Consider the entanglement of formation as defined by (6.5); it can be interpreted as the minimal averaged entanglement cost of $\rho_{12}$. In fact, given a convex decomposition of $\rho_{12} = \sum_j \lambda_j |\,\psi^j_{12}\,\rangle\langle\,\psi^j_{12}\,|$, the entropies $S\left(\rho^{(1)}_{\psi^j_{12}}\right)$ are the entanglement cost of the pure states that decompose it. However, in line of principle, it could be more advantageous to create the tensor product $\rho_{12}^{\otimes n}$ instead of the $n$ copies of $\rho_{12}$ one by one. One is thus led to define the so-called *regularized entanglement of formation*

$$\mathrm{E_F^\infty}[\rho_{12}] := \lim_{n\to+\infty} \frac{1}{n}\,\mathrm{E_F}[\rho_{12}^{\otimes n}]\ . \tag{6.8}$$

Such a limit exists because the entanglement of formation is subadditive. Indeed, consider the state $\rho^{(1)}_{12} \otimes \rho^{(2)}_{12}$ of two copies of the bipartite system $S_1 + S_2$ and suppose $\mathrm{E_F}[\rho^{(i)}_{12}]$ is achieved at the (optimal) decompositions $\rho^{(i)}_{12} = \sum_j \nu^{(i)}_j |\,\phi^{(i)}_j\,\rangle\langle\,\phi^{(i)}_j\,|$. Since the decomposition

$$\rho^{(1)}_{12} \otimes \rho^{(2)}_{12} = \sum_{j,k} \nu^{(1)}_j\,\nu^{(2)}_k\,|\,\phi^{(1)}_j\,\rangle\langle\,\phi^{(1)}_j\,| \otimes |\,\phi^{(2)}_k\,\rangle\langle\,\phi^{(2)}_k\,|$$

need not in general be optimal for $\mathrm{E_F}[\rho^{(1)}_{12} \otimes \rho^{(2)}_{12}]$, it follows that

$$\mathrm{E_F}[\rho^{(1)}_{12} \otimes \rho^{(2)}_{12}] \leq \mathrm{E_F}[\rho^{(1)}_{12}] + \mathrm{E_F}[\rho^{(2)}_{12}]\ .$$

[5] One can achieve entanglement increase by *LOCC* only probabilistically for certain states of a mixture, for instance in some of the states in (6.6), but not on the average for the whole mixture.

In [137] it is proved that the regularized entanglement of formation equals the entanglement cost: $\mathrm{E}_{\mathrm{C}}[\rho_{12}] = \mathrm{E}_{\mathrm{F}}^{\infty}[\rho_{12}]$. Moreover (see P. Horodecki's contribution in [71]), it has been proved that $\mathrm{E}_{\mathrm{C}}[\rho_{12}] > 0$ for all entangled $\rho_{12}$.

As a consequence of the fact that, if $\rho_{12}$ is *PPT* entangled, no entanglement can be distilled from it, it thus turns out that a non-zero non-retrievable amount of entanglement (of singlets) is always necessary to create *PPT* entangled states.

### Concurrence

We shall now elaborate a little bit more in detail on the entanglement of formation of two *qubit* states.

Let $S_{A,B}$ be two *qubits* , $|0\rangle = \begin{pmatrix} 1 \\ 0 \end{pmatrix}$, $|1\rangle = \begin{pmatrix} 0 \\ 1 \end{pmatrix}$ the standard basis in $\mathbb{C}^2$ and consider a generic two *qubit* vector state of $S_A + S_B$ of the form

$$|\,\Psi_{AB}\,\rangle = C_{00}|\,00\,\rangle + C_{01}^{*}|\,01\,\rangle + C_{10}|\,10\,\rangle + C_{11}|\,11\,\rangle\ .$$

Then, the marginal state $\rho_A = CC^{\dagger}$, $C = \begin{pmatrix} C_{00} & C_{01} \\ C_{10} & C_{11} \end{pmatrix}$ has eigenvalues

$$\frac{1 \pm \sqrt{1 - \mathcal{C}(\Psi_{AB})^2}}{2}\ , \qquad \mathcal{C}(\Psi_{AB}) := 2\,|C_{00}C_{11} - C_{01}C_{10}|\ . \tag{6.9}$$

This expression can be recast as follows. Let $|\Psi_{AB}^{*}\rangle$ denote the complex conjugate of $|\,\Psi_{AB}\,\rangle$ with respect to the standard product basis $\{|ij\rangle\}_{i,j=0,1}$ and denote

$$|\,\widetilde{\Psi}_{AB}\,\rangle := \sigma_2 \otimes \sigma_2 |\,\Psi_{AB}\,\rangle = -C_{00}^{*}|\,11\,\rangle + C_{01}^{*}|\,10\,\rangle + C_{10}^{*}|\,01\,\rangle - C_{11}^{*}|\,00\,\rangle\ , \tag{6.10}$$

for $\sigma_2 = \begin{pmatrix} 0 & -i \\ i & 0 \end{pmatrix}$ is such that $\sigma_2|\,0\,\rangle = i|\,1\,\rangle$, $\sigma_2|\,1\,\rangle = -i|\,0\,\rangle$. Then,

$$\left|\langle\,\widetilde{\Psi}_{AB}\,|\,\Psi_{AB}\,\rangle\right| = \mathcal{C}(\Psi_{AB})\ . \tag{6.11}$$

Since $\sigma_2 \begin{pmatrix} \alpha^{*} \\ \beta^{*} \end{pmatrix} \perp \begin{pmatrix} \alpha \\ \beta \end{pmatrix}$, it turns out that $\mathcal{C}(\Psi_{AB}) = 0$ when $\Psi_{AB}$ is separable, while $\mathcal{C}(\Psi_{AB})$ reaches its maximum $\mathcal{C}(\Psi_{AB}) = 1$ when $\Psi_{AB}$ is maximally entangled and (only) two coefficients $C_{ij}$ are proportional to $2^{-1/2}$.

Therefore, for two *qubit* vector states, the entanglement of formation reads

$$\mathrm{E}_{\mathrm{F}}[|\,\Psi_{AB}\,\rangle\langle\,\Psi_{AB}\,|] = \mathcal{E}(\Psi_{AB}) := H_2\left(\frac{1 + \sqrt{1 - \mathcal{C}(\Psi_{AB})^2}}{2}\right)\ , \tag{6.12}$$

where $H_2(x) := -x \log x - (1 - x)\log(1 - x)$.

The variational problem embodied in Definition 6.2.2 is in general extremely difficult to solve and a general closed expression of $E(\rho)$ as a function of $\rho$ has been found only in the case of two *qubits* ; it is based upon the notion of *concurrence* [320].

Given a two *qubit* density matrix $\rho \in M_4(\mathbb{C})$, one first constructs the density matrix

$$\tilde{\rho} = \sigma_2 \otimes \sigma_2 \rho^* \sigma_2 \otimes \sigma_2 \ , \tag{6.13}$$

obtained via the operation (6.10), where $\rho^*$ denotes complex conjugation with respect to the the standard basis $\{|ij\rangle\}_{i,j=0,1}$. Then, the quantity $\mathcal{C}(\Psi_{AB})$ in (6.9) generalizes to density matrices as follows.

**Definition 6.2.3 (Concurrence).** *Let $\lambda_i$, $i = 1,2,3,4$, be the positive eigenvalues of $\sqrt{\sqrt{\rho}\tilde{\rho}\sqrt{\rho}}$ in decreasing order. The* concurrence *of $\rho$ is*

$$\mathcal{C}(\rho) := \max\{\lambda_1 - \lambda_2 - \lambda_3 - \lambda_4 \, , 0\} \ . \tag{6.14}$$

**Examples 6.2.2.**

1. **Pure states** Let $\rho = |\psi\rangle\langle\psi|$, $|\psi\rangle \in \mathbb{C}^4$. Then,

$$\sqrt{\rho}\tilde{\rho}\sqrt{\rho} = \left|\langle\psi|\tilde{\psi}\rangle\right|^2 \, |\psi\rangle\langle\psi| \ ,$$

whence $\mathcal{C}(\rho) = \left|\langle\psi|\tilde{\psi}\rangle\right|$.

2. **Werner states** Setting $d = 2$ in (6.2)

$$|0\rangle\langle 0| = \frac{1+\sigma_3}{2} \ , \quad |0\rangle\langle 1| = \frac{\sigma_1 + i\sigma_2}{2}$$
$$|1\rangle\langle 1| = \frac{1-\sigma_3}{2} \ , \quad |1\rangle\langle 0| = \frac{\sigma_1 - i\sigma_2}{2} \ ,$$

it follows that

$$\widehat{P}^2_+ = \frac{1}{4}\Big(1 \otimes 1 + \sigma_1 \otimes \sigma_1 - \sigma_2 \otimes \sigma_2 + \sigma_3 \otimes \sigma_3\Big)$$
$$V = \mathrm{id} \otimes T[P^2_+] = \frac{1}{2}\Big(1 \otimes 1 + \sigma_1 \otimes \sigma_1 + \sigma_2 \otimes \sigma_2 + \sigma_3 \otimes \sigma_3\Big)$$
$$\rho_W = \frac{1}{4}\Big(1 \otimes 1 + \frac{2W-1}{3}(\sigma_1 \otimes \sigma_1 + \sigma_2 \otimes \sigma_2 + \sigma_3 \otimes \sigma_3)\Big) \ .$$

Since $W \in \mathbb{R}$, the algebraic relations among the Pauli matrices yield $\widetilde{\rho_W} = \rho_W$, whence the eigenvalues of $\sqrt{\sqrt{\rho_W}\tilde{\rho}_W\sqrt{\rho_W}}$ are those of $\rho_W$, namely $\frac{1+W}{6}$ (thrice degenerate) and $\frac{1-W}{2}$. It then follows that

$$\mathcal{C}(\rho_W) = \begin{cases} \max\{-W, 0\} & -1 \leq W \leq 1/2 \\ \max\{(W-2)/3, 0\} & 1/2 \leq W \leq 1 \end{cases} \ ,$$

whence $\mathcal{C}(\rho_W) > 0$ and $\rho_W$ is entangled if and only if $W < 0$, in agreement with Example 6.2.1.1

3. **Isotropic States** Setting $d = 2$ in (6.3) and arguing as in the previous example, the isotropic states read

$$\rho_F = \frac{1}{4}\Big(1 \otimes 1 + \frac{4F-1}{3}(\sigma_1 \otimes \sigma_1 - \sigma_2 \otimes \sigma_2 + \sigma_3 \otimes \sigma_3)\Big) .$$

Again, it turns out that $\tilde{\rho}_F = \rho_F$ so that the eigenvalues of $\sqrt{\sqrt{\rho_F}\tilde{\rho}_F\sqrt{\rho_F}}$ are those of $\rho_F$ itself, namely $\frac{1-F}{3}$ thrice degenerate and $F$. Thus,

$$C(\rho_F) = \begin{cases} \max\{2F-1, 0\} & 1/4 \leq F \leq 1 \\ \max\{-(1+2F)/3, 0\} & 0 \leq F \leq 1/4 \end{cases} ,$$

whence $\rho_F$ is entangled if and only if $F > 1/2$, in agreement with Example 6.2.1.2.

By direct inspection, the function (see (6.12))

$$\mathcal{E}(C(\psi)) = H_2\left(\frac{1 + \sqrt{1 - C(\psi)^2}}{2}\right) , \qquad (6.15)$$

is monotonically increasing ($\mathcal{E}'(x) \geq 0$, $0 \leq x \leq 1$) and *convex* ($\mathcal{E}''(x) \geq 0$, $0 \leq x \leq 1$) in the concurrence. As $0 \leq C(\psi) \leq 1$, the entanglement increases from $\mathcal{E}(C(\psi)) = 0$ for separable vector states to $\mathcal{E}(C(\psi)) = 1$ for maximally entangled states and

$$\mathcal{E}(\lambda x_1 + (1-\lambda)x_2) \leq \lambda\mathcal{E}(x_1) + (1-\lambda)\mathcal{E}(x_2) , \quad 0 \leq \lambda \leq 1 , \quad 0 \leq x_{1,2} \leq 1 .$$

Further, given a decomposition $\rho = \sum_j p_j |\psi_j\rangle\langle\psi_j|$, let

$$\langle C\rangle_{\rho=\sum_j p_j|\psi_j\rangle\langle\psi_j|} := \sum_j p_j\, C(\psi_j) \qquad (6.16)$$

$$\langle \mathcal{E}\rangle_{\rho=\sum_j p_j|\psi_j\rangle\langle\psi_j|} := \sum_j p_j\, \mathcal{E}(C(\psi_j)) \qquad (6.17)$$

denote the corresponding average concurrence, respectively the average entanglement. Because of convexity, it turns out that

$$\mathcal{E}\Big(\langle C\rangle_{\rho=\sum_j p_j|\psi_j\rangle\langle\psi_j|}\Big) \leq \langle \mathcal{E}\rangle_{\rho=\sum_j p_j|\psi_j\rangle\langle\psi_j|} . \qquad (6.18)$$

**Theorem 6.2.1.** *[320] The entanglement of formation (6.2.2) of any state $\rho$ of a two qubit system is given by $\mathrm{E_F}[\rho] = \mathcal{E}(C(\rho))$ and is thus a monotonically increasing function of the concurrence (6.14).*

**Proof:** The right hand side of (6.18) is the argument of the minimum in (6.5) (see (6.12)), whereas the left hand side is an increasing function of its argument. It thus follows that $\mathrm{E_F}[\rho]$ cannot be smaller than $\mathcal{E}(\mathcal{C}_{min})$ where $\mathcal{C}_{min}$ is the smallest average concurrence. Therefore, if $\mathcal{C}_{min}$ is attained at a suitable decomposition, then the same decomposition yields $\mathrm{E_F}[\rho] = \mathcal{E}(\mathcal{C}_{min})$. We will construct a density matrix $\rho = \sum_j p_j |\,\psi_j\,\rangle\langle\,\psi_j\,|$ such that $\langle\mathcal{C}\rangle_{\rho=\sum_j p_j |\,\psi_j\,\rangle\langle\,\psi_j\,|} = \mathcal{C}(\rho)$ and show that no smaller average concurrence can be achieved, namely $\mathcal{C}_{min} = \mathcal{C}(\rho)$.

In order to arrive at such decomposition, we first consider the expansion $\rho = \sum_{i=1}^n |\,v_i\,\rangle\langle\,v_i\,|$, $n \leq 4$ being the rank of $\rho$, and $|v_j\rangle$ its (non-normalized) eigenvectors such that $\langle v_i|v_j\rangle = r_j\delta_{ij}$, with $r_j$ the eigenvalues of $\rho$.

The $n \times n$ matrix $\tau$ with entries $\tau_{ij} := \langle v_i|\tilde{v}_j\rangle$, where $|\tilde{v}_i\rangle := \sigma_2 \otimes \sigma_2 |v_i^*\rangle$, is symmetric, $\langle v_i|\tilde{v}_j\rangle = \langle v_j|\tilde{v}_i\rangle$, but not hermitian and

$$(\tau\tau^\dagger)_{ij} = (\tau\tau^*)_{ij} = \sum_{k=1}^n \langle v_i|\tilde{v}_k\rangle\langle\tilde{v}_k|v_j\rangle = \langle r_i|\,\sqrt{\rho}\tilde{\rho}\sqrt{\rho}\,|r_j\rangle\ .$$

Thus the eigenvalues of $\tau\tau^*$ are the squares of the eigenvalues $\lambda_j$ of $\sqrt{\rho}\tilde{\rho}\sqrt{\rho}$ in decreasing order (see (6.14)). Let $Z$ be the $n \times n$ unitary matrix that diagonalizes $\tau\tau^*$,

$$Z\tau\tau^*Z^\dagger = (Z\tau Z^T)(Z\tau Z^T)^* = \mathrm{diag}(\lambda_1^2, \lambda_2^2, \lambda_3^2, \lambda_4^2)\ ,$$

then $Z$ can be chosen such that $Z\tau Z^T = \mathrm{diag}(\lambda_1, \lambda_2, \lambda_3, \lambda_4)$ is diagonal with the $\lambda_j$'s as eigenvalues. Setting $|w_i\rangle := \sum_{j=1}^n Z_{ij}^*|v_j\rangle$ gives $\rho = \sum_{i=1}^n |w_i\rangle\langle w_i|$, with decomposers such that

$$\langle w_i|\tilde{w}_j\rangle = \sum_{k,\ell=1}^n Z_{ik}Z_{j\ell}\,\langle v_k|\tilde{v}_\ell\rangle = (Z\tau Z^T)_{ij} = \lambda_i\delta_{ij}\ .$$

**Case 1:** $\lambda_1 < \lambda_2 + \lambda_3 + \lambda_4$.

Because of the ordering of the $\lambda_j$'s, this case is possible if $n \geq 3$. Consider the quantity $f(\boldsymbol{\theta}) := \sum_{i=1}^4 \lambda_i \mathrm{e}^{2i\theta_i}$: since $f(0, \pi/2, \pi/2, \pi/2) < 0$ while $f(0,0,0,0) > 0$, by continuity $f(\boldsymbol{\varphi}) = 0$ at some $\boldsymbol{\varphi}$. Using the vectors $|\,w_i\,\rangle$ introduced above, let

$$|\,z_i\,\rangle = \sum_{j=1}^4 C_{ij}\mathrm{e}^{i\varphi_j}|\,w_j\,\rangle,\quad C := [C_{ij}] = \frac{1}{2}\begin{pmatrix} 1 & 1 & 1 & 1 \\ 1 & 1 & -1 & -1 \\ 1 & -1 & 1 & -1 \\ 1 & -1 & -1 & 1 \end{pmatrix},$$

where $\mathrm{e}^{2i\varphi_4}$ and $|\,z_4\,\rangle$ do not appear if $\lambda_4 = 0$.

Introducing the normalized vectors $|\,\psi_j\,\rangle := |\,z_j\,\rangle/\|z_j\|$, from $C^\dagger C = 1$ it turns out that

$$\rho = \sum_{i=1}^{4} \|z_j\|^2 \,|\,\psi_j\,\rangle\langle\,\psi_j\,| \;, \quad \text{and} \quad |\langle\,\psi_i\,|\,\widetilde{\psi_i}\,\rangle| = \frac{|f(\varphi)|}{4\|z_i\|^2} = 0 \;,$$

for all $i = 1, 2, 3, 4$. Then, the vectors $|\,\psi_j\,\rangle$ and thus $\rho$ are separable.

**Case 2:** $\lambda_1 > \lambda_2 + \lambda_3 + \lambda_4$.

Set $|\,y_1\,\rangle = |\,w_1\,\rangle$, $|\,y_j\,\rangle = i|\,w_j\,\rangle$, if $j \geq 2$. Then $\rho = \sum_{j=1}^{n} |\,y_j\,\rangle\langle\,y_j\,|$; further, consider the diagonal matrix $Y = \mathrm{diag}(\lambda_1, -\lambda_2, -\lambda_3, -\lambda_4)$ with entries $Y_{ij} := \langle\,y_i\,|\,\tilde{y}_j\,\rangle$.

Because of Example 5.5.4, any other decomposition $\rho = \sum_{j=1}^{n} |\,z_j\,\rangle\langle\,z_j\,|$ is such that $|\,z_j\,\rangle = \sum_{i=1}^{n} V_{ji}^*|\,y_i\,\rangle$, with $V$ a unitary matrix on $\mathbb{C}^n$. Therefore, for orthogonal $V$, the quantity

$$\begin{aligned} c_{\rho=\sum_{j=1}^{n}|\,z_j\,\rangle\langle\,z_j\,|} := \sum_{j=1}^{n} \langle\,z_j\,|\,\tilde{z}_j\,\rangle &= \sum_{j,i,k=1}^{n} V_{ji}V_{jk}Y_{ij} \\ &= \mathrm{Tr}(VYV^T) = \mathrm{Tr}(Y) = \lambda_1 - \lambda_2 - \lambda_3 - \lambda_4 = \mathcal{C}(\rho) \end{aligned}$$

is independent of $V$. By using this invariance property, one can find a decomposition $\rho = \sum_{j=1}^{n} |\,z_j\,\rangle\langle\,z_j\,|$ such that

$$\langle\,z_j\,|\,\tilde{z}_j\,\rangle = \mathcal{C}(\rho) = |\langle\,z_j\,|\,\tilde{z}_j\,\rangle| = \|z_j\|^2\, \mathcal{C}\left(\frac{z_j}{\|z_j\|}\right) \;,$$

for all $1 \leq j \leq n$. Thus, its average concurrence (6.16) equals $\mathcal{C}(\rho)$,

$$\langle \mathcal{C} \rangle = \sum_{j=1}^{n} \|z_j\|^2\, \mathcal{C}\left(\frac{z_j}{\|z_j\|}\right) = \sum_{j=1}^{n} \langle\,z_j\,|\,\tilde{z}_j\,\rangle = \mathcal{C}(\rho) \;.$$

The $|\,z_j\,\rangle$ are constructed as follows: unless all the $Y_{ii}$ are already equal to $\mathcal{C}(\rho)$, there must be one decomposer, $y_1$ say, with $Y_{11} > \mathcal{C}(\rho)$, and another one, $y_2$, with $Y_{22} < \mathcal{C}(\rho)$. Choosing $V$ that exchanges $y_1$ with $y_2$ and leaves the other decomposers fixed, we obtain a decomposition with the same average concurrence and $Y_{11}$, $Y_{22}$ exchanged. By continuity there must exist an orthogonal matrix $V$ such that $\langle z_1|\tilde{z}_1\rangle = \langle z_2|\tilde{z}_2\rangle = \mathcal{C}(\rho)$, with $|z_j\rangle = \sum_{i=1}^{n} V_{ji}^*|y_i\rangle$. Iteration of this argument for the remaining decomposers yields the result.

The proof of the theorem is then concluded by showing that no decomposition can achieve a smaller average concurrence than $\mathcal{C}(\rho)$. Indeed, using again Example 5.5.4, a generic decomposition has average concurrence

$$\langle \mathcal{C} \rangle_{\rho=\sum_{j=1}^{Q}|\,z_q\,\rangle\langle\,z_q\,|} = \sum_{q=1}^{Q} |\langle\,z_q\,|\,\tilde{z}_q\,\rangle| = \sum_{q=1}^{Q} \left| \sum_{i=1}^{n} (Z_{qj})^2 Y_{ii} \right| \;,$$

where $Z : \mathbb{C}^n \mapsto \mathbb{C}^Q$ is an isometry: $\sum_{q=1}^{Q} \left|(Z_{qi})^2\right| = 1$ for all $1 \leq i \leq n$. Since one can always adjust the phases of the $Z_{qi}$ in such a way that $(Z_{q1})^2 = |(Z_{q1})^2| \geq 0$, for all $1 \leq q \leq Q$, then

$$\langle \mathcal{C} \rangle_{\rho=\sum_{q=1}^{Q} |z_q\rangle\langle z_q|} \geq \left| \sum_{q=1}^{Q} \sum_{i=1}^{n} (Z_{qi})^2 Y_{ii} \right| = \left| \lambda_1 - \sum_{q=1}^{Q} \sum_{i=2}^{n} (Z_{qi})^2 \lambda_j \right|$$

$$\geq \left| \lambda_1 - \left| \sum_{q=1}^{Q} \sum_{i=2}^{n} (Z_{qi})^2 \lambda_i \right| \right| \geq \mathcal{C}(\rho) \ .$$

Indeed, by assumption,

$$\left| \sum_{q=1}^{Q} \sum_{i=2}^{n} (Z_{qi})^2 \lambda_i \right| \leq \sum_{i=2}^{n} \sum_{q=1}^{Q} |(Z_{qi})^2| \, \lambda_i = \lambda_2 + \lambda_3 + \lambda_4 < \lambda_1 \ .$$

□

## Two-Mode Gaussian States

Let $S$ be a bipartite continuous variable system consisting of two subsystems $A$ and $B$ described by annihilation and creation operators $a_i^{\#}$, $i = 1, 2, \ldots, p$, and $b_i^{\#}$, $i = 1, 2, \ldots, q$, $p + q = f$, satisfying the *CCR* (5.92), and arranged, as in (5.95), into a vector

$$\widehat{\boldsymbol{X}} = (\boldsymbol{a}, \boldsymbol{b}, \boldsymbol{a}^\dagger, \boldsymbol{b}^\dagger) \ , \ \boldsymbol{a} = (a_1, \ldots, a_p) \ , \ \boldsymbol{b} := (b_1, \ldots, b_q) \ .$$

We know that a state of $S$ described by a density matrix $\rho$ is specified by the characteristic function (5.117) which now reads

$$F_\rho^V(\boldsymbol{z}) = \mathrm{Tr}\left(\rho\, \mathrm{e}^{\boldsymbol{Z}^* \cdot \widehat{\boldsymbol{X}}}\right) = \mathrm{Tr}\left(\rho\, \mathrm{e}^{\boldsymbol{Z}_a^* \cdot \widehat{\boldsymbol{A}}} \otimes \mathrm{e}^{\boldsymbol{Z}_b^* \cdot \widehat{\boldsymbol{B}}}\right) \ , \tag{6.19}$$

where $\boldsymbol{A} := (\boldsymbol{a}, \boldsymbol{a}^\dagger)$, $\boldsymbol{B} := (\boldsymbol{b}, \boldsymbol{b}^\dagger)$, $\boldsymbol{Z}_{a,b} := (\boldsymbol{z}_{a,b}, -\boldsymbol{z}_{a,b}^*)$ with $\boldsymbol{z}_a := (z_{a1}, z_{a2}, \ldots, z_{ap})$ and $\boldsymbol{z}_b := (z_{b1}, z_{b2}, \ldots, z_{bq})$. Let $T_a$ denote the transposition with respect to the orthonormal basis of the occupation number states $|\,\boldsymbol{k}_a\,\rangle = |\,k_{a1} k_{a2} \ldots k_{ap}\,\rangle$, $k_{ai} \in \mathbb{N}$, of the subsystem $A$ (see (5.93)); then, using the number state basis $\{|\,\boldsymbol{k}_a \boldsymbol{k}_b\,\rangle\}_{\boldsymbol{k}_a, \boldsymbol{k}_b}$, one calculates

$$\begin{aligned}
\mathrm{Tr}\left(\rho^{T_a}\, \mathrm{e}^{\boldsymbol{Z}^* \cdot \widehat{\boldsymbol{X}}}\right) &= \sum_{\substack{\boldsymbol{k}_a, \boldsymbol{k}_b \\ \boldsymbol{j}_a, \boldsymbol{j}_b}} \langle\, \boldsymbol{k}_a \boldsymbol{k}_b \,|\, \rho^{T_a} \,|\boldsymbol{j}_a \boldsymbol{j}_b\,\rangle \langle\, \boldsymbol{j}_a \boldsymbol{j}_b \,|\, \mathrm{e}^{\boldsymbol{Z}_a^* \cdot \widehat{\boldsymbol{A}}} \otimes \mathrm{e}^{\boldsymbol{Z}_b^* \cdot \widehat{\boldsymbol{B}}} \,|\boldsymbol{k}_a \boldsymbol{k}_b\,\rangle \\
&= \sum_{\substack{\boldsymbol{k}_a, \boldsymbol{k}_b \\ \boldsymbol{j}_a, \boldsymbol{j}_b}} \langle\, \boldsymbol{j}_a \boldsymbol{k}_b \,|\, \rho \,|\boldsymbol{k}_a \boldsymbol{j}_b\,\rangle \langle\, \boldsymbol{j}_a \,|\, \mathrm{e}^{\boldsymbol{Z}_a^* \cdot \widehat{\boldsymbol{A}}} \,|\boldsymbol{k}_a\,\rangle \langle\, \boldsymbol{j}_b \,|\, \mathrm{e}^{\boldsymbol{Z}_b^* \cdot \widehat{\boldsymbol{B}}} \,|\boldsymbol{k}_b\,\rangle \\
&= \sum_{\substack{\boldsymbol{k}_a, \boldsymbol{k}_b \\ \boldsymbol{j}_a, \boldsymbol{j}_b}} \langle\, \boldsymbol{k}_a \boldsymbol{k}_b \,|\, \rho \,|\boldsymbol{j}_a \boldsymbol{j}_b\,\rangle \langle\, \boldsymbol{k}_a \,|\, \mathrm{e}^{\boldsymbol{Z}_a^* \cdot \widehat{\boldsymbol{A}}} \,|\boldsymbol{j}_a\,\rangle \langle\, \boldsymbol{j}_b \,|\, \mathrm{e}^{\boldsymbol{Z}_b^* \cdot \widehat{\boldsymbol{B}}} \,|\boldsymbol{k}_b\,\rangle \\
&= \mathrm{Tr}\left(\rho\, \mathrm{e}^{\boldsymbol{Z}_a^* \cdot \widehat{\boldsymbol{A}}'} \otimes \mathrm{e}^{\boldsymbol{Z}_b^* \cdot \widehat{\boldsymbol{B}}}\right) \ ,
\end{aligned}$$

where $\widehat{\boldsymbol{A}}' := (\boldsymbol{a}^\dagger, \boldsymbol{a})$. The last equality easily follows from

$$\langle \boldsymbol{k}_a \,|\, \mathrm{e}^{\boldsymbol{Z}_a^* \cdot \widehat{\boldsymbol{A}}} \,|\boldsymbol{j}_a \rangle = \prod_{i=1}^{p} \langle k_{ai} \,|\, \mathrm{e}^{z_{ai} a_i - z_{ai}^* a_i^\dagger} \,|j_{ai} \rangle$$

and

$$\langle k \,|\, \mathrm{e}^{za - z^* a^\dagger} \,|j \rangle = (\langle j \,|\, \mathrm{e}^{-za + z^* a^\dagger} \,|k \rangle)^* = \langle j \,|\, \mathrm{e}^{-z^* a + z a^\dagger} \,|k \rangle \,.$$

Partial transposition thus amounts to changing annihilation operators of a chosen subsystem into creation operators within the Weyl operator appearing into the characteristic function of a bipartite state.

In terms of position and momentum operators this means keeping fixed $\widehat{\boldsymbol{q}}_a = (\boldsymbol{a} + \boldsymbol{a}^\dagger)/\sqrt{2}$, $\widehat{\boldsymbol{q}}_b = (\boldsymbol{b} + \boldsymbol{b}^\dagger)/\sqrt{2}$ and $\widehat{\boldsymbol{p}}_b = (\boldsymbol{b} - \boldsymbol{b}^\dagger)/(i\sqrt{2}$ while changing $\widehat{\boldsymbol{p}}_a = (\boldsymbol{a} - \boldsymbol{a}^\dagger)/(i\sqrt{2})$ into $-\widehat{\boldsymbol{p}}_a$. This observation identifies partial transposition as a local *mirror reflection* [278]. Also in the continuous variable case, separable bipartite states must remain positive, hence well-defined states, under partial transposition. Then, if the correlation matrix associated with $\rho^{T_a}$ fails to satisfy (5.113) the state $\rho$ is surely entangled. In view of the fact that positivity under partial transposition fails to be equivalent to separability already for two 3-level systems, one may suspect this to be the case for all continuous variable systems as well. Surprisingly it turns out that partial transposition is an exhaustive entanglement witness also for two-mode Gaussian states [278] (see also [103, 102, 198]).

We shall use the notation of Examples 5.5.3 and start by noting that one can always consider Gaussian states $\rho$ with characteristic function

$$G(\boldsymbol{R}) = \mathrm{e}^{-\frac{1}{2}\boldsymbol{R}\cdot(\widehat{\Sigma}_1 \mathcal{V} \widehat{\Sigma}_1 \boldsymbol{R})}$$

as in (5.131). Indeed, as local operations that do not alter the entanglement properties, the displacement operators $D(\boldsymbol{u}) = D(\boldsymbol{u}_1) \otimes D(\boldsymbol{u}_2)$ can be used to set the mean values $\mathrm{Tr}(\rho\,(\widehat{\boldsymbol{q}}, \widehat{\boldsymbol{p}})) = 0$. Partial transposition on the first mode amounts to replacing $\widehat{p}_1$ with $-\widehat{p}_1$ in $\mathcal{V}$ thus $I_3$ with $-I_3$ in (5.138) (see (5.131)– (5.135)). Thus $\rho$ and $\rho^{T_a}$ are well-defined states if and only if both the following inequalities hold

$$\frac{1}{4} + I_4 \geq I_1 + I_2 + 2\,I_3 \tag{6.20}$$

$$\frac{1}{4} + I_4 \geq I_1 + I_2 - 2\,I_3 \,. \tag{6.21}$$

Also, the operations leading from a generic $\mathcal{V}$ to the standard form $\mathcal{V}_0$ in (5.137) are local ones, acting independently on the two subsystems $A$ and $B$; this means that if a two-mode Gaussian state $\rho$ with correlation matrix $\mathcal{V}$ is separable the same is true of the two-mode Gaussian state $\rho_0$ with correlation matrix $\mathcal{V}_0$. In [278] it is showed that

**Lemma 6.2.1.** *All two-mode Gaussian states with $I_3 \geq 0$ are separable.*

Notice that, when $I_3 \geq 0$, (6.20) implies (6.21) whence $\rho^{T_a} \geq 0$ in agreement with its being separable. Suppose instead that a two-mode Gaussian state $\rho$ with $I_3 < 0$ be *PPT* , then its mirror reflected $\rho^{T_a}$ has $I_3 > 0$ and is thus separable by the Lemma, whence by a second mirror reflection also $\rho$ is separable.

**Proof of Lemma 6.2.1** The strategy of the proof is to show that if a two-mode Gaussian state $\rho$ has a correlation matrix $\mathcal{V}$ with $I_3 \geq 0$, then $\mathcal{V} \geq \mathbb{1}_4/2$ whence, because of Example 5.5.3.3, $\rho$ is separable. Because of the possibility of reducing $\mathcal{V}$ to the standard form

$$\mathcal{V}_0 = \begin{pmatrix} \alpha & 0 & \gamma_1 & 0 \\ 0 & \alpha & 0 & \gamma_2 \\ \gamma_1 & 0 & \beta & 0 \\ 0 & \gamma_2 & 0 & \beta \end{pmatrix} ,$$

by local operations, one can equivalently show that $I_3 = \gamma_1\gamma_2 \geq 0$ implies $\mathcal{V}_0 \geq \mathbb{1}/2$. Analogously, since matrices of the form $O(x) = \mathrm{diag}(x, x^{-1})$, $0 \neq x \in \mathbb{R}$, implement local scalings of positions and momenta which preserve the symplectic matrix $J = \begin{pmatrix} 0 & 1 \\ -1 & 0 \end{pmatrix}$, one can focus upon

$$\begin{pmatrix} O(y)O(x) & 0 \\ 0 & O(y)O(x^{-1}) \end{pmatrix} \mathcal{V}_0 \begin{pmatrix} O(x)O(y) & 0 \\ 0 & O(x^{-1})O(y) \end{pmatrix} =$$

$$= \begin{pmatrix} \alpha(xy)^2 & 0 & \gamma_1 y^2 & 0 \\ 0 & \alpha(xy)^{-2} & 0 & \gamma_2 y^{-2} \\ \gamma_1 y^2 & 0 & \beta x^{-2} y^2 & 0 \\ 0 & \gamma_2 y^{-2} & 0 & \beta x^2 y^{-2} \end{pmatrix} =: \mathcal{V}_0' .$$

Consider the $2 \times 2$ matrices

$$X := \begin{pmatrix} \alpha x^2 & \gamma_1 \\ \gamma_1 & \beta x^{-2} \end{pmatrix} , \quad Y := \begin{pmatrix} \alpha x^{-2} & \gamma_2 \\ \gamma_2 & \beta x^2 \end{pmatrix}$$

and notice that, according to (5.132)– (5.135), their entries are correlations involving position ($\widehat{q}_{1,2}$), respectively momentum operators ($\widehat{p}_{1,2}$). Their eigenvalues are

$$x_\pm := \frac{1}{2}\Big(a\,x^2 + b\,x^{-2} \pm \sqrt{4\,\gamma_1 + (a\,x^2 - b\,x^{-2})^2}\Big)$$

$$x_\pm := \frac{1}{2}\Big(a\,x^{-2} + b\,x^2 \pm \sqrt{4\,\gamma_2 + (a\,x^{-2} - b\,x^2)^2}\Big)$$

with eigenvectors $|\,x_\pm\,\rangle = \begin{pmatrix} x^1_\pm \\ x^2_\pm \end{pmatrix}$, respectively $|\,y_\pm\,\rangle = \begin{pmatrix} y^1_\pm \\ y^2_\pm \end{pmatrix}$ such that

$$\frac{x^2_\pm}{x^1_\pm} = \frac{c_1}{x_\pm - ax^2} = 2\Big(-(ax^2 - bx^{-2})c_1^{-1} \pm \sqrt{4 + (ax^2 - bx^{-2})^2 c_1^{-2}}\Big)^{-1}$$

$$\frac{y^2_\pm}{y^1_\pm} = \frac{c_2}{y_\pm - ax^{-2}} = 2\Big(-(ax^{-2} - bx^2)c_2^{-1} \pm \sqrt{4 + (ax^{-2} - bx^2)^2 c_2^{-2}}\Big)^{-1} .$$

Since $|\,x_\pm\,\rangle$, respectively $|\,y_\pm\,\rangle$ are the rows of the orthogonal rotation matrices $O_X$ and $O_Y$ which diagonalize $X$, respectively $Y$, one can make $O_X = O_Y = 0$ by choosing the scaling parameter $x$ such that

$$\frac{\gamma_1}{ax^2 - bx^{-2}} = \frac{\gamma_2}{ax^{-2} - bx^2} ,$$

namely $x^2 = \sqrt{\dfrac{a\gamma_1 + b\gamma_2}{a\gamma_2 + b\gamma_1}}$. Since the diagonalization of the two sub-matrices $X$ and $Y$ of $\mathcal{V}_0'$ is obtained by means of a same orthogonal rotation, the overall transformation is symplectic (that is it preserves $\Omega = \begin{pmatrix} \mathbb{J}_2 & \mathbb{O}_2 \\ \mathbb{O}_2 & \mathbb{J}_2 \end{pmatrix}$). Therefore, one can study the diagonal matrix

$$\mathcal{V}_0'' := \begin{pmatrix} y^2\,x_+ & 0 & 0 & 0 \\ 0 & y^{-2}\,y_+ & 0 & 0 \\ 0 & 0 & y^2\,x_- & 0 \\ 0 & 0 & 0 & y^{-2}\,y_- \end{pmatrix} ,$$

which must satisfy $\mathcal{V}_0'' \pm \frac{i}{2}\Omega \geq 0$ whence $x_+y_+ \geq 1/4$ and $x_-y_- \geq 1/4$. By choosing the remaining scaling parameter $y$ such that $y^2\,x_- = y^{-2}\,y_-$ one gets that all four eigenvalues are $\geq 1/2$ and thus that $\mathcal{V}_0'' \geq \mathbb{1}_4/2$. This means that the two-mode Gaussian state $\rho_0''$ corresponding to such a correlation matrix has a $P$-representation (5.111) with a positive phase-space function $R_0''(\boldsymbol{r}_0'')$ and is thus separable according to Example 5.5.3.3. Observe that this fact does not allows one to directly infer that also the state $\rho_0'$ with correlation matrix $\mathcal{V}_0'$ is separable; indeed, the diagonalization of $\mathcal{V}_0'$ has been obtained by non-local rotations involving both sub-systems. However, using (5.141) in Remark 5.5.3, the positive phase-space distribution $R_0''(\boldsymbol{r}_0'')$ is obtained from the function $R_0'(\boldsymbol{r}_0')$ relative to the $P$-representation of $\rho'$ by means of a symplectic matrix $S$ composed of a same rotation $O$ in the $q_{1,2}$ and $p_{1,2}$ planes, so that $\|\boldsymbol{r}_0''\| = \|\widetilde{S}^T\boldsymbol{r}_0'\|$. It thus follows that $R_0''(\boldsymbol{r}_0'') = R_0'(S^{-1}\boldsymbol{r}_0') > 0$, whence $\rho_0$ and thus $\rho$ are separable.

If $I_3 = 0$, let $\gamma_1 > \gamma_2 = 0$ and choose $x^2 = \sqrt{\alpha/\beta}$, $y^2 = 2\sqrt{\alpha\beta}$ so that

$$\mathcal{V}_0' = \begin{pmatrix} 2\alpha^2 & 0 & 2\gamma_1\sqrt{\alpha\beta} & 0 \\ 0 & 1/2 & 0 & 0 \\ 2\gamma_1\sqrt{\alpha\beta} & 0 & 2\beta^2 & 0 \\ 0 & 0 & 0 & 1/2 \end{pmatrix} .$$

Similarly as before, one checks that $\mathcal{V}_0' \geq \pm\dfrac{i}{2}\Omega \Longrightarrow \mathcal{V}_0' \geq \pm\dfrac{\mathbb{1}_4}{2}$. □

## Positive Maps and Semigroups

We have seen in Section 6.2 that positive but not completely positive maps cannot be directly used as mathematical descriptions of fully consistent state-transformations; however, they play a major role as entanglement witnesses (see Proposition 6.2.1). Unfortunately, since there are no general rules that allows one to identify positive maps, only particular instances of them can be provided [179, 84, 85, 86, 114, 67]. The one which follows assumes the existence of just one negative eigenvalue in decompositions as in (5.41) that is smaller in absolute value than all the other ones [35].

**Proposition 6.2.1.** *Let* $\{L_k\}_{k=1}^{d^2}$ *be a Hilbert-Schmidt ONB in* $M_d(\mathbb{C})$ *and* $\Lambda : M_d(\mathbb{C}) \mapsto M_d(\mathbb{C})$ *a positive map with a decomposition*

$$\Lambda[X] = \sum_{k=1}^{d^2} \ell_k \, L_k^\dagger \, X \, L_k \ , \quad X \in M_d(\mathbb{C}) \ ,$$

*where* $0 \leq \ell_i \leq \ell_{i+1}$ *for* $i \geq 2$ *while* $\ell_1 = -|\ell_1| < 0$, *with* $|\ell_1| \leq \ell_2$. *If* $|\ell_1| \leq \ell_2 \dfrac{1 - \|L_1\|^2}{\|L_1\|^2}$, *then* $\Lambda$ *is positive.*

**Proof:** The matrices $L_k$ form a Hilbert-Schmidt *ONB*, thus, using (5.30), it turns out that, for all normalized $\psi, \phi \in \mathbb{C}^d$,

$$\sum_{k=1}^{d^2} \left| \langle \psi | L_k^\dagger | \phi \rangle \right|^2 = \sum_{k=1}^{d} \langle \psi \, | L_k^\dagger \, | \, \phi \, \rangle \langle \, \phi \, | \, L_k | \, \psi \, \rangle = \langle \, \psi \, | \mathrm{Tr}(| \, \phi \, \rangle \langle \, \phi \, |) | \, \psi \, \rangle = 1 \ .$$

Then, since $\|L_j\|^2 = \|L_j^\dagger\|^2 \leq \mathrm{Tr}(L_j^\dagger L_j) = 1$ (see Remark 5.2.4), it follows that

$$\langle \psi | \left( \sum_{k=1}^{d^2} \ell_k \, L_k^\dagger | \phi \rangle \langle \phi | L_k \right) | \psi \rangle \geq \ell_2 \sum_{k=2}^{d^2} \left| \langle \psi | L_k^\dagger | \phi \rangle \right|^2 - |\ell_1| \left| \langle \psi | L_1^\dagger | \phi^* \rangle \right|^2$$
$$= \ell_2 - (\ell_2 + |\ell_1|) \left| \langle \psi | L_1^\dagger | \phi \rangle \right|^2 \geq \ell_2 (1 - \|L_1\|^2) - |\ell_1| \|L_1\|^2 \ .$$

If $|\ell_1| \leq \ell_2 \dfrac{1 - \|L_1\|^2}{\|L_1\|^2}$, then it follows that $\langle \, \psi \, | \, \Lambda[| \, \phi \, \rangle \langle \, \phi \, |] \, \rangle \psi \geq 0$ for all normalized $\psi, \phi \in \mathbb{C}^d$, whence $\Lambda$ is positive. □

In Remark 5.6.4.6 it has been stressed that, apart from the fact that the Kossakowski matrix $C = [C_{ij}]$ cannot be positive, there are no general prescriptions on $C$ such that the corresponding semigroup surely consist of positive, but not *CP* maps; as well as for positive maps, one can however seek sufficient conditions.

In the following, we shall consider a system consisting of two $d$-level systems $S_d$ and construct [35] a semigroup of positive, but not *CP* maps $\Gamma_t = \gamma_t^1 \otimes \gamma_t^2$ on $M_d(\mathbb{C}) \otimes M_d(\mathbb{C})$, where $\gamma_t^{(1)} = \exp(tL_1)$ is a semigroup of *CP* maps, while $\gamma_t^{(2)}$ is a semigroup of positive, but not *CP* maps [6]. The construction will also provide non-decomposable positive maps able to witness bound entangled states within a particular class of bipartite states with $d = 4$ [36]. We shall consider generators as in Proposition 5.6.1 without asking for the positivity of the Kossakowski matrix.

**Proposition 6.2.2.** *Suppose* $\gamma_t^{(1,2)} : M_d(\mathbb{C}) \mapsto M_d(\mathbb{C})$ *to be semigroups with generators*

$$L_i[X] = i[H^{(i)}\,,X] + \sum_{\ell=1}^{d^2-1} c_\ell^{(i)}\Big(G_\ell^{(i)}\,X\,G_\ell^{(i)} - \frac{1}{2}\Big\{(G_\ell^{(i)})^2\,,X\Big\}\Big)\,,\quad c_i^\ell \in \mathbb{R}\,,$$

*for* $i = 1, 2$, *where* $G_\ell^{(i)} = (G_\ell^{(i)})^\dagger \in M_d(\mathbb{C})$, *together with* $G_{d^2}^{(i)} = 1/\sqrt{d}$, *form two Hilbert-Schmidt ONBs in* $M_d(\mathbb{C})$.

*Assume* $c_\ell^{(1)} > 0$, $\ell = 1, 2, \ldots, d^2-1$, *and* $c_k^{(2)} = -|c_k^{(2)}| < 0$, *for one index* $k$, *while* $c_\ell^{(2)} > 0$ *for* $\ell \neq k$. *Then, the semigroups of maps* $\Gamma_t = \gamma_t^{(1)} \otimes \gamma_t^{(2)}$ *on* $M_d(\mathbb{C}) \otimes M_d(\mathbb{C})$ *preserves positivity if* $c_\ell^{(1)} \geq |c_k^{(2)}|$, $\ell = 1, 2, \ldots, d^2-1$ *and* $c_\ell^{(2)} \geq |c_k^{(2)}|$, $\ell = 1, 2, \ldots, d^2-1$, $\ell \neq k$.

**Proof:** According to [177, 178] (see also [64]), in order to show that the semigroup $\{\Gamma_t\}_{t\geq 0}$, with generator $L = L^{(1)} \otimes \mathrm{id}_d + \mathrm{id}_d \otimes L^{(2)}$, consists of positive maps, it is sufficient to prove that

$$I(\psi,\phi) := \langle\,\psi\,|\,L[|\,\phi\,\rangle\langle\,\phi\,|]\,|\psi\,\rangle \geq 0$$

for all orthogonal $\psi, \phi \in \mathbb{C}^d \otimes \mathbb{C}^d$. Since $\langle\,\psi\,|\,\phi\,\rangle = 0$, it follows that

$$I(\psi,\phi) = \sum_{\ell=1}^{d^2-1} c_\ell^{(1)}\,\Big|\langle\,\psi\,|\,G_\ell^{(1)} \otimes \mathbb{1}_2\,|\phi\,\rangle\Big|^2 + \sum_{\ell=1}^{d^2-1} c_\ell^{(2)}\,\Big|\langle\,\psi\,|\,\mathbb{1}_1 \otimes G_\ell^{(2)}\,|\phi\,\rangle\Big|^2\,.$$

It proves convenient to define the following $d^2 \times d^2$ matrices $\Psi = [\psi_{ij}]$ and $\Phi = [\phi_{ij}]$ where $\psi_{ij}$ and $\phi_{ij}$ are the components of the vectors $\psi$ and $\phi$ with respect to a fixed *ONB* $\{|\,i,j\,\rangle\}_{i,j=1}^d$ in $\mathbb{C}^d \otimes \mathbb{C}^2$. Then, one rewrites

$$\begin{aligned}
I(\psi,\phi) &= \sum_{\ell=1}^{d^2-1} c_\ell^{(1)}\,\Big|\mathrm{Tr}(G_\ell^{(1)}\Phi\Psi^\dagger)\Big|^2 + \sum_{\ell=1}^{d^2-1} c_\ell^{(2)}\,\Big|\mathrm{Tr}(G_\ell^{(2)}(\Psi^\dagger\Phi)^T)\Big|^2 \\
&= \sum_{\ell=1}^{d^2-1} (c_\ell^{(1)} - |c_k^{(2)}|)\,\Big|\mathrm{Tr}(G_\ell^{(1)}\Phi\Psi^\dagger)\Big|^2 + \sum_{k\neq\ell=1}^{d^2-1} c_\ell^{(2)}\,\Big|\mathrm{Tr}(G_\ell^{(2)}(\Psi^\dagger\Phi)^T)\Big|^2 \\
&+ |c_k^{(2)}|\,\left(\sum_{\ell=1}^{d^2-1}\Big|\mathrm{Tr}(G_\ell^{(1)}\Phi\Psi^\dagger)\Big|^2 - \Big|\mathrm{Tr}(G_\ell^{(2)}(\Psi^\dagger\Phi)^T)\Big|^2\right)\,.
\end{aligned}$$

---

[6] If the two semigroups $\gamma_t^{(1,2)}$ were the same, then, according to Proposition 5.6.1 and the successive remark, $\Gamma_t$ positive would mean $\Gamma_t$ *CP*.

As $\langle \psi | \phi \rangle = 0$, the matrices $\Phi\Psi^\dagger$ and $\Psi^\dagger\Phi$ are traceless; using the *ONB*s consisting of the matrices $G^{(i)}_\ell = (G^{(i)}_\ell)^\dagger$, $\ell = 1, 2, \ldots, d^2 - 1$, one thus gets

$$\begin{aligned}\sum_{\ell=1}^{d^2-1} (\mathrm{Tr}(G^{(1)}_\ell \Phi\Psi^\dagger))^2 &= \mathrm{Tr}\Big(\Big(\sum_{\ell=1}^{d^2-1} \mathrm{Tr}(G^{(1)}_\ell \Phi\Psi^\dagger)\, G^{(1)}_\ell\Big)\Phi\Psi^\dagger\Big) \\ &= \mathrm{Tr}(\Phi\Psi^\dagger)^2 = \mathrm{Tr}(\Psi^\dagger\Phi)^2 = \mathrm{Tr}((\Psi^\dagger\Phi)^T)^2 \\ &= \sum_{\ell=1}^{d^2-1} (\mathrm{Tr}(G^{(2)}_\ell (\Psi^\dagger\Phi)^T)^2 \,.\end{aligned}$$

This yields

$$\Big|\mathrm{Tr}(G^{(2)}_k (\Psi^\dagger\Phi)^T)^2\Big| \le \sum_{\ell=1}^{d^2-1} \Big|\mathrm{Tr}(G^{(1)}_\ell (\Phi\Psi^\dagger)\Big|^2 + \sum_{k\neq\ell=1}^{d^2-1} \Big|\mathrm{Tr}(G^{(2)}_\ell (\Psi^\dagger\Phi)^T)\Big|^2 \,,$$

whence one concludes

$$\begin{aligned} I(\psi, \phi) \ge &\sum_{\ell=1}^{d^2-1} (c^{(1)}_\ell - |c^{(2)}_k|) \Big|\mathrm{Tr}(G^{(1)}_\ell \Phi\Psi^\dagger)\Big|^2 \\ &+ \sum_{\ell=1}^{d^2-1} (c^{(2)}_\ell - |c^{(2)}_k|) \Big|\mathrm{Tr}(G^{(2)}_\ell (\Psi^\dagger\Phi)^T)\Big|^2 \ge 0 \,. \end{aligned}$$

□

**Example 6.2.3.** [35] Let $d = 2$ and $\sigma_\alpha$, $\alpha = 0, 1, 2, 3$, be the Pauli matrices plus the $2 \times 2$ identity matrix $\sigma_0$. Let $S_a : M_2(\mathbb{C}) \mapsto M_2(\mathbb{C})$ be the completely positive map $X \mapsto S_\alpha[X] = \sigma_\alpha X \sigma_\alpha$, and set

$$X \mapsto \frac{1}{2}\sum_{\alpha=0}^{3} S_\alpha[X] \,, \quad X \mapsto \frac{1}{2}\sum_{\alpha=0}^{3} \varepsilon_\alpha S_\alpha[X] \,,$$

where $\varepsilon_\alpha = 1$ when $\alpha \neq 2$, whereas $\varepsilon_2 = -1$. The first map amounts to the trace map $\mathrm{Tr}_2$ (see (5.30)), while the second one corresponds to the transposition $\mathsf{T}_2$ with respect to the basis of eigenvectors of $\sigma_3$: indeed, it changes $\sigma_2$ into $-\sigma_2$ and leaves all other Pauli matrices unchanged. According to Proposition 6.2.1, it is positive but not *CP* , for $\Lambda_{\alpha\beta} = \mathrm{diag}(1, 1, -1, 1)$ and $\|\sigma_\alpha\|^2 = 1$.

Consider generators $L_{1,2}$ as in Proposition 6.2.2 with $d = 2$, $F_i = \sigma_i/\sqrt{2}$ and choose as Kossakowski matrices

$$C^{(1)} = \begin{pmatrix} 1 & 0 & 0 \\ 0 & 1 & 0 \\ 0 & 0 & 1 \end{pmatrix} \,, \qquad C^{(2)} = \begin{pmatrix} 1 & 0 & 0 \\ 0 & -1 & 0 \\ 0 & 0 & 1 \end{pmatrix} \,.$$

Then, the corresponding master equations read

$$\partial_t \gamma_t^{(1)}[\rho] = L_1[\rho] := \frac{1}{2}\Big(\sum_{i=1}^{3} S_i[\rho] \, - \, 3\,\rho\Big)$$

$$\partial_t \gamma_t^{(2)}[\rho] = L_2[\rho] := \frac{1}{2}\Big(\sum_{i=1}^{3} \varepsilon_i \, S_i[\rho] \, - \, \rho\Big) \, .$$

The second one has been considered in Example 5.6.5 and generates a positive semigroup such that

$$\gamma_t^{(2)}[\rho] = \frac{1}{2}\Big(\mathbb{1} + \rho_1\sigma_1 + \mathrm{e}^{-2t}\rho_2\sigma_2 + \rho_3\sigma_3\Big) = \rho + \frac{\mathrm{e}^{-2t}-1}{2}\rho_2\sigma_2 \, .$$

Since $L_1[\sigma_i] = -2\sigma_i$ while $L_1[\sigma_0] = 0$, the solutions of the first master equation are the following *CPU* maps,

$$\gamma_t^{(1)}[\rho] = \frac{\mathbb{1} + \mathrm{e}^{-2t}\boldsymbol{\rho}\cdot\boldsymbol{\sigma}}{2} = \frac{1-\mathrm{e}^{-2t}}{2} + \mathrm{e}^{-2t}\,\rho \, .$$

Let $\mathrm{id}_n$, $\mathrm{Tr}_n$ and $\mathsf{T}_n$ denote identity, trace and transposition operations on $(\mathbb{C}^2)^{\otimes n}$; since $1 = \mathrm{Tr}(\rho)$ and $\rho - \mathsf{T}[\rho] = \rho_2\sigma_2$, one rewrites

$$\gamma_t^{(1)} = \mathrm{e}^{-2t}\mathrm{id}_2 + \frac{1-\mathrm{e}^{-2t}}{2}\mathrm{Tr}_2 \, , \quad \gamma_t^{(2)} = \frac{1+\mathrm{e}^{-2t}}{2}\mathrm{id}_2 + \frac{1-\mathrm{e}^{-2t}}{2}\mathsf{T}_2 \, .$$

As $\mathsf{T}_4 = \mathsf{T}_2 \otimes \mathsf{T}_2$ and $\mathrm{Tr}_2 \circ \mathsf{T}_2 = \mathrm{Tr}_2$, the tensor product maps $\Gamma_t = \gamma_t^1 \otimes \gamma_t^2$ can be recast in the form

$$\Gamma_t = \underbrace{\mathrm{e}^{-2t}\frac{1+\mathrm{e}^{-2t}}{2}\mathrm{id}_4 + \frac{1-\mathrm{e}^{-4t}}{4}\mathrm{Tr}_2\otimes\mathrm{id}_2}_{\Gamma_t^1}$$
$$+ \underbrace{\frac{1-\mathrm{e}^{-2t}}{2}\Big(\mathrm{e}^{-2t}\mathsf{T}_2\otimes\mathrm{id}_2 + \frac{1-\mathrm{e}^{-2t}}{2}\mathrm{Tr}_2\otimes\mathrm{id}_2\Big)}_{\Gamma_t^2}\circ\mathsf{T}_4 \, . \qquad (6.22)$$

The semigroup $\{\Gamma_t\}_{t\geq 0}$ consists of positive maps because the chosen Kossakowski matrices satisfy the sufficient condition of Proposition 6.2.2; moreover, the maps $\Gamma_t$ are of the form $\Gamma_t = \Gamma_t^1 + \Gamma_t^2 \circ \mathsf{T}_4$. It turns out that $\Gamma_t^1$ is completely positive for all $t \geq 0$ for it is the sum of tensor products of completely positive maps. If $\Gamma_t^2$ were also *CP*, each map $\Gamma_t$ would then be decomposable; in order to check whether this is true or not, consider the Choi matrix $\mathrm{id}_4 \otimes \Lambda_t[P_+^4]$, where

$$\Lambda_t := \mathrm{e}^{-2t}\mathsf{T}_2\otimes\mathrm{id}_2 + \frac{1-\mathrm{e}^{-2t}}{2}\mathrm{Tr}_2\otimes\mathrm{id}_2 \, .$$

Fixing a basis $\{|0\rangle, |1\rangle\} \in \mathbb{C}^2$ and writing $|\widehat{\Psi}^4_+\rangle = \frac{1}{2}\sum_{a,b=0}^{1} |ab\rangle \otimes |ab\rangle$, one explicitly computes

$$\mathrm{id}_4 \otimes \Lambda_t[P^4_+] =$$
$$\begin{pmatrix} (1+\mathrm{e}^{-2t})\widehat{P}^2_+ & 0 & 0 & 0 \\ 0 & (1-\mathrm{e}^{-2t})\widehat{P}^2_+ & 2\mathrm{e}^{-2t}\widehat{P}^2_+ & 0 \\ 0 & 2\mathrm{e}^{-2t}\widehat{P}^2_+ & (1-\mathrm{e}^{-2t})\widehat{P}^2_+ & 0 \\ 0 & 0 & 0 & (1+\mathrm{e}^{-2t})\widehat{P}^2_+ \end{pmatrix} .$$

This $16 \times 16$ matrix has eigenvalue 0 with eigenvectors

$$(|\hat{\Psi}_j\rangle, 0, 0, 0)\ ,\ (0, |\hat{\Psi}_j\rangle, 0, 0)\ ,\ (0, 0, |\hat{\Psi}_j\rangle, 0)\ ,\ (0, 0, 0, |\hat{\Psi}_j\rangle)\ ,\ j = 2, 3, 4\ ,$$

where $|\hat{\Psi}_j\rangle$ are the Bell states orthogonal to $|\hat{\Psi}_{00}\rangle$, while

$$(|\hat{\Psi}_{00}\rangle, 0, 0, 0)\ ,\ (0, 0, 0, |\hat{\Psi}_{00}\rangle)\ ,\ (0, |\hat{\Psi}_{00}\rangle, |\hat{\Psi}_{00}\rangle, 0)$$

are eigenvectors relative to the positive eigenvalue $(1+\mathrm{e}^{-2t})/\sqrt{2}$. More interesting is the last eigenvalue $\dfrac{1-3\mathrm{e}^{-2t}}{\sqrt{2}}$ with eigenvector $(0, |\hat{\Psi}_{00}\rangle, -|\hat{\Psi}_{00}\rangle, 0)$: it is positive only if $t \geq t^* = (\log 3)/2$. It follows that $\Gamma_t$ is surely decomposable for $t = 0$ ($\Gamma_0 = \mathrm{id}_{16}$) and for $t \geq t^*$.

In order to ascertain whether the positive maps $\Gamma_t$ constructed in the previous example are not decomposable for $0 < t < t^*$, we need some further insight. Indeed, the decomposition (6.22) need not be unique and there might be other decompositions revealing that $\Gamma_t$ is decomposable for all $t \geq 0$. In order to proceed, we use the following result.

**Lemma 6.2.2.** *[35] Let $\Lambda : M_{d_1}(\mathbb{C}) \mapsto M_{d_2}(\mathbb{C})$ be a positive map and $\rho$ a PPT state of a bipartite system $S_1 + S_2$. If $\mathrm{Tr}\Big(\mathrm{id}_{d_1} \otimes \Lambda[P^{d_1}_+]\,\rho\Big) < 0$, the state $\rho$ is bound-entangled and $\Lambda$ not decomposable.*

**Proof:** If $\Lambda$ is decomposable, so is its dual $\Lambda^T : M_{d_2}(\mathbb{C}) \mapsto M_{d_1}(\mathbb{C})$; indeed,

$$\Lambda = \Lambda_1 + \Lambda_2 \circ \mathsf{T}_{d_1} \Longrightarrow \Lambda^T = \Lambda_1^T + \mathsf{T}_{d_1} \circ \Lambda_2^T = \Lambda_1^T + \tilde{\Lambda}_2 \circ \mathsf{T}_{d_2}\ ,$$

where $\Lambda_1^T$ is *CP* for it is the dual of a *CP* map, while $\tilde{\Lambda}_2 := \mathsf{T}_{d_1} \circ \Lambda_2^T \circ \mathsf{T}_{d_2}$ is also *CP* as the corresponding Choi matrix is positive. In fact,

$$\begin{aligned} \mathrm{id}_{d_2} \otimes \tilde{\Lambda}[\tilde{E}^{d_2}] &= \sum_{i,j=1}^{d_2} E^{d_2}_{ij} \otimes (\mathsf{T}_{d_1} \circ \Lambda_2^T)[E^{d_2}_{ji}] = \mathsf{T}_{d_1 d_2} \circ \sum_{i,j=1}^{d_2} E^{d_2}_{ji} \otimes \Lambda_2^T[E^{d_2}_{ji}] \\ &= \mathsf{T}_{d_1 d_2} \circ (\mathrm{id}_{d_2} \otimes \Lambda^T)[\tilde{E}^{d_2}] \geq 0\ , \end{aligned}$$

where $\mathsf{T}_{d_1d_2} = \mathsf{T}_{d_2} \circ \mathsf{T}_{d_1}$ is the transposition on $M_{d_1d_2}(\mathbb{C}) = M_{d_1}(\mathbb{C}) \otimes M_{d_2}(\mathbb{C})$. It preserves the positivity of the Choi matrix $\mathrm{id}_{d_2} \otimes \Lambda^T[\tilde{E}^{d_2}]$ associated with the *CP* map $\Lambda_2^T$. Since $\rho$ is assumed to be *PPT*, if $\Lambda$ is decomposable or $\rho$ separable, then $\mathrm{id}_{d_2} \otimes \Lambda^T[\rho] \geq 0$. □

To make good use of the above result, it is convenient to introduce a particular class [34, 33] of 16-dimensional density matrices as states of a bipartite system consisting of two pairs of *qubits*; their structure is simple, yet flexible enough to represent an interesting setting where to test the decomposability of a wider range of positive maps $\Lambda : M_4(\mathbb{C}) \mapsto M_4(\mathbb{C})$.

**Example 6.2.4.** Let $S = S_1 + S_2$ be a bipartite system where $S_{1,2}$ are each a two *qubit* system; consider the sub-class of $16 \times 16$ density matrices constructed by associating to the pairs of the set $L_{16} := \{(\alpha, \beta)\}_{\alpha,\beta=0}^3$ the vectors $|\Psi_{\alpha\beta}\rangle := \left(\mathbb{1}_4 \otimes \sigma_{\alpha\beta}\right) |\widehat{\Psi}_+^4\rangle$, where $|\widehat{\Psi}_+^4\rangle$ is the Bell state $|\,\Psi_{00}\,\rangle$ in (5.164) and $\sigma_{\alpha\beta} := \sigma_\alpha \otimes \sigma_\beta$ are tensor products of Pauli matrices with $\sigma_0 = \mathbb{1}_2$. The vectors $|\,\Psi_{\alpha\beta}\,\rangle$ form an *ONB* in $\mathbb{C}^{16}$,

$$\langle\,\Psi_{\alpha\beta}\,|\,\Psi_{\gamma\delta}\,\rangle = \langle\widehat{\Psi}_+^4|\mathbb{1}_4 \otimes \sigma_{\alpha\beta}\sigma_{\gamma\delta}|\widehat{\Psi}_+^4\rangle = \frac{1}{4}\mathrm{Tr}(\sigma_\alpha\sigma_\gamma)\mathrm{Tr}(\sigma_\beta\sigma_\delta) = \delta_{\alpha\gamma}\delta_{\beta\delta}\ .$$

Given the corresponding orthogonal projections

$$P_{\alpha\beta} := |\Psi_{\alpha\beta}\rangle\langle\Psi_{\alpha\beta}| = \left(\mathrm{id}_4 \otimes \sigma_{\alpha\beta}\right) \widehat{P}_+^4 \left(\mathrm{id}_4 \otimes \sigma_{\alpha\beta}\right)\ , \quad P_{\alpha\beta}\, P_{\gamma\varepsilon} = \delta_{\alpha\gamma}\, \delta_{\beta\varepsilon}\, P_{\alpha\beta}\ ,$$

consider the states consisting of all equidistributed convex combinations

$$\rho_I := \frac{1}{N_I} \sum_{(\alpha,\beta)\in I} P_{\alpha\beta}\ ,$$

where $I$ is a subset of $L_{16}$ and $N_I$ its cardinality.

The behavior of such states under partial transposition can be deduced by means of the fact that the flip operator $V = d\, \mathrm{id}_d \otimes \mathsf{T}_d[\widehat{P}_+^d]$ (5.32) is such that $V|\widehat{\Psi}_+^d\rangle = |\widehat{\Psi}_+^d\rangle$ and $V(A \otimes B)V = B \otimes A$ for all $A, B \in M_d(\mathbb{C})$; while

$$\begin{aligned} A \otimes B|\widehat{\Psi}_+^d\rangle &= \frac{1}{\sqrt{d}} \sum_{i=1}^d A|\,i\,\rangle \otimes B|\,i\,\rangle = \frac{1}{\sqrt{d}} \sum_{i,j=1}^d \langle\,j\,|\,A\,|i\,\rangle\,|\,j\,\rangle \otimes B|\,i\,\rangle \\ &= \frac{1}{\sqrt{d}} \sum_{j=1}^d |\,j\,\rangle \otimes \sum_{i=1}^d B|\,i\,\rangle\langle\,i\,|\,A^T\,|j\,\rangle = \mathbb{1}_d \otimes BA^T|\widehat{\Psi}_+^d\rangle\ . \end{aligned}$$

Then, setting $\widetilde{P}_{\alpha\beta} := \mathrm{id}_4 \otimes \mathsf{T}_4[P_{\alpha\beta}] = \frac{1}{4}\mathbb{1}_4 \otimes \sigma_{\alpha\beta}\, V\, \mathbb{1}_4 \otimes \sigma_{\alpha\beta}$, it turns out that

$$\begin{aligned} \widetilde{P}_{\alpha\beta}|\,\Psi_{\gamma\delta}\,\rangle &= \frac{1}{4}\mathbb{1}_4 \otimes \sigma_{\alpha\beta}\, V\, \mathbb{1}_4 \otimes \sigma_{\alpha\beta}\sigma_{\gamma\delta}V|\widehat{\Psi}_+^4\rangle = \frac{1}{4}\sigma_{\alpha\beta}\sigma_{\gamma\delta} \otimes \sigma_{\alpha\beta}|\widehat{\Psi}_+^4\rangle \\ &= \frac{1}{4}\mathbb{1}_4 \otimes \sigma_{\alpha\beta}(\sigma_{\alpha\beta}\sigma_{\gamma\delta})^T|\widehat{\Psi}_+^4\rangle = \varepsilon_\alpha\varepsilon_\gamma\varepsilon_\beta\varepsilon_\delta\, \mathbb{1}_4 \otimes \sigma_{\alpha\beta}\sigma_{\gamma\delta}\sigma_{\alpha\beta}|\widehat{\Psi}_+^4\rangle \\ &= \frac{1}{4}\eta_{\alpha\gamma}\eta_{\beta\delta}|\,\Psi_{\gamma\delta}\,\rangle\ , \end{aligned}$$

where it has been used that $\sigma_\alpha^T = \varepsilon_\alpha \sigma_\alpha$ with $\varepsilon_\alpha = 1$ if $\alpha \neq 2$, $= -1$ otherwise, that the algebra of the Pauli matrices implies

$$\sigma_\alpha \sigma_\gamma \sigma_\alpha = \tilde{\eta}_{\alpha\gamma}\, \sigma_\gamma \ , \quad \tilde{\eta} := [\tilde{\eta}_{\alpha\gamma}] = \begin{pmatrix} 1 & 1 & 1 & 1 \\ 1 & 1 & -1 & -1 \\ 1 & -1 & 1 & -1 \\ 1 & -1 & -1 & 1 \end{pmatrix}$$

and it has been set

$$\eta_{\alpha\gamma} := \varepsilon_\alpha \varepsilon_\gamma\, \tilde{\eta}_{\alpha\gamma} \ , \quad \eta := [\eta_{\alpha\beta}] = \begin{pmatrix} 1 & 1 & -1 & 1 \\ 1 & 1 & 1 & -1 \\ -1 & 1 & 1 & 1 \\ 1 & -1 & 1 & 1 \end{pmatrix} .$$

The vectors $|\Psi_{\gamma\delta}\rangle$ are thus eigenvectors of $\tilde{P}_{\alpha\beta}$ with eigenvalues $\eta_{\alpha\gamma}\eta_{\beta\delta}$ and

$$\tilde{\rho}_I := \mathrm{id}_4 \otimes \mathsf{T}_4[\rho_I] = \sum_{(\gamma,\delta)\in L_{16}} \Big( \frac{1}{4\,N_I} \sum_{(\alpha,\beta)\in I} \eta_{\alpha\gamma}\eta_{\beta\delta} \Big)\, P_{\gamma\delta} \ .$$

Since the matrix $\eta_{\alpha\gamma}$ is symmetric, the eigenvalues of $\tilde{\rho}_I$ can be recast as

$$\frac{1}{4\,N_I} \sum_{(\alpha,\beta)\in I} \eta_{\alpha\gamma}\eta_{\beta\delta} = (\eta\, X^I\, \eta)_{\gamma\delta}$$

where $X^I$ is a sort of characteristic matrix of the sublattice $I$ with entries $X^I_{\mu\nu} = \dfrac{1}{4\,N_I}$ if $(\mu,\nu) \in I$, $= 0$ otherwise. Concretely, consider

$$\rho = \frac{1}{6}\Big(P_{02} + P_{11} + P_{23} + P_{31} + P_{32} + P_{33}\Big) \ ;$$

then, $\tilde{\rho} = \dfrac{1}{6}(P_{01} + P_{02} + P_{20} + P_{22} + P_{32} + P_{33})$ for

$$X^I = \frac{1}{6}\begin{pmatrix} 0 & 0 & 1 & 0 \\ 0 & 1 & 0 & 0 \\ 0 & 0 & 0 & 1 \\ 0 & 1 & 1 & 1 \end{pmatrix} , \quad \eta\, X^I\, \eta = \frac{1}{6}\begin{pmatrix} 0 & 1 & 1 & 0 \\ 0 & 0 & 0 & 0 \\ 1 & 0 & 1 & 0 \\ 0 & 0 & 1 & 1 \end{pmatrix} .$$

Thus, $\tilde{\rho}$ is positive, hence $\rho$ is *PPT*.

We now show that $\mathrm{Tr}(\mathrm{id}_4 \otimes \Gamma_t[P_{00}]\rho) < 0$ for $0 < t < (\log 3)/2$, where $\Gamma_t$ is the positive semigroup of Example 6.2.3; by Lemma 6.2.2 it thus follows that $\rho$ is *PPT* entangled and $\Gamma_t$ indecomposable in that time-interval.

Since $\mathrm{Tr}_2(\,\cdot\,) = \dfrac{1}{2}\sum_{\mu=0}^{3} S_\mu[\,\cdot\,]$ and $\mathsf{T}_2[\,\cdot\,] = \dfrac{1}{2}\sum_{\mu=0}^{3} \varepsilon_\mu\, S_\mu[\,\cdot\,]$, the two semigroups in Example 6.2.3 can be recast in the form

$$\gamma_t^{(1)} = \frac{1+3\lambda_t}{4} S_0 + \frac{1-\lambda_t}{4} \sum_{i=1}^3 S_i \ , \quad \gamma_t^{(2)} = \frac{3+\lambda_t}{4} S_0 + \frac{1-\lambda_t}{4} \sum_{i=1}^3 \varepsilon_i S_i \ ,$$

where $\lambda_t := \mathrm{e}^{-2t}$, so that, with $S_{\alpha\beta}[\cdot] := \sigma_{\alpha\beta} \cdot \sigma_{\alpha\beta}$,

$$\begin{aligned}
\Gamma_t &= \frac{(1+3\lambda_t)(3+\lambda_t)}{16} S_{00} + \frac{(1-2\lambda_t)(3+\lambda_t)}{16} \sum_{i=1}^3 S_{i0} \\
&+ \frac{(1+3\lambda_t)(1-2\lambda_t)}{16} \sum_{i=1}^3 \varepsilon_i S_{0i} + \frac{(1-2\lambda_t)^2}{16} \sum_{i,j=1}^3 \varepsilon_j S_{ij} \\
\mathrm{id}_4 \otimes \Gamma_t[P_{00}] &= \frac{(1+3\lambda_t)(3+\lambda_t)}{16} P_{00} + \frac{(1-2\lambda_t)(3+\lambda_t)}{16} \sum_{i=1}^3 P_{i0} \\
&+ \frac{(1+3\lambda_t)(1-2\lambda_t)}{16} \sum_{i=1}^3 \varepsilon_i P_{0i} + \frac{(1-2\lambda_t)^2}{16} \sum_{i,j=1}^3 \varepsilon_j P_{ij} .
\end{aligned}$$

It then turns out that

$$0 < t < (\log 3)/2 \Longrightarrow \mathrm{Tr}\Big(\mathrm{id}_4 \otimes \Gamma_t[P_{00}]\,\rho\Big) = \frac{(1-\lambda_t)(1-3\lambda_t)}{48} < 0 \ .$$

## 6.3 Relative Entropy

A notion directly related to the von Neumann entropy with several useful applications in quantum information and of great importance for the topics discussed later in the book, is the *quantum relative entropy.* It is the quantum counterpart of the Kullbach-Leibler distance (see (2.94)) and has already been introduced in the proof of some properties of the von Neumann entropy (see Proposition (5.5.6)).

**Definition 6.3.1 (Relative Entropy).** *Let $S$ be a quantum system described by a d-dimensional Hilbert space $\mathbb{H}$ and $\rho, \sigma \in \mathbb{B}_1^+(\mathbb{H})$ two density matrices acting on $\mathbb{H}$. The relative entropy of $\rho$ with respect to $\sigma$ is*

$$S(\rho\,;\,\sigma) := \begin{cases} \mathrm{Tr}\Big(\rho\,\Big(\log\rho - \log\sigma\Big)\Big) & \text{if } Ker(\sigma) \subset Ker(\rho) \\ +\infty & \text{otherwise} \ , \end{cases} \tag{6.23}$$

*$Ker(\sigma)$ and $Ker(\rho)$ being the subspaces where $\sigma$ and $\rho$ vanish.*

**Example 6.3.1 (Relative modular operator).** [225, 237, 261, 262]
Given the Hilbert space $\mathbb{H} = \mathbb{C}^d$, equip the algebra $M_d(\mathbb{C})$ with the Hilbert-Schmidt scalar product (5.26) and denote it as $<< \cdot, \cdot >>$. Then, consider the following linear operators on $M_d(\mathbb{C})$:

$$L_X[Y] = X\,Y\ , \quad R_X[Y] = Y\,X\ .$$

They commute with each other, $L_X R_Z[Y] = X\,Y\,Z = R_Z L_X[Y]$; moreover, if $X = X^\dagger$ they are self-adjoint. Indeed, the ciclycity of the trace operation yields

$$<< Y\,,\, L_X[Z] >> = \mathrm{Tr}\Big(Y^\dagger\,X\,Z\Big) = \mathrm{Tr}\Big((XY^\dagger)^\dagger\,Z\Big) =<< L_X[Y], Z >>$$

$$<< Y\,,\, R_X[Z] >> = \mathrm{Tr}\Big(Y^\dagger\,Z\,X\Big) = \mathrm{Tr}\Big((Y^\dagger X)^\dagger\,Z\Big) =<< R_X[Y], Z >>\ .$$

Further, if $X \geq 0$ the operators $L_X$ and $R_X$ turn out to be positive; in fact,

$$<< Z\,,\, L_X[Z] >> = \mathrm{Tr}\Big(Z^\dagger\,X\,Z\Big) \geq 0$$

$$<< Z\,,\, R_X[Z] >> = \mathrm{Tr}\Big(Z^\dagger\,Z\,X\Big) = \mathrm{Tr}\Big(Z\,X\,Z^\dagger\Big) \geq 0\ .$$

Let $P_i$ and $P_j$ be orthogonal projections; then, $L_{P_i} L_{P_j} = \delta_{ij} L_{P_i}$ and $R_{P_i} R_{P_j} = \delta_{ij} R_{P_i}$. As a consequence, from the spectral representation $X = X^\dagger = \sum_{i=1}^d x_i\,|\,x_i\,\rangle\langle\,x_i\,|$, one derives the spectral representations

$$L_X = \sum_{i=1}^d x_i\,L_{|\,x_i\,\rangle\langle\,x_i\,|}\ , \qquad R_X = \sum_{i=1}^d x_i\,R_{|\,x_i\,\rangle\langle\,x_i\,|}\ ,$$

with orthogonal projections $\{L_{|\,x_i\,\rangle\langle\,x_i\,|}\}_{i=1}^d$, respectively $\{R_{|\,x_i\,\rangle\langle\,x_i\,|}\}_{i=1}^d$. Suppose $\rho \in \mathbb{B}_1^+(\mathbb{H})$ is strictly positive, then $R_{\rho^{-1}} = R_\rho^{-1}$ is well defined as well as the *relative modular operator* of $\rho$ and $\sigma \in \mathbb{B}_1^+(\mathbb{H})$,

$$\Delta_{\rho,\sigma} := L_\sigma\,R_\rho^{-1} = R_\rho^{-1}\,L_\sigma = \sum_{i,j=1}^d s_i r_j^{-1}\,L_{|\,s_i\,\rangle\langle\,s_i\,|} R_{|\,r_j\,\rangle\langle\,r_j\,|}\ , \qquad (6.24)$$

where the spectralizations $\rho = \sum_{j=1}^d r_j|\,r_j\,\rangle\langle\,r_j\,|$ and $\sigma = \sum_{i=1}^d s_i|\,s_i\,\rangle\langle\,s_i\,|$ have been used. If both $\rho$ and $\sigma$ are strictly positive, the same is true of their relative modular operator: for all $X \in |mdd$,

$$<< X\,,\, \Delta_{\rho,\sigma} X >> = \mathrm{Tr}\Big(\sum_{i,j=1}^d s_i r_j^{-1}\,X^\dagger|\,s_i\,\rangle\langle\,s_i\,|X|\,r_j\,\rangle\langle\,r_j\,|\Big)$$

$$= \sum_{i,j=1}^d s_i r_j^{-1}\,\left|\langle\,r_j\,|\,X^\dagger\,|s_i\,\rangle\right|^2 \geq 0$$

Also,

$$\log \Delta_{\rho,\sigma} = \log L_\sigma + \log R_\rho^{-1} = \sum_{i=1}^d \Big(\log s_i\,L_{|\,s_i\,\rangle\langle\,s_i\,|} - \log r_i\,R_{|\,r_i\,\rangle\langle\,r_i\,|}\Big)\ .$$

This yields the following expression for the relative entropy

$$<< \sqrt{\rho}\,,\, -(\log \Delta_{\rho,\sigma})[\sqrt{\rho}] >>= S\,(\rho\,,\,\sigma)\ .$$

Of the following properties of the relative entropy, *joint convexity* and *monotonicity under CPU maps* have already been used in the proof of the properties (5.162) and (5.163) of the von Neumann entropy in Proposition 5.5.6.

**Proposition 6.3.1.** *The relative entropy of $\rho, \sigma \in \mathbb{B}_1(\mathbb{H})$ is*

1. positive: $S(\rho;\sigma) \geq 0$, $S(\rho;\sigma) = 0$ *iff* $\rho = \sigma$;
2. jointly convex: *given weights* $\lambda_i \geq 0$, $i \in I$, $\sum_{i\in I}\lambda_i = 1$, *and density matrices* $\rho_i, \sigma_i \in \mathbb{B}_1(\mathbb{H})$, $i \in I$,
$$S\left(\sum_{i\in I}\lambda_i\rho_i\,,\,\sum_{i\in I}\lambda_i\sigma_i\right) \leq \sum_{i\in I}\lambda_i\,S\left(\rho_i\,,\,\sigma_i\right)\ ; \tag{6.25}$$
3. invariant under unitary maps: *let* $\rho, \sigma \in \mathbb{B}_1^+(\mathbb{H})$ *and* $U : \mathbb{H} \mapsto \mathbb{H}$ *a unitary map, then*
$$S\left(U\rho U^\dagger\,,\,U\sigma U^\dagger\right) = S\left(\rho\,,\,\sigma\right)\ . \tag{6.26}$$
4. monotonically decreasing under trace-preserving CP maps: *let* $\rho, \sigma \in \mathbb{B}_1^+(\mathbb{H})$ *and* $\mathbb{F} : \mathbb{B}_1^+(\mathbb{H}) \mapsto \mathbb{B}_1^+(\mathbb{H})$ *be a CP map such that* $\mathrm{Tr}(\mathbb{F}[\rho]) = \mathrm{Tr}(\rho)$, *then*
$$S\left(\mathbb{F}[\rho]\,,\,\mathbb{F}[\sigma]\right) \leq S\left(\rho\,,\,\sigma\right)\ . \tag{6.27}$$

Writing $\mathbb{F}[\rho] = \rho \circ \mathbb{E}$, where $\mathbb{E} : \mathbb{B}(\mathbb{H}) \mapsto \mathbb{B}(\mathbb{H})$ is the dual *CPU* map of $\mathbb{F}$, monotonicity reads

$$S\left(\rho \circ \mathbb{E}\,,\,\sigma \circ \mathbb{E}\right) \leq S\left(\rho\,,\,\sigma\right)\ . \tag{6.28}$$

Also, joint convexity is equivalently expressed by the inequality

$$S\left(\sum_{i\in I}\widetilde{\rho}_i\,,\,\sum_{i\in I}\widetilde{\sigma}_i\right) \leq \sum_{i\in I}\lambda_i\,S\left(\widetilde{\rho}_i\,,\,\widetilde{\sigma}_i\right)\ , \tag{6.29}$$

where $\widetilde{\rho}_i := \lambda_i\rho_i$ and $\widetilde{\sigma}_i := \lambda_i\sigma_i$.

**Proof:**
• **Positivity**: by means of the eigenvalues $r_j$, $s_k$ (repeated according to their multiplicities) and of the eigenbases $|\,r_j\,\rangle$, $|\,s_k\,\rangle$ of $\rho$, respectively $\sigma$, one computes

$$\begin{aligned}
S\left(\rho\,,\,\sigma\right) &= \sum_i r_i(\log r_i - \langle\, r_j\,|\,\log\sigma\,|r_j\,\rangle) \\
&= \sum_i r_i\Big(\log r_i - \sum_k |\langle\, r_j\,|\,s_k\,\rangle)|^2\log s_k\Big) \\
&\geq \sum_i r_i\Big(\log r_i - \log(\sum_k s_k\,|\langle\, r_i\,|\,s_k\,\rangle|^2)\Big)
\end{aligned}$$

$$= \sum_i r_i(\log r_i - \log(\langle r_i \,|\, \sigma \,| r_i \rangle)$$
$$\geq \sum_i (r_i - \langle r_i \,|\, \sigma \,| r_i \rangle) = 0 \ .$$

The first inequality follows from $\sum_k |\langle r_i \,|\, s_k \rangle|^2 = 1$ and the concavity of the logarithm, while the second one is a consequence of the concavity of $\eta(x) = -x \log x$, $0 \leq x \leq 1$ (see (2.85) in Section 2.4.3), which also implies that equality only holds when the eigenvalues and thus the density matrices coincide.

- **Joint convexity**: we shall establish (6.29). Notice that, for all $w > 0$,

$$\begin{aligned} -\log w &= \int_0^\infty \mathrm{d}t \left(\frac{1}{w+t} - \frac{1}{1+t}\right) \\ &= (1-w) \int_0^\infty \mathrm{d}t \left(\frac{1}{(w+t)(1+t)} - \frac{1}{(1+t)^2} + \frac{1}{(1+t)^2}\right) \\ &= (1-w) + \int_0^\infty \frac{\mathrm{d}t}{(1+t)^2} \frac{(w-1)^2}{w+t} \ . \end{aligned}$$

Then, use the spectral representation of the relative modular operator (6.24) to insert $\Delta_{\rho,\sigma}$ in the place of $w$, act with $-\log \Delta_{\rho,\sigma}$ on $\rho$ and take the trace of the resulting matrix. Since $\mathrm{Tr}\Big((\mathbb{1} - \Delta_{\rho,\sigma})[\rho]\Big) = \mathrm{Tr}(\rho - \sigma) = 0$, it follows

$$S(\rho\,,\sigma) = -\mathrm{Tr}\Big(\log \Delta_{\rho,\sigma}[\rho]\Big) = \int_0^\infty \frac{\mathrm{d}t}{(1+t)^2} \mathrm{Tr}\Big((\mathbb{1} - \Delta_{\rho,\sigma}) \frac{1}{\Delta_{\rho,\sigma} + t\mathbb{1}}[\rho - \sigma]\Big) \ .$$

Further, setting $Y = \mathbb{1}$ and $X = (\Delta_{\rho,\sigma} + t\mathbb{1})^{-1}[\rho - \sigma]$ in

$$\begin{aligned} << Y\,, (\mathbb{1} - \Delta_{\rho,\sigma})[X] >> &= << Y\,, (R_\rho - L_\sigma) R_\rho^{-1}[X] >> \\ &= << (R_\rho - L_\sigma)[Y]\,, R_\rho^{-1}[X] >> \ , \end{aligned}$$

yields

$$S(\rho\,,\sigma) = \int_0^\infty \frac{\mathrm{d}t}{(1+t)^2} \mathrm{Tr}\Big((\rho - \sigma) \frac{1}{R_\rho + tL_\sigma}[\rho - \sigma]\Big) \ . \tag{6.30}$$

Let now $\tilde{\rho}_j$ and $\tilde{\sigma}_j$ be as in (6.29) and

$$X_j := (L_{\tilde{\sigma}_j} + tR_{\tilde{\rho}_j})^{-1/2}[\tilde{\rho}_j - \tilde{\sigma}_j] - (L_{\tilde{\sigma}_j} + tR_{\tilde{\rho}_j})^{1/2}[B] \ ,$$

with $B = B^\dagger \in M_d(\mathbb{C})$ to be defined later. Then, since the various operators are self-adjoint with respect to the Hilbert-Schmidt scalar product, by observing that $\sum_j (L_{\tilde{\sigma}_j} + tR_{\tilde{\rho}_j}) = L_\sigma + tR_\rho$, one obtains

$$\begin{aligned} 0 \leq \sum_j << X_j\,, X_j >> &= \sum_j << \tilde{\rho}_j - \tilde{\sigma}_j\,, (L_{\tilde{\sigma}_j} + tR_{\tilde{\rho}_j})^{-1}[\tilde{\rho}_j] >> \\ &- << \rho - \sigma\,, B >> - << B\,, \rho - \sigma >> + << B\,, (L_\sigma + tR_\rho)[B] >> \ . \end{aligned}$$

By choosing $B = (L_\sigma + tR_\rho)^{-1}[\rho]$, one gets the inequality

$$\sum_j << \tilde{\rho}_j - \tilde{\sigma}_j \,,\, (L_{\tilde{\sigma}_j} + tR_{\tilde{\rho}_j})^{-1}[\tilde{\rho}_j - \tilde{\sigma}_j] >>=$$

$$= \sum_j \mathrm{Tr}\Big((\tilde{\rho}_j - \tilde{\sigma}_j)(L_{\tilde{\sigma}_j} + tR_{\tilde{\rho}_j})^{-1}[\tilde{\rho}_j - \tilde{\sigma}_j]\Big)$$

$$\geq \mathrm{Tr}\Big((\rho - \sigma)(L_{\tilde{\sigma}_j} + tR_{\tilde{\rho}_j})^{-1}[\tilde{\rho}_j]\Big) \,,$$

which, once inserted in (6.30), yields the result.

- **Invariance:** it follows from the fact that $\log(U\rho U^\dagger) = U(\log\rho)U^\dagger$ and that the same holds for $\sigma$.
- **Monotonicity:** it is implied by joint convexity. We shall first show that

$$S(\rho_1 \,,\, \sigma_1) \leq S(\rho_{12} \,,\, \sigma_{12}) \,,$$

where $\rho_{12}, \sigma_{12} \in \mathbb{B}_1^+(\mathbb{H}_{12})$ with marginal states $\rho_{1,2} = \mathrm{Tr}_{2,1}\rho_{12}$, respectively $\sigma_{1,2} = \mathrm{Tr}_{2,1}\sigma_{12}$. Let $d_i = \dim(\mathbb{H}_i)$, fix an *ONB* $\{|\,j\,\rangle\}_{j=1}^{d_2}$ in $\mathbb{H}_2$ and define the unitary matrices $U_\ell \in M_{d_2}(\mathbb{C})$, $\ell = 1, 2, \ldots, d_2$, with entries $(U_\ell)_{jk} = \delta_{jk} \exp(\frac{2\pi i}{d_2} j\ell)$. Then, for all $X \in M_{d_2}(\mathbb{C})$,

$$\Phi[X] := \frac{1}{d_2}\sum_{\ell=1}^{d_2} U_\ell\, X\, U_\ell^\dagger = \frac{1}{d_2}\sum_{\ell,j,k=1}^{d_2} \mathrm{e}^{2\pi i\ell(j-k)}\langle\, j\,|\,X\,|k\,\rangle\,|\,j\,\rangle\langle\, k\,|$$

$$= \sum_{j=1}^{d_2} |\,j\,\rangle\langle\, j\,|\,\langle\, j\,|\,X\,|j\,\rangle \,;$$

whence $\rho_1 = \mathrm{id}_1 \otimes \Phi[\rho_{12}]$ and similarly for $\sigma_1$. Furthermore, using the basis $\{|\,j\,\rangle\}_{j=1}^{d_2}$ to write $\rho_{12} = \sum_{j,k=1}^{d_2} \rho_{jk}^{(1)} \otimes |\,j\,\rangle\langle\, k\,|$, then

$$S(\rho_1 \,,\, \sigma_1) = S\left(\sum_{j=1}^{d_2} \rho_{jj}^{(1)} \,,\, \sum_{j=1}^{d_2} \sigma_{jj}^{(1)}\right) \leq \sum_{j=1}^{d_2} S\left(\rho_{jj}^{(1)} \,,\, \sigma_{jj}^{(1)}\right)$$

$$= S\left(\sum_{j=1}^{d_2} \rho_{jj}^{(1)} \otimes |\,j\,\rangle\langle\, j\,| , \sum_{j=1}^{d_2} \sigma_{jj}^{(1)} \otimes |\,j\,\rangle\langle\, j\,|\right)$$

$$= S(\mathrm{id}_1 \otimes \Phi[\rho_{12}] \,,\, \mathrm{id}_1 \otimes \Phi[\sigma_{12}])$$

$$\leq \frac{1}{d_2}\sum_{\ell=1}^{d_2} S\left((\mathbb{1}_1 \otimes Z_\ell)\rho_{12}(\mathbb{1}_1 \otimes Z_\ell)^\dagger \,,\, (\mathbb{1}_1 \otimes Z_\ell)\sigma_{12}(\mathbb{1}_1 \otimes Z_\ell)^\dagger\right)$$

$$= S(\rho_{12} \,,\, \sigma_{12}) \,,$$

where the second equality follows from the orthogonality of the matrices contributing to the sums, the last equality follows since the matrices $\mathbb{1}_1 \otimes Z_\ell$

are unitary and because of the invariance of the relative entropy, while the two inequalities are a consequence of its joint convexity.

Monotonicity under trace preserving *CP* maps thus results from Remark 5.6.3, by writing $\mathbb{E}[\rho] = \mathrm{Tr}_E\Big(U\,(\rho\otimes\rho_E)\,U^\dagger\Big)$:

$$\begin{aligned} S\,(\mathbb{E}[\rho]\,,\,\mathbb{E}[\sigma]) &\leq S\,\big(U\,(\rho\otimes\rho_E)U^\dagger\,,\,U\,(\sigma\otimes\rho_E)U^\dagger\big) \\ &= S\,(\rho\otimes\rho_E\,,\,\sigma\otimes\rho_E) = S\,(\rho\,,\,\sigma)\ . \end{aligned}$$

□

**Example 6.3.2.** As an application of joint convexity, let $\rho \in \mathbb{B}_1(\mathbb{H})$ be a density matrix describing a statistical mixture $\{\lambda_{ij},\rho_{ij}\}$: $\rho = \sum_{ij}\lambda_{ij}\rho_{ij}$, $\sum_{ij}\lambda_{ij} = 1$. Setting $\widetilde{\rho}_{ij} := \lambda_{ij}\rho_{ij}$, $\widetilde{\rho}^1_i := \sum_j \widetilde{\rho}_{ij}$ and $\widetilde{\rho}^2_j := \sum_i \widetilde{\rho}_{ij}$, one derives

$$\begin{aligned} \sum_j \lambda_{ij} S\,(\rho_{ij}\,,\,\rho) &= \sum_j \mathrm{Tr}\widetilde{\rho}_{ij}\log\widetilde{\rho}_{ij} - \mathrm{Tr}\widetilde{\rho}^1_i\log\rho - \sum_j \lambda_{ij}\log\lambda_{ij} \\ &= \sum_j S\,(\widetilde{\rho}_{ij}\,,\,\widetilde{\rho}^1_i) + S\,(\widetilde{\rho}^1_i\,,\,\rho) - \sum_j \lambda_{ij}\log\lambda_{ij}\ , \end{aligned}$$

whence (6.29) applied with reference to the sum over the index $i$ and the fact that $\sum_i \widetilde{\rho}^1_i = \rho$ yield

$$\begin{aligned} \sum_{ij}\lambda_{ij}S\,(\rho_{ij}\,,\,\rho) &\geq \sum_j S\,(\widetilde{\rho}^2_j\,,\,\rho) + \sum_i S\,(\widetilde{\rho}^1_i\,,\,\rho) - \sum_{ij}\lambda_{ij}\log\lambda_{ij} \\ &= \sum_j S\,(\rho^2_j\,,\,\rho) + \sum_i S\,(\rho^1_i\,,\,\rho) \\ &\quad + \sum_i \lambda^1_i\log\lambda^1_i + \sum_j \lambda^2_j\log\lambda^2_j - \sum_{ij}\lambda_{ij}\log\lambda_{ij}\ , \end{aligned}$$

$$\text{where}\qquad \lambda^1_i := \sum_j \lambda_{ij}\ ,\ \lambda^2_j := \sum_i \lambda_{ij}\ ,\ \rho^1_i := \frac{\widetilde{\rho}^1_i}{\lambda^1_i}\ ,\ \rho^2_j := \frac{\widetilde{\rho}^2_j}{\lambda^2_j}\ .$$

The physical interpretation of the relative entropy comes from thermodynamics: there, it amounts to *free energy*. As such, it can only decrease under dissipative time-evolutions [286, 189]. Let $\sigma$ in (6.23) be the Gibbs state at inverse temperature $\beta = T^{-1}$ with respect to a Hamiltonian operator $H \in \mathbb{B}(\mathbb{H})$ (see (5.177)):

$$\sigma = \rho_\beta := Z_\beta \exp(-\beta H)\ ,\quad \mathcal{Z}_\beta^{-1} = \mathrm{Tr}(\exp(-\beta H))\ .$$

The free energy of a state $\rho \in \mathbb{B}_1(\mathbb{H})$ is

$$F(\rho) := T\,S(\rho) - \langle H\rangle_\rho\ ,\quad \langle H\rangle_\rho := \mathrm{Tr}(\rho\,H)\ .$$

From $F(\rho_\beta) = -T \log Z_\beta$ it follows that

$$S(\rho\,;\,\rho_\beta) = -S(\rho) - \log Z_\beta + \beta\langle H\rangle_\rho = \beta\Big(F(\rho_\beta) - F(\rho)\Big)\,.$$

Let $S$ undergo an irreversible evolution described by a quantum dynamical semigroup $\gamma_t : \mathcal{S}(S) \mapsto \mathcal{S}(S)$, $t \geq 0$, with $\rho_\beta$ an equilibrium state, namely $\gamma_t[\rho_\beta] = \rho_\beta$. Then, as seen in Chapter 3, the dynamical maps $\gamma_t$ are completely positive and fulfill $\gamma_t = \gamma_{t-s} \circ \gamma_s$, $t \geq s$. Thus, monotonicity (6.27) yields

$$\begin{aligned} S\Big(\gamma_t[\rho]\,;\,\gamma_t[\rho_\beta]\Big) &= S\Big(\gamma_t[\rho]\,;\,\rho_\beta\Big) = \beta\Big(F(\rho_\beta) - F(\gamma_t[\rho])\Big) \\ &= S\Big(\gamma_{t-s} \circ \gamma_s[\rho]\,;\,\gamma_{t-s}[\rho_\beta]\Big) \\ &\leq S\Big(\gamma_s[\rho]\,;\,\rho_\beta\Big) = \beta\Big(F(\rho_\beta) - F(\gamma_s[\rho])\Big)\,. \end{aligned} \tag{6.31}$$

While the relative entropy behaves monotonically, this is not true of the von Neumann entropy (see Example 5.6.3). For instance, if the quantum dynamical semigroup mentioned before represents the reduced dynamics of a quantum open system $S$ interacting with a reservoir, the free energy of an initial state may decrease in time, showing tendency to equilibrium, while but its von Neumann entropy may in some cases decrease (for more details see [41]). The following example provide a class of dynamical operations on the states of $S$ which always increase the entropy of its states (or keep it constant).

**Examples 6.3.3.**

1. **Bistochastic maps** [302, 303] Completely positive unital maps $\mathbb{E} : \mathbb{B}(\mathbb{H}) \mapsto \mathbb{B}(\mathbb{H})$ are called *bistochastic* if their dual maps $\mathbb{F} : \mathcal{S}(S) \mapsto \mathcal{S}(S)$ preserve the tracial state: $\mathbb{F}\left[\frac{\mathbb{1}}{d}\right] = \frac{\mathbb{1}}{d}$.
The most natural bistochastic maps are those associated with projective *POVMs*, $\mathbb{F}_{\mathcal{P}}[\rho] = \sum_i P_i \rho P_i$, $P_i P_j = \delta_{ij} P_j$, $\sum_i P_i = \mathbb{1}$ (see Section 5.6.1). These maps always increase the von Neumann entropy; indeed, from (6.27)
$$S\left(\mathbb{F}[\rho]\,,\,\mathbb{F}\left[\frac{\mathbb{1}}{d}\right]\right) = \log d\, -\, S\left(\mathbb{F}[\rho]\right) \leq S\left(\rho\,,\,\frac{\mathbb{1}}{d}\right) = \log d\, -\, S\left(\rho\right)\,.$$
2. Let $\mathcal{P}$ be a projective *POVM* as in the previous point. The linear span of the orthogonal projectors $P_i$, $i \in I$ (not necessarily one-dimensional, so that $\mathrm{card}(I) \leq d$) is an Abelian subalgebra $\boldsymbol{A}_{\mathcal{P}} \subset \mathbb{B}(\mathbb{H})$ with identity, whose typical elements have the form $a = \sum_{i\in I} a_i\, P_i$. The space of states $\mu$ over $\boldsymbol{A}_{\mathcal{P}}$ consists of normalized, positive linear expectations

$$A_{\mathcal{P}} \ni a \mapsto \mu(a) = \sum_{i\in I} a_i\,\mu(P_i)\ .$$

They thus correspond to all possible discrete probability distributions with $\mathrm{card}(I)$ elements. Given a state $\rho \in \mathcal{S}(S)$ its restriction to $\boldsymbol{A}_{\mathcal{P}}$, denoted by $\rho\lceil \boldsymbol{A}_{\mathcal{P}}$, corresponds to the discrete probability distribution $\mu_\rho := \{\mathrm{Tr}(\rho\, P_i)\}_{i\in I}$. It thus follows that $\rho\lceil A_{\mathcal{P}} = \mathbb{F}_{\mathcal{P}}[\rho] = \sum_{i\in I} P_i\rho\, P_i$, indeed

$$\mathrm{Tr}\Big(\mathbb{F}_{\mathcal{P}}[\rho]\, a\Big) = \sum_{i,j\in I} a_j \mathrm{Tr}(P_i\rho P_i\, P_j) = \sum_{i\in I} a_i\,\mathrm{Tr}(\rho\, P_i)\ .$$

From the previous point, it then follows that

$$S(\rho) = \min\Big\{S(\rho\lceil \boldsymbol{A})\ :\ \boldsymbol{A}\subset \mathbb{B}(\mathbb{H})\ \text{Abelian with identity}\Big\}\ ,$$

the minimum being achieved at any Abelian subalgebra $\boldsymbol{A}$ generated by the eigenprojectors $P_i = |\,r_i\,\rangle\langle\, r_i\,|$ of $\rho$, for in this case $\mu_\rho(P_i) = \mathrm{Tr}(\rho\,|\,r_i\,\rangle\langle\, r_i\,|) = r_i$.

The following result emphasizes the connections between von Neumann entropy and relative entropy [213]; the idea is to exploit the (infinitely many) convex decompositions of mixed states.

**Proposition 6.3.2.** *Let $S$ be a quantum system described by a Hilbert space $\mathbb{H}$ and let $\rho = \sum_{i\in I}\lambda_i\rho_i$, $\lambda_i \geq 0$, $\sum_{i\in I}\lambda_i = 1$, be any convex decomposition of a mixed state $\rho \in \mathcal{S}(S)$ in terms of other density matrices $\rho_i \in \mathcal{S}(S)$. Then,*

$$S(\rho) = \min\Big\{\sum_{i\in I}\lambda_i\, S(\rho_i\,;\,\rho)\ :\ \rho = \sum_{i\in I}\lambda_i\rho_i\Big\}\ .$$

**Proof:** From (6.23), $\sum_{i\in I}\lambda_i\, S(\rho_i\,;\,\rho) = S(\rho) - \sum_{i\in I}\lambda_i S(\rho_i) \leq S(\rho)$, while the spectral eigenprojectors of $\rho$ give the upper bound. □

### 6.3.1 Holevo's Bound and the Entropy of a Subalgebra

As seen in Example 6.1.2, by encoding classical information into non-orthogonal quantum states one may always detect the presence of eavesdroppers during transmission. However, the non-orthogonality of the quantum code-words does not allow for perfect retrieval of the encoded classical information, for no measurement can perfectly distinguish between non-orthogonal states.

In fact, let $|\,\psi_1\,\rangle$ and $|\,\psi_2\,\rangle = \alpha|\,\psi_1\,\rangle + \beta|\,\psi_1^\perp\,\rangle$, $\alpha \neq 0$ be two vector states in some Hilbert space $\mathbb{H}$. Suppose there exists a set of orthogonal projections

$P_j$ on $\mathbb{H}$, $j \in J = J_1 \cup J_2$, such that if $j \in J_1$, respectively $j \in J_2$, is measured then the state is $\psi_1$, respectively $\psi_2$, with probability 1. Then, the orthogonal projections $Q_{1,2} := \sum_{j \in J_{1,2}} P_j$, would be such that $\langle \psi_{1,2} | Q_{1,2} | \psi_{1,2} \rangle = 1$, whereas $\langle \psi_{1,2} | Q_{2,1} | \psi_{1,2} \rangle = 0$. Therefore,

$$1 = \langle \psi_2 | Q_2 | \psi_2 \rangle = |\beta|^2 \langle \psi_1^\perp | Q_2 | \psi_1^\perp \rangle \leq |\beta|^2 \leq 1 \Longrightarrow |\beta| = 1$$

namely $\psi_1 \perp \psi_2$.

More in general, the symbols $i \in I_A = \{1, 2, \ldots, a\}$ of a classical alphabet, emitted with probabilities $p_1, p_2, \ldots, p_a$, might be encoded by means of mixed states $\rho_i \in \mathbb{B}_1(\mathbb{H})$. Or, from a more realistic viewpoint, given an encoding of the classical symbols into pure states $\psi_i \in \mathbb{H}$, a noisy transmission channel might transform them into mixed states $\rho_i := \mathbb{F}[| \psi_i \rangle\langle \psi_i |]$, where $\mathbb{F}$ is the dual of a *CPU* map $\mathbb{E} : \mathbb{B}(\mathbb{H}) \to \mathbb{B}(\mathbb{H})$. The receiver must then reconstruct the encoded classical message with the least possible error; practically speaking, he must seek a *POVM* $\mathcal{B} = \{B_i\}_{i \in I_B} \subset \mathbb{B}(\mathbb{H})$, $I_B = \{1, 2, \ldots, b\}$, such that, when measured on the statistical mixture $\rho = \sum_{a \in I_A} p_a \rho_a$, it maximizes the accessible information.

In such a context, three random variables appear: $A$, $B$, and $A \vee B$, with probability distributions $\pi_A$, $\pi_B$ and $\pi_{A \vee B}$:

1. the outcomes of $A$ correspond to the indices $i \in I_A$ of the incoming states and $\pi_A = \{p_a\}_{a \in I_A}$;
2. the outcomes of $B$ correspond to the indices $i \in I_B$ of the *POVM* and $\pi_B = \{\mathrm{Tr}(\rho B_i\}_{i \in I_B}$;
3. the outcomes of $A \vee B$ correspond to the joint events consisting of an incoming state $\rho_a$ and a measured index $i$: $\pi_{A \vee B} = \Big\{p_a \mathrm{Tr}(\rho_a B_i)\Big\}_{a \in I_A, i \in I_B}$.

According to Section 2.4.5, the mutual information $I(A, B)$ measures how much knowledge one gains about $A$, that is about which state $\rho_i$ has reached Bob, from measuring on $B$ the *POVM* $\mathcal{B} = \{B_i\}_{i \in I_B}$:

$$\begin{aligned}
I(A, B) &= H(A) + H(B) - H(A \vee B) \\
&= -\sum_{a \in I_A}^{a} p_a \log p_a - \sum_{i \in I_B} (\mathrm{Tr}(\rho B_i) \log(\mathrm{Tr}(\rho B_i)) \\
&\qquad + \sum_{a \in I_A} \sum_{i \in I_B} p_a(\mathrm{Tr}(\rho_a B_i)) \log(p_a(\mathrm{Tr}(\rho_a B_i))) \\
&= -\sum_{i \in I_B} (\mathrm{Tr}(\rho B_i)) \log(\mathrm{Tr}(\rho B_i)) \\
&\qquad + \sum_{a \in I_A} p_a \sum_{i \in I_B} (\mathrm{Tr}(\rho_a B_i)) \log(\mathrm{Tr}(\rho_a B_i)) . \qquad (6.32)
\end{aligned}$$

In the classical case, perfect knowledge of $A$ from knowing $B$ can be achieved by choosing $B$ such that $H(A|B) = 0$; in the quantum case, there is a more stringent upper bound on $I(A,B)$ that depends on the given decomposition $\rho = \sum_{a\in I_A} p_a\rho_a$ and is denoted by $\chi(\rho, \{p_a\rho_a\}_{a\in I_A})$.

**Proposition 6.3.3 (Holevo's Bound).** *Given* $\rho = \sum_{a\in I_A} p_a\,\rho_a \in \mathbb{B}_1^+(\mathbb{H})$ *and the POVM* $\mathcal{B} = \{B_i\}_{i\in I_B} \subseteq \mathbb{B}(\mathbb{H})$,

$$I(A,B) \leq \chi(\rho, \{\rho, p_a\rho_a\}_{a\in I_A}) := S(\rho) - \sum_{a\in I_A} p_a\, S(\rho_a) \ . \tag{6.33}$$

**Proof:** [23] Given the *POVM* $\mathcal{B} = \{B_i\}_{i\in I_B} \subseteq \mathbb{B}(\mathbb{H})$, let $\boldsymbol{B} = \{\widehat{b}_j\}_{j\in I_B}$ be an Abelian algebra with minimal projections $\widehat{b}_j$, $\widehat{b}_i\widehat{b}_j = \delta_{ij}\widehat{b}_i$, $\sum_{j\in I_B}\widehat{b}_j = \mathbb{1}_{\boldsymbol{B}}$. It can be embedded into $\mathbb{B}(\mathbb{H})$ (as a linear space) by means of the linear maps $\gamma_{\mathcal{B}} : \boldsymbol{B} \mapsto \mathbb{B}(\mathbb{H})$ such that $\gamma_{\mathcal{B}}[\widehat{b}_i] = B_i$. Positive operators in $\mathcal{B}$ are of the form $b = \sum_{i\in I_B} \beta_i\,\widehat{b}_i$, $\beta_i \geq 0$, therefore $\gamma_{\mathcal{B}}(b) = \sum_{i\in I_B} \beta_i\, B_j \geq 0$ so that $\gamma_{\mathcal{B}}$ is a positive map and, because of Example 5.2.6.7, completely positive. Also, $\gamma_{\mathcal{B}}(\mathbb{1}_B) = \sum_{i\in I_B} \gamma(\widehat{b}_i) = \sum_{i\in I_B} B_i = \mathbb{1}_A$, whence $\gamma_{\mathcal{B}}$ is a *CPU* map. Furthermore, the states $\rho\circ\gamma_{\mathcal{B}}$ and $\rho_i\circ\gamma_{\mathcal{B}}$ on $\boldsymbol{B}$ are diagonal density matrices with eigenvalues $\{\mathrm{Tr}(\rho\, B_i\}_{i\in I_B}$, respectively $\{\mathrm{Tr}(\rho_a\, B_i\}_{i\in I_B}$. Thus, (6.32) and the monotonicity of the relative entropy (6.27) yield

$$I(A,B) = \sum_{a\in I_A} p_a\, S(\rho\circ\gamma_{\mathcal{B}}\ \rho_a\circ\gamma_{\mathcal{B}}) \leq \sum_{a\in I_A} p_a\, S(\rho,\rho_a) \ .$$

□

**Remarks 6.3.1.**

1. Using (5.156) one derives that

$$\chi(\rho, \{p_a\rho_a\}_{a\in I_A}) = S(\rho) - \sum_{a\in I_A} p_a\, S(\rho_i) \leq -\sum_{a\in I_A} p_a \log p_a = H(A) \ .$$

   Thus, if (6.33) is a strict inequality, perfect reconstruction of $A$ upon knowledge of $B$ is not possible; on the other hand, the inequality is strict unless the states $\rho_a$ are orthogonal to each other and thus perfectly distinguishable.
2. A consequence of the Holevo's bound is that any quantum encoding of $n$ *bits* into $n$ non-orthogonal *qubits* states $|\,\psi_i\,\rangle \in \mathbb{C}^{2^n}$ achieves secure transmission, but cannot transfer more than $H(A) \leq n$ *bits* of information (when the entropy is expressed in base 2).
3. Whether the upper bound is achieved or not depends on the ability on the part of the receiver to find one or more optimal detection strategies, namely those *POVM* 's $\mathcal{B}$ that maximize $I(A,B)$. As we shall see this is a remarkably difficult analytical problem even in low dimension.

4. By taking the supremum over all possible *POVM* 's $\mathcal{B}$ that the receiver may devise as detection strategies, one defines the *maximal accessible information* as

$$I(A) := \sup_{\mathcal{B}} I(A, B) \leq \chi\left(\rho, \{p_a \rho_a\}_{a \in I_A}\right) \,. \tag{6.34}$$

**Example 6.3.4.** Suppose $A$ transmits the *bits* 0 and 1 to $B$ by encoding them into the non-orthogonal states

$$|\pm\rangle = \sqrt{p}\,|\,\psi_1\,\rangle \pm \sqrt{1-p}\,|\,\psi_2\,\rangle \in \mathbb{C}^N \;,\; 0 \leq p \leq 1 \;,\; \langle\,\psi_1\,|\,\psi_2\,\rangle = 0 \;,$$

chosen with equal probability. The statistics of the encoded quantum signals is thus described by

$$M_d(\mathbb{C}) \ni \rho = \frac{1}{2}|+\rangle\langle+| + \frac{1}{2}|-\rangle\langle-| = p\,|\,\psi_1\,\rangle\langle\,\psi_1\,| + (1-p)\,|\,\psi_2\,\rangle\langle\,\psi_2\,| \;,$$

and $\chi(\{\rho, \lambda \rho_j\}) = H_2(p) = -p\log_2 p + -(1-p)\log_2(1-p)$ reaches its maximum of 1 *bit* of transmissible information only when $p = 1/2$ so that $\langle\,+\,|\,-\,\rangle = 1 - 2p = 0$.

The problem of achieving the maximal accessible information $I(A)$ appeared earlier than in quantum information, in relation to finding *optimal decompositions* achieving the so-called *entropy of a subalgebra* [213, 88], the building block for constructing a particular quantum extension of the *KS* entropy to be discussed later in Chapter 8.

Let $\boldsymbol{M} \subseteq \mathbb{B}(\mathbb{H})$ be a finite-dimensional subalgebra with identity, $\rho \in \mathbb{B}_1^+(\mathbb{H})$ a state on $\mathbb{B}(\mathbb{H})$ and $\rho\lceil\boldsymbol{M}$ the state on $\boldsymbol{M}$ which results from restricting $\rho$ to act (as an expectation) on the observables in $\boldsymbol{M}$, only.

**Example 6.3.5.** Let $\boldsymbol{A} \subset \mathbb{B}(\mathbb{H})$ be an Abelian subalgebra with $k \leq \dim(\mathbb{H})$ minimal projectors $\widehat{a}_i$ and $\rho \in \mathbb{B}_1^+(\mathbb{H})$ a density matrix; then, $\rho\lceil\boldsymbol{A}$ amounts to the classical probability distribution $\pi_{\boldsymbol{A}} = \{\mathrm{Tr}(\rho\,\widehat{a}_i)\}_{i=1}^k$.

**Definition 6.3.2 (Entropy of a Subalgebra).** *Let* $\boldsymbol{M} \subseteq \mathbb{B}(\mathbb{H})$ *be a subalgebra and* $\rho \in \mathbb{B}_1^+(\mathbb{H})$ *a state; the entropy of* $\boldsymbol{M}$ *relative to* $\rho$ *is*

$$\mathrm{H}_\rho(\boldsymbol{M}) := \sup_{\rho=\sum_{i\in I}\lambda_i\rho_i} \sum_{i\in I} \lambda_i\, S\left(\rho\lceil\boldsymbol{M}\,,\,\rho_i\lceil\boldsymbol{M}\right) \tag{6.35}$$

$$= S\left(\rho\lceil\boldsymbol{M}\right) - \inf_{\rho=\sum_{i\in I}\lambda_i\rho_i} \sum_{i\in I} \lambda_i S\left(\rho_i\lceil\boldsymbol{M}\right) \,, \tag{6.36}$$

*where the* sup *in (6.35) and the* inf *in (6.36) are taken with respect to all possible linear convex decompositions* $\rho = \sum_{i\in I}\lambda_i\rho_i$.

It must be stressed that, unlike in Proposition 6.3.2, the decomposition comes first and the restriction to the subalgebra only afterwards; this is what makes the explicit computation of the entropy of a subalgebra a complicated variational problem, in general. Luckily, there are particular instances of states and subalgebras where things are easier.

**Example 6.3.6.** Consider the Abelian subalgebra $\boldsymbol{A}$ of Example 6.3.5 and let $\rho \in \mathbb{B}_1^+(\mathbb{H})$ be a state which commutes with all elements of $\boldsymbol{A}$. The following decomposition of $\rho$ is optimal,

$$\rho = \sum_{i=1}^{k} \mathrm{Tr}(\rho\widehat{a}_i)\,\rho_i\ , \quad \rho_i := \frac{\sqrt{\rho}\,\widehat{a}_i\,\sqrt{\rho}}{\mathrm{Tr}(\rho\,\widehat{a}_i)} = \frac{\rho\,\widehat{a}_i}{\mathrm{Tr}(\rho\,\widehat{a}_i)}\ .$$

Indeed, the restrictions $\rho\lceil\boldsymbol{A}$ and $\rho_i\lceil\boldsymbol{A}$ are probability distributions

$$\pi_{\boldsymbol{A}} = \Big\{\mathrm{Tr}(\rho\,\widehat{a}_j)\Big\}_{j=1}^{k}\ , \ \pi^i_{\boldsymbol{A}} = \left\{\frac{\mathrm{Tr}(\rho\,\widehat{a}_i\,\widehat{a}_j)}{\mathrm{Tr}(\rho\,\widehat{a}_i)} = \delta_{ij}\right\}_{j=1}^{k}\ ,$$

such that $S\left(\rho_i\lceil\boldsymbol{A}\right) = 0$ and

$$\mathrm{H}_\rho\left(\boldsymbol{A}\right) = S\left(\rho\lceil\boldsymbol{A}\right) = -\sum_{i=1}^{k} \mathrm{Tr}(\rho\,\widehat{a}_i)\log\mathrm{Tr}(\rho\,\widehat{a}_i)\ .$$

The simplest context in which the above argument applies is when, instead of $\mathbb{B}(\mathbb{H})$, one deals with an Abelian von Neumann algebra. Then, as seen in Section 5.3.2, via the Gelfand transform, any finite subalgebra $\boldsymbol{A} \subset \mathcal{A}$ has minimal projections which correspond to the characteristic functions of suitable measurable subsets of a measure space and thus identify a finite partition of the latter or, equivalently a random variable $A$. Moreover, the state $\omega$ becomes a probability measure $\mu$ and gives a probability distribution over $\boldsymbol{A}$ such that the entropy of $\boldsymbol{A}$ yields the Shannon entropy of $A$: $\mathrm{H}_\omega\left(\boldsymbol{A}\right) = H(A)$.

Instead, the simplest non-commutative application of the previous argument is when $\mathbb{B}(\mathbb{H}) = M_d(\mathbb{C})$ and $\rho = \mathbb{1}_d/d$ is the tracial state; then,

$$\mathrm{H}_\rho\left(\boldsymbol{A}\right) = S\left(\rho\lceil\boldsymbol{A}\right) = H(A) = -\sum_{i=1}^{k}\frac{\mathrm{Tr}(\widehat{a}_i)}{d}\log\frac{\mathrm{Tr}(\widehat{a}_i)}{d}\ .$$

We now relate the variational problem in (6.35) to the one in (6.34). The clue is that *POVMs* as $\mathcal{B} = \{B_i\}_{i\in I_B}$ give rise to decompositions and vice versa, while the main technical tools are provided by the *GNS* construction $\pi_\rho(\mathbb{B}(\mathbb{H}))$ based on the state $\rho$. Of particular importance is the possibility of dealing with decompositions by means of positive elements in the commutant $\pi_\rho(\mathbb{B}(\mathbb{H}))'$ or even in $\pi_\rho(\mathbb{B}(\mathbb{H}))$ itself (see Remark 5.3.2.3 and the relation in (5.146)), an instance of which already appears in the previous example. It follows that the probabilities in (6.32) can be recast as

$$\mathrm{Tr}(\rho_a B_i) = \frac{\langle \Omega_\rho \,|\, \pi_\rho(B_i)\, X'_a \,|\Omega_\rho \rangle}{p_a} = \frac{\mathrm{Tr}(\rho\, B_i)}{p_a}\, \rho'_i(X'_a) \tag{6.37}$$

$$\rho'_i(X'_a) := \frac{\langle \Omega_\rho \,|\, \pi_\rho(B_i)\, X'_a \,|\Omega_\rho \rangle}{\mathrm{Tr}(\rho\, B_i)} \;, \; p_a := \langle \Omega_\rho \,|\, X'_a \,|\Omega_\rho \rangle \;, \tag{6.38}$$

where $|\,\Omega_\rho\,\rangle$ is the *GNS* cyclic vector and the $X'_a \in \pi_\rho(\mathbb{B}(\mathbb{H}))'$, $a \in I_A$, are positive operators in the commutant such that $\sum_{a\in I_A} X'_a = \mathbb{1}$.

It turns out that the linear functionals $\pi_\rho(\mathbb{B}(\mathbb{H}))' \ni X' \mapsto \rho'_i(X')$ are positive and normalized, hence states on the commutant, as well as

$$\rho'(X') := \sum_{i\in I_B} (\mathrm{Tr}(\rho\, B_i))\, \rho'_i(X') = \langle \Omega_\rho \,|\, X' \,|\Omega_\rho \rangle \;. \tag{6.39}$$

In analogy with the proof of the Holevo's bound, let $\boldsymbol{A} = \{\widehat{a}_j\}_{j\in I_A}$ be an Abelian algebra (with identity) generated by minimal projections $\widehat{a}_j$ and introduce the *CPU* map $\gamma'_A : \boldsymbol{A} \mapsto \pi_\rho(\mathbb{B}(\mathbb{H}))'$, $\gamma'_A[\widehat{a}_j] = X'_j$ that sends $\boldsymbol{A}$ into the commutant $\pi_\rho(\mathbb{B}(\mathbb{H}))'$. Then, using (6.37) and (6.38), one sees that the states $\rho' \circ \gamma'_A$ and $\rho'_i \circ \gamma'_A$, $i \in I_B$, on $\boldsymbol{A}$ correspond to the probability distributions $\pi'_{\boldsymbol{A}} = \{p_a\}_{a\in I_A}$ and $(\pi'_{\boldsymbol{A}})^i = \{\mathrm{Tr}(\rho'_i(X'_a)\}_{a\in I_A}$, whence (6.32) can be rewritten as

$$\begin{aligned}
I(A,B) = &- \sum_{i\in I_B} (\mathrm{Tr}(\rho\, B_i)) \log(\mathrm{Tr}(\rho\, B_i)) \\
&\qquad + \sum_{a\in I_A} \sum_{i\in I_B} p_a\, (\mathrm{Tr}(\rho_a\, B_i)) \log(\mathrm{Tr}(\rho_a\, B_i)) \\
= &- \sum_{i\in I_B} \Big( \sum_{a\in I_A} (\mathrm{Tr}(\rho\, B_i))\, \rho'_i(X'_a) \Big) \log p_a \\
&\qquad + \sum_{a\in I_A} (\mathrm{Tr}(\rho\, B_i)) \sum_{i\in I_B} \rho'_i(X'_j)\, \log \rho'_i(X'_j) \\
= &- \sum_{i\in I_B} p_a \log p_a + \sum_{i\in I_B} (\mathrm{Tr}(\rho\, B_i)) \sum_{a\in I_A} \rho'_i(X'_a)\, \log \rho'_i(X'_a) \\
= &\; S(\rho' \circ \gamma'_A) - \sum_{i\in I_B} (\mathrm{Tr}(\rho\, B_i))\, S(\rho'_i \circ \gamma'_A) \;.
\end{aligned} \tag{6.40}$$

Therefore, the maximal accessible information relative to the encoding $\{\rho\,,\, p_a \rho_a\}$ equals the entropy of the *CPU* map $\gamma'_A : \boldsymbol{A} \mapsto \pi_\rho(\mathbb{B}(\mathbb{H}))'$ relative to the state $\rho'$ on the commutant: $I(A) = H'_\rho(\gamma'_A)$. If $\rho$ is a faithful state, one can use (5.146) to substitute the $X'_j$ with elements of a *POVM* in $\mathbb{B}(\mathbb{H})$ and $\gamma'_A$ with a *CPU* map $\gamma_A : \boldsymbol{A} \mapsto \mathbb{B}(\mathbb{H})$, so that $I(A) = \mathrm{H}_\rho(\gamma_A)$.

The natural embedding $\imath_{\boldsymbol{M}}$ of a subalgebra $\boldsymbol{M} \subseteq \mathbb{B}(\mathbb{H})$ into $\mathbb{B}(\mathbb{H})$ is a *CPU* map (see Examples 5.2.3.7 and 8) such that $\rho\restriction\boldsymbol{M} = \rho \circ \imath_{\boldsymbol{M}}$. This observation suggests the following extension of Definition 6.3.2.

**Definition 6.3.3 (Entropy of *CPU* maps).** *Given a completely positive unital map* $\gamma : \boldsymbol{M} \mapsto \mathbb{B}(\mathbb{H})$, *where* $\boldsymbol{M}$ *is a finite-dimensional algebra, its entropy relative to a state* $\rho \in \mathbb{B}_1^+(\mathbb{H})$ *is*

$$\mathrm{H}_\rho(\gamma) := \sup_{\rho=\sum_{i\in I}\lambda_i\rho_i} \sum_{i\in I} \lambda_i\, S(\rho\circ\gamma\,,\,\rho_i\circ\gamma) \tag{6.41}$$

$$= S(\rho\circ\gamma) - \inf_{\rho=\sum_{i\in I}\lambda_i\rho_i} \sum_{i\in I} \lambda_i S(\rho_i\circ\gamma)\ . \tag{6.42}$$

**Lemma 6.3.1.**

1. *Given a CPU map* $\gamma : \boldsymbol{M} \mapsto \mathbb{B}(\mathbb{H})$ *from a finite dimensional algebra* $\boldsymbol{M}$ *into* $\mathbb{B}(\mathbb{H})$, *one has*

$$0 \le \mathrm{H}_\rho(\gamma) \le S(\rho\circ\gamma) \le \log\dim(\boldsymbol{M})\ , \tag{6.43}$$

*where* $\dim(\boldsymbol{M})$ *is the dimension of any maximally Abelian subalgebra contained in* $\boldsymbol{M}$.

2. *If* $\rho$ *is a faithful state, then* $\mathrm{H}_\rho(\boldsymbol{M}) > 0$ *unless* $\boldsymbol{M}$ *is the trivial algebra, consisting only of multiples of the identity.*
3. *Consider two finite dimensional algebras* $\boldsymbol{M}_{1,2}$ *and two CPU maps* $\gamma_1 : \boldsymbol{M}_1 \mapsto \boldsymbol{M}_2$, $\gamma_2 : \boldsymbol{M}_2 \mapsto \mathbb{B}(\mathbb{H})$,

$$\mathrm{H}_\rho(\gamma_2\circ\gamma_1) \le \mathrm{H}_\rho(\gamma_2)\ . \tag{6.44}$$

*In particular, if* $\boldsymbol{N} \subseteq \boldsymbol{M} \subseteq \mathbb{B}(\mathbb{H})$ *are two finite dimensional subalgebras,*

$$\mathrm{H}_\rho(\boldsymbol{N}) \le \mathrm{H}_\rho(\boldsymbol{M})\ . \tag{6.45}$$

**Proof:** Positivity and boundedness are evident, monotonicity under *CPU* maps follows from (6.28) applied to Definition 6.3.3, while monotonicity under algebraic embeddings follows from considering the *CPU* maps consisting of the natural inclusions $\imath_{\boldsymbol{M}}$ of $\boldsymbol{M}$ into $\mathbb{B}(\mathbb{H})$ and $\imath_{\boldsymbol{NM}}$ of $\boldsymbol{N}$ into $\boldsymbol{M}$:

$$\mathrm{H}_\rho(\boldsymbol{N}) = \mathrm{H}_\rho(\imath_{\boldsymbol{M}}\circ\imath_{\boldsymbol{NM}}) \le \mathrm{H}_\rho(\imath_{\boldsymbol{M}}) = \mathrm{H}_\rho(\boldsymbol{M})\ .$$

As regards the second property, suppose that $\mathrm{H}_\rho(\boldsymbol{M}) = 0$, then, the first property of the relative entropy in Proposition 6.3.1 yields $\rho\lceil\boldsymbol{M} = \rho_i\lceil\boldsymbol{M}$ for all decompositions $\rho = \sum_{i\in I}\lambda_i\rho_i$. Then, consider the *GNS* representation of $\mathbb{B}(\mathbb{H})$ based on $\rho$ and set $\rho(M) := \mathrm{Tr}(\rho M) = \langle\,\Omega_\rho\,|\,\pi_\rho(M)\,|\,\Omega_\rho\,\rangle$, for all $M \in \boldsymbol{M}$. It follows that

$$\langle\,\Omega_\rho\,|\,X_i'\,\pi_\rho(M)\,|\,\Omega_\rho\,\rangle = \rho(M)\,\langle\,\Omega_\rho\,|\,X'\,|\,\Omega_\rho\,\rangle \quad \text{equivalently}$$
$$\langle\,\Omega_\rho\,|\,X_i'\Big(\pi_\rho(M) - \rho(M)\,\mathbb{1}\Big)\,|\,\Omega_\rho\,\rangle = 0\ ,$$

for all $0 \leq X_i' \leq \mathbb{1}$ in the commutant $\pi_\rho(\mathbb{B}(\mathbb{H}))'$. Notice that any such $X'$ can be written as a sum of positive $\mathbb{1} \geq X_{1,2}' \in \pi_\rho(\mathbb{B}(\mathbb{H}))'$; then, since $\rho$ faithful on $\mathbb{B}(\mathbb{H})$ implies $|\,\Omega_\rho\,\rangle$ separating for $\pi_\rho(\mathbb{B}(\mathbb{H}))$ and thus cyclic for $\pi_\rho(\mathbb{B}(\mathbb{H}))'$ (see Lemma 5.3.1), it follows that $(\pi_\rho(M) - \rho(M))|\,\Omega_\rho\,\rangle$ is orthogonal to a dense subset of the *GNS* Hilbert space $\mathbb{H}_\rho$ whence, again from the faithfulness of $\rho$, $M = \rho(M)\,\mathbb{1}$ for all $M \in \boldsymbol{M}$. □

**Example 6.3.7.** The last result in Example 6.3.6 extends to subalgebras $\boldsymbol{M} \subseteq \mathbb{B}(\mathbb{H})$ which are not Abelian but commute with the state $\rho$ [7]; then

$$\mathrm{H}_\rho\,(\boldsymbol{M}) = S\,(\rho\restriction\boldsymbol{A}) \ , \tag{6.46}$$

where $\boldsymbol{A}$ is any maximally Abelian subalgebra contained in $\boldsymbol{M}$. Indeed, from Example 6.3.3.2 and the first case discussed in Example 6.3.6 it follows that $S\,(\rho\restriction\boldsymbol{M}) = S\,(\rho\restriction\boldsymbol{A}) = \mathrm{H}_\rho\,(\boldsymbol{A})$, where $\boldsymbol{A} \subseteq \boldsymbol{M}$ is maximally Abelian; on the other hand, from Lemma 6.3.1, one deduces that

$$S\,(\rho\restriction\boldsymbol{M}) = S\,(\rho\restriction\boldsymbol{A}) = \mathrm{H}_\rho\,(\boldsymbol{A}) \leq \mathrm{H}_\rho\,(\boldsymbol{M}) \leq S\,(\rho\restriction\boldsymbol{M}) \ .$$

Apart for the simple cases discussed in Examples 6.3.6 and 6.3.7, the minimization of the linear convex combination of von Neumann entropies in (6.42) is in general an extremely difficult task. At first sight, one might even suspect to be forced to consider more than discrete convex decompositions of the state $\rho$; luckily, the following result ensures that $\mathrm{H}_\rho\,(\boldsymbol{M})$ can be reached within $\varepsilon > 0$, by means of discrete decompositions [88, 222]. We shall denote by $H_\rho^\gamma\Big(\{\lambda_i, \rho_i\}\Big)$ the argument of the supremum in (6.41) evaluated at a given decomposition $\rho = \sum_{i\in I} \lambda_i \rho_i$, namely

$$H_\rho^\gamma\Big(\{\lambda_i, \rho_i\}_{i\in I}\Big) := \sum_{i\in I} \lambda_{i\in I}\, S\,(\rho\circ\gamma\,,\,\rho_i\circ\gamma) \ . \tag{6.47}$$

**Proposition 6.3.4.** *Let $\gamma : \boldsymbol{M} \mapsto \mathbb{B}(\mathbb{H})$ be a CPU map from a finite dimensional algebra $\boldsymbol{M}$ into $\mathbb{B}(\mathbb{H})$ and $\rho \in \mathbb{B}_1^+(\mathbb{H})$ a density matrix. Given a decomposition $\rho = \sum_{i\in I} \lambda_i \rho_i$ and $\varepsilon > 0$, there exists a decomposition $\rho = \sum_{j\in J} \lambda_j' \rho_j'$ where* card$(J)$ *depends on* $\dim(\boldsymbol{M})$ *and $\varepsilon$, such that*

$$\left| H_\rho^\gamma\Big(\{\lambda_i, \rho_i\}_{i\in I}\Big) - H_\rho^\gamma\Big(\{\lambda_j', \rho_j'\}_{j\in J}\Big)\right| \leq \varepsilon \ . \tag{6.48}$$

**Proof:** Consider a finite partition $\mathcal{Z} = \{Z_j\}_{j\in J}$ of the state-space $\mathcal{S}(\boldsymbol{M})$ of $\boldsymbol{M}$ into subsets $Z_j$ such that

---

[7] In such a case, one says that such $\boldsymbol{M}$ are contained in the centralizer of $\rho$.

$$\sigma_{1,2} \in Z_j \Longrightarrow \|\sigma_1 - \sigma_2\| \leq \delta \quad \forall Z_j \in \mathcal{Z} \ .$$

For instance, card($J$) can be chosen not larger than the least number of balls of radius $\delta$ that are necessary to cover $\mathcal{S}(\boldsymbol{M})$. Define

$$\rho'_j := \sum_{\substack{i \in I \\ \rho_i \circ \gamma \in Z_j}} \frac{\lambda_i}{\lambda'_j} \rho_i \ , \quad \lambda'_j := \sum_{\substack{i \in I \\ \rho_i \circ \gamma \in Z_j}} \lambda_i \ .$$

By construction, $\rho = \sum_{j \in J} \lambda'_j \rho'_j$ and

$$\begin{aligned} \left| H^\gamma_\rho\left(\{\lambda_i, \rho_i\}_{i\in I}\right) - H^\gamma_\rho\left(\{\lambda'_j, \rho'_j\}_{j\in J}\right) \right| &\leq \left| \sum_{i\in I} \lambda_i \, S\left(\rho_i \circ \gamma\right) - \sum_{j \in J} \lambda'_j \, S\left(\rho'_j\right) \right| \\ &\leq \sum_{j\in J} \sum_{\substack{i \in I \\ \rho_i \circ \gamma \in Z_j}} \lambda_i \left| S\left(\rho_i \circ \gamma\right) - S\left(\rho'_j\right) \right| \ . \end{aligned}$$

By choosing $\delta$ appropriately, the result follows from the Fannes inequality (see (5.157)). □

From this result, it follows that, for any $\varepsilon > 0$, there exists a decomposition $\rho = \sum_{i \in I} \lambda_i \, \rho_i$ with card($I$) depending on dim($\boldsymbol{M}$) and $\varepsilon$, such that

$$H^\gamma_\rho\left(\{\lambda_i, \rho_i\}_{i\in I}\right) \geq \mathrm{H}_\rho\left(\gamma\right) - \varepsilon \ . \tag{6.49}$$

We shall call $\varepsilon$-*optimal* for $\gamma$ the decompositions which achieve $\mathrm{H}_\rho(\gamma)$ within $\varepsilon > 0$ and *optimal for* $\gamma$ those decompositions $\rho = \sum_i \lambda_j \rho_i$ such that

$$\mathrm{H}_\rho\left(\gamma\right) = S\left(\rho \circ \gamma\right) - \sum_i \lambda_j S\left(\rho_i \circ \gamma\right) \ .$$

### 6.3.2 Entropy of a Subalgebra and Entanglement of Formation

In this section, we shall consider some techniques developed in [45, 46, 47] that are of help in calculating the entropy of a subalgebra $\mathrm{H}_\rho(\boldsymbol{A})$ where $\boldsymbol{A}$ is a maximally Abelian ($n$-dimensional) subalgebra of a full matrix algebra $M_n(\mathbb{C})$. The first step is to extract from (6.36) the expression

$$\mathrm{E}_\rho[\mathcal{M}, \boldsymbol{M}] := \inf_{\rho = \sum_{i\in I} \lambda_i \rho_i} \sum_{i \in I} \lambda_i S\left(\rho_i \restriction \boldsymbol{M}\right) \ , \tag{6.50}$$

where we have specified the state of the system, the total algebra of its observables $\mathcal{M}$ and the selected subalgebra $\boldsymbol{M} \subseteq \mathcal{M}$. Then, one notices that the variational problem can be solved by restricting to decompositions of $\rho$ in terms of pure states; this is so for the von Neumann entropy is concave (see (5.156)). In fact, assume $\rho = \sum_j \lambda_j \rho_j$ optimal for $\boldsymbol{M}$ (so that $\mathrm{E}_\rho[\mathcal{M},\boldsymbol{M}] = \sum_{i\in I}\lambda_i S(\rho_i \restriction \boldsymbol{M}))$, with non-pure decomposers $\rho_j$. Then, by further decomposing $\rho_j = \sum_k \lambda_{jk}\rho_{jk}$, one gets another decomposition $\rho = \sum_{j,k}\lambda_j\lambda_{jk}\rho_{jk}$; thence, $S(\rho_j \restriction \boldsymbol{M}) \geq \sum_k \lambda_{jk} S(\rho_{jk} \restriction \boldsymbol{M})$ yields

$$\mathrm{E}_\rho[\mathcal{M},\boldsymbol{M}] \leq \sum_{j,k}\lambda_j\lambda_{jk} S(\rho_{jk} \restriction \boldsymbol{M}) \leq \sum_j \lambda_j S(\rho_j \restriction \boldsymbol{M}) = \mathrm{E}_\rho[\mathcal{M},\boldsymbol{M}] .$$

Notice that, since pure states $P_j$ cannot be decomposed, for them it holds that

$$\mathrm{E}_{P_j}[\mathcal{M},\boldsymbol{M}] = S(P_j \restriction \boldsymbol{M}) . \tag{6.51}$$

In a similar way, one shows that the functional $\mathrm{E}_\rho[M_n(\mathbb{C}),\boldsymbol{M}]$ is convex over the state space $\mathbb{B}_1^+(\mathbb{C}^n)$: given a convex combination $\rho = \sum_j \nu_j\,\rho_j$, the optimal decompositions $\rho_j = \sum_k \lambda_{jk}\rho_{jk}$ that achieve

$$\mathrm{E}_{\rho_j}[M_n(\mathbb{C}),\boldsymbol{M}] = \sum_k \lambda_k S(\rho_{jk} \restriction \boldsymbol{M})$$

for each $j$, provide a decomposition $\rho = \sum_{j,k}\nu_j\lambda_{jk}\rho_{jk}$ which need not be optimal, whence

$$\mathrm{E}_{\sum_j \nu_j\rho_j}[M_n(\mathbb{C}),\boldsymbol{M}] \leq \sum_{j,k}\nu_j\lambda_{jk} S(\rho_{jk} \restriction \boldsymbol{M}) \leq \sum_j \nu_j\, S(\rho_j \restriction \boldsymbol{M}) . \tag{6.52}$$

We shall fix $\mathcal{M} = M_n(\mathbb{C})$ for some $n$; the following results turn out to be useful [47].

**Proposition 6.3.5.** *For a fixed density matrix $\rho \in M_n(\mathbb{C})$ and $\boldsymbol{M} \subseteq M_n(\mathbb{C})$,*

1. *there is an optimal decomposition consisting of no more than $n^2$ decomposers;*
2. *the functional $\mathrm{E}_\rho[M_n(\mathbb{C}),\boldsymbol{M}]$ is linear on the convex hull of the optimal decomposers of $\rho$; namely, if $\rho = \sum_i \lambda_i\,P_i$ is an optimal decomposition for $\mathrm{E}_\rho[M_n(\mathbb{C}),\boldsymbol{M}]$, where the $P_i$ are projections, then any other convex combination $\widetilde{\rho} = \sum_j \nu_j\,P_j$, with weights $\nu_j > 0$, $\sum_j \nu_j = 1$, is also optimal in the sense that,*

$$\mathrm{E}_{\widetilde{\rho}}[M_n(\mathbb{C}),\boldsymbol{M}] = \sum_j \nu_j\, S(P_j \restriction \boldsymbol{M}) .$$

**Proof:** The first statement results from a theorem of Caratheodory [20] since $M_n(\mathbb{C})$ is $n^2$ dimensional as a linear space and the set of pure states is compact [304] (see Remark 5.3.2.5).

The second statement is a consequence of (6.51) and (6.52); indeed, as a convex functional, $\mathrm{E}_\rho[M_n(\mathbb{C}),\boldsymbol{M}]$ can be expressed as

$$\mathrm{E}_\rho[M_n(\mathbb{C}),\boldsymbol{M}] = \sup\left\{\Lambda[\rho]\,:\,\Lambda \quad \text{affine functional on } \mathbb{B}_1^+(\mathbb{C}^n)\right\}\,.$$

Let $\mathrm{E}_\rho[M_n(\mathbb{C}),\boldsymbol{M}] = \Lambda[\rho]$, then $\Lambda[\sigma] \leq \mathrm{E}_\sigma[M_n(\mathbb{C}),\boldsymbol{M}]$ for a different state $\sigma$; thus, given an optimal decomposition $\rho = \sum_i \lambda_i\, P_i$, where $\lambda_i > 0$ and the $P_i$ are projections,

$$\begin{aligned}\mathrm{E}_\rho[M_n(\mathbb{C}),\boldsymbol{M}] &= \sum_i \lambda_i\, S\,(P_i\restriction\boldsymbol{M}) = \sum_i \lambda_i \mathrm{E}_{P_i}[M_n(\mathbb{C}),\boldsymbol{M}]\\ &\geq \sum_i \lambda_i \Lambda[P_i] = \Lambda[\sum_i \lambda_i\, P_i] = \mathrm{E}_\rho[M_n(\mathbb{C}),\boldsymbol{M}]\,.\end{aligned}$$

Therefore, $\Lambda[P_i] = \mathrm{E}_{P_i}[M_n(\mathbb{C}),\boldsymbol{M}]$ for all $i$; consequently, if $\widetilde{\rho} = \sum_j \nu_j\, P_j$ is any convex combination of these optimal projections, then

$$\begin{aligned}\mathrm{E}_{\widetilde{\rho}}[M_n(\mathbb{C}),\boldsymbol{M}] &\leq \sum_j \nu_j S\,(P_j\restriction\boldsymbol{M}) = \sum_j \nu_j \mathrm{E}_{P_j}[M_n(\mathbb{C}),\boldsymbol{M}] = \sum_j \nu_j \Lambda[P_j]\\ &= \Lambda[\widetilde{\rho}] \leq \mathrm{E}_{\widetilde{\rho}}[M_n(\mathbb{C}),\boldsymbol{M}]\,.\end{aligned}$$

□

Calculating $\mathrm{E}_\rho[M_n(\mathbb{C}),\boldsymbol{M}]$ can be simplified if the state $\rho$ enjoys symmetries that leave the subalgebra $\boldsymbol{M}$ invariant as a set; namely, suppose there exists a unitary matrix $U\,:\,\mathbb{C}^n \mapsto \mathbb{C}^n$ such that $\Gamma_u[\rho] = U\,\rho\,U^\dagger = \rho$ and $\Gamma_u^T[\boldsymbol{M}] = \boldsymbol{M}$, where $\Gamma_u^T : M_n(\mathbb{C}) \mapsto M_n(\mathbb{C})$ is the dual map of $\Gamma_u$. Then,

**Proposition 6.3.6.** *Let $\mathrm{E}_\rho[M_n(\mathbb{C}),\boldsymbol{M}]$ be achieved at the optimal decomposition $\rho = \sum_i \lambda_i\, P_i$; then, the symmetry map $\Gamma_u$ gives other optimal decompositions.*

**Proof:** From $\rho = \Gamma_u[\rho] = \sum_i \lambda_i\, \Gamma_u[P_i]$ and $\Gamma_u[P_i]\restriction\boldsymbol{M} = P_i\restriction\Gamma_u^T[\boldsymbol{M}] = P_i\restriction\boldsymbol{M}$ it follows that

$$\mathrm{E}_\rho[M_n(\mathbb{C}),\boldsymbol{M}] \leq \sum_i \lambda_i\, S\,(\Gamma_u[P_i]\restriction\boldsymbol{M}) = \sum_i \lambda_i\, S\,(P_i\restriction\boldsymbol{M}) = \mathrm{E}_\rho[M_n(\mathbb{C}),\boldsymbol{M}]\,.$$

□

Particularly suggestive instances of states $\rho \in \mathbb{B}_1^+(\mathbb{C}^d)$ with symmetries are those that are permutation invariant with respect to a given *ONB* $\{|\,i\,\rangle\}_{i=1}^d$; they are of the form

$$\rho_x^{(d)} = \frac{1}{d}\mathbb{1}_d + \frac{x}{d}\sum_{i\neq j=1}^{d} |\,i\,\rangle\langle\,j\,| = \frac{1-x}{d}\mathbb{1}_n + x|\,\psi_+\,\rangle\langle\,\psi_+\,|\,, \tag{6.53}$$

where $|\,\psi_+\,\rangle := \frac{1}{d}\sum_{i=1}^{d}|\,i\,\rangle$ so that $-\frac{1}{d-1} \leq x \leq 1$.

By defining $F := \langle \psi_+ \,|\, \rho_x \,|\psi_+ \rangle$, $0 \le F \le 1$, one can rewrite $\rho_x$ in a way which is directly comparable with the isotropic states (6.3):

$$\rho_F^{(d)} = \frac{1-F}{d-1} \mathbb{1}_d + \frac{dF-1}{d-1} |\, \psi_+ \rangle\langle \psi_+ \,| \,, \tag{6.54}$$

which we shall denote in the following by $\rho_F^{(d^2)}$ as they are obtained from (6.3) by changing $d$ into $d^2$; notice that

$$\langle \widehat{\Psi}_+^d | \rho_F^{(d^2)} | \widehat{\Psi}_+^d \rangle = \langle \psi_+ \,|\, \rho_F^{(d)} \,|\psi_+ \rangle = F \,. \tag{6.55}$$

Let $\pi$ denote the $d!$ permutations $i \mapsto \pi(i)$, $1 \le i \le d$; it turns out that

$$\rho_F^{(d)} = \frac{1}{d!} \sum_\pi \underbrace{U_\pi \,|\, \phi \rangle\langle \phi \,|\, U_\pi^{-1}}_{P_\phi^\pi} \,, \tag{6.56}$$

where $|\, \phi \rangle \in \mathbb{C}^d$ is any vector such that $|\langle \psi_+ \,|\, \phi \rangle|^2 = F$ and $U_\pi$ unitarily implements the permutation of the chosen *ONB* corresponding to $\pi$.

Let $\boldsymbol{A}$ denote the maximally Abelian subalgebra generated by the projections $\{|\, i \rangle\langle i \,|\}_{i=1}^d$; the decomposition (6.56) is such that

$$\begin{aligned} \mathrm{E}_{\rho_F^{(d)}}[M_d(\mathbb{C}), \boldsymbol{A}] &\le \frac{1}{d!} \sum_\pi S\left(P_\phi^\pi \restriction \boldsymbol{A}\right) = S\left(P_\phi^\pi \restriction \boldsymbol{A}\right) \\ &= -\sum_{j=1}^d |\langle \phi \,|\, j \rangle|^2 \log |\langle \phi \,|\, j \rangle|^2 =: r(F) \,. \end{aligned} \tag{6.57}$$

**Proposition 6.3.7.** *If $\rho_F^{(d)}$ is a permutation invariant state on $M_d(\mathbb{C})$ and $r(F)$ is a convex function of $F \in [0,1]$, the decomposition (6.56) achieves $\mathrm{E}_{\rho_F^{(d)}}[M_{d^2}(\mathbb{C}), \boldsymbol{A}]$.*

**Proof:** Let $\rho_F^{(d)} = \sum_i \lambda_i P_i$, $P_i = |\, \phi_i \rangle\langle \phi_i \,|$, achieve $\mathrm{E}_{\rho_F^{(d)}}[M_d(\mathbb{C}), \boldsymbol{A}]$ and consider

$$\rho_F^{(d)} = \frac{1}{d!} \sum_\pi U_\pi^\dagger \rho_F^{(d)} U_\pi = \sum_i \lambda_i \underbrace{\frac{1}{d!} \sum_\pi U_\pi^\dagger P_i U_\pi}_{P_i^u} \,.$$

The states $P_i^u$ are permutation invariant; according to (6.56) they are completely characterized by parameters $F_i$ that satisfy

$$F = \langle \psi_+ \,|\, \rho_F^{(d)} \,|\psi_+ \rangle = \sum_i \lambda_i \langle \psi_+ \,|\, P_i^u \,|\psi_+ \rangle = \sum_i \lambda_i F_i \,.$$

Thus, Proposition 6.3.6, the assumed convexity of $r(F)$ and (6.57) yield

$$\mathrm{E}_{\rho_F^{(d)}}[M_d(\mathbb{C}),\boldsymbol{A}] = \sum_i \lambda_i\, S\left(P_i^u \restriction \boldsymbol{A}\right) = \sum_i \lambda_i\, r(F_i) \geq r(F)$$
$$\geq \mathrm{E}_{\rho_F^{(d)}}[M_d(\mathbb{C}),\boldsymbol{A}] .$$

□

**Examples 6.3.8.**

1. For $d=2$, $\rho_F^{(2)} = \frac{1}{2}\begin{pmatrix} 1 & 2F-1 \\ 2F-1 & 1\end{pmatrix}$ can be written as

$$\rho_F^{(2)} = \frac{1}{2}\begin{pmatrix} \frac{1+a}{2} & \frac{\sqrt{1-a^2}}{2} \\ \frac{\sqrt{1-a^2}}{2} & \frac{1-a}{2}\end{pmatrix} + \frac{1}{2}\begin{pmatrix} \frac{1-a}{2} & \frac{\sqrt{1-a^2}}{2} \\ \frac{\sqrt{1-a^2}}{2} & \frac{1+a}{2}\end{pmatrix},$$

where $a := 2\sqrt{F(1-F)}$; then, with $\eta(x) = -x\log x$ and the notation of (6.12),

$$r(F) = \eta\left(\frac{1+a}{2}\right) + \eta\left(\frac{1-a}{2}\right) = H_2\left(\frac{1+2\sqrt{F(1-F)}}{2}\right) . \tag{6.58}$$

In order to use the previous proposition, we need show that $r(F)$ is convex on $[0,1]$; for this we calculate

$$\frac{\mathrm{d}^2 r(F)}{\mathrm{d}F^2} = \frac{2}{a^3}\left(\log\frac{1+a}{1-a} - 2a\right) .$$

The function within the parenthesis is monotonically increasing from 0 to $+\infty$; the second derivative is thus non-negative and the function $r(F)$ is convex. Then,

$$\mathrm{E}_{\rho_F^{(2)}}[M_2(\mathbb{C}),\boldsymbol{A}] = H_2\left(\frac{1+2\sqrt{F(1-F)}}{2}\right)$$
$$\mathrm{H}_{\rho^{(2)}}(\boldsymbol{A}) = \log 2 - H_2\left(\frac{1+2\sqrt{F(1-F)}}{2}\right),$$

where $\boldsymbol{A}$ is the Abelian subalgebra of diagonal $2\times 2$ matrices. Notice that this is the only Abelian subalgebra in the $d=2$ case: in [45] $\mathrm{H}_\rho(\boldsymbol{A})$ has been computed for all states $\rho \in M_2(\mathbb{C})$.

2. Given a fixed *ONB* $\{|\,i\,\rangle\}_{i=1}^d$ in $\mathbb{C}^d$, consider the *doubling map*

$$M_d(\mathbb{C}) \ni X = \sum_{i,j=1}^d x_{ij}\,|\,i\,\rangle\langle\,j\,| \mapsto \mathbb{D}[X] = \sum_{i,j=1}^d x_{ij}\,|\,ii\,\rangle\langle\,jj\,| . \tag{6.59}$$

It is a homomorphism from $M_d(\mathbb{C})$ onto a subalgebra $\mathcal{M}_0 \subset M_{d^2}(\mathbb{C})$,

$$\mathbb{D}[X]\,\mathbb{D}[Y] = \sum_{i,j\,;\,k,\ell=1}^{d} x_{ij}\, y_{k\ell} |\, ii\,\rangle\langle\, jj \,|\, kk\,\rangle|\, \ell\ell\,\rangle$$
$$= \sum_{i,j=1}^{d} \left( \sum_{k=1}^{d} x_{ik}\, y_{k\ell} \right) |\, ii\,\rangle\langle\, \ell\ell \,| = \mathbb{D}[XY] \;. \qquad (6.60)$$

It is thus a positive linear map from $M_d(\mathbb{C})$ onto $\mathcal{M}_0 \subset M_{d^2}(\mathbb{C})$ where it is invertible:

$$\mathcal{M}_0 \ni X_0 = \sum_{i,j=1}^{d} x_{i,j} |\, ii\,\rangle\langle\, jj \,| \mapsto \mathbb{D}^{-1}[X_0] = \sum_{i,j=1}^{d} x_{ij} |\, i\,\rangle\langle\, j \,| \in M_d(\mathbb{C}) \;. \qquad (6.61)$$

Let $d = 2$ and $|\,0\,\rangle, |\,1\,\rangle$ be the fixed *ONB* in $\mathbb{C}^2$; when applied to the permutation invariant state in the previous example, the doubling map gives the state

$$R_F^{(2)} := \mathbb{D}[\rho_F^{(2)}] = \frac{|\,00\,\rangle 00 + |\,11\,\rangle\langle\,11\,|}{2} + \frac{2F-1}{2}\Big(|\,00\,\rangle\langle\,11\,| + |\,11\,\rangle\langle\,00\,|\Big)$$
$$= \frac{1}{4}\Big(1\!\!1_4 + (2F-1)(\sigma_1 \otimes \sigma_1 - \sigma_2 \otimes \sigma_2) + \sigma_3 \otimes \sigma_3\Big)$$
$$= \frac{1}{2} \begin{pmatrix} 1 & 0 & 0 & 2F-1 \\ 0 & 0 & 0 & 0 \\ 0 & 0 & 0 & 0 \\ 2F-1 & 0 & 0 & 1 \end{pmatrix} .$$

If thus turns out that $\widetilde{R}_F^{(2)}$ as defined in (6.13) equals $R_F^{(2)}$ so that, for $F \neq 1/2$, $R_F^{(2)}$ is entangled with concurrence $\mathcal{E}(R_F^{(2)}) = |1 - 2F|$ (see (6.14)) and entanglement of formation (see (6.15) and Theorem 6.2.1) given by

$$\mathrm{E}_{\mathrm{F}}[R_F^{(2)}] = H_2 \left( \frac{1 + 2\sqrt{F(1-F)}}{2} \right) - \mathrm{E}_{\rho_F^{(2)}}[M_d(\mathbb{C}), \boldsymbol{A}] \;.$$

3. In [46], Proposition 6.3.7 has been used to compute $\mathrm{E}_{\rho_F^{(3)}}[M_3(\mathbb{C}), \boldsymbol{A}]$ where

$$\rho_F^{(3)} = \frac{1}{3} \begin{pmatrix} 1 & \frac{3F-1}{2} & \frac{3F-1}{2} \\ \frac{3F-1}{2} & 1 & \frac{3F-1}{2} \\ \frac{3F-1}{2} & \frac{3F-1}{2} & 1 \end{pmatrix} ,$$

and $\boldsymbol{A}$ consists of diagonal matrices in this representation. While for $d = 2$ there is only one optimal decomposition achieving $\mathrm{E}_{\rho_F^{(2)}}[M_2(\mathbb{C}), \boldsymbol{A}]$, when $d = 3$ more optimal decompositions appear. Indeed, one can decompose $\rho_F^{(3)}$ by means of the unitary operator $U : \mathbb{C}^2 \mapsto \mathbb{C}^3$ that implements the permutation $(1, 2, 3) \mapsto (3, 1, 2)$:

$$\rho_F^{(3)} = \frac{1}{3} |\,\phi\,\rangle\langle\,\phi\,| + \frac{1}{3} U |\,\phi\,\rangle\langle\,\phi\,| U^{-1} + \frac{1}{3} U^2 |\,\phi\,\rangle\langle\,\phi\,| U^{-2} \,, \qquad (6.62)$$

where

$$| \phi \rangle = \begin{pmatrix} a + 2b\cos\theta \\ a - 2b\cos(\theta - \pi/3) \\ a - 2b\cos(\theta + \pi/3) \end{pmatrix} , \; a := \sqrt{3F} \; , \; b := \sqrt{\frac{3}{2}(1 - F)} \; .$$

It turns out that, for $0 \leq F \leq 8/9$, the function $r(F) = S\left(| \phi \rangle\langle \phi | \restriction \boldsymbol{A}\right)$ is convex, whence

$$\mathrm{E}_{\rho_F^{(3)}}[M_3(\mathbb{C}), \boldsymbol{A}] = \eta\left(\frac{2 - F + 2\sqrt{2F(1-F)}}{3}\right) + 2\,\eta\left(\frac{1 + F - 2\sqrt{2F(1-F)}}{6}\right) \; . \tag{6.63}$$

There exists a value $0 < F^* \leq 8/9$ such that $\mathrm{E}_{\rho_F^{(3)}}[M_3(\mathbb{C}), \boldsymbol{A}]$ is achieved at a unique decompositions of the form (6.62) given by

$$| \phi \rangle = \frac{1}{\sqrt{3}} \begin{pmatrix} \sqrt{F} + \sqrt{2F(1-F)} \\ \sqrt{F} - \sqrt{F(1-F)/2} \\ \sqrt{F} - \sqrt{F(1-F)/2} \end{pmatrix} ,$$

for $F^* \leq F \leq 8/9$; while, for $0 < F < F^*$, two optimal decompositions of the form (6.62) appear with

$$| \phi_F^{\pm} \rangle = \frac{1}{\sqrt{3}} \begin{pmatrix} a + 2b\cos\theta_f \\ a - 2b\cos(\pi/3 \mp \theta_F) \\ a - 2b\cos(\pi/3 \pm \theta_F) \end{pmatrix} ,$$

where the angle $\theta_F$ varies with $F$. According to Proposition 6.3.5, all linear convex combinations of the projections onto these vectors also provide optimal decompositions. When $8/9 \leq F \leq 1$, the function $r(F) = S\left(| \phi \rangle\langle \phi | \restriction \boldsymbol{A}\right)$ is no longer convex and one cannot use Proposition 6.3.7; in this case it is the close relation of $\mathrm{H}_\rho\left(\boldsymbol{M}\right)$ with the entanglement of formation which is of help. Indeed, (6.63) coincides with the entanglement of formation of the $d = 3$ isotropic states (6.3) for $1/3 \leq F \leq 8/9$ as calculated in [298].

In order to expose the relation between the entanglement of formation (6.5) and $\mathrm{E}_\rho[\mathcal{M}, \boldsymbol{M}]$ in (6.50), set $\mathcal{M} = M_{d^2}(\mathbb{C}) := M_d(\mathbb{C}) \otimes M_d(\mathbb{C})$, $\boldsymbol{M} = M_d(\mathbb{C})$ embedded as $M_d(\mathbb{C}) \otimes 1\!\!1_d$ into $M_{d^2}(\mathbb{C})$ and $\rho_j = | \psi_j \rangle\langle \psi_j |$. Since the marginal density matrices $\rho^{(1)}_{\psi_j} = \rho_j \restriction \boldsymbol{M}$, it turns out that

$$\mathrm{E_F}[\rho] = \mathrm{E}_\rho[M_{d^2}(\mathbb{C}), M_d(\mathbb{C})] \; . \tag{6.64}$$

Further insights into the connections between these two notions, with particular reference to Examples (6.3).2,3, come from [47]

**Proposition 6.3.8.** *Let $\boldsymbol{A} \subset M_d(\mathbb{C})$ be the maximally Abelian subalgebra corresponding to a fixed ONB $\{|\,i\,\rangle\}_{i=1}^d$ and $\mathbb{D}[\cdot]$ is the doubling map (6.59); then,*

$$\mathrm{E}_\rho[M_d(\mathbb{C}), \boldsymbol{A}] = \mathrm{E}_{\mathbb{D}[\rho]}[M_{d^2}(\mathbb{C}), M_d(\mathbb{C})] \ . \tag{6.65}$$

**Proof:** Suppose $\rho = \sum_i \lambda_i P_i$ achieves $\mathrm{E}_\rho[M_d(\mathbb{C}), \boldsymbol{A}]$; then, with $\rho = \sum_{i,j=1}^d r_{ij} |\,i\,\rangle\langle\,j\,|$,

$$\begin{aligned}\mathbb{D}[\rho] \upharpoonright M_d(\mathbb{C}) = \mathrm{Tr}_2(\mathbb{D}[\rho]) = \sum_{i=1}^d r_{ii} |\,i\,\rangle\langle\,i\,| = \rho \upharpoonright \boldsymbol{A} \quad \text{implies}& \\ \mathrm{E}_{\mathbb{D}[\rho]}[M_{d^2}(\mathbb{C}), M_d(\mathbb{C})] \le \sum_i \lambda_i \, S\left(\mathbb{D}[P_i] \upharpoonright M_d(\mathbb{C})\right) = \sum_i \lambda_i S\left(P_i \upharpoonright \boldsymbol{A}\right)& \\ = \mathrm{E}_\rho[M_d(\mathbb{C}), \boldsymbol{A}] \ .&\end{aligned}$$

Vice versa, let $\mathbb{D}[\rho] = \sum_j \nu_j \, Q_j$ achieve $\mathrm{E}_{\mathbb{D}[\rho]}[M_{d^2}(\mathbb{C}), M_d(\mathbb{C})]$; if the optimal decomposers $Q_j$ were of the form $Q_j = \sum_{k,\ell} q^j_{k\ell} |\,kk\,\rangle\langle\,\ell\ell\,|$, by the inverse doubling map (6.61) one would get a decomposition of $\rho$ that could be used to reverse the previous inequality and thus prove the result. The decomposers $Q_j$ are indeed of the claimed form as they are one-dimensional projections that can always be recast as follows

$$Q_j = \frac{\sqrt{\mathbb{D}[\rho]}\,|\,\Psi_j\,\rangle\langle\,\Psi_j\,|\,\sqrt{\mathbb{D}[\rho]}}{\langle\,\Psi\,|\,\mathbb{D}[\rho]\,|\Psi\,\rangle} \ .$$

Because of (6.60), it turns out that $\mathbb{D}[\rho]^n = \mathbb{D}[\rho^n]$ whence, by power series expansion, $\sqrt{\mathbb{D}[\rho]} = \mathbb{D}[\sqrt{\rho}]$. □

**Example 6.3.9.** In [298], the entanglement of formation of an isotropic state $\rho_F^{(d^2)}$ was computed by 1) considering the twirling (2) of suitable vectors of diagonal form, $|\,\Phi\,\rangle = \sum_{i=1}^d \sqrt{\mu_i} |\,ii\,\rangle$, with respect to the chosen *ONB*, and by 2) minimizing the von Neumann entropy $S\left(\rho_\Phi^{(1)}\right) = -\sum_{i=1}^d \mu_i \log \mu_i$ of the marginal density matrix.

Choose one such $|\,\Phi\,\rangle$ from an optimal decomposition for $\mathrm{E_F}[\rho_F^{(d^2)}]$ and construct the density matrix

$$R_F^{(d^2)} := \frac{1}{d!} \sum_\pi U_\pi \otimes U_\pi \,|\,\Phi\,\rangle\langle\,\Phi\,|\, U_\pi^{-1} \otimes U_\pi^{-1}$$

by using the permutation operators $U_\pi$. Since, by definition, $U_\pi \otimes U_\pi$ are symmetries for the isotropic state $\rho_F^{(d^2)}$, using Proposition 6.3.6 one deduces that $\mathrm{E}_{R_F^{(d^2)}}[M_{d^2}(\mathbb{C}), M_d(\mathbb{C})] = S\left(\rho_\Phi^{(1)}\right)$. On the other hand, in terms of the doubling map (6.59) and using (6.55),

$$R_F^{(d^2)} = \mathbb{D}[\rho_F^{(d)}] = \frac{1}{d!}\sum_{\pi} U_\pi \,|\,\phi\,\rangle\langle\,\phi\,|\, U_\pi^{-1} ,$$

where $\rho_F^{(d)}$ is as in (6.56) and $|\,\phi\,\rangle = \sum_{i=1}^{d} \sqrt{\mu_i}\,|\,i\,\rangle$. Finally, from Proposition 6.3.8, it results

$$\mathrm{E}_{\rho_F^{(d)}}[M_d(\mathbb{C}), \boldsymbol{A}] = \mathrm{E}_{R_F^{(d^2)}}[M_{d^2}(\mathbb{C}), M_d(\mathbb{C})] = S\left(\rho_\Phi^{(1)}\right) .$$

In this way one can use the results of [298] to extend the computation of $\mathrm{E}_{\rho^{(3)}}[M_3(\mathbb{C}), \boldsymbol{A}]$ to those values of $F \in [8/9, 1]$, where the methods employed in Example 6.3.8.3 are useless.

### Trace-distance and Fidelities

In this section we review some mathematical techniques that are used to compare two quantum states of a system $S$; the importance of such an issue will become apparent in the next chapter when we shall deal with the compression and retrieval of strings of *qubits* . We shall assume $S$ to be an $N$-level system.

**Definition 6.3.4.** *Given $\rho_{1,2} \in \mathcal{S}(S)$, their* trace-distance *is given by*

$$D(\rho_1, \rho_2) := \frac{1}{2}\,\mathrm{Tr}|\rho_1 - \rho_2| \ . \tag{6.66}$$

Namely, the trace distance of two density matrices is defined as half the trace-norm $\|\rho_1 - \rho_2\|_{tr}$ of their difference: $D(\rho_1, \rho_2)$ is a proper distance on the state-space $\mathcal{S}(S)$.

**Proposition 6.3.9.** *The trace distance enjoys the following properties:*

1. *Let $P \in M_N(\mathbb{C})$ be any orthogonal projector, then*

$$D(\rho_1, \rho_2) = \max_P \mathrm{Tr}(P(\rho_1 - \rho_2)) \ . \tag{6.67}$$

2. *The trace-distance monotonically decreases under completely positive trace-preserving maps $\mathbb{F} : \mathcal{S}(S) \mapsto \mathcal{S}(S)$:*

$$D(\mathbb{F}[\rho_1], \mathbb{F}[\rho_2]) \ \leq \ D(\rho_1, \rho_2) \ . \tag{6.68}$$

3. *The trace-distance is jointly convex:*

$$D(\sum_j \lambda_j \rho_j, \sum_j \lambda_j \sigma_j) \leq \sum_j \lambda_j D(\rho_j, \sigma_j) \ , \tag{6.69}$$

*where $\lambda_j \geq 0$ and $\sum_j \lambda_j = 1$.*

**Proof:** As seen in Example 5.5.6, $\rho_1 - \rho_2 = A - B$ and $|\rho_1 - \rho_2| = A + B$, with $A, B$ positive orthogonal matrices $A, B \geq 0$, $AB = 0$, so that $\mathrm{Tr}A = \mathrm{Tr}B$ since $\mathrm{Tr}\rho_{1,2} = 1$. Thus $D(\rho_1, \rho_2) = \mathrm{Tr}A = \mathrm{Tr}B$. Let $P$ be any projector, then

$$\mathrm{Tr}(P(A - B)) \leq \mathrm{Tr}(PA) \leq \mathrm{Tr}A = D(\rho_1, \rho_2) \ ,$$

for $\mathrm{Tr}(PB)$ is a positive quantity and $P$ projects onto a subspace. Further, if this subspace supports $A$, it annihilates $B$ and the maximum is achieved.

The second property is proved as follows: let $P$ be the projector which achieves the trace distance $D(\mathbb{F}[\rho_1], \mathbb{F}[\rho_2])$, then, because of the assumed trace-preserving character of $\mathbb{F}$,

$$\begin{aligned} D(\rho, \sigma) &= \mathrm{Tr}A = \mathrm{Tr}\mathbb{F}[A] \geq \mathrm{Tr}(P\mathbb{F}[A]) \geq \mathrm{Tr}(P\mathbb{F}[A]) - \mathrm{Tr}(P\mathbb{F}[B]) \\ &= \mathrm{Tr}(P\mathbb{F}[A - B]) = \mathrm{Tr}(P(\mathbb{F}[\rho_1] - \mathbb{F}[\rho_2])) = D(\mathbb{F}[\rho], \mathbb{F}[\rho_2]) \ . \end{aligned}$$

□

In order to introduce some useful notions of fidelity, let us begin with a simple observation: the closer two vector states $\psi_{1,2} \in \mathbb{H} = \mathbb{C}^N$ to each other, the closer to 1 is $|\langle \psi_1 \,|\, \psi_2 \rangle|$. Indeed, the latter quantity is 1 iff $\psi = \phi$ (a part for an overall multiplicative phase) and vanishes when $\psi \perp \phi$. This idea extends to density matrices of an $N$ level system as follows.

**Definition 6.3.5 (Fidelity).** *The fidelity of two density matrices $\rho_{1,2} \in \mathcal{S}(S)$ is*

$$\begin{aligned} F(\rho_1, \rho_2) &:= \mathrm{Tr}\sqrt{\sqrt{\rho_1}\rho_2\sqrt{\rho_1}} = \mathrm{Tr}\sqrt{(\sqrt{\rho_2}\sqrt{\rho_1})^\dagger(\sqrt{\rho_2}\sqrt{\rho_1})} \\ &= \mathrm{Tr}\left|\sqrt{\rho_2}\sqrt{\rho_1}\right| \ . \end{aligned} \tag{6.70}$$

If $\rho_1 = |\,\psi_1\,\rangle\langle\,\psi_1\,| =: P_1$ then $\sqrt{P_1} = P_1$ so that

$$F(P_1, \rho_2) = \sqrt{\langle\,\psi_1\,|\,\rho_2\,|\psi_1\,\rangle} \ . \tag{6.71}$$

Thus if $\rho_2 = |\,\psi_2\,\rangle\langle\,\psi_2\,| =: P_2$, then $F(P_1, P_2) = |\langle\,\psi_1\,|\,\psi_2\,\rangle|$.

**Proposition 6.3.10.** *The fidelity enjoys the following properties:*

1. *Let $|\,\Psi_\rho^{UV}\,\rangle$ be a purification of $\rho \in \mathcal{S}(S)$ of the form*

$$|\,\Psi_\rho^{UV}\,\rangle = \sum_{i=1}^N \sqrt{\rho}\, U|\,i\,\rangle \otimes V|\,i\,\rangle \ , \tag{6.72}$$

*where $\{|\,i\,\rangle\}_{i=1}^N$ is an orthonormal basis in $\mathbb{H} = \mathbb{C}^N$ and $U$ any partial isometry such that $U^\dagger U$ projects onto the orthogonal complement of $Ker(\rho)$ and $V$ is any unitary matrix. Then,*

$$F(\rho_1, \rho_2) = \max_{U_2\,,\,V_2} \left|\langle\,\Psi_{\rho_2}^{U_2V_2}\,|\,\Psi_{\rho_1}^{U_1V_1}\,\rangle\right| \ , \tag{6.73}$$

*that is the fidelity is the largest such scalar product achievable by fixing one purification and varying the other.*

2. *The fidelity does not depend on the order of its arguments. Further,* $F(\rho_1, \rho_2) = 1$ *if and only if* $\rho_1 = \rho_2$, *otherwise* $0 \leq F(\rho_1, \rho_2) < 1$.
3. *Let* $\mathcal{E} = \{E_i\}_{i\in I}$ *denote any POVM with elements in* $M_N(\mathbb{C})$, *then*

$$F(\rho_1, \rho_2) = \max\Big\{\sum_{i\in I} \mathrm{Tr}(\rho_1 E_i)\,(\mathrm{Tr}(\rho_2\, E_i) \ :\ E_i \in \mathcal{E}\Big\} . \tag{6.74}$$

4. *The fidelity is jointly concave, namely if* $\rho_{1,2} = \sum_i \lambda_i \sigma^i_{1,2}$ *with* $0 < \lambda_i < 1$, $\sum_i \lambda_i = 1$ *and* $\sigma_i^{1,2} \in \mathcal{S}(S)$,

$$F(\rho_1, \rho_2) \geq \sum_i \lambda_i\, F(\sigma_i^1, \sigma_i^2) . \tag{6.75}$$

5. *The fidelity monotonically increases under the action of trace-preserving completely positive maps* $\mathbb{F} : \mathcal{S}(S) \mapsto \mathcal{S}(S)$:

$$F\Big(\mathbb{F}[\rho_1], \mathbb{F}[\rho_2]\Big) \geq F(\rho_1, \rho_2) . \tag{6.76}$$

**Proof:** It is easy to check that $\mathrm{Tr}_1 |\Psi_\rho^{UV}\rangle\langle\Psi_\rho^{UV}| = \rho$, so that (6.72) is a purification of the mixed state $\rho$. One computes,

$$\begin{aligned} |\langle \Psi_{\rho_2}^{U_2V_2} \,|\, \Psi_{\rho_1}^{U_1V_2}\rangle| &= \sum_{i,j=1}^N \langle i \,|\, U_2^\dagger \sqrt{\rho_2}\sqrt{\rho_1} U_1 \,|j\rangle \langle i \,|\, V_2^\dagger V_1 \,|j\rangle \\ &= \mathrm{Tr}\Big(U_2^\dagger \sqrt{\rho_2}\sqrt{\rho_1} U_1 (V_1 V_2^\dagger)^T\Big) \leq \|\sqrt{\rho_2}\sqrt{\rho_1}\|_{tr} = F(\rho_1, \rho_2) , \end{aligned}$$

where $T$ means transposition. Further, the upper bound is achieved by choosing $V_2 = V_1$ and $U_2 = W^\dagger U_1$ with $W$ such that $\sqrt{\rho_2}\sqrt{\rho_1} = W|\sqrt{\rho_2}\sqrt{\rho_1}|$.

From the previous point, the second point follows at once. □

One expects a relation between trace-distance and fidelity of the kind: the smaller the trace-distance, the closer to 1 the fidelity. That this is indeed so is the content of the following

**Proposition 6.3.11.** *Given* $\rho_{1,2} \in \mathcal{S}(S)$, *the following bounds hold*

$$1 - F(\rho_1, \rho_2) \leq D(\rho_1, \rho_2) \leq \sqrt{1 - F^2(\rho_1, \rho_2)} . \tag{6.77}$$

The following proposition establishes that if $\rho_1 \in \mathcal{S}(S)$ is close to $\rho_2$ in the sense that $F(\rho_1, \rho_2) \simeq 1$ while while $F(\rho_2, \rho_3) \simeq 0$, then also $F(\rho_1, \rho_3) \simeq 0$.

**Proposition 6.3.12.** *[19] Let* $\rho_{1,2,3} \in \mathcal{S}(S)$, *then* $F_{ij} := F^2(\rho_i, \rho_j)$ *satisfy*

$$F_{13} \leq F_{23} + 2\,(1 - F_{12}) + 2\sqrt{(1 - F_{12})F_{23}} , \tag{6.78}$$

*where* $\mathrm{Tr}\rho_3 = 1$, *but* $\mathrm{Tr}\rho_{1,2} < 1$ *(subnormalization).*

**Proof:** Notice that subnormalization does not alter either the definition of fidelity or the first property in Proposition (6.3.10). Let then $|\Psi_1\rangle$ be a fixed purification of $\rho_1$, choose $|\Psi_{2,3}\rangle$ in order to achieve $F_{12}$ and $F_{13}$. Further, adjust the phases of the three vectors so that

$$F_{12} = \langle\Psi_1|\Psi_2\rangle^2 \ , \ F_{13} = \langle\Psi_1|\Psi_3\rangle^2 \ , \ F_{23} \geq \langle\Psi_2|\Psi_3\rangle^2 \ .$$

Setting $|\Psi\rangle := |\Psi_2\rangle - |\Psi_1\rangle$, one estimates

$$\langle\Psi|\Psi\rangle = \langle\Psi_1|\Psi_1\rangle + \langle\Psi_2|\Psi_2\rangle - 2\langle\Psi_1|\Psi_2\rangle \leq 2\left(1-\sqrt{F_{12}}\right) ,$$

for subnormalization gives $\langle\Psi_{1,2}|\Psi_{1,2}\rangle < 1$. Then, from $\langle\Psi_3|\Psi_3\rangle = \mathrm{Tr}\rho_3 = 1$ and the bound

$$\begin{aligned}\sqrt{F_{13}} &= \langle\Psi_1|\Psi_3\rangle = \langle\Psi_2|\Psi_3\rangle + \langle\Psi|\Psi_3\rangle \leq F_{23} + |\langle\Psi|\Psi_3\rangle| \\ &\leq F_{23} + \sqrt{\langle\Psi|\Psi\rangle} \leq F_{23} + \sqrt{2(1-\sqrt{F_{12}})} \ ,\end{aligned}$$

the result follows. □

Let $\rho \in \mathcal{S}(S)$ correspond to a mixture $\{\lambda_j, \rho_j\}$, $\rho = \sum_j \lambda_j \rho_j$, subjected to the action of a trace-preserving completely positive map $\mathbb{F} : \mathcal{S}(S) \mapsto \mathcal{S}(S)$. Then would like to keep track of how much $\mathbb{F}[\rho]$ differs form $\rho$ in the mean: this is well described by

**Definition 6.3.6 (Ensemble Fidelity).** *The ensemble fidelity relative to a mixture $\{\lambda_j, \rho_j\}$ and a completely positive action $\mathbb{F}$ is defined as the ensemble average of square fidelities,*

$$F_{av}\Big(\{\lambda_j, \rho_j\}, \mathbb{F}\Big) := \sum_j \lambda_j \, F^2(\rho_j, \mathbb{F}[\rho_j]) \ . \tag{6.79}$$

*We shall also denote by*

$$F_s(\rho, \mathbb{F}) := \sup\Big\{F_{av}(\{\lambda_j, P_j\}, \mathbb{F}) \ : \ P_j^2 = P_j = P_j^\dagger\Big\} \ , \tag{6.80}$$

*the supremum of the ensemble fidelities over all possible decompositions of $\rho$ as a mixture of pure states.*

**Example 6.3.10.** [19] Let $|i\rangle \in \mathbb{H} = \mathbb{C}^3$, $i = 1, 2, 3$, be an othonormal basis and consider the mixture represented by

$$\begin{aligned}&\rho = p_1|\psi_1\rangle\langle\psi_1| + p_2|\psi_1\rangle\langle\psi_1| + p_3|\psi_3\rangle\langle\psi_3| \ , \quad \text{where} \\ &|\psi_1\rangle := \cos\alpha|1\rangle + \sin\alpha|2\rangle \ , \ |\psi_2\rangle := \sin\alpha|1\rangle + \cos\alpha|2\rangle\end{aligned}$$

and $|\psi_3\rangle$ is such that

$$\langle \psi_3 | \psi_1 \rangle = \langle \psi_3 | \psi_2 \rangle = \langle \psi_1 | \psi_2 \rangle = \sin 2\alpha \ .$$

Suppose the system is subjected to a trace-preserving completely positive map such that

$$\begin{aligned} &\mathbb{F}[| \psi_1 \rangle\langle \psi_1 |] = | 1 \rangle\langle 1 | \ , \ \mathbb{F}[| \psi_2 \rangle\langle \psi_2 |] = | 2 \rangle\langle 2 | \\ &\mathbb{F}[| \psi_3 \rangle\langle \psi_3 |] = \frac{1}{2} | \psi_1 \rangle\langle \psi_1 | + \frac{1}{2} | \psi_2 \rangle\langle \psi_2 | \ . \end{aligned} \tag{6.81}$$

The purification of a state $\rho \in \mathcal{S}(S)$ actually couples $S$ to an ancilla and this coupling is embodied by an entangled pure state $| \Psi_\rho \rangle$. The following fidelity reflects how much the action of an operation on $S$ described by a completely positive trace-preserving map $\mathbb{F} : \mathcal{S}(S) \mapsto \mathcal{S}(S)$ preserves this entanglement.

**Definition 6.3.7 (Entanglement Fidelity).** *The entanglement fidelity of $\rho$ relative to $\mathbb{F}$ is defined by the square fidelity of the states $| \Psi_\rho \rangle\langle \Psi_\rho |$ and $\mathbb{F} \otimes \mathrm{id}[| \Psi_\rho \rangle\langle \Psi_\rho |]$, where $| \Psi_\rho \rangle$ is any purification of $\rho$:*

$$F_{ent}(\rho, \mathbb{F}) := F^2 \Big( | \Psi_\rho \rangle\langle \Psi_\rho | , \mathbb{F} \otimes \mathrm{id}[| \Psi_\rho \rangle\langle \Psi_\rho |] \Big) \ . \tag{6.82}$$

Relations between these various fidelities are as follows [224].

**Proposition 6.3.13.**

$$0 \leq F_{ent}(\rho, \mathbb{F}) \leq F_{av} \leq F(\rho, \mathbb{F}[\rho]) \leq 1 \ , \tag{6.83}$$

*where $F_{av}$ is any ensemble fidelity corresponding to a decomposition of $\rho$.*

## Bibliographical Notes

The literature on quantum information, communication and computation is vast and still growing. There are nowadays various introductions to quantum information and related topics: the most extended ones can be found in the books [48, 49, 224] and the lecture notes [242].

The book [165] offers a detailed introduction to quantum computation, while more mathematically oriented presentations are to be found in [145, 239].

As far as entanglement theory is concerned the review [152] provides an exhaustive and up-to-date overview of the subject in its manifold aspects together with a complete list of references to the relevant literature.

The book [71] collects a series of reviews from various experts in the field where one can find relevant information about many of the topics discussed or just touched upon in this chapter; among others, Gaussian states, entanglement measures, bound entanglement, continuous variable systems, quantum algorithms, quantum cryptography and one-way quantum computation. Furthermore, the second half of the book provides an overview of the state-of-the-art concerning experimental and technological implementations.

A thorough collection of recent results of quantum information and entanglement theory concerning continuous variable atomic and optical systems can be found in [76, 2].

As regards the role and use of Gaussian states in quantum information theory see also [115].

Rapid introductions to the basic tools of quantum information are provided in [100, 188]

# 7 Quantum Mechanics of Infinite Degrees of Freedom

Quantum systems with infinite degrees of freedom exhibit properties, like relaxation to equilibrium, phase-transitions and the existence of inequivalent representations of the *CAR* and *CCR*, that can satisfactorily be dealt with by means of the methods and techniques of algebraic quantum statistical mechanics [108, 64, 65]. The point of departure from standard quantum mechanics, is that in an infinite dimensional context one is usually provided with the algebraic properties of the relevant observables, but, in general, not with an a priori given representation on a Hilbert space; the latter rather depends on to the physical properties of the systems under consideration [274, 290, 291].

## Relaxation to Equilibrium

As discussed in Remark 2.1.3.4, by discretizing chaotic classical systems, properties like the exponential growth of errors or a constant entropy production can survive only over times that scale logarithmically with respect to the discretization parameter. Indeed, beyond this time-scale, due to the finite number of allowed states, quasi-periodicity and recursion appear. While in classical dynamical systems, recursion can be eliminated by going to the continuum, this is impossible in quantum mechanics because of the intrinsic discretization of phase-space, due to $\hbar > 0$ and to the Heisenberg uncertainty relations. However, recursion times can be made longer and longer by letting the number of degrees of freedom go to infinity [212, 300].

If we let $N \to \infty$ in Example 5.6.1.2, the recurrence time diverges and, unlike for finitely many spins, infinite spin chains may exhibit relaxation to equilibrium. Indeed, observe that

$$\begin{aligned} f_N(0,t) := \prod_{i=1}^{(N-1)/2} \cos\frac{t}{2^i} &= \prod_{i=1}^{(N-1)/2} \cos\frac{2t}{2^{i+1}} \\ &= \prod_{i=2}^{(N+1)/2} \cos\frac{2t}{2^i} = \frac{\cos 2^{-(N+1)/2} t}{\cos t}\, f_N(0,2t) \ ; \end{aligned}$$

then, when $N \to \infty$, $f_N(0,t)$ tends to a function $f_\infty(t)$ which satisfies $f_\infty(t)\cos t = f_\infty(2t)$ together with $f_\infty(0) = 1$. By expanding both members of the first equality and comparing equal powers in $t$, it turns out that

F. Benatti, *Dynamics, Information and Complexity in Quantum Systems*, Theoretical and Mathematical Physics, DOI 10.1007/978-1-4020-9306-7_7,

$f_\infty(t) = \frac{\sin t}{t}$; (5.173) thus becomes

$$\rho^{\infty\otimes}(\sigma_+^j(t)) = \mathrm{e}^{2itB\mu_j}\,\frac{sint}{2t} \ .$$

As a consequence, when $t \to \infty$, the time-dependent state $\rho_t^\infty$ defined on the infinite spin array by the expectations

$$\sigma_\#^j \mapsto \rho_t^{\otimes\infty}(\sigma_\#^j) := \rho^{\otimes\infty}(\sigma_\#^j(t)) \ ,$$

tends to the state $\omega_\infty$ such that $\omega_\infty(\sigma^j) = 1/2$ and $\omega_\infty(\sigma_\pm^j) = 0$ for all $j$.

### Inequivalent representations

One of the most relevant aspects of quantum mechanics with infinite degrees of freedom is the existence of inequivalent irreducible representations of a same algebra; this fact explains physical phenomena such as symmetry breaking and phase-transitions [290, 291, 274, 275].

We let $N \to \infty$ in Example 5.4.1 and set

$$|\,vac\,\rangle_\uparrow := |\,0\,\rangle^{\otimes\infty} \ , \qquad |\,\boldsymbol{i}^{(n)}\,\rangle_\uparrow := \prod_{j=1}^n \sigma_-^{i_j}|\,vac\,\rangle_\uparrow \tag{7.1}$$

$$|\,vac\,\rangle_\downarrow := |\,1\,\rangle^{\otimes\infty} \ , \qquad |\,\boldsymbol{i}^{(n)}\,\rangle_\downarrow := \prod_{j=1}^n \sigma_+^{i_j}|\,vac\,\rangle_\downarrow \ , \tag{7.2}$$

where $\boldsymbol{i}^{(n)} = i_1 i_2 \cdots i_n$ with $i_j \in \mathbb{N}$.

The physical interpretation is straightforward: $|\,vac\,\rangle_{\uparrow,\downarrow}$ represent configurations consisting of infinitely many spins all pointing up, respectively down, whereas the vector states $|\,\boldsymbol{i}^{(n)}\,\rangle_{\uparrow,\downarrow}$ describe local configurations that are obtained from $|\,vac\,\rangle_{\uparrow,\downarrow}$ by flipping the spins at the sites specified by $i_1, i_2, \ldots, i_n$ by means of the raising and lowering operators $\sigma_\pm$. By defining the scalar product of infinite tensor products of vectors as infinite products of scalar products of vectors at single sites [300], one gets

$${}_k\langle\,\boldsymbol{i}^{(n)}\,|\,\boldsymbol{j}^{(m)}\,\rangle_k = \delta_{n,m}\prod_{\ell=1}^n \delta_{i_\ell j_\ell} \ , \ k = \uparrow, \downarrow, \quad {}_\uparrow\langle\,\boldsymbol{i}^{(n)}\,|\,\boldsymbol{j}^{(m)}\,\rangle_\downarrow = 0 \ .$$

Indeed, in the first scalar product, outside a local region where the spins may be flipped, there are infinitely many spins all pointing up (down). Therefore, the value of the scalar product is determined by the spins within the region where they are flipped: it is 0 unless the flipped spins on both sides of the scalar product match each other. On the contrary, in the second scalar product there are always (infinitely many) spins in $|\,\boldsymbol{j}^{(m)}\,\rangle$ that are orthogonal to the ones at the corresponding sites in $|\,\boldsymbol{i}^{(n)}\,\rangle$.

It thus follows that the completions of the linear spans of all vectors of the form $|\,\boldsymbol{i}^{(n)}\,\rangle_\uparrow$, respectively $|\,\boldsymbol{i}^{(n)}\,\rangle_\downarrow$ give rise to two orthogonal Hilbert spaces $\mathbb{H}_\uparrow$, respectively $\mathbb{H}_\downarrow$. Furthermore, let $\mathcal{A}$ be the (*local*) algebra generated by all products of Pauli matrices as in (5.61); the two Hilbert spaces then provide two irreducible, inequivalent representations $\pi_{\uparrow,\downarrow}(\mathcal{A})$. In fact, suppose $X \in \mathbb{B}(\mathbb{H}_\uparrow)$ commutes with $\pi_\uparrow(\mathcal{A})$, then

$$
\begin{aligned}
{}_\uparrow\langle\,\boldsymbol{i}^{(n)}\,|\,X\,|\boldsymbol{j}^{(m)}\,\rangle_\uparrow &= {}_\uparrow\langle\,vac\,|\prod_{r=1}^{p}\sigma_+^{i_r}\prod_{s=1}^{q}\sigma_-^{j_s}\,X\,|vac\,\rangle_\uparrow \\
&= {}_\uparrow\langle\,vac\,|\,X\,|vac\,\rangle_\uparrow\;{}_\uparrow\langle\,\boldsymbol{i}^{(n)}\,|\,\boldsymbol{j}^{(m)}\,\rangle_\uparrow\ ,
\end{aligned}
$$

for $\prod_{s=1}^{q}\sigma_+^{j_s}\prod_{r=1}^{p}\sigma_-^{i_r}|\,vac\,\rangle_\uparrow = 0$ unless raising and lowering operators match each other. Thus, $X$ acts as ${}_\uparrow\langle\,vac\,|\,X\,|vac\,\rangle_\uparrow\,\mathbb{1}$ on a dense set of $\mathbb{H}_\uparrow$, whence $X = {}_\uparrow\langle\,vac\,|\,X\,|vac\,\rangle_\uparrow\,\mathbb{1}$ and the commutant of $\pi_\uparrow(\mathcal{A})$ is trivial.

Consider now the magnetization $\boldsymbol{m}(N)$ relative to the first $N$ spins, that is the operator-valued vector $\boldsymbol{m}(N)$ of components

$$
m_i(N) := \mu\sum_{n=1}^{N}\sigma_i^n\ , \qquad i = 1,2,3\ ,
$$

and the average magnetization $\boldsymbol{m} = (m_1, m_2, m_3)$ given by the formal limits

$$
m_i := \lim_{N\to+\infty}\frac{m_i(N)}{N}\ . \tag{7.3}
$$

Choose $N > j_n \geq i_m$, $1 \leq i_1 \leq j_1$, then ($k = \uparrow,\downarrow$)

$$
{}_k\langle\,\boldsymbol{i}^{(n)}\,|\,m_3(N)\,|\boldsymbol{j}^{(m)}\,\rangle_k = \pm\mu(N + i_n - j_m) + \mu\sum_{\ell=i_n}^{j_m}{}_k\langle\,\boldsymbol{i}^{(n)}\,|\,\sigma_3^n\,|\boldsymbol{j}^{(m)}\,\rangle_k
$$

$$
{}_k\langle\,\boldsymbol{i}^{(n)}\,|\,m_{1,2}(N)\,|\boldsymbol{j}^{(m)}\,\rangle_k = \mu\sum_{\ell=i_n}^{j_m}{}_k\langle\,\boldsymbol{i}^{(n)}\,|\,\sigma_{1,2}^n\,|\boldsymbol{j}^{(m)}\,\rangle_k\ ,
$$

for $\sigma_3|\,0\,\rangle = |\,0\,\rangle$, $\sigma_3|\,1\,\rangle = -|\,1\,\rangle$, while $\langle\,0\,|\,\sigma_{1,2}\,|0\,\rangle = \langle\,1\,|\,\sigma_{1,2}\,|1\,\rangle = 0$. Then,

$$
\begin{aligned}
&\lim_{N\to+\infty}{}_k\langle\,\boldsymbol{i}^{(n)}\,|\,\frac{m_3(N)}{N}\,|\boldsymbol{j}^{(m)}\,\rangle_k = \begin{cases}\mu & k=\uparrow\\ -\mu & k=\downarrow\end{cases}\\
&\lim_{N\to+\infty}{}_k\langle\,\boldsymbol{i}^{(n)}\,|\,\frac{m_{1,2}(N)}{N}\,|\boldsymbol{j}^{(m)}\,\rangle_k = 0\ .
\end{aligned}
$$

Therefore, the mean magnetization $\boldsymbol{m}$ exists as a weak-limit (see (5.8)), that is with respect to the weak-operator topology determined by the representations $\pi_{\uparrow,\downarrow}(\mathcal{A})$ on $\mathbb{H}_{\uparrow,\downarrow}$. It thus depends on the representation with respect to which the limit is calculated and belongs to the bicommutant $\pi_{\uparrow,\downarrow}(\mathcal{A})''$ (see Definition 5.3.1) where $\boldsymbol{m}_0 = (0,0,\mu)$, $\boldsymbol{m}_1 = (0,0,-\mu)$.

If $\pi_{\uparrow,\downarrow}(\mathcal{A})$ were unitarily equivalent, then there would exist a unitary operator $U : \mathbb{H}_\uparrow \mapsto \mathbb{H}_\downarrow$ such that $U^\dagger \pi_\downarrow(\mathcal{A}) U = \pi_\uparrow(\mathcal{A})$; if so, it could be extended by continuity to the weak closures $\pi_{\uparrow,\downarrow}(\mathcal{A})''$. Then, its action on the mean magnetization would give rise to a contradiction:

$$-\mu = m_{13} = U^\dagger m_{03} U = \mu U^\dagger U = \mu \ .$$

The vectors $|\, vac \,\rangle_{\uparrow,\downarrow}$ behave as vacuum vectors for the spin algebra $\mathcal{A}$; indeed, $|\, vac \,\rangle_{\uparrow,\downarrow}$ and the representations $\pi_{\uparrow,\downarrow}(\mathcal{A})$ are unitarily equivalent to the *GNS* representations based on the expectation functionals

$$\omega_{\uparrow,\downarrow}(\prod_{r=1}^p \sigma_+^{i_r} \prod_{s=1}^q \sigma_-^{j_s}) := {}_{\uparrow,\downarrow}\langle\, vac \,|\, \prod_{r=1}^p \sigma_+^{i_r} \prod_{s=1}^q \sigma_-^{j_s} \,|vac\,\rangle_{\uparrow,\downarrow} \ .$$

Other interesting representations of $\mathcal{A}$ can be obtained by means of the following expectation functional

$$\omega_s\left(\prod_{\ell=1}^n \sigma_{j_\ell}^{i_\ell}\right) = \prod_{\ell=1}^n \mathrm{Tr}(\rho\, \sigma_{j_\ell}) \ , \tag{7.4}$$

where $\rho$ is the spin density matrix of Example 5.5.5 and $\sigma_{j_\ell}^{i_\ell}$ is the $j_\ell$ Pauli matrix at site $i_\ell$. The corresponding *GNS* vector $|\, \Omega_s \,\rangle$ can be identified with the infinite tensor product of the vector states resulting from purifying $\rho$:

$$|\, \Omega_s \,\rangle = \bigotimes_n |\, \sqrt{\rho}\,\rangle = \bigotimes_{n=1}^\infty \left(\sqrt{\frac{1-s}{2}}\,|\, 0 \,\rangle \otimes |\, 0 \,\rangle + \sqrt{\frac{1+s}{2}}\,|\, 1 \,\rangle \otimes |\, 1 \,\rangle\right) .$$

Further, the *GNS* representation $\pi_\omega(\mathcal{A})$ can be identified with the infinite tensor product

$$\pi_s(\mathcal{A}) = \bigotimes_n \pi_\rho(M_2(\mathbb{C})) = \bigotimes_n \Big(M_2(\mathbb{C}) \otimes \mathbb{1}_2\Big) \ .$$

The von Neumann algebra $\pi_s(\mathcal{A})''$ is not irreducible, but it is a factor since the commutant is $\pi_s(\mathcal{A})' = \bigotimes_n \Big(\mathbb{1}_2 \otimes M_2(\mathbb{C})\Big)$. Since $\frac{\mathbb{1}+\sigma_3}{2}\sigma_-|\, 0 \,\rangle = 0$, the projection $\mathcal{A} \ni P_N := \prod_{i=N}^{2N} \frac{\mathbb{1}+\sigma_3^i}{2}$ is such that $P_N|\, \boldsymbol{i}^{(n)} \,\rangle_0 = 0$ if the sites $i_1, i_2, \ldots, i_n \in [N, 2N]$, else $P_N|\, \boldsymbol{i}^{(n)} \,\rangle_0 = |\, \boldsymbol{i}^{(n)} \,\rangle_0$.

Given $|\, \psi \,\rangle_\uparrow \in \mathbb{H}_\uparrow$ and $\varepsilon > 0$, one can find a vector $|\, \phi \,\rangle_\uparrow$ in the subset linearly spanned by $|\, \boldsymbol{i}^{(n)} \,\rangle_\uparrow$ indexed by the sites within a suitable finite interval $I_\varepsilon$ such that $\||\, \psi \,\rangle_\uparrow - |\, \phi \,\rangle_\uparrow\| \leq \varepsilon$; then, by choosing $N$ such that $I_\varepsilon \cap [N, 2N] = \emptyset$, one estimates

$$\|(P_N - \mathbb{1})|\psi\rangle_\uparrow\| \leq \|(P_N - \mathbb{1})|\phi\rangle_\uparrow\| + 2\||\psi\rangle_\uparrow - |\phi\rangle_\uparrow\| \leq 2\varepsilon \,.$$

Therefore, $P_N \to \mathbb{1}$ strongly on $\mathbb{H}_\uparrow$. Instead, $\mathrm{Tr}(\rho\sigma_3) = s$ yields

$$\lim_{N\to+\infty} \omega_s(P_N) = \lim_{N\to+\infty} \left(\frac{1+s}{2}\right)^N = \begin{cases} 1 & s = 1 \\ 0 & 0 \leq s < 1 \end{cases}$$

Therefore, for $0 \leq s < 1$, the *GNS* representation $\pi_s(\mathcal{A})$ cannot be unitarily equivalent to $\pi_\uparrow(\mathcal{A})$. If so, there would exist an isometry $U : \mathbb{H}_s \mapsto \mathbb{H}_\uparrow$ such that

$$0 = \lim_{N\to+\infty} \langle\, \Omega_s \,|\, \pi_s(P_N) \,|\Omega_s\,\rangle = \lim_{N\to+\infty} \langle\, \Omega_s \,|\, U^\dagger \pi_\uparrow(P_N) U \,|\Omega_s\,\rangle = 1 \,.$$

In fact, $U|\,\Omega_s\,\rangle \in \mathbb{H}_\uparrow$ and $P_N$ converges strongly and thus weakly on $\mathbb{H}_\uparrow$.

### Factor Types

According to Example 5.6.2, the states $\omega_s$ on the spin algebra $\mathcal{A}$ may be interpreted as thermal spin states at inverse temperature

$$\beta_s = \frac{1}{2\omega} \log \frac{1+s}{1-s} \;;$$

- the zero temperature state $\omega_1$ is equivalent to the vacuum state $\omega_\uparrow$;
- $\omega_0$ is an infinite temperature state with the properties of a tracial state (compare (5.55)) such that $\omega_0(XY) = \omega_0(YX)$ for all $X, Y \in \mathcal{A}$;
- for $0 < s < 1$, $\omega_s$ is a thermal state with no specific properties.

Correspondingly, the von Neumann algebras $\pi_s''(\mathcal{A})$ that arise from the strong closures of the spin algebras $\pi_s(\mathcal{A})$ on the *GNS* Hilbert spaces $\mathbb{H}_s$ are instances of the so-called factors of *type I*, *II* and *III* [300].

The classification of von Neumann algebras starts by considering different possible classes of their projections. A projection $p = p^\dagger = p^2$ of a von Neumann algebra $\mathcal{A}$ is called an *Abelian projection* if $p\,\mathcal{M}\,p$ is Abelian. Typical examples in this class are the minimal projections of Example 5.3.4.1: if $p \in \mathcal{A}$ is a minimal projection and $q \in \mathcal{A}$ is another projection, then $0 \leq pqp \leq p$. Therefore, $pqp = \lambda p$ as all spectral projections of $pqp$ are $\leq p$ and must then be equal to $p$. Since $\mathcal{A}$ is generated by its projections $q$, $p\,\mathcal{A}\,p$ is Abelian.

Two projectors $p, q$ are said to be *equivalent*, $p \simeq q$, if there exists $U \in \mathcal{A}$ such $U^\dagger U = P$ and $UU^\dagger = q$. This is the case for the initial and range projections in the polar decomposition (see Remark 5.2.2). A projection $p \in \mathcal{A}$ is said to be *finite* if $\mathcal{A} \ni q = q^* = q^2 \leq p$ and $q \simeq p$ imply $\Longrightarrow q = p$. Any Abelian projection $p \in \mathcal{A}$ is finite; in fact, let $q \leq p$ and $p = U^\dagger U$, $q = UU^\dagger$. Then, since $q \leq p \Longrightarrow qp = pq = q$ (see Example 5.3.4.1), it turns out that $V := pUp = pqU = qU$ and $V^\dagger$ commute so that

$$V^\dagger V = U^\dagger q U = p^2 = p = VV^\dagger = qUU^\dagger q = q \,.$$

1. A unital von Neumann algebra $\mathcal{A}$ is said to be of type $I$ if its identity $\mathbb{1}$ can be decomposed into an orthogonal sum of Abelian projections $p_n \in \mathcal{A}$. Typical examples are $\mathcal{A} := \mathbb{L}^\infty_\mu(\mathcal{X})$ and $\mathbb{B}(\mathbb{H})$ with $\mathbb{H}$ a separable Hilbert space. In the first case, the characteristic functions of the atoms of any partition of $\mathcal{X}$ into disjoint measurable atoms are the required Abelian projections. In the second case, the projections $p_n = |n\rangle\langle n|$ onto the orthonormal vectors $\{|n\rangle\}_{n\in\mathbb{N}}$ of any *ONB* are Abelian and sum up to the identity. Then, also $\mathcal{A}\otimes\mathbb{B}(\mathbb{H})$ is of type $I$, the required Abelian projections being given by $\mathbb{1}_\mathcal{A}\otimes p_n$.
2. A unital von Neumann algebra is said to be *finite* if its identity is a finite projection, *semi-finite* if its identity can be decomposed into an orthogonal sum of finite projections. Since the projections $p_n$ in the previous point are minimal, type $I$ von Neumann algebras on infinite dimensional Hilbert spaces are semi-finite, finite if $\dim(\mathbb{H}) = n$.
3. A unital von Neumann algebra is said to be of type $II$ if it is semi-finite, but does not contain any non-zero Abelian projection; of type $III$ if it does not contain any finite projection.

The trace for finite dimensional systems (see (5.19)) is a particular realization of the following general notion.

**Definition 7.0.8 (Traces).** *[300] A* trace *on a von Neumann algebra $\mathcal{A}$ is a map $\Phi : \mathcal{A}_+ \mapsto \mathbb{R}_+$ from its positive elements into the positive reals $\mathbb{R}_+$ such that*

$$\Phi(\sum_i \lambda_i A_i) = \sum_i \lambda_i \Phi(A_i) \qquad \forall \lambda_i \in \mathbb{R}_+\,,,\ A_i \in \mathcal{A}_+$$

$$\Phi(A) = \Phi(U^\dagger A\, A) \qquad \forall A \in \mathcal{A}_+\,,\ U \in \mathcal{A} \quad unitary\,.$$

*The trace is*

- faithful *if $\Phi(A) = 0 \Longleftrightarrow A = 0$ for $A \in \mathcal{A}_+$;*
- finite *if $\Phi(A) < +\infty$ for all $\mathcal{A} \in \mathcal{A}_+$;*
- semi-finite *if for all $A \in \mathcal{A}_+$ there exists $\mathcal{A}_+ \ni B \leq A$ with $\Phi(B) < +\infty$;*
- normal *if $\sup \Phi(A_\alpha) = \Phi(\sup A_\alpha)$ for every increasing net $\{A_\alpha\} \subset \mathcal{A}_+$.*

Analogously to what has been proved for states (see Example 5.3.2.3), it turns out that if $\Phi \leq \Psi$ for two faithful, semi-finite traces on $\mathcal{A}$, then there exists $0 \leq X' \leq \mathbb{1}$ in the center $\mathcal{Z} = \mathcal{A}\cap\mathcal{A}'$ such that $\Phi(A) = \Psi(X'A)$ for all $A \in \mathcal{A}_+$ [300]. Then, if $\mathcal{A}$ is a factor, $\mathcal{Z} = \{\lambda\mathbb{1}\}$ and all traces on it are proportional; in fact, given any two traces $\Phi_i$, $i = 1, 2$,

$$\Phi_1 \leq \Phi_1 + \Phi_2 \Longrightarrow \Phi_1 = \lambda_1(\Phi_1 + \Phi_2)\,, \quad \Phi_2 \leq \Phi_1 + \Phi_2 \Longrightarrow \Phi_2 = \lambda_2(\Phi_1 + \Phi_2)\,,$$

whence $\Phi_1 = \lambda_1\lambda_2^{-1}\Phi_2$.

Consequently, any chosen trace on a factor von Neumann algebra can be used to assign its projections an intrinsic dimension, thus providing a characterization of types [162, 117, 300]:

1. Factors of type $I$ have a semi-finite, faithful, normal trace given by (5.19) whose range on projections is discrete and finite for finite type $I_n$ factors, or countable for infinite type $I_\infty$ factors: an example of the latter case is the spin algebra $\Pi_s(\mathcal{A})''$ with respect to the zero temperature state $\omega_1$;
2. factors of type $II$ have a semi-finite, faithful, normal trace whose range on projections is the whole interval $[0,1]$ for finite type $II_1$ factors or the whole of $\mathbb{R}_+$ for infinite type $II_\infty$ factors. An instance of the first occurrence is the spin algebra $\pi_s(\mathcal{A})''$ with respect to the infinite temperature state $\omega_s$ when $s=0$;
3. finally, type $III$ factors have no semi-finite, faithful, normal traces: this is the case of the spin algebra $\pi_s(\mathcal{A})''$ when $0<s<1$.

## 7.1 Observables, States and Dynamics

The physical scenario in the examples discussed in the previous section is a common one in quantum statistical mechanics. Indeed, the limit of infinitely many degrees of freedom is in general achieved in the so-called *thermodynamical limit*, where one starts with $N$ particles in a finite volume $V \subset \mathbb{R}^3$ (or $\mathbb{Z}^3$ in the case of a lattice system) and lets $N, V \to \infty$ in such a way that $N/V \mapsto \rho$, where $\rho \geq 0$ is a given spatial density.

Each $V \subset \mathbb{R}^3$ has its own Hilbert space $\mathbb{H}_V = \mathbb{L}^2_{\mathrm{d}r}(V)$ of Lebesgue square-summable functions and the corresponding $C^*$ algebra $\mathcal{A}_V = \mathbb{B}(\mathbb{H}_V)$ of bounded operators. Instead, in the case of a lattice system, each $\boldsymbol{x} \in V$ carries a Hilbert space $\mathbb{H}_{\boldsymbol{x}}$ and the $C^*$ algebra $\mathcal{A}_{\boldsymbol{x}} := \mathbb{B}(\mathbb{H}_{\boldsymbol{x}})$, so that $\mathbb{H}_V = \bigotimes_{\boldsymbol{x}\in V} \mathbb{H}_{\boldsymbol{x}}$ and $\mathcal{A}_V = \bigotimes_{\boldsymbol{x}\in V} \mathcal{A}_{\boldsymbol{x}}$. Notice that in the continuous case each Hilbert space $\mathbb{H}_V$ is infinite dimensional, while in the discrete case it depends on whether the Hilbert spaces $\mathbb{H}_{\boldsymbol{x}}$ at the lattice sites are finite dimensional or not. If $V_1 \subseteq V_2$, set $V_{12}^c := V_2 \setminus V_1$, then $\mathbb{H}_{V_1} = \mathbb{H}_{V_2} \otimes \mathbb{H}_{V_{12}^c}$ and $\mathcal{A}_{V_1}$ becomes a subalgebra of $\mathcal{A}_{V_2}$ by embedding any $A_1 \in \mathcal{A}_{V_1}$ into $\mathcal{A}_{V_2}$ as $A_1 \otimes 1\!\!1_{V_{12}^c}$ where $1\!\!1_{V_{12}^c}$ denotes the identity operator on the Hilbert space $\mathbb{H}_{V_{12}^c}$. It follows that the set $\mathcal{A}_0 := \bigcup_V \mathcal{A}_V$ is a $*$-algebra, namely it is closed under addition and multiplication of its elements; also, it is naturally endowed with the norm $\mathcal{A}_V \ni X \mapsto \|X\|$ for all $V \subset \mathbb{R}^3$.

**Definition 7.1.1 (Quasi-Local $C^*$ algebras).** *The normed $*$-algebra $\mathcal{A}_0$ is the algebra of* local observables, *while its norm-closure* $\mathcal{A} := \overline{\bigcup_V \mathcal{A}_V}^{\|\cdot\|}$ *is known as a* quasi-local $C^*$ algebra.

**Remark 7.1.1.** The notion of quasi-local algebra is physically motivated by the fact that the only experimentally accessible observables of infinitely extended quantum systems are the local ones. These can then be used to approximate as much as one desires the non-local ones. From a mathematical point of view, the construction is an instance of *inductive limit* [117] of a directed net of $C^*$ algebras. In the case of an increasing sequence $\{\mathcal{A}_{n_i}\}_{n_i\in\mathbb{N}}$ of finite-dimensional $C^*$ algebras $\mathcal{A}_{n_i} \subseteq \mathcal{A}_{n_{i+1}}$ the generated quasi-local $C^*$ algebra is called *Almost Finite* (*AF*), if the algebras $\mathcal{A}_{n_i}$ are full matrix algebras $M_{n_i}(\mathbb{C})$ then $\mathcal{A}$ is known as *Uniformly Hyperfinite* (*UHF*) [277, 244, 245].

Suppose an increasing sequence of finite-dimensional unital $C^*$ algebras $\{\mathcal{A}_{n_i}\}_{n_i}$ is represented on a Hilbert space $\mathbb{H}$, then the strong closure of $\bigcup_{n_i} \mathcal{A}_{n_i}$ is a von Neumann algebra $\mathcal{M}$ which is called *Hyperfinite.* An Abelian instance of such an algebra is the von Neumann algebra of essentially bounded functions, $\mathbb{L}^\infty_\mu(\mathcal{X})$, which one can generate by means of the characteristic functions of finer and finer finite partitions of $\mathcal{X}$ as explained in Remark 2.2.3.4.

**Example 7.1.1.** [10] Let $M_{n_1}(\mathbb{C}) \subseteq M_{n_2}(\mathbb{C})$ be two matrix algebras; given a system of matrix units $\{E^{(1)}_{jk}\}_{j,k=1}^{n_1}$ for the smaller one (see (5.12)), the orthogonal projections $E^{(1)}_{kk}$ sum up to the identity $\sum_{k=1}^{n_1} E^{(1)}_{kk} = \mathbb{1}_2 \in M_{n_2}(\mathbb{C})$, whence $\sum_{k=1}^{n_1} \mathrm{Tr}_2(E^{(1)}_{kk}) = n_2$, where $\mathrm{Tr}_2$ denotes the trace computed with respect to the Hilbert space $\mathbb{C}^{n_2}$. But then, using the cyclicity of the trace,

$$\mathrm{Tr}_2(E^{(1)}_{kk}) = \mathrm{Tr}_2(E^{(1)}_{kp}\, E^{(1)}_{pk}) = \mathrm{Tr}_2(E^{(1)}_{pk}\, E^{(1)}_{kp}) = \mathrm{Tr}_2(E^{(1)}_{pp}) ,$$

for all $k, p = 1, 2, \ldots, n_1$, whence $n_2 = n_1 \times d$, where $d := \mathrm{Tr}_2(E^{(1)}_{kk})$ for all $k = 1, 2, \ldots, n_1$. Let $\{|\, f_i\,\rangle \in \mathbb{C}^{n_2}\}_{i=1}^d$ be an *ONB* in the subspace projected out by $E^{(1)}_{11}$ and set

$$E^{(2)}_{(k_1,k_2);(j_1,j_2)} := E^{(1)}_{k_1 1}|\, f_{k_2}\,\rangle\langle\, f_{j_2}\,|E^{(1)}_{1 j_2} . \tag{7.5}$$

Since $1 \leq k_1, j_1 \leq n_1$ while $1 \leq k_2, j_2 \leq d$, these are $n_1^2 \times d^2 = n_2^2$ matrices in $M_{n_2}(\mathbb{C})$; moreover, from (5.12) and $E^{(1)}_{11}|\, f_p\,\rangle = |\, f_p\,\rangle$ it follows that

$$\begin{aligned} E^{(2)}_{(k_1,k_2);(j_1,j_2)}\, E^{(2)}_{(p_1,p_2);(q_1,q_2)} &= E^{(1)}_{k_1 1}\,|\, f_{k_2}\,\rangle\langle\, f_{j_2}\,|\, E^{(1)}_{1 j_1}\, E^{(1)}_{p_1 1}\,|\, f_{p_2}\,\rangle\langle\, f_{q_2}\,|\, E^{(1)}_{1 q_1} \\ &= \delta_{j_1 p_1}\, E^{(1)}_{k_1 1}\,|\, f_{k_2}\,\rangle\langle\, f_{j_2}\,|\, E^{(1)}_{11}\,|\, f_{p_2}\,\rangle\langle\, f_{q_2}\,|\, E^{(1)}_{1 q_1} \\ &= \delta_{j_1 p_1}\delta_{j_2 p_2}\, E^{(1)}_{k_1 1}\,|\, f_{k_2}\,\rangle\langle\, f_{q_2}\,|\, E^{(1)}_{1 q_1} \\ &= \delta_{j_1 p_1}\, \delta_{j_2 p_2}\, E^{(2)}_{(k_1,k_2);(q_1,q_2)} . \end{aligned}$$

Thus, (7.5) defines a set of matrix units in $M_{n_2}(\mathbb{C})$. Set $E^d_{k_2 j_2} := |\, f_{k_2}\,\rangle\langle\, f_{j_2}\,|$; then, $E^{(2)}_{(k_1,k_2);(j_1,j_2)}$ can be isomorphically represented by $E^{(1)}_{k_1 j_1} \otimes E^d_{k_2 j_2}$ on $\mathbb{C}^{n_2} = \mathbb{C}^{n_1} \otimes \mathbb{C}^d$. Therefore $M_{n_2}(\mathbb{C})$ is isomorphic to $M_{n_1}(\mathbb{C}) \otimes M_d(\mathbb{C})$. The matrix algebras $M_{n_i}(\mathbb{C}) \subseteq M_{n_{i+1}}(\mathbb{C})$ that generate a *UHF* algebra $\mathcal{A}$ must be such that any $n_i$ must divide the subsequent one so that $\mathcal{A}$ is isomorphic to an infinite tensor product of matrix algebras. The simplest instance of *UHF* algebra $\mathcal{A}$ is a *quantum spin chain* (see Section 7.1.5). In the case of the previously discussed infinite spin system, $\mathcal{A}$ is the quasi-local algebra $\mathcal{A}$ generated by the local algebras $\mathcal{A}_{[-k,k]} = \bigotimes_{\ell=-k}^{k}(M_2(\mathbb{C}))_\ell$ which are tensor products of $2 \times 2$ matrix algebras at each lattice site.

### 7.1.1 Bosons and Fermions

Physical systems of quantum statistical mechanics usually consist of indistinguishable particles and are described by operators of creation and annihilation satisfying either the *CAR* (5.62) or the *CCR* (5.92). More precisely, one considers the *Fock representation* built upon the existence of a distinguished vacuum vector $|\,vac\,\rangle$ (which was considered in Examples 5.6.2.1,2 for finitely many degrees of freedom).

Let $\mathbb{H}$ be the Hilbert space describing a single Fermion or Boson and let $\{|\,\psi_i\,\rangle\}_{i\in\mathbb{N}}$ be an *ONB*. Then, one introduces operators

$$\begin{aligned} a_i &:= a(\psi_i) \qquad \text{such that} \quad a(\psi_i)|\,vac\,\rangle = 0 \quad \forall i \in \mathbb{N} \\ a_i^\dagger &:= a^\dagger(\psi_i) \qquad \text{such that} \quad a_i^\dagger|\,vac\,\rangle = |\,\psi_i\,\rangle\ . \end{aligned}$$

They are required to satisfy the *CAR* (5.62) if the particles are Fermions, the *CCR* (5.92) if Bosons.

By expanding any $|\,\psi\,\rangle \in \mathbb{H}$ along the chosen *ONB* , $|\,\psi\,\rangle = \sum_i c_i|\,\psi_i\,\rangle$, one can consistently define creation and annihilation operators of generic $|\,\psi\,\rangle \in \mathbb{H}$:

$$a^\dagger(\psi) = \sum_{i\in\mathbb{N}} c_i\, a_i^\dagger\ , \quad a(\psi) = \sum_{i\in\mathbb{N}} c_i^*\, a_i\ .$$

This yields $a(\psi)|\,vac\,\rangle = 0$, $a^\dagger(\psi)|\,vac\,\rangle = |\,\psi\,\rangle$ and

$$\begin{aligned} \Big[a(\psi)\,,\,a(\phi)\Big] &= \Big[a^\dagger(\psi)\,,\,a^\dagger(\phi)\Big] = 0\ ,\ \Big[a(\psi)\,,\,a^\dagger(\phi)\Big] = \langle\,\psi\,|\,\phi\,\rangle\ (CCR\,) \\ \Big\{a(\psi)\,,\,a(\phi)\Big\} &= \Big\{a^\dagger(\psi)\,,\,a^\dagger(\phi)\Big\} = 0\ ,\ \Big\{a(\psi)\,,\,a^\dagger(\phi)\Big\} = \langle\,\psi\,|\,\phi\,\rangle\ (CAR\,)\ , \end{aligned}$$

for all $|\,\psi\,\rangle, |\,\phi\,\rangle \in \mathbb{H}$. Furthermore, by using these algebraic relations one gets

$$a(\psi)|\,\phi\,\rangle = a(\psi)a^\dagger\phi|\,vac\,\rangle = \langle\,\psi\,|\,\phi\,\rangle|\,vac\,\rangle\ . \tag{7.6}$$

For both Fermions and Bosons the number operator is defined by

$$N := \sum_{i\in\mathbb{N}} a_i^\dagger\, a_i\ .$$

Directly for Bosons and by means of

$$[a_i^\dagger a_i\,,\,a^\dagger(f)] = a_i^\dagger\{a_i\,,\,a^\dagger(f)\} - \{a_i^\dagger\,,\,a^\dagger(f)\}a_i = f_i a_i^\dagger\;, \tag{7.7}$$

where $f_i := \langle\,\psi_i\,|\,f\,\rangle$, for Fermions, one finds that

$$[N\,,\,a^\dagger(f)] = a^\dagger(f)\;, \qquad [N\,,\,a(f)] = -a(f)\;. \tag{7.8}$$

The Fock space $\mathbb{H}_{F,B}$ for Fermions, respectively Bosons is generated by the completion of the linear span of vectors of the form $P(a(f), a^\dagger(g))|\,vac\,\rangle$ where $P(a(f), a^\dagger(g))$ is any polynomial in Fermi, respectively Bose annihilation and creation operators. The Fermi operators $a^\#(f)$ are bounded on the Fock space; indeed, from the *CAR* it follows that, for any normalized $|\,\Psi\,\rangle \in \mathbb{H}_F$,

$$\|a^\dagger(f)|\,\Psi\,\rangle\|^2 + \|a^\dagger(f)|\,\Psi\,\rangle\|^2 = \|f\|^2\;.$$

The polynomials $P(a(f), a^\dagger(g)$ with $f, g \in \mathbb{H}_V$, where $V$ is a finite volume, generate, by norm completion, a local $C^*$-algebra, $\mathcal{A}_V^F$.

**Remark 7.1.2.** Given two volumes $V_1 \subset V_2 \subset \mathbb{R}^3$, the local Fermi algebra cannot be isomorphic to $\mathcal{A}_{V_2}^F \neq \mathcal{A}_{V_1}^F \otimes \mathcal{A}_{V_3}^F$, where $V_3 := V_2 \setminus V_2$. In fact, if $|\,f\,\rangle \in \mathbb{H}_{V_1}$ and $|\,g\,\rangle \in \mathbb{H}_{V_3}$, despite the fact that $\langle\,f\,|\,g\,\rangle = 0$, commutators of the form $[a^\#(f)\,,\,a^\#(g)]$ need not vanish. However, because of (5.63), commutators vanish if one considers polynomial with even numbers of creation and annihilation operators: the quasi-local $C^*$ algebra they generate is denoted by $\mathcal{A}^G$. The quasi-local algebra $C^*$ generated by polynomial with a same number of creation and annihilation operators is denoted by $\mathcal{A}^E$; it is known as *even* Fermi algebra and commutes with the number operator. Indeed, using (7.8), it turns out that the number operator generates the *gauge-transformation*

$$\begin{aligned}
\mathrm{e}^{i\alpha N}\,a^\dagger(f)\,\mathrm{e}^{-i\alpha N} &= \sum_{k=0}^{+\infty}\frac{(i\alpha)^k}{k!}\underbrace{[N\,,[N\,,\cdots[N}_{k\ times}\,,\,a^\#(f)]\cdots]] \\
&= \sum_{k=0}^{+\infty}\frac{(i\alpha)^k}{k!}\sum_j f_j\underbrace{[N\,,[N\,,\cdots[N}_{k-1\ times}\,,\,a_j^\dagger]\cdots]] \\
&= \sum_j f_j\sum_{k=0}^{+\infty}\frac{(i\alpha)^k}{k!} = \mathrm{e}^{i\alpha}\,a^\dagger(f) = a^\dagger(\mathrm{e}^{i\alpha}f) \qquad (7.9)\\
\mathrm{e}^{i\alpha N}\,a^\dagger(f)\,\mathrm{e}^{-i\alpha N} &= \mathrm{e}^{-i\alpha}\,a(f) = a(\mathrm{e}^{i\alpha}f)\;. \qquad (7.10)
\end{aligned}$$

Therefore, the various phases compensate each other in polynomials with equal numbers of $a$ and $a^\dagger$; these are thus left invariant by the gauge-transformation for any $\alpha \in \mathbb{R}$ and must therefore commute with the number operator $N$.

For Bosons, $[a^\#(f_1)\,,\,a^\#(f_2)] = 0$ if $f_i \in \mathbb{H}_{V_i}$ and $V_1 \cap V_2 = \emptyset$; however, the operators $a^\#(f)$ cannot be bounded (see (5.69)) In order to construct local $C^*$ algebras generating a Bose quasi-local algebra $\mathcal{A}^B$, one associates to $|\,\psi\,\rangle \in \mathbb{H}$ the bounded operators [108]

$$W(\psi) := \exp\Big(i\frac{a(\psi) + a^\dagger(\psi)}{2}\Big) \tag{7.11}$$

which generalize the Weyl operators (5.96) and linearly generate a local Bosonic $C^*$ subalgebra $\mathcal{A}^B_V$ by choosing $\psi$ supported within the volume $V$.

The $C^*$ algebras $\mathcal{A}_{B,F}$ are irreducibly represented on the Fock spaces $\mathbb{H}_{B,F}$. In order to show this one can use a similar argument as for the spin algebras $\pi_{\uparrow,\downarrow}(\mathcal{A})$ discussed in the previous section. If $X$ belongs to the commutant, $X \in \mathcal{A}'_{B,F}$, then (7.6) yields

$$\begin{aligned}\langle\, vac \,|\, a(g_n)\cdots a^\dagger(g_1)\, X\, a^\dagger(f_1)\cdots a^\dagger(f_m)\,|vac\,\rangle &= \\ &= \langle\, vac \,|\, X\, a(g_n)\cdots a(g_1)a^\dagger(f_1)\cdots a^\dagger(f_m)\,|vac\,\rangle \\ &= \langle\, vac \,|\, X\,|vac\,\rangle\,\langle\, vac \,|\, a(g_n)\cdots a(g_1)a^\dagger(f_1)\cdots a^\dagger(f_m)\,|vac\,\rangle\ ,\end{aligned}$$

where (see (5.185) and (5.183))

$$\langle\, vac \,|\, a(g_n)\cdots a(g_1)a^\dagger(f_1)\cdots a^\dagger(f_m)\,|vac\,\rangle = \begin{cases}\mathrm{per}([\langle\, g_i \,|\, f_j\,\rangle]) & CCR \\ \det([\langle\, g_i \,|\, f_j\,\rangle]) & CAR\end{cases}\ . \tag{7.12}$$

Therefore, $X = \langle\, vac \,|\, X\,|vac\,\rangle\, \mathbb{1}$ whence the commutant is trivial; this means (see Lemma 5.3.2) that $\mathcal{A}_{B,F}$ are irreducibly represented.

### Quasi-free Automorphisms and Quasi-free States

The gauge-transformation (7.9) is a particularly simple example of *quasi-free automorphism.*

**Definition 7.1.2.** *Every single particle unitary transformation* $U : \mathbb{H} \mapsto \mathbb{H}$ *gives rise to a* quasi-free automorphism *on* $\mathcal{A}^{B,F}$ *given by*

$$\Theta_U[a^\#(f)] = a^\#(Uf)\ . \tag{7.13}$$

Quasi-free automorphisms are typical time-evolutions of non-interacting particles possibly subjected to external potentials. They preserve the number operator; indeed, by expanding $U|\,f_i\,\rangle = \sum_j c_{ij}|\,f_j\,\rangle$ with respect to the chosen *ONB*, it turns out that

$$\Theta[N] = \sum_{i\in\mathbb{N}} a^\dagger(Uf_i)a(Uf_i) = \sum_{i,j,k} c_{ij}c^*_{ik}a^\dagger_j a_k = \sum_{j\in\mathbb{N}} a^\dagger_j a_j\ ,$$

for the matrix $C = [c_{ij}]$ is unitary.

**Examples 7.1.2.**

1. Let $\{U_{\boldsymbol{x}}\}_{\boldsymbol{x}\in\mathbb{R}^3}$ be the unitary group of *space-translations*,

$$|\, f\,\rangle \mapsto U_{\boldsymbol{x}}|\, f\,\rangle = |\, f_{\boldsymbol{x}}\,\rangle\ , \qquad \langle\, \boldsymbol{r}\,|\, f_{\boldsymbol{x}}\,\rangle = f(\boldsymbol{r}-\boldsymbol{x})\ ,$$

for all $f \in \mathbb{L}^2_{\mathrm{d}\boldsymbol{r}}(\mathbb{R}^3)$; then, $\{\Theta_{\boldsymbol{x}}\}_{\boldsymbol{x}\in\mathbb{R}^3}$ is the group of space-translation automorphisms of $\mathcal{A}^{B,F}$:

$$\Theta_{\boldsymbol{x}}[a^{\#}(f)] = a^{\#}(f_{\boldsymbol{x}})\ . \tag{7.14}$$

2. Let $h : \mathbb{H} \mapsto \mathbb{H}$ be a single-particle Hamiltonian with discrete spectrum, $h = \sum_i \varepsilon_i |\,\psi_i\,\rangle\langle\,\psi_i\,|$ being its spectral decomposition, and set $a_i^{\#} := a^{\#}(\psi_i)$. Therefore, the basic annihilation and creation operators annihilate and create single-particle energy eigenvectors. Consider the second-quantized Hamiltonian $H = \sum_i \varepsilon_i\, a_i^{\dagger} a_i$ and the generated one-parameter group of automorphisms $\Theta := \{\Theta_t\}_{t\in\mathbb{R}}$,

$$a^{\#}(f) \mapsto \Theta_t[a^{\#}(f)] := \mathrm{e}^{itH}\, a^{\#}(f)\, \mathrm{e}^{-itH}\ .$$

By expanding and summing as in Remark 7.1.2 one finds that, for both Bosons and Fermions,

$$\mathrm{e}^{\alpha H}\, a(f)\, \mathrm{e}^{-\alpha H} = a(\mathrm{e}^{-\alpha^* h} f)\ , \qquad \mathrm{e}^{\alpha H}\, a^{\dagger}(f)\, \mathrm{e}^{-\alpha H} = a(\mathrm{e}^{\alpha h} f)\ , \tag{7.15}$$

for all $\alpha \in \mathbb{C}$. Therefore,

$$\Theta_t[a^{\#}(f)] = a^{\#}(\mathrm{e}^{ith} f)\ , \tag{7.16}$$

whence the group $\Theta$ is a quasi-free time-evolution.

3. Let us consider a single-particle Hamiltonian $h$ with an absolutely continuous spectrum and the corresponding quasi-free time-evolution (7.16). For instance, the free-time evolution given by $\langle\, \boldsymbol{p}\,|\, h\,|\, f\,\rangle = \boldsymbol{p}^2/(2m) f(\boldsymbol{p})$ in momentum representation.
In this cases, one can use the so-called *Riemann-Lebesgue Lemma* [258]. For an integrable function $f : \mathbb{R} \mapsto \mathbb{C}$ with integrable first derivative, it follows from integration by parts:

$$\begin{aligned}\int_{\mathbb{R}} \mathrm{d}\nu\ f(\nu)\, \mathrm{e}^{i\nu t} &= \frac{f(\nu)\mathrm{e}^{i\nu t}}{it}\Big|_{-\infty}^{+\infty} + \frac{i}{t}\int_{\mathbb{R}} \mathrm{d}\nu\ f'(\nu)\, \mathrm{e}^{i\nu t} \\ &= \frac{i}{t}\int_{\mathbb{R}} \mathrm{d}\nu\ f'(\nu)\, \mathrm{e}^{i\nu t} \longrightarrow 0\end{aligned}$$

when $t \to \pm\infty$. Because of the assumed absolute continuity of the spectrum of the single-particle Hamiltonian $h$, this lemma ensures that

$$\lim_{t\to\pm\infty} \|[a(f)\,,\, a^{\dagger}(\mathrm{e}^{iht} g)]\| = \lim_{t\to\pm\infty} |\langle\, f\,|\, \mathrm{e}^{iht} g\,\rangle| = 0 \tag{7.17}$$

for all $f, g \in \mathbb{H}$ for Bosons and

$$\lim_{t\to\pm\infty} \|\{a(f)\,,\, a^\dagger(\mathrm{e}^{iht}g)\}\| = \lim_{t\to\pm\infty} |\langle\, f\,|\,\mathrm{e}^{iht}g\,\rangle| = 0 \tag{7.18}$$

for Fermions. While Bosonic annihilation and creation operators commute asymptotically in time, Fermionic ones anticommute. However, by means of (7.7), one computes

$$[a(f_1)a(f_2)\,,\, a^\dagger(\mathrm{e}^{iht}g)] = a(f_1)\,\langle\, f_2\,|\,\mathrm{e}^{iht}g\,\rangle \;-\; \langle\, f_1\,|\,\mathrm{e}^{iht}g\,\rangle\, a(f_2)\;.$$

Then, $\|[a(f_1)a(f_2)\,,\, a^\dagger(\mathrm{e}^{iht}g)]\| \to 0$ when $t \to \pm\infty$; furthermore, the same asymptotic commutativity in time holds for $[X\,,\,\Theta_t[Y]]$ where $X$ is an even polynomial in $a, a^\dagger$ and $Y$ any polynomial. By continuity, it extends to all $X$ belonging to the even Fermi algebra $\mathcal{A}^G$ and all $Y \in \mathcal{A}^F$. Moreover, this result holds for all quasi-free automorphisms $\Theta_t(a^\#(f) = a(U_t f)$ over $\mathcal{A}^G$ consisting of a discrete or continuous group $\{U_t\}_{t\in\mathbb{R},\mathbb{Z}}$ of single-particle unitaries $U_t : \mathbb{H} \mapsto \mathbb{H}$ such that $\lim_{t\to\pm\infty}\langle\, f\,|\,U_t g\,\rangle = 0$ for all $f, g \in \mathbb{H}$. This phenomenon is known as *asymptotic Abelianess.*

**Definition 7.1.3.** *[65, 108] A quasi-free state on $\mathcal{A}^B$ is any linear functional $\omega_A$ such that $\omega(\mathbb{1}) = 1$ and*

$$\omega_A(W(\psi)) = \exp\left(-\frac{1}{4}\langle\,\psi\,|\,(\mathbb{1}+2A)\,|\psi\,\rangle\right)\,, \tag{7.19}$$

*where $0 \le A \in \mathbb{B}(\mathbb{H})$ is a positive bounded operator on the single particle Hilbert space $\mathbb{H} = \mathbb{L}^2_{\mathrm{d}r}(\mathbb{R}^3)$.*

*A quasi-free state on $\mathcal{A}^F$ is any linear functional $\omega_A$ such that $\omega_A(\mathbb{1}) = 1$ and*

$$\omega_A(a^\dagger(f_m)\cdots a^\dagger(f_1)a(g_1)\cdots a(g_n)) = \delta_{nm}\,\mathrm{Det}[\langle\, g_i\,|\,A\,|f_j\,\rangle]\,, \tag{7.20}$$

*where $0 \le A \le \mathbb{1} \in \mathbb{B}(\mathbb{H})$ is a single particle operator on $\mathbb{H} = \mathbb{L}^2_{\mathrm{d}r}(\mathbb{R}^3)$.*

The Fock vacuum satisfying (7.12) is the simplest instance of a *quasi-free state*; the one with $A = 0$. Like classical Gaussian states, quasi-free states also can be reconstructed from their two-point correlation functions

$$\omega_A(a^\dagger(f)a(g)) = \langle\, g\,|\,A\,|f\,\rangle \qquad \forall f, g \in \mathbb{H}\;. \tag{7.21}$$

This property results directly from the determinant in (7.20) for Fermions, while for Bosons it can be proved by showing that (7.19) leads to

$$\omega_A(a^\dagger(f_m)\cdots a^\dagger(f_1)a(g_1)\cdots a(g_n)) = \delta_{nm}\,\mathrm{Per}[\langle\, g_i\,|\,A\,|f_j\,\rangle]\,, \tag{7.22}$$

where the so-called permanent is as in (5.185); indeed, (7.19) is a generalization of (5.186) to the infinite dimensional case.

**Example 7.1.3.** [65] Suppose a quasi-free state satisfies the *KMS* conditions (5.179) with respect to the quasi-free time-evolution (7.16); using the commutation relations and (7.15), it turns out that

$$\begin{aligned}\omega_A(a^\dagger(f)a(g)) &= \langle\, g \,|\, A \,|\, f\,\rangle = \omega_A(a(g)\Theta_{i\beta}[a^\dagger(f)]) = \omega_A(a(g)a^\dagger(\mathrm{e}^{-\beta\, h} f)) \\ &= \langle\, g \,|\, \mathrm{e}^{-\beta\, h} \,|\, f\,\rangle \pm \omega_A(a^\dagger(\mathrm{e}^{-\beta\, h} f)a(g)) \\ &= \langle\, g \,|\, \mathrm{e}^{-\beta\, h} \pm A\mathrm{e}^{-\beta\, h} \,|\, f\,\rangle \ .\end{aligned}$$

Since this is true for all $f, g \in \mathbb{H}$ it turns out that (compare (5.182) and (5.184)) $A_\mp = \dfrac{\mathrm{e}^{-\beta\, h}}{1\!\!1 \mp \mathrm{e}^{-\beta\, h}}$, where the plus sign holds for Fermions and the minus sign for Bosons.

## *KMS* States and Modular Theory

We have seen that Gibbs states of finite dimensional quantum systems satisfy the *KMS* relations (5.179). These relations can be extended to infinitely many degrees of freedom where they identify equilibrium states at a given temperature [131] (a simple instance of this fact was offered in the previous example). Unlike with finitely many degrees of freedom (see Remark 5.6.1.1), there can be more than one equilibrium state at inverse temperature $\beta$. An equilibrium state is called *extremal* when it cannot be decomposed into a linear convex combination of other equilibrium states at the same temperature; extremal equilibrium states give rise to factor representations and can be in some cases rightly identified as pure thermodynamical phases [300, 65].

Given a triplet $(\mathcal{A}, \Theta, \omega)$, with a faithful state $\omega$. The latter is said to be a *KMS state at inverse temperature* $\beta$ with respect to the automorphism $\Theta$ if the functions (compare (5.178))

$$F_{XY}(t) := \omega(\Theta_t[X]Y) \ , \quad G_{XY}(t) := \omega(Y\Theta_t[X]) \qquad \forall X, Y \in \mathcal{A} \ ,$$

can be extended to analytic functions $F_{XY}(z)$, respectively $G_{XY}(z)$, on the strips $-\beta < \Im(z) < 0$, respectively $0 < \Im(z) < \beta$, and continuous on their borders, where they satisfy

$$\omega(\Theta_t[X]Y) = \omega(Y\Theta_{t+i\beta}[X]) \ . \tag{7.23}$$

We outline a few of the many properties of *KMS* states [300] (for a more detailed analysis see [65, 108]). These properties involve the *GNS* cyclic representation $\pi_\omega(\mathcal{A})$ on the *GNS* Hilbert space $\mathbb{H}_\omega$ and the *GNS* implementation of $\Theta$ by a unitary operator $U_\omega$.

**Remarks 7.1.3.**

1. When extended to the von Neumann algebra $\pi_\omega(\mathcal{A})''$, a *KMS* state $\omega$ remain *KMS* in the sense that
$$\langle\, \Omega_\omega \,|\, X\, U_\omega(t) Y \,|\Omega_\omega\,\rangle = \langle\, \Omega_\omega \,|\, Y\, U^\dagger_\omega(t+i\beta) X \,|\Omega_\omega\,\rangle \quad \forall X, Y \in \pi_\omega(\mathcal{A})''\ ,$$
2. *KMS* states are $\Theta$-invariant: indeed, (7.23) implies
$$f_X(t) := \omega(\Theta_t[X]) = \omega(\Theta_{t+i\beta}[X]) = f_X(t+i\beta)$$
for all $X \in \mathcal{A}$. Thus, $f_X(t)$ can be periodically extended over the whole of $\mathbb{C}$ where it defines a bounded analytic function for $f(t)$ is bounded on the strip $-\beta < \Im(z) < 0$. Therefore, this function must be constant: $f_X(t) = f_X(0)$, whence $\Theta_t[X] = X$ for all $X \in \mathcal{A}$.
3. The center $\mathcal{Z}_\omega = \pi_\omega(\mathcal{A})'' \cap \pi_\omega(\mathcal{A})'$ of the *GNS* representation based on a *KMS* state $\omega$ consists of $\Theta$-invariant global observables. Indeed, if $T \in \mathcal{Z}_\omega$, by the same argument as in the previous point, the function
$$\begin{aligned} f_{X,T}(t) &:= \langle\, \Omega_\omega \,|\, \pi_\omega(\Theta_t[X])\, T \,|\Omega_\omega\,\rangle = \langle\, \Omega_\omega \,|\, T\, \Theta_{t+i\beta}[X] \,|\Omega_\omega\,\rangle \\ &= \langle\, \Omega_\omega \,|\, \pi_\omega(\Theta_{t+i\beta}[X])\, T \,|\Omega_\omega\,\rangle = f_{X,T}(t+i\beta) \end{aligned}$$
can be extended to a bounded analytic function over $\mathbb{C}$, for all $X \in \mathcal{A}$. Then, it must be $f_{X,T}(t) = f_{X,T}(0)$. Since $t \in \mathcal{Z}_\omega$, choosing $X = Y^\dagger Z$, $Y, Z \in \mathcal{A}$, yields
$$\begin{aligned} f_{Y^\dagger Z,T}(t) &= \langle\, \Omega_\omega \,|\, \pi_\omega(Y)^\dagger\, U_\omega(t)\, T\, U^\dagger_\omega(t)\, \pi_\omega[Z] \,|\Omega_\omega\,\rangle \\ &= f_{Y^\dagger Z,T}(0) = \langle\, \Omega_\omega \,|\, \pi_\omega(Y)^\dagger\, T\, \pi_\omega(Z) \,|\Omega_\omega\,\rangle \end{aligned}$$
on a dense set, whence $U_\omega(t)\, T\, U^\dagger_\omega(t) = T$ for all $t \in \mathbb{R}$.
4. For fixed inverse temperature, the *KMS* states form a convex set which is compact in the $w^*$-topology [300].
5. A *KMS* state is said to be *extremal KMS* if it cannot be written as a convex combination of other *KMS* states (at the same inverse temperature). The *GNS* representation based on an extremal *KMS* state is a factor: $\mathcal{Z}_\omega = \{\lambda \mathbb{1}\}$. If not, there would exist $0 \leq T_{1,2} \in \mathcal{Z}_\omega$ with $T_1 + T_2 = \mathbb{1}$ which could be used to construct the states
$$\omega_i(X) = \frac{\langle\, \Omega_\omega \,|\, T_i \pi_\omega(X) \,|\Omega_\omega\,\rangle}{\langle\, \Omega_\omega \,|\, T_i \,|\Omega_\omega\,\rangle}$$
which turns out to be a *KMS* state with respect to $\Theta$ at inverse temperature $\beta$. Indeed,
$$\begin{aligned} \omega_i(\Theta_t[X]Y) &= \frac{\langle\, \Omega_\omega \,|\, T_i \pi_\omega(\Theta_t[X]) \pi_\omega(Y) \,|\Omega_\omega\,\rangle}{\langle\, \Omega_\omega \,|\, T_i \,|\Omega_\omega\,\rangle} \\ &= \frac{\langle\, \Omega_\omega \,|\, \pi_\omega(\Theta_t[X]) T_i \pi_\omega(Y) \,|\Omega_\omega\,\rangle}{\langle\, \Omega_\omega \,|\, T_i \,|\Omega_\omega\,\rangle} \\ &= \frac{\langle\, \Omega_\omega \,|\, T_i \pi_\omega(Y) \pi_\omega(\Theta_{t+i\beta}[X]) \,|\Omega_\omega\,\rangle}{\langle\, \Omega_\omega \,|\, T_i \,|\Omega_\omega\,\rangle} = \omega_i(Y\Theta_{t+i\beta}[X])\ . \end{aligned}$$

The modular theory or Tomita-Takesaki theory, which has been introduced in its simplified finite-dimensional version in Section 5.5.1, extend to generic von Neumann algebras $\mathcal{M} \subseteq \mathbb{B}(\mathbb{H})$ with a faithful state (see Definition 5.3.2) [64]. More precisely, given a quantum triplet $(\mathcal{A}, \Theta, \omega)$, if the *GNS* state is such that

$$X|\,\Omega_\omega\,\rangle = 0 \Longrightarrow X = 0 \quad \forall X \in \pi_\omega(\mathcal{A})'' \ ,$$

then there exists a *modular conjugation* $J : \mathbb{H}_\omega \mapsto \mathbb{H}_\omega$, such that

$$J^2 = 1\!\!1 \ , \quad J_\omega|\,\Omega_\omega\,\rangle = |\,\Omega_\omega\,\rangle \ , \quad J_\omega\,\pi_\omega(\mathcal{A})''\,J_\omega \ = \ \pi_\omega(\mathcal{A})' \ , \tag{7.24}$$

and a *modular operator* $\Delta_\omega : \mathbb{H}_\omega \mapsto \mathbb{H}_\omega$ such that

$$J_\omega\sqrt{\Delta_\omega}X|\,\Omega_\omega\,\rangle = X^\dagger|\,\Omega_\omega\,\rangle \quad \forall\, X \in \pi_\omega(\mathcal{A})'' \ . \tag{7.25}$$

Furthermore, the maps

$$\sigma^t_\omega : \pi_\omega(\mathcal{A})'' \ni X \mapsto \Delta^{it}_\omega\, X\, \Delta^{-it}_\omega \tag{7.26}$$

form a group $\{\sigma^t_\omega\}_{t\in\mathbb{R}}$ of automorphisms, called *modular group*; moreover, they satisfy the *KMS* conditions

$$\langle\,\Omega_\omega\,|\,X\,Y\,|\Omega_\omega\,\rangle \ = \ \langle\,\Omega_\omega\,|\,Y\,\sigma^{-i}_\omega(X)\,|\Omega_\omega\,\rangle \quad \forall X, Y \in \pi_\omega(\mathcal{A})'' \ , \tag{7.27}$$

that we will shortly write as $\omega(XY) = \omega(Y\sigma^{-i}_\omega(X))$.

**Example 7.1.4.** If $\mathcal{M} \subseteq \mathbb{B}(\mathbb{H})$ is an Abelian von Neumann algebra ($\mathcal{M} \subseteq \mathcal{M}'$) with a cyclic vector $|\,\Omega\,\rangle$, then it is maximally Abelian. In fact, $|\,\Omega\,\rangle$ is necessarily cyclic also for the commutant $\mathcal{M}'$, thence separating for the bicommutant $\mathcal{M}'' = \mathcal{M}$ (see Lemma 5.3.1). Then, (7.25) gives

$$\begin{aligned}\|J\sqrt{\Delta}X|\,\Omega\,\rangle\|^2 &= \|X^\dagger|\,\Omega\,\rangle\|^2 = \langle\,\Omega\,|\,XX^\dagger\,|\Omega\,\rangle \\ &= \langle\,\Omega\,|\,X^\dagger X\,|\Omega\,\rangle = \|X|\,\Omega\,\rangle\|^2 \ ,\end{aligned}$$

for all $X \in \mathcal{M}$ since $\mathcal{M}$ is Abelian. Therefore, $\Delta = 1\!\!1$ and $JXJ = X^\dagger$, whence (7.24) yields $\mathcal{M} = \mathcal{M}'$.

**Example 7.1.5.** A most used *GNS* representation [300], is the so-called *thermal representation* whose cyclic and separating vector is the tensor product of two vacuum states, $|\,\Omega_\beta\,\rangle = |\,vac\rangle \otimes |\,vac\rangle$, so that the *GNS* Hilbert space is isomorphic to the tensor product of two Fock Hilbert spaces.

We shall consider the framework of Examples 5.6.2.2,3 without restrictions on the dimensionality of the single particle Hilbert space and on the cardinality of the spectrum of the single particle Hamiltonian $h$. We shall denote by $a^{\#}$, respectively $b^{\#}$, Bose, respectively Fermi, creation and annihilation operators. In the Bose case, their action on $|\,\Omega_\beta\,\rangle$ is

$$\begin{aligned}\pi_\beta(a(f)) &=: a_\beta(f) = a(\sqrt{1\!\!1 + A_-}f)\otimes 1\!\!1 \,+\, 1\!\!1\otimes a^\dagger(j\sqrt{A_-}f)\\ \pi_\beta(a^\dagger(f)) &=: a^\dagger_\beta(f) = a^\dagger(\sqrt{1\!\!1 + A_-}f)\otimes 1\!\!1 \,+\, 1\!\!1\otimes a(j\sqrt{A_-}f)\ ,\end{aligned}$$

where the single particle operator $j:\mathbb{H}\mapsto\mathbb{H}$ is antilinear and satisfies $\langle\, jf\,|\,jg\,\rangle = \langle\, g\,|\,f\,\rangle$, while in the Fermi case

$$\begin{aligned}\pi_\beta(b(f)) &=: b_\beta(f) = b(\sqrt{1\!\!1 - A_+}f)\otimes 1\!\!1 \,+\, \theta\otimes a^\dagger(j\sqrt{A_+}f)\\ \pi_\beta(b^\dagger(f)) &=: b^\dagger_\beta(f) = b^\dagger(\sqrt{1\!\!1 - A_+}f)\otimes 1\!\!1 \,+\, \theta\otimes a(j\sqrt{A_+}f)\ ,\end{aligned}$$

where $\theta$ is an operator on the Fock space such that $\theta\, b^{\#} = -b^{\#}\,\theta$ and $\theta|\,vac\,\rangle = |\,vac\,\rangle$. Then,

$$\begin{aligned}a^\dagger_\beta(f)|\,\Omega_\beta\,\rangle &= |\,\sqrt{1\!\!1 + A_-}f\,\rangle\otimes|\,vac\,\rangle\ , &\quad a_\beta(f)|\,\Omega_\beta\,\rangle &= |\,vac\,\rangle\otimes|\,j\sqrt{A_-}f\,\rangle\\ b^\dagger_\beta(f)|\,\Omega_\beta\,\rangle &= |\,\sqrt{1\!\!1 - A_+}f\,\rangle\otimes|\,vac\,\rangle\ , &\quad b_\beta(f)|\,\Omega_\beta\,\rangle &= |\,vac\,\rangle\otimes|\,j\sqrt{A_+}f\,\rangle\ ,\end{aligned}$$

whence

$$\begin{aligned}\langle\,\Omega_\beta\,|\,a^\dagger_\beta(f)a_\beta(g)\,|\Omega_\beta\,\rangle &= \langle\, j\sqrt{A_-}f\,|\,j\sqrt{A_-}g\,\rangle = \langle\, g\,|\,A_-\,|f\,\rangle\\ \langle\,\Omega_\beta\,|\,b^\dagger_\beta(f)b_\beta(g)\,|\Omega_\beta\,\rangle &= \langle\, j\sqrt{A_+}f\,|\,j\sqrt{A_+}g\,\rangle = \langle\, g\,|\,A_+\,|f\,\rangle\ .\end{aligned}$$

The modular operators read

$$\Delta^-_\beta = \mathrm{e}^{-\beta\sum_i\varepsilon_i a^\dagger_i a_i}\otimes\mathrm{e}^{+\beta\sum_i\varepsilon_i a^\dagger_i a_i}\ ,\ \Delta^+_\beta = \mathrm{e}^{-\beta\sum_i\varepsilon_i b^\dagger_i b_i}\otimes\mathrm{e}^{+\beta\sum_i\varepsilon_i b^\dagger_i b_i}\ ,$$

where $a^{\#}_i$ and $b^{\#}_i$ create or annihilate eigenstates $|\,\varepsilon_i\rangle$ of the single-particle Hamiltonian $h$. By means of calculations similar to those that led to (7.9) and (7.10), one explicitly calculates

$$\begin{aligned}\Delta^-_\beta a_\beta(f)|\,\Omega_\beta\,\rangle &= |\,vac\,\rangle\otimes|\,j\mathrm{e}^{\beta h}\sqrt{A_-}f\,\rangle\\ \Delta^+_\beta b_\beta(f)|\,\Omega_\beta\,\rangle &= |\,vac\,\rangle\otimes|\,j\mathrm{e}^{\beta h}\sqrt{A_+}f\,\rangle\ .\end{aligned}$$

One can thus explicitly evaluate the action of the modular conjugation; from (7.25),

$$\begin{aligned}J_\beta\, a_\beta(f)|\,\Omega_\beta\,\rangle &= \sqrt{\Delta^-_\beta}a^\dagger_\beta(f)|\,\Omega_\beta\,\rangle = |\,\mathrm{e}^{-\beta h/2}\sqrt{1\!\!1 + A_-}f\,\rangle\otimes|\,vac\,\rangle\\ &= |\,\sqrt{A_-}f\,\rangle\otimes|\,vac\,\rangle\\ J_\beta\, a^\dagger_\beta(f)|\,\Omega_\beta\,\rangle &= \sqrt{\Delta^-_\beta}a_\beta(f)|\,\Omega_\beta\,\rangle = |\,vac\,\rangle\otimes|\,j\mathrm{e}^{\beta h/2}\sqrt{A_-}f\,\rangle\\ &= |\,vac\,\rangle\otimes|\,j\sqrt{1\!\!1 + A_-}f\,\rangle\ ,\end{aligned}$$

in the case of Bosons, while for Fermions one obtains

$$J_\beta\, b_\beta(f)|\,\Omega_\beta\,\rangle = \sqrt{\Delta_\beta^{+}}b_\beta^\dagger(f)|\,\Omega_\beta\,\rangle = |\,\mathrm{e}^{-\beta\,h/2}\sqrt{\mathbb{1}-A_+}f\,\rangle\otimes|\,vac\rangle$$
$$= |\,\sqrt{A_+}f\,\rangle\otimes|\,vac\rangle$$
$$J_\beta\, b_\beta^\dagger(f)|\,\Omega_\beta\,\rangle = \sqrt{\Delta_\beta^{+}}b_\beta(f)|\,\Omega_\beta\,\rangle = |\,vac\rangle\otimes|\,j\mathrm{e}^{\beta\,h/2}\sqrt{A_+}f\,\rangle$$
$$= |\,vac\rangle\otimes|\,j\sqrt{\mathbb{1}-A_+}f\,\rangle\ .$$

Since thermal states are faithful, $|\,\Omega_\beta\,\rangle$ is cyclic and separating, therefore

$$J_\beta\, a_\beta(f)\, J_\beta = a^\dagger(\sqrt{A_-}f)\otimes\mathbb{1}+\mathbb{1}\otimes a(j\sqrt{\mathbb{1}+A_-}f)$$
$$J_\beta\, a_\beta^\dagger(f)\, J_\beta = a(\sqrt{A_-}f)\otimes\mathbb{1}+\mathbb{1}\otimes a^\dagger(j\sqrt{\mathbb{1}+A_-}f)$$

for Bosons and, for Fermions,

$$J_\beta\, b_\beta(f)\, J_\beta = b^\dagger(\sqrt{A_+}f)\otimes\mathbb{1}+\theta\otimes b(j\sqrt{\mathbb{1}-A_+}f)$$
$$J_\beta\, b_\beta^\dagger(f)\, J_\beta = b(\sqrt{A_+}f)\otimes\mathbb{1}+\theta\otimes b^\dagger(j\sqrt{\mathbb{1}-A_+}f)\ .$$

The thermal representation has been used in [211] to implement the transposition in an infinite dimensional context and study the entanglement properties of infinitely extended quantum systems (see also [308]), the starting point being (5.149) in the finite-dimensional case. Let $V$ the flip operator which exchange vectors in tensor products $V|\,\phi\otimes\psi\,\rangle = |\,\psi\otimes\phi\,\rangle$, then

$$V^\dagger\, J_\rho\, X^\dagger\ \otimes\mathbb{1}_N\, J_\rho\, V = X^T\ \otimes\ \mathbb{1}\ .$$

Analogously, in the thermal representation one may represent the transposition as follows

$$\gamma_T^-[a_\beta(f)] = V_-^\dagger\, J_\beta\, a_\beta^\dagger(f)\, J_\beta\, V_- = a^\dagger(j\sqrt{\mathbb{1}+A_-}f)\otimes\mathbb{1}+\mathbb{1}\otimes\ a^\dagger(\sqrt{A_-}f)$$
$$\gamma_T^+[b_\beta(f)] = V_+^\dagger\, J_\beta\, b_\beta^\dagger(f)\, J_\beta\, V_+ = b^\dagger(j\sqrt{\mathbb{1}-A_+}f)\otimes\mathbb{1}+\theta\otimes\ b(\sqrt{A_+}f)\ .$$

Among the *CPU* maps on a $C^*$ algebra $\mathcal{A}$, a special role is played by the conditional expectations (see Definition 5.2.3). Suppose the orthogonal projections $P_i$ in Example 5.2.9.1 commute with a given density matrix $\rho\in\mathbb{B}_1^+(\mathbb{H})$, it then follows that

$$\rho\circ\mathbb{E}(X) = \mathrm{Tr}(\rho\mathbb{E}[X]) = \sum_{i\in I}\mathrm{Tr}(P_i\,\rho\,P_i\,X) = \mathrm{Tr}(\sum_{i\in I}P_i\rho\,X) = \rho(X)$$

for all $X\in\mathbb{B}(\mathbb{H})$ for $\sum_{i\in I}P_i = \mathbb{1}$. One says that the conditional expectation from $\mathbb{B}(\mathbb{H})$ onto the Abelian subalgebra $\mathcal{P}\subset\mathbb{B}(\mathbb{H})$ generated by the $P_j$ respects the state $\rho$. Also, notice that if $\rho$ is faithful then $\mathcal{P}$ is left invariant by the modular automorphism (5.180), that is $\sigma_\rho^t[\mathcal{P}] = \mathcal{P}$. This is the key point how to extends these considerations to the case of general von Neumann algebras with faithful normal states, where conditional expectations are identified with normal projections of norm one (see Remark (5.2.7)). We state the result, for a proof see [293].

**Proposition 7.1.1.** *Let $\mathcal{A}$ be a von Neumann algebra, $\omega$ a faithful normal state with associated modular group of automorphisms $\sigma^t_\omega$, $t \in \mathbb{R}$. Moreover, let $\mathcal{A}_0 \subset \mathcal{A}$ be a von Neumann subalgebra and $\omega_0$ the restriction $\omega \lceil \mathcal{A}_0$ with associated modular automorphisms $\sigma^t_{\omega_0}$. Then, the following conditions are equivalent:*

1. *there exists a normal conditional expectation $\mathbb{E} : \mathcal{A} \mapsto \mathcal{A}_0$ that respects the state, $\omega \circ \mathbb{E} = \omega$;*
2. *$\sigma^t_\omega[\mathcal{A}_0] \subseteq \mathcal{A}_0$ for all $t \in \mathbb{R}$;*
3. *$\sigma^t_{\omega_0}[A] = \sigma^t_\omega[A]$ for all $A \in \mathcal{A}_0$.*

### 7.1.2 *GNS* Representation and Dynamics

Quantum dynamical systems will be identified as non-commutative algebraic triplets (compare the analogous commutative Definition 2.2.4).

**Definition 7.1.4.** *Quantum dynamical systems are triplets $(\mathcal{A}, \Theta, \omega)$, where $\mathcal{A}$ is a $C^*$ algebra with identity $\mathbb{1}$, the dynamics $\Theta$ corresponds to a group of automorphisms $\Theta_t : \mathcal{A} \mapsto \mathcal{A}$, $t \in G$, such that*

$$\Theta_t \circ \Theta_s = \Theta_s \circ \Theta_t = \Theta_{t+s} \ , \quad \omega \circ \Theta_t = \omega \ , \quad \forall\, s,t \in G \ ,$$

*where $G = \mathbb{Z}$ or $G = \mathbb{R}$ and the state $\omega : \mathcal{A} \mapsto \mathbb{C}$ is a normalized, positive, $\Theta$-invariant expectation, namely $\omega \circ \Theta_t = \omega$ for all $t \in G$.*

Given an algebraic triplet $(\mathcal{A}, \Theta, \omega)$, a natural Hilbert space formulation is based on the *GNS* construction (see Definition 5.3.7); it does provide not only a representation $\pi_\omega(\mathcal{A})$ on a Hilbert space $\mathbb{H}_\omega$ with a cyclic invariant vector $|\,\Omega\,\rangle$, but also an implementation of the dynamics by a group of unitary operators.

**Proposition 7.1.2.** *[107, 300] Let $(\mathcal{A}, \Theta, \omega)$ be a $C^*$ dynamical system and $(\mathbb{H}_\omega, \pi_\omega, \Omega_\omega)$ the associated GNS triplet, then, the $C^*$ automorphism $\Theta$ is implemented by a unique unitary operator $U_\omega : \mathbb{H}_\omega \mapsto \mathbb{H}_\omega$,*

$$\pi_\omega(\Theta(X)) = U^\dagger_\omega\, \pi_\omega(X)\, U_\omega \qquad \forall\, X \in \mathcal{A} \ . \tag{7.28}$$

**Proof:** Given the *GNS* representation $\pi_\omega$, $\pi := \pi_\omega \circ \Theta$ is another representation of $\mathcal{A}$ on $\mathbb{H}_\omega$ such that

$$\langle\, \Omega_\omega \,|\, \pi(X) \,|\Omega_\omega \,\rangle = \langle\, \Omega_\omega \,|\, \pi_\omega(\Theta(X)) \,|\Omega_\omega \,\rangle = \omega(\Theta(X)) = \omega(X) \ .$$

Therefore, Remark 5.3.2.1 ensures the existence of a unitary operator $U_\omega$ such that (7.28) holds. If another unitary operator $W$ with the same properties exists, then $\Big[W^\dagger U_\omega \,,\, \pi_\omega(X)\Big]\pi_\omega(Y)|\,\Omega_\omega\,\rangle = 0$ for all $Y \in \mathcal{A}$, namely on a dense set; therefore, $W^\dagger U_\omega$ belongs to $\pi_\omega(\mathcal{A})'$ for which $|\,\Omega_\omega\,\rangle$ is separating (see Lemma 5.3.1). Then, $W = U_\omega$, since $(W^\dagger U_\omega - \mathbb{1})|\,\Omega_\omega\,\rangle = 0$. □

**Remarks 7.1.4.**

1. If the dynamics is specified by a one-parameter group of $C^*$ automorphisms $\{\Theta_t\}_{t\in\mathbb{R}}$ which is weakly-continuous in the *GNS* representation, then the group $U_\omega(G) := \{U_\omega(t)\}_{t\in G}$, $G = \mathbb{R}$, is strongly continuous on $\mathbb{H}_\omega$. In fact, the previous Proposition asserts that each $\Theta_t$ is implemented by a unitary operator $U_\omega(t)$; furthermore,

$$\begin{aligned} U^\dagger_\omega(t)U^\dagger_\omega(s)\pi_\omega(X)|\,\Omega_\omega\,\rangle &= U^\dagger_\omega(t)\pi_\omega(\Theta_s(X))|\,\Omega_\omega\,\rangle = \pi_\omega(\Theta_{t+s}(X))|\,\Omega_\omega\,\rangle \\ &= U^\dagger_\omega(t+s)\pi_\omega(X)|\,\Omega_\omega\,\rangle \end{aligned}$$

on a dense set, whence the family $\{U_\omega(t)\}_{t\in\mathbb{R}}$ forms a one-parameter group of unitaries on $\mathbb{H}_\omega$. Strong continuity follows from weak continuity and

$$\|(U_\omega(t) - \mathbb{1})\pi_\omega(X)|\,\Omega_\omega\,\rangle\|^2 = 2\Big(\omega(X^\dagger X) - \Re(\omega(X^\dagger\Theta_t(X)))\Big)\ .$$

2. The Fock representation is unitarily equivalent to the *GNS* representation based on the vacuum state: $\omega(a(f)) = \langle\, vac\,|\, a(f)\,|vac\rangle = 0$ for all $|\, f\,\rangle$ in the single-particle Hilbert space $\mathbb{H}$. Suppose $\Theta$ is a quasi-free Fermi automorphism as in Example 7.1.2.3; let $V : \mathbb{H}_F \mapsto \mathbb{H}_F$ be the unitary operator that implements it on the Fock space. Since the number operator is left invariant by quasi-free automorphisms, if $V$ belonged to $\mathcal{A}_F$, then it should also belong to the even Fermi algebra $\mathcal{A}_E$. Further, from asymptotic Abelianess, the invariance of the norm under unitary transformations and the fact that the various $V_t$ commute, it turns out that, for any $\varepsilon > 0$ and $X \in \mathcal{A}_F$,

$$\begin{aligned} \|[\,V_t\,,\,\Theta_s[X]\,]\| &= \|V_s^\dagger\,V_t\,X\,V_s\,-\,V_s^\dagger\,X\,V_t\,V_s\| \\ &= \|V_t\,X\,-\,X\,V_t\| = \|X\,-\,V_t^\dagger\,X\,V_t\| \le \varepsilon \end{aligned}$$

for all $t \in \mathbb{R}$. This cannot be true for all $X \in \mathcal{A}_F$ so that the unitary operator $V \in \mathbb{B}(\mathbb{H}_F)$ does not belong to $\mathcal{A}_F$. However, since $\mathcal{A}_F$ is irreducible (see (7.12)), $V$ belongs to the bicommutant $\mathcal{A}''_{B,F} = \{\lambda\mathbb{C}\}' = \mathbb{B}(\mathbb{H}_F)$.
3. Very rarely, starting from the Hamiltonian of a system of $N$ interacting particles and going to the thermodynamic limit, one obtains a norm-continuous dynamics at the $C^*$ algebraic level, that is independently of a given time-invariant state. Usually, the dynamics exists only in the *GNS* representation provided by that state; however, an instance of Galilei invariant interaction which gives rise to a norm-continuous group of automorphisms of the *CAR* algebra can be found in [300].

**Example 7.1.6 (Infinite Dimensional Quantum Cat Maps).**
The finite dimensional quantization of the torus $\mathbb{T}^2$ studied in Example 5.4.2 can be turned into an infinite dimensional one by lifting the condition (5.86), namely the quantum counterpart of the folding constraint (2.15) in Example 2.1.3. Concretely, the Weyl relations (5.83) become

$$U_\theta^{n_1} V_\theta^{n_2} = \mathrm{e}^{4\pi\, i\,\theta\, n_1 n_2}\, V_\theta^{n_2}\, U_\theta^{n_1} \ , \qquad \theta \in [0,1) \ ,$$

where $U$ and $V$ are two abstract unitary operators and $\boldsymbol{n} = (n_1, n_2) \in \mathbb{Z}^2$. Notice that $2\theta$ plays the role of $1/N$ in Example 5.4.2 and is a continuous *deformation parameter*: when $\theta = 0$, the commutation relations are those that hold for the exponential functions (2.21), namely

$$e_{\boldsymbol{n}} e_{\boldsymbol{m}} = e_{\boldsymbol{m}} e_{\boldsymbol{n}} \ .$$

Then, as in (5.84), we define the unitary Weyl-like operators

$$W_\theta(\boldsymbol{n}) := \mathrm{e}^{-2i\,\pi\,\theta\, n_1 n_2}\, U_\theta^{n_1}\, V_\theta^{n_2} \ , \qquad \boldsymbol{n} \in \mathbb{Z}^2 \ ,$$

that satisfy relations similar to those in (5.85),

$$W_\theta(\boldsymbol{n})\, W_\theta(\boldsymbol{m}) = \mathrm{e}^{2i\,\pi\,\theta\,\sigma(\boldsymbol{n},\boldsymbol{m})}\, W_\theta(\boldsymbol{n}+\boldsymbol{m}) \ , \quad \forall \boldsymbol{n},\boldsymbol{m} \in \mathbb{Z}^2 \ , \tag{7.29}$$

with symplectic form $\sigma(\boldsymbol{n},\boldsymbol{m}) := n_1 m_2 - n_2 m_1$. Also, in analogy with (7.11), one sets

$$W_\theta(f) = \sum_{\boldsymbol{n}} f(\boldsymbol{n})\, W_\theta(\boldsymbol{n}) \ , \tag{7.30}$$

where $|\, f\,\rangle = \{f(\boldsymbol{n})\}_{\boldsymbol{n}\in\mathbb{Z}^2}$ belongs to the subspace $\ell_*(\mathbb{Z}^2) \subset \ell^2(\mathbb{Z}^2)$ of square-summable sequences with finitely many non-zero components. We shall call *support* of $f$ the set

$$\mathrm{Supp}(f) := \Big\{\boldsymbol{n} \in \mathbb{Z} \ : \ f(\boldsymbol{n}) \neq 0\Big\} \tag{7.31}$$

The following properties hold for all $f, g \in \ell_*(\mathbb{Z}^2)$,

$$W_\theta(f)^\dagger = W_\theta(f^T) \ , \qquad f^T(\boldsymbol{n}) = f(-\boldsymbol{n})^*$$

$$W_\theta(f) W_\theta(g) = W_\theta(f * g) \ , \quad \text{with} \tag{7.32}$$

$$(f * g)(\boldsymbol{n}) := \sum_{\boldsymbol{m}\in\mathbb{Z}^2} \mathrm{e}^{2i\,\pi\,\theta\,\sigma(\boldsymbol{n},\boldsymbol{m})}\, f(\boldsymbol{n}-\boldsymbol{m}) g(\boldsymbol{m}) \ . \tag{7.33}$$

Consider the $*$-algebra $\mathcal{A}_\theta^* := \big\{W_\theta(f) \ : \ f \in \ell^*(\mathbb{Z}^2)\big\}$ generated by all possible linear combinations of Weyl operators $W_\theta(f)$ with $f \in \ell^*(\mathbb{Z}^2)$. Let then $\omega$ denote the linear functional $\omega : \mathcal{A}_\theta^* \mapsto \mathbb{C}$ such that

$$\omega(W_\theta(\boldsymbol{n})) = \delta_{\boldsymbol{n}\boldsymbol{0}} \ . \tag{7.34}$$

Using (7.32) with (7.33) one checks that

$$\omega\left(W_\theta(f)^\dagger\, W_\theta(g)\right) = (f^T * g)(\mathbf{0}) = \sum_{\boldsymbol{m}\in\mathbb{Z}^2} f^*(\boldsymbol{m})g(\boldsymbol{m}) = \langle\, f\,|\, g\,\rangle\,. \qquad (7.35)$$

Thus $\omega$ is a positive normalized functional, namely a state on $\mathcal{A}_\theta^*$ similar to $\omega_N$ in (5.187). Consider the associated *GNS* representation $\pi_\omega(\mathcal{A}_\theta^*)$ and set

$$|\, f\,\rangle_\theta := \pi_\omega(W_\theta(f))|\,\Omega\,\rangle \qquad \forall\,|\, f\,\rangle \in \ell^*(\mathbb{Z}^2)\ ,$$

where $|\,\Omega_\omega\,\rangle$ is the cyclic *GNS* vector. Then, the vectors $|\,\boldsymbol{n}\,\rangle_\theta$ form an *ONB* in $\mathbb{H}_\omega$ and ${}_\theta\langle\,\boldsymbol{n}\,|\, f\,\rangle_\theta = \langle\,\boldsymbol{n}\,|\, f\,\rangle = f(\boldsymbol{n})$. Therefore, $\mathbb{H}_\omega = \ell^2(\mathbb{Z}^2) = \mathbb{L}^2_{d\boldsymbol{r}}(\mathbb{T}^2)$, independently of the deformation parameter $\theta$; also

$$\pi_\omega(W_\theta(f))|\, g\,\rangle = |\, f * g\,\rangle\,. \qquad (7.36)$$

The $*$-algebra $\mathcal{A}_\theta^*$ can be equipped with a $*$-automorphism which extends to the present case the dynamics discussed in Example 5.6.1.4: it is defined on the Weyl operators by

$$\Theta_{\mathbb{A}}[W_\theta(\boldsymbol{n})] = W_\theta(\mathbb{A}^T\boldsymbol{n})\ , \qquad (7.37)$$

where $\mathbb{A}$ is a $2\times 2$ integer matrix as in (5.176). The state $\omega$ is left invariant by $\Theta_{\mathbb{A}}$ which is then implemented by a same unitary operator $U_\omega$ for all $\theta \in [0,1)$ that coincides with the Koopman operator $U_{\mathbb{A}}$ of Example 2.1.3. Indeed,

$$\Theta_{\mathbb{A}}[W_\theta(f)] = \sum_{\boldsymbol{n}} f(\boldsymbol{n})\, W_\theta(\mathbb{A}^T\boldsymbol{n}) = \sum_{\boldsymbol{p}} f(\mathbb{A}^{-T}\boldsymbol{p})\, W_\theta(\boldsymbol{p}) = W_\theta(U_{\mathbb{A}}\, f)\ , \qquad (7.38)$$

for $f(\mathbb{A}^{-T}\boldsymbol{n}) = (U_{\mathbb{A}}\, f)(\boldsymbol{n})$ (compare (2.1.3)). Consequently,

$$\begin{aligned}\omega\left(W_\theta(f)\Theta_{\mathbb{A}}[W_\theta(g)]\right) &= \langle\,\Omega\,|\,\pi_\omega(W_\theta(f))^\dagger\, U_\omega\,\pi_\omega(W_\theta(g))\,|\Omega\,\rangle \\ &= \omega(W_\theta(f)W_\theta(U_{\mathbb{A}}\, g)) = \langle\, f\,|\, U_{\mathbb{A}}\,|g\,\rangle\,. \qquad (7.39)\end{aligned}$$

The dependence on $\theta \in [0,1)$ emerges when considering the closure of $\pi_\omega(\mathcal{A}_\theta^*)$ with respect to strong-operator topology thus obtaining von Neumann subalgebras $\mathcal{M}_\theta \subseteq \mathbb{B}(\mathbb{H}_\omega)$.

While the *GNS* Hilbert space and the unitary implementation of the dynamics are the same for all von Neumann dynamical triplets $(\mathcal{M}_\theta, \Theta_{\mathbb{A}}, \omega)$ [1], for $\theta = 0$ $\mathcal{M}_\theta$ is isomorphic to the maximally Abelian von Neumann algebra $\mathbb{L}^\infty_{d\boldsymbol{r}}(\mathbb{T}^2)$ of essentially bounded functions on $\mathbb{T}^2$ (see Section 5.3.2), for $\theta$ irrational $\mathcal{M}_\theta$ is a hyperfinite $II_1$ factor, while $\mathcal{M}_\theta$ is not a factor and of finite type $I_n$ when $\theta$ is rational.

Let us consider the case $\theta = 0$; clearly, $\mathcal{M}_0$ is an Abelian von Neumann subalgebra of $\mathbb{B}(\mathbb{H}_\omega)$, actually, maximally Abelian since it has a cyclic vector $|\,\Omega_\omega\,\rangle$ (see Example 7.1.4.1); therefore, it is isomorphic to $\mathbb{L}^\infty_{d\boldsymbol{r}}(\mathbb{T}^2)$ via the argument of Theorem 5.3.3 and a one-to-one mapping

[1] $\Theta_{\mathbb{A}}$ and $\omega$ denote the extensions of the automorphism $\Theta_{\mathbb{A}}$ and of the state $\omega$ from $\mathcal{A}_\theta^*$ to the strong-operator closures $\mathcal{M}_\theta$.

$$e_{\boldsymbol{n}} \mapsto \Phi[e_{\boldsymbol{n}}] = W_0(\boldsymbol{n}) \tag{7.40}$$

between the Weyl operators $W_0(\boldsymbol{n})$ and the exponential functions (2.21). Indeed, one observes that the multiplication of vectors in $\mathbb{L}^2_{\mathrm{d}\boldsymbol{r}}(\mathbb{T}^2)$ by $e_{\boldsymbol{n}}$ and the action (7.36) of $W_0(\boldsymbol{n})$ on vectors in $\mathbb{H}_\omega$ do coincide. We shall thus identify $\mathcal{M}_0 = \mathbb{L}^\infty_{\mathrm{d}\boldsymbol{r}}(\mathbb{T}^2)$.

Consider now the case of a rational deformation parameter, $\theta = p/q$, $p, q \in \mathbb{N}$; then, (7.29) yields

$$W_{p/q}(q\boldsymbol{n})W_{p/q}(\boldsymbol{m}) = \mathrm{e}^{2\pi\, i\, p\, \sigma(\boldsymbol{n},\boldsymbol{m})} W_{p/q}(q\boldsymbol{n}+\boldsymbol{m}) = W_{p/q}(\boldsymbol{m})W_{p/q}(q\boldsymbol{n}) \ ,$$

for all $\boldsymbol{n}, \boldsymbol{m} \in \mathbb{Z}^2$. Further, set

$$\mathbb{Z}^2 \ni \boldsymbol{n} = (n_1, n_2) = [\boldsymbol{n}] + <\boldsymbol{n}> := ([n_1] + <n_1>, [n_2] + <n_2>) \ ,$$

where, for any $n \in \mathbb{Z}$, $[n] = q\, m$ denotes the unique multiple of $q$ such that $0 \leq n - [n] =: <n> \leq q-1$. Then, one gets

$$W_{p/q}(\boldsymbol{n}) = W_{p/q}([\boldsymbol{n}])\, W_{p/q}(<\boldsymbol{n}>) \ .$$

As a consequence, when $\theta$ is rational, every $W_{p/q}(f)$ can be written as

$$\begin{aligned} W_{p/q}(f) &= \sum_{<\boldsymbol{n}>\in J(q)} \Big(\sum_{[\boldsymbol{n}]} f([\boldsymbol{n}] + <\boldsymbol{n}>)\, W_{p/q}([\boldsymbol{n}])\Big)\, W_{p/q}(<\boldsymbol{n}>) \\ &= \sum_{\boldsymbol{s}\in J(q)} X_f(\boldsymbol{s})\, W_{p/q}(\boldsymbol{s}) \qquad \text{with} \qquad (7.41) \\ X_f(\boldsymbol{s}) &:= \sum_{\boldsymbol{n}\in\mathbb{Z}^2} f(q\boldsymbol{n}+\boldsymbol{s})\, W_{p/q}(q\boldsymbol{n}) \in \mathcal{M}^{(q)} \ , \qquad (7.42) \end{aligned}$$

$J(q) := \Big\{\boldsymbol{s} = (s_1, s_2) \ : \ 0 \leq s_i \leq q-1\Big\}$ and where

$$\mathcal{M}^{(q)} := \Big\{\sum_{\boldsymbol{n}\in\mathbb{Z}^2} f(\boldsymbol{n})\, W_{p/q}(q\boldsymbol{n})\Big\} \tag{7.43}$$

denotes the von Neumann subalgebra of $\mathcal{M}_{p/q}$ linearly generated by the Weyl operators of the form $W_{p/q}(q\boldsymbol{n})$, $\boldsymbol{n} \in \mathbb{Z}^2$. Because of (7.41), they commute with $\mathcal{M}_{p/q}$ whence $\mathcal{M}^{(q)}$ belongs to the center of $\mathcal{M}_{p/q}$.

Moreover, the exponential functions of the form $W_0(q\boldsymbol{n})$ fulfil

$$W_0(q\boldsymbol{n})(\boldsymbol{r}) = W_0(q\boldsymbol{n})(\boldsymbol{r} + \boldsymbol{s}/q) \quad \forall\, \boldsymbol{s} \in J(q) \ . \tag{7.44}$$

They generate a $*$-algebra whose strong-closure is a von Neumann subalgebra $\mathcal{M}_0^{(q)} \subset \mathcal{M}_0$ of essentially bounded functions $f$ on $\mathbb{T}^2$ such that

$$f(\boldsymbol{r}) = \gamma^{(q)}_{\boldsymbol{s}}[f](\boldsymbol{r}) := f(\boldsymbol{r} + \boldsymbol{s}/q) \ , \tag{7.45}$$

for all $s \in J(q)$. Let $\Pi_s : \mathcal{M}_0 \mapsto \mathcal{M}_0^{(q)}$ be defined by

$$\mathcal{M}_0 \ni f \mapsto \Pi_s[f] := \sum_{n\in\mathbb{Z}} f(qn+s) W_0(qn) \in \mathcal{M}_0^{(q)} , \tag{7.46}$$

where $s \in J(q)$. By decomposing

$$f = \sum_n f(n)\, W_0(n) = \sum_{s\in J(q)} \Big(\sum_m f(m+s)\, W_0(qm)\Big)\, W_0(s) ,$$

it follows that $\Pi_s$ can be recast as

$$\Pi_s[f] = W_0(-s)\, \frac{1}{q^2} \sum_{t\in J(q)} \gamma_t^{(q)}[f]\, \mathrm{e}^{-2\pi i\, s\cdot t/q} . \tag{7.47}$$

Furthermore, a map similar to the one in (7.40),

$$W_0(qn) \mapsto \Phi_q[W_0(qn)] = W_{p/q}(qn) \qquad \forall n \in \mathbb{Z}^2 , \tag{7.48}$$

makes $\mathcal{M}^{(q)}$ and $\mathcal{M}_0^{(q)}$ isomorphic so that (7.42), respectively (7.41) read $X_f(s) = \Phi_q[\Pi_s[f]]$, respectively

$$W_{p/q}(f) = \sum_{s\in J(q)} \Phi_q[\Pi_s[f]]\, W_{p/q}(s) . \tag{7.49}$$

Concluding: 1) $\mathcal{M}_{p/q}$ is not a factor, 2) due to the finitely many non-commuting $W_{p/q}(s)$ with $s \in J(q)$, the type of $\mathcal{M}_{p/q}$ is finite $I_n$ (see the discussion of types preceding Section 7.1) and 3) $\mathcal{M}_{p/q}$ is hyperfinite for such is $\mathcal{M}_0$ (and thus $\mathcal{M}_0^{(q)}$) according to Remark 7.1.1.

For $\theta$ irrational, $\mathcal{M}_\theta$ is a factor; indeed, from (7.29),

$$\Big[W_\theta(n) , W_\theta(m)\Big] = 2i \sin(2\pi\, \theta\sigma(n,m))\, W_\theta(n+m) \tag{7.50}$$

cannot vanish for $n \neq m$, whence the center $\mathcal{Z} = \mathcal{M}_\theta \cap \mathcal{M}'_\theta$ is trivial, that is it consists of multiples of the identity only. Since the state (7.34) is a trace on $\mathcal{M}_\theta$, according to the discussion following Definition 7.34, $\mathcal{M}_\theta$ is a type $II_1$ factor and also hyperfinite [263].

In the commutative setting of Example 2.2.3, the unitary $U_\omega$ corresponds to the Koopman-von Neumann operator which cannot belong to the commutative von Neumann algebra $\mathcal{M}_0 = \mathbb{L}^\infty_\mu(\mathcal{X})$.

As much as in this case and unlike for finite level quantum systems, the quantum dynamics is typically implemented by unitary operators which map the algebra of observables into itself, indeed

$$U_\omega^\dagger(t)\, \pi_\omega(X)\, U_\omega(t) = \pi_\omega(\Theta_t(X)) \in \pi_\omega(\mathcal{A}) , \tag{7.51}$$

without $U_\omega(t)$ itself belonging to $\pi_\omega(\mathcal{A})$. Given $\pi_\omega(\mathcal{A})$ and $U_\omega(G)$, one can however consider all linear combinations of products of operators in $\pi_\omega(\mathcal{A})$ and elements $U_\omega(t)$ which can always be reduced to the form (compare Example 5.3.2.7 for a similar structure)

$$\sum_{i\in I} \pi_\omega(X_i)\, U_\omega(t_i) \ . \tag{7.52}$$

These elements form an algebra, denoted by $\{\pi_\omega(\mathcal{A}), U_\omega(G)\}$ whose bi-commutant, namely its strong closure on $\mathbb{H}_\omega$, turns out to be a useful tool to discuss ergodicity and mixing in quantum dynamics.

**Definition 7.1.5 (Covariance Algebra).** *Given a quantum dynamical system $(\mathcal{A}, \Theta, \omega)$ and the GNS implementation of the dynamics, the associated covariance algebra is the von Neumann algebra $\mathcal{R}_\omega := \{\pi_\omega(\mathcal{A}), U_\omega(G)\}''$.*

As we have seen in Section 5.3, besides the von Neumann algebra $\mathcal{R}_\omega$ itself, what is also important is its commutant; in particular, in the framework of the *GNS* construction, for what concerns the convex decompositions of the reference state $\omega$ (see Remark 5.3.2.3). As regards the covariance algebra $\mathcal{R}_\omega$ and its commutant $\mathcal{R}'_\omega$, notice that if $X \in \mathbb{B}(\mathbb{H}_\omega)$ commutes with $\mathcal{R}_\omega$, it must commute with both $\pi_\omega(\mathcal{A})$ and $U_\omega(G)$. Vice versa, if $X \in \mathbb{B}(\mathbb{H}_\omega)$ commutes with $\pi_\omega(\mathcal{A})$ and $U_\omega(G)$, by continuity, it also commutes with the von Neumann algebra generated by them. Therefore,

$$\mathcal{R}'_\omega = \pi_\omega(\mathcal{A})' \cap U_\omega(G)' \ , \tag{7.53}$$

where $U_\omega(G)'$ is the algebra of the bounded constants of the motion, that is the algebra of all $X \in \mathbb{B}(\mathbb{H}_\omega)$ such that $U_\omega(t)^\dagger\, X\, U_\omega(t) = X$.

**Example 7.1.7.** For the case of Example 5.6.2 the covariance algebra is

$$\begin{aligned}\mathcal{R}_\rho &= \Big(\pi_\rho(M_2(\mathbb{C})) \cup U_\rho(\mathbb{R})\Big)'' = \Big(M_2(\mathbb{C}) \otimes \mathbb{1}_2 \cup \rho^{it} \otimes \rho^{-it}\Big)'' \\ &= M_2(\mathbb{C}) \otimes \{\mathbb{1}_2, \sigma_3\} \ ,\end{aligned}$$

where $\{\mathbb{1}_2, \sigma_3\}$ stands for the commutative algebra of $2 \times 2$ matrices which are diagonal in the eigenbasis of $\rho$. Furthermore, the constants of the motion are contained in $U_\rho(\mathbb{R})' = \{\mathbb{1}_2, \sigma_3\} \otimes \{\mathbb{1}_2, \sigma_3\}$ and

$$\mathcal{R}'_\rho = \pi_\rho(M_2(\mathbb{C}))' \cap U_\rho(\mathbb{R})' = \mathbb{1}_2 \otimes \{\mathbb{1}_2, \sigma_3\} \ .$$

### 7.1.3 Quantum Ergodicity and Mixing

As seen in Section 2.3, ergodicity corresponds to a specific behavior of the time-averages of two-point correlation functions. Given a quantum dynamical triplet $(\mathcal{A}, \Theta, \omega)$, two-point correlation functions have the form $\omega(A\Theta_t(B))$ where $A, B \in \mathcal{A}$ and $t \in G$, where $G = \mathbb{R}$ or $\mathbb{Z}$. For sake of concreteness [2], we will consider averages or *invariant means* as in (7.3), namely of the form (compare Definition 2.3.1)

$$\eta_t\left[\omega(A^\dagger\Theta_t(B)C)\right] = \lim_{T\to\infty}\frac{1}{2T+1}\sum_{t=-T}^{T-1}\omega(A^\dagger\Theta_t(B)C) \qquad (7.54)$$

in discrete time, or else

$$\eta_t\left[\omega(A^\dagger\Theta_t(B)C)\right] = \lim_{T\to\infty}\frac{1}{2T}\int_{-T}^{T}\mathrm{d}t\;\omega(A^\dagger\Theta_t(B)C) \quad \text{if} \quad G=\mathbb{R}\,. \qquad (7.55)$$

Because of (5.49), it turns out that these averages are bounded,

$$\left|\eta_t\left[\omega(A^\dagger\Theta_t(B)C)\right]\right| \leq \|A\|\,\|B\|\,\|C\|\,.$$

Also, in the *GNS* construction based on the invariant state $\omega$, the averages

$$\eta_t\left[\omega(A^\dagger\Theta_t(B)C)\right] = \eta_t\left[\langle\,\Omega_\omega\,|\,\pi_\omega(A)^\dagger U_\omega^\dagger(t)\,\pi_\omega(B)\,U_\omega(t)\,\pi_\omega(C)\,|\Omega_\omega\,\rangle\right]\;,$$

provide bounded sesquilinear forms on a dense subset of the *GNS* Hilbert space $\mathbb{H}_\omega$. This observation together with an argument similar to that in Example 5.3.2.3 lead one to introduce

1. an operator $\eta_\omega\left[U_\omega\right] \in \mathbb{B}(\mathbb{H}_\omega)$ with matrix elements

$$\langle\,\psi\,|\,\eta_\omega\left[U_\omega\right]\,|\phi\,\rangle = \eta_t\left[\langle\,\psi\,|\,U_\omega(t)\,|\phi\,\rangle\right] \quad \forall\psi,\phi\in\mathbb{H}_\omega\;; \qquad (7.56)$$

2. a linear map $\eta_\omega : \mathcal{A} \mapsto \mathbb{B}(\mathbb{H}_\omega)$ defined by

$$\langle\,\psi\,|\,\eta_\omega\left[A\right]\,|\phi\,\rangle = \eta_t\left[\langle\,\psi\,|\,U_\omega^\dagger(t)\,\pi_\omega(A)\,U_\omega(t)\,|\phi\,\rangle\right] \quad \forall\psi,\phi\in\mathbb{H}_\omega\,. \qquad (7.57)$$

Notice that, because of $\Theta$-invariance, $\omega(\eta_\omega\left[X\right]) = \omega(X)$.

**Examples 7.1.8.**

1. Consider a finite-dimensional dynamical system described by a finite-dimensional Hilbert space $\mathbb{H} = \mathbb{C}^d$, by observables that are matrices in $M_d(\mathbb{C})$ and by a Hamiltonian $H$ assumed to have non-degenerate eigenvalues $e_d \geq e_{d-1} \geq \dots e_0 = 0$ and eigenvectors $|\,j\,\rangle$, $j = 0, 2, \dots, d-1$. The dynamics is thus given by (see (5.170))

[2] For more details on the existence of invariant means see [107].

$$U_t = e^{-itH} = |0\rangle\langle 0| + \sum_{j=1}^{d-1} e^{-ie_j t} |j\rangle\langle j|$$

$$X \mapsto X_t = U_t^\dagger X U_t = \sum_{j,k=0}^{d-1} e^{it(e_j - e_k)} \langle k|X|j\rangle |k\rangle\langle j| ,$$

for all $X \in M_d(\mathbb{C})$. Then, the time-average $\eta$ yields

$$\eta(U) = |0\rangle\langle 0| , \quad \eta(X) = \sum_{j=0}^{d-1} \langle j|X|j\rangle |j\rangle\langle j| .$$

Thus, $\eta : M_d(\mathbb{C}) \mapsto M_d(\mathbb{C})$ amounts to the conditional expectation (see Example 5.2.9.1) onto the Abelian subalgebra of $M_d(\mathbb{C})$ generated by the minimal projections $|j\rangle\langle j|$.

2. All the eigenvectors $|j\rangle$ in the previous example provide, $\mathcal{U}_t$-invariant expectation functionals $\omega_j$ on $M_d(\mathbb{C})$. The corresponding irreducible *GNS* representations $\pi_{\omega_j}(M_d(\mathbb{C}))$ (see Section 5.5.1) act on a Hilbert space isomorphic to $\mathbb{C}^d$ with *GNS* cyclic vectors of the form $|\Omega_j\rangle = |j\rangle \otimes |j\rangle$. The dynamics is implemented by unitary operators of the form

$$U_{\omega_j}(t) = e^{-itH} \otimes e^{ite_j} \mathbb{1}_d ,$$

so that they preserve $|\Omega_j\rangle$. It follows that $\eta_{\omega_j}\left[U_{\omega_j}\right] = |\Omega_j\rangle\langle\Omega_j|$, while $\eta_{\omega_j}[X] = \eta(X) \otimes |j\rangle\langle j|$ for all $X \in M_d(\mathbb{C})$.

3. Consider the projection $P_\omega$ onto the subspace of vectors $|\psi\rangle \in \mathbb{H}_\omega$ such that $U_\omega|\psi\rangle = |\psi\rangle$ (the *GNS* vector $|\Omega_\omega\rangle$ is certainly one of them). Then, since $\psi, \phi \in \mathbb{H}_\omega$ are arbitrary,

$$\langle\psi|\eta_\omega[U_\omega] P_\omega|\phi\rangle - \eta_t[\langle\psi|U_\omega(t) P_\omega|\phi\rangle] = \langle\psi|P_\omega|\phi\rangle$$

implies $\eta_\omega[U_\omega] P_\omega = P_\omega$; analogously, $P_\omega \eta_\omega[U_\omega] = P_\omega$. Moreover,

$$\langle\psi|U_\omega(s)\eta_\omega[U_\omega]|\phi\rangle = \eta_t[\langle\psi|U_\omega(t+s)|\phi\rangle] = \langle\psi|\eta_\omega[U_\omega]|\phi\rangle ,$$

whence $\eta_\omega[U_\omega]|\phi\rangle$ is invariant. Thus, $P_\omega\eta_\omega[U_\omega]|\phi\rangle = \eta_\omega[U_\omega]|\phi\rangle$ for all $\phi \in \mathbb{H}_\omega$ and then $\eta_\omega[U_\omega] = P_\omega \eta_\omega[U_\omega] = P_\omega$.
Notice that $\eta_\omega[U_\omega]$ belongs to $\mathcal{R}_\omega$, for it arises from averaging correlation functions; furthermore, since it equals $P_\omega$, it does not depend on the specific invariant mean used.

While the average of the time-evolution $U_\omega(t)$ gives rise to the projector onto $U_\omega(t)$-invariant vectors (compare Proposition 2.3.4), the average of quasi-local observables $A \in \mathcal{A}$ transforms them into global constants of the motion, namely into observables that belong to the strong-closure of $\mathcal{A}$ and that are left invariant by the dynamics.

**Proposition 7.1.3.** $\eta_\omega[\mathcal{A}] \subseteq \pi_\omega(\mathcal{A})'' \cap U_\omega(G)'$.

**Proof:** We check this on the dense subset $\pi(\mathcal{A})|\,\Omega_\omega\,\rangle \subseteq \mathbb{H}_\omega$. Let $X$ belong to $\mathcal{A}$ and $X'$ to the commutant $\pi_\omega(\mathcal{A})'$, then, using (7.51),

$$\begin{aligned}
&\langle\,\Omega_\omega\,|\,\pi_\omega(A)^\dagger\eta_\omega[X]\;X'\pi_\omega(C)\,|\Omega_\omega\,\rangle = \\
&\qquad = \eta_t\left[\langle\,\Omega_\omega\,|\,\pi_\omega(A)^\dagger\pi_\omega(\Theta_t(X))\,X'\pi_\omega(C)\,|\Omega_\omega\,\rangle\right] \\
&\qquad = \eta_t\left[\langle\,\Omega_\omega\,|\,\pi_\omega(A)^\dagger\,X'\,\pi_\omega(\Theta_t(X))\,\pi_\omega(C)\,|\Omega_\omega\,\rangle\right] \\
&\qquad = \langle\,\Omega_\omega\,|\,\pi_\omega(A)^\dagger\,X'\,\eta_\omega[X]\;\pi_\omega(C)\,|\Omega_\omega\,\rangle\ .
\end{aligned}$$

Thus, $\eta_\omega[X]$ commutes with $\pi_\omega(\mathcal{A})'$ whence $\eta_\omega[\mathcal{A}] \subseteq \pi_\omega(\mathcal{A})''$. Similarly,

$$\begin{aligned}
&\langle\,\Omega_\omega\,|\,\pi_\omega(A)^\dagger\eta_\omega[X]\;U_\omega(s)\pi_\omega(C)\,|\Omega_\omega\,\rangle = \\
&\quad = \eta_t\left[\langle\,\Omega_\omega\,|\,\pi_\omega(A)^\dagger\,U_\omega^\dagger(t)\pi_\omega(X)U_\omega(t+s)\,\pi_\omega(C)\,|\Omega_\omega\,\rangle\right] \\
&\quad = \eta_t\left[\langle\,\Omega_\omega\,|\,\pi_\omega(A)^\dagger\,U_\omega(s)U_\omega^\dagger(t+s)\pi_\omega(X)U_\omega(t+s)\,\pi_\omega(C)\,|\Omega_\omega\,\rangle\right] \\
&\quad = \langle\,\Omega_\omega\,|\,\pi_\omega(A)^\dagger\,U_\omega(s)\,\eta_\omega[X]\;\pi_\omega(C)\,|\Omega_\omega\,\rangle\ ,
\end{aligned}$$

for all $X \in \mathcal{A}$ and $s \in G$, whence $\pi_\omega(X) \in U_\omega(G)'$. □

**Example 7.1.9.** We have just showed that $X \in \mathcal{A} \Longrightarrow \eta_\omega[X] \in U_\omega(G)'$; also, by construction (see Example 7.1.8.3), $P_\omega = \eta_\omega[U_\omega] \in U_\omega(G)''$. Then, it follows that $[\eta_\omega[X]\ ,\ P_\omega] = 0$, whence

$$\begin{aligned}
&\langle\,\Omega_\omega\,|\,\pi_\omega(A)^\dagger\eta_\omega[X]\;P_\omega\,\pi_\omega(B)\,|\Omega_\omega\,\rangle = \\
&\qquad = \eta_t\left[\langle\,\Omega_\omega\,|\,\pi_\omega(A)^\dagger\,P_\omega\,\pi_\omega(\Theta_t(X))\,P_\omega\,\pi_\omega(B)\,|\Omega_\omega\,\rangle\right] \\
&\qquad = \langle\,\Omega_\omega\,|\,\pi_\omega(A)^\dagger\,P_\omega\,\pi_\omega(X)\,P_\omega\,\pi_\omega(B)\,|\Omega_\omega\,\rangle
\end{aligned}$$

on a dense set. Therefore, for all $X \in \mathcal{A}$, it holds that

$$\eta_\omega[X]\,P_\omega = P_\omega\,\eta_\omega[X] = P_\omega\,\pi_\omega(X)\,P_\omega \quad \forall X \in \mathcal{A}\ .$$

In the case of classical ergodic systems, these latter systems are equivalently identified by the clustering properties of their two-point correlation functions (Propositions (2.3.2) and 2.3.9), by the spectral properties of the corresponding Koopman operator (Corollary 2.3.1) and by the extremality of their invariant states (Proposition 2.3.8). Clustering in the mean as in (2.65) or (2.75) and extremality are notions that readily extend to the non-commutative setting.

**Definition 7.1.6.** *A quantum dynamical system* $(\mathcal{A},\Theta,\omega)$ *is* $\eta$*-clustering if*

$$\eta_t\left[\omega(A\Theta_t(B))\right] = \omega(A)\omega(B) \quad \forall A,B \in \mathcal{A}\ , \tag{7.58}$$

*clustering if*

$$\lim_{t\to\pm\infty} \omega(A\,\Theta_t(B)) = \omega(A)\,\omega(B) \quad \forall X, Y \in \mathcal{A}\;. \tag{7.59}$$

*A state $\nu$ on $\mathcal{A}$ is* extremal *if $\nu = \lambda\nu_1 + (1-\lambda)\nu_2$ with $0 < \lambda < 1$ and $\nu_{1,2}$ states on $\mathcal{A}$ implies $\nu_{1,2} = \nu$; $\nu$ is an* extremal $\Theta$-invariant *if it cannot be written as a convex sum of other $\Theta$-invariant states on $\mathcal{A}$.*

**Remarks 7.1.5.**

1. If $\pi_\omega(\mathcal{A})'' \cap U_\omega(G)' = \{\lambda\,\mathbb{1}\}$, where $\{\lambda\,\mathbb{1}\}$ denotes the trivial algebra consisting only of multiples of the identity, then the global constants of the motion are trivial. It follows that $(\mathcal{A}, \Theta, \omega)$ is $\eta$-clustering. In fact, in such a case $\eta_\omega[B] = \omega(B)\,\mathbb{1}$ for all $B \in \mathcal{A}$, whence
$$\eta_t\left[\omega(A\Theta_t(B))\right] = \langle\, \Omega_\omega \,|\, \pi_\omega(A)\eta_\omega[B]\,|\Omega_\omega\,\rangle = \omega(A)\omega(B)\;.$$
2. Suppose the invariant state $\omega$ in $(\mathcal{A}, \Theta, \omega)$ is not extremal invariant; then, there exists a state $\nu$ on $\mathcal{A}$ such that $\lambda\nu \le \omega$, for some $0 < \lambda < 1$, and $\nu \circ \Theta = \nu$. Thus, from Remark 5.3.2.3, $\lambda\nu(X) = \langle\, \Omega_\omega \,|\, T'\pi_\omega(X)\,|\Omega_\omega\,\rangle$ for all $X \in \mathcal{A}$, with $0 \le T' \in \pi_\omega(\mathcal{A})'$. It turns out that $T' \in U_\omega(G)'$, too; indeed, $\Theta$-invariance yields
$$\begin{aligned}\langle\, \Omega_\omega \,|\, \pi_\omega(A)^\dagger T'\, U_\omega(t)\pi_\omega(B)\,|\Omega_\omega\,\rangle &= \langle\, \Omega_\omega \,|\, \pi_\omega(A)^\dagger T'\, \pi_\omega(\Theta_{-t}(B))\,|\Omega_\omega\,\rangle \\ &= \lambda\,\nu(A^\dagger\,\Theta_{-t}(B)) = \lambda\,\nu(\Theta_t(A)^\dagger\,B) \\ &= \langle\, \Omega_\omega \,|\, \pi_\omega(A)^\dagger\, U_\omega(t)\,T'\,\pi_\omega(B)\,|\Omega_\omega\,\rangle\;,\end{aligned}$$
on a dense set. Therefore, $T' \in \mathcal{R}'_\omega$.

Consider now the following list (we shall refer to it as ergodic list in the following) of statements [107] concerning clustering, extremal $\Theta$-invariance and the spectral properties of the dynamics of quantum dynamical systems.

$$\omega \quad \text{is extremal invariant} \tag{7.60}$$
$$\mathcal{R}'_\omega = \{\lambda\mathbb{1}\} \tag{7.61}$$
$$\mathcal{R}_\omega \cap \mathcal{R}'_\omega = \{\lambda\mathbb{1}\} \tag{7.62}$$
$$(\mathcal{A}, \Theta, \omega) \quad \text{is } \eta\text{-clustering} \tag{7.63}$$
$$P_\omega \quad \text{is a one-dimensional projector} \tag{7.64}$$
$$\pi_\omega(\mathcal{A})'' \cap \mathcal{R}'_\omega = \{\lambda\mathbb{1}\} \tag{7.65}$$
$$\eta_\omega[X] = \omega(X)\,\mathbb{1} \qquad \forall X \in \mathcal{A} \tag{7.66}$$
$$\omega \quad \text{is the only normal invariant state on} \quad \pi_\omega(\mathcal{A})'' \tag{7.67}$$
$$|\,\Omega_\omega\,\rangle \quad \text{is the only invariant vector state in} \quad \pi_\omega(\mathcal{A})|\,\Omega_\omega\,\rangle\;. \tag{7.68}$$

From Remark 7.1.5.2 it follows that (7.61)$\Longrightarrow$(7.60); also, if $\mathcal{R}'_\omega$ is not trivial, then, as in Remark 5.3.2.4, $\omega$ can be decomposed into a convex combination of $\Theta$-invariant states. Therefore, (7.60)$\Longleftrightarrow$ (7.61). If $\omega$ is not extremal invariant, the corresponding operator $\mathbb{1} \neq T' \in \mathcal{R}'_\omega$ provides an invariant vector state $U_\omega(t)T'|\,\Omega_\omega\,\rangle = T'|\,\Omega_\omega\,\rangle$, thus $|\,\Omega_\omega\,\rangle$ is not the only invariant vector state and the projector $P_\omega$ onto the invariant subspace of $\mathbb{H}_\omega$ cannot be one-dimensional, whence (7.68)$\Longrightarrow$ (7.60) and (7.64)$\Longrightarrow$ (7.60). Furthermore, using Example 7.1.8, one gets

$$\begin{aligned}\eta_t\left[\omega(A\Theta_t(B)\right] &= \langle\,\Omega_\omega\,|\,\pi_\omega(A)\,\eta_\omega\left[U_\omega\right]\,\pi_\omega(B)\,|\Omega_\omega\,\rangle\\ &= \langle\,\Omega_\omega\,|\,\pi_\omega(A)\,P_\omega\,\pi_\omega(B)\,|\Omega_\omega\,\rangle\ ;\end{aligned}$$

together with $P_\omega|\,\Omega_\omega\,\rangle = |\,\Omega_\omega\,\rangle$ and

$$\omega(A)\omega(B) = \langle\,\Omega_\omega\,|\,\pi_\omega(A)\,|\,\Omega_\omega\,\rangle\langle\,\Omega_\omega\,|\,\pi_\omega(B)\,|\Omega_\omega\,\rangle\ ,$$

this yields (7.64)$\Longleftrightarrow$ (7.63).

From Proposition 5.5.2 it follows that the normal invariant states $\rho$ on $\pi_\omega(\mathcal{A})''$ must correspond to density matrices $\rho \in \mathbb{B}_1^+(\mathbb{H}_\omega)$ that commute with $U_\omega(G)$; thus, $\rho(\eta_\omega\left[X\right]) = \rho(\pi_\omega(X)))$ for all $X \in \mathcal{A}$ and, if $\eta_\omega\left[X\right] = \omega(X)\,\mathbb{1}$, then $\rho(\eta_\omega\left[X\right]) = \rho(\pi_\omega(X)) = \omega(X)$, whence (7.66)$\Longrightarrow$ (7.67). Also, from Remark 7.1.5.1, (7.66)$\Longrightarrow$ (7.63).

Other implications are (7.61)$\Longrightarrow$ (7.62)$\Longrightarrow$ (7.65) and (7.67)$\Longrightarrow$ (7.68). Summarizing,

**Proposition 7.1.4.** *With reference to the previous list of ergodic properties, the following implications hold*

$$\begin{array}{ccccccccc} & & (7.64) & \Longleftrightarrow & (7.63) & \Longleftarrow & (7.66) & \Longrightarrow & (7.67)\\ & & & & \Downarrow & & & & \Downarrow\\ (7.65) & \Longleftarrow & (7.62) & \Longleftarrow & (7.61) & \Longleftrightarrow & (7.60) & \Longleftarrow & (7.68)\end{array}\quad .$$

While in a commutative setting the properties (7.60)– (7.68) are equivalent and each of them identifies an ergodic system, it is not so in the quantum realm, unless $(\mathcal{A},\Theta,\omega)$ is *asymptotic Abelian* [107, 300].

**Definition 7.1.7 (Asymptotic Abelianess).** *Suppose $(\mathcal{A},\Theta,\omega)$ enjoys one of the following properties*

$$\eta_t\left[\omega(A^\dagger\left[B,\Theta_t(C)\right]D)\right] = 0 \quad \forall\, A,B,C,D \in \mathcal{A} \tag{7.69}$$

$$\lim_{t\pm\infty}\omega(A^\dagger\left[B,\Theta_t(C)\right]D) = 0 \quad \forall\, A,B,C,D \in \mathcal{A} \tag{7.70}$$

$$\lim_{t\pm\infty}\omega(A^\dagger\left[B,\Theta_t(C)\right]^\dagger\left[B,\Theta_t(C)\right]A) = 0 \quad \forall\, A,B \in \mathcal{A} \tag{7.71}$$

$$\lim_{t\pm\infty}\|[B,\Theta_t(C)]\| = 0 \quad \forall\, B,C \in \mathcal{A}\ . \tag{7.72}$$

*In the first case, $(\mathcal{A},\Theta,\omega)$ is called $\eta$-Abelian, in the second one weakly asymptotic Abelian, in the third case strongly Asymptotic Abelian and in the last one norm asymptotic Abelian.*

Physically speaking, asymptotic Abelianess corresponds to the fact that the non-commutativity of any pair of local observables is a property which dies out asymptotically under the action of the dynamics on one of them.

**Example 7.1.10.** Quantum spin chains (see Example 7.1.1) are the simplest instance of norm-asymptotic Abelianess with respect to lattice-translations on the quasi-local $C^*$ algebra $\mathcal{A}$ [277]. In the case of spins 1/2 at each site, lattice-translations correspond to the shift-automorphism $\Theta_\sigma : \mathcal{A} \mapsto \mathcal{A}$ defined by

$$\Theta_\sigma \left[\prod_{\ell=1}^{n} \sigma_{j_\ell}^{i_\ell}\right] = \prod_{\ell=1}^{n} \sigma_{j_\ell}^{i_\ell+1} .$$

Given $A, B \in \mathcal{A}$, for any $\varepsilon > 0$ we can find strictly local $A_\varepsilon, B_\varepsilon \in \mathcal{A}_{[-k,k]}$ such that $\|A - A_\varepsilon\| \leq \varepsilon$ and $\|B - B_\varepsilon\| \leq \varepsilon$. If $\mathbb{N} \ni n > 2k$, $\Theta_\sigma^n[B] \in \mathcal{A}_{[n-k,n+k]}$ commutes with $A_\varepsilon$, $\left[A\,,\,\Theta_\sigma^n[B_\varepsilon]\right] = 0$, whence

$$\begin{aligned}\left\|\left[A\,,\,\Theta_\sigma^n[B]\right]\right\| &\leq \left\|\left[A - A_\varepsilon\,,\,\Theta_\sigma^n[B]\right]\right\| + \left\|\left[A_\varepsilon\,,\,\Theta_\sigma^n[B - B_\varepsilon]\right]\right\| \\ &\leq 2\varepsilon\|B\| + 2\varepsilon(\varepsilon + \|A\|) .\end{aligned}$$

The mean magnetization $\boldsymbol{m}$ in (7.3) is not a quasi-local observable for the spatial-average collects contributions from all local regions: nevertheless, it exists in the *GNS* representations $\pi_{\uparrow,\downarrow}(\mathcal{A})''$ corresponding to the translation-invariant states $\omega_{\uparrow,\downarrow}$ in (7.1) and (7.2). Furthermore, local non-commutativity is suppressed by dividing by larger and larger number of sites with the result that the mean magnetization commutes with all local observables. By its very construction it also commutes with the *GNS* unitary operator $U_\omega$ which implements the space-translations on the *GNS* Hilbert space. Therefore, $\boldsymbol{m} \in \mathcal{R}'_{\uparrow,\downarrow} = \pi_{\uparrow,\downarrow}(\mathcal{A})' \cap U_\omega(\mathbb{Z})'$. Since $\pi_{\uparrow,\downarrow}(\mathcal{A})' = \{\lambda\mathbb{1}\}$ it thus follows that $\boldsymbol{m}$ is a scalar multiple of the identity in the two representations.

The fact that the mean magnetization commutes with all local observables holds, more in general, as a consequence of $\eta$-Abelianess; namely, $\eta_\omega[\mathcal{A}] \subseteq \mathcal{R}'_\omega$ follows from observing that (7.69) yields

$$\begin{aligned}&\eta_t\left[\langle\,\Omega_\omega\,|\,\pi_\omega(A)^\dagger\left[\pi_\omega(X)\,,\,\pi_\omega(\Theta_t(Y))\right]\pi_\omega(B)\,|\Omega_\omega\,\rangle\right] = \\ &\quad = \langle\,\Omega_\omega\,|\,\pi_\omega(A)^\dagger\left[\pi_\omega(X)\,,\,\eta_\omega\,[Y]\right]\pi_\omega(B)\,|\Omega_\omega\,\rangle = 0\,.\end{aligned}$$

**Remark 7.1.6.** Proposition 7.1.3 states that $\eta_\omega[\mathcal{A}] \subseteq \pi_\omega(\mathcal{A})'' \cap U_\omega(G)'$; if $(\mathcal{A}, \Theta, \omega)$ is $\eta$-Abelian then also $\eta_\omega[\mathcal{A}] \subseteq \pi_\omega(\mathcal{A})'$, whence $\eta_\omega[\mathcal{A}] \subseteq \mathcal{R}_\omega \cap \mathcal{R}'_\omega$. Since the latter is an Abelian von Neumann algebra, actually the center of $\mathcal{R}_\omega$

(see Definiton 5.3.4), it turns out that $\Big[\eta_\omega\,[X]\,,\,\eta_\omega\,[Y]\Big] = 0$ for all $X, Y \in \mathcal{A}$. Therefore, using Example 7.1.9, one proves that on a dense set,

$$
\begin{aligned}
&\langle\, \Omega_\omega \,|\, \pi_\omega(A)^\dagger \Big[P_\omega\, \pi_\omega(X)\, P_\omega\,,\, P_\omega, \pi_\omega(Y)\, P_\omega\Big]\, \pi_\omega(B)\,|\Omega_\omega\,\rangle = \\
&\quad = \eta_t \Big[\langle\, \Omega_\omega \,|\, \pi_\omega(A)^\dagger \Big[P_\omega\, \pi_\omega(X)\, P_\omega\,,\, U_\omega(t)\, \pi_\omega(Y)\, P_\omega\Big]\, \pi_\omega(B)\,|\Omega_\omega\,\rangle\Big] \\
&\quad = \eta_t \Big[\langle\, \Omega_\omega \,|\, \pi_\omega(A)^\dagger \Big[P_\omega\, \pi_\omega(X)\, P_\omega\,,\, \pi_\omega(\Theta_t(Y))\, P_\omega\Big]\, \pi_\omega(B)\,|\Omega_\omega\,\rangle\Big] \\
&\quad = \langle\, \Omega_\omega \,|\, \pi_\omega(A)^\dagger \Big[P_\omega \pi_\omega(X)\, P_\omega\,,\, \eta_\omega\,[Y]\;P_\omega\Big]\, \pi_\omega(B)\,|\Omega_\omega\,\rangle \\
&\quad = \eta_t \Big[\langle\, \Omega_\omega \,|\, \pi_\omega(A)^\dagger \Big[\pi_\omega(\Theta_t(X))\, P_\omega\,, \eta_\omega\,[Y]\;P_\omega\Big]\, \pi_\omega(B)\,|\Omega_\omega\,\rangle\Big] \\
&\quad = \langle\, \Omega_\omega \,|\, \pi_\omega(A)^\dagger\, P_\omega \Big[\eta_\omega\,[X]\,,\, \eta_\omega\,[Y]\Big]\, P_\omega\, \pi_\omega(B)\,|\Omega_\omega\,\rangle = 0\;,
\end{aligned}
$$

for all $X, Y \in \mathcal{A}$. Therefore, $\eta$-Abelianess implies Abeliannes of $P_\omega\, \pi_\omega(\mathcal{A})\, P_\omega$ as a set (in general it is not an algebra).

If $(\mathcal{A}, \Theta, \omega)$ corresponds to a classical dynamical system, then (7.70), (7.71) and (7.72) are equivalent statements implying (7.69). In a genuinely quantum setting, however, (7.71), (7.70) and (7.72) take into account the differences between convergence in the weak, strong and norm topologies, whereby, in general, (7.72)$\Longrightarrow$ (7.71)$\Longrightarrow$ (7.70)$\Longrightarrow$ (7.69). Interestingly, the weakest sort of Abelianess is sufficient to make properties (7.60)– (7.68) equivalent to each other. Indeed, the consequences of $\eta$-Abeliannes are as follows.

**Proposition 7.1.5.** *If $(\mathcal{A}, \Theta, \omega)$ is $\eta$-Abelian, then $\mathcal{R}'_\omega = \pi_\omega(\mathcal{A})'' \cap U_\omega(G)'$.*

**Corollary 7.1.1.** *If $(\mathcal{A}, \Theta, \omega)$ is $\eta$-Abelian, the properties (7.60)– (7.68) in the ergodic list are equivalent.*

**Proof:** Because of $\eta$-Abelianess, applying the previous proposition one gets that (7.65)$\Longrightarrow$ (7.66), whence the claimed equivalence follows from Proposition 7.1.4. □

Then, the following definition makes sense.

**Definition 7.1.8.** *An $\eta$-Abelian quantum dynamical system $(\mathcal{A}, \Theta, \omega)$ is ergodic if $\omega$ is extremal invariant.*

**Remark 7.1.7.** In the case of Example 7.1.8.2, the eigenvectors of the Hamiltonian are extremal as functionals on $M_d(\mathbb{C})$ and thus ergodic according to the definition above. However, finite-dimensional systems cannot be asymptotic Abelian; indeed, while condition (7.64) holds, condition (7.66) does not.

**Proof of Proposition 7.1.5** The proof is based on 3 steps.

- **Step 1**: If $(\mathcal{A}, \Theta, \omega)$ is $\eta$-Abelian then $P_\omega \mathcal{R}_\omega P_\omega$ is Abelian.

Indeed, from Remark 7.1.6 we know that $P_\omega \pi_\omega(\mathcal{A}) P_\omega$ is Abelian, thus, by continuity, also $P_\omega \pi_\omega(\mathcal{A})'' P_\omega$. On the other hand, since the elements of $\mathcal{R}_\omega$ are strong limits of operators of the form (7.52), it turns out that $P_\omega \pi_\omega(\mathcal{A})'' P_\omega = P_\omega \mathcal{R}_\omega P_\omega$, whence the latter is Abelian.

- **Step 2**: If $P_\omega \mathcal{R}_\omega P_\omega$ is Abelian, then $\mathcal{R}'_\omega$ is Abelian ($\mathcal{R}'_\omega \subseteq \mathcal{R}''_\omega = \mathcal{R}_\omega$).

Since $P_\omega \in \mathcal{R}_\omega$, we can use Example 5.3.2.6 to deduce that

$$P_\omega \mathcal{R}_\omega P_\omega \subseteq (P_\omega \mathcal{R}_\omega P_\omega)' = P_\omega \mathcal{R}'_\omega P_\omega \ .$$

Further, since $|\,\Omega_\omega\,\rangle$ is cyclic for $P_\omega \mathcal{R}_\omega P_\omega$ with respect to the Hilbert space $P_\omega \mathbb{H}_\omega$, using Example 7.1.4.1 we conclude that $P_\omega \mathcal{R}_\omega P_\omega = P_\omega \mathcal{R}'_\omega P_\omega$, whence the latter algebra is Abelian. In order to prove the statement, we show that $\mathcal{R}'_\omega$ and $P_\omega \mathcal{R}'_\omega P_\omega$ are isomorphic; for any $X' \in \mathcal{R}'_\omega$ set $\lambda(X') = P_\omega X' P_\omega$. This map is obviously linear and surjective; further, since $P_\omega \in \mathcal{R}_\omega$, $\lambda(X'Y') = \lambda(X')\lambda(Y')$ for all $X', Y' \in \mathcal{R}'_\omega$. Finally, suppose $\lambda(Z') = 0$ for some $Z' \in \mathcal{R}'_\omega$, then, since $P_\omega|\,\Omega_\omega\,\rangle = |\,\Omega_\omega\,\rangle$,

$$Z' \pi_\omega(X)|\,\Omega_\omega\,\rangle = \pi_\omega(X)\, Z' P_\omega|\,\Omega\,\rangle = \pi_\omega(X)\, P_\omega Z' P_\omega|\,\Omega_\omega\,\rangle = 0$$

on a dense set. Thus, $Z' = 0$ and $\lambda$ is also injective.

- **Step 3**: If $X \in \mathcal{A}$ and $\pi_\omega(X) \in \pi_\omega(\mathcal{A}) \cap U_\omega(G)'$ then $\eta_\omega[X] = \pi_\omega(X) \in \pi_\omega(\mathcal{A})'$.

This fact can be extended by continuity to the constants of the motion in the strong closure $\pi_\omega(\mathcal{A})'' \cap U_\omega(G)'$ so that

$$\pi_\omega(\mathcal{A})'' \cap U_\omega(G)' \subseteq \pi_\omega(\mathcal{A})' \Longrightarrow \pi_\omega(\mathcal{A})'' \cap U_\omega(G)' \subseteq \mathcal{R}'_\omega \ . \tag{7.73}$$

If $X \in \pi_\omega(\mathcal{A})'' \cap U_\omega(G)'$, it commutes with the projector onto the invariant vectors $P_\omega \in \mathcal{R}_\omega$, whence $X P_\omega = P_\omega X P_\omega$ implies

$$\left(\pi_\omega(\mathcal{A})'' \cap U_\omega(G)'\right) P_\omega \subseteq P_\omega \pi_\omega(\mathcal{A})'' P_\omega \ .$$

On the other hand, Example 7.1.9 and Proposition 7.1.3 yield

$$\eta_\omega[\mathcal{A}] = P_\omega \pi_\omega(\mathcal{A}) P_\omega \subseteq \pi_\omega(\mathcal{A})'' \cap U_\omega(G)' \ ,$$

whence by continuity

$$P_\omega \pi_\omega(\mathcal{A})'' P_\omega = \left(\pi_\omega(\mathcal{A})'' \cap U_\omega(G)'\right) P_\omega \ . \tag{7.74}$$

Finally, from the first two steps, (7.73) and (7.74), we derive

$$\begin{aligned}\mathcal{R}'_\omega\, P_\omega &= P_\omega\, \mathcal{R}'_\omega\, P_\omega \subseteq P_\omega\, \mathcal{R}_\omega\, P_\omega = P_\omega\, \pi_\omega(\mathcal{A})''\, P_\omega \\ &= \Big(\pi_\omega(\mathcal{A})'' \cap U_\omega(G)'\Big)\, P_\omega = \Big(\pi_\omega(\mathcal{A})'' \cap U_\omega(G)'\Big)\, P_\omega\ .\end{aligned}$$

Consequently, given $X' \in \mathcal{R}'_\omega$ there exists $Y' \in \pi_\omega(\mathcal{A})'' \cap U_\omega(G)'$ such that $(X' - Y')|\,\Omega_\omega\,\rangle$. As $|\,\Omega_\omega\,\rangle$ is cyclic for $\mathcal{R}_\omega$, it is separating for $\mathcal{R}'_\omega$, thence $X' = Y'$ so that $\mathcal{R}'_\omega \subseteq \pi_\omega(\mathcal{A})''$ or, equivalently, $\mathcal{R}'_\omega \subseteq \pi_\omega(\mathcal{A})'' \cap U_\omega(G)'$ which, together with (7.73) completes the proof. □

**Remark 7.1.8.** In case of $\eta$-Abelianess, from $\mathcal{R}'_\omega = \pi_\omega(\mathcal{A})'' \cap U_\omega(G)'$ it follows that $\mathcal{R}'_\omega$ is contained in the center $\mathcal{Z}_\omega = \pi_\omega(\mathcal{A})'' \cap \pi_\omega(\mathcal{A})'$ of $\pi_\omega(\mathcal{A})''$ and is thus Abelian. Actually, $\mathcal{R}'_\omega$ coincides with the commutative algebra of $\Theta$-invariant classical observables of the quantum system $(\mathcal{A}, \Theta, \omega)$. Therefore, if $(\mathcal{A}, \Theta, \omega)$ is $\eta$-Abelian and $\omega$ is a factor state (see Remark 5.3.2.2), then $\mathcal{R}'_\omega = \{\lambda \mathbb{1}\}$ and $\omega$ is extremal invariant and thus ergodic, according to Definition 7.1.8.

If $(\mathcal{A}, \Theta, \omega)$ is $\eta$-Abelian, but the state $\omega$ is not extremal invariant, that is not ergodic with respect to $\Theta$, then $\mathcal{R}'_\omega$ cannot be trivial. However, it is Abelian and thus generated by a unique set of minimal projections $P'_j$ (see Section 5.3.2). Each of these projections are such that $(U^\dagger_\omega)^t\, P_j\, U^t_\omega = P_j$ and thus provides a $\Theta$-invariant state $\omega_j$ on $\mathcal{A}$:

$$\omega_j(X) := \frac{\langle\, \Omega_\omega \,|\, P_j\, \pi_\omega(X)\, |\Omega_\omega\,\rangle}{\omega(P_j)}\ .$$

Since the $P_j$ are minimal projections in $\mathcal{R}'_\omega$, they cannot be further decomposed in $\mathcal{R}'_\omega$, whence they are extremal invariant and yield a decomposition $\omega = \sum_j \omega(P_j)\, \omega_j$ of $\omega$ into its ergodic components.

**Example 7.1.11.** [220] Let $\omega_\pm$ be states of a one-dimensional spin chain of the form (7.4), where (the upper indices label the lattice sites, the lower ones the Pauli matrices)

$$\begin{aligned}\omega_+\left(\prod_{\ell=1}^n \sigma^{i_\ell}_{j_\ell}\right) &= \prod_{\ell=1}^n \mathrm{Tr}\left(\frac{\mathbb{1} + (-1)^{i_\ell}\sigma_3}{2}\,\sigma_{j_\ell}\right) = \prod_{\ell=1}^n (-1)^{i_\ell}\delta_{j_\ell 3} \\ \omega_-\left(\prod_{\ell=1}^n \sigma^{i_\ell}_{j_\ell}\right) &= \prod_{\ell=1}^n \mathrm{Tr}\left(\frac{\mathbb{1} - (-1)^{i_\ell}\sigma_3}{2}\,\sigma_{j_\ell}\right) = \prod_{\ell=1}^n (-1)^{i_\ell+1}\delta_{j_\ell 3}\ .\end{aligned}$$

Unlike the states (7.1) and (7.2) which are characterized by infinitely many spins pointing up, respectively down along the $z$ axis, these states are antiferromagnetic alternating spins up, at even sites $\omega_+$, at odd sites $\omega_-$, and spins down. These are pure states and give rise to factor GNS representations $\pi_\omega(\mathcal{A})''$: in fact, as for the states (7.1) and (7.2), one can show that the commutants $\pi_\pm(\mathcal{A})'$ are trivial. Also, by considering the shift-automorphism of Example 7.1.10, it turns out that $\omega_\pm \circ \Theta_\sigma = \omega_\mp$ so that the state

$$\omega := \frac{\omega_+ + \omega_-}{2} \tag{7.75}$$

is translation-invariant and can be decomposed in terms of pure states:

$$\omega_\pm(A) = \frac{\langle\, \Omega_\omega \,|\, Q_\pm \pi_\omega(A) \,|\Omega_\omega\,\rangle}{\langle\, \Omega_\omega \,|\, Q_\pm \,|\Omega_\omega\,\rangle} \ ,$$

with $Q_\pm \in \pi_\omega(\mathcal{A})'$. If $\omega$ could be decomposed into $\Theta_\sigma$-invariant states $\omega_i$, these would correspond to $0 \leq Q_i \in \mathcal{R}'_\omega = \pi_\omega(\mathcal{A})' \cap U_\omega(G)'$ that provide decompositions of the pure states $\omega_\pm$ as well

$$\omega_\pm(A) = \frac{\langle\, \omega_\omega \,|\, Q_\pm \, Q_i \pi_\omega(A) \,|\Omega_\omega\,\rangle}{\langle\, \Omega_\omega \,|\, Q_\pm \,|\Omega_\omega\,\rangle} + \frac{\langle\, \omega_\omega \,|\, Q_\pm \,(\mathbb{1} - Q_i)\pi_\omega(A) \,|\Omega_\omega\,\rangle}{\langle\, \Omega_\omega \,|\, Q_\pm \,|\Omega_\omega\,\rangle} \ ,$$

which is impossible. Therefore $\omega$ is extremal invariant, but not a factor state. As for the spin system considered at the beginning of this chapter, the following even and odd magnetizations $\boldsymbol{m}^{e,o} = (m_1^{e,o}, m_2^{e,o}, m_3^{e,o})$ (see (7.3)) exist as strong operator limits in the *GNS* representations $\pi_\pm$ and $\pi_\omega$:

$$m_i^e := \lim_{N\to+\infty} \frac{\mu}{2N+1} \sum_{\ell=-N}^{N} \sigma_i^{2i} \ , \quad m_i^o := \lim_{N\to+\infty} \frac{\mu}{2N+1} \sum_{\ell=-N}^{N} \sigma_i^{2i+1} \ .$$

They commute with all local observables and thus belong to the trivial centers $\mathcal{Z}_\pm$ and the non-trivial one $\mathcal{Z}_\omega$. In the first two ones they are multiples of the identity, $\boldsymbol{m}_\pm^e = (0, 0, \pm\mu)$ and $\boldsymbol{m}_\pm^o = (0, 0, \mp\mu)$, while in the representation $\pi_\omega$ which can be conveniently split as $\pi_\omega(\mathcal{A}) = \pi_+(\mathcal{A}) \bigoplus \pi_-(\mathcal{A})$,

$$\pi_\omega(m_3^e) = \mu \begin{pmatrix} \mathbb{1} & 0 \\ 0 & -\mathbb{1} \end{pmatrix} \ , \quad \pi_\omega(m_3^0) = \mu \begin{pmatrix} -\mathbb{1} & 0 \\ 0 & \mathbb{1} \end{pmatrix} \ .$$

Furthermore, while enjoying clustering in the mean, the representation $\pi_\omega$, which is not a factor, is not clustering; indeed,

$$\omega(\sigma_3^i \sigma_3^{i+\ell}) = \frac{(-1)^\ell + (-1)^\ell}{2} = (-1)^\ell \text{ while } \omega(\sigma_3^k) = \frac{(-1)^k + (-1)^{1+k}}{2} = 0 \ .$$

From the previous discussion, it turns out that if $(\mathcal{A}, \Theta, \omega)$ is $\eta$-Abelian and $\omega$ extremal invariant (property (7.60) in the ergodic list), then two-point correlation functions factorize in the mean (property (7.63) in the ergodic list). Concerning the extension of the classical property of mixing (see Proposition 2.3.3, (2.66) and (2.76)) to quantum dynamical systems, due to non commutativity, one distinguishes between various way of clustering beside (7.59).

**Definition 7.1.9.** *A quantum dynamical system* $(\mathcal{A}, \Theta, \omega)$ *is weakly mixing if [42]*

$$\lim_{t\to\pm\infty} \omega(A\Theta_t(B)C) = \omega(AC)\,\omega(B) \quad \forall\, A, B, C \in \mathcal{A}\ ; \tag{7.76}$$

*strongly mixing [42] (or hyperclustering in [215]) if*

$$\lim_{t\to\pm\infty} \omega(A\Theta_t(B)C\Theta_t(D)E) = \omega(ACE)\,\omega(BD) \quad \forall\, A, B, C, D, E \in \mathcal{A}\ . \tag{7.77}$$

Clearly, strong mixing implies weak-mixing and weak-mixing implies $\eta$-Abelianess; it also implies weak asymptotic Abelianess, whereas strong-mixing implies strong-asymptotic Abelianess. The proof of the latter statement (the proof of the former one is similar) comes from (7.77) applied to

$$\begin{aligned}
&\omega(A^\dagger[\Theta_t(B)\,,\,C]^\dagger[\Theta_t(B), C]A) = \\
&\quad = \omega((CA)^\dagger\,\Theta_t(B^\dagger\,B)\,CA) \;-\; \omega((CA)^\dagger\,\Theta_t(B)^\dagger\,C\,\Theta_t(B)\,A) \\
&\quad -\omega(A^\dagger\,\Theta_t(B)^\dagger\,C^\dagger\,\Theta_t(B)\,CA) \;+\; \omega(A^\dagger\,\Theta_t(B)^\dagger\,C^\dagger\,C\,\Theta_t(B)\,A)\ ,
\end{aligned}$$

which yields

$$\begin{aligned}
&\lim_{t\to\pm\infty} \omega(A^\dagger[\Theta_t(B), C]^\dagger[\Theta_t(B), C]A) = \\
&\quad = \omega((CA)^\dagger\,CA)\,\omega(B^\dagger\,B) \;-\; \omega((CA)^\dagger\,CA)\,\omega(B^\dagger\,B) \\
&\quad -\omega(A^\dagger\,C^\dagger\,CA)\,\omega(B^\dagger\,B) \;+\; \omega(A^\dagger\,C^\dagger\,CA)\,\omega(B^\dagger\,B) = 0\ .
\end{aligned}$$

Also, weak-mixing and strong-asymptotic Abelianess together imply strong-mixing; this can be seen by rewriting

$$\begin{aligned}
&\omega(A^\dagger[\Theta_t(B), C]^\dagger[\Theta_t(B), C]A) = \\
&= \omega((CA)^\dagger\,\Theta_t(B^\dagger\,B)\,C\,A) \;-\; \omega((CA)^\dagger\,\Theta_t(B^\dagger\,B)\,C\ A) \\
&-\omega(A^\dagger\,\Theta_t(B^\dagger\,B)\,C^\dagger\,C\,A) \;+\; \omega(A^\dagger\,C^\dagger\,C\,\Theta_t(B^\dagger\,B)\,A) \\
&-\omega\Big((CA)^\dagger\,\Theta_t(B^\dagger)\,\Big[C^\dagger\,,\,\Theta_t(B)\Big]\,CA\Big) \;-\; \omega\Big(A^\dagger\,\Theta_t(B)^\dagger\,\Big[C^\dagger\,,\,\Theta_t(B)\Big]CA\Big) \\
&+\omega\Big(A^\dagger\,\Theta_t(B)^\dagger\,\Big[C^\dagger\,C\,,\,\Theta_t(B)\Big]\,A\Big)\ .
\end{aligned}$$

Now, weak-mixing means that the first four terms factorize in the same way and thus cancel each other, asymptotically in time; on the other hand, each of the terms with the commutators vanish asymptotically if the system is strongly asymptotic Abelian. This comes out from the Cauchy-Schwartz inequality (5.49) which gives upper bounds of the form

$$\begin{aligned}
&\left|\omega\Big((CA)^\dagger\,\Theta_t(B)^\dagger\,\Big[C^\dagger\,,\,\Theta_t(B)\Big]\,CA\Big)\right|^2 \le \\
&\quad \le \omega((CA)^\dagger\,\Theta_t(B^\dagger\,B)\,CA)\;\omega\Big((CA)^\dagger\,\Big[C^\dagger\,,\,\Theta_t(B)\Big]^\dagger\,\Big[C^\dagger\,,\,\Theta_t(B)\Big]\,CA\Big)\ .
\end{aligned}$$

**Proposition 7.1.6.** *Given a quantum dynamical system $(\mathcal{A}, \Theta, \omega)$, we have the following implications:*

1. *strong-mixing (7.77) implies weak-mixing (7.76);*
2. *strong-mixing (7.77) implies strong asymptotic Abelianess (7.71):*
3. *weak-mixing (7.76) implies weak asymptotic Abelianess (7.70);*
4. *weak-mixing (7.76) and strong-asymptotic Abelianess (7.71) imply strong-mixing (7.77).*

**Remark 7.1.9.** If the state $\omega$ is faithful then weak-mixing is equivalent to the factorization of two-point correlation functions (see (7.59)). Of course, if $(\mathcal{A}, \Theta\,\omega)$ is weakly mixing, then $\omega(X\,\Theta_t(Y))$ asymptotically split into $\omega(X)\omega(Y)$. Vice versa, using the *KMS* conditions (7.27), if (7.59) holds, then

$$\begin{aligned}\lim_t \omega(A\,\Theta_t(B)\,C) &= \lim_{t\to\pm\infty} \omega(\sigma_\omega(i)(C)\,A\,\Theta_t(B) \\ &= \omega(\sigma_\omega(i)(C)\,A)\,\omega(B) = \omega(A\,C)\,\omega(B)\ .\end{aligned}$$

The same conclusion that (7.59) implies weak-mixing follows if one knows $(\mathcal{A}, \Theta, \omega)$ to be weakly asymptotic Abelian; one uses

$$\omega(A\,\Theta_t(B)\,C) = \omega(A\,C\,\Theta_t(B)) + \omega\Big(A\,\Big[\Theta_t(B)\,,\,C\Big]\Big)\ .$$

The following proposition establishes a link, similar to the classical one, between mixing and the spectral properties of the time-evolution group $U_\omega(G)$ in the *GNS* construction.

**Proposition 7.1.7.** *If $(\mathcal{A}, \Theta, \omega)$ is weakly mixing the following equivalent properties hold,*

$$w - \lim_{to\pm\infty} \pi_\omega(\Theta_t(X)) = \omega(X)\,1\!\!1 \qquad \forall\, X \in \mathcal{A} \tag{7.78}$$

$$w - \lim_{t\to\pm\infty} U_\omega(t) = |\,\Omega_\omega\,\rangle\langle\,\Omega_\omega\,|\ . \tag{7.79}$$

**Proof:** Weak-mixing asserts that $\lim_{t\to\pm\infty} \pi_\omega(\Theta_t(X)) = \omega(X)\,1\!\!1$ on a dense set, whence (7.78). Furthermore, from

$$\begin{aligned}\omega(A^\dagger\,\Theta_t(B)) &= \langle\,\Omega_\omega\,|\,\pi_\omega(A)^\dagger\,U_\omega(t)\pi_\omega(B)\,|\Omega_\omega\,\rangle \quad \text{and} \\ \omega(A^\dagger)\,\omega(B) &= \langle\,\Omega_\omega\,|\,\pi_\omega(A)^\dagger|\,\Omega_\omega\,\rangle\langle\,\Omega_\omega\,|\pi_\omega(B)\,|\Omega_\omega\,\rangle\ ,\end{aligned}$$

for all $A, B \in \mathcal{A}$, it follows that (7.78) and (7.79) are equivalent. □

**Remarks 7.1.10.**

1. If $(\mathcal{A}, \Theta, \omega)$ is norm-asymptotic Abelian and $\omega$ is a factor state, then $\omega$ is clustering as in (7.59) [64]. Indeed, $\mathcal{Z}_\omega = \{\lambda \mathbb{1}\}$ says that the $C^*$ algebra $\mathcal{B}$ generated by $\pi_\omega(\mathcal{A})$ and $\pi_\omega(\mathcal{A})'$ has trivial commutant $\mathcal{B}' = \{\lambda \mathbb{1}\}$, namely $\mathcal{B}'' = \mathbb{B}(\mathbb{H}_\omega)$. Let $X \in \mathcal{A}$ and consider the vector
$$\mathbb{H}_\omega \ni |\,\Psi_X\,\rangle := (\pi_\omega(X) - \omega(X)\mathbb{1})|\,\Omega_\omega\,\rangle\ .$$
It is orthogonal to $|\,\Omega_\omega\,\rangle$: thus, there exists $\mathbb{B}(\mathbb{H}_\omega) \ni T = T^\dagger$ such that
$$T|\,\Psi_X\,\rangle = 0\ , \qquad T|\,\Omega_\omega\,\rangle = |\,\Omega_\omega\,\rangle\ .$$
Actually, $T$ can be chosen in $\mathcal{B}$ [64]: the operators
$$C_1 := T\,(\pi_\omega(X) - \omega(X)\mathbb{1})\ , \quad C_2 := (\mathbb{1} - T)\,(\pi_\omega(X) - \omega(X)\mathbb{1})$$
are in $\mathcal{B}$. Further, $C_1|\,\Omega_\omega\,\rangle = C^\dagger|\,\Omega_\omega\,\rangle = 0$ and $\pi_\omega(X) = C_1 + C_2 = \omega(X)\mathbb{1}$; consequently,
$$\begin{aligned}\omega(X\Theta_t[Y]) - \omega(X)\omega(Y) &= \langle\,\Omega_\omega\,|\,C_1\pi_\omega(\Theta_t[X])\,|\Omega_\omega\,\rangle \\ &= \langle\,\Omega_\omega\,|\,\Big[C_1\,,\,\pi_\omega(\Theta_t[X])\Big]\,|\Omega_\omega\,\rangle\ .\end{aligned}$$
Now, for any $\varepsilon > 0$ one can approximate $C_1 \in \mathcal{B}$ by a finite sum in $\pi_\omega(\mathcal{A}) \cup \pi_\omega(\mathcal{A})'$:
$$\Big\|C_1 - \sum_{j=1}^n \pi_\omega(X_j)Y_j'\Big\| \le \frac{\varepsilon}{\|Y\|}\ ,$$
so that
$$\begin{aligned}|\omega(X\Theta_t[Y]) - \omega(X)\omega(Y)| &\le \sum_{j=1}^n \Big|\langle\,\Omega_\omega\,|\,\Big[\pi_\omega(C_1)\,,\,\pi_\omega(A_j)\Big]B_j\,|\Omega_\omega\,\rangle\Big| + \varepsilon \\ &\le \sum_{j=1}^n \|B_j\|\,\Big\|\Big[\pi_\omega(C_1)\,,\,\pi_\omega(A_j)\Big]\Big\| + \varepsilon\ .\end{aligned}$$
Thus, since $(\mathcal{A}, \Theta, \omega)$ is assumed to be norm-asymptotic Abelian, then it turns out to be clustering whence, according to Remark 7.1.9, also weakly mixing.
2. If $(\mathcal{A}, \Theta, \omega)$ is norm-asymptotic Abelian and $\omega$ is an extremal *KMS* state, then it is a factor state (see Remark 7.1.3.4) and the system is weakly mixing.

**Example 7.1.12.** The infinite dimensional systems of Example 7.1.6 provide an interesting framework to apply the previous abstract considerations [44].

Since the $\Theta_{\mathbb{A}}$-invariant state is tracial, as explained in Remark 7.1.9, correlation functions as those appearing in (7.76) can be reduced to two-point correlation functions as in (7.59). Then, (7.39) yields

$$\omega\left(W_\theta(f)\Theta^t_{\mathbb{A}}[W_\theta(g)]\right) = \langle\, f \,|\, U^t_{\mathbb{A}}\,| g\,\rangle\ .$$

Therefore, if the underlying classical system is mixing, that is if the Koopman operator $U_A$ has absolutely continuous spectrum on the subspace orthogonal to the constant function ( namely to the *GNS* cyclic vector $|\,\Omega_\omega\,\rangle$), then, for all $f, g \in \mathbb{L}^2_{\mathrm{d}r}(\mathbb{T}^2)$,

$$\begin{aligned}\lim_{t\to\pm\infty}\omega\left(W_\theta(f)\Theta^t_{\mathbb{A}}[W_\theta(g)]\right) &= \lim_{t\to\pm\infty}\langle\, f \,|\, U^t_{\mathbb{A}}\,| g\,\rangle\\ &= \langle\, f^* \,|\, \Omega\,\rangle\langle\, \Omega \,|\, g\,\rangle = \omega(W_\theta(f))\,\omega(W_\theta(g))\ .\end{aligned}$$

This means that, independently of the deformation parameter $\theta$, the quantum dynamical systems $(\mathcal{M}_\theta, \Theta_{\mathbb{A}}, \omega)$ are mixing when such is the classical dynamical system of which they represent a quantization.

As regards strong mixing (7.77), observe that (7.50) implies

$$\begin{aligned}\omega\left(\Big[W_\theta(\boldsymbol{n})\,,\,\Theta^t_{\mathbb{A}}[W_\theta(\boldsymbol{m})]\Big]^\dagger\Big[W_\theta(\boldsymbol{n})\,,\,\Theta^t_{\mathbb{A}}[W_\theta(\boldsymbol{m})]\Big]\right) &=\\ = 4\,\sin^2(2\pi\theta\sigma(\boldsymbol{n},(\mathbb{B})^t\boldsymbol{m}))\ ,\end{aligned} \tag{7.80}$$

where we have set $\mathbb{B} = \mathbb{A}^T$, the transposed of $\mathbb{A}$. Since $\sigma(\boldsymbol{n}, \mathbb{B}^t\boldsymbol{m})$ is an integer, when $\theta \in \mathbb{Q}$ is rational, the right hand side of (7.80) is periodic in $t$ and cannot vanish when $t \to \pm\infty$. Therefore, the quantum dynamical systems $(\mathcal{M}_{p/q}, \Theta_{\mathbb{A}}, \omega)$ cannot be strong asymptotic Abelian, and thus not strongly mixing, because of Proposition 7.1.6.

If $\theta$ is irrational, then, as proved in [44], the right hand side of (7.80) vanishes asymptotically at most for a countable set of $\theta \in [0, 1]$. A concrete construction of a countable set of $\theta$ is as follows [209].

Let $t \geq 0$; using (2.17)– (2.19) in Example 2.1.3 with $b$ and $c$ exchanged, one explicitly computes

$$\begin{aligned}\sigma(\boldsymbol{n}\,,\,\mathbb{B}^t\boldsymbol{m}) &= C_+(\boldsymbol{m})\alpha^t(n_1a_{2+} - n_2a_{1+}) + C_-(\boldsymbol{m})\alpha^{-t}(n_1a_{2-} - n_2a_{1-})\\ &= \alpha^t\frac{m_1a_{2-} - m_2a_{1-}}{\Delta}(n_1a_{2+} - n_2a_{1+}) + O(\alpha^{-t})\\ &= \frac{\alpha^{t+1}}{b(1-\alpha^2)}\Big(m_1n_1(\alpha^{-1} - a)(\alpha - a)\\ &\qquad + m_2n_2b^2 - m_1n_2\,(\alpha^{-1} - a) - m_2n_2\,(\alpha - a)\Big)\ .\end{aligned}$$

The transposed matrix $\mathbb{B} = \mathbb{A}^T$ has eigenvalues $\alpha^{\pm1}$ as $\mathbb{A}$; therefore, $\mathrm{Tr}(\mathbb{B}^t) = \alpha^t + \alpha^{-t} \in \mathbb{Z}$ for all $t \geq 0$. It thus follows that

$$\begin{aligned}\sigma(\boldsymbol{n}\,,\,\mathbb{B}^t\boldsymbol{m}) &= \frac{\alpha^{t+1}}{\alpha^2-1}\Big(m_1n_1\,c - m_2n_2\,b - (m_1n_2+m_2n_1)\,a \\ &\qquad + m_1n_2\alpha^{-1} + m_2n_1\,\alpha\Big) + O(\alpha^{-t}) \\ &= \frac{1}{\alpha^2-1}\sum_{k=0}^{2} r_k\alpha^{t+k} + O(\alpha^{-t}) \\ &= \frac{1}{\alpha^2-1}\sum_{k=0}^{2} r_k\Big(\alpha^{t+k} + \alpha^{-(t+k)}\Big) + O(\alpha^{-t})\,,\end{aligned}$$

where the coefficient $r_k \in \mathbb{Z}$ and thus also the sum is an integer.

Choose $\theta = \alpha^2 s \mod (1)$, $s \in \mathbb{Z}$, then

$$\theta\,\sigma(\boldsymbol{n}\,,\,\mathbb{B}^t\boldsymbol{m}) = s(\alpha^2-1)\sigma(\boldsymbol{n}\,,\,\mathbb{B}^t\boldsymbol{m}) \mod (1) = O(\alpha^{-t}) \mod (1)\,.$$

Therefore, when $t \to +\infty$, the function in (7.80) vanishes and norm-asymptotic Abelianess holds. Indeed, consider $W_\theta(f)$ and $W_\theta(g)$ where $f$ and $g$ have compact supports, namely, there exists $K > 0$ such that $f(\boldsymbol{n}) = g(\boldsymbol{n}) = 0$ when $\|\boldsymbol{n}\| \geq K$; then,

$$\begin{aligned}\Big\|\Big[W_\theta(f)\,,\,\Theta^t_{\mathbb{A}}[W_\theta(g)]\Big]\Big\| &\leq \sum_{\boldsymbol{n},\boldsymbol{m}} |f(\boldsymbol{n})|\,|g(\boldsymbol{m})|\,\Big\|\Big[W_\theta(\boldsymbol{n})\,,\,W_\theta(\mathbb{B}^t\boldsymbol{m})\Big]\Big\| \\ &\leq \alpha^{-t}\sum_{\boldsymbol{n},\boldsymbol{m}} |f(\boldsymbol{n})|\,|g(\boldsymbol{m})|\,C_{\boldsymbol{n},\boldsymbol{m}}\end{aligned}\tag{7.81}$$

for a suitable constant $C_{\boldsymbol{n},\boldsymbol{m}}$. Because of the assumption on $\mathrm{Supp}(f)$ and $\mathrm{Supp}(g)$, (7.81) goes to 0 with $t \to +\infty$.

Thus, for $\theta = s\alpha^2 \mod (1)$, $s \in \mathbb{Z}$, $(\mathcal{M}_\theta, \Theta_{\mathbb{A}}, \omega)$ are weakly mixing and norm-asymptotic Abelian; whence, according to Proposition 7.1.6, strongly mixing.

### 7.1.4 Algebraic Quantum $K$-Systems

The notion of $K$-systems is naturally extended by removing the Abelian constraint from Definition 2.3.6.

**Definition 7.1.10.** *Let $(\mathcal{A}, \Theta, \omega)$ be a quantum dynamical system; if $\mathcal{A}$ is a $C^*$ algebra it is called a $C^*$* algebraic quantum $K$-system *if there exists a $C^*$ subalgebra $\mathcal{A}_0 \subset \mathcal{A}$ such that*

1. $\mathcal{A}_t : \Theta_t[\mathcal{A}_0] \subset \mathcal{A}_{t+1}$ *for all* $t \in \mathbb{Z}$;
2. $\bigvee_{t\in\mathbb{Z}} \mathcal{A}_t = \mathcal{A}$;
3. $\bigwedge_{t\in\mathbb{Z}} \mathcal{A}_t = \{\lambda 1\!\!1\}$,

*where* $\bigwedge_{t\in\mathbb{Z}} \mathcal{A}_t$ *denotes the set theoretic intersection of the* $C^*$ *subalgebras* $\mathcal{A}_t$ *and* $\bigvee_{t\in\mathbb{Z}} \mathcal{A}_t$ *the* $C^*$ *they generate by norm closure. The nested sequence will be called a* quantum $K$-sequence.

*If* $\mathcal{A}$ *is a von Neumann algebra acting on a Hilbert space, then* $(\mathcal{A}, \Theta, \omega)$ *is a* von Neumman algebraic $K$-system *if there exists a quantum* $K$*-sequence of von Neumann subalgebras* $\mathcal{A}_n$*, with* $\bigvee_{t\in\mathbb{Z}} \mathcal{A}_t$ *denoting the von Neumann algebra obtained by strong-operator closure.*

**Remark 7.1.11.** Typically [220], starting from a $C^*$ algebraic quantum $K$-system with $K$-sequence $\mathcal{A}_t$, one considers the *GNS* representation $\pi_\omega(\mathcal{A})$ and the sequence of von Neumann subalgebras $\pi_\omega(\mathcal{A}_t)''$ and checks whether it is a (von Neumann) $K$-sequence for $(\pi_\omega(\mathcal{A})'', \Theta, \omega)$.

We have seen in Section 2.3 that classical $K$-systems enjoy the strongest possible clustering properties corresponding to $K$-mixing; to some extent this notion extends to von Neumann quantum $K$-systems. Let a quantum dynamical system $(\mathcal{A}, \Theta, \omega)$ posses a sequence $\{\mathcal{A}_t\}_{t\in\mathbb{Z}}$ of $C^*$ subalgebras such that, setting $\mathcal{M} := \pi_\omega(\mathcal{A})''$ and $\mathcal{M}_t := \pi_\omega(\mathcal{A}_t)''$, $\{\mathcal{M}_t\}_{t\in\mathbb{Z}}$ is a $K$-sequence for the von Neumann triplet $(\mathcal{M}, \Theta, \omega)$. We assume $\omega$ to be a faithful state and the $\mathcal{M}_0$ to be invariant under the modular automorphism $\sigma_\omega$, so that Proposition (7.1.1) ensures the existence of a normal conditional expectation $E_0 : \mathcal{M} \mapsto \mathcal{M}_0$ which respects the state. It thus follows that the *CPU* maps $E_t := \Theta_t \circ E_0 \circ \Theta_{-t} : \mathcal{M} \mapsto \mathcal{M}_t$ are $\omega$-preserving conditional expectations. Setting

$$P_t \pi_\omega(A) | \Omega_\omega := E_t[\pi_\omega(A)] | \,\Omega_\omega\,\rangle \qquad \forall A \in \mathcal{A} \,,$$

one obtains a bounded linear operator which can be extended to a bounded operator $P_t : \mathbb{H}_\omega \mapsto \mathbb{H}_t$, where $\mathbb{H}_t$ is the closure of the linear span $\pi_\omega(\mathcal{A})|\,\Omega_\omega\,\rangle$ and $\mathbb{H}_\omega$ is the *GNS* Hilbert space corresponding to $\omega$.

**Proposition 7.1.8.** *The* $P_t$ *are projectors such that* $P_t = U^\dagger_\omega(t)\, P_0\, U_\omega(t)$*, where* $U_\omega(t)$ *is the GNS unitary operator which implements* $\Theta$ *on* $\mathbb{H}_\omega$*. If* $\{\mathcal{M}_t\}_{t\in\mathbb{Z}}$ *is a* $K$*-sequence, then [217]*

1. $P_t \leq P_{t+1}$ *for all* $t \in \mathbb{Z}$*;*
2. $s - \lim_{t\to+\infty} P_t = 1\!\!1$*;*
3. $s - \lim_{t\to+\infty} P_t = |\,\Omega_\omega\,\rangle\langle\,\Omega_\omega\,|$*.*

**Proof:** From the properties of the conditional expectations (see (5.44) in Proposition 5.2.2), it follows that $P_t^2 = P_t$. That $P_t^\dagger = P_t$ follows from the assumption that $\omega \circ E_t = \omega$; indeed, using (5.42) and (5.43), one gets

$$\begin{aligned}\langle\, \pi_\omega(A)\Omega_\omega \,|\, P_t\pi_\omega(B)\Omega_\omega \,\rangle &= \omega(A^\dagger\, E_t[B]) = \omega(E_t[A]^\dagger\, E_t[B])) = \omega(E_t[A]^\dagger)\, B) \\ &= \langle\, P_t\pi_\omega(A)\Omega_\omega \,|\, \pi_\omega(B)\Omega_\omega \,\rangle \,,\end{aligned}$$

for all $A \in \mathcal{A}$ (thus on a dense subset of $\mathbb{H}_\omega$). Also, on a dense subset, it holds that

$$\begin{aligned} P_t \pi_\omega(A)|\,\Omega_\omega\,\rangle &= \Theta_t \circ E_0 \circ \Theta_{-t} \pi_\omega(A)|\,\Omega_\omega\,\rangle \\ &= U^\dagger_\omega(t)\, E_0[U_\omega(t)\pi_\omega(A) U^\dagger_\omega(t)]|\,\Omega_\omega\,\rangle \\ &= U^\dagger_\omega(t)\, P_0\, U_\omega(t)\pi_\omega(A)]|\,\Omega_\omega\,\rangle \ . \end{aligned}$$

This proves the second statement of the Proposition, while, of the last assertions, the first one is a consequence of $\mathcal{M}_t \subset \mathcal{M}_{t+1}$ and the last two relations follow from

$$\begin{aligned} \lim_{t\to+\infty} \langle\, \pi_\omega(A)^\dagger \,|\, (P_t - \mathbb{1})\, |\pi_\omega(B)\,\rangle &= \lim_{t\to+\infty} \omega(A^\dagger\, (E_t[B] - B)) = 0 \\ \lim_{t\to-\infty} \langle\, \pi_\omega(A)^\dagger \,|\, (P_t - |\,\Omega_\omega\,\rangle\langle\,\Omega_\omega\,|)\, |\pi_\omega(B)\,\rangle &= \lim_{t\to+\infty} \omega(A^\dagger\, E_t[B]) - \\ &\quad -\omega(A)\omega(B) = 0 \ . \end{aligned}$$

In fact, these two limits imply weak-operator convergence of projections to projections which is equivalent to strong-operator convergence. Notice that the second limit holds since $E_t$ maps onto the trivial algebra when $t \to -\infty$ and $\omega(E_t[B]) = \omega(B)$. □

**Corollary 7.1.2.** *Let $(\mathcal{M}, \Theta, \omega)$ be a von Neumann algebraic quantum $K$-system as specified above; then, for any $\varepsilon > 0$, $A_0 \in \mathcal{M}_0$ and $A \in \mathcal{M}$ there exists $T > 0$ such that*

$$\Big|\omega(A_0 \Theta_t[A]) - \omega(A_0)\omega(A)\Big| \leq \varepsilon \sqrt{\omega(A_0\, A_0^\dagger)}$$

*for all $t \leq -T$.*

**Proof:** The result is a consequence of the second strong-operator limit in the previous proposition and of [217, 300]

$$\omega(A_0 \Theta_t[A]) = \omega(A_0\, P_0\, U^\dagger_\omega(t) A) = \omega(A_0\, U^\dagger_\omega(t)\, P_{-t}\, A) \ ,$$

whence

$$\begin{aligned} \Big|\omega(A_0 \Theta_t[A]) - \omega(A_0)\omega(A)\Big| &= \Big|\omega\Big(A_0\, U^\dagger_\omega(t)\, (P_t - |\,\Omega_\omega\,\rangle\langle\,\Omega_\omega\,|)\, A\Big)\Big| \\ &\leq \sqrt{\omega(A_0\, A_0^\dagger)}\ \|(P_t - |\,\Omega_\omega\,\rangle\langle\,\Omega_\omega\,|)\, A|\,\Omega_\omega\,\rangle\| \ . \end{aligned}$$

□

**Corollary 7.1.3.** *Let $(\mathcal{M}, \Theta, \omega)$ be a von Neumann algebraic quantum $K$-system as specified above; then, it is weakly-mixing.*

**Proof:** Because of Remark 7.1.9, one need show

$$\lim_{t\pm\infty} \omega(A\Theta_t[B]) = \omega(A)\,\omega(B) \qquad \forall A, B \in \mathcal{M}\ .$$

Since $\{\mathcal{M}_t\}_{t\in\mathbb{Z}}$ is a $K$-sequence, let $\varepsilon > 0$ and choose $A_\varepsilon \in \mathcal{M}_0$ and $s \in \mathbb{Z}$ large enough such that

$$\|(A - \Theta_s[A_\varepsilon])|\,\Omega_\omega\,\rangle| \leq \varepsilon\ .$$

Then, for $t$ sufficiently large,

$$\Big|\omega(A\Theta_{-t}[B]) - \omega(A)\omega(B)\Big| \leq \Big|\omega((A - \Theta_s[A_\varepsilon])\Theta_{-t}[B])\Big| + \\ +\Big|\omega(A_\varepsilon\Theta_{-(t+s)}[B]) - \omega(A)\omega(B)\Big| \leq 2\varepsilon\,\|B\|\ .$$

This shows clustering when $t \to -\infty$; when $t \to +\infty$, one uses the modular automorphism to rewrite

$$\omega(A\Theta_t[B]) = \omega(\Theta_{-t}[A]B) = \omega(\sigma^i_\omega[B]\theta_{-t}[A])\ ,$$

and then applies the previous argument. □

**Remarks 7.1.12.**

1. The result in Corollary 7.1.2 is the maximum of uniformity one can achieve in clustering; indeed, if
$$\Big|\omega(B\Theta_t[A]) - \omega(B)\omega(A)\Big| \leq \varepsilon\,\sqrt{\omega(B\,B^\dagger)}$$
for all $t \leq -T$ and $B \in \mathcal{M}$, then $B = \Theta_t[A^\dagger]$ would yield
$$0 = \omega(A^\dagger\,A) - |\omega(A)|^2 = \omega\Big((A^\dagger - \omega(A)^*)(A - \omega(A))\Big)\ .$$
As $\omega$ was assumed faithful, this gives $A = \omega(A)\mathbb{1}$ for all $A$.
2. By substituting $A_0$ with $\Theta_s[A_0]$ for fixed $s$, the uniformity in Corollary 7.1.2 holds with respect to any fixed $\mathcal{M}_s$.
3. The structure of the nested sequence of Hilbert subspaces $\{\mathbb{H}_t\}_{t\in\mathbb{Z}}$ corresponding to the projections $P_t$ very much resembles that arising from the Lebesgue spectrum of classical $K$-systems (see the discussion after Remark 2.3.5). However, the projections $\{P_t\}_{t\in\mathbb{Z}}$ have been constructed by relying on the existence of $\omega$-preserving conditional expectations $E_t : \mathcal{M} \mapsto \mathcal{M}_t$. For a state like the tracial state which has trivial modular automorphism, they surely exist; however, this need not be true in general. Luckily, the previous results can also be proved without referring to the existence of a $K$-sequence of projections [220].

**Example 7.1.13.** As sketched in Example 2.3.3.4, the classical hyperbolic automorphisms of the torus are $K$-systems. Aided by this fact we shall show that, for rational values of the deformation parameter $\theta = p/q$, their quantized versions $(\mathcal{M}_\theta, \Theta, \omega)$ are algebraic von Neumann quantum $K$-systems [44].

According to Example 7.1.6, we shall identify the classical automorphisms of the torus as triplets $(\mathcal{M}_0, \Theta_{\mathbb{A}}, \omega)$, where $\mathcal{M}_0$ is the von Neumann Abelian algebra of essentially bounded functions on $\mathbb{T}^2$, $\Theta_{\mathbb{A}}$ is such that $\Theta_{\mathbb{A}}[f](\boldsymbol{r}) = f(\mathbb{A}\,\boldsymbol{r})$, $f \in \mathcal{M}_0$, and $\omega$ is the integration on $\mathbb{T}^2$ with respect to the uniform measure $\mathrm{d}\boldsymbol{r}$. Therefore, the $K$-partition characterizing them as $K$-systems amounts to the existence of a $K$-sequence $\{\mathcal{N}_t\}_{t\in\mathbb{Z}}$ of von Neumann subalgebras of $\mathcal{M}$ (see Definition 2.3.6).

Because of the decomposition (7.49), let $\mathcal{M}_t \subset \mathcal{M}_{p/q}$ be defined, with obvious use of the notation, as

$$\mathcal{M}_t := \sum_{\boldsymbol{s}\in J(q)} \Phi_q[\Pi_{\boldsymbol{s}}[\mathcal{N}_t]]\, W_{p/q}(\boldsymbol{s}) \ . \tag{7.82}$$

In this way, the $K$-properties of the classical $K$-sequence $\{\mathcal{N}_t\}_{t\in\mathbb{Z}}$ would make the characterizing properties in Definition 7.1.10 also hold for the non-commutative sequence $\{\mathcal{M}_t\}_{t\in\mathbb{Z}}$ of subalgebras of $\mathcal{M}_{p/q}$. The first two conditions are in fact immediate, while the third one comes from the fact that (7.46) implies $\Pi_{\boldsymbol{s}}[\mathbb{1}] = \delta_{\boldsymbol{s},\boldsymbol{0}}$.

Unfortunately, one has first to ensure that the $\mathcal{M}_t$ are subalgebras, namely that, if $f, g \in \mathcal{N}_t$, also

$$\begin{aligned}
&\sum_{\boldsymbol{s},\boldsymbol{t}\in J(q)} \Phi_q[\Pi_{\boldsymbol{s}}[f]\, \Phi_q[\Pi_{\boldsymbol{t}}[g]]\, W_{p/q}(\boldsymbol{s})\, W_{p/q}(\boldsymbol{t}) = \\
&= \sum_{\boldsymbol{s},\boldsymbol{t}\in J(q)} \Phi_q[\Pi_{\boldsymbol{s}}[f]\, \Phi_q[\Pi_{\boldsymbol{t}}[g]]\, W_{p/q}(q[\boldsymbol{s}+\boldsymbol{t}])\, W_{p/q}(<\boldsymbol{s}+\boldsymbol{t}>) \\
&= \sum_{\boldsymbol{s},\boldsymbol{t}\in J(q)} \Phi_q\Big[\Pi_{\boldsymbol{s}}[f]\, \Phi_q[\Pi_{\boldsymbol{t}}[g]\, W_0(q[\boldsymbol{s}+\boldsymbol{t}])\Big]\, W_{p/q}(<\boldsymbol{s}+\boldsymbol{t}>)
\end{aligned} \tag{7.83}$$

belongs to $\mathcal{M}_t$.

To this purpose, consider the von Neumann subalgebra $\mathcal{M}_0^{(q)} \subset \mathcal{M}_0$ consisting of those essentially bounded functions on $\mathbb{T}^2$ that satisfy (7.45). Since $\mathcal{M}_0^{(q)}$ is mapped into itself by $\Theta_{\mathbb{A}}$, the $K$-properties of the sequence $\{\mathcal{N}_t\}_{t\in\mathbb{Z}}$ extend to the sequence $\{\mathcal{N}_t^{(q)} := \mathcal{N}_t \cap \mathcal{M}_0^{(q)}\}_{t\in\mathbb{Z}}$, whence the triplets $(\mathcal{M}_0^{(q)}, \Theta_{\mathbb{A}}, \omega)$ are also $K$-systems [3]. Further, let $\mathcal{B}$ denote the (von Neumann) subalgebra generated by the characteristic functions $\chi_{\Delta(\boldsymbol{s})}$ of the partition of the torus into atoms

$$\Delta(\boldsymbol{s}) := \left\{\boldsymbol{r} \,:\, \frac{s_i}{q} \le x_i \le \frac{s_i+1}{q} \ , \ i = 1,2\right\} ,$$

[3] Here, $\Theta_{\mathbb{A}}$ and $\omega$ denote the restrictions of the dynamics and the state to $\mathcal{M}_0$

where $s$ in $\in J(q)$; set $\mathcal{B}_t := \Theta_{\mathbb{A}}(\mathcal{B})$, $\mathcal{B}_{t]} := \bigvee_{s=-\infty}^{t} \mathcal{B}_s$. Finally, construct the von Neumann subalgebras

$$\widetilde{\mathcal{N}}_t := \mathcal{N}_t^{(q)} \vee \mathcal{B}_{t]} ,$$

and consider the sequence $\{\widetilde{\mathcal{N}}_t\}_{t\in\mathbb{Z}}$.

This is a $K$-sequence for $\mathcal{M}_0$; indeed, $\widetilde{\mathcal{N}}_t \subset \widetilde{\mathcal{N}}_{t+1}$ directly follows from the analogous property of the $K$-sequence $\{\mathcal{N}^{(q)}\}_{t\in\mathbb{Z}}$, while $\bigvee_{t\in\mathbb{Z}} \widetilde{\mathcal{N}}_t = \mathcal{M}$ is a consequence of the fact that $\bigvee_{t\in\mathbb{Z}} \mathcal{N}_t^{(q)} = \mathcal{M}_0^{(q)}$ together with the observation that $\mathcal{M}_0 = \mathcal{M}_0^{(q)} \vee \mathcal{B} \subset \bigvee_{t\in\mathbb{Z}} \widetilde{\mathcal{N}}_t$. Finally, $\bigwedge_{t\in\mathbb{Z}} \widetilde{\mathcal{N}}_t = \{\lambda \mathbb{1}\}$ follows from the fact that, according to Proposition 2.3.5, $\mathrm{Tail}\,(\mathcal{B}) = \{\lambda \mathbb{1}\}$.

Let us now insert the subalgebras $\widetilde{\mathcal{N}}_t$ in the place of $\mathcal{N}_t$ in (7.82); using (7.47), for $f \in \widetilde{\mathcal{N}}_t$, $\boldsymbol{s} \in J(q)$, let us consider

$$\Pi_{\boldsymbol{s}}[f]\, W_0(\boldsymbol{s}) = \frac{1}{q^2} \sum_{\boldsymbol{t}\in J(q)} \gamma_{\boldsymbol{t}}^{(q)}[f]\, \mathrm{e}^{-2\pi i \boldsymbol{s}\cdot\boldsymbol{t}/q} .$$

Now, it turns out that the map in (7.45) fulfils

$$\begin{aligned}\gamma_{\boldsymbol{s}}^{(q)}\left[\Theta_{\mathbb{A}}\left[f\right]\right](\boldsymbol{r}) &= \Theta_{\mathbb{A}}[f]\left(\boldsymbol{r} + \frac{\boldsymbol{s}}{q}\right) = f\left(\mathbb{A}\boldsymbol{r} + \frac{\mathbb{A}\boldsymbol{s}}{q}\right) = f\left(\mathbb{A}\boldsymbol{r} + \frac{<\mathbb{A}\boldsymbol{s}>}{q}\right) \\ &= \Theta_{\mathbb{A}}\left[\gamma_{<\mathbb{A}\boldsymbol{s}>}^{(q)}[f]\right](\boldsymbol{r}) .\end{aligned}$$

Since $\gamma_{\boldsymbol{s}}^{(q)}$ maps the subalgebra $\mathcal{B}$ into itself for all $\boldsymbol{s} \in J(q)$, it turns out that, when $f \in \widetilde{\mathcal{N}}_t$, all the components in (7.82) also belong to $\widetilde{\mathcal{N}}_t$. Therefore, with $f, g \in \widetilde{\mathcal{N}}_t$, it follows that

$$\widetilde{\mathcal{N}}_t \ni \Pi_{\boldsymbol{s}}[f]\, W_0(\boldsymbol{s})\, \Pi_{\boldsymbol{t}}[g]\, W_0(\boldsymbol{t}) = \underbrace{\Big(\Pi_{\boldsymbol{s}}[f]\, \Pi_{\boldsymbol{t}}[g]\, W_0(q[\boldsymbol{s}+\boldsymbol{t}])\Big)}_{\in \Pi_{<\boldsymbol{s}+\boldsymbol{t}>}(\widetilde{\mathcal{N}}_t)} W_0(<\boldsymbol{s}+\boldsymbol{t}>) .$$

This shows that the linear sets in (7.83) are subalgebras of $\mathcal{M}_{p/q}$ and completes the proof that the quantum dynamical triplets $(\mathcal{M}_{p/q}, \Theta_{\mathbb{A}}, \omega)$ are algebraic von Neumann quantum $K$-systems.

**Remark 7.1.13.** The reason why all quantized hyperbolic automorphisms of the torus are algebraic $K$-systems for rational deformation parameters is that the properties of their classical counterparts are inherited by the commutative subsystems (the centers) contained in $(\mathcal{M}_{p/q}, \Theta_{\mathbb{A}}, \omega)$. When the deformation parameter is irrational, this is no longer true and indeed one can prove that a part from countable sets of $\theta \in [0,1]$ $(\mathcal{M}_\theta, \Theta_{\mathbb{A}}, \omega)$ cannot be algebraic $K$-systems [44].

### 7.1.5 Quantum Spin Chains

As sketched in Example 7.1.1, the algebraic structure of quantum spin chains is as in Definition 2.2.5, the difference being that at each site of the one-dimensional lattice indexed by the integers $k \in \mathbb{Z}$, instead of diagonal matrix algebras, there remain assigned copies $\mathcal{A}_k$ of a same non-commutative algebra $\mathcal{A}$. In the following, we shall consider $\mathcal{A} = M_d(\mathbb{C})$, namely chains consisting of linear arrays of $d$-level quantum systems (or spins).

The algebra $\mathcal{A}_{\mathbb{Z}}$ associated with the infinite chain is the quasi-local $C^*$-algebra which arises by taking the norm closure of the $*$-algebra consisting of operators from all *local algebras* $\mathcal{A}_{[-\ell,\ell]} := \bigotimes_{k=-\ell}^{\ell} \mathcal{A}_k = M_d(\mathbb{C})^{\otimes(2\ell+1)}$, supported by the lattice sites in the interval $[-\ell,\ell]$. If $\mathcal{A}_0$ denotes the strictly local $*$ algebra $\bigcup_{\ell\in\mathbb{N}} \mathcal{A}_{[-\ell,\ell]}$, then $\mathcal{A}_{\mathbb{Z}} = \overline{\mathcal{A}_0}^{\|\cdot\|}$.

The local algebras $\mathcal{A}_{[-\ell,\ell]} = \bigotimes_{k=-\ell}^{\ell} \mathcal{A}_k$ describe spin arrays located at finitely many lattice sites $-\ell \leq k \leq \ell$. Their elements are linear combinations of tensor products $\otimes_{k=-\ell}^{\ell} A_k$, $A_k \in \mathcal{A}_k$. If $0 \leq \ell \leq p$, the local algebra $\mathcal{A}_{[-\ell,\ell]}$ can be embedded into $\mathcal{A}_{[-p,p]}$ as follows; we shall denote by

$$\mathbb{1}_{[i,j]} := \otimes_{k=i}^{j} \mathbb{1}_k \ , \quad \mathbb{1}_{i-1]} := \otimes_{k=-\infty}^{i-1} \mathbb{1}_k \ , \quad \mathbb{1}_{[j+1} := \otimes_{k=j+1}^{\infty} \mathbb{1}_k \tag{7.84}$$

the tensor products of identities at sites from $i$ to $j$, from $-\infty$ to $i-1$ and from $j+1$ to $+\infty$, respectively. Then, $\mathcal{A}_{[-\ell,\ell]}$ is embedded into $\mathcal{A}_{[-p,p]}$ as $\mathbb{1}_{[-p,-\ell-1]} \otimes \mathcal{A}_{[-\ell,\ell]} \otimes \mathbb{1}_{[\ell+1,p]}$. Analogously, $\mathcal{A}_{[-\ell,\ell]}$ is embedded into $\mathcal{A}_0$ as $\mathbb{1}_{-\ell-1]} \otimes \mathcal{A}_{[-\ell,\ell]} \otimes \mathbb{1}_{[\ell+1}$. In the following, for sake of simplicity, we shall often identify local algebras $\mathcal{A}_{[-\ell,\ell]}$ with their embeddings, as well as their elements as elements of $\mathcal{A}_0$.

The dynamics over $\mathcal{A}_{\mathbb{Z}}$ is the shift automorphism $\Theta_\sigma : \mathcal{A}_{\mathbb{Z}} \to \mathcal{A}_{\mathbb{Z}}$

$$\Theta_\sigma(\mathcal{A}_{[-\ell,\ell]}) = \mathcal{A}_{[-\ell+1,\ell+1]}$$
$$\Theta_\sigma\Big(\mathbb{1}_{-\ell-1]} \otimes \Big(\otimes_{k=-\ell}^{\ell} A_k\Big) \otimes \mathbb{1}_{[\ell+1}\Big) = \mathbb{1}_{-\ell]} \otimes \Big(\otimes_{k=-\ell+1}^{\ell+1} A_k\Big) \otimes \mathbb{1}_{[\ell+2} \ .$$

In order to complete the description of quantum spin chains as quantum dynamical systems we need provide $\mathcal{A}_{\mathbb{Z}}$ with translation invariant states, that is with positive functionals $\omega : \mathcal{A}_{\mathbb{Z}} \mapsto \mathbb{C}$ such that $\omega \circ \Theta_\sigma = \omega$. Given any such state, its restrictions $\omega_{[i,j]} := \omega \lceil \mathcal{A}_{[i,j]}$ to a local subalgebras $\mathcal{A}_{[i,j]} := \bigotimes_{k=i}^{j} \mathcal{A}_k$ are density matrices $\rho_{[i,j]} \in M_d(\mathbb{C})^{\otimes(j-i+1)}$. Since it originates from the global state $\omega$, the family of $\rho_{[i,j]}$ is automatically compatible with the embedding of $\mathcal{A}_{[i,j]} \subset \mathcal{A}_{[i,j+1]}$, that is they fulfil the condition

$$\begin{aligned} \omega\left(A_i \otimes \cdots A_j \otimes \mathbb{1}_{j+1}\right) &= \mathrm{Tr}_{[i,j+1]}\left(\rho_{[i,j+1]} A_i \otimes \cdots A_j \otimes \mathbb{1}_{j+1}\right) \\ &= \mathrm{Tr}_{[i,j]}\left(\left(\mathrm{Tr}_{\{j+1\}} \rho_{[i,j+1]}\right) A_i \otimes \cdots A_j\right) \\ &= \mathrm{Tr}_{[i,j]}\left(\rho_{[i,j]} A_i \otimes \cdots A_j\right) \ , \end{aligned}$$

where $\mathrm{Tr}_{[i,j]}$ indicates that the trace has to be performed with respect to an orthonormal basis of the Hilbert space $(\mathbb{C}^d)^{\otimes(j-i+1)}$ associated with the spins at sites $i \leq k \leq j$. In other words,

$$\mathrm{Tr}_{\{j+1\}}(\rho_{[i,j+1]}) = \rho_{[i,j]} , \qquad \forall i,j \in \mathbb{Z} . \tag{7.85}$$

Further, translation invariance implies

$$\begin{aligned}\omega(\Theta_\sigma(A_i \otimes \cdots A_j)) &= \omega(\mathbb{1}_i \otimes A_i \otimes \cdots A_j) \\ &= \mathrm{Tr}_{[i,j+1]}\left(\rho_{[i,j+1]}\mathbb{1}_i \otimes A_i \otimes \cdots A_j \otimes \mathbb{1}_{j+1}\right) \\ &= \mathrm{Tr}_{[i,j]}\left(\left(\mathrm{Tr}_{\{i\}}\rho_{[i,j+1]}\right) A_i \otimes \cdots A_j\right) \\ &= \omega(A_i \otimes \cdots A_j) = \mathrm{Tr}_{[i,j]}\left(\rho_{[i,j]}A_i \otimes \cdots A_j\right) ,\end{aligned}$$

whence the local states satisfy

$$\mathrm{Tr}_{\{i\}}(\rho_{[i,j+1]}) = \rho_{[i+1,j+1]} = \rho_{[i,j]} , \qquad \forall i,j \in \mathbb{Z} . \tag{7.86}$$

Vice versa, if $\mathcal{A}_\mathbb{Z}$ is equipped with a family of local states $\rho_{[i,j]}$, $i,j \in \mathbb{Z}$, satisfying (7.85) and (7.86), then they define a translation invariant state $\omega$ on $\mathcal{A}_\mathbb{Z}$ such that its restrictions to local subalgebras satisfy $\omega\lceil\mathcal{A}_{[i,j]} = \rho_{[i,j]}$ [10].

**Definition 7.1.11 (Quantum Spin Chains).**

*Quantum spin chains are dynamical systems represented by algebraic triplets $(\mathcal{A}_\mathbb{Z}, \Theta_\sigma, \omega)$ where*

1. *$\mathcal{A}_\mathbb{Z}$ is a quasi-local algebra with a d-level system at each site;*
2. *$\Theta_\sigma : \mathcal{A}_\mathbb{Z} \mapsto \mathcal{A}_\mathbb{Z}$ is the shift-automorphism over $\mathcal{A}_\mathbb{Z}$;*
3. *$\omega : \mathcal{A}_\mathbb{Z} \mapsto \mathbb{C}$ is a translation invariant state over $\mathcal{A}_\mathbb{Z}$*

**Example 7.1.14.** Quantum spin chains turn out to be $C^*$ algebraic quantum $K$-systems; indeed, one argues in the same way as for classical spin chains (see Remark 2.3.5). The $K$-sequence consist of the quasi-local algebras $\mathcal{A}_t := \Theta_\sigma^t(\mathcal{A}_0)$ where $\mathcal{A}_0 \subset \mathcal{A}_\mathbb{Z}$ is the quasi-local algebra which arises as the $C^*$ inductive limit of the local matrix algebras $\mathcal{A}_{[p,q]}$, with $p \leq q \leq 0$.

If $\omega$ is a factor state, $(\mathcal{A}_\mathbb{Z}, \Theta_\sigma, \omega)$ is also a von Neumann algebraic quantum $K$-system. This can be seen as follows: denote by $\mathcal{A}_{[t}$ the quasi-local algebra generated by all matrix algebras of the form $\mathcal{A}_{[p,q]}$ with $t \leq p \leq q$ and set $\mathcal{M}_t := \pi_\omega(\mathcal{A}_t)''$, $\mathcal{M}_{[t} := \pi_\omega(\mathcal{A}_{[t})''$, $\mathcal{M}_t' := (\mathcal{M}_t)'$ for the various commutants. Clearly, the first two conditions in the second part of Definition 7.1.10 are satisfied. The third one is obtained as follows: since $\mathcal{M}_{[t+1} \subset \mathcal{M}_t'$, one finds [220]

$$\begin{aligned}\left(\bigcap_{t\in\mathbb{Z}}\mathcal{M}_t\right)' &= \bigcup_{t\in\mathbb{Z}}\mathcal{M}_t' = \bigcup_{t\in\mathbb{Z}}(\mathcal{M}_t' \cup \mathcal{M}_{[t+1}) = \left(\bigcup_{t\in\mathbb{Z}}\mathcal{M}_t'\right) \cup \left(\bigcup_{t\in\mathbb{Z}}\mathcal{M}_{[t+1}\right) \\ &= \mathcal{M}' \cup \mathcal{M} = (\mathcal{M} \cap \mathcal{M}')' = \lambda\mathbb{1}'\end{aligned}$$

for $\omega$ has been assumed to be a factor whence the center $\mathcal{Z}_\omega = \mathcal{M} \cap \mathcal{M}'$ is trivial.

This is not true in general; for instance, in the case of Example 7.1.11, the state (7.75) is not a factor and the von Neumann system is not clustering. Therefore, according to Corollary 7.1.3, $(\mathcal{M}, \theta_\sigma, \omega)$ cannot be a von Neummann algebraic quantum $K$-system.

### Ergodic Quantum Spin Chains

In Remark 7.1.8, we have seen that asymptotic Abelianess allows to uniquely decompose non-extremal invariant states into their ergodic components.

As a concrete application, consider a quantum spin-chain $(\mathcal{A}_{\mathbb{Z}}, \Theta_\sigma, \omega)$ whose state $\omega$ is extremal invariant (see Definition 7.1.8) with respect to the lattice translation by one site, $\Theta_\sigma$, but not with respect to lattice translations by $\ell$ sites, $\Theta_\sigma^\ell$, for some $\ell \in \mathbb{N}$. We prove the following result [58].

**Proposition 7.1.9.** *Let $(\mathcal{A}_{\mathbb{Z}}, \Theta_\sigma, \omega)$ be an ergodic quantum spin-chain. For any $\ell \in \mathbb{N}$ the state $\omega$ can be written as a convex decomposition*

$$\omega = \frac{1}{n_\ell} \sum_{j=0}^{n_\ell - 1} \omega_j \ , \tag{7.87}$$

*where $n_\ell$ divides $\ell$ and the $\omega_j$ are shift-invariant states over the spin-chain which are ergodic with respect to $\Theta_\sigma^\ell$.*

**Proof:** Consider the commutant $(\mathcal{R}_\omega^\ell)' := \pi_\omega(\mathcal{A}_{\mathbb{Z}})' \cap \{U_\omega^\ell)\}'$ of the covariance algebra (see Definition 7.1.5) $\mathcal{R}_\omega^\ell := \Big(\pi_\omega(\mathcal{A}) \cup U_\omega^\ell(\mathbb{Z})\Big)''$ which is built by means of the $C^*$ algebra $\pi_\omega(\mathcal{A}_{\mathbb{Z}})$ and the group of unitary *GNS* operators $U_\omega^{n\ell}$, $n \in \mathbb{N}$, instead of $U_\omega(\mathbb{Z})$.

Since $\mathcal{A}_{\mathbb{Z}}$ is norm-asymptotic Abelian and $\omega$ is assumed not to be $\Theta_\sigma^\ell$-ergodic, because of Corollary 7.1.1, it turns out that $\mathcal{R}_\omega^\ell \neq \{\lambda \mathbb{1}\}$. Let $\{Q_i\}_{i\in I}$ be a decomposition of the identity by orthogonal projections in $(\mathcal{R}_\omega^\ell)'$; then, the cardinality of $I$, $n_\ell$, must fulfil $n_\ell \leq \ell$.

Indeed, let $P$ denote any of the $Q_i$, $i \in I$, and set $P_j := U_\omega^j P (U_\omega^\dagger)^j$, $0 \leq j \leq \ell - 1$. Since $P$ commutes with $\pi_\omega(\mathcal{A}_{\mathbb{Z}})$, one derives

$$\pi_\omega(X) P_j = U_\omega^j \pi_\omega(\Theta^j[X]) P (U_\omega^\dagger)^j = U_\omega^j P \pi_\omega(\Theta^j[X]) (U_\omega^\dagger)^j = P_j \pi_\omega(X) \ ,$$

for all $X \in \mathcal{A}_{\mathbb{Z}}$. Moreover, $P_j$ commutes with $U_\omega^\ell$, whence $P_j \in (\mathcal{R}_\omega^\ell)'$ for all $0 \leq j \leq \ell - 1$ and so does $\overline{P} := \bigvee_{j=0}^{\ell-1} P_j$, namely the smallest one among the projections $Q$ such that $Q \geq P_j$ for all $0 \leq j \leq \ell - 1$.

By decomposing $n \in \mathbb{Z}$ as $n = m\ell + r$, $0 \leq r \leq \ell - 1$, it follows that

$$U_\omega^n \{P_j\}_{j=0}^{\ell-1} (U_\omega^\dagger)^n = \{P_j\}_{j=0}^{\ell-1} \ .$$

Therefore, $\overline{P}$ is $\Theta_\sigma$-invariant, whence $\overline{P} = \mathbb{1}$ as $\omega$ is ergodic with respect to $\Theta_\sigma$-ergodicity of $\omega$. Then, (7.61) in the ergodic list yields

$$1 = \langle \Omega | \overline{P} | \Omega \rangle \leq \sum_{j=0}^{n_\ell - 1} \langle \Omega | P_j | \Omega \rangle = \ell \langle \Omega | P | \Omega \rangle \ .$$

Applying this argument to each of the $Q_j$, one obtains

$$1 = \langle \Omega | \sum_{i\in I} Q_i | \Omega \rangle \geq \frac{n_\ell}{\ell} \, .$$

Since $\mathcal{A}$ is norm-asymptotic Abelian, from the second step in the proof of Proposition 7.1.5, one deduces that $(\mathcal{R}^\ell_\omega)'$ is Abelian. Let then $Q_j$, $0 \leq j \leq n_\ell - 1 \leq \ell$, be its minimal projections (see Example 5.3.4.1) with $Q_0$ such that

$$q_0 := \langle \Omega | Q_0 | \Omega \rangle \leq q_j := \langle \Omega | Q_j | \Omega \rangle$$

for all $j > 0$. Then, introduce the set

$$\mathcal{S}_0 := \left\{ j \in \mathbb{Z} \, : \, U^j_\omega \, Q_0 \, (U^\dagger_\omega)^j = Q_0 \right\} .$$

It follows that $\mathcal{S}_0 \supseteq \ell\,\mathbb{Z}$; further, set $k_0 := \min\{0 < j \in \mathcal{S}_0\}$. Then, $\ell \propto k_0$; otherwise, $\ell = p\,k_0 + q$, for some $p \geq 0$ and $0 < q < k_0$, so that $q \in \mathcal{S}_0$ thus contradicting the minimality of $k_0$.

Furthermore, set $Q_{0,j} := U^j_\omega \, Q_0 \, (U^\dagger_\omega) j$, $0 \leq j \leq k_\ell$; $Q_{0,j}$ belongs to $(\mathcal{R}^\ell_\omega)'$. If it is not a minimal projector $Q_{i(j)}$, then, $Q_{0,j} = \sum_i Q_i$ thus contradicting $q_0 \leq q_j$ when $j > 0$. Thus, $Q_{0,j} = Q_{i(j)}$. If $\overline{Q}_0 := \bigvee_{j=0}^{k_0-1} Q_j = \sum_{j=0}^{k_0-1} Q_j$ (the projectors are now orthogonal), it follows that, as shown before, $\overline{Q}_0 \in (\mathcal{R}^\ell_\omega)'$ and hence $\overline{Q}_0 = \mathbb{1}$. Consequently, because of the uniqueness of the orthogonal decomposition of the identity in an Abelian algebra, it turns out that $k_0 = n_\ell$ whence $Q_j = U^j_\omega \, Q_0 \, (U^\dagger_\omega)^j$ and $q_0 = q_j = n_\ell^{-1}$.

One can now introduce the states on $\mathcal{A}_\mathbb{Z}$ defined by

$$\begin{aligned} \mathcal{A}_\mathbb{Z} \ni X \mapsto \omega_j(X) &:= n_\ell \, \langle \Omega | Q_j \pi_\omega(X) | \Omega \rangle = \langle \Omega | Q_0 \pi_\omega \left(\Theta_\sigma^{-j}(X)\right) | \Omega \rangle \\ &= \omega_0(\Theta^j_\sigma[X]) \, , \end{aligned}$$

for all $X \in \mathcal{A}_\mathbb{Z}$. It turns out that the $\omega_j$'s are all $\Theta^\ell_\sigma$-ergodic, otherwise it would be possible to further convexly decompose them (hence $\omega$) into $\Theta^\ell_\sigma$-invariant components:

$$\omega_j = \sum_i \lambda_{ji} \omega_{ji} \, , \quad \omega = \sum_{j=0}^{n_\ell - 1} \sum_i \frac{\lambda_{ji}}{n_\ell} \omega_{ji} \, .$$

As explained in Remark 7.1.5.2, the decomposers $\omega_{ji}$ correspond to projectors $P_{ji} \in (\mathcal{R}^\ell_\omega)'$ such that $\dfrac{\lambda_{ji}}{n_\ell} = \langle \Omega | P_{ji} | \Omega \rangle$ and

$$\begin{aligned} \lambda_{ji} \omega_{ji}(X) &= n_\ell \, \langle \Omega | P_{ji} \pi_\omega(X) | \Omega \rangle \\ &\leq \omega_j(X) = n_\ell \langle \Omega | P_j \pi_\omega(X) | \Omega \rangle \, , \quad \forall X \in \mathcal{A}_\mathbb{Z} \, . \end{aligned}$$

As the projections $P_{ji}$ and $P_j$ belong to the commutant $\pi_\omega(\mathcal{A}_\mathbb{Z})'$, by choosing $X = Y^\dagger Z$ one gets

$$\langle \Omega \,|\, \pi_\omega(Y)^\dagger \, P_{ji} \, \pi_\omega(Z) \,|\Omega\rangle \leq \langle \Omega \,|\, \pi_\omega(Y)^\dagger \, P_j \, \pi_\omega(Z) \,|\Omega\rangle \ .$$

Since $Y$ and $Z$ are arbitrary elements of $\mathcal{A}_\mathbb{Z}$ and the vectors $\pi_\omega(X)|\,\Omega\,\rangle$ are dense in the *GNS* Hilbert space, it turns out that $P_{ji} \leq P_j$. But the $P_j$'s are minimal projections, thus $P_{ji} = P_j$. □

### Finitely Correlated States

An interesting class of translation-invariant states on a quantum spin-chain $\mathcal{A}_\mathbb{Z}$ is constructed as follows [113]. Let $(\mathcal{B}, \rho, \mathbb{E})$ be an auxiliary triplet, where

1. $\mathcal{B}$ is a finite dimensional algebra that we shall fix to be the algebra $M_b(\mathbb{C})$ of $b \times b$ matrices acting on $\mathbb{C}^b$ ;
2. $\rho$ is a state on $\mathcal{B}$ identified by a density matrix: $\rho(B) = \mathrm{Tr}(\rho\, B)$.
3. $\mathbb{E} : \mathcal{A} \mapsto \mathcal{B}$ is a completely positive map such that

$$\mathbb{E}(\mathbb{1}_\mathcal{A} \otimes \mathbb{1}_\mathcal{B}) = \mathbb{1}_\mathcal{B} \tag{7.88}$$
$$\rho \circ \mathbb{E}(\mathbb{1}_\mathcal{A} \otimes B) = \rho(B) \ , \tag{7.89}$$

where $\mathbb{1}_{\mathcal{A},\mathcal{B}}$ denote the identities of the algebras $\mathcal{A}$, respectively $\mathcal{B}$.

Since they result from iteratively composing *CPU* maps, the following maps

$$\mathbb{E}^{(n)} := \mathbb{E} \circ \left(\mathrm{id}_\mathcal{A} \otimes \mathbb{E}^{(n-1)}\right) \ : \ \mathcal{A}^{\otimes n} \mapsto \mathcal{B} \ , \ n \geq 1 \ , \quad \mathbb{E}^{(0)} := \mathbb{E} \ , \tag{7.90}$$

are also *CPU* . Consequently, the functionals $\rho \circ \mathbb{E}^{(n)}$ on $\mathcal{A}^{\otimes n}$ are positive and normalized. They are thus states on the local algebras $\mathcal{A}^{\otimes n}$: moreover, the corresponding density matrices in $M_d(\mathbb{C})^{\otimes n} \bigotimes M_b(\mathbb{C})$ can be obtained by duality: in fact,

$$\mathrm{Tr}_\mathcal{B}\Big(\rho\, \mathbb{E}[A \otimes B]\Big) = \mathrm{Tr}_{\mathcal{A}\otimes\mathcal{B}}\Big(\mathbb{F}[\rho]\, A \otimes B\Big) \ , \tag{7.91}$$

where $\mathbb{F} : \mathcal{S}(\mathcal{B}) \mapsto \mathcal{S}(\mathcal{A} \otimes \mathcal{B})$ is the trace-preserving dual map of $\mathbb{E}$ which transforms states over $\mathcal{B}$ into states over $\mathcal{A} \otimes \mathcal{B}$. Analogously, to the *CPU* maps $\mathbb{E}^{(n)}$ there correspond the dual maps $\mathbb{F}^{(n)} \ : \ \mathcal{S}(\mathcal{B}) \mapsto \mathcal{S}(\mathcal{A}^{\otimes(n+1)} \otimes \mathcal{B})$ given by

$$\mathbb{F}^{(n)} := (\mathrm{id}_{\mathcal{A}^{\otimes n}} \otimes \mathbb{F}) \circ \mathbb{F}^{(n-1)} \ , \ n \geq 1 \ , \qquad \mathbb{F}^{(0)} := \mathbb{F} \ .$$

Consider the states $\omega_{[-\ell,\ell]}$ defined on the local subalgebras $\mathcal{A}_{[-\ell,\ell]}$ by

$$\begin{aligned}\omega_{[-\ell,\ell]}(\otimes_{k=-\ell}^{\ell} A_k) &:= \mathrm{Tr}_\mathcal{B}\Big(\rho\, \mathbb{E}^{(n)}[(\otimes_{k=-\ell}^{\ell} A_k) \otimes \mathbb{1}_\mathcal{B}]\Big) \\ &= \mathrm{Tr}_\mathcal{B}\Big(\mathbb{F}^{(n-1)}[\rho]\, (\otimes_{k=-\ell}^{\ell} A_k) \otimes \mathbb{1}_\mathcal{B}\Big) \ .\end{aligned} \tag{7.92}$$

As a consequence of (7.88) and (7.89), they satisfy the compatibility relations (7.85), that is $\omega_{[-\ell-1,\ell+1]} \lceil \mathcal{A}_{[-\ell,\ell]} = \omega_{[-\ell,\ell]}$, and the translation-invariance conditions (7.86), namely $\omega_{[-\ell,\ell]} = \omega_{[-\ell+1,\ell+1]}$. We illustrate these properties by means of the simplest non trivial case and choose $\mathcal{A}_{[1,2]}$; then

$$\omega(A \otimes \mathbb{1}_2) = \mathrm{Tr}_{\mathcal{B}}\Big(\rho\, \mathbb{E}[A \otimes \mathbb{E}[\mathbb{1}_{\mathcal{A}} \otimes \mathbb{1}_{\mathcal{B}}]]\Big) = \mathrm{Tr}_{\mathcal{B}}\Big(\rho\, \mathbb{E}[A \otimes \mathbb{1}_{\mathcal{B}}]\Big) = \omega(A)$$
$$\omega(\mathbb{1}_1 \otimes A) = \mathrm{Tr}_{\mathcal{B}}\Big(\rho\, \mathbb{E}[\mathbb{1}_{\mathcal{A}} \otimes \mathbb{E}[A \otimes \mathbb{1}_{\mathcal{B}}]]\Big) = \mathrm{Tr}_{\mathcal{B}}\Big(\rho\, \mathbb{E}[A \otimes \mathbb{1}_{\mathcal{B}}]\Big) = \omega(A) \ .$$

Therefore, the family of local states $\omega_{[-\ell,\ell]}$ defines a global invariant state overt the quantum spin chain $\mathcal{A}_{\mathbb{Z}}$.

**Definition 7.1.12 (Finitely Correlated States).** *Given a triplet $(\mathcal{B}, \rho, \mathbb{E})$ as specified before, all functionals $\omega$ on $\mathcal{A}_{\mathbb{Z}}$ locally defined on $\mathcal{A}_{[i,j]}$ by*

$$\omega(\otimes_{k=i}^{j} A_k) = \mathrm{Tr}_{\mathcal{B}}\Big(\rho\ \mathbb{E}^{(j-i)}[\otimes_{k=i}^{j} A_k]\Big) \ ,$$

*are translation invariant states called finitely correlated (FCS).*

**Remark 7.1.14.** The specification *finitely correlated* refers to the finite dimensionality of the auxiliary algebra $\mathcal{B}$. Without such a restriction, every translation-invariant state over $\mathcal{A}_{\mathbb{Z}}$ would be given as in the previous definition. Indeed, one could then choose $\mathcal{B} := \mathcal{A}_{[1,+\infty]}$, $\rho := \omega \lceil \mathcal{A}_{[0,+\infty]}$ and as $\mathbb{E}$ the natural embedding of any $\mathcal{A}_{[i,j]}$ into $\mathcal{A}_{[1,+\infty]}$.

Because of translation-invariance, $\omega \lceil \mathcal{A}_{[i,j]} = \omega \lceil \mathcal{A}_{[1,j-i+1]}$; therefore, the local structure of $\omega$ is determined by the density matrices $\rho_{[1,n]}$ corresponding to $\omega \lceil \mathcal{A}_{[1,n]}$. They are recursively obtained by means of the dual maps (7.92),

$$\rho_{[1,n]} := \mathrm{Tr}_{\mathcal{B}}\Big(\mathbb{F}^{(n-1)}[\rho]\Big) \ . \tag{7.93}$$

In order to take a closer look at the recursive structure of *FCS*, we make use of the Kraus-Stinespring representation (5.195); concretely,

$$\mathbb{E}(A \otimes B) = \sum_{j \in J} V_j^\dagger\, A \otimes B\, V_j \ , \quad \sum_{j \in J} V_j^\dagger V_j = \mathbb{1}_{\mathcal{B}} \tag{7.94}$$
$$V_j^\dagger : \mathbb{C}^b \mapsto \mathbb{C}^d \otimes \mathbb{C}^b \ , \quad V_j^\dagger : \mathbb{C}^d \otimes \mathbb{C}^b \mapsto \mathbb{C}^b \ , \tag{7.95}$$

where $J$ is an index set of finite cardinality. With $|\psi_i^A\rangle$, $i = 1, , \dots, d$ and $|\psi_k^B\rangle$, $k = 1, 2, \dots, b$ two *ONB*s in $\mathbb{C}^d$, respectively $\mathbb{C}^b$, the action of $V_j$ can be represented in the following two ways,

$$V_j |\, \psi_i^B \,\rangle = \sum_{k=1}^{b} |\, \Psi_{j,ik}^A \,\rangle \otimes |\, \psi_k^B \,\rangle \tag{7.96}$$
$$V_j |\, \psi_i^B \,\rangle = \sum_{\ell=1}^{d} |\, \psi_\ell^A \,\rangle \otimes |\, \Psi_{j,i\ell}^B \,\rangle \ , \tag{7.97}$$

where the $\Psi^A_{j,ik}$, $\Psi^B_{j,i\ell} \in \mathbb{C}^d$ are in general neither orthogonal nor normalised. From (7.96) it follows that

$$V_j = \sum_{i,k=1}^{b} |\,\Psi^A_{j,ik} \otimes \psi^B_k\,\rangle\langle\,\psi^B_k\,|\,, \quad V_j^\dagger = \sum_{i,k=1}^{b} |\,\psi^B_k\,\rangle\langle\,\Psi^A_{j,ik} \otimes \psi^B_k\,|\,.$$

Thus, $\sum_{j\in J} V_j^\dagger V_j = \mathbb{1}$ whence $\sum_{j\in J}\sum_{k=1}^b \langle\,\Psi^A_{j,pk}\,|\,\Psi^A_{j,qk}\,\rangle = \delta_{pq}$. On the other hand, from (7.91) one gets

$$\begin{aligned}\mathbb{F}[\rho] &= \sum_{j\in J}\sum_{\substack{i,k=1\\p,q=1}}^{b} \langle\,\psi^B_p\,|\,\rho\,|\psi^B_k\,\rangle\,|\,\Psi^A_{j,pq}\,\rangle\langle\,\Psi^A_{j,ki}\,|\otimes|\,\psi^B_q\,\rangle\langle\,\psi^B_i\,| \\ &= \sum_{j\in J}\sum_{\ell,p=1}^{b}\sum_{i,q=1}^{b} r_\ell\,|\,\Psi^A_{j,\ell q}\,\rangle\langle\,\Psi^A_{j,\ell i}\,|\otimes|\,\psi^B_q\,\rangle\langle\,\psi^B_i\,|\,, \end{aligned} \tag{7.98}$$

where we have conveniently chosen the eigenprojections of $\rho$ as *ONB* in $\mathbb{C}^b$, that is $\rho = \sum_{\ell=1}^b r_\ell |\,\psi^B_\ell\,\rangle\langle\,\psi^B_\ell\,|$. As condition (7.89) amounts to $\mathrm{Tr}_{\mathcal{A}}\mathbb{F}[\rho] = \rho$, the vectors $\Psi^A_{j,ik}$ must also satisfy $\sum_{j\in J}\sum_{\ell=1}^b r_\ell\langle\,\Psi^A_{j,\ell i}\,|\,\Psi^A_{j,\ell q}\,\rangle = \delta_{iq}\,r_q$.

By recursively inserting (7.98) into (7.93), one gets the following expression for the local density matrices $\rho_{[1,n]}$,

$$\rho_{[1,n]} = \sum_{\boldsymbol{j}^{(n)}\in I_J^n}\sum_{\ell,p=1}^{b} r_\ell\,|\,\Psi^{\boldsymbol{j}^{(n)}}_{\ell p}\,\rangle\langle\,\Psi^{\boldsymbol{j}^{(n)}}_{\ell p}\,|\,, \tag{7.99}$$

$$\begin{aligned}|\,\Psi^{\boldsymbol{j}^{(n)}}_{\ell p}\,\rangle := \sum_{\boldsymbol{i}^{(n-1)}\in\Omega_b^{(n-1)}} &|\,\Psi^A_{j_1,\ell i_1}\,\rangle\otimes|\,\Psi^A_{j_2,i_1 i_2}\,\rangle\otimes|\,\Psi^A_{j_2,i_2 i_3}\,\rangle\otimes\cdots \\ &\cdots\otimes|\,\Psi^A_{j_{n-1},i_{n-2}i_{n-1}}\,\rangle\otimes|\,\Psi^A_{j_n,i_{n-1}p}\,\rangle\,. \end{aligned} \tag{7.100}$$

**Remark 7.1.15.** Notice that, despite the recursive structure involving more and more factor components, for each $n$-tuple $\boldsymbol{j}^{(n)} = j_1 j_2\cdots j_n \in I_J^n$ there are at most $b^2$ vectors $\Psi^{\boldsymbol{j}^{(n)}}_{\ell p} \in (\mathbb{C}^d)^{\otimes n}$.

**Example 7.1.15.** The vectors $\Psi^{\boldsymbol{j}^{(n)}}_{\ell p}$ need not be normalized, $\|\Psi^{\boldsymbol{j}^{(n)}}_{\ell p}\| \neq 1$; taking this fact into account, (7.99) provides the following natural decomposition of the local restrictions of *FCS* states,

$$\rho_{[1,n]} = \sum_{\boldsymbol{j}^{(n)}\in I_J^n} p(\boldsymbol{j}^{(n)})\,\rho^{\boldsymbol{j}^{(n)}}_{[1,n]}\,, \quad \rho^{\boldsymbol{j}^{(n)}}_{[1,n]} = \sum_{\ell,p=1}^{b}\frac{r_\ell\|\Psi^{\boldsymbol{j}^{(n)}}_{\ell p}\|^2}{\underbrace{\sum_{\ell,p=1}^{b} r_\ell\|\Psi^{\boldsymbol{j}^{(n)}}\|^2}_{p(\boldsymbol{j}^{(n)})}}\,P^{\boldsymbol{j}^{(n)}}_{\ell p}\,, \tag{7.101}$$

where $P^{j^{(n)}}_{\ell p}$ projects onto $|\Psi^{j^{(n)}}_{\ell p}\rangle/\|\Psi^{j^{(n)}}_{\ell p}\|$. It follows that the support of each $\rho^{j^{(n)}}_{[1,n]}$ has dimension at most $b^2$. By defining the completely positive (non-unital) maps $\mathcal{A}\otimes\mathcal{B}\ni\mathcal{A}\otimes\mathbb{1}\mapsto\mathbb{E}_j[A\otimes B]:=V_j^\dagger\, A\otimes B\, V_j\in\mathcal{B}$, the weights $p(\boldsymbol{j}^{(n)})$, $\boldsymbol{j}^{(n)}=j_1j_2\cdots j_n\in I_J^n$, can be rewritten as

$$p(\boldsymbol{j}^{(n)})=\sum_{\ell=1}^{b} r_\ell\,\|\Psi^{\boldsymbol{j}^{(n)}}\|^2=\mathrm{Tr}\Big(\rho_{[1,n]}\,\mathbb{E}_{j_1}\circ\mathbb{E}_{j_2}\circ\cdots\mathbb{E}_{j_n}[\mathbb{1}_{\mathcal{A}^{\otimes n}}\otimes\mathbb{1}_{\mathcal{B}}]\Big)\ .$$

Since $\sum_{j\in J}\mathbb{E}_j=\mathbb{E}$, from (7.88) and (7.89) it follows that the probabilities $\pi^{(n)}=\{p(\boldsymbol{j}^{(n)}\}_{\boldsymbol{j}^{(n)}\in I^n_J}$ define a shift invariant global state $\omega_\pi$ over the Abelian algebra of generated by tensor products of infinitely many $\mathrm{card}(J)\times\mathrm{card}(J)$ diagonal matrices, thus a classical spin chain $(\mathfrak{D}_J^{\otimes\infty},\Theta_\sigma,\omega_\pi)$.

As regards the action of $V_j$ in (7.97), we proceed as follows. Given the vectors $|\Psi^B\rangle_{j,i\ell}\in\mathbb{C}^b$, where $j\in J$, $i=1,2,\ldots,b$ and $\ell=1,2,\ldots,d$, let $v^\dagger_{j\ell}\in M_b(\mathbb{C})$ be the matrix such that $\langle\psi^B_k\,|\,v^\dagger_{j\ell}\,|\psi^B_i\rangle=\langle\psi^B_k\,|\,\Psi^B_{j,k\ell}\rangle$. Then,

$$V_j=\sum_{\ell=1}^{d}\sum_{i=1}^{b}\Big(|\,\psi^A_\ell\,\rangle\otimes\ v^\dagger_{j\ell}|\,\psi^B_i\,\rangle\Big)\langle\,\psi^B_i\,| \tag{7.102}$$

$$V_j^\dagger=\sum_{\ell=1}^{d}\sum_{i=1}^{b}|\,\psi^B_i\,\rangle\Big(\langle\,\psi^A_\ell\,|\otimes\ \langle\,\psi^B_i\,|v_{j\ell}\Big)\ . \tag{7.103}$$

Then, (7.88) implies $\mathbb{1}_{\mathcal{B}}=\sum_{j\in J}V_j^\dagger V_j=\mathbb{1}_{\mathcal{B}}=\sum_{j\in J}\sum_{\ell=1}^d v_{j\ell}v^\dagger_{j\ell}$. Further, the dual map $\mathbb{F}$ reads

$$\mathbb{F}[\rho]=\sum_{j\in J}\sum_{p,q=1}^{d}|\,\psi^A_p\,\rangle\langle\,\psi^A_q\,|\otimes v^\dagger_{jp}\,\rho\, v_{jq}\ . \tag{7.104}$$

It then turns out that, in terms of the $v_{j\ell}$s, the translation-invariant condition (7.89) amounts to $\sum_{j\in J}\sum_{\ell=1}^d v^\dagger_{j\ell}\,\rho\, v_{j\ell}=\rho$. Finally, using (7.93), the local density matrices $\rho_{[1,n]}$ exhibit the following recursive structure,

$$\rho_{[1,n]}=\sum_{\substack{\boldsymbol{j}^{(n)}\in I^n_J\\ \boldsymbol{k}^{(n)},\boldsymbol{i}^{(n)}\in\Omega^{(n)}_d}}|\,\Psi^A_{\boldsymbol{k}^{(n)}}\,\rangle\langle\,\Psi^A_{\boldsymbol{i}^{(n)}}\,|\ \mathrm{Tr}_{\mathcal{B}}\Big(v^\dagger_{\boldsymbol{j}^{(n)}\boldsymbol{k}^{(n)}}\,\rho\, v_{\boldsymbol{j}^{(n)}\boldsymbol{i}^{(n)}}\Big)\ , \tag{7.105}$$

where $|\,\Psi^A_{\boldsymbol{i}^{(n)}}\,\rangle:=|\,\psi^A_{k_1}\,\rangle\otimes|\,\psi^A_{k_2}\,\rangle\otimes\cdots|\,\psi^A_{k_n}\,\rangle$ and $v_{\boldsymbol{j}^{(n)}\boldsymbol{i}^{(n)}}:=v_{j_1i_1}v_{j_2i_2}\cdots v_{j_ni_n}$. The above expression is particularly suited to deal with

**Definition 7.1.13 (Purely Generated FCS).** *A FCS $\omega$ is called purely generated if the defining CPU $\mathbb{E}$ consists of only one Kraus operator [113]:*

$$\mathbb{E}(A\otimes B)=V^\dagger\, A\otimes B\, V\ .$$

In terms of (7.102) and (7.103), the map $\mathbb{E}$ and its dual $\mathbb{F}$ read

$$\mathbb{E}[A\otimes B]=\sum_{i,k=1}^{a}\langle\,\psi_i^A\,|\,A\,|\psi_j^A\,\rangle\,v_i\,B\,v_k^\dagger\ ,\ \mathbb{F}[\rho]=\sum_{i,k=1}^{a}|\,\psi_k^A\,\rangle\langle\,\psi_i^A\,|\otimes v_k^\dagger\,\rho\,v_i\ ,$$

whereby compatibility (7.88) and translation-invariance (7.89) impose

$$\sum_{i=1}^{a}v_i\,v_i^\dagger=\mathbb{1}_B\ ,\quad \sum_{i=1}^{a}v_i^\dagger\,\rho\,v_i=\rho\ . \tag{7.106}$$

The states $\rho_{\mathcal{A}\otimes B}:=\mathbb{F}[\rho]$ on $\mathcal{A}\otimes\mathcal{B}$ and $\rho_{[1,2]}=\mathrm{Tr}_{\mathcal{B}}(\mathrm{id}_{\mathcal{A}}\otimes\mathbb{F}\circ\mathbb{F}[\rho])$ on $\mathcal{A}_{[1,2]}$ can be explicitly written out. Notice that, because of translation-invariance, $\rho_{[1,2]}$ describes any two nearest neighbor spins:

$$\rho_{\mathcal{A}\otimes\mathcal{B}}=\sum_{i,j=1}^{a}|\,\psi_i^A\,\rangle\langle\,\psi_j^A\,|\otimes v_i\,\rho\,v_j^\dagger=\begin{pmatrix} v_1\,\rho\,v_1^\dagger & \dots & v_1\,\rho\,v_a^\dagger\\ \cdot & \cdots & \cdot\\ \cdot & \cdots & \cdot\\ \cdot & \cdots & \cdot\\ v_a\,\rho\,v_1^\dagger & \dots & v_a\,\rho\,v_a^\dagger\end{pmatrix} \tag{7.107}$$

$$\rho_{[1,2]}=\sum_{i,j=1}^{a}|\,\psi_i^A\,\rangle\langle\,\psi_j^A\,|\otimes\begin{pmatrix} R_{1ij1} & \dots & R_{1ija}\\ \cdot & \cdots & \cdot\\ \cdot & \cdots & \cdot\\ \cdot & \cdots & \cdot\\ R_{aij1} & \dots & R_{aija}\end{pmatrix}\ , \tag{7.108}$$

where $R_{ij\ell k}:=\mathrm{Tr}(v_i^\dagger v_j^\dagger\,\rho\,v_\ell v_k)$.

**Example 7.1.16 (AKLT Model).** A typical instance of the recursive finitely correlated structure is provided by the AKLT-model [4, 5], a spin-chain consisting of spin 1 particles with nearest-neighbor interactions described by the Hamiltonian

$$H=\sum_{k=1}^{N-1}\left\{\frac{1}{2}\boldsymbol{S}_k\cdot\boldsymbol{S}_{k+1}+\frac{1}{6}\Big(\boldsymbol{S}_k\cdot\boldsymbol{S}_{k+1}\Big)^2+\frac{1}{3}\right\}\ , \tag{7.109}$$

where $\boldsymbol{S}_k=(S_{1k},S_{2k},S_{3k})$ represents the spin operator for the $k$-th spin along the chain.

The possible values of the total spin of two nearest neighbors are 0, 1 and 2 with corresponding orthogonal projectors $P_0^{(k)}$, $P_1^{(k)}$, respectively $P_2^{(k)}$. Therefore, since

$$\boldsymbol{S}_k\cdot\boldsymbol{S}_{k+1}=-2P_0^{(k)}-P_1^{(k)}+P_2^{(k)}\ ,\ \Big(\boldsymbol{S}_k\cdot\boldsymbol{S}_{k+1}\Big)^2=4P_0^{(k)}+P_1^{(k)}+P_2^{(k)}$$

and $P_0^{(k)}+P_1^{(k)}+P_2^{(k)}=1$, it follows that the interaction between sites $k$ and $k+1$ amounts to the projection $P_2^{(k)}$ onto the subspace with total spin 2.

The spin 1 at site $k$ can be described by means of two spins 1/2, labeled by $k$, $\overline{k}$, by projecting with $P_{k\overline{k}}$ from $\mathbb{C}^4$ onto the 3-dimensional subspace orthogonal to the singlet state $|\Psi^{(-)}_{k,\overline{k}}\rangle$ of the pair of spins 1/2 at $k$ and $\overline{k}$. Further, after associating the spins 1 at $k$, $k+1$ with the pairs $k$, $\overline{k}$, respectively $k+1$, $\overline{k+1}$ of spins 1/2, one imposes *valence bonds* between the pairs, by requiring that the spins 1/2 at $\overline{k}$ and $k+1$ be in a singlet state $|\Psi^{(-)}_{\overline{k},k+1}\rangle$. It follows that the common state of the pairs $k$, $\overline{k}$ and $k+1$, $\overline{k+1}$, namely of two neighboring spins 1 is eigenstate of $P_2^{(k)}$ with eigenvalue 0.

Further, by appending two spins 1/2 at the opposite ends, $\overline{0}$ and $N+1$, of the spin 1 chain, it thus follows that the vector state

$$\left(\otimes_{k=1}^{N-1} P_{k,k+1}\right)|\Psi^{(-)}_{\overline{0}1}\rangle \otimes |\Psi^{(-)}_{\overline{1}2}\rangle \otimes \cdots |\Psi^{(-)}_{\overline{N-1}N}\rangle \otimes |\Psi^{(-)}_{\overline{N}N+1}\rangle \tag{7.110}$$

is the unique ground state for the Hamiltonian

$$H = \sum_{k=1}^{N-1} P_k^{(2)} + \frac{2}{3}\Big(1 + \boldsymbol{s}_{\overline{0}} \cdot \boldsymbol{S}_1\Big) + \frac{2}{3}\Big(1 + \boldsymbol{s}_{N+1} \cdot \boldsymbol{S}_N\Big) \tag{7.111}$$

which is obtained from 7.109 by adding two boundary interactions involving the boundary spin 1/2 operators $\boldsymbol{s}_{\overline{0}}$ and $\boldsymbol{s}_{N+1}$. In the limit of an infinitely long spin-chain, the above valence-bond construction provides a unique, translation-invariant ground state of the AKLT-model, known as valence-bond solid, which exhibits short-range correlations and an energy gap.

In the limit of an infinite spin-chain, its ground state, the valence-bond solid, corresponds to the triplet $(\mathcal{B}, \rho, \mathbb{E})$ with $\mathcal{B} = M_2$, $\rho(B) = \frac{1}{2}\mathrm{Tr}(B)$ and

$$\mathbb{E} : M_3 \otimes M_2 \ni A \otimes B \mapsto V^\dagger (A \otimes B) V \in M_2 \ , \tag{7.112}$$

where, with $|b_{1,2}\rangle \in \mathbb{C}^2$ the eigenvectors of the Pauli matrix $\sigma_3$ relative to the eigenvalues $1, -1$ and $|a_{1,2,3}\rangle$ the eigenvectors of $S_z$ relative to the eigenvalues $-1, 0, 1$,

$$V|b_1\rangle = \sqrt{\frac{2}{3}}\,|a_3, b_1\rangle - \frac{1}{\sqrt{3}}\,|a_2, b_1\rangle\,, \qquad V|b_2\rangle = \frac{1}{\sqrt{3}}\,|a_2, b_2\rangle - \sqrt{\frac{2}{3}}\,|a_1, b_1\rangle\ . \tag{7.113}$$

From (7.102), with $\sigma_\pm := (\sigma_1 \pm i\sigma_2)/2$, it thus follows that

$$v_1 = -\sqrt{\frac{2}{3}}\,\sigma_+\ , \quad v_2 = -\frac{1}{\sqrt{3}}\sigma_3\ , \quad v_3 = \sqrt{\frac{2}{3}}\,\sigma_-\ . \tag{7.114}$$

One can thus check that the conditions (7.106) are satisfied and, moreover, that the identity matrix $1_2 \in M_2$ is the only solution of the second relation in (7.106) in agreement with the translation invariance and purity of the valence bond solid. Further, from (7.107) one explicitly computes

$$\rho_{\mathcal{A}\otimes\mathcal{B}} = \frac{1}{6}\begin{pmatrix} 1-\sigma_3 & \sqrt{2}\sigma_- & 0 \\ \sqrt{2}\sigma_+ & 1 & \sqrt{2}\sigma_- \\ 0 & \sqrt{2}\sigma_+ & 1+\sigma_3 \end{pmatrix} = \frac{1}{6}\begin{pmatrix} 0 & 0 & 0 & 0 & 0 & 0 \\ 0 & 2 & \sqrt{2} & 0 & 0 & 0 \\ 0 & \sqrt{2} & 1 & 0 & 0 & 0 \\ 0 & 0 & 0 & 1 & \sqrt{2} & 0 \\ 0 & 0 & 0 & \sqrt{2} & 1 & 0 \\ 0 & 0 & 0 & 0 & 0 & 0 \end{pmatrix}, \tag{7.115}$$

while (7.108) gives the nearest neighbor states

$$\rho_{[1,2]} = \frac{1}{9}\begin{pmatrix} 0 & 0 & 0 & 0 & 0 & 0 & 0 & 0 & 0 \\ 0 & 1 & 0 & -1 & 0 & 0 & 0 & 0 & 0 \\ 0 & 0 & 2 & 0 & -1 & 0 & 0 & 0 & 0 \\ 0 & -1 & 0 & 1 & 0 & 0 & 0 & 0 & 0 \\ 0 & 0 & -1 & 0 & 1 & 0 & -1 & 0 & 0 \\ 0 & 0 & 0 & 0 & 0 & 1 & 0 & -1 & 0 \\ 0 & 0 & 0 & 0 & -1 & 0 & 2 & 0 & 0 \\ 0 & 0 & 0 & 0 & 0 & -1 & 0 & 1 & 0 \\ 0 & 0 & 0 & 0 & 0 & 0 & 0 & 0 & 0 \end{pmatrix}. \tag{7.116}$$

**Remark 7.1.16.** Finitely correlated states are a useful arena for investigating the behavior of entanglement in quantum spin chains for these are determined by the triplet $(\mathcal{B}, \rho, \mathbb{E})$ (see [38, 204]). Furthermore, they are particular important as ground states of certain solid state Hamiltonians whereby one is interested in either the relations between entanglement and long-range order effects [227] or in the possibility to create entanglement between distant sites by suitable local measurements [241].

## Price-Powers Shifts

The so-called Price-Powers shift [246, 247] are quantum dynamical systems described by an infinite-dimensional $C^*$ algebra $\mathcal{A}_g$, whose building blocks are the identity operator $\mathbb{1}$ and operators $e_i$, $i = \mathbb{N}$, satisfying

$$e_i = e_i^* \;, \quad e_i^2 = \mathbb{1} \;. \tag{7.117}$$

Their algebraic properties are determined by a function

$$g : \mathbb{N}_0 \mapsto \{0, 1\} \;, \quad g(0) = 0 \;, \tag{7.118}$$

called *bitstream*: according to its values, different $e_i$'s commute or anticommute

$$e_i\, e_j = (-)^{g(|i-j|)}\, e_j\, e_i \;, \quad \forall i, j \in \mathbb{N} \;. \tag{7.119}$$

By means of the above relations, every product of finitely many $e_i$'s can be reduced (up to a sign) to an operator of the form

$$W_{\boldsymbol{i}} = e_{i_1} e_{i_2} \cdots e_{i_n} \ , \tag{7.120}$$

where $\boldsymbol{i} = i_1 i_2 \cdots i_n \in \mathbb{N}^*$ stands for any choice of finitely many indices such that $i_1 < i_2 < \ldots < i_n$: $\boldsymbol{i}$ will be called the *support* of $W_{\boldsymbol{i}}$.

It is convenient for later purposes to explicitly compute the commutator of two operators $W_{\boldsymbol{i}}$ and $W_{\boldsymbol{j}}$ with supports $\boldsymbol{i} = i_1 i_2 \cdots i_n$ and $\boldsymbol{j} = j_1 j_2 \cdots j_m$:

$$[W_{\boldsymbol{i}}\,,\,W_{\boldsymbol{j}}] = W_{\boldsymbol{i}}\,W_{\boldsymbol{j}}\left(1 - (-)^{\sum_{p=1}^{n}\sum_{q=1}^{m} g(|i_p - j_q|)}\right) . \tag{7.121}$$

By means of the operators $W_{\boldsymbol{i}}$ one constructs finite-dimensional local subalgebras. Indeed, again by means of (7.117) and (7.119), products of $W_{\boldsymbol{i}}$, with $\boldsymbol{i}$'s consisting of indices from a same interval $[p,q]$, $p \leq q$, reduce up to a sign to some other $W_{\boldsymbol{i}}$ from the same interval. Thus, the algebra $\mathcal{A}_{[p,q]}$ generated by $W_{\boldsymbol{i}}$ with $\boldsymbol{i}$ from $[p,q]$ is a finite-dimensional unital $C^*$ algebra that can be embedded into the spin algebra $M_2(\mathbb{C})^{\otimes(q-p+1}$. This becomes apparent by representing the operators $e_j$ by means of tensor products of Pauli matrices:

$$e_j = \bigotimes_{i=1}^{j-1} (\sigma_z^{g(j-i)})_i \otimes (\sigma_x)_j \otimes \mathbb{1}_{[j+1} \ . \tag{7.122}$$

Since $\sigma_z \sigma_x = -\sigma_x \sigma_z$, one can check that the relations (7.119) are indeed satisfied. The local $C^*$ algebras generate the $*$-algebra

$$\mathcal{A}_* = \bigcup_{n \geq 1} \mathcal{A}_{[1,n]} \ ,$$

and by norm closure (for instance as a subalgebra of the quantum spin chain $\bigotimes_{n=1}^{\infty} M_2(\mathbb{C})$) the quasi local algebra

$$\mathcal{A}_g := \overline{\mathcal{A}_*}^{\|\cdot\|} \ . \tag{7.123}$$

As for quantum spin chains the dynamics on $\mathcal{A}_g$ is given by the shift to the right of the support of the operators $W_{\boldsymbol{i}}$:

$$W_{\boldsymbol{i}} \mapsto \Theta_\sigma^t[W_{\boldsymbol{i}}] =: W_{\boldsymbol{i}+t} = e_{i_1+t} e_{i_2+t} \cdots e_{i_n+t} \ , \quad t \in \mathbb{N} \ . \tag{7.124}$$

**Proposition 7.1.10.** *Let $W_\emptyset := \mathbb{1}$; the linear functional $\omega : \mathcal{A}_g \mapsto \mathbb{C}$ obtained by setting*

$$\omega(W_{\boldsymbol{i}}) = \delta_{\boldsymbol{i},\emptyset} \tag{7.125}$$

*and by linearly extending it to $\mathcal{A}_*$ defines a tracial $\Theta_\sigma$-invariant state on $\mathcal{A}_g$. This is the only tracial state on $\mathcal{A}_g$ if and only if the following property holds: for all finite supports $\boldsymbol{i} = i_1 i_2 \cdots i_n \in \mathbb{N}^*$ there exists $k \in \mathbb{N}$ such that $\sum_{\ell=1}^{n} g(|k - i_\ell|)$ is odd.*

**Proof:** The positivity of $\omega$ can be checked by setting $W = \sum_i c_i W_i$, $c_i \in \mathbb{C}$ and considering

$$\omega(W^\dagger W) = \sum_{i,j} c_i^* \, c_j \, \omega(W_i^\dagger \, W_j) \ .$$

The expectations $\omega(W_i^\dagger \, W_j)$ vanish unless $W_i^\dagger \, W_j = \mathbb{1}$ which can be true only if each $e_{i_k}$ in $W_i$ is matched by an $e_{j_\ell}$ in $W_j$. Thus, $\omega(W_i^\dagger \, W_j) = \delta_{i,j}$ whence $\omega(W^\dagger W) \geq 0$. Therefore, $\omega$ is positive, thus continuous (see (5.49)) and can be extended by continuity to the whole of $\mathcal{A}_g$. That $\omega \circ \Theta_\sigma = \omega$ follows directly from (7.124) and (7.125). Such a state has the tracial property $\omega(W_i \, W_j) = \omega(W_j \, W_i)$.

Suppose that a state $\widetilde{\omega}$ on $\mathcal{A}_g$ has the tracial property and that for all finite supports $\boldsymbol{i} = i_1 i_2 \cdots i_n \in \mathbb{N}^*$ there exists $k$ such that $\sum_{\ell=1}^n g(|k - i_\ell|)$ is odd, then [10] using (7.117) and (7.119), one gets

$$\begin{aligned}\widetilde{\omega}(W_i) &= \widetilde{\omega}(e_k^2 \, W_i) = \widetilde{\omega}(e_k \, W_i \, e_k) \\ &= (-)^{\sum_{\ell=1}^n g(|k-i_\ell|)} \, \widetilde{\omega}(e_k^2 \, W_i) = -\widetilde{\omega}(W_i) \ .\end{aligned}$$

Therefore, $\widetilde{\omega}(W_i) = 0$ for all $W_i \neq \mathbb{1}$ whence $\widetilde{\omega} = \omega$.

Viceversa, if there exists a support $\boldsymbol{i} = i_1 i_2 \cdots i_n$ such that for all $k \in \mathbb{N}$ $\sum_{\ell=1}^n g(|k - i_\ell|)$ is even, then (7.121) implies that $W_i$ commutes with $\mathcal{A}_*$ and thus with $\mathcal{A}_g$, whence, setting $W := \mathbb{1} + W_i$ ($\omega(W) = 1$)), it turns out that

$$\mathcal{A}_g \ni W_j \mapsto \widetilde{\omega}(W_j) := \omega(W \, W_j) \ , \qquad W_j \in \mathcal{A}_g \ ,$$

defines another state on $\mathcal{A}_g$ with the tracial property. □

**Remark 7.1.17.** The property that ensures the uniqueness of the tracial state $\omega$ defined by (7.125) is guaranteed by non-periodic bitstreams [222]; we shall assume this in the following.

**Definition 7.1.14 (Price-Powers Shifts).** *We shall call Price-Powers shifts the dynamical triplets $\left(\mathcal{A}_g, \Theta_\sigma, \omega\right)$ constructed as above with a unique invariant tracial state $\omega$.*

**Examples 7.1.17.** [13]

1. The von Neumann algebras $\mathcal{M}_g := \pi_\omega(\mathcal{A}_g)''$ that arise from the strong closure of $\mathcal{A}_g$ in the *GNS* construction based on $\omega$ are hyperfinite. By extending $\Theta$ and $\omega$ to $\mathcal{M}_g$ one gets von Neumann triplets $(\mathcal{M}_g, \Theta, \omega)$ with $\omega$ still a unique tracial state.
2. If $g \equiv 0$ then $(\mathcal{M}_g, \Theta_\sigma, \omega)$ is an algebraic version of the classical balanced two-valued Bernoulli shift $(\Omega_2, T_\sigma, \mu)$. The von Neumann algebra $\mathcal{M}_0$ is generated by the projections $p_i := \dfrac{\mathbb{1} \pm e_i}{2}$ which are orthogonal for a same index $i$ and otherwise commute.

3. If $g \neq 0$, because of the existence of a unique normalized trace, all $\mathcal{M}_g$ are hyperfinite factors of type $II_1$. Indeed, their center $\mathcal{Z}_g := \mathcal{M}_g \cap \mathcal{M}'_g = \emptyset$, otherwise there would be a positive $Z \in \mathcal{Z}_g$ with $\omega(Z) > 0$ such that $\omega_Z(W_i) := \dfrac{\omega(Z\,W_i)}{\omega(Z)}$ gives a different tracial state on $\mathcal{M}_g$ contradicting the uniqueness of $\omega$:
$$\omega_Z(W_i\,W_j) = \frac{\omega(Z\,W_i\,W_j)}{\omega(Z)} = \frac{\omega(W_j\,Z\,W_i)}{\omega(Z)} = \frac{\omega(Z\,W_j\,W_i)}{\omega(Z)} = \omega_Z(W_j\,W_i)\ .$$
4. If $g(1) \equiv 1$ then $\mathcal{A}_g$ amounts to a discrete Fermi algebra endowed with an infinite temperature state. Indeed, $e_i e_j + e_j e_i = 0$ if $i \neq j$ so that the operators
$$a_i := \frac{e_{2i-1} + i\,e_{2i}}{2}\ , \quad a_i^\dagger := \frac{e_{2i-1} - i\,e_{2i}}{2}\ ,$$
$i \geq 1$, satisfy the *CAR* (5.62). Moreover, the expectations
$$\omega(a_i^\dagger a_j) = \frac{\delta_{ij}}{2}$$
are those of a Fermionic *KMS* state at infinite temperature (see Example (7.1.3)).
5. The bitstream can be chosen such that the von Neumann dynamical triplets $(\mathcal{A}_g, \Theta_\sigma, \omega)$ are *asymptotically highly anti-commutative* [222]. This means the following: there exists a subset $\mathcal{S} \subset \mathcal{A}_g$ such that 1) the set $1\!\!1 \cup \mathcal{S}$ is dense in $\mathcal{A}_g$ and 2) for any $S \in \mathcal{S}$, $\varepsilon > 0$ and $N \in \mathbb{N}$ there exist $0 < n_1 < n_2 < \cdots < n_N \in \mathbb{N}$ such that the anti-commutators satisfy
$$\left\|\left\{\Theta_\sigma^{n_i}[S^\dagger]\,,\,\Theta_\sigma^{n_j}[S]\right\}\right\| \leq \varepsilon$$
for all $n_i \neq n_j$. In this case the tracial state $\omega$ turns out to be the only state which is invariant under the shift $\Theta_\sigma$. Indeed, choose $S \in \mathcal{S}$ and set $X := \dfrac{1}{N}\sum_{i=1}^{N}\Theta_\sigma^{n_i}[S]$, then
$$\left\|X^\dagger\,X + X\,X^\dagger\right\| \leq \frac{2}{N}\|S\|^2 + \frac{1}{N^2}\sum_{i\neq j}\left\|\left\{\Theta^{n_i}[S^\dagger]\,,\,\Theta^{n_j}[S]\right\}\right\|$$
$$\leq \frac{2}{N}\|S\|^2 + \frac{\varepsilon(N-1)}{N}\ .$$
Further, if $\nu$ is a translation-invariant state on $\mathcal{A}_g$, then $\nu(X) = \nu(S)$; thus, by applying (5.49),
$$|\nu(S)| = |\nu(X)| \leq \frac{1}{2}\left(\sqrt{\nu(X^\dagger\,X)} + \sqrt{\nu(X\,X^\dagger)}\right) \leq \sqrt{\frac{\omega(X^\dagger\,X + X\,X^\dagger)}{2}}$$
$$\leq \sqrt{\frac{\|S\|^2}{N} + \frac{\varepsilon(N-1)}{2N}}\ .$$

Because of the arbitrariness of $N$ and $\varepsilon > 0$, it follows that $\nu(S) = 0$ for all $S \in \mathcal{S}$ and because of the assumed density of the set $\mathbb{1} \cup \mathcal{S}$, $\nu$ coincides with $\omega$ as defined in (7.125).

Like in Example 7.1.12, Price-Powers shifts are weakly mixing for all bitstreams; indeed, because of the quasi-local structure of the algebra and the of the tracial property of $\omega$, one need only study the asymptotic behavior of

$$\omega(W_{\boldsymbol{i}}\Theta_\sigma[W_{\boldsymbol{j}}]) = \omega(W_{\boldsymbol{i}}\, W_{\boldsymbol{j}+t}) \ .$$

Clearly, for sufficiently large $t$, $\boldsymbol{i} \cap (\boldsymbol{j}+t) = \emptyset$, then

$$\lim_{t\to+\infty} \omega(W_{\boldsymbol{i}}\Theta_\sigma[W_{\boldsymbol{j}}]) = 0 \ ,$$

unless $\boldsymbol{i} = \boldsymbol{j} = \emptyset$.

As regards strong-mixing, it is convenient to consider strong-asymptotic Abelianess first; namely, at its simplest, using (7.121), it turns out that

$$\omega\left([e_i\,,\, e_{j+t}]^\dagger [e_i\,,\, e_{j+t}]\right) = \left(1 - (-1)^{g(|j+t-i|)}\right)^2 \ .$$

Therefore, unlike for weak-asymptotic Abelianess, the possibility of strong-asymptotic Abelianess depend on the asymptotic behavior of the bitstream; for instance, highly anti-commutative Price-Powers shifts cannot be strongly asymptotic Abelian.

## 7.2 von Neumann Entropy Rate

As seen in the introduction to this chapter, the usual setting of quantum statistical mechanics consists of a quasi-local algebra $\mathcal{A}$ which is the $C^*$ inductive limit of local $C^*$-algebras $\mathcal{A}_V \subseteq \mathbb{B}(\mathbb{H}_V)$ of operators localized in finite volumes $V \subset \mathbb{R}^3$; also, $\mathcal{A}$ is equipped with a locally normal state, namely with a state whose local restriction to $\mathcal{A}_V$, $\omega\lceil\mathcal{A}_V$ is a density matrix $\rho_V \in \mathbb{B}_1^+(\mathbb{H}_V)$. Usually, $\omega$ is translation-invariant, that is $\rho_{V+\boldsymbol{a}} = \rho_V$, where $V + \boldsymbol{a}$ denotes the volume $V$ rigidly translated by $\boldsymbol{a} \in \mathbb{R}^3$ (or by $\boldsymbol{a} \in \mathbb{Z}^3$ in the case of a lattice system).

Consider two disjoint volumes $V_1$ and $V_2$ and let $V := V_1 \cup V_2$; then, $\mathcal{A}_V = \mathcal{A}_{V_1} \otimes \mathcal{A}_{V_2}$ and $\rho_{V_{1,2}} = \mathrm{Tr}_{\mathbb{H}_{V_{2,1}}} \rho_V$, namely the states localized within $V_{1,2}$ are obtained as marginal states of $\rho_V$ localized within the larger volume $V$. Each local state $\rho_V$ has von Neumann entropy

$$S(V) := S(\rho_V) = -\mathrm{Tr}(\rho_V \log \rho_V) \ ;$$

then, the subadditivity of the von Neumann entropy, that is the upper bound in (5.161) reads

$$S(V) \leq S(V_1) + S(V_2) \ , \tag{7.126}$$

where the equality holds if and only if $\rho_V = \rho_{V_1} \otimes \rho_{V_1}$. In order to understand the meaning of strong subadditivity in this setting, consider two volumes $V$ and $U$ such that $V_2 := V \cap U \neq \emptyset$ and set $V_1 := V \setminus V_2$, $V_3 := U \setminus V_2$, $W = V_1 \cup V_2 \cup V_3 = U \cup V$. Since $V_{1,2,3}$ are disjoint volumes, it follows that $\mathcal{A}_W = \mathcal{A}_{V_1} \otimes \mathcal{A}_{V_2} \otimes \mathcal{A}_{V_3}$, $\mathcal{A}_V = \mathcal{A}_{V_1} \otimes \mathcal{A}_{V_2}$ and $\mathcal{A}_U = \mathcal{A}_{V_2} \otimes \mathcal{A}_{V_3}$; further

$$\rho_{V_2} = \mathrm{Tr}_{\mathbb{H}_{V_1} \otimes \mathbb{H}_{V_3}}(\rho_W) \ , \ \rho_V = \mathrm{Tr}_{\mathbb{H}_{V_3}}(\rho_W) \ , \ \rho_U = \mathrm{Tr}_{\mathbb{H}_{V_1}}(\rho_W) \ .$$

Then (5.162) reads

$$S(U \cup V) + S(V_2) \leq S(U) + S(V) \ . \tag{7.127}$$

In general, the von Neumann entropy of $\rho_V$ diverges when $V \uparrow \mathbb{R}^3$ (or $V \uparrow \mathbb{Z}^2$); on thus wonders whether the rate $S(V)/|V|$ exists when the $V \uparrow \mathbb{R}^3, \mathbb{Z}^3$, where $|V| = \int_V \mathrm{d}\boldsymbol{r}$. Among the many ways a sequence of volumes may fill the whole space $\mathbb{R}^3$ (or $\mathbb{Z}^3$), a convenient one [314] is to consider a family of parallelepipeds $V(\boldsymbol{a}) := \Big\{\boldsymbol{x} = (x_1, x_2, x_3) \in \mathbb{R}^3 \ : \ -a_i \leq x_i \leq a_i\Big\}$, where $\boldsymbol{a} \in \mathbb{R}^3_+$ and then to let each $a_i \to +\infty$ so that $V(\boldsymbol{a}) \to \mathbb{R}^3$.

**Proposition 7.2.1 (Mean von Neumann Entropy).** *[314]*
*If $(\mathcal{A}, \omega)$ is a quasi-local shift-dynamical system with a locally normal translation invariant state $\omega$, its mean von Neumann entropy is given by*

$$s(\omega) := \lim_{V(\boldsymbol{a}) \to \mathbb{R}^3} \frac{S(V(\boldsymbol{a}))}{|V(\boldsymbol{a})|} = \inf_{V(\boldsymbol{a})} \frac{S(V(\boldsymbol{a}))}{|V(\boldsymbol{a})|} \ . \tag{7.128}$$

**Proof:** [314] Because of translation invariance, in (7.128) we can consider parallelepipeds of the form $V(\boldsymbol{a}) = \Big\{\boldsymbol{x} \in \mathbb{R}^3 : 0 \leq x_i \leq a_i\Big\}$. Choose $\varepsilon > 0$ and a parallelepiped $V(\boldsymbol{a}_0)$ in such a way that

$$s(\omega) = \inf_{V(\boldsymbol{a})} \frac{S(V(\boldsymbol{a}))}{|V(\boldsymbol{a})|} \geq \frac{S(V(\boldsymbol{a}_0))}{|V(\boldsymbol{a}_0)|} - \varepsilon \ . \qquad (*)$$

By decomposing $\mathbb{R}_+ \ni a_i = n_i a_0^i + b_i$ with $n_i \in \mathbb{N}$ and $0 \leq b_i \leq a_0^i$, any other $V(\boldsymbol{a})$ can be written as the union of disjoint parallelepipeds

$$V(\boldsymbol{a}) = \bigcup_{\substack{0 \leq k_1 \leq n_1 - 1 \\ 0 \leq k_2 \leq n_2 - 1 \\ 0 \leq k_3 \leq n_3 - 1}} V_{\boldsymbol{k}}(\boldsymbol{a}_0) \cup V_{\boldsymbol{b}}(\boldsymbol{a}_0)$$

$$V_{\boldsymbol{k}}(\boldsymbol{a}_0) := \Big\{\boldsymbol{x} \in \mathbb{R}^3 : k_i a_0^i \leq x_i \leq (k_i + 1) a_0^i\Big\}$$

$$V_{\boldsymbol{b}}(\boldsymbol{a}_0) := \Big\{\boldsymbol{x} \in \mathbb{R}^3 : n_i a_0^i \leq x_i \leq n_i a_0^i + b_i\Big\} \ .$$

Then, (7.126) yields

$$S(V(\boldsymbol{a})) \leq \underbrace{\left(\prod_{i=1}^{3} n_i\right)}_{|V(\boldsymbol{a})|} S(V(\boldsymbol{a}_0)) + S(V_{\boldsymbol{b}}(\boldsymbol{a}_0)) \qquad (**) .$$

By translation, $V_{\boldsymbol{b}}(\boldsymbol{a}_0)$ can be embedded within $V(\boldsymbol{a}_0)$; furthermore, since $0 \leq b_i \leq a_0^i$, each of them is the intersection of the interval $[0, a_0^i]$ with an interval $[-c_i, a_0^i - c_i]$, $c_i \geq 0$, of the same length. It thus follows that $V_{\boldsymbol{b}}(\boldsymbol{a}_0)$ can be written as the intersection $V_2$ of $V := V(\boldsymbol{a}_0)$ with a suitably translated $V(\boldsymbol{a}_0)$ denoted by $U$. Then, from (7.127) and translation-invariance one derives the upper bound

$$S(V_{\boldsymbol{b}}(\boldsymbol{a}_0)) = S(V_2) \leq S(U \cup V) + S(V_2) \leq S(U) + S(V) = 2\, S(V(\boldsymbol{a}_0)) .$$

Finally, dividing $(**)$ by $V(\boldsymbol{a})$ and going to the limit, similarly as in the proof of the existence of the Shannon entropy rate in (3.2), using $(*)$ one gets

$$\limsup_{V(\boldsymbol{a})} \frac{S(V(\boldsymbol{a}))}{|V(\boldsymbol{a})|} \leq \frac{S(V(\boldsymbol{a}_0))}{|V(\boldsymbol{a}_0)|} \leq s(\omega) + \varepsilon \leq \liminf_{V(\boldsymbol{a})} \frac{S(V(\boldsymbol{a}))}{|V(\boldsymbol{a})|} + \varepsilon ,$$

whence the result follows from the arbitrariness of $\varepsilon > 0$. □

**Examples 7.2.1.**

1. For quantum spin chains $(\mathcal{A}_{\mathbb{Z}}, \Theta_\sigma, \omega)$, the mean entropy is given by

$$s(\omega) = \lim_{n \to +\infty} \frac{1}{n} S\left(\rho_{[1,n]}\right) = \inf_n \frac{1}{n} S\left(\rho_{[1,n]}\right) , \qquad (7.129)$$

   where $\rho_{[1,n]}$ is the density matrix corresponding to the restriction $\omega \upharpoonright \mathcal{A}_{[1,n]}$ of the translation invariant state $\omega$ to the local subalgebra $\mathcal{A}_{[1,n]}$.
2. Because of their structure (see Remark 7.1.15), purely generated *FCS* $\omega$ have $s(\omega) = 0$. Indeed, the support of local states $\rho_{[1,n]}$ is at most $b^2$-dimensional where $M_b(\mathbb{C}) = \mathcal{B}$ is the auxiliary algebra in the triplet $(\mathcal{B}, \rho, \mathbb{E})$; then $S\left(\rho_{[1,n]}\right) \leq 2 \log_2 b$.
3. Consider the Bosonic (7.22) and Fermionic (7.20) quasi-free states $\omega_A$ and assume the action of the operator $A$ on $\mathbb{L}^2_{\mathrm{d}\boldsymbol{r}}(\mathbb{R}^3)$ to be given by

$$\langle \boldsymbol{r} \,|\, A\psi \rangle = \int_{\mathbb{R}^3} \mathrm{d}\boldsymbol{x}\; K_A(\boldsymbol{r} - \boldsymbol{x}) \psi(\boldsymbol{x}) ,$$

   where the kernel $K_A$ has Fourier transform

$$\widehat{K}_A(\boldsymbol{k}) := \frac{1}{(2\pi)^3} \int_{\mathbb{R}^3} \mathrm{d}\boldsymbol{x}\; \mathrm{e}^{-i\boldsymbol{k}\cdot\boldsymbol{x}}\, K_A(\boldsymbol{x})$$

   such that $0 \leq \widehat{K}_A(\boldsymbol{k}) \leq 1$ for Fermions, $0 \leq \widehat{K}_A(\boldsymbol{k}) \leq M < +\infty$ for Bosons. These quasi-free states are translation-invariant and their mean entropies can be explicitly calculated [233, 110],

$$s(\omega_A) = \frac{1}{(2\pi)^3}\int_{\mathbb{R}^3} \mathrm{d}\boldsymbol{k}\,\Big(\eta(\widehat{K}_A(\boldsymbol{k})) + \eta(1-\widehat{K}_A(\boldsymbol{k}))\Big)\ \text{(Fermions)}$$

$$s(\omega_A) = \frac{1}{(2\pi)^3}\int_{\mathbb{R}^3} \mathrm{d}\boldsymbol{k}\,\Big(\eta(\widehat{K}_A(\boldsymbol{k})) - \eta(1+\widehat{K}_A(\boldsymbol{k}))\Big)\ \text{(Bosons)}\ .$$

**Remarks 7.2.1.**

1. The entropy density of quantum spin-chains scales as the power of the shift-automorphism, that is the entropy production per length $\ell$ time-step is $\ell$ times the entropy production per unit time-step:

$$s_\ell(\omega) := \lim_{n\to\infty}\frac{1}{n}S\left(\rho^{(n\ell)}\right) = \ell\, s(\omega)\ , \tag{7.130}$$

where $\rho^{(n\ell)} = \omega\lceil\mathcal{A}_{0,n\ell-1}$. Indeed, since the limit in (7.129) exists, it can be computed as

$$s(\omega) = \lim_{n\to+\infty}\frac{1}{n\ell}S\left(\rho_{[1,n\ell]}\right) = \frac{1}{\ell}\, s_\ell(\omega)\ .$$

2. From (5.156), it follows that the entropy density is affine over all convex decompositions of $\Theta_\sigma$-invariant states $\omega$ of quantum spin-chains into $\Theta_\sigma$-invariant components $\omega_j$. Namely, if $\omega = \sum_j \lambda_j\omega_j$, with $0\leq\lambda_j\leq 1$, $\sum_j\lambda_j = 1$, then

$$s(\omega) = \sum_j \lambda_j\, s(\omega_j)\ ,\qquad \forall\ell\in\mathbb{N}\ . \tag{7.131}$$

3. In the case of a decomposition of a translation-invariant state $\omega$ of a quantum spin-chain into $\theta^\ell_\sigma$-invariant components $\omega_j$, the previous two points give

$$s_\ell(\omega) = \sum_j \lambda_j s_\ell(\omega_j) = \ell\sum_j\lambda_j s(\omega_j)\ ,\qquad \forall\ell\in\mathbb{N}\ . \tag{7.132}$$

Let us consider the decomposition of a $\Theta_\sigma$-invariant state $\omega$ over a quantum spin-chain which is not $\Theta^\ell_\sigma$-ergodic into $n_\ell$ $\Theta^\ell_\sigma$-ergodic states (see Proposition 7.1.9).

**Lemma 7.2.1.** *Given the decomposition $\omega = \frac{1}{n_\ell}\sum_{j=0}^{n_\ell-1}\omega_j$, using the notation of Remark 7.2.1, it turns out that*

1. *all states $\omega_j$ have the same entropy density with respect to $\Theta^\ell_\sigma$: $s_\ell(\omega_j) = s_\ell(\omega)$, $0\leq j\leq n_\ell-1$.*

2. *Set* $s_j^{(\ell)} := \frac{1}{\ell} S\left(\rho_j^{(l)}\right)$, $s^{(\ell)} := \frac{1}{\ell} S\left(\rho^{(l)}\right)$ *and fix* $\eta > 0$*; it turns out that the subsets of states*

$$A_{\ell,\eta} := \left\{\omega_j \,:\, s_j^{(\ell)} \geq s(\omega) + \eta\right\} \tag{7.133}$$

*has asymptotically zero density. Namely, if* $\#(A_{\ell,\eta})$ *denotes its cardinality, then*

$$\lim_{n_\ell \to \infty} \frac{\#(A_{\ell,\eta})}{n_\ell} = 0 \ . \tag{7.134}$$

**Proof:**

**Part 1** Because of (7.132), $s_\ell(\omega_j) = s_\ell(\omega_0)$, for $1 \leq j \leq n_\ell - 1$. This fact follows from subadditivity (5.161) and the fact that

$$\rho_j^{(n\ell)} = \omega_0 \circ \Theta_\sigma^{-j} \lceil \mathcal{A}_{[0,n\ell-1]} = \underbrace{\omega_0 \lceil \mathcal{A}_{[-j,n\ell-j-1]}}_{:=\rho_0^{[-j,n\ell-j-1]}} \ .$$

Indeed, split the intervals $[-j, n\ell - j - 1]$, $0 \leq j \leq n_\ell - 1$ into disjoint pieces (notice that, according to Proposition 7.1.9, $n_\ell \leq \ell$),

$$[-j, n\ell - j - 1] = \underbrace{[-j, \ell - 1]}_{I_1} \cup \underbrace{[\ell, n\ell - \ell - 1]}_{I_2} \cup \underbrace{[n\ell - \ell, n\ell - j - 1]}_{I_3} \ ;$$

then, apply (5.161) to the density matrices $\rho_j^{(n\ell)}$, respectively $\rho_0^{I_1 \cup I_3} \otimes \rho_0^{I_2}$, and use translation-invariance together with the bound (5.155). It then follows

$$S\left(\rho_j^{(n\ell)}\right) \leq S\left(\rho_0^{[\ell,n\ell-\ell-1]}\right) + S\left(\rho_0^{I_1 \cup I_3}\right) \leq S\left(\rho_0^{(n\ell-2\ell)}\right) + 2\ell \log_2 d \ .$$

Vice versa, if instead of subdividing the interval $[-j, n\ell - j - 1]$ of interest, we include it as a disjoint piece in a larger one

$$[-\ell, n\ell + \ell - 1] = \underbrace{[-\ell, -1]}_{I_1} \cup \underbrace{[-j, n\ell - j - 1]}_{I_2} \cup \underbrace{[n\ell - j, n\ell + \ell - 1]}_{I_3} \ ,$$

then, subadditivity and boundedness give

$$S\left(\rho_j^{(n\ell)}\right) \geq S\left(\rho_0^{[-\ell,n\ell+\ell-1]}\right) - S\left(\rho_0^{I_1 \cup_3}\right) \geq S\left(\rho_0^{(n\ell+2\ell)}\right) - 2\ell \log_2 d \ .$$

Dividing by $n$ and taking the limit $n \to \infty$ yield the result.

**Part 2** If there were $\eta_0$ such that $\limsup_{n_\ell \to \infty} \frac{\#(A_{\ell,\eta_0})}{n_\ell} = a > 0$, then there would be a subsequence $\ell_j$ such that $\lim_{j \to \infty} \frac{\#(A_{\ell_j,\eta_0})}{n_{\ell_j}} = a$. Then, since $\rho^{(\ell_j)} = \sum_{k=0}^{n_{\ell_j}-1} \rho_k^{\ell_j}$, subadditivity implies

$$
\begin{aligned}
n_{\ell_j}\, s^{(\ell_j)} &= \frac{n_{\ell_j}}{\ell_j} S\left(\rho^{(\ell_j)}\right) \geq \frac{1}{\ell_j} \sum_{k=0}^{n_{\ell_j}-1} S\left(\rho_k^{(\ell_j)}\right) = \sum_{k=0}^{n_{\ell_j}-1} s_k^{(\ell_j)} \\
&\geq \#(A_{\ell_j,\eta_0})\,(s(\omega)+\eta_0) + \#(A^c_{\ell_j,\eta_0}) \min_{k\in A^c_{\ell_j,\eta_0}} s_k^{(\ell_j)} .
\end{aligned}
$$

The previous point, (7.130) and (7.129) obtain

$$
\begin{aligned}
\ell_j\, s(\omega) &= s_{\ell_j}(\omega_j) = \inf_m \frac{1}{m} S\left(\rho_j^{m\ell_j}\right) \leq \ell_j\, s_j^{(\ell_j)} \quad \text{whence} \\
s^{(\ell_j)} &\geq \frac{\#(A_{\ell_j,\eta_0})}{n_{\ell_j}} (s(\omega)+\eta_0) + \frac{\#(A^c_{\ell_j,\eta_0})}{n_{\ell_j}} s(\omega) .
\end{aligned}
$$

When $n_{\ell_j} \to \infty$, a contradiction arises:

$$
s(\omega) \geq (s(\omega)+\eta_0)a + s(\omega)(1-a) > s(\omega) .
$$

□

## 7.3 Quantum Spin Chains as Quantum Sources

Quantum spin chains $(\mathcal{A}_{\mathbb{Z}}, \Theta_\sigma, \omega)$, with $\mathcal{A} = M_d(\mathbb{C})$, provide useful algebraic descriptions of quantum sources whose signals consist of quantum states acting on Hilbert spaces of increasing dimension. The local states $\rho^{(n)}$ obtained as restrictions of $\omega$ to the local subalgebras $\mathcal{A}^{(n)} := \mathcal{A}_{[1,n]}$ describe ensembles of quantum strings of length $n$ emitted by these sources.

Quantum sources are one of the two ends of quantum transmission channels; like their classical counterparts, these consist of a source, a sender who encodes, a channel which transmits and a receiver which decodes. Channel inputs and outputs are generic quantum states and the encoding and decoding procedures, as well as the channel action are quantum operations described by trace-preserving *CP* maps on the state-space.

In analogy with Figure 2.2, a quantum transmission scheme can be pictorially represented as in Figure 7.1.

1. At each stroke of time, a source $A$ emits quantum states, represented by density matrices $\rho_i \in \mathbb{B}_1^+(\mathbb{H})$, $i = 1, 2, \ldots, a$, $\mathbb{H} = \mathbb{C}^a$, with weights $p(i)$. The statistical description of a single use of the source is given by means of the density matrix $\rho = \sum_{i=1}^a p(i)\, \rho_i$.
2. As a result of $n$ uses of the source, the sender would collect generic density matrices $\rho^{(n)}_{\boldsymbol{i}^{(n)}} \in \mathbb{B}_1^+(\mathbb{H}^{\otimes n})$, $\boldsymbol{i}^{(n)} = i_1 i_2 \cdots i_n \in \Omega_a^{(n)}$, with weights $p^{(n)}(\boldsymbol{i}^{(n)})$. Consequently, the statistics of $n$ uses of the source is described by the density matrix

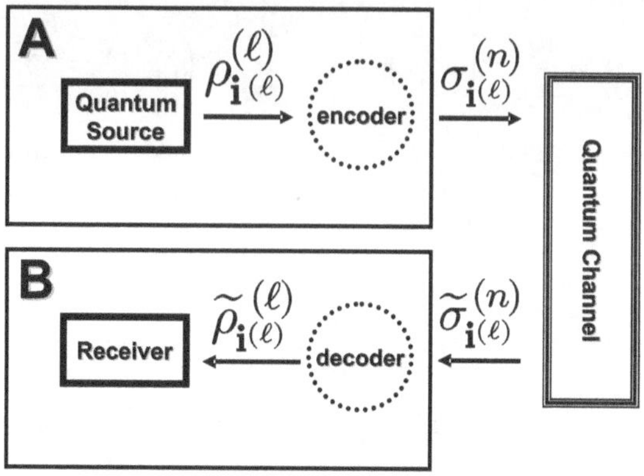


**Fig. 7.1.** Quantum Transmission Channel

$$\rho^{(n)} := \sum_{\boldsymbol{i}^{(n)} \in \Omega_a^{(n)}} p^{(n)}(\boldsymbol{i}^{(n)}) \, \rho^{(n)}_{\boldsymbol{i}^{(n)}} \,, \tag{7.135}$$

which embodies purely classical correlations (the weights) and quantum correlations due to the states $\rho^{(n)}_{\boldsymbol{i}^{(n)}}$.

3. The encoding is a trace-preserving *CP* map $\mathcal{E}^{(n)} : \mathbb{B}_1^+(\mathbb{H}^{\otimes n}) \mapsto \mathbb{B}_1^+(\mathbb{K}^{(n)}_{in})$ such that $\mathcal{E}^{(n)}[\rho^{(n)}_{\boldsymbol{i}^{(n)}}] = \sigma^{(n)}_{\boldsymbol{i}^{(n)}}$ are density matrices that can all be considered as acting on a same (finite dimensional) Hilbert space $\mathbb{K}^{(n)}_{in}$.
4. The code-states $\sigma^{(n)}_{\boldsymbol{i}^{(n)}}$ go through the (lossless) channel that transform them as a trace-preserving *CP* map $\mathbb{F}^{(n)} : \mathbb{B}_1^+(\mathbb{K}^{(n)}_{in}) \mapsto \mathbb{B}_1(\mathbb{K}^{(n)}_{out})$ such that $\mathbb{F}^{(n)}[\sigma^{(n)}_{\boldsymbol{i}^{(n)}}] = \widetilde{\sigma}^{(n)}_{\boldsymbol{i}^{(n)}}$, the latter being a, possibly not-normalized, positive matrix acting on a (finite dimensional) Hilbert space $\mathbb{K}^{(n)}_{out}$.
5. The channel output $\widetilde{\sigma}^{(n)}_{\boldsymbol{i}^{(n)}}$ finally undergoes a decompressing procedure corresponding to the action of a *CP* map $\mathcal{D}^{(n)} : \mathbb{B}_1(\mathbb{K}^{(n)}_{out}) \mapsto \mathbb{B}_1^+(\mathbb{H}^{\otimes n})$ such that $\mathcal{D}^{(n)}[\widetilde{\sigma}^{(n)}_{\boldsymbol{i}^{(n)}}] = \widetilde{\rho}^{(n)}_{\boldsymbol{i}^{(n)}}$.
6. The efficiency of the encoding-decoding procedures with respect to the channel action $\mathbb{F}$ is measured by how faithfully the decompressed states $\widetilde{\rho}^{(n)}_{\boldsymbol{i}^{(n)}}$ reproduce the input states $\rho^{(n)}_{\boldsymbol{i}^{(n)}}$.

The simplest instance of quantum source is the generalization of a classical Bernoulli process: at each use of the source, vector states $|\,\psi_i\,\rangle \in \mathbb{H} := \mathbb{C}^d$, $i = 1, 2, \ldots, a$, (not necessarily orthogonal) are independently emitted with weights $p(i)$. The quantum statistics of $n$ uses of the source is thus described by the density matrix

$$\rho^{\otimes n} := \bigotimes_{j=1}^{n} \rho = \sum_{\boldsymbol{i}^{(n)} \in \Omega_a^{(n)}} p^{(n)}(\boldsymbol{i}^{(n)}) \bigotimes_{j=1}^{n} \rho_{i_j} \,, \quad \rho_{i_j} = |\,\psi_{i_j}\,\rangle\langle\,\psi_{i_j}\,| \,, \tag{7.136}$$

where $\rho = \sum_{i=1}^{a} p(i) | \psi_i \rangle\langle \psi_i |$ and $p^{(n)}(\boldsymbol{i}^{(n)}) = \prod_{j=1}^{n} p(i_j)$.

**Remarks 7.3.1.**

1. Quantum spin chains as they appear in quantum statistical mechanics provide fairly general models of quantum sources. Their local states over $n$ successive chain sites correspond to density matrices $\rho^{(n)}$ that describe a variety of possible quantum strings of length $n$ consisting of separable and entangled states that can in turn be pure and mixed.
2. Like classical strings, quantum strings emitted from quantum sources of Bernoulli type can be chained together by tensorizing them; this is not anymore so obvious for generic quantum strings [60].
3. Two classical strings can always be told apart, for instance by a non-zero value of the Hamming distance that counts by how many symbols they differ. Instead, there are uncountably many quantum strings that can be arbitrarily close to one another, for instance with respect to the trace-distance (6.66), and which cannot then be perfectly distinguished.

### 7.3.1 Quantum Compression Theorems

In analogy with classical coding, the idea how to compress quantum information in absence of noise is to consider quantum strings acting on Hilbert spaces $\mathbb{H}^{\otimes n}$ with $n$ large and to map them into quantum strings acting on Hilbert spaces $\mathbb{H}^{(n)}$ of smaller dimension in a way that allows for faithful decompression.

Concretely, the procedure consists in a *coding operation* corresponding to a trace-preserving *CP* compression map $\mathcal{E}^{(n)} : \mathbb{B}_1^+(\mathbb{H}^{\otimes n}) \mapsto \mathbb{B}_1^+(\mathbb{H}^{(n)})$ and a *decoding operation* described by a trace-preserving *CP* decompression map $\mathcal{D}^{(n)} : \mathbb{B}_1^+(\mathbb{H}^{(n)}) \mapsto \mathbb{B}_1^+(\mathbb{H}^{\otimes n})$ that tries to retrieve the source signals. If the task is to compress the information contained in $n$ uses of a quantum source, then each of the quantum strings in (7.136)) is subjected to the chain of maps

$$\rho^{(n)}_{\boldsymbol{i}^{(n)}} \mapsto \sigma^{(n)}_{\boldsymbol{i}^{(n)}} := \mathcal{E}^{(n)}[\rho^{(n)}_{\boldsymbol{i}^{(n)}}] \longmapsto \tilde{\rho}^{(n)}_{\boldsymbol{i}^{(n)}} := \mathcal{D}^{(n)}[\sigma^{(n)}_{\boldsymbol{i}^{(n)}}] \ . \tag{7.137}$$

Any sequence $\{\mathcal{E}^{(n)}, \mathcal{D}^{(n)}\}_n$ will be referred to as a *compression scheme* and denoted by $(\mathcal{E}, \mathcal{D})$.

In the following, we shall first focus on quantum Bernoulli sources emitting *qubits*, that is we shall consider Hilbert spaces $\mathbb{H}^{\otimes n} = \mathbb{C}^{2^n}$ and local algebras $\mathcal{A}^{(n)} = M_{2^n}(\mathbb{C})$. For them, the compression rate of a scheme $(\mathcal{C}, \mathcal{D})$ is defined as follows.

**Definition 7.3.1 (Compression Rate).** *The compression rate of $(\mathcal{E}, \mathcal{D})$ for a qubit quantum source $(\mathcal{A}_{\mathbb{Z}}, \omega)$ is given by*

$$R(\mathcal{E}) := \limsup_{n \to +\infty} \frac{1}{n} \log_2 dim(\mathbb{H}^{(n)}) \ .$$

*where* $\mathbb{H}^{(n)}$ *is the minimal support subspace of all quantum code-words* $\sigma^{(n)}_{\boldsymbol{i}^{(n)}}$.

According to the previous definition, for large $n$, $2^{n\,R(\mathcal{E})}$ estimates the dimension of the subspace supporting the encoded signals, with $R(\mathcal{E})$ roughly being the used number of *qubits* per encoded *qubit* . Clearly, one looks for compression schemes $(\mathcal{E},\mathcal{D})$ such that $R(\mathcal{E})<1$ with $\mathcal{D}^{(n)}\circ\mathcal{E}^{(n)}$ asymptotically approximating the identity map in a suitable topology.

### Compression of *qubit* Bernoulli Sources

In the case of a Bernoulli quantum source, Shannon's noiseless coding theorem 3.2.2 has a natural quantum extension whereby the von Neumann entropy plays the role of the Shannon entropy as optimal compression rate.

A convenient fidelity is the ensemble fidelity introduced in Definition 6.3.6; using (6.71) it reads:

$$F_{av}\Big(\Big\{p^{(n)}_{\boldsymbol{i}^{(n)}}\rho^{(n)}_{\boldsymbol{i}^{(n)}}\Big\},\mathcal{D}^{(n)}\circ\mathcal{E}^{(n)}\Big)=\sum_{\boldsymbol{i}^{(n)}\in\Omega_2^{(n)}}p^{(n)}_{\boldsymbol{i}^{(n)}}\,\mathrm{Tr}\Big(\rho^{(n)}_{\boldsymbol{i}^{(n)}}\,\tilde{\rho}^{(n)}_{\boldsymbol{i}^{(n)}}\Big)\ . \tag{7.138}$$

It is positive, bounded by 1 and equal to 1 if and only if $\tilde{\rho}^{(n)}_{\boldsymbol{i}^{(n)}}=\rho^{(n)}_{\boldsymbol{i}^{(n)}}$. Useful upper and lower bounds to $F_{av}$ are obtained as follows.

If $\rho=\sum_{j=1}^{2}r_j|\,r_j\,\rangle\langle\,r_j\,|\in\mathcal{S}(\mathbb{C}^2)$ is the spectral decomposition of the state describing a single use of the source, the eigenvalues $r^{(n)}_{\boldsymbol{j}^{(n)}}$ of $\rho^{\otimes n}$ are of the form $r^{(n)}_{\boldsymbol{j}^{(n)}}=:\prod_{i=1}^{n}r_{j_i}$. Let $\mathbb{H}^{(n)}\subseteq\mathbb{H}^{\otimes n}$ be the smallest subspace, of dimension $d(n)$, supporting all code-words $\tilde{\rho}^{(n)}_{\boldsymbol{i}^{(n)}}$ and let $\Gamma^{(n)}:\mathbb{H}^{\otimes n}\mapsto\mathbb{H}^{(n)}$ denote the corresponding orthogonal projection. Then,

$$\begin{aligned}F_{av}\Big(\Big\{p^{(n)}_{\boldsymbol{i}^{(n)}}\rho^{(n)}_{\boldsymbol{i}^{(n)}}\Big\},\mathcal{D}^{(n)}\circ\mathcal{E}^{(n)}\Big)&\leq\sum_{\boldsymbol{i}^{(n)}\in\Omega_2^{(n)}}p^{(n)}_{\boldsymbol{i}^{(n)}}\,\mathrm{Tr}\Big(\rho^{(n)}\,\Gamma^{(n)}\Big)\\&\leq\sum_{j=1}^{d(n)}e_j(\rho^{\otimes n})\ .\end{aligned} \tag{7.139}$$

Indeed, the first inequality is implied by the fact that $\tilde{\rho}^{(n)}_{\boldsymbol{i}^{(n)}}\leq\Gamma^{(n)}$, while the second one is the Ky Fan inequality (5.158), where $e_j(\rho^{\otimes n})$, $j=1,2,\cdots,d(n)$, are the first $d(n)$ largest eigenvalues of $\rho^{\otimes n}$.

Vice versa, given $\Gamma^{(n)}:\mathbb{H}^{\otimes n}\mapsto\mathbb{H}^{(n)}$, consider the trace-preserving *CP* map $\mathcal{E}^{(n)}:\mathbb{B}_1^+(\mathbb{H}^{\otimes n})\mapsto\mathbb{B}_1^+(\mathbb{H}^{(n)})$ defined by

$$\mathcal{E}^{(n)}[\rho]=\Gamma^{(n)}\,\rho\,\Gamma^{(n)}+\underbrace{\sum_{|\,\Phi_k\,\rangle\perp\mathbb{K}^{(n)}}|\,0\,\rangle\langle\,\Phi_k\,|\,\rho\,|\,\Phi_k\,\rangle\langle\,0\,|}_{|\,0\,\rangle\langle\,0\,|\,\mathrm{Tr}\big((\mathbb{1}-P)\,\rho\big)}\ , \tag{7.140}$$

where $|\,0\,\rangle \in \mathbb{H}^{(n)}$ is a suitable reference state. As a decompression map $\mathcal{D}^{(n)}$, choose the identity map on $\mathbb{H}^{(n)}$ which embeds it into $\mathbb{H}^{\otimes n}$. Then,

$$\tilde{\rho}^{(n)}_{\boldsymbol{i}^{(n)}} = \Gamma^{(n)}\,\rho^{(n)}_{\boldsymbol{i}^{(n)}}\,\Gamma^{(n)} + |\,0\,\rangle\langle\,0\,|\,\mathrm{Tr}\Big((1\!\!1 - P)\,\rho^{(n)}_{\boldsymbol{i}^{(n)}}\Big)\ ,$$

whence, since $\rho^{(n)}_{\boldsymbol{i}^{(n)}}$ is a pure state,

$$\begin{aligned}
&F_{av}\Big(\Big\{p^{(n)}_{\boldsymbol{i}^{(n)}}\rho^{(n)}_{\boldsymbol{i}^{(n)}}\Big\}, \mathcal{D}^{(n)}\circ\mathcal{E}^{(n)}\Big) \geq \sum_{\boldsymbol{i}^{(n)}\in\Omega_2^{(n)}} p^{(n)}_{\boldsymbol{i}^{(n)}}\,\mathrm{Tr}\Big(\Big(\rho^{(n)}_{\boldsymbol{i}^{(n)}}\,\Gamma^{(n)}\Big)^2\Big)\\
&= \sum_{\boldsymbol{i}^{(n)}\in\Omega_2^{(n)}} p^{(n)}_{\boldsymbol{i}^{(n)}}\Big(\mathrm{Tr}\Big(\rho^{(n)}_{\boldsymbol{i}^{(n)}}\,\Gamma^{(n)}\Big)\Big)^2 \geq \sum_{\boldsymbol{i}^{(n)}} \lambda_{\boldsymbol{i}^{(n)}}\Big(2\,\mathrm{Tr}\Big(\rho^{(n)}_{\boldsymbol{i}^{(n)}}\,\Gamma^{(n)}\Big) - 1\Big)\\
&\geq 2\,\mathrm{Tr}\Big(\rho^{\otimes n}\,\Gamma^{(n)}\Big) - 1\ .
\end{aligned} \tag{7.141}$$

Exactly as the Shannon entropy in the classical case, a theorem of Schumacher [267, 159] shows that, for quantum sources of Bernoulli type, the von Neumann entropy $S\,(\rho)$ is the optimal compression rate. Namely, this rate can be achieved by suitable compression and decompression schemes with high-fidelity retrieval of increasingly long *qubit* strings; on the other hand, compression and decompression schemes with rates exceeding $S\,(\rho)$ perform poorly with long *qubit* strings.

**Theorem 7.3.1 (Schumacher Theorem).**

*Let $(\mathcal{A}_{\mathbb{Z}}, \rho^{\otimes\infty})$ be a qubit Bernoulli source with entropy density $S(\rho)$. If $R \geq S(\rho)$ there exists a compression scheme $(\mathcal{E}, \mathcal{D})$ with rate $R(\mathcal{E}) = R$ and ensemble fidelity $F_{av}$ tending to 1. On the contrary, if $R < S(\rho)$, then for every compression scheme such that $R(\mathcal{E}) = R$ the ensemble fidelity tends to 0 in the limit $n \to \infty$.*

**Proof:** The eigenvalues $r^{(n)}_{\boldsymbol{j}^{(n)}}$ of $\rho^{\otimes n}$ provide a probability distribution $\pi^{(n)} = \{r^{(n)}_{\boldsymbol{j}^{(n)}}\}_{\boldsymbol{j}^{(n)}\in\Omega_2^{(n)}}$ on the strings $\boldsymbol{j}^{(n)} \in \Omega_2^{(n)}$ with Shannon entropy $n\,S(\rho)$. According to Proposition 3.2.2, for any $\delta > 0$, $\epsilon > 0$ and $n$ large enough, there exists a subset $A^{(n)}_\epsilon$ of probability

$$\mathrm{Prob}(A^{(n)}_\epsilon) = \sum_{\boldsymbol{j}\in A^{(n)}_\epsilon} r^{(n)}_{\boldsymbol{j}^n} = \mathrm{Tr}\Big(\rho^{\otimes n}\,\Gamma^{(n)}\Big) \geq 1 - \delta$$

and cardinality $d(n)$ satisfying

$$(1-\delta)2^{n(S(\rho)-\epsilon)} \leq d(n) \leq 2^{n(S(\rho)+\epsilon)}\ .$$

Let $\Gamma^{(n)}$ project onto the subspace linearly spanned by the eigenvectors $|\,r^{(n)}_{\boldsymbol{j}^{(n)}}\,\rangle = |\,r_{j_1}\,\rangle \otimes |\,r_{j_2}\,\rangle \cdots \otimes |\,r_{j_n}\,\rangle$ corresponding to the eigenvalues $r^{(n)}_{\boldsymbol{j}^{(n)}}$,

$\boldsymbol{j}^{(n)} \in A_\varepsilon^{(n)}$. Such $\Gamma^{(n)}$ can be used to construct the compression map (7.140); hence, from (7.141), $F_{av} \geq 1-2\delta$. Also, the bounds on $d(n)$ ensures that any rate $R < S(\rho)$ is achievable.

Vice versa, if $d(n) \leq 2^{n(S(\rho)-\varepsilon)}$, then, according to Theorem 3.2.2, given the probability distribution $\pi^{(n)} = \{r^{(n)}_{\boldsymbol{j}^{(n)}}\}_{\boldsymbol{j}^{(n)}\in\Omega_2^{(n)}}$, any subset $B_{d(n)}$ with $d(n)$ strings has vanishingly small probability,

$$\mathrm{Prob}(B_{d(n)}) = \sum_{\boldsymbol{j}\in B_{d(n)}} r^{(n)}_{\boldsymbol{j}^n} = \mathrm{Tr}\Big(\rho^{\otimes n}\,\Gamma_{dn)}\Big) \leq \epsilon$$

for $n$ large enough, where $\Gamma_{d(n)}$ projects onto the subset spanned by the eigenvectors relative to the eigenvalues indexed by $\boldsymbol{j}^{(n)} \in B_{d(n)}$. It then follows that also the sum of the first $d(n)$ largest eigenvalues of $\rho^{\otimes n}$ must be smaller than $\epsilon$ and so also $F_{av} \leq \varepsilon$ because of (7.139). □

**Example 7.3.1.** [159] In a single use, a Bernoulli *qubit* source emits the non-orthogonal states

$$|\,\psi_0\,\rangle := \sqrt{1-\varepsilon}|\,0\,\rangle + \sqrt{\varepsilon}|\,1\,\rangle\ ,\ |\,\psi_1\,\rangle := \sqrt{1-\varepsilon}|\,0\,\rangle - \sqrt{\varepsilon}|\,1\,\rangle$$

where $0 < \varepsilon < 1/2$, with probability $1/2$ each; the corresponding statistical ensemble is described by

$$\rho = \frac{1}{2}|\,\psi_0\,\rangle\langle\,\psi_0\,| + \frac{1}{2}|\,\psi_1\,\rangle\langle\,\psi_1\,| = (1-\varepsilon)|\,0\,\rangle\langle\,0\,| + \varepsilon|\,1\,\rangle\langle\,1\,|\ .$$

Suppose that, given the 3-*qubit* strings $|\,\psi_{i_1i_2i_3}\,\rangle := |\,\psi_{i_1}\,\rangle \otimes |\,\psi_{i_2}\,\rangle \otimes |\,\psi_{i_3}\,\rangle$, only two *qubits* can be transmitted; how can the transmission of quantum information be optimized?

Since $\langle\,0\,|\,\psi_0\,\rangle = \langle\,0\,|\,\psi_1\,\rangle = \sqrt{1-\varepsilon}$, $\langle\,1\,|\,\psi_0\,\rangle = -\langle\,1\,|\,\psi_1\,\rangle = \varepsilon$ and $\varepsilon < 1/2$, the sender may encode and decode each 3-*qubit* string by tracing over the third *qubit* and appending the high probability state $|\,0\,\rangle\langle\,0\,|$ in its place:

$$\begin{aligned}|\,\psi_{i_1i_2i_3}\,\rangle\langle\,\psi_{i_1i_2i_3}\,| \mapsto \tilde{\rho}^{(3)}_{i_1i_2i_3} &= \mathcal{D}^{(3)}_1 \circ \mathcal{E}^{(3)}_1[|\,\psi_{i_1i_2i_3}\,\rangle\langle\,\psi_{i_1i_2i_3}\,|] \\ &= |\,\psi_{i_1i_2}\,\rangle\langle\,\psi_{i_1i_2}\,| \otimes |\,0\,\rangle\langle\,0\,|\ .\end{aligned}$$

The average fidelity then results

$$F^{(1)}_{av} = \frac{1}{8}\sum_{i_1,i_2,i_3} \langle\,\psi_{i_1i_2i_3}\,|\,\tilde{\rho}^{(3)}_{i_1i_2i_3}\,|\psi_{i_1i_2i_3}\,\rangle = 1-\varepsilon\ .$$

A better strategy arises from considering the components of a 3-*qubit* string along the eigenvectors of $\rho^{\otimes 3}$:

$$\begin{array}{ll} |\langle\,000\,|\,\psi_{i_1i_2i_3}\,\rangle| = (1-\varepsilon)^{3/2} & |\langle\,110\,|\,\psi_{i_1i_2i_3}\,\rangle| = \varepsilon\sqrt{1-\varepsilon} \\ |\langle\,001\,|\,\psi_{i_1i_2i_3}\,\rangle| = (1-\varepsilon)\sqrt{\varepsilon} & |\langle\,101\,|\,\psi_{i_1i_2i_3}\,\rangle| = \varepsilon\sqrt{1-\varepsilon} \\ |\langle\,010\,|\,\psi_{i_1i_2i_3}\,\rangle| = (1-\varepsilon)\sqrt{\varepsilon}\ , & |\langle\,011\,|\,\psi_{i_1i_2i_3}\,\rangle| = \varepsilon\sqrt{1-\varepsilon}\ . \\ |\langle\,100\,|\,\psi_{i_1i_2i_3}\,\rangle| = (1-\varepsilon)\sqrt{\varepsilon} & |\langle\,111\,|\,\psi_{i_1i_2i_3}\,\rangle| = \varepsilon^{3/2} \end{array}$$

Since $\varepsilon < 1/2$, the eigenvectors $|\,000\,\rangle$, $|\,001\,\rangle$, $|\,010\,\rangle$ and $|\,100\,\rangle$ provide higher probabilities than the second four eigenvectors; let $P$ project onto the linear span. Observe that the unitary permutation

$$U : \begin{cases} |\,000\,\rangle \mapsto |\,000\,\rangle & |\,111\,\rangle \mapsto |\,001\,\rangle \\ |\,001\,\rangle \mapsto |\,010\,\rangle & |\,110\,\rangle \mapsto |\,011\,\rangle \\ |\,010\,\rangle \mapsto |\,100\,\rangle & |\,101\,\rangle \mapsto |\,101\,\rangle \\ |\,100\,\rangle \mapsto |\,110\,\rangle & |\,011\,\rangle \mapsto |\,111\,\rangle \end{cases} ,$$

is such that $UPU^\dagger = \mathbb{1}_{12} \otimes |\,0\,\rangle\langle\,0\,|$, where $\mathbb{1}_{12} = \sum_{i,j=0}^{1} |\,ij\,\rangle\langle\,ij\,|$ is the identity matrix of the first two *qubits* . Therefore, one can construct a compression map as follows; first, introduce the trace-preserving *CP* maps

$$\begin{aligned} \mathbb{E}[\rho] &= P\,\rho\,P + |\,000\,\rangle\langle\,000\,|\,\mathrm{Tr}\Big((\mathbb{1} - P)\,\rho\Big) \\ U\,\mathbb{E}[\rho]\,U^\dagger &= \mathbb{1}_{12} \otimes |\,0\,\rangle\langle\,0\,|\Big(U^\dagger\,\rho\,U\Big)\mathbb{1}_{12} \otimes |\,0\,\rangle\langle\,0\,| \\ &\qquad + |\,000\,\rangle\langle\,000\,|\,\mathrm{Tr}\Big((\mathbb{1} - P)\,\rho\Big) . \end{aligned}$$

Then, define $\mathcal{E}_2^{(3)} : \mathbb{B}_1^+(\mathbb{C}^3) \mapsto \mathbb{B}_1^+(\mathbb{C}^2)$ as $\mathcal{E}_2^{(3)}[\rho] := \mathcal{E}_1^{(3)}[U\,\mathbb{E}[\rho]\,U^\dagger]$ and $\mathcal{D}_2^{(3)} : \mathbb{B}_1^+(\mathbb{C}^2) \mapsto \mathbb{B}_1^+(\mathbb{C}^3)$ as $\mathcal{D}_2^{(3)}[\sigma] = U^\dagger\,\sigma \otimes |\,0\,\rangle\langle\,0\,|\,U$. It follows that

$$\mathcal{D}_2^{(3)} \circ \mathcal{E}_2^{(3)}[\rho] = P\,\rho\,P + |\,000\,\rangle\langle\,000\,|\,\mathrm{Tr}\Big((\mathbb{1} - P)\,\rho\Big) .$$

Thus, with $\tilde{\rho}^{(3)}_{i_1 i_2 i_3} = \mathcal{D}_2^{(3)} \circ \mathcal{E}_2^{(3)}[|\,\psi_{i_1 i_2 i_3}\,\rangle\langle\,\psi_{i_1 i_2 i_3}\,|]$,

$$\begin{aligned} \langle\,\psi_{i_1 i_2 i_3}\,|\,\tilde{\rho}^{(3)}_{i_1 i_2 i_3}\,|\psi_{i_1 i_2 i_3}\,\rangle &= |\langle\,\psi_{i_1 i_2 i_3}\,|\,P\,|\psi_{i_1 i_2 i_3}\,\rangle|^2 \\ &\quad + |\langle\,000\,|\,\psi_{i_1 i_2 i_3}\,\rangle|^2\,\langle\,\psi_{i_1 i_2 i_3}\,|\,(\mathbb{1} - P)\,|\psi_{i_1 i_2 i_3}\,\rangle \\ &= 1 - 9\varepsilon^3 + 15\varepsilon^4 - 9\varepsilon^5 + 2\varepsilon^6 . \end{aligned}$$

As the right end side of the last inequality is the same for all 3-*qubit* strings considered, this is also the value of the fidelity $F^{(2)}_{av}$. As shown in the figure below, the latter turns out to be larger than $F^{(1)}_{av}$ for $0 < \varepsilon < 1/2$.

### Compression of Ergodic Quantum Sources

As showed in Section 3.2.1, Proposition 3.2.2, Theorem 3.2.1 and Theorem 3.2.2 establish the role the entropy rate as the optimal compression rate of classical ergodic sources. Theorem 7.3.1 assigns the same role to the von Neumann entropy in the case of quantum sources of Bernoulli type.

For this particular family of quantum chains, the von Neumann entropy coincides with their entropy density as defined in Section 7.2; it is thus expected that a kind of general Quantum Shannon-Mc Millan-Breiman Theorem should hold for generic ergodic quantum sources. The relevance for quantum

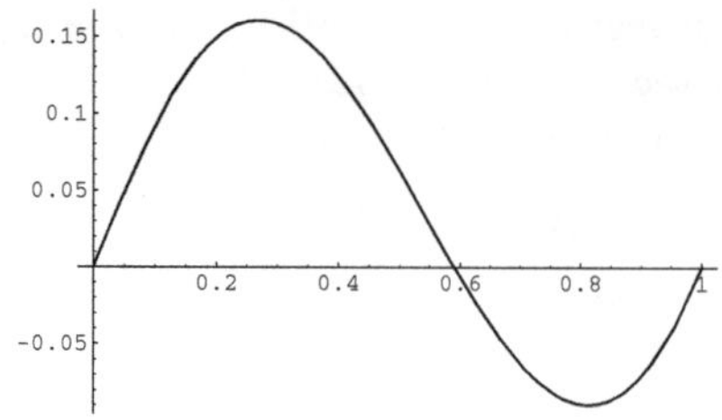


**Fig. 7.2.** $F_{av}^{(2)} - F_{av}^{(1)}$ against $0 \leq \varepsilon \leq 1$.

information of quantum spin chains endowed with states with a more general structure than a tensor product has been emphasized in Section 7.3, where analogies and differences between classical and quantum contexts have also been outlined.

In particular, in Remark 7.3.1.2 it was pointed out that, unlike for classical *bit* strings, one cannot profit from any natural "chaining together" *qubit* -strings. Though ideas how to circumvent such a problem have been put forward [60], this fact represents an obstruction to a full quantum generalization of the classical Breiman theorem. The latter is an almost everywhere statement regarding single sequences, while the Shannon-Mc Millan formulation is concerned with statistical ensembles; of this theorem there exist a number of extensions to particular non-commutative settings [223, 240, 169, 94] and a full quantum extension [58]. This general result has then been used [59] to devise compression protocols for ergodic sources consisting of encoding and decoding procedures similarly to what outlined in the previous section.

**Theorem 7.3.2.** *Let* $(\mathcal{A}_{\mathbb{Z}}, \Theta_\sigma, \omega)$, *with* $\mathcal{A} = M_d(\mathbb{C})$ *as site-algebras, be an ergodic quantum spin-chain with mean entropy* $s(\omega)$. *Then, for all* $\delta > 0$ *there is* $N_\delta \in \mathbb{N}$ *such that for all* $n \geq N_\delta$ *there exists an orthogonal projection* $p_n(\delta) \in \mathcal{A}_n$ *such that*

*1.* $\omega(p_n(\delta)) = \mathrm{Tr}_n(\rho^{(n)}\, p_n(\delta)) \geq 1 - \delta$,

*2. for all minimal projections* $0 \neq p_n \in \mathcal{A}_n$ *dominated by* $p_n(\delta)$ $(p \leq p_n(\delta))$

$$(1-\delta)2^{-n(s(\omega)+\delta)} < \omega(p_n(\delta)) < 2^{-n(s(\omega)-\delta)} ,$$

*3.* $2^{n(s(\omega)-\delta)} < \mathrm{Tr}_n(p_n(\delta)) < 2^{n(s(\omega)+\delta)}$.

That the above results extend Proposition 3.2.2 is apparent: classical high probability subsets are replaced by orthogonal projections $p_n(\delta)$ whose statistical weight with respect to the translation-invariant state $\omega$ is nearly 1; further, the typical subsets correspond to orthogonal projections whose normalization (dimension of the associated Hilbert subspaces), $\mathrm{Tr}(p_n(\delta))$ goes as $2^{ns(\omega)}$ for large $n$.

Like in the case of a Bernoulli quantum source, the proof of Theorem 7.3.2 hinges upon considering the discrete probability distributions $\pi^{(\ell)} = \{r_i^{(\ell)}\}_{i=1}^{d^\ell}$ consisting of the eigenvalues of the density matrices $\rho^{(\ell)}$ describing the restrictions $\omega$ to the local subalgebra $\mathcal{A}_\ell = M_d(\mathbb{C})^{\otimes \ell}$ with spectral decomposition $\rho^{(\ell)} = \sum_{i=1}^{d^\ell} r_i^{(\ell)} \,|\, r_i^{(\ell)} \rangle\langle r_i^{(\ell)} \,|$. The Shannon entropy $H(\pi^{(\ell)})$ equals the von Neumann entropy $S\left(\rho^{(\ell)}\right)$; further, from the definition of entropy density $s(\omega)$, given $\eta > 0$, for infinitely many $\ell$ one has

$$s(\omega) = \inf_n \frac{1}{n} S\left(\rho^{(n)}\right) \leq \frac{1}{\ell} S\left(\rho^{(\ell)}\right) = \frac{1}{\ell} H(\pi^{(\ell)}) \leq s(\omega) + \eta \,. \tag{7.142}$$

For Bernoulli quantum sources, the products of eigenvalues of single site density matrices provide a natural Bernoulli stochastic process, whose entropy density $s(\omega)$ is exactly the von Neumann entropy of $\rho$. Such a structure is missing in the case of generic ergodic quantum source. However, from (7.142), one observes that choosing $\ell$ large enough, $S\left(\rho^{(\ell)}\right) \simeq \ell\, s(\omega)$. Moreover, the eigenvectors $|\, r_i^{(\ell)} \rangle\langle r_i^{(\ell)} \,|$ are minimal projections generating a maximally Abelian subalgebra $\mathfrak{D} \subset \mathcal{A}_\ell$ and the eigenvalues $r_i^{(\ell)}$ define a probability $\pi^{(\ell)}$ over the symbols $i \in I := \{1, 2, \dots d^\ell\}$. By tensorizing copies of the Abelian subalgebra $\mathfrak{D}$, one can embed the Abelian subalgebras $\mathfrak{D}_n := \mathfrak{D}^{\otimes n}$ into the local algebras $\mathcal{A}_{n\ell}$ and $C^*$-induction yields a quasi-local Abelian algebra $\mathfrak{D}^\infty$ embedded into the quantum spin-chain $\mathcal{A}_\mathbb{Z}$.

The Abelian algebra $\mathfrak{D}^\infty$ is clearly associated to a triplet, or symbolic model $\left(\widetilde{\Omega}_I, T_\sigma, \mu^{(\ell)}\right)$ (see Definition 2.2.5 and the preceding discussion) where $\widetilde{\Omega}_I$ is the space of sequences of symbols from $I$, $T_\sigma$ is the shift along these sequences and $\mu^{(\ell)}$ is the measure on $\widetilde{\Omega}_I$ that arises from $\pi^{(\ell)}$. Further, from (7.130), the (classical) entropy rate is

$$h(\mu^{(\ell)}) = \lim_{n\to\infty} \frac{1}{n} S\left(\rho^{(n\ell)}\right) = s_\ell(\omega) = \ell\, s(\omega) \,. \tag{7.143}$$

As the automorphism over $\mathfrak{D}^\infty$ corresponding to the shift $T_\sigma$ on $\widetilde{\Omega}_I$ is not $\Theta_\sigma$, but its $\ell$-th power $\Theta_\sigma^\ell$, the Abelian spin-chain associated to the symbolic model of above is $\left(\mathfrak{D}_I^\infty, \Theta_\sigma^\ell, \omega \restriction \mathfrak{D}_I^\infty\right)$ (see Defintion 2.2.5 and the preceding discussion). The state $\omega$ is $\Theta_\sigma$-ergodic, but not in general $\Theta_\sigma^\ell$-ergodic. If it were $\Theta_\sigma^\ell$-ergodic, $\left(\mathfrak{D}_I^\infty, \Theta_\sigma^\ell, \omega \restriction \mathfrak{D}_I^\infty\right)$ would amount to an ergodic process and we could use the classical techniques as in Proposition 3.2.2 with the mean entropy $h(\omega \restriction \mathfrak{D}_I^\infty) = \ell\, s(\omega)$ in the place of the Shannon entropy as follows from the classical Shannon-Mc Millan-Breiman result in Theorem 3.2.1.

**Remarks 7.3.2.**

1. The ergodicity of the embedded Abelian spin-chain $\left(\mathfrak{D}_I^\infty, \Theta_\sigma^\ell, \omega\lceil\mathfrak{D}_I^\infty\right)$ would follow from that of the quantum spin-chain $\left(\mathcal{A}_{\mathbb{Z}}, \Theta_\sigma^\ell, \omega\right)$, since otherwise the resulting decomposition of $\omega\lceil\mathfrak{D}_I^\infty$ into ergodic components would provide a decomposition of $\omega$ as well.
2. In [240], the quantum Shannon-Mc Millan theorem was proved under the assumption of $\Theta_\sigma^{(\ell)}$-ergodicity of the spin-chain state $\omega$: such a property is known as *complete ergodicity*. This restriction has been removed in [58].

The possible lack of $\Theta_\sigma^\ell$-ergodicity can be overcome by means of Proposition 7.1.9 and of Lemma 7.2.1. Indeed, the argument of above can be developed for the $\Theta_\sigma^\ell$-ergodic components $\omega_j$ indexed by $j \in A_{\ell,\eta}^c$ for which $s_\ell(\omega_j) = \ell s(\omega)$ and $s_j^{(\ell)} := \frac{1}{\ell} S\left(\rho_j^{(\ell)}\right) \leq s(\omega) + \eta$, for some fixed $\eta > 0$. For each of these $\omega_j$, one considers the local states $\rho_j^{(\ell)}$ over $\mathcal{A}_\ell$, the probability distributions $\pi_j^{(\ell)}$ corresponding to their spectra, the Abelian subalgebras $\mathfrak{D}^j$ generated by their spectral projections $p_{j,i}^{(\ell)}$, $i \in I_j$, and the associated ergodic Abelian spin-chains $\left(\mathfrak{D}_{I_j}^\infty, \Theta_\sigma^\ell, \omega_j\lceil\mathfrak{D}_{I_j}^\infty\right)$. Because of the bound (3.7) in Remark 3.1.1.1 and of the choice of indices $j \in A_{\ell,\eta}^c$, these chains have entropy rates

$$h_j \leq H(\pi_j^{(\ell)}) = S\left(\rho_j^{(\ell)}\right) \leq \ell\,(s(\omega) + \eta)\ . \tag{7.144}$$

After identifying strings $\boldsymbol{i}^{(n)}$ of symbols from $I_j$ with minimal projections $p_{\boldsymbol{i}^{(n)}} \in \mathfrak{D}_n^j \subset \mathcal{A}_{n\ell}$, so that $\pi_{\boldsymbol{i}^{(n)}}^{(n\ell)} = \mathrm{Tr}_{[0,n\ell-1]}(\rho^{(n\ell)} p_{\boldsymbol{i}^{(n)}})$, one can choose positive $\varepsilon, \delta$ and select subsets of minimal projections,

$$\mathcal{C}_j^{(n)} := \left\{p_{\boldsymbol{i}^{(n)}} \in \mathfrak{D}_n^j \,:\, 2^{-n(h_j+\delta)} < \mathrm{Tr}_{[0,n\ell-1]}(\rho^{(n\ell)} p_{\boldsymbol{i}^{(n)}}) < 2^{-n(h_j-\delta)}\right\} \tag{7.145}$$

such that, by using Proposition 3.2.2, Theorem 3.2.1 and (7.144), for $n$ large enough

$$\#(\mathcal{C}_j^{(n)}) = \mathrm{Tr}_{[0,n\ell-1]}(p_{\boldsymbol{i}^{(n)}}) \leq 2^{n(h_j+\delta)} \leq 2^{n(\ell\, s(\omega)+\eta)+\delta)} \tag{7.146}$$

$$\pi_j^{(n)}(\mathcal{C}_j^{(n)}) = \sum_{p_{\boldsymbol{i}^{(n)}} \in \mathcal{C}_j^{(n)}} \mathrm{Tr}_{[0,n\ell-1]} p_j^{(n)}) \geq 1 - \frac{\varepsilon}{2}\ , \tag{7.147}$$

where $p_j^{(n)} := \sum_{p_{\boldsymbol{i}^{(n)}} \in \mathcal{C}_j^{(n)}} p_{\boldsymbol{i}^{(n)}}$.

In order to use these arguments and conclude the proof of the quantum Shannon-Mc Millan theorem, some further results are needed. The first one deals with discrete subsets equipped with (not necessarily compatible) probability distributions and with the asymptotic behavior of their minimal cardinality.

**Lemma 7.3.1.** *Let $D > 0$ and $\left\{\left(I_n, \pi_n\right)\right\}_{n\in\mathbb{N}}$ be a countable family of finite sets $I_n$ of cardinality $\#(I_n)$ with associated probability distributions $\pi_n = \{p_n(i)\}_{i\in I_n}$. Suppose $\frac{1}{n}\log_2\#(I_n) \leq D$ for all $n$ and define*

$$\alpha_{\varepsilon,n}(\pi_n) := \min\left\{\log_2\#(\Omega)\ :\ \Omega\subset I_n\,,\ \pi_n(\Omega)\geq 1-\varepsilon\right\}. \tag{7.148}$$

*If $\left\{\left(I_n, \pi_n\right)\right\}_{n\in\mathbb{N}}$ satisfies*

$$\lim_{n\to\infty}\frac{1}{n}H(\pi_n) = h < \infty \ (1) \quad \text{and} \quad \limsup_{n\to\infty}\frac{1}{n}\alpha_{\varepsilon,n}(\pi_n) \leq h \quad (2)\,,$$

*for all $\varepsilon\in(0,1)$, then*

$$\lim_{n\to\infty}\frac{1}{n}\alpha_{\varepsilon,n}(\pi_n) \leq h\,, \qquad \forall\varepsilon\in(0,1)\,. \tag{7.149}$$

**Proof:** Let $\delta > 0$ be arbitrarily chosen and distinguish the following disjoint subsets of $I_n$:

$$I_n^1(\delta) := \left\{i\in I_n\ :\ \pi_n(i) > 2^{-n(h-\delta)}\right\},$$
$$I_n^2(\delta) := \left\{i\in I_n\ :\ 2^{-n(h+\delta} < \pi_n(i) < 2^{-n(h-\delta)}\right\},$$
$$I_n^3(\delta) := \left\{i\in I_n\ :\ \pi_n(i) < 2^{-n(h+\delta)}\right\}.$$

Suppose $1 \geq \limsup_{n\to\infty}\pi_n(I_n^3(\delta)) - b > 0$; then, there exists $n$ such that $\pi_n(I_n^3(\delta) \geq b$ and $\pi_n(I_n^1(\delta)\cup I_n^2(\delta)) < 1-b$. Choose $0<\varepsilon<b$; if $\pi_n(\Omega)\geq 1-\varepsilon$,

$$1-\varepsilon \leq \pi_n(\Omega)\ <\ 1-b+\pi_n\left(\Omega\cap I_n^3(\delta)\right) \quad \text{implies}$$
$$b-\varepsilon \leq \pi_n\left(\Omega\cap I_n^3(\delta)\right)\ <\ \#\left(\Omega\cap I_n^3(\delta)\right)2^{-n(h+\delta} \quad \text{and}$$
$$\log_2\#\left(\Omega\cap I_n^3(\delta)\right)\ >\ \log_2(b-\varepsilon)\ +\ n(h+\delta)\,,$$

whence $\lim_{n\to\infty}\frac{1}{n}\alpha_{\varepsilon,n}(\pi_n)\ >\ h+\delta$ contradicting the second condition in the statement of the lemma. Thus, $\lim_{n\to\infty}\pi_n(I_n^3(\delta) = 0$.

It also follows that $I_n^3(\delta)$ cannot asymptotically contribute to the Shannon entropy $H(\pi_n)$; indeed, applying inequality (2.85) to the (non-normalized) distributions $\{p_n(i)\}_{i\in I_n^3(\delta)}$ and $\left\{q_n(i) := \frac{\pi_n(I_n^3(\delta)}{\#(I_n^3(\delta)}\right\}_{i\in I_n^3(\delta)}$, which are such that $\sum_{i\in I_n^3(\delta)}p_n(i) = \sum_{i\in I_n^3(\delta)}q_n(i)$, yields

$$\begin{aligned}
h_n^3 &:= -\frac{1}{n}\sum_{i\in I_n^3(\delta)} p_n(i)\log_2 p_n(i) \le -\frac{1}{n}\sum_{i\in I_n^3(\delta)} p_n(i)\log_2 q_n(i) \\
&= -\frac{1}{n}\sum_{i\in I_n^3(\delta)} p_n(i)\Big(\log_2 \pi_n(I_n^3(\delta)) - \log_2 \#(I_n^3(\delta))\Big) \\
&\le -\pi_n(I_n^3(\delta))\frac{1}{n}\log_2 \pi_n(I_n^3(\delta) + D\,\pi_n(I_n^3(\delta)) .
\end{aligned}$$

The right hand side of the last inequality goes to 0 with $n \to \infty$ due to $\pi_n(I_n^3(\delta) \mapsto 0$ with $n \to \infty$ and because $\log_2 \#(I_n) \le nD$ by assumption. Further, this very same fact implies $\lim_{n\to\infty}\pi_n(I_n^1(\delta) = 0$ for all $\delta > 0$, otherwise

$$\begin{aligned}
\frac{1}{n}H(\pi_n) &= -\frac{1}{n}\sum_{i\in I_n^1(\delta)} p_n(i)\log_2 p_n(i) - \frac{1}{n}\sum_{i\in I_n^2(\delta)} p_n(i)\log_2 p_n(i) - h_n^3 \\
&< \pi_n(I_n^1(\delta))\,(h-\delta) + \pi_n(I_n^2(\delta)) + h_n^3 \\
&< h + \delta\Big(\pi_n(I_n^1(\delta)) - \pi_n(I_n^2(\delta))\Big) + h_n^3
\end{aligned}$$

would contradict the first condition of the lemma for sufficiently small $\delta$.

Consequently, $\lim_{n\to\infty}\pi_n(I_n^2(\delta)) = 1$ for sufficiently small $\delta$. Thus, choosing $n$ so that $\pi_n(\Omega) \ge 1-\varepsilon$ and $\pi_n(I_n^2(\delta)) \ge 1-\eta$, it follows that $\pi_n\Big(\Omega\cap I_n^2(\delta)\Big) \ge 1-\varepsilon-\eta$, whence

$$\begin{aligned}
&\#\Big(\Omega\cap I_n^2(\delta)\Big) \ge (1-\varepsilon-\eta)\,2^{n(h-\delta)} \quad \text{implies} \\
&\frac{1}{n}\alpha_{\varepsilon,n}(\pi_n) \ge \frac{1}{n}\log_2(1-\varepsilon-\eta) + h - \delta .
\end{aligned}$$

Since $\delta$ can be chosen arbitrarily small, the result follows. □

Returning to the probability distribution $\pi^{(\ell)}$ associated with the ordered spectrum of $\rho^{(\ell)}$, fix $\varepsilon \in (0,1)$ and set

$$N_{\varepsilon,\ell} := \min\Big\{1 \le k \le d^\ell \ :\ \sum_{i=1}^{k} r_i^{(\ell)} \ge 1-\varepsilon\Big\} , \tag{7.150}$$

so that $\alpha_{\varepsilon,\ell}(\pi^{(\ell)}) = \log_2 N_{\varepsilon,\ell}$. If

$$\limsup_{\ell\to\infty}\frac{1}{\ell}N_{\varepsilon,\ell} \le s(\omega) , \tag{7.151}$$

then, together with (7.142), this allows using Lemma 7.3.1. As follows: let $I_\ell$ be the set of indices labeling the eigenprojections $|\,r_i^{(\ell)}\,\rangle\langle\,r_i^{(\ell)}\,|$. Then, choose $0 < \delta' < \delta$ and consider the subset $I_\ell^2(\delta')$ as constructed in the lemma and set

$$P_\ell(\delta) := \sum_{i\in I^2_\ell(\delta')} |\, r^{(\ell)}_i \,\rangle\langle\, r^{(\ell)}_i \,| \ .$$

It turns out that, for $\ell$ sufficiently large,

$$\mathrm{Tr}\Big(\rho^{(\ell)}\, P_\ell(\delta)\Big) = \sum_{i\in I^2_\ell(\delta')} r^{(\ell)}_i = \pi^{(\ell)}(I^2_n(\delta')) \geq 1-\delta \ .$$

Further, every minimal projection $p \leq P_\ell(\delta)$ dominated by $P_\ell(\delta)$ projects onto a vector of $(\mathbb{C}^2)^{\otimes \ell}$,

$$p = |\,\psi\,\rangle\langle\,\psi\,| \ , \quad |\,\psi\,\rangle = \sum_{i\in I^2_\ell(\delta')} c_i\,|\, r^{(\ell)}_i\,\rangle \ , \qquad \sum_{i\in I^2_\ell(\delta')} |c_i|^2 = 1 \ ,$$

whence, by the definition of the subset $I^2_\ell(\delta')$

$$2^{-\ell(s(\omega)+\delta)} < 2^{-\ell(s(\omega)+\delta')} < \mathrm{Tr}(\rho^{(\ell)}\, p) < 2^{-\ell(s(\omega)-\delta')} < 2^{-\ell(s(\omega)-\delta)} \ .$$

Finally, from $\mathrm{Tr}(\rho^{(\ell)}\, P_\ell(\delta)) \geq 1-\delta$ and

$$\#(I^2_\ell(\delta'))\, 2^{-\ell(s(\omega)+\delta)} \leq \mathrm{Tr}(\rho^{(\ell)}\, P_\ell(\delta)) = \sum_{i\in I^2_\ell(\delta')} r^{(\ell)}_i \leq \#(I^2_\ell(\delta'))\, 2^{-\ell(s(\omega)-\delta)}$$

it follows that $(1-\delta)\, 2^{\ell(s(\omega)-\delta)} \leq \mathrm{Tr}(P_\ell(\delta)) = \#(I^2_\ell(\delta')) \leq 2^{\ell(s(\omega)+\delta)}$, thus concluding the proof of Theorem 7.3.2.

Of course, it remains to be showed that (7.151) really holds true. The proof of this fact hinges upon a *per se* interesting result concerning the minimal dimension of the so-called high probability subspaces. Practically speaking, these are the relevant subspaces: as already seen in the case of Bernoulli quantum sources and as it will be showed at the end of this section, they allow for quantum compression with reliable retrieval.

**Definition 7.3.2 (Typical Subspaces).**

1. *Given a quantum spin chain $(\mathcal{A}_\mathbb{Z}, \omega)$, projectors $p_n \in \mathcal{A}_n$ such that $\omega(p_n) = \mathrm{Tr}(\rho^{(n)}\, p_n) \geq 1-\varepsilon$ will be termed $\omega$-typical projectors and $\omega$-typical subspaces the subspaces of $(\mathbb{C}^d)^{\otimes n}$ onto which they project.*
2. *For any $\varepsilon > 0$, let*

$$\beta_{\varepsilon,n}(\omega) := \min\Big\{\log_2 \mathrm{Tr}(q) \ : \ \mathcal{A}_n \ni q = q^\dagger = q^2 \ , \ \mathrm{Tr}(\rho^{(n)}\, q) \geq 1-\varepsilon\Big\} \ . \tag{7.152}$$

The following result relates the spectrum of local states to the dimension of high probability subspaces [58, 140, 141].

**Lemma 7.3.2.** *$\beta_{\varepsilon,n}(\omega)$ equals $N_{\varepsilon,n}$ in (7.150).*

**Proof:** By definition,

$$\omega\Big(\sum_{i=1}^{N_{\varepsilon,n}} |r_i^{(n)}\rangle\langle r_i^{(n)}|\Big) = \mathrm{Tr}(\rho^{(n)} \sum_{i=1}^{N_{\varepsilon,n}} |r_i^{(n)}\rangle\langle r_i^{(n)}|) \geq 1-\varepsilon ,$$

whence $\beta_{\varepsilon,n}(\omega) \leq N_{\varepsilon,n}$. If the inequality is strict, then there exists a projection $q \in \mathcal{A}_n$ such that $m := \mathrm{Tr}(q) < N_{\varepsilon,n}$ and $\mathrm{Tr}(\rho^{(n)}\, q) \geq 1-\varepsilon$. Then, using Ky Fan inequality (5.158), a contradiction emerges:

$$1-\varepsilon \leq \mathrm{Tr}(\rho^{(n)}\, q) = \sum_{i=1}^{m} \langle q_i \,|\, \rho^{(n)} \,| q_i \rangle \leq \sum_{i=1}^{m} r_i^{(n)} < 1-\varepsilon ,$$

where $|\, q_i \rangle\langle\, q_i\,|$ are minimal projections such that $q = \sum_{i=1}^m |\, q_i \rangle\langle\, q_i\,|$. □

For showing (7.151), the key result is

**Lemma 7.3.3.** *For an ergodic quantum spin-chain* $\Big(\mathcal{A}_{\mathbb{Z}}, \Theta_\sigma, \omega\Big)$

$$\limsup_{n\to\infty} \frac{1}{n} \beta_{\varepsilon,n}(\omega) \leq s(\omega) , \quad \forall \varepsilon \in (0,1) .$$

Before proving it, observe that the previous two lemmas imply a quantum counterpart to the *AEP* (see Theorem 3.2.2).

**Proposition 7.3.1 (Quantum *AEP* (*QAEP*)).**
*Let* $(\mathcal{A}_{\mathbb{Z}}, \Theta_\sigma, \omega)$ *be an ergodic quantum source with entropy rate* $s(\omega)$. *Then, for every* $0 < \varepsilon < 1$,

$$\lim_{n\to\infty} \frac{1}{n} \beta_{\varepsilon,n}(\omega) = s(\omega) . \tag{7.153}$$

**Remark 7.3.3.** Operatively, the previous proposition states that any sequence of typical projections must project onto subspaces whose dimension goes as $2^{ns(\omega)}$, asymptotically.

**Proof of Proposition 7.3.3** Lemma 7.2.1 ensures that for any $\varepsilon > 0$ and fixed $\eta > 0$ there exists $L \in \mathbb{N}$ such that $\ell > L$ implies

$$0 \leq \frac{\#(A_{\ell,\eta})}{n_\ell} \leq \frac{\varepsilon}{2} , \quad \frac{\#(A^c_{\ell,\eta})}{n_\ell} = 1 - \frac{\#(A_{\ell,\eta})}{n_\ell} \geq 1 - \frac{\varepsilon}{2} .$$

Consider the $\Theta^\ell_\sigma$-ergodic components $\omega_j$ of $\omega$, the sets $\mathcal{C}_j^{(n)}$ in (7.146) and the smallest projector $q^{(n\ell} := \bigvee_{j\in A^c_{\ell,\eta}} p_j^{(n)}$ larger than all $p_j^{(n)}$, $j \in A^c_{\ell,\eta}$. If

$m = n\ell + r$, $0 \leq r < \ell$, set $q_m := q^{(n\ell)} \otimes \mathbb{1}_{[n\ell, n\ell+r-1]}$ and, by means of (7.147), estimate

$$\begin{aligned}
\mathrm{Tr}_{[0,m-1]}(\rho^{(m)} q_m) &= \frac{1}{n_\ell} \sum_{j=0}^{n_\ell - 1} \mathrm{Tr}_{[0,m-1]}(\rho_j^{(m)} q_m) \\
&= \frac{1}{n_\ell} \sum_{j=0}^{n_\ell - 1} \mathrm{Tr}_{[0,n\ell-1]}(\rho_j^{(n\ell)} q^{(n\ell)}) \\
&\geq \frac{1}{n_\ell} \sum_{j=0}^{n_\ell - 1} \mathrm{Tr}_{[0,n\ell-1]}(\rho_j^{(n\ell)} p_j^{(n)}) \geq \frac{\#(A^c_{\ell,\eta})}{n_\ell}(1 - \frac{\varepsilon}{2}) \geq 1 - \varepsilon \,.
\end{aligned}$$

Then, definition (7.152) and (7.146) imply

$$\begin{aligned}
\beta_{\varepsilon,m}(\omega) &\leq \log_2 \mathrm{Tr}_{[0,m-1]}(q_m) = \log_2 \mathrm{Tr}_{[0,n\ell-1]}(q^{(n\ell)}) \,+\, r \log_2 d \\
&\leq \log_2\Big( \sum_{j \in A^c_{\ell,\eta}} \mathrm{Tr}_{[0,n\ell-1]}(p_j^{(n)})\Big) \,+\, r \log_2 d \\
&\leq \log_2 \#(A^c_{\ell,\eta}) \,+\, n(\ell(s(\omega) + \eta) + \delta) \,+\, r \log_2 d \,,
\end{aligned}$$

whence the result follows from the arbitrariness of $\eta$ and $\delta$ and from

$$\limsup_{m\to\infty} \frac{1}{m} \beta_{\varepsilon,m}(\omega) \leq \limsup_{m\to\infty} \frac{1}{m} \log_2 \#(A^c_{\ell,\eta}) \,+\, s(\omega) \,+\, \eta \,+\, \frac{\delta}{\ell} \,.$$

□

## Universal Quantum Compression

Based on the classical construction of universal codes [325, 168], of which a particular instance has been given in Section 3.2.1, one may disengage the compression from its explicit dependence on the quantum source statistics by resorting to Universal Quantum Compression Schemes [163].

In the following, the construction in [163] will be slightly modified. We shall consider a quantum spin chain $\mathcal{A}_\mathbb{Z}$ with an ergodic translation-invariant state $\omega$, an increasing sequence of local subalgebras $\mathcal{A}_n$ as defined in Section 7.3 and local states $\omega \restriction \mathcal{A}_n$ described by density matrices $\rho^{(n)}$.

The idea is to construct the analogous of the typical subsets $A^{(n)}$ as in the proof of Proposition 3.2.3, with cardinality growing as $2^{nR}$ and probability $\pi_A^{(n)}(A^{(n)})$ tending to 1 with $n$ for all sources $A$ (Bernoulli in that case) with entropy (rate) $H(A) < R$. As always when passing from classical to quantum sources typical subsets will be replaced by typical subspaces and by the associated orthogonal projectors.

**Theorem 7.3.3 (Universal Typical Subspaces).**

*Let $s > 0$ and $\varepsilon > 0$. There exists a sequence of projectors $Q^{(n)}_{s,\varepsilon} \in \mathcal{A}_n$, $n \in \mathcal{N}$, such that for $n$ large enough*

$$\mathrm{Tr}\Big(Q^{(n)}_{s,\varepsilon}\Big) \leq 2^{n(s+\varepsilon)} \tag{7.154}$$

*and for every ergodic quantum state $\omega$ on $\mathcal{A}_{\mathbb{Z}}$ with entropy rate $s(\omega) \leq s$ it holds that*

$$\lim_{n\to\infty} \omega^{(n)}(Q^{(n)}_{s,\varepsilon}) = \mathrm{Tr}(\rho^{(n)} Q^{(n)}_{s,\varepsilon}) = 1 \ . \tag{7.155}$$

**Definition 7.3.3.** *The orthogonal projectors $Q^{(n)}_{s,\varepsilon}$ in the above theorem will be called universal typical projectors at level $s$.*

We subdivide the proof of Theorem 7.3.3 in various steps.

**Step 1** Let $\ell \in \mathbb{N}$ and $R > 0$. Any Abelian quasi-local subalgebra $\mathcal{C}^\infty_\ell \subseteq \mathcal{A}_{\mathbb{Z}}$ constructed from a maximal Abelian $\ell-$block subalgebra $\mathcal{C}_\ell \subseteq \mathcal{A}_\ell$, together with the probability distribution $\omega\lceil\mathcal{C}^\infty_\ell$ corresponds to a classical ergodic stochastic process.

The results in [168] imply that, independently of the latter, there exists a universal sequence of projectors (corresponding to classical universal typical subspaces) $p^{(n)}_{\ell,R} \in \mathcal{C}^{(n)}_\ell \subseteq \mathcal{A}_{\ell n}$ with

$$\frac{1}{n}\log \mathrm{Tr}(p^{(n)}_{\ell,R}) \leq R \ , \quad \text{such that} \quad \lim_{n\to\infty} \pi^{(n)}(p^{(n)}_{\ell,R}) = 1$$

for any ergodic state $\pi$ on the Abelian algebra $\mathcal{C}^\infty_\ell$ with entropy rate $s(\pi) < R$. Notice that ergodicity and entropy rate of $\pi$ are defined with respect to the shift on $\mathcal{C}^\infty_\ell$, which corresponds to the $\ell$-shift on $\mathcal{A}_{\mathbb{Z}}$.

One then applies unitary operators of the factorized form $U^{\otimes n}$, with $U \in \mathcal{A}_\ell$ unitary, to the $p^{(n)}_{\ell,R}$ and introduces the projectors

$$w^{(\ell n)}_{\ell,R} := \bigvee_{U\in\mathcal{A}_\ell \ \text{unitary}} U^{\otimes n} p^{(n)}_{\ell,R} U^{*\otimes n} \in \mathcal{A}^{(\ell n)} \ . \tag{7.156}$$

These are, by definition, the smallest projectors such that, for all $U$,

$$U^{\otimes n} p^{(n)}_{\ell,R} U^{*\otimes n} \leq w^{(\ell n)}_{\ell,R} \ .$$

Let $p^{(n)}_{\ell,R} = \sum_{i\in I} |\, i^{(n)}_{\ell,R}\,\rangle\langle\, i^{(n)}_{\ell,R}\,|$ be a spectral decomposition of $p^{(n)}_{\ell,R}$ (with $I \subset \mathcal{N}$ some index set), and let $\mathbf{P}(V)$ denote the orthogonal projector onto a given subspace $V$. Then, $w^{(\ell n)}_{\ell,R}$ can also be written as

$$w^{(\ell n)}_{\ell,R} = \mathbf{P}\left(\mathrm{span}\Big\{U^{\otimes n} |\, i^{(n)}_{\ell,R}\,\rangle \ : \ i \in I, U \in \mathcal{A}_\ell \ \text{unitary}\Big\}\right) \ .$$

It proves convenient to consider the projectors

$$W^{(\ell n)}_{\ell,R} := \mathbf{P}\left(\mathrm{span}\left\{A^{\otimes n}|\,i^{(n)}_{\ell,R}\rangle\ :\ i \in I, A \in \mathcal{A}_\ell\right\}\right)\ ,\quad w^{(\ell n)}_{\ell,R} \leq W^{(\ell n)}_{\ell,R}\ . \tag{7.157}$$

Given $m = n\ell + k$ with $n \in \mathbb{N}$ and $k \in \{0,\ldots,\ell-1\}$, let

$$w^{(m)}_{\ell,R} := w^{(\ell n)}_{\ell,R} \otimes \mathbf{1}^{\otimes k} \in \mathcal{A}_m\ ,\qquad W^{(m)}_{\ell,R} := W^{(\ell n)}_{\ell,R} \otimes \mathbf{1}^{\otimes k} \in \mathcal{A}_m\ .$$

These are projectors and, as in [160], one estimates the trace of $W^{(m)}_{\ell,R} \in \mathcal{A}_m$ as follows. By an argument similar to that used in the proof of Lemma 3.2.2, the dimension of the symmetric subspace $SYM(\mathcal{A}_{\ell n} := \mathrm{span}\{A^{\otimes n} : A \in \mathcal{A}_\ell\}$ is upper bounded by $(n+1)^{\dim(\mathcal{A}_\ell)}$, thus

$$\mathrm{Tr}W^{(m)}_{\ell,R} = \mathrm{Tr}W^{(\ell n)}_{\ell,R}\cdot\mathrm{Tr}\mathbb{1}^{\otimes k} \leq (n+1)^{2^{2\ell}}\mathrm{Tr}p^{(n)}_{\ell,R}\cdot 2^\ell \leq (n+1)^{2^{2\ell}}\cdot 2^{Rn}\cdot 2^\ell\ . \tag{7.158}$$

**Step 2.** Consider a stationary ergodic state $\omega$ on the spin-chain $\mathcal{A}_\mathbb{Z}$ with entropy rate $s(\omega) \leq s$. Let $\varepsilon,\delta > 0$. If $\ell$ is chosen large enough, then the projectors $w^{(m)}_{\ell,R}$, where $R := \ell(s + \frac{\varepsilon}{2})$, are $\delta-$typical for $\omega$ i.e.

$$\mathrm{Tr}\left(\rho^{(m)} w^{(m)}_{\ell,R})\right) \geq 1 - \delta\ ,$$

for $m \in \mathcal{N}$ sufficiently large. This follows from the result in Proposition 7.1.9 concerning the convex decomposition of the ergodic state $\omega$ into $k(\ell) \leq \ell$ states $\omega^{(\ell)}_{i,l}$, $\omega = \dfrac{1}{k(l)}\sum_{i=1}^{k(l)} \omega^{(\ell)}_{i,l}$, that are ergodic with respect to the $\ell-$shift on $\mathcal{A}_\mathbb{Z}$ and have an entropy rate (with respect to the $\ell-$shift) equal to $\ell\, s(\omega)$.

Moreover, according to Lemma 7.2.1, for every $\Delta > 0$, if one defines the set of integers $A_{\ell,\Delta} := \{i \in \{1,\ldots,k(\ell)\}\ :\ S(\omega^{(\ell)}_{i,\ell}) \geq \ell(s(\omega) + \Delta)\}$, then these states enjoy the following property with respect to the von Neumann entropy: $\lim_{\ell\to\infty} \dfrac{\#(A_{\ell,\Delta})}{k(\ell)} = 0$.

Let $\mathcal{C}_{i,\ell}$ be the maximal Abelian subalgebra of $\mathcal{A}_\ell$ generated by the one-dimensional eigenprojectors of the density matrices corresponding to $\omega^{(\ell)}_{i,\ell} \in \mathcal{A}_\ell$. The restriction of $\omega_{i,\ell}$ to the Abelian quasi-local algebra $\mathcal{C}^\infty_{i,\ell}$ generated by $\mathcal{C}_{i,\ell}$ is again an ergodic state. From the properties of the entropy density and of the von Neumann entropy one derives the chain of bounds

$$\ell\cdot s(\omega) = s(\omega^{(\ell)}_{i,\ell}) \leq s(\omega^{(\ell)}_{i,\ell}\lceil\mathcal{C}^\infty_{i,\ell}) \leq S(\omega^{(\ell)}_{i,l}\lceil\mathcal{C}_{i,\ell}) = S(\omega^{(\ell)}_{i,\ell})\ .$$

Further, with $\Delta := \frac{R}{\ell} - s(\omega)$, if $i \in A^c_{\ell,\Delta}$ one has the upper bound $S(\omega^{(l)}_{i,l}) < R$.

Let $U_i \in \mathcal{A}_\ell$ be a unitary operator such that $U_i^{\otimes n} p^{(n)}_{\ell,R} U_i^{*\otimes n} \in \mathcal{C}^{(n)}_{i,\ell}$. For every $i \in A^c_{\ell,\Delta}$, it holds that

$$\omega^{(\ell n)}_{i,\ell}(w^{(\ell n)}_{\ell,R}) \geq \omega^{(\ell n)}_{i,\ell}(U_i^{\otimes n} p^{(n)}_{\ell,R} U_i^{*\otimes n}) \longrightarrow 1\ . \tag{7.159}$$

We can thus fix an $\ell \in \mathcal{N}$ large enough to fulfill $\frac{\#(A^c_{\ell,\Delta})}{k(\ell)} \geq 1 - \frac{\delta}{2}$ and use the ergodic decomposition to obtain the lower bound

$$\omega^{(\ell n)}(w^{(\ell n)}_{\ell,R}) \geq \frac{1}{k(\ell)} \sum_{i \in A^c_{\ell,\Delta}} \omega^{(\ell n)}_{\ell,i}(w^{(\ell n)}_{\ell,R}) \geq \left(1 - \frac{\delta}{2}\right) \min_{i \in A^c_{\ell,\Delta}} \omega^{(\ell n)}_{i,\ell}(w^{(\ell n)}_{\ell,R}) .$$

Then (7.159) yields

$$\omega^{(\ell n)}(W^{(\ell n)}_{\ell,R}) \geq \omega^{(\ell n)}(w^{(\ell n)}_{\ell,R}) \geq 1 - \delta .$$

**Step 3.** One can now proceed as in [163] and introduce a sequence of integers $\ell_m$, $m \in \mathcal{N}$, where each $\ell_m$ is a power of 2 fulfilling the inequality

$$\ell_m 2^{3 \cdot \ell_m} \leq m < 2\ell_m 2^{3 \cdot 2\ell_m} . \tag{7.160}$$

Let the integer sequence $n_m$ and the real-valued sequence $R_m$ be defined by $n_m := \lfloor \frac{m}{\ell_m} \rfloor$, respectively $R_m := \ell_m \cdot \left(s + \frac{\varepsilon}{2}\right)$ and set

$$Q^{(m)}_{s,\varepsilon} := \begin{cases} W^{(\ell_m n_m)}_{\ell_m,R_m} & \text{if } m = \ell_m 2^{3 \cdot \ell_m} , \\ W^{(\ell_m n_m)}_{\ell_m,R_m} \otimes \mathrm{id}^{\otimes(m - \ell_m n_m)} & \text{otherwise} . \end{cases} \tag{7.161}$$

Observe that

$$\begin{aligned} \frac{1}{m} \log \mathrm{Tr}\, Q^{(m)}_{s,\varepsilon} &\leq \frac{1}{n_m \ell_m} \log \mathrm{Tr} Q^{(m)}_{s,\varepsilon} \leq \frac{4^{\ell_m}}{\ell_m} \frac{\log(n_m + 1)}{n_m} + \frac{R_m}{\ell_m} + \frac{1}{n_m} \\ &\leq \frac{4^{\ell_m}}{\ell_m} \frac{6\ell_m + 2}{2^{3\ell_m} - 1} + s + \frac{\varepsilon}{2} + \frac{1}{2^{3\ell_m} - 1} , \end{aligned}$$

where the second inequality follows from (7.158) and the last one from the bounds on $n_m$

$$2^{3\ell_m} - 1 \leq \frac{m}{\ell_m} - 1 \leq n_m \leq \frac{m}{\ell_m} \leq 2^{6\ell_m + 1} .$$

Thus, for large $m$, it holds

$$\frac{1}{m} \log \mathrm{Tr} Q^{(m)}_{s,\varepsilon} \leq s + \varepsilon. \tag{7.162}$$

By the special choice (7.160) of $\ell_m$ it is ensured that the sequence of projectors $Q^{(m)}_{s,\varepsilon} \in \mathcal{A}_m$ is indeed typical for any quantum state $\omega$ with entropy rate $s(\omega) \leq s$. This means that $\{Q^{(m)}_{s,\varepsilon}\}_{m \in \mathbf{N}}$ is a sequence of universal typical projectors at level $s$.

### 7.3.2 Quantum Capacities

The $\chi$-quantity in Holevo's bound 6.33 limits the amount of classical information that can be retrieved by a *POVM* measurement from encoding classical symbols $i \in I_A = \{1, 2, \ldots, a\}$ by quantum (mixed) states, $i \mapsto \rho_i$ coming from a mixture $\rho = \sum_{i \in I_A} p_i \rho_i \in \mathbb{B}_1^+(\mathbb{H})$ with a priori probabilities $p_i$. In particular, the bound is in general hardly reachable; however, like in classical capacity theory, the amount of retrievable classical information per transmitted quantum state can be made arbitrarily close to the Holevo bound by means of suitable encodings of longer and longer strings $\boldsymbol{i}^{(n)} = i_1 i_2 \cdots i_n \in I_A^{(n)} := \underbrace{I_A \times \cdots I_A}_{n \text{ times}}$. Also, the Holevo bound is a limit to the classical information per letter that can be encoded into quantum states and retrieved with negligible errors. As we shall see, this state of affairs will lead to different definitions of *quantum capacities.*

In order to prepare the ground for a detailed discussion, appropriate notations must be introduced.

We spectralize $\rho = \sum_{\alpha \in I_\rho} r(\alpha) |\, r(\alpha)\,\rangle\langle\, r(\alpha)\,|$, set $I_\rho^n := \underbrace{I_\rho \times \cdots I_\rho}_{n \text{ times}}$, denote $\boldsymbol{\alpha}^{(n)} = \alpha_1 \alpha_2 \cdots \alpha_n$, $\alpha_i \in I_\rho$, and write

$$r(\boldsymbol{\alpha}^{(n)}) := \prod_{i=1}^n r(\alpha_i) \,, \qquad |\,\boldsymbol{\alpha}^{(n)}\,\rangle := \bigotimes_{j=1}^n |\, r(\alpha_j)\,\rangle \tag{7.163}$$

$$\rho(\boldsymbol{i}^{(n)}) := \rho_{i_1} \otimes \rho_{i_2} \otimes \cdots \rho_{i_n} \,, \qquad P(\boldsymbol{i}^{(n)}) := \prod_{j=1}^n p_{i_j} \tag{7.164}$$

$$\rho^{\otimes n} = \sum_{\boldsymbol{i}^{(n)} \in I_A^{(n)}} P(\boldsymbol{i}^{(n)})\, \rho(\boldsymbol{i}^{(n)}) = \sum_{\boldsymbol{\alpha}^{(n)} \subset I_\rho^n} r(\boldsymbol{\alpha}^{(n)}) |\,\boldsymbol{\alpha}^{(n)}\,\rangle\langle\,\boldsymbol{\alpha}^{(n)}\,| \tag{7.165}$$

On the other hand, the spectral decompositions of the quantum code-words

$$\rho_i = \sum_{k \in J_i} p(k|i)\; |\, p(k|i)\,\rangle\langle\, p(k|i)\,| \,, \tag{7.166}$$

with eigenvalues $0 \le p(k|i) \le 1$ and eigenprojectors $|\, p(k|i)\,\rangle\langle\, p(k|i)\,|$, provide conditional and joint probabilities.

Denote by $A^{(n)}$, respectively $K^{(n)}$, the stochastic variables with outcomes $\boldsymbol{i}^{(n)}$, respectively $\boldsymbol{k}^{(n)} = k_{i_1 k_{i_2}} \cdots k_{i_n} \in I_K^n$, where $I_K^n := \bigcup_{\boldsymbol{i}^{(n)} \in I_A^n} J(\boldsymbol{i}^{(n)})$ with $J(\boldsymbol{i}^{(n)}) := \times_{j=1}^n J_{i_j}$. Finally assign to $A^{(n)}$ and $K^{(n)}$ conditional and joint probabilities defined by $\pi_{K^{(n)}|A^{(n)}} := \{P(\boldsymbol{k}^{(n)}|\boldsymbol{i}^{(n)}\}_{\boldsymbol{i}^{(n)} \in I_A^n, \boldsymbol{k}^{(n)} \in J(\boldsymbol{i}^{(n)})}$, where

$$P(\boldsymbol{k}^{(n)}|\boldsymbol{i}^{(n)}) := \prod_{j=1}^n p(k_j|i_j) \,, \tag{7.167}$$

and by $\pi_{A^{(n)}\vee K^{(n)}} = \{P(\boldsymbol{i}^{(n)}, \boldsymbol{k}^{(n)})\}_{\boldsymbol{i}^{(n)}\in I_A^n, \boldsymbol{k}^{(n)}\in I_K^n}$, where

$$P(\boldsymbol{i}^{(n)}, \boldsymbol{k}^{(n)}) := P(\boldsymbol{i}^{(n)})\, P(\boldsymbol{k}^{(n)}|\boldsymbol{i}^{(n)}) \,. \tag{7.168}$$

Shannon entropies are computed using (7.164) and additivity of von Neumann entropy, as follows:

$$H(A^{(n)}) = n\, H(A) = n \sum_{i=1}^{a} p_i \log_2 p_i \tag{7.169}$$

$$\begin{aligned} H(A^{(n)} \vee K^{(n)}) &= H(A^{(n)}) + \sum_{\boldsymbol{i}^{(n)}\in I_A^n} P(\boldsymbol{i}^{(n)})\, H(K^{(n)}|\boldsymbol{i}^{(n)}) \\ &= H(A^{(n)}) + \sum_{\boldsymbol{i}^{(n)}\in I_A^n} P(\boldsymbol{i}^{(n)})\, S(\rho(\boldsymbol{i}^{(n)})) \\ &= n\Big(H(A) + \sum_{i=1}^{a} p_i\, S(\rho_i)\Big) \,. \end{aligned} \tag{7.170}$$

According to the *AEP* (see Proposition 3.2.2), for any fixed $\varepsilon > 0$, we can distinguish a subset of $\pi_{A^{(n)}}$-typical strings $\boldsymbol{i}^{(n)} \in \mathcal{U}_\varepsilon^{(n)} \subseteq I_A^{(n)}$ and a subset of $\pi_{A^{(n)}\vee K^{(n)}}$-typical strings $(\boldsymbol{i}^{(n)}, \boldsymbol{k}^{(n)}) \in \mathcal{V}_\varepsilon^{(n)} \subseteq I_A^{(n)} \times I_K^{(n)}$. These subsets are such that $\pi_{A^{(n)}}(\mathcal{U}_\varepsilon^{(n)}) \geq 1-\varepsilon$ and $\pi_{A^{(n)}\vee B^{(n)}}(\mathcal{V}_\varepsilon^{(n)}) \geq 1-\varepsilon$. Furthermore, if $\boldsymbol{i}^{(n)} \in \mathcal{U}_\varepsilon^{(n)}$ then

$$2^{-n(H(A)+\varepsilon)} \leq P(\boldsymbol{i}^{(n)}) \leq 2^{-n(H(A)-\varepsilon)} \,, \tag{7.171}$$

while if $(\boldsymbol{i}^{(n)}, \boldsymbol{k}^{(n)}) \in \mathcal{V}_\varepsilon^{(n)}$, then

$$2^{-n\left(H(A)+\sum_{i=1}^{a} p_i S(\rho_i)+\varepsilon\right)} \leq P(\boldsymbol{i}^{(n)}, \boldsymbol{k}^{(n)}) \leq 2^{-n\left(H(A)+\sum_{i=1}^{a} p_i S(\rho_i)-\varepsilon\right)} \,. \tag{7.172}$$

As for classical capacity (see Section 3.2.2), we distinguish one more typical subset, $\mathcal{W}_\varepsilon^{(n)} \subseteq I_A^{(n)} \times I_B^{(n)}$, consisting of all jointly typical pairs, $(\boldsymbol{i}^{(n)}, \boldsymbol{k}^{(n)}) \in \mathcal{V}_\varepsilon^{(n)}$, where also $\boldsymbol{i}^{(n)} \in \mathcal{U}^{(n)}$. From (7.168), (7.171) and (7.172),

$$2^{-n\left(S(\rho)-\chi+2\varepsilon\right)} \leq P(\boldsymbol{k}^{(n)}|\boldsymbol{i}^{(n)}) \leq 2^{-n\left(S(\rho)-\chi-2\varepsilon\right)} \,, \tag{7.173}$$

where (6.33) has been used and $\chi$ is Holevo's $\chi$-quantity for $\rho$ and its decomposition $\rho = \sum_{i=1}^{a} p_i \rho_i$. Finally, in terms of (7.164) and (7.165),

$$\rho(\boldsymbol{i}^{(n)}) = \sum_{\boldsymbol{k}^{(n)}\in J(\boldsymbol{i}^{(n)})} P(\boldsymbol{k}^{(n)}|\boldsymbol{i}^{(n)})\; |\, P(\boldsymbol{k}^{(n)}|\boldsymbol{i}^{(n)})\,\rangle\langle\, P(\boldsymbol{k}^{(n)}|\boldsymbol{i}^{(n)})\,| \tag{7.174}$$

$$\rho^{\otimes n} = \sum_{\substack{\boldsymbol{i}^{(n)}\in I_A^n \\ \boldsymbol{k}^{(n)}\in J(\boldsymbol{i}^{(n)})}} P(\boldsymbol{i}^{(n)}, \boldsymbol{k}^{(n)})\; |\, P(\boldsymbol{k}^{(n)}|\boldsymbol{i}^{(n)})\,\rangle\langle\, P(\boldsymbol{k}^{(n)}|\boldsymbol{i}^{(n)})\,| \,, \tag{7.175}$$

where (see (7.166)) $|P(\boldsymbol{k}^{(n)}|\boldsymbol{i}^{(n)})\rangle := \bigotimes_{j=1}^{n}|p(k_j|i_j)\rangle$.

As showed in the proof of Theorem 7.3.1, for any $\varepsilon > 0$ there exists an orthogonal projector $\Pi_\varepsilon \in \mathbb{B}(\mathbb{H}^{\otimes n})$ that commutes with $\rho^{\otimes n}$ and $\mathrm{Tr}(\rho^{\otimes n}\,\Pi_\varepsilon) \geq 1-\varepsilon$. Analogously, if the sum in (7.175) is restricted to $\mathcal{W}_\varepsilon^{(n)}$, one gets a positive operator $\tilde{\rho}_n \leq \rho^{\otimes n} \in \mathbb{B}_1^+(\mathbb{H}^{\otimes n})$ with

$$\begin{aligned}\mathrm{Tr}\tilde{\rho}_n &= \sum_{(\boldsymbol{i}^{(n)},\boldsymbol{k}^{(n)})\in\mathcal{W}_\varepsilon^{(n)}} P(\boldsymbol{i}^{(n)},\boldsymbol{k}^{(n)}) \\ &= \Bigg(\sum_{(\boldsymbol{i}^{(n)},\boldsymbol{k}^{(n)})\in\mathcal{V}_\varepsilon^{(n)}} - \sum_{\substack{(\boldsymbol{i}^{(n)},\boldsymbol{k}^{(n)})\in\mathcal{V}_\varepsilon^{(n)} \\ \boldsymbol{i}^{(n)}\notin\mathcal{U}_\varepsilon^{(n)}}}\Bigg) P(\boldsymbol{i}^{(n)},\boldsymbol{k}^{(n)}) \geq 1-2\varepsilon\,. \qquad (7.176)\end{aligned}$$

**Theorem 7.3.4.** *[136, 269] Let $\rho = \sum_{i\in I_A} p_i\rho_i \in \mathbb{B}_1^+(\mathbb{H})$ provide a statistical mixture of quantum states available for encoding symbols $i \in I_A$ emitted by a classical source; let $\chi := \chi\left(\rho, \{p_i\rho_i\}_{i\in I_A}\right)$. For any fixed $\delta > 0$ and sufficiently large $n$, there exists an encoding $\boldsymbol{i}^{(n)} \mapsto \mathcal{E}(\boldsymbol{i}^{(n)}) = \rho(\boldsymbol{i}^{(n)})$ on a subset $\tilde{I}_A \subseteq I_A^n$ consisting of $M$ strings and a decoding POVM*

$$\begin{aligned}\mathbb{B}(\mathbb{H}^{\otimes n}) \supset \mathcal{B}^{(n)} := \Bigg\{\Big\{|\Psi(\boldsymbol{k}^{(n)}|\boldsymbol{i}^{(n)})\rangle\langle\Psi(\boldsymbol{k}^{(n)}|\boldsymbol{i}^{(n)})|\Big\}_{\substack{\boldsymbol{i}^{(n)}\in\tilde{I}_A \\ (\boldsymbol{i}^{(n)},\boldsymbol{k}^{(n)})\in\mathcal{W}_\varepsilon^{(n)}}} \\ \cup\ \mathbb{1} - \sum_{\substack{\boldsymbol{i}^{(n)}\in\tilde{I}_A \\ (\boldsymbol{i}^{(n)},\boldsymbol{k}^{(n)})\in\mathcal{W}_\varepsilon^{(n)}}} |\Psi(\boldsymbol{k}^{(n)}|\boldsymbol{i}^{(n)})\rangle\langle\Psi(\boldsymbol{k}^{(n)}|\boldsymbol{i}^{(n)})|\Bigg\}\,,\end{aligned}$$

*such that* $\left|\frac{M}{n} - \chi\right| \leq \delta$ *and* $\boldsymbol{e}_n \leq \delta$, *where* $\boldsymbol{e}_n$ *is the decoding error*

$$\boldsymbol{e}_n := 1 - \frac{1}{M}\sum_{\substack{\boldsymbol{i}^{(n)}\in\tilde{I}_A \\ \boldsymbol{k}^{(n)}\in J(\boldsymbol{i}^{(n)})}} P(\boldsymbol{k}^{(n)}|\boldsymbol{i}^{(n)})\,\Big|\langle\Psi(\boldsymbol{k}^{(n)}|\boldsymbol{i}^{(n)})\,|\,P(\boldsymbol{k}^{(n)}|\boldsymbol{i}^{(n)})\rangle\Big|^2\,. \qquad (7.177)$$

The strategy of the proof consists of the following steps.

1. choose an equidistributed set $\tilde{I}_A \subseteq I_A^n$ of $M$ sequences $\boldsymbol{i}^{(n)}$ and to encode each of them by a density matrix $\mathcal{E}(\boldsymbol{i}^{(n)}) = \rho(\boldsymbol{i}^{(n)})$. The encoding thus provides the density matrix

$$\begin{aligned}\rho_{\mathcal{E}}^{(n)} &:= \frac{1}{M}\sum_{\boldsymbol{i}^{(n)}\in\tilde{I}_A} \mathcal{E}(\boldsymbol{i}^{(n)}) \\ &= \frac{1}{M}\sum_{\substack{\boldsymbol{i}^{(n)}\in\tilde{I}_A \\ \boldsymbol{k}^{(n)}\in J(\boldsymbol{i}^{(n)})}} P(\boldsymbol{k}^{(n)}|\boldsymbol{i}^{(n)})\,|P(\boldsymbol{k}^{(n)}|\boldsymbol{i}^{(n)})\rangle\langle P(\boldsymbol{k}^{(n)}|\boldsymbol{i}^{(n)})|\,. \qquad (7.178)\end{aligned}$$

2. Choose random encodings $\mathcal{E}$ as in the proof of Theorem 3.2.3: the argument does ensure that the required encoding and decoding procedures exist, but does not provide concrete instances of them (for a similar result see [144, 146]).
3. As regards the decoding protocol, the idea is to try to identify by means of (possibly of norm less than 1) vectors $|\,\Psi(\boldsymbol{k}^{(n)}|\boldsymbol{i}^{(n)})\,\rangle$ only those $|\,P(\boldsymbol{k}^{(n)}|\boldsymbol{i}^{(n)})\,\rangle$ in (7.178) which are labeled by pairs $(\boldsymbol{i}^{(n)},\boldsymbol{k}^{(n)}) \in \mathcal{W}_\varepsilon^{(n)}$ after the same have been projected by $\Pi_\varepsilon$ onto the chosen high probability subspace of $\rho^{(n)}$. All other $|\,P(\boldsymbol{k}^{(n)}|\boldsymbol{i}^{(n)})\,\rangle$ will be made correspond to $|\,\Psi(\boldsymbol{k}^{(n)}|\boldsymbol{i}^{(n)})\,\rangle = 0$.
4. The non-trivial decoding vectors are constructed as follows. (For sake of simplicity we shall denote multi-indices $\boldsymbol{i}^{(n)}$, $\boldsymbol{k}^{(n)}$ as $i$ and $k$. Set $|\,\Phi(i,k)\,\rangle := \Pi_\varepsilon|\,P(i,k)\,\rangle$, consider the matrix $S$ with entries

$$S_{(i,k),(\overline{i},\overline{k})} := \langle\,\Phi(i,k)\,|\,\Phi(\overline{i},\overline{k})\,\rangle = \langle\,P(k|i)\,|\,\Pi_\varepsilon\,|P(\overline{k}|\overline{i})\,\rangle\ , \tag{7.179}$$

where $i,\overline{i} \in \tilde{I}_A$, $k,\overline{k} \in J(i), J(\overline{i})$ and $(i,k),(\overline{i},\overline{k}) \in \mathcal{W}_\varepsilon^{(n)}$. This matrix is positive and its square root defines vectors $|\,\Psi(i,k)\,\rangle$ such that [136]

$$\sqrt{S}_{(i,k),(\overline{i},\overline{k})} =: \langle\,\Psi(k|i)\,|\,\Phi(\overline{k}|\overline{i})\,\rangle = \langle\,\Psi(k|i)\,|\,P(\overline{k}|\overline{i})\,\rangle\ .$$

Namely, $Q^{-1/2}|\,\Phi(k|i)\,\rangle$ where $Q^{-1/2}$ is the (positive) inverse square-root of $Q := \sum^*_{(i,k)} |\,\Phi(k|i)\,\rangle\langle\,\Phi(k|i)\,|$ (defined only on the range of $Q$), where $\sum^*_{(i,k)}$ denotes the sum restricted to the pairs $(i,k) \in \mathcal{W}_\varepsilon^{(n)}$.
5. The error (7.177) is complementary to the ensemble fidelity (see Definition 6.3.6) that has been used in Theorem 7.3.1. Using the previous definitions and that $Q^{-1/2} \geq 0$, it can be bounded as follows:

$$\begin{aligned} e_n &\leq \frac{2}{M} \sum_{\substack{i\in\tilde{I}_A\\ k\in J(i)}} P(k|i)\Big(1 - \langle\,\Psi(k|i)\,|\,P(k|i)\,\rangle\Big) \\ &\leq 2 - \frac{2}{M} \sum_{i\in\tilde{I}_A} \sum^*_{(i,k)} P(k|i)\,\sqrt{S}_{(i,k);(i,k)}\ . \end{aligned} \tag{7.180}$$

**Proof of Theorem 7.3.4** Since $2(x-\sqrt{x}) \leq x - x^2$ for all $x \geq 0$, the same inequality holds by substituting $x$ with the positive matrix $S$ in (7.179); taking diagonal values:

$$\sqrt{S}_{(i,k);(i,k)} \geq \frac{3}{2}\,S_{(i,k);(i,k)} - \frac{1}{2}\sum^*_{(j,k)} S_{(i,k);(j,\ell)}\,S_{(j,\ell);(i,k)}\ .$$

Then, using the first inequality in (7.173), the bound (7.180) can conveniently be recast as follows

$$\begin{aligned} e_n \leq 2 &- \frac{3}{M} \sum_{i \in \tilde{I}_A} \sum_{(i,k)}^{*} P(k|i)\, S_{(i,k);(i,k)} \\ &+ \frac{1}{M} \sum_{i \in \tilde{I}_A} \sum_{(i,k)}^{*} P(k|i)\, S^2_{(i,k);(j,\ell)} \\ &+ \frac{2^{n(S(\rho)-\chi+2\varepsilon)}}{M} \sum_{i,j \in \tilde{I}_A} \sum_{(i,k)\neq(j,\ell)}^{*} P(k|i)\, P(\ell|j)\, S_{(i,k);(j,\ell)}\, S_{(j,\ell);(i,k)} \,. \end{aligned}$$

Suppose now to choose the $M$ words $\boldsymbol{i}^{(n)} \in \tilde{I}_A$ randomly according to the probability distribution $P(i)$; we obtain in this way a statistical ensemble of random codes and, as much as in the classical case, by averaging over the contributions of the randomly chosen $\boldsymbol{i}^{(n)}$ one eliminates the dependence on $\tilde{I}_A$ and remains with a sum over all $\boldsymbol{i}^{(n)} \in I^n_A$. Therefore, the average error can be estimated from above by

$$e^{av}_n \leq 2 - 3 \underbrace{\sum_{i \in I_A^{(n)}} \sum_{(i,k)}^{*} P(i)\, P(k|i)\, S_{(i,k);(i,k)}}_{L_1} \tag{7.181}$$

$$+ \underbrace{\sum_{i \in I_A^{(n)}} \sum_{(i,k)}^{*} P(i)\, P(k|i)\, S^2_{(i,k);(i,k)}}_{L_{2a}} \tag{7.182}$$

$$\begin{aligned} &+ \frac{M(M-1)}{M}\, 2^{n(S(\rho)-\chi+2\varepsilon)} \\ &\times \underbrace{\sum_{i,j \in I_A^{(n)}} \sum_{(i,k)\neq(j,\ell)}^{*} P(i)\, P(j)\, P(k|i)\, P(\ell|j)\, S_{(i,k);(j,\ell)}\, S_{(j,\ell);(i,k)}}_{L_{2b}} \end{aligned} \tag{7.183}$$

From (7.179) and (7.176),

$$L_1 = \mathrm{Tr}\Big(\tilde{\rho}_n \Pi_\varepsilon\Big) = \mathrm{Tr}\Big(\rho^{\otimes n} \Pi_\varepsilon\Big) - \mathrm{Tr}\Big((\tilde{\rho}_n - \rho^{\otimes n})\Pi_\varepsilon\Big) \geq 1 - 3\varepsilon \quad (1) \,.$$

Further, $S_{(i,k);(i,k)} \leq 1$, thus $L_{2a} \leq 1$ $(2a)$. Finally, the last sum can be bounded from above by observing that if $0 \leq A \leq B$ and $0 \leq C \leq D$, then by means of cyclicity under the trace operation,

$$\mathrm{Tr}(AC) = \mathrm{Tr}(\sqrt{A} C \sqrt{A}) \leq \mathrm{Tr}(AD) = \mathrm{Tr}(\sqrt{D} A \sqrt{D}) \leq \mathrm{Tr}(BD) \,.$$

Therefore, since $\Pi_\varepsilon$ commutes with $\rho^{\otimes n}$, $L_{2b} \leq \mathrm{Tr}\Big(\Pi_\varepsilon (\rho^{\otimes n})^2\Big)$; on the other hand, from the quantum *AEP* we know that the dimension of the subspace projected out by $\Pi_\varepsilon$ is $\leq 2^{n(S(\rho)+\varepsilon)}$ with eigenvalues $\leq 2^{-n(S(\rho)-\varepsilon)}$, whence $L_{2b} \leq 2^{-n(S(\rho)-3\varepsilon)}$ (2*b*). Altogether, inequalities (1), (2*a*) and (2*b*) yield

$$e_n^{av} \leq 2 - 3(1-3\varepsilon) + 1 + M\,2^{-n(\chi-5\varepsilon)} = 9\varepsilon + 2^{-n(\chi-R-5\varepsilon)} ,$$

where we have put the growth rate $R$ into evidence $M = 2^{n\,R}$. The latter can be chosen arbitrarily close to the Holevo $\chi$ quantity and still the average error becomes negligible with $n \to \infty$. Therefore, for any $\delta \geq 0$ and $n$ large enough there is an $\tilde{I}_A \subseteq I_A^n$ with $R \geq \chi - \delta$ and $\boldsymbol{e}_n \leq \delta$. □

**Example 7.3.2.** Suppose the sender encodes classical symbols $i \in I$ into states that she obtains by acting locally with unitary operators $U_i$ on her system in a state $\rho_{12} \in M_{d_1}(\mathbb{C}) \otimes M_{d_2}(\mathbb{C})$ which she shares with the receiver. She selects the unitary operators $U_i$ with probabilities $p_i$, and after changing $\rho_{12}$ into $\rho_i = U_i \otimes \mathbb{1}_2\, \rho_{12}\, U_i^\dagger \otimes \mathbb{1}$ she sends her system to the receiver. The sender tries to maximize the information accessible to the receiver by optimizing the Holevo bound, thus seeking [71]

$$C_M := \max_{p_i, U_i} \left\{ S(\rho) - \sum_i p_i S(\rho_i) \right\} , \quad \rho = \sum_i p_i\, \rho_i .$$

Note that $S(\rho_i) = S(\rho_{12})$ for unitary transformations do not change the von Neumann entropy; in order to maximize $S(\rho)$, consider the marginal states

$$\rho^{(1)} = \mathrm{Tr}_2(\rho) , \quad \rho_1^{(2)} = \mathrm{Tr}_1(\rho) = \rho_2 \, (= \mathrm{Tr}_1(\rho_{12})) .$$

By subadditivity (5.160) and (5.155),

$$C_M \leq S\left(\rho^{(1)}\right) + S(\rho_2) - S(\rho_{12}) \leq \log d_1 + S(\rho_2) - S(\rho_{12}) .$$

Choose as unitary operators the $d_1^2$ Weyl operators $W_{d_1}(\boldsymbol{n})$ of Example 5.4.2 with equal probabilities $1/d_1^2$; then, using (5.88) and (5.30), one gets

$$\rho = \frac{1}{d_1^2} \sum_{\boldsymbol{n}=(n_1,n_2)} W_{d_1}(\boldsymbol{n}) \otimes \mathbb{1}_2\, \rho_{12}\, W_{d_1}(\boldsymbol{n}) \otimes \mathbb{1}_2 = \frac{\mathbb{1}_1}{d_1} \otimes \rho_2 ,$$

so that $C_M \geq \log d_1 + S(\rho_2) - S(\rho_{12})$. This transmission protocol can thus achieve an optimal quantum transmission rate $C_M = \log d_1 + S(\rho_2) - S(\rho_{12})$.

In general, like in classical transmission, the quantum states that have been used to code and transmit information are subjected to perturbing effects of the transmission channel which is being used. Concretely, if the the quantum code-words are projections $P_i \in M_d(\mathbb{C})$ chosen with probabilities $p_i$, thus making a statistical ensemble described by the density matrix

$\rho = \sum_i p_i P_i$, a probability preserving channel acts on them as a trace-preserving *CP* map $\Lambda$. In the light of the previous theorem, the *channel capacity* is defined by [145, 276]

$$C_M[\Lambda] := \max_{p_i, P_i} \left\{ S(\Lambda[\rho]) - \sum_i p_i S(\Lambda[P_i]) \right\} .$$

As much as for the entanglement cost (see (6.8)), in order to improve the capacity, one may consider $n$ uses of the channel, thus a *CP* map $\Lambda^{\otimes n}$ acting on states on $M_d(\mathbb{C})^{\otimes n}$ which may carry entanglement between different uses. Then one introduces the *regularized capacity*

$$C_\infty[\Lambda] := \lim_{n\to+\infty} \frac{1}{n} C_M[\Lambda^{\otimes n}] .$$

Such a limit exists because the capacity is superadditive; indeed, consider $C_M[\Lambda_1 \otimes \Lambda_2]$ and two statistical ensembles $\{p_j^{(1)}, P_i^{(1)}\}$ and $\{p_j^{(1)}, P_i^{(1)}\}$ that achieve $C_M[\Lambda_1]$, respectively $C_M[\Lambda_2]$. The additivity of the von Neumann entropy over tensor products states implies that, for the not necessarily optimal statistical ensemble, $\{p_i^{(1)} p_j^{(2)}, P_i^{(1)} \otimes P_j^{(2)}\}$,

$$\begin{aligned} C_M[\Lambda_1 \otimes \Lambda_2] &\geq S\left(\Lambda_1[\rho^{(1)}]\right) - \sum_i p_i^{(1)} S\left(\Lambda_1[P_i^{(1)}]\right) \\ &\quad + S\left(\Lambda_2[\rho^{(2)}]\right) - \sum_i p_i^{(2)} S\left(\Lambda_2[P_i^{(2)}]\right) = C_M[\Lambda_1] + C_M[\Lambda_2] . \end{aligned}$$

Were the capacity additive, the regularized capacity would coincide with $C_M[\Lambda]$. This is another important open question in quantum information which is actually equivalent to the additivity of the entanglement of formation [276] (see also [43] for an approach to this problem based on the relations between the entanglement of formation and the the entropy of a subalgebra.)

## Bibliographical Notes

A most exhaustive and complete review of the algebraic approach to quantum statistical mechanics is provided by [64]; in [65] one finds plenty of applications to spin and continuous systems. A fully developed mathematical theory of the canonical commutation and anti-commutation relations, quasi-free states and quasi-free automorphisms can be found in [237, 200, 201, 202, 255, 256, 257].

Quantum ergodicity and mixing are presented in [260, 277, 300, 107, 64] in increasing order of mathematical sophistication; the second reference has provided most of the material of this book concerning these topics, the first one concerning mixing. In [64] one also finds a detailed discussion of decomposition theory, while in [300] more recent developments are presented.

In [215, 221] what has been called mixing in this book is termed clustering, the qualification mixing being assigned to a stronger clustering behavior which is discussed in connection with Galilei-invariant interactions [218]. In [107, 300] one finds enlightening discussions about the physical meaning of the different algebraic factor types and of Tomita-Takesaki modular theory.

Applications of quantum mechanics with infinite degrees of freedom to collective phenomena and thermodynamics can be found in [274], while in [290, 291] the emphasis is more on symmetry breaking phenomena and on the existence of inequivalent representations of the *CCR* and *CCR* with applications to physically relevant models.

Quantum information related issues involving infinitely many degrees of freedom and the necessary mathematical tools like quantum compression theorems and quantum capacities are discussed in [145, 239, 250]. For a review of different formulation of quantum capacity related quantities and their relations see [182].

# Part III

# Quantum Dynamical Entropies and Complexities

The last part of the book first deals with two extensions of the Kolmogorov dynamical entropy to quantum systems and with their applications. Then, it discusses some recent generalizations to quantum systems of classical algorithmic complexity.

# 8 Quantum Dynamical Entropies

The first part of this book has been devoted to illustrate some of the many properties of the classical dynamical entropy of Kolmogorov and Sinai; in particular, it has been showed that it provides the optimal compression rate of ergodic sources (Shannon-Mc Millan-Breiman Theorem 3.2.1), while, through the positive Lyapounov exponents (Pesin's Theorem), it measures the dynamical instability of classical dynamical systems; finally, it gives the complexity rate of almost all trajectories of ergodic dynamical systems (Brudno's theorem 4.2.1).

Several extensions of the *KS* entropy to quantum dynamical systems can be found in the mathematical and physical literature (see the bibliographical notes at the end of this chapter). All of them predated or were developed independently of quantum information; due to its rapid growth, the latter more and more appears as an ideal ground for testing the physical meaning and the technical usefulness of these proposals.

One of the aims behind the attempts at defining quantum dynamical entropies was the possibility of classifying quantum dynamical systems, as much as it had been done for classical dynamical systems by means of the *KS* entropy (see Remark 3.1.2). Afterwards, the quantum dynamical entropies have been applied to the study of quantum chaotic phenomena and the quantum/classical correspondence; recently, they have been used to shed light on certain foundational aspects of quantum information, like quantum capacity and *quantum algorithmic complexity*.

Of the various quantum dynamical entropies that have been proposed in recent years, we shall mainly focus upon two of them; namely, the entropy of Connes, Narnhofer and Thirring [88] (*CNT* entropy) and of Alicki and Fannes [9] (*AFL* entropy) [1]. These quantum dynamical entropies embody two radically different ways of approaching the notion of information production in quantum mechanics; indeed, they may behave differently on a same quantum dynamical system.

In Section 2.4, we have seen that partitions of the phase-space into finitely many, disjoint measurable atoms provide classical dynamical systems $(\mathcal{X}, T, \mu)$ (see Definition 2.2.2) with symbolic models: finite measurable partitions $\mathcal{P} = \{P_i\}_{i \in I}$ can be used to successively localize the moving phase-point

[1]The $L$ in *AFL* stands for Lindblad who introduced the notion in [194, 195, 196].

F. Benatti, *Dynamics, Information and Complexity in Quantum Systems*, Theoretical and Mathematical Physics, DOI 10.1007/978-1-4020-9306-7_8,

within their atoms $P_i$ and to quantify the predictability of the dynamics via the information relative to the next time-step that is gained by observing the evolving system.

In Chapter 3, partitions have been interpreted as *POVMs* taken from a commutative dynamical triple $(\mathbb{L}^\infty_\mu(\mathcal{X}), \Theta_T, \omega_\mu)$, where atoms have been identified with their characteristic functions and thus with orthogonal projections summing up to the identity (see Definition 2.2.3.2). Thus, partitions $\mathcal{P}$ define partitions of unit (see Definition 5.6.1) and *CPU* maps $\mathbb{E}_\mathcal{X}$ on the $C^*$-algebra $\mathbb{B}(\mathbb{L}^2_\mu(\mathcal{X}))$. However, because of commutativity, the action of $\mathbb{E}$ on $\mathbb{L}^\infty_\mu(\mathcal{X})$ reduces to the identity map; indeed, for all $f \in \mathbb{L}^\infty_\mu(\mathcal{X})$,

$$\mathbb{E}_\mathcal{X}[f](x) = \sum_{i\in I} \chi_{P_i}(x)\, f(x)\, \chi_{P_i}(x) = \sum_{i\in I} \chi_{P_i}(x)\, f(x) = f(x) \ .$$

The dynamics is thus insensitive to measurements, $\Theta_T \circ \mathbb{E}_\mathcal{X} = \mathbb{E}_\mathcal{X} \circ \Theta_T = \Theta_T$, as well as the states on $\mathbb{L}^\infty_\mu(\mathcal{X})$: $\mathbb{F}_\mathcal{X}[\omega_\mu] = \omega_\mu$, where $\mathbb{F}_\mathcal{X}$ is the dual *CP* map such that $\mathbb{F}_\mathcal{X}[\omega_\mu](f) = \omega_\mu(\mathbb{E}_\mathcal{X}[f])$.

Given a quantum dynamical triplet $(\mathcal{A}, \Theta, \omega)$, if one wants to extend the notion of partition to such a non-commutative context, a natural step is to substitute commuting projections with non-commuting ones or, more in general, with non-projective *POVMs* . Differently from the classical case, the *CPU* maps $\mathbb{E} : \mathcal{A} \mapsto \mathcal{A}$ associated with them do not in general commute with the quantum dynamics, $\Theta \circ \mathbb{E} \neq \Theta \neq \mathbb{E} \circ \Theta$, nor do the dual maps $\mathbb{F}$ preserve the quantum state, $\mathbb{F}[\omega] \neq \omega$. Both $\mathbb{E}$ and $\mathbb{F}$ act as external perturbations; therefore, a preliminary question arises whether one should or not incorporate measurement processes into the very construction of quantum dynamical entropies.

If the answer is **yes**, then, beside the dynamics itself, measurement processes themselves may act as a source of randomness; on the other hand, if the answer is **no**, the regular and irregular features of the dynamics refer to the system only, but are insensitive to the typical quantum phenomenon that getting information about quantum systems in general perturbs them. In other words, a perturbation-free quantifier of quantum dynamical randomness might not measure the actual information production that always comes from observations of the time-evolving system; on the other hand, a quantifier of quantum dynamical randomness that takes into account acquisition of information through measurement processes would add the randomness coming from the latter ones to that proper to the quantum dynamics itself.

Purpose of this and the last chapter is to convey the idea that, unlike in classical dynamics, randomness in quantum dynamics has more than one facet and that choosing one of the two answers above just means exploring two of these inequivalent aspects.

## 8.1 *CNT* Entropy: Decompositions of States

The non-commutative algebraic structure which more closely resembles a commutative one is that of type $II_1$ factor von Neumann algebras $\mathcal{A}$ (see point 2 after Definition 7.0.8). Indeed, the state $\omega$ which makes $\mathcal{A}$ a type $II_1$ factor is a normalized trace such $\omega(X\,Y) = \omega(Y\,X)$ for all $X, Y \in \mathcal{A}$. The *CNT* entropy [88] generalizes to generic von Neumann algebras previous extensions of the *KS* entropy to type $II_1$ factors that were based on the above commutativity with respect to the state [107, 89].

The *CNT* entropy quantifies the information rate in quantum dynamical systems described by algebraic triplets $(\mathcal{A}, \Theta, \omega)$ and it does it independently of external measurement processes, by relying only on the algebraic properties of $\mathcal{A}$, $\Theta$ and $\omega$. The basic idea of the whole construction is a clever use of the relation between entropy and relative entropy that has been discussed in Section 6.3.1 in relation to the entropy of a subalgebra (more in general of a *CPU* map: see Definitions 6.3.2 and 6.3.3). Before getting to that, we indicate why, in general, the steps that in Section 3.1 led to the *KS* entropy are not practicable in a quantum setting.

Suppose $\boldsymbol{M} \subset \mathcal{A}$ is a finite-dimensional subalgebra; if $(\mathcal{A}, \Theta, \omega)$ is a classical dynamical triplet, the *KS* entropy is constructed by considering

- the finite partition corresponding to the subalgebra $\boldsymbol{M}$;
- the partition $\boldsymbol{M}^{(n)} = \bigvee_{j=0}^{n-1} \Theta^k(\boldsymbol{M})$ generated by the time-evolved partitions $\{\Theta^k(\boldsymbol{M})\}_{k=0}^{n-1}$;
- the Shannon entropy of the state $\omega$ restricted to $\boldsymbol{M}^{(n)}$: $H(\omega\restriction\boldsymbol{M}^{(n)})$,
- the asymptotic rate $\lim_{n\to\infty} \frac{1}{n} H(\omega\restriction\boldsymbol{M}^{(n)})$.

In the non-commutative context, by sheer analogy one might consider

- any finite-dimensional subalgebra $\boldsymbol{M} \subseteq \mathcal{A}$;
- the finite-dimensional subalgebras $\boldsymbol{M}^{(n)} := \bigvee_{k=0}^{n-1} \Theta^k(\boldsymbol{M})$ generated by the $n$ (finite-dimensional) subalgebras $\Theta^k(\boldsymbol{M})_{k=0}^{n-1}$;
- the von Neumann entropy of the restricted state $\omega\restriction\boldsymbol{M}^{(n)}$: $S\left(\omega\restriction\boldsymbol{M}^{(n)}\right)$;
- the asymptotic rate $\lim_{n\to\infty} \frac{1}{n} S\left(\omega\restriction\boldsymbol{M}^{(n)}\right)$.

However, this argument generally fails and the reason why it does fail is non-commutativity [301]: despite being finite-dimensional, the subalgebras at different times $\Theta^k(\boldsymbol{M})$ need not commute and can thus generate an infinite-dimensional subalgebra $\boldsymbol{M}^{(n)}$ so that $S\left(\omega\restriction\boldsymbol{M}^{(n)}\right)$ cannot in general be controlled.

In order to overcome these difficulties, the idea is to extend the entropy of a *CPU* map $\mathrm{H}_\rho(\gamma)$ (see Definition 6.3.3) to the entropy $\mathrm{H}_\omega(\gamma_1, \gamma_2, \ldots, \gamma_n)$ of $n$ *CPU* maps $\gamma_i : \boldsymbol{M}_i \mapsto \mathcal{A}$ from finite-dimensional $C^*$ algebras (that we shall always suppose with identity) $\boldsymbol{M}_i$ into $\mathcal{A}$.

**Remark 8.1.1.** As in Section 6.3.1, when $\boldsymbol{M}$ is a subalgebra of $\mathcal{A}$, then one chooses as *CPU* map $\gamma$ the natural embedding $\iota_{\boldsymbol{M}}$ of $\boldsymbol{M}$ into $\mathcal{A}$. Therefore, when dealing with the set of subalgebras $\{\Theta^j(\boldsymbol{M})\}_{j=0}^{n-1}$, the $n$ *CPU* maps are given by $\gamma_j := \Theta^j \circ \iota_{\boldsymbol{M}}$.

Like in the case of $\mathrm{H}_\rho(\gamma)$, we shall consider linear convex decompositions of the state $\omega$ in terms of states $\omega_{\boldsymbol{i}^{(n)}}$. These states will now be indexed by strings $\boldsymbol{i}^{(n)} = i_1 i_2 \cdots i_n$, $i_j \in I_j$, each *CPU* map $\gamma_j$ being associated with a generic index set $I_j$, carrying a total weight $\lambda(\boldsymbol{i}^{(n)})$. Fixing $i_j \in I_j$, after summing over the all other indices and after renormalization, one obtains auxiliary decompositions of $\omega$ associated to each $\gamma_j$. Concretely, from the multi-index decomposition

$$\omega = \sum_{\boldsymbol{i}^{(n)} \in I^{(n)}} \lambda_{\boldsymbol{i}^{(n)}} \, \omega_{\boldsymbol{i}^{(n)}} \ , \qquad I^{(n)} := I_1 \times I_2 \times \cdots I_n \ , \tag{8.1}$$

one obtains subdecompositions $\omega = \sum_{i_j \in I_j} \lambda^j_{i_j} \, \omega^j_{i_j}$, $j = 1, 2, \ldots, n$, where

$$\omega^j_{i_j} := \sum_{\substack{\boldsymbol{i}^{(n)} \\ i_j \text{ fixed}}} \frac{\lambda_{\boldsymbol{i}^{(n)}}}{\lambda^j_{i_j}} \, \omega_{\boldsymbol{i}^{(n)}} \ , \qquad \lambda^j_{i_j} := \sum_{\substack{\boldsymbol{i}^{(n)} \\ i_j \text{ fixed}}} \lambda_{\boldsymbol{i}^{(n)}} \ . \tag{8.2}$$

Let $\Lambda^{(n)} := \left\{ \lambda^{(n)}_{\boldsymbol{i}^{(n)}} \right\}_{\boldsymbol{i}^{(n)} \in I^{(n)}}$ be the probability distribution associated with the weights in (8.1) and $\Lambda_j := \left\{ \lambda^j_{i_j} \right\}_{i_j \in I_j}$ the marginal probability distributions consisting of the weights in (8.2). The generalization of (6.3.3) is as follows.

**Definition 8.1.1 (*n-CPU* Entropies).** *Given a $C^*$ algebra $\mathcal{A}$ equipped with a state $\omega$, let $\gamma_i : \boldsymbol{M}_i \subset \mathcal{A}$, $i = 1, 2, \ldots, n$, be CPU maps from finite-dimensional $C^*$ algebras into $\mathcal{A}$. Their entropy with respect to $\omega$ is:*

$$\begin{aligned}
\mathrm{H}_\omega(\gamma_1, \gamma_2, \ldots, \gamma_n) := \sup_{\omega = \sum_{\boldsymbol{i}^{(n)}} \lambda_{\boldsymbol{i}^{(n)}} \omega_{\boldsymbol{i}^{(n)}}} & \left\{ H(\Lambda^{(n)}) - \sum_{j=1}^n H(\Lambda_j) \right. \\
& \left. + \sum_{j=1}^n \sum_{i_j \in I_j} \lambda^j_{i_j} \, S\left( \omega^j_{i_j} \restriction \gamma_j \, , \, \omega \restriction \gamma_j \right) \right\}
\end{aligned} \tag{8.3}$$

$$\begin{aligned}
= \sup_{\omega = \sum_{\boldsymbol{i}^{(n)}} \lambda_{\boldsymbol{i}^{(n)}} \omega_{\boldsymbol{i}^{(n)}}} & \left\{ H(\Lambda^{(n)}) - \sum_{j=1}^n H(\Lambda_j) \right. \\
& \left. + \sum_{j=1}^n \left( S(\omega \restriction \boldsymbol{M}_j) - \sum_{i_j \in I_j} \lambda^j_{i_j} S\left( \omega^j_{i_j} \restriction \gamma_j \right) \right) \right\} \ ,
\end{aligned} \tag{8.4}$$

*where, with* $\eta(x) := -x\log x$, $x \in [0,1]$,

$$H(\Lambda^{(n)}) = \sum_{\boldsymbol{i}^{(n)}} \eta(\lambda_{\boldsymbol{i}^{(n)}}) \ , \quad H(\Lambda_j) = \sum_{i_j \in I_j} \eta(\lambda^j_{i_j}) \ .$$

As in Section 6.3.1, a concrete way to construct decompositions of $\omega$ is to use the *GNS* construction based on the state $\omega$; it follows that the states $\omega_{\boldsymbol{i}^{(n)}}$ contributing to $\omega = \sum_{\boldsymbol{i}^{(n)}} \lambda_{\boldsymbol{i}^{(n)}} \omega_{\boldsymbol{i}^{(n)}}$ are in one-to-one correspondence with the positive elements of the commutant of $\pi_\omega(\mathcal{A})$:

$$\lambda_{\boldsymbol{i}^{(n)}} \omega_{\boldsymbol{i}^{(n)}}(X) = \langle \Omega_\omega \,|\, Y'_{\boldsymbol{i}^{(n)}} \pi_\omega(X) \,|\Omega_\omega \rangle \ , \tag{8.5}$$

where $0 \leq Y'_{\boldsymbol{i}^{(n)}} \in \pi_\omega(\mathcal{A})'$ and $\sum_{\boldsymbol{i}^{(n)}} Y'_{\boldsymbol{i}^{(n)}} = \mathbb{1}$. Also, if $\omega$ is faithful, one can express the decomposing states in terms of $0 \leq X \in \pi_\omega(\mathcal{A})''$ by means of the modular automorphism $\sigma_\omega$ (see (5.181) in Remark 5.6.1.3):

$$\lambda_{\boldsymbol{i}^{(n)}} \omega_{\boldsymbol{i}^{(n)}}(X) = \langle \Omega_\omega \,|\, \sigma_\omega^{i/2}\Big(\pi_\omega(Y_{\boldsymbol{i}^{(n)}})\Big) \pi_\omega(X) \,|\Omega_\omega \rangle \ , \tag{8.6}$$

where $0 \leq Y_{\boldsymbol{i}^{(n)}} \in \mathcal{A}$ and $\sum_{\boldsymbol{i}^{(n)}} Y_{\boldsymbol{i}^{(n)}} = \mathbb{1}$.

In analogy with (6.47), we shall denote by

$$H_\omega^{\{\gamma_j\}_{j=1}^n}\Big(\{\lambda_{\boldsymbol{i}^{(n)}}, \omega_{\boldsymbol{i}^{(n)}}\}\Big) := H(\Lambda^{(n)}) - \sum_{j=1}^n H(\Lambda_j)$$
$$+ \sum_{j=1}^n \sum_{i_j \in I_j} \lambda^j_{i_j} \, S\left(\omega^j_{i_j} \restriction \gamma_j \, , \, \omega \restriction \gamma_j\right) \ , \tag{8.7}$$

the contribution to the $n$-subalgebra entropy coming from a chosen decomposition of the state $\omega$.

The $n$-*CPU* entropies enjoy a number of very useful properties that can luckily be proved without being obliged to know the optimal decompositions; the first one of these is a generalization of Proposition 6.3.4.

**Proposition 8.1.1.** *Given a* $C^*$ *algebra* $\mathcal{A}$, *a state* $\omega$ *on it and CPU maps* $\gamma_j : \boldsymbol{M}_j \mapsto \mathcal{A}$, $j = 1, 2, \ldots, n$, *from finite-dimensional* $C^*$ *algebras* $\boldsymbol{M}_j$ *(*$\dim \boldsymbol{M}_j \leq d$*) into* $\mathcal{A}$, *consider a decomposition* $\omega = \sum_{\boldsymbol{i}^{(n)} \in I^{(n)}} \lambda^{(n)}_{\boldsymbol{i}^{(n)}} \omega_{\boldsymbol{i}^{(n)}}$ *and* $\varepsilon > 0$. *Then, there exists a decomposition* $\omega = \sum_{\boldsymbol{j} \in J^{(n)}} \lambda'^{(n)}_{\boldsymbol{j}^{(n)}} \omega'_{\boldsymbol{j}^{(n)}}$ *where* $J^{(n)} := J_1 \times J_2 \times \cdots J_n$ *is a multi-index set with* card$(J^{(n)})$ *depends on d and* $\varepsilon$ *and* $\boldsymbol{j}^{(n)} := j_1 j_2 \cdots j_n$, $j_k \in J_k$, *such that*

$$H_\omega^{\{\gamma_j\}_{j=1}^n}\left(\left\{\lambda^{(n)}_{\boldsymbol{i}^{(n)}}, \omega_{\boldsymbol{i}^{(n)}}\right\}_{\boldsymbol{i}^{(n)} \in I^{(n)}}\right) \leq H_\omega^{\{\gamma_j\}_{j=1}^n}\left(\left\{\lambda'_{\boldsymbol{j}^{(n)}}, \omega'_{\boldsymbol{j}^{(n)}}\right\}_{\boldsymbol{j}^{(n)} \in J^{(n)}}\right) + \varepsilon \ .$$

**Proof:** Following the steps in the proof of Proposition 8.1.1, consider partitions $\mathcal{Z}_j = \{Z^j_{k_j}\}_{k_j=1}^{n_j}$ of the state-spaces $\mathcal{S}(\boldsymbol{M}_j)$ into atoms $Z^j_{k_j}$ such that if $\nu^j_{1,2}$ are two states on $\boldsymbol{M}_j$ belonging to any $Z^j_{k_j}$, then $\|\nu_1 - \nu_2\| \leq \delta$. The cardinality $n_j$ of each of these partitions can be chosen not larger than the smallest number, $r$, of balls of radius $\leq \delta$ needed to cover each $\mathcal{S}(\boldsymbol{M}_j)$, a number which depends on $\dim \boldsymbol{M}_j$ and thus on $d$. Given the decompositions

$$\omega = \sum_{\boldsymbol{i}^{(n)} \in I^{(n)}} \lambda^{(n)}_{\boldsymbol{i}^{(n)}} \omega_{\boldsymbol{i}^{(n)}} \quad \text{and} \quad \omega = \sum_{i_k} \lambda^k_{i_k} \omega^k_{i_k} \,, \qquad k = 1, 2, \ldots, n$$

one constructs the states

$$\omega'_{j_k} := \sum_{\substack{i_k \in I_k \\ \omega^k_{i_k} \in Z^k_{j_k}}} \frac{\lambda^k_{i_k}}{\lambda'_{j_k}} \omega^k_{i_k} \,, \qquad \lambda'_{j_k} := \sum_{\substack{i_k \in I_k \\ \omega^k_{i_k} \in Z^k_{j_k}}} \lambda^k_{i_k}$$

$$\omega'_{\boldsymbol{j}^{(n)}} := \sum_{\substack{\boldsymbol{i}^{(n)} \in I^{(n)} \\ \omega^k_{i_k} \in Z^k_{j_k}, k=1,2,\ldots,n}} \frac{\lambda^{(n)}_{\boldsymbol{i}^{(n)}}}{\lambda'_{\boldsymbol{j}^{(n)}}} \omega_{\boldsymbol{i}^{(n)}} \,, \qquad \lambda'_{\boldsymbol{j}^{(n)}} := \sum_{\substack{\boldsymbol{i}^{(n)} \in I^{(n)} \\ \omega^k_{i_k} \in Z^k_{j_k}, k=1,2,\ldots,n}} \lambda^{(n)}_{\boldsymbol{i}^{(n)}} \,,$$

and the corresponding decompositions

$$\omega = \sum_{\boldsymbol{j}^{(n)} \in J^{(n)}} \lambda'_{\boldsymbol{j}^{(n)}} \omega'_{\boldsymbol{j}^{(n)}} \quad \text{and} \quad \omega = \sum_{j_k \in J_k} \lambda'_{j_k} \omega'_{j_k} \,.$$

Then, introducing the probability distributions $\Lambda^{(n)} := \left\{\lambda^{(n)}_{\boldsymbol{i}^{(n)}}\right\}_{\boldsymbol{i}^{(n)} \in I^{(n)}}$ and $\Lambda' := \left\{\lambda'_{\boldsymbol{j}^{(n)}}\right\}_{\boldsymbol{j}^{(n)} \in J^{(n)}}$, together with the respective marginal distributions $\Lambda_k := \left\{\lambda^k_{i_k}\right\}_{i_k \in I_k}$ and $\Lambda'_k := \left\{\lambda'_{j_k}\right\}_{j_k \in J_k}$, one estimates

$$\begin{aligned}
&H^{\{\gamma_j\}_{j=1}^n}_\omega\left(\left\{\lambda^{(n)}_{\boldsymbol{i}^{(n)}}, \omega_{\boldsymbol{i}^{(n)}}\right\}_{\boldsymbol{i}^{(n)} \in I^{(n)}}\right) - H^{\{\gamma_j\}_{j=1}^n}_\omega\left(\left\{\lambda'_{\boldsymbol{j}^{(n)}}, \omega'_{\boldsymbol{j}^{(n)}}\right\}_{\boldsymbol{j}^{(n)} \in J^{(n)}}\right) = \\
&= H(\Lambda^{(n)}) - H(\Lambda') + \sum_{k=1}^n \Big(H(\Lambda'_k) - H(\Lambda_k)\Big) \\
&+ \sum_{k=1}^n \left( \sum_{i_k \in I_k} \lambda^k_{i_k} S\left(\omega^k_{i_k} \lceil \gamma_k \,, \omega \lceil \gamma_k\right) - \sum_{j_k \in J_k} \lambda'_{j_k} S\left(\omega'_{j_k} \lceil \gamma_k \,, \omega \lceil \gamma_k\right) \right) \\
&\leq n\varepsilon \,.
\end{aligned}$$

Indeed, according to the proof of Proposition 8.1.1, each term in the second sum over $k$ is $\leq \varepsilon$, while the first line after the equality is $\leq 0$. This can be seen by considering the $\Lambda_k$ as probability distributions over partitions $\mathcal{P}_k$ with atoms $P^k_{i_k}$ and $\Lambda^{(n)}$ as a probability distribution over the finite partition $\mathcal{P}^{(n)} := \bigvee_{k=1}^n \mathcal{P}_k$. Then (compare Remarks 2.4.2), the $\Lambda'_k$ and $\Lambda'_k$

are probability distributions over coarser partitions $\mathcal{Q}_k \preceq \mathcal{P}_k$, respectively $\mathcal{Q}^{(n)} := \bigvee_{k=1}^n \mathcal{Q}_k \preceq \mathcal{P}^{(n)}$, whence, using the conditional entropy (2.90),

$$\begin{aligned} H(\mathcal{P}^{(n)} \vee \mathcal{Q}^{(n)}) &= H(\mathcal{P}^{(n)}) = H(\mathcal{Q}^{(n)}) + H(\mathcal{P}^{(n)}|\mathcal{Q}^{(n)}) \\ H(\mathcal{P}_k \vee \mathcal{Q}_k) &= H(\mathcal{P}_k) = H(\mathcal{Q}_k) + H(\mathcal{P}_k|\mathcal{Q}_k) \ . \end{aligned}$$

Thus, from Lemma 2.4.3 and Corollary 2.4.2,

$$H(\mathcal{P}^{(n)}) - H(\mathcal{Q}^{(n)}) \leq \sum_{k=1}^n H(\mathcal{P}_k|\mathcal{Q}_k) = \sum_{k=1}^n \Big(H(\mathcal{P}_k) - H(\mathcal{Q}_k)\Big) \ .$$

□

From proposition 8.1.1 it follows that for any $\varepsilon > 0$, there exists a finite decomposition $\omega = \sum_{\boldsymbol{i}^{(n)} \in I^{(n)}} \lambda^{(n)}_{\boldsymbol{i}^{(n)}} \, \omega_{\boldsymbol{i}^{(n)}}$ such that

$$H_\omega^{\{\gamma_j\}_{j=1}^n}\left(\left\{\lambda^{(n)}_{\boldsymbol{i}^{(n)}}, \omega_{\boldsymbol{i}^{(n)}}\right\}_{\boldsymbol{i}^{(n)} \in I^{(n)}}\right) \geq \mathrm{H}_\omega\left(\gamma_1, \gamma_2, \ldots, \gamma_n\right) - n\,\varepsilon \ , \tag{8.8}$$

while, from Definition 8.1.1,

$$\mathrm{H}_\omega\left(\gamma_1, \gamma_2, \ldots, \gamma_n\right) \geq H_\omega^{\{\gamma_j\}_{j=1}^n}\left(\left\{\lambda^{(n)}_{\boldsymbol{i}^{(n)}}, \omega_{\boldsymbol{i}^{(n)}}\right\}_{\boldsymbol{i}^{(n)} \in I^{(n)}}\right) \ .$$

We shall call these decompositions $\varepsilon$-optimal and remark that their cardinality $r := \#(I^{(n)})$ depends on $\varepsilon$ and on the maximal dimension $d$ of the finite-dimensional $C^*$ algebras on which the $\gamma_j$ act. Then, the $n$-*CPU* map entropies result equicontinuous in the maps $\gamma_j : \boldsymbol{M}_j \mapsto \mathcal{A}$ with respect to the topology defined on their linear space by the norm

$$\|\gamma_1 - \gamma_2\|_\omega := \sup_{\substack{X \in \mathcal{A} \\ \|X\| \leq 1}} \|(\gamma_1 - \gamma_2)(X)\|_\omega \ , \tag{8.9}$$

where $\|X\|_\omega^2 := \omega(X^\dagger X)$ for all $X \in \mathcal{A}$.

**Proposition 8.1.2.** *Let $\gamma_j$ and $\gamma'_j$, $j = 1, 2, \ldots, n$, be CPU maps from finite-dimensional $C^*$ algebras $\boldsymbol{M}_j$ with $\dim \boldsymbol{M}_j \leq d$ into $\mathcal{A}$. Then, for any $\varepsilon > 0$ there can be found $\delta > 0$ depending on $\varepsilon$ and $d$ such that*

$$\Big|\mathrm{H}_\omega\left(\gamma_1, \gamma_2, \ldots, \gamma_n\right) - \mathrm{H}_\omega\left(\gamma'_1, \gamma'_2, \ldots, \gamma'_n\right)\Big| \leq n\,\varepsilon$$

*when $\|\gamma_j - \gamma'_j\|_\omega \leq \delta$.*

**Proof:** Consider a finite decomposition of cardinality $r$ as in (8.8) with $\varepsilon/2$ in the place of $\varepsilon$; then,

$$\mathrm{H}_\omega(\gamma_1,\gamma_2,\ldots,\gamma_n) - \mathrm{H}_\omega(\gamma'_1,\gamma'_2,\ldots,\gamma'_n) \leq H_\omega^{\{\gamma_j\}_{j=1}^n}\left(\left\{\lambda^{(n)}_{\boldsymbol{i}^{(n)}},\omega_{\boldsymbol{i}^{(n)}}\right\}_{\boldsymbol{i}^{(n)}\in I^{(n)}}\right)$$
$$-H_\omega^{\{\gamma'_j\}_{j=1}^n}\left(\left\{\lambda^{(n)}_{\boldsymbol{i}^{(n)}},\omega_{\boldsymbol{i}^{(n)}}\right\}_{\boldsymbol{i}^{(n)}\in I^{(n)}}\right) + \frac{n\varepsilon}{2} = \sum_{j=1}^n\Big(S(\omega\circ\gamma_j) - S(\omega\circ\gamma'_j) +$$
$$+\sum_{i_j\in I_j}\lambda^j_{i_j}\left(S\left(\omega^j_{i_j}\circ\gamma_j\right) - S\left(\omega^j_{i_j}\circ\gamma'_j\right)\right)\Big) + \frac{n\varepsilon}{2}\ .$$

Choose $0<\delta<\delta_0$ and $\delta_0$ such that the Fannes inequality 5.157 implies

$$\Big|S(\nu_1\lceil\boldsymbol{M}) - S(\nu_2\lceil\boldsymbol{M})\Big| \leq \frac{\varepsilon}{6}$$

when $\boldsymbol{M}$ is a finite-dimensional $C^*$ algebra with $\dim\boldsymbol{M}\leq d$ and $\nu_{1,2}\in\mathcal{S}(\boldsymbol{M})$ are states on it such that $\|\nu_1-\nu_2\|\leq\delta_0$. Since $|\omega(X)|\leq\|X\|_\omega$ (see (5.49)), it follows that

$$\|\omega\circ(\gamma_j-\gamma'_j)\| = \sup_{M\in\boldsymbol{M},\|M\|\leq 1}|\omega\circ(\gamma_j-\gamma'_j)(M)| \leq \|\gamma_-\gamma'_j\|_\omega \leq \delta < \delta_0\ ,$$

whence $\sum_{j=1}^n\Big(S(\omega\circ\gamma_j) - S(\omega\circ\gamma'_j)\Big) \leq n\,\frac{\varepsilon}{6}$.

In order to estimate the remaining sums, let us divide each index set $I_j$ into two disjoint subsets, namely

$$I_j^0 := \left\{i_j\in I_J\,:\,\lambda^j_{i_j}\geq\frac{\delta^2}{\delta_0^2}\right\}$$

and its complement. Since $\omega = \sum_{i_j\in I_j}\lambda^j_{i_j}\omega^j_{i_j}$, from (8.9), the state-based distances are such that, for all $i_j\in I_j^0$,

$$\|\gamma_j-\gamma'_j\|_{\omega^j_{i_j}} \leq \frac{1}{\sqrt{\lambda^j_{i_j}}}\|\gamma_j-\gamma'_j\|_\omega \leq \delta_0\ ,$$

so that $\sum_{j=1}^n\sum_{i_j\in I_j^0}\lambda^j_{i_j}\Big(S(\omega\circ\gamma_j) - S(\omega\circ\gamma'_j)\Big) \leq n\,\frac{\varepsilon}{6}$. Finally, using that $\mathrm{card}(I_j)\leq\mathrm{card}(I^{(n)})\leq r$ and $\dim\boldsymbol{M}_j\leq d$ together with (5.155), one derives

$$\sum_{i_j\notin I_j^0}\lambda^j_{i_j}\left(S\left(\omega^j_{i_j}\circ\gamma_j\right) - S\left(\omega^j_{i_j}\circ\gamma'_j\right)\right) \leq r\frac{\delta^2}{\delta_0^2}\log d\ .$$

Therefore, $\delta<\delta_0$ such that $r\dfrac{\delta^2}{\delta_0^2}\log d\leq\dfrac{\varepsilon}{6}$, yields

$$\mathrm{H}_\omega(\gamma_1,\gamma_2,\ldots,\gamma_n) - \mathrm{H}_\omega(\gamma'_1,\gamma'_2,\ldots,\gamma'_n) \leq n(\frac{\varepsilon}{2}+3\frac{\varepsilon}{6}) = n\,\varepsilon\ ,$$

whence the result follows by exchanging the sets $\{\gamma_j\}_{j=1}^n$ and $\{\gamma'_j\}_{j=1}^n$. □

Other properties that are important for applications to concrete quantum dynamical systems are the following ones.

**Proposition 8.1.3 (Properties of *n-CPU* Entropies).**
*Given a $C^*$ algebra $\mathcal{A}$ equipped with a state $\omega$ and $n$ CPU maps $\gamma_i : \boldsymbol{M}_j \mapsto \mathcal{A}$ from unital $C^*$ algebras $\boldsymbol{M}_j$, $j = 1, 2, \ldots, n$, with $\dim \boldsymbol{M}_j \leq d$, into $\mathcal{A}$, it holds that:*

1. *the n-CPU map entropies are positive and bounded,*

$$0 \leq \mathrm{H}_\omega(\gamma_1, \gamma_2, \ldots, \gamma_n) \leq \sum_{j=1}^n \mathrm{H}_\omega(\gamma_j) \leq \sum_{j=1}^n S(\omega \circ \gamma_j) \leq n \log d \; ; \tag{8.10}$$

2. *the n-CPU map entropies do not depend on the order of their arguments:*

$$\mathrm{H}_\omega(\gamma_1, \gamma_2, \ldots, \gamma_n) = \mathrm{H}_\omega\left(\gamma_{\pi(1)}, \gamma_{\pi(2)}, \ldots, \gamma_{\pi(n)}\right) \; , \tag{8.11}$$

   *with $\pi(i)$ any permutation of $1, 2, \ldots, n$;*

3. *the n-CPU map entropies are not sensitive to repetitions of their arguments:*

$$\mathrm{H}_\omega(\gamma_1, \ldots \gamma_{j-1}, \gamma_j, \gamma_j, \gamma_{j+1}, \ldots, \gamma_n) = \mathrm{H}_\omega(\gamma_1, \gamma_{j-1}, \gamma_j, \gamma_{j+1}, \ldots, \gamma_n) \; ; \tag{8.12}$$

4. *if $\Theta : \mathcal{A} \to \mathcal{A}$ is an automorphism such that $\omega \circ \Theta = \omega$, then*

$$\mathrm{H}_\omega(\Theta \circ \gamma_1, \Theta \circ \gamma_2, \ldots, \Theta \circ \gamma_n) = \mathrm{H}_\omega(\gamma_1, \gamma_2, \ldots, \gamma_n) \; ; \tag{8.13}$$

5. *the n-CPU map entropies are subadditive:*

$$\begin{aligned}\mathrm{H}_\omega(\gamma_1, \ldots, \gamma_p, \gamma_{p+1}, \ldots, \gamma_n) &\leq \mathrm{H}_\omega(\gamma_1, \gamma_2, \ldots, \gamma_p) \\ &+ \mathrm{H}_\omega(\gamma_{p+1}, \gamma_{p+2}, \ldots, \gamma_n) \; ;\end{aligned} \tag{8.14}$$

6. *the n-CPU map entropies are monotonic under composition of CPU maps; namely, if $\gamma'_j : \boldsymbol{N}_j \mapsto \boldsymbol{M}_j$ are CPU maps from finite-dimensional $C^*$ algebras $\boldsymbol{N}_j$ into finite-dimensional $C^*$-algebras $\boldsymbol{M}_j$, $j = 1, 2, \ldots, n$, which are in turn mapped into $\mathcal{A}$ by CPU maps $\gamma_j$, then*

$$\mathrm{H}_\omega(\gamma_1 \circ \gamma'_1, \gamma_2 \circ \gamma'_2, \ldots, \gamma_n \circ \gamma'_n) \leq \mathrm{H}_\omega(\gamma_1, \gamma_2, \ldots, \gamma_n) \; ; \tag{8.15}$$

7. *the n-CPU map entropies increase by non-trivially increasing the number of their arguments:*

$$\mathrm{H}_\omega(\gamma_1, \gamma_2, \ldots, \gamma_n) \leq \mathrm{H}_\omega(\gamma_1, \gamma_2, \ldots, \gamma_n, \gamma_{n+1}) \; . \tag{8.16}$$

Before proving the previous properties we examine some simple cases where, in analogy with Example 6.3.6, the *n-CPU* map entropies can explicitly be computed.

**Examples 8.1.1.**

1. If in (8.15) one considers as *CPU* maps $\gamma'_j$ and $\gamma_j$ the natural embeddings of subalgebras $\boldsymbol{N}_j \subseteq \boldsymbol{M}_j \subseteq \mathcal{A}$, then that property asserts that the $n$-subalgebra entropies increases under embeddings into larger subalgebras. Suppose the finite-dimensional $C^*$ subalgebras $\{\boldsymbol{M}_j\}_{j=1}^n$ to be such that they together generate a finite-dimensional subalgebra $\boldsymbol{M}^{(n)} := \bigvee_{j=1}^n \boldsymbol{M}_j$. This is the case, for instance, when each $\boldsymbol{M}_j$ is a spin-algebra at site $j$ on a lattice so that $\boldsymbol{M}^{(n)}$ is the algebra of $n$ spins at sites $1, 2, \ldots, n$. Then, $\boldsymbol{M}_j \subseteq \boldsymbol{M}^{(n)}$ whence property (8.15) together with property (8.12) and property (8.10) give

$$\begin{aligned} \mathrm{H}_\omega\left(\boldsymbol{M}_1, \boldsymbol{M}_2, \ldots, \boldsymbol{M}_n\right) &\leq \mathrm{H}_\omega\left(\boldsymbol{M}^{(n)}, \boldsymbol{M}^{(n)}, \ldots, \boldsymbol{M}^{(n)}\right) \\ &= \mathrm{H}_\omega\left(\boldsymbol{M}^{(n)}\right) \leq S\left(\omega \upharpoonright \boldsymbol{M}^{(n)}\right) . \qquad (8.17) \end{aligned}$$

2. Suppose $\mathcal{A}$ is an Abelian von Neumann algebra and $\{\boldsymbol{A}_j\}_{j=1}^n$ are finite dimensional subalgebras generated by minimal projectors $\{\widehat{a}_{ji}\}_{i=1}^{d_j}$. Then, consider the products $\widehat{a}_{\boldsymbol{i}^{(n)}} := \widehat{a}_{1i_1}\widehat{a}_{2i_2}\cdots\widehat{a}_{ni_n}$, $I^{(n)} = \times_{j=1}^n I_j$, where $\boldsymbol{i}^{(n)} = i_1 i_2 \cdots i_n$, $i_j \in I_j$ and $I_j = \{1, 2, \ldots, d_j\}$. Because of commutativity, these are projectors that one can use to decompose $\omega$ as follows

$$\omega = \sum_{\boldsymbol{i}^{(n)} \in I^{(n)}} \lambda_{\boldsymbol{i}^{(n)}}\, \omega_{\boldsymbol{i}^{(n)}} , \qquad \omega_{\boldsymbol{i}^{(n)}}(a) = \frac{\omega(\widehat{a}_{\boldsymbol{i}^{(n)}} a)}{\omega(\widehat{a}_{\boldsymbol{i}^{(n)}})} \qquad \forall a \in \mathcal{A} .$$

Further, the various probability distributions and elements of the subdecompositions amounts to

$$\Lambda^{(n)} = \{\omega(\widehat{a}_{\boldsymbol{i}^{(n)}})\}_{\boldsymbol{i}^{(n)} \in I^{(n)}} , \qquad \omega^j_{i_j}(a) = \frac{\omega(\widehat{a}_{ji_j} a)}{\omega(\widehat{a}_{ji_j})} , \qquad \Lambda_j = \{\omega(\widehat{a}_{ji_j})\}_{i_j \in I_j} .$$

It thus follows that 1) the states $\omega \upharpoonright \boldsymbol{A}_j = \Lambda_j$ and the states $\omega^j \upharpoonright \boldsymbol{A}_j$ are pure states because the orthogonality of the minimal projectors implies

$$\omega^j_{i_j}(\widehat{a}_{jk}) = \frac{\omega(\widehat{a}_{ji_j}\widehat{a}_{jk})}{\omega(\widehat{a}_{ji_j})} = \delta_{i_j k} .$$

Since the minimal projections $\widehat{a}^{(n)}_{\boldsymbol{i}^{(n)}}$ generate the Abelian subalgebra $\boldsymbol{A}^{(n)} := \bigvee_{j=1}^n \boldsymbol{A}_j$, this yields

$$H_\omega^{\{\boldsymbol{A}_j\}_{j=1}^n}\left(\left\{\lambda^{(n)}_{\boldsymbol{i}^{(n)}}, \omega_{\boldsymbol{i}^{(n)}}\right\}\right) = -\sum_{\boldsymbol{i}^{(n)} \in I^{(n)}} \lambda^{(n)}_{\boldsymbol{i}^{(n)}} \log \lambda^{(n)}_{\boldsymbol{i}^{(n)}} = S\left(\omega \upharpoonright \boldsymbol{A}^{(n)}\right) .$$

Because of (8.17) this result is optimal; thus,

$$\mathrm{H}_\omega\left(\boldsymbol{A}_1, \boldsymbol{A}_2, \ldots, \boldsymbol{A}_n\right) = S\left(\omega \restriction \boldsymbol{A}^{(n)}\right) . \tag{8.18}$$

Notice that the latter von Neumann entropy is the Shannnon entropy of the random variable $\bigvee_{j=1}^n A_j$ distributed with probability $\Lambda^{(n)}$, where the random variables $A_j$ are distributed with marginal probabilities $\Lambda_j$. The outcomes of these random variables correspond to the minimal projections $\widehat{a}_{ji}$; according to Section 5.3.2, via the Gelfand transform these can be turned into characteristic functions of the atoms of suitable measurable partitions.

3. The result in (8.18) also holds when the $\boldsymbol{A}_j$ are commuting Abelian finite-dimensional subalgebras of a non-commutative $\mathcal{A}$, but $\omega$ is the tracial state. Indeed, the minimal projectors of the $\boldsymbol{A}_j$ provide an optimal decomposition as in the previous example. The reason is that the modular automorphism of the tracial state is trivial; thus, (8.6) yields

$$\begin{aligned} \lambda_{\boldsymbol{i}^{(n)}}\omega_{\boldsymbol{i}^{(n)}}(a) &= \langle\, \Omega_\omega \,|\, \sigma_\omega^{i/2}\left(\pi_\omega(\widehat{a}_{\boldsymbol{i}^{(n)}})\right)\pi_\omega(a)\,|\Omega_\omega\,\rangle \\ &= \langle\, \Omega_\omega \,|\, \pi_\omega(\widehat{a}_{\boldsymbol{i}^{(n)}}a)\,|\Omega_\omega\,\rangle = \omega(\widehat{a}_{\boldsymbol{i}^{(n)}}a) . \end{aligned}$$

4. Suppose $\{\boldsymbol{M}_j\}_{j=1}^n$ are finite-dimensional $C^*$ subalgebras that generate a finite-dimensional subalgebra $\boldsymbol{M}^{(n)} := \bigvee_{j=1}^n \boldsymbol{M}_j \subseteq \mathcal{A}$. Further, suppose they contain pairwise commuting Abelian subalgebras $\boldsymbol{A}_j \subseteq \boldsymbol{M}_j$ each belonging to the centralizer of $\omega$ [2] and such that the algebra $\boldsymbol{A}^{(n)} := \bigvee_{j=1}^n \boldsymbol{A}_j$ they generate is maximally Abelian in $\boldsymbol{M}^{(n)}$. Since the $\boldsymbol{A}_j$ pairwise commute the products of their minimal projectors $\widehat{a}_{ji_j}$ provide the minimal projectors $\widehat{a}_{\boldsymbol{i}^{(n)}}$ of $\boldsymbol{A}_j^{(n)}$. By assumption, they are left invariant by the modular automorphism of $\omega$ and can thus be used to decompose $\omega$ as in the previous two examples. Then, from the second example above and from Example 6.3.3.2 one derives

$$\begin{aligned} \mathrm{H}_\omega\left(\boldsymbol{M}_1, \boldsymbol{M}_2, \ldots, \boldsymbol{M}_n\right) &\geq H_\omega^{\{\boldsymbol{M}_j\}_{j=1}^n}\left(\{\omega(\widehat{a}_{\boldsymbol{i}^{(n)}}), \omega_{\boldsymbol{i}^{(n)}}\}_{\boldsymbol{i}^{(n)}\in I^{(n)}}\right) \\ &= S\left(\omega \restriction \boldsymbol{A}^{(n)}\right) = S\left(\omega \restriction \boldsymbol{M}^{(n)}\right) . \end{aligned}$$

Therefore, the first example yields

$$\mathrm{H}_\omega\left(\boldsymbol{M}_1, \boldsymbol{M}_2, \ldots, \boldsymbol{M}_n\right) = S\left(\omega \restriction \boldsymbol{M}_1 \vee \boldsymbol{M}_2 \vee \cdots \boldsymbol{M}_n\right) . \tag{8.19}$$

## Proof of Proposition 8.1.3

1. Positivity comes from choosing not to decompose $\omega$ at all, in which case the argument of the supremum in (8.3) vanishes. Further, the first line in the argument of the supremum equals minus the relative entropy (see (2.94)) $S\left(\Lambda^{(n)}, \widetilde{\Lambda}^{(n)}\right)$ of the two probability distributions

[2] They are therefore left pointwise invariant by the modular automorphism of $\omega$.

$\Lambda^{(n)} = \left\{\lambda^{(n)}_{\boldsymbol{i}^{(n)}}\right\}_{\boldsymbol{i}^{(n)}\in I^{(n)}}$, respectively $\widetilde{\Lambda}^{(n)} := \left\{\prod_{j=1}^n \lambda^j_{i_j}\right\}_{\boldsymbol{i}^{(n)}\in I^{(n)}}$ on the strings set of strings $\boldsymbol{i}^{(n)}$. Since the relative entropy is non-negative, the upper bound to the $n$-$CPU$ map entropies follows from Lemma 6.3.1.

2. In order to show (8.11), let $\omega = \sum_{\boldsymbol{i}^{(n)}\in I^{(n)}} \lambda^{(n)}_{\boldsymbol{i}^{(n)}}\omega_{\boldsymbol{i}^{(n)}}$ be an $\varepsilon$-optimal decomposition such that, as in (8.8),

$$\mathrm{H}_\omega\left(\gamma_1,\gamma_2,\ldots,\gamma_n\right) \le H_\omega^{\{\gamma_j\}_{j=1}^n}\left(\left\{\lambda^{(n)}_{\boldsymbol{i}^{(n)}},\omega_{\boldsymbol{i}^{(n)}}\right\}_{\boldsymbol{i}^{(n)}\in I^{(n)}}\right) + n\,\varepsilon\ .$$

Since $H_\omega^{\{\gamma_{\pi(j)}\}_{j=1}^n}\left(\left\{\lambda^{(n)}_{\pi(\boldsymbol{i}^{(n)})},\omega_{\pi(\boldsymbol{i}^{(n)})}\right\}_{\boldsymbol{i}^{(n)}\in I^{(n)}}\right)$, where $\pi(\boldsymbol{i}^{(n)})$ denotes the string $i_{\pi(1)}i_{\pi(2)}\cdots i_{\pi(n)}$, equals $H_\omega^{\{\gamma_j\}_{j=1}^n}\left(\left\{\lambda^{(n)}_{\boldsymbol{i}^{(n)}},\omega_{\boldsymbol{i}^{(n)}}\right\}_{\boldsymbol{i}^{(n)}\in I^{(n)}}\right)$, it follows that

$$\begin{aligned}\mathrm{H}_\omega\left(\gamma_1,\gamma_2,\ldots,\gamma_n\right) &\le H_\omega^{\{\gamma_{\pi(j)}\}_{j=1}^n}\left(\left\{\lambda^{(n)}_{\pi(\boldsymbol{i}^{(n)})},\omega_{\pi(\boldsymbol{i}^{(n)})}\right\}_{\boldsymbol{i}^{(n)}\in I^{(n)}}\right) + n\,\varepsilon\\ &\le \mathrm{H}_\omega\left(\gamma_{\pi(1)},\gamma_{\pi(2)},\ldots,\gamma_{\pi(n)}\right) + n\,\varepsilon\ .\end{aligned}$$

Equality follows from the arbitrariness of $\varepsilon > 0$ by exchange of the sets $\{\gamma_j\}_{j=1}^n$ and $\{\gamma_{\pi(j)}\}_{j=1}^n$.

3. In view of (8.11), to prove (8.12) we show that $\mathrm{H}_\omega\left(\gamma_1,\gamma_2,\ldots,\gamma_n\right)$ does not change if the argument $\gamma_n$ appears twice. Consider an $\varepsilon$-optimal decomposition for the right hand side of (8.12) and, according to (8.5), the corresponding positive decomposition of unity in the commutant $\pi(\mathcal{A})'$, $\{Y_{\boldsymbol{i}^{(n)}}\}_{\boldsymbol{i}^{(n)}\in I^{(n)}}$. Then, construct a new decomposition

$$\omega = \sum_{\boldsymbol{j}^{(n+1)}\in J^{(n+1)}} \widetilde{\lambda}^{(n+1)}_{\boldsymbol{j}^{(n+1)}}\widetilde{\omega}_{\boldsymbol{j}^{(n+1)}}$$

based on a decomposition of unit consisting of $\pi(\mathcal{A})' \ni \widetilde{Y}_{\boldsymbol{j}^{(n+1)}} := Y_{\boldsymbol{i}^{(n)}}\,\mathbb{1}$, where $\boldsymbol{j}^{(n+1)} := \boldsymbol{i}^{(n)}j_{n+1}$, $\boldsymbol{i}^{(n)}\in I^{(n)}$ and $j_{n+1}\in J_{n+1} = \{1\}$. Then,

$$\widetilde{\lambda}^{(n+1)}_{\boldsymbol{j}^{(n+1)}} = \omega(\widetilde{Y}_{\boldsymbol{j}^{(n+1)}}) = \lambda^{(n)}_{\boldsymbol{i}^{(n)}}\ ,\quad \widetilde{\omega}^k_{j_k} = \omega^k_{i_k}\quad \forall k\neq n+1\ ,$$

while $\widetilde{\lambda}^{n+1}_{j_{n+1}} = 1$ and $\widetilde{\omega}^{n+1}_{j_{n+1}} = \omega$. Therefore,

$$\begin{aligned}\mathrm{H}_\omega\left(\gamma_1,\gamma_2,\ldots,\gamma_n\right) &\le H_\omega^{\{\gamma_j\}_{j=1}^n}\left(\left\{\lambda^{(n)}_{\boldsymbol{i}^{(n)}},Y_{\boldsymbol{i}^{(n)}}\right\}_{\boldsymbol{i}^{(n)}\in I^{(n)}}\right) + n\,\varepsilon\\ &= H_\omega^{\{\gamma_j\}_{j=1}^n\cup\gamma_n}\left(\left\{\widetilde{\lambda}^{(n+1)}_{\boldsymbol{j}^{(n+1)}},\widetilde{Y}_{\boldsymbol{j}^{(n+1)}}\right\}_{\boldsymbol{j}^{(n+1)}\in J^{(n+1)}}\right) + n\,\varepsilon\\ &\le \mathrm{H}_\omega\left(\gamma_1,\gamma_2,\ldots,\gamma_n,\gamma_n\right) + n\,\varepsilon\ .\end{aligned}$$

In order to invert this inequality, let

$$\omega = \sum_{\boldsymbol{i}^{(n+1)}\in I^{(n+1)}} \lambda^{(n+1)}_{\boldsymbol{i}^{(n+1)}}\ \omega_{\boldsymbol{i}^{(n+1)}}$$

be an $\varepsilon$-optimal decomposition for the left hand side of (8.12) and consider

$$\omega = \sum_{\boldsymbol{j}^{(n)} \in J^{(n)}} \widetilde{\lambda}^{(n)}_{\boldsymbol{j}^{(n)}} \, \widetilde{\omega}_{\boldsymbol{j}^{(n)}} \; ,$$

where $\boldsymbol{j}^{(n)} = j_1 j_2 \cdots j_n$ with $j_k = i_k$ for $1 \leq k \leq n-1$, while $j_n \in J_n := I_n \times I_{n+1}$ enumerates the pairs $(i_n i_{n+1})$ so that $J^{(n)} = I_1 \times \cdots I_{n-1} \times J_n$, $\widetilde{\lambda}^{(n)}_{\boldsymbol{j}^{(n)}} = \lambda^{(n)}_{\boldsymbol{i}^{(n)}}$, $\widetilde{\omega}_{\boldsymbol{j}^{(n)}} = \omega_{\boldsymbol{i}^{(n)}}$. If $1 \leq k \leq n-1$, it also turns out that $\widetilde{\lambda}^k_{i_k} = \lambda^k_{i_k}$ and $\widetilde{\omega}^k_{j_k} = \omega^k_{i_k}$, while

$$\widetilde{\lambda}^n_{j_n} = \sum_{\substack{\boldsymbol{i}^{(n+1)} \\ i_n, i_{n+1} \text{ fixed}}} \lambda^{(n+1)}_{\boldsymbol{i}^{(n+1)}} \; , \quad \widetilde{\omega}^n_{j_n} = \frac{1}{\widetilde{\lambda}^n_{j_n}} \sum_{\substack{\boldsymbol{i}^{(n+1)} \\ i_n, i_{n+1} \text{ fixed}}} \lambda^{(n+1)}_{\boldsymbol{i}^{(n+1)}} \omega_{\boldsymbol{i}^{(n+1)}} \; .$$

Then, $\widetilde{\Lambda}^{(n)} := \left\{ \widetilde{\lambda}^{(n)}_{\boldsymbol{j}^{(n)}} \right\}_{\boldsymbol{j}^{(n)} \in J^{(n)}} = \Lambda^{(n+1)}$, and $\widetilde{\Lambda}_k := \left\{ \lambda^k_{i_k} \right\}_{i_k \in I_k} = \Lambda_k$ for $1 \leq k \leq n-1$, whereas $\widetilde{\Lambda}_n := \left\{ \widetilde{\lambda}^n_{j_n} \right\}$. Finally, one can estimate

$$\begin{aligned}
\mathrm{H}_\omega(\gamma_1, \gamma_2, \ldots, \gamma_{n-1}, \gamma_n) &\geq H^{\{\gamma_i\}_{i=1}^n}_\omega \left( \left\{ \widetilde{\lambda}^{(n)}_{\boldsymbol{j}^{(n)}}, \widetilde{\omega}_{\boldsymbol{j}^{(n)}} \right\}_{\boldsymbol{j}^{(n)} \in J^{(n)}} \right) \\
&= H(\widetilde{\Lambda}^{(n)}) - \sum_{j=1}^{n-1} H(\widetilde{\Lambda}_j) + \sum_{j=1}^{n-1} \sum_{i_j \in I_j} \widetilde{\lambda}^j_{i_j} S\left( \widetilde{\omega}^j_{i_j} \circ \gamma_j \, , \, \omega \circ \gamma_j \right) \\
&\quad - H(\widetilde{\Lambda}_n) + \sum_{j_n \in J_n} \widetilde{\lambda}_{j_n} S\left( \widetilde{\omega}^n_{j_n} \circ \gamma_n \, , \, \omega \circ \gamma_n \right) . \qquad (*)
\end{aligned}$$

From (2.88) it follows that $H(\widetilde{\Lambda}_n) \leq H(\Lambda_n) + H(\Lambda_{n+1})$. On the other hand, with $j_n = (i_n, i_{n+1})$,

$$\omega^n_{i_n} = \frac{1}{\lambda^n_{i_n}} \sum_{i_{n+1}} \widetilde{\lambda}^n_{j_n} \widetilde{\omega}^n_{j_n} \; , \quad \omega^{n+1}_{i_{n+1}} = \frac{1}{\lambda^{n+1}_{i_{n+1}}} \sum_{i_n} \widetilde{\lambda}^n_{j_n} \widetilde{\omega}^n_{j_n}$$

and (6.31) imply

$$\sum_{j_n \in J_n} \widetilde{\lambda}^n_{j_n} S\left( \widetilde{\omega}^n_{j_n} \circ \gamma_n \, , \, \omega \circ \gamma_n \right) \geq \sum_{k=n}^{n+1} \sum_{i_k \in I_k} \lambda^k_{i_k} \, S\left( \omega^k_{i_k} \circ \gamma_k \, , \, \omega \circ \gamma_k \right) .$$

Together with $(*)$ this yields

$$\begin{aligned}
\mathrm{H}_\omega(\gamma_1, \gamma_2, \ldots, \gamma_{n-1}, \gamma_n) &\geq H^{\{\gamma_j\}_{j=1}^n \cup \gamma_n}_\omega \left( \left\{ \lambda^{(n+1)}_{\boldsymbol{i}^{(n+1)}}, \omega_{\boldsymbol{i}^{(n+1)}} \right\}_{\boldsymbol{i}^{(n+1)} \in I^{(n+1)}} \right) \\
&\geq \mathrm{H}_\omega(\gamma_1, \gamma_2, \ldots, \gamma_n, \gamma_n) \; - \; n\,\varepsilon \; .
\end{aligned}$$

4. Property (8.13) is a consequence of $\omega \circ \Theta = \omega$. In fact, given an $\varepsilon$-optimal decomposition for the right (left) hand side of (8.13) and the corresponding positive decomposition of unit in the commutant, $\left\{ Y^{(n)}_{\boldsymbol{i}^{(n)}} \right\}_{\boldsymbol{i}^{(n)} \in I^{(n)}}$,

then $\left\{U_\omega^\dagger Y_{\boldsymbol{i}^{(n)}}^{(n)} U_\omega\right\}_{\boldsymbol{i}^{(n)}\in I^{(n)}}$ $\left(\left\{U_\omega Y_{\boldsymbol{i}^{(n)}}^{(n)} U_\omega^\dagger\right\}_{\boldsymbol{i}^{(n)}\in I^{(n)}}\right)$, where $U_\omega$ is the unitary *GNS* implementation of $\Theta$, provides a decomposition for the left (right) side.

5. In order to prove subadditivity, let $\omega = \sum_{\in^{(n)}} \lambda_{\boldsymbol{i}^{(n)}}^{(n)} \omega_{\boldsymbol{i}^{(n)}}$ be an $\varepsilon$-optimal decomposition for $\mathrm{H}_\omega(\gamma_1, \gamma_2, \ldots, \gamma_n)$ and construct from it the following two decompositions:

$$\omega = \sum_{\boldsymbol{j}^{(p)}} \lambda_{\boldsymbol{j}^{(p)}}^1 \omega_{\boldsymbol{j}^{(p)}}^1 \ , \qquad \omega = \sum_{\boldsymbol{k}^{(n-p+1)}} \lambda_{\boldsymbol{k}^{(n-p+1)}}^2 \omega_{\boldsymbol{k}^{(n-p+1)}}^2$$

where $\boldsymbol{j}^{(p)} = i_1 i_2 \cdots i_p \in I^{(p)} := I_1 \times I_2 \times \cdots \times I_p$, while the indexes $\boldsymbol{k}^{(n-p+1)} = i_{p+1} i_{p+2} \cdots i_n \in I^{(n-p+1)} := I_{p+1} \times I_{p+2} \times \cdots \times I_n$ and

$$\omega_{\boldsymbol{j}^{(p)}}^1 := \sum_{\boldsymbol{k}^{(n-p+1)}\in I^{(n-p+1)}} \frac{\lambda_{\boldsymbol{i}^{(n)}}^{(n)}}{\lambda_{\boldsymbol{j}^{(p)}}^1} \omega_{\boldsymbol{i}^{(n)}} \ , \qquad \lambda_{\boldsymbol{j}^{(p)}}^1 := \sum_{\boldsymbol{k}^{(n-p+1)}\in I^{(n-p+1)}} \lambda_{\boldsymbol{i}^{(n)}}^{(n)}$$

$$\omega_{\boldsymbol{k}^{(n-p+1)}}^2 := \sum_{\boldsymbol{j}^{(p)}\in I^{(p)}} \frac{\lambda_{\boldsymbol{i}^{(n)}}^{(n)}}{\lambda_{\boldsymbol{k}^{(n-p+1)}}^2} \omega_{\boldsymbol{i}^{(n)}} \ , \qquad \lambda_{\boldsymbol{k}^{(n-p+1)}}^2 := \sum_{\boldsymbol{j}^{(p)}\in I^{(p)}} \lambda_{\boldsymbol{i}^{(n)}}^{(n)} \ .$$

Since $\Lambda^1 := \left\{\lambda_{\boldsymbol{j}^{(p)}}^1\right\}_{\boldsymbol{j}^{(p)}\in I^{(p)}}$ and $\Lambda^2 := \left\{\lambda_{\boldsymbol{k}^{(n-p+1)}}^2\right\}_{\boldsymbol{k}^{(n-p+1)}\in I^{(n-p+1)}}$ are marginal distributions of $\Lambda^{(n)} = \left\{\lambda_{\boldsymbol{i}^{(n)}\in I^{(n)}}^{(n)}\right\}$, (2.88) yields

$$\begin{aligned}
&H(\Lambda^{(n)}) \leq H(\Lambda^1) + H(\Lambda^2) \qquad \text{whence} \\
&\mathrm{H}_\omega(\gamma_1, \gamma_2, \ldots, \gamma_p) + \mathrm{H}_\omega(\gamma_{p+1}, \gamma_{p+2}, \ldots, \gamma_n) \geq \\
&\quad \geq H_\omega^{\{\gamma_j\}_{j=1}^p}\left(\left\{\lambda_{\boldsymbol{j}^{(p)}}^1, \omega_{\boldsymbol{j}^{(p)}}^1\right\}_{\boldsymbol{j}^{(p)}\in I^{(p)}}\right) + \\
&\qquad + H_\omega^{\{\gamma_j\}_{j=p+1}^n}\left(\left\{\lambda_{\boldsymbol{k}^{(n-p+1)}}^2, \omega_{\boldsymbol{k}^{(n-p+1)}}^2\right\}_{\boldsymbol{k}^{(n-p+1)}\in I^{(n-p+1)}}\right) \\
&\quad \geq H_\omega^{\{\gamma_j\}_{j=1}^n}\left(\left\{\lambda_{\boldsymbol{i}^{(n)}}^{(n)}, \omega_{\boldsymbol{i}^{(n)}}\right\}_{\boldsymbol{i}^{(n)}\in I^{(n)}}\right) \geq \mathrm{H}_\omega(\gamma_1, \gamma_2, \ldots, \gamma_n) - n\varepsilon \ .
\end{aligned}$$

6. Property (8.15) follows from the monotonicity of the relative entropy under *CPU* maps.
7. Finally, property (8.16) is a consequence of (8.15) and of the fact that, given an $\varepsilon$-optimal decomposition for the left hand side, one may construct a decomposition for the right hand side as in the first part of point 3 above.

□

**Proposition 8.1.4.** *[216] Given a C* algebra $\mathcal{A}$ equipped with a state $\omega$, let $\{\gamma_j\}_{j=1}^n$, $\gamma_n'$ be CPU maps from finite-dimensional $C^*$-algebras $\{\boldsymbol{M}_j\}_{j=1}^n$ into $\mathcal{A}$, then*

$$\mathrm{H}_\omega(\gamma_1,\gamma_2,\ldots,\gamma_{n-1},\gamma_n) - \mathrm{H}_\omega(\gamma_1,\gamma_2,\ldots,\gamma_{n-1},\gamma_n') \leq \mathrm{H}_\omega(\gamma_n \mid \gamma_n') \quad (8.20)$$

$$\mathrm{H}_\omega(\gamma_1 \mid \gamma_2) := \sup_{\omega=\sum_{i\in I}\lambda_i\omega_i} \mathrm{H}_\omega^{\gamma_1,\gamma_2}\left(\{\lambda_i,\omega_i\}_{i\in I}\right) \quad (8.21)$$

$$\mathrm{H}_\omega^{\gamma_1,\gamma_2}\left(\{\lambda_i,\omega_i\}_{i\in I}\right) := \sum_{i\in I}\lambda_i\Big(S\left(\omega_i\circ\gamma_1\,,\,\omega\circ\gamma_1\right) - S\left(\omega_i\circ\gamma_2\,,\,\omega\circ\gamma_2\right)\Big)\,. \quad (8.22)$$

**Proof:** Let $\omega = \sum_{\boldsymbol{i}^{(n)}\in I^{(n)}}\lambda^{(n)}_{\boldsymbol{i}^{(n)}}\omega_{\boldsymbol{i}^{(n)}}$ be an $\varepsilon$-optimal decomposition such that, as in (8.8),

$$\mathrm{H}_\omega(\gamma_1,\gamma_2,\ldots,\gamma_n) \leq H_\omega^{\{\gamma_j\}_{j=1}^n}\left(\left\{\lambda^{(n)}_{\boldsymbol{i}^{(n)}},\omega_{\boldsymbol{i}^{(n)}}\right\}_{\boldsymbol{i}^{(n)}\in I^{(n)}}\right) + n\,\varepsilon\,,$$

then, according to (8.7),

$$\begin{aligned}
&\mathrm{H}_\omega(\gamma_1,\gamma_2,\ldots,\gamma_{n-1},\gamma_n) - \mathrm{H}_\omega(\gamma_1,\gamma_2,\ldots,\gamma_{n-1},\gamma_n') \leq \\
&\quad\leq H_\omega^{\{\gamma_j\}_{j=1}^n}\left(\left\{\lambda^{(n)}_{\boldsymbol{i}^{(n)}},\omega_{\boldsymbol{i}^{(n)}}\right\}_{\boldsymbol{i}^{(n)}\in I^{(n)}}\right) - \\
&\quad - H_\omega^{\{\gamma_j\}_{j=1}^{n-1}\cup\gamma_n'}\left(\left\{\lambda^{(n)}_{\boldsymbol{i}^{(n)}},\omega_{\boldsymbol{i}^{(n)}}\right\}_{\boldsymbol{i}^{(n)}\in I^{(n)}}\right) + n\,\varepsilon \\
&\quad\leq \sum_{i_n\in I_n}\lambda^n_{i_n}\Big(S\left(\omega^n_{i_n}\circ\gamma_1\,,\,\omega\circ\gamma_1\right) - S\left(\omega^n_{i_n}\circ\gamma_2\,,\,\omega\circ\gamma_2\right)\Big) + n\varepsilon\,.
\end{aligned}$$

Since $n$ is fixed and $\varepsilon$ is arbitrary the result follows. □

**Example 8.1.2.** If $\mathcal{A}$ is an Abelian $C^*$ algebra and $\gamma_{1,2}$ are the natural embeddings of two finite-dimensional Abelian $C^*$ algebras $\boldsymbol{A}_{1,2} \subseteq \mathcal{A}$, then $\mathrm{H}_\omega(\boldsymbol{A}_1 \mid \boldsymbol{A}_2)$ reduces to the conditional entropy of the random variables $A_{1,2}$ associated with the minimal projections $\{\widehat{a}_{1i}\}_{i\in I_1}$, $I_1 = 1,2,\ldots,d_1$ and $\{\widehat{a}_{2j}\}_{j\in I_2}$, $I_2 = 2,\ldots,d_2$, of $\boldsymbol{A}_{1,2}$. These projections give rise to probability distributions $\omega\lceil\boldsymbol{A}_1 = \{\omega(\widehat{a}_{1i}\}_{i=1}^{d_1}$ and $\omega\lceil\boldsymbol{A}_2 = \{\omega(\widehat{a}_{1j}\}_{j=1}^{d_2}$ and can be considered as the outcomes of $A_{1,2}$. Accordingly, the set of expectations $\omega(\widehat{a}_{1i}\widehat{a}_{2j})$ corresponds to the probability distribution of the joined random variable $A_1 \vee A_2$. Consequently, by using the minimal projections of $\boldsymbol{A}_1$ to decompose

$$\omega = \sum_{i\in I_1}\omega(\widehat{a}_{1i})\omega_i\,,\quad \omega_i(a) := \frac{\omega(\widehat{a}_i\,a)}{\omega(\widehat{a}_i)}\ \forall a\in\mathcal{A}\,,$$

it turns out that $\omega_i\lceil\boldsymbol{A}_1 = \{\omega_i(\widehat{a}_{1k}) = \delta_{ik}\}_{k=1}^{d_1}$ and $\omega_i\lceil\boldsymbol{A}_2 = \left\{\dfrac{\omega(\widehat{a}_{1i}\widehat{a}_{2j})}{\omega(\widehat{a}_{1i})}\right\}_{j=1}^{d_2}$.

Then, by means of (8.22) and of (2.91) one gets

$$\begin{aligned}
\mathrm{H}_\omega^{\boldsymbol{A}_1,\boldsymbol{A}_2}\left(\{\omega(\widehat{a}_{1i}),\omega_i\}_{i\in I_1}\right) &= H(A_1) - H(A_2) - H(A_1) + H(A_1\vee A_2) \\
&= H(A_1\vee A_2)\,,
\end{aligned}$$

where $H(A_{1,2}) := S(\omega \upharpoonright \boldsymbol{A}_{1,2})$ are the Shannon entropies of the random variables $A_{1,2}$. We now show that no decomposition can do better; indeed, in the Abelian case at hands, decompositions of $\omega$ correspond to partitions of unit with positive elements $\{\widehat{c}_k\}_{k\in K}$ in $\mathcal{A}$ such that

$$\omega = \sum_{k\in K} \omega(\widehat{c}_k)\,\omega_k \ , \quad \omega_k(a) = \frac{\omega(\widehat{c}_k\, a)}{\omega(\widehat{c}_k)} \quad \forall a \in \mathcal{A} \ .$$

Finally, Corollary 2.4.1 and strong subadditivity (see Proposition 2.4.1) yield

$$\begin{aligned}
\mathrm{H}_\omega^{\boldsymbol{A}_1,\boldsymbol{A}_2}(\{\omega(\widehat{c}_k),\omega_k\}_{k\in K}) &= H(A_1) - H(A_1 \vee C) - H(A_2) + H(A_2 \vee C) \\
&\leq H(A_1) - H(A_1 \vee C) - H(A_2) + H(A_1 \vee A_2 \vee C) \\
&\leq H(A_1 \vee A_2) - H(A_2) = H(A_1|A_2) \ ,
\end{aligned}$$

where $C$, $A_1 \vee C$ and $A_2 \vee C$ are random variables with probability distributions $\{\omega(\widehat{c}_k)\}_{k\in K}$, $\{\omega(\widehat{c}_k\,\widehat{a}_{1i})\}_{i\in I_1,k\in K}$ and $\{\omega(\widehat{c}_k\,\widehat{a}_{2j})\}_{j\in I_2,k\in K}$.

### *CNT* Entropy Rate and *CNT* Entropy

Apart from the relation (8.13), all other properties in Proposition 8.1.3 regard $n$-tuples of arbitrary maps $\gamma$ without reference to the dynamics. Since the purpose of the *CNT* entropy is to quantify the information production in a given quantum dynamical triplet $(\mathcal{A},\Theta,\omega)$, we set $\gamma_j := \Theta^j \circ \gamma$, where $j = 0,1,\ldots,n-1$ and $\gamma : \boldsymbol{M} \mapsto \mathcal{A}$ is a *CPU* map from a finite-dimensional $C^*$ algebra into $\mathcal{A}$. The first step is to ensure the existence of the rate

$$\lim_{n\to\infty} \frac{1}{n} \mathrm{H}_\omega\left(\gamma,\Theta\circ\gamma,\ldots,\Theta^{n-1}\circ\gamma\right) \ .$$

This limit exists since (8.14) together with (8.13) yield

$$\begin{aligned}
&\mathrm{H}_\omega\left(\gamma,\Theta\circ\gamma,\ldots,\Theta^{n-1}\circ\gamma\right) \leq \mathrm{H}_\omega\left(\gamma,\Theta\circ\gamma,\ldots,\Theta^{p-1}\circ\gamma\right) + \\
&\quad + \mathrm{H}_\omega\left(\Theta^p\circ\gamma,\Theta^{p+1}\circ\gamma,\ldots,\Theta^{n-1}\circ\gamma\right) \\
&\quad = \mathrm{H}_\omega\left(\gamma,\Theta\circ\gamma,\ldots,\Theta^{p-1}\circ\gamma\right) + \mathrm{H}_\omega\left(\gamma,\Theta\circ\gamma,\ldots,\Theta^{n-p-1}\circ\gamma\right) \ .
\end{aligned}$$

Thus, one can apply the same argument already used to show the existence of the classical entropy rate (3.2) or of the mean von Neumann entropy in quantum spin chains: actually,

$$\lim_{n\to\infty} \frac{1}{n} \mathrm{H}_\omega\left(\gamma,\Theta\circ\gamma,\ldots,\Theta^{n-1}\circ\gamma\right) = \inf_n \frac{1}{n} \mathrm{H}_\omega\left(\gamma,\Theta\circ\gamma,\ldots,\Theta^{n-1}\circ\gamma\right) \ . \tag{8.23}$$

**Definition 8.1.2.** *Given a quantum dynamical triplet $(\mathcal{A}, \Theta, \omega)$, where $\mathcal{A}$ is a $C^*$ or a von Neumann algebra and a CPU map $\gamma : \boldsymbol{M} \mapsto \mathcal{A}$ from a finite-dimensional $C^*$ algebra $\boldsymbol{M}$ into $\mathcal{A}$, the CNT* **entropy rate** *of $\gamma$ is*

$$\mathrm{h}_\omega^{\mathrm{CNT}}(\Theta, \gamma) := \lim_{n\to\infty} \frac{1}{n} \mathrm{H}_\omega\left(\gamma, \Theta \circ \gamma, \ldots, \Theta^{n-1} \circ \gamma\right) , \tag{8.24}$$

*while the CNT* **entropy** *of $(\mathcal{A}, \Theta, \omega)$ is defined by*

$$\mathrm{h}_\omega^{\mathrm{CNT}}(\Theta) = \sup_\gamma \mathrm{h}_\omega^{\mathrm{CNT}}(\Theta, \gamma) . \tag{8.25}$$

There are a few properties of the *CNT* entropy that easily follows from the above construction.

**Proposition 8.1.5.** *The following bounds hold for the CNT entropy rate of a quantum dynamical triplet $(\mathcal{A}, \Theta, \omega)$; given any CPU map $\gamma$ from a finite dimensional $C^*$ algebra $\boldsymbol{M}$ into $\mathcal{A}$, one has:*

$$0 \le \mathrm{h}_\omega^{\mathrm{CNT}}(\Theta, \gamma) \le \mathrm{H}_\omega(\gamma) \tag{8.26}$$

$$\frac{1}{n}\mathrm{h}_\omega^{\mathrm{CNT}}(\Theta^n, \gamma) \le \mathrm{h}_\omega^{\mathrm{CNT}}(\Theta, \gamma) \le \mathrm{h}_\omega^{\mathrm{CNT}}(\Theta^n, \gamma) . \tag{8.27}$$

**Proof:** The bounds in (8.26) come from the properties (8.10) and (8.13) of the *n-CPU* map entropies. The upper bound in (8.27) follows from subadditivity (8.14) together with property (8.13); indeed, since the limit (8.24) exists, one can fix $\mathbb{N} \ni n > 0$ and compute

$$\begin{aligned}
\mathrm{h}_\omega^{\mathrm{CNT}}(\Theta, \gamma) &= \lim_{k\to+\infty} \frac{1}{kn} \mathrm{H}_\omega\left(\gamma, \Theta \circ \gamma, \ldots, \Theta^{nk-1} \circ \gamma\right) \\
&\le \frac{1}{n} \lim_{k\to+\infty} \frac{1}{k} \sum_{j=0}^{n-1} \mathrm{H}_\omega\left(\Theta^j \circ \gamma, \Theta^{n+j} \circ \gamma, \ldots, \Theta^{n(k-1)+j} \circ \gamma\right) \\
&= \lim_{k\to+\infty} \frac{1}{k} \mathrm{H}_\omega\left(\gamma, \Theta^n \circ \gamma, \ldots, \Theta^{n(k-1)} \circ \gamma\right) = \mathrm{h}_\omega^{\mathrm{CNT}}(\Theta^n, \gamma) .
\end{aligned}$$

For the lower bound, first notice that *n-CPU* entropies remain unchanged by adding to the maps $\gamma_j$ in their arguments any number of *CPU* maps $\gamma_j'$ from trivial finite dimensional $C^*$ algebras $\{c\mathbb{1}_j\}$ [3] into $\mathcal{A}$. Indeed, for any such *CPU* map $\mathrm{H}_\omega(\gamma') = 0$, thus by subadditivity,

$$\mathrm{H}_\omega(\gamma_1, \gamma_2, \ldots, \gamma_n, \gamma_1', \ldots, \gamma_m') \le \mathrm{H}_\omega(\gamma_1, \gamma_2, \ldots, \gamma_n) .$$

However, any optimal decomposition for the right hand side of the previous inequality can always be used to decompose $\omega$ in the left hand side, too. This

[3] That is algebras consisting only of multiples of an identity operator $\mathbb{1}_j$

decomposition can then be used to invert the previous inequality; then, one expands $\Theta^{jn} \circ \gamma$ into the set

$$\Gamma^{(j)} := \left\{ \Theta^{jn} \circ \gamma \,,\, \Theta^{jn+1} \circ \gamma' \,,\, \ldots\, \Theta^{jn+n-1} \circ \gamma' \right\} ,$$

where $\gamma = \gamma \circ \gamma'$ and $\gamma'$ embeds the trivial subalgebra of $\boldsymbol{M}$ into $\boldsymbol{M}$. Using (8.15), one finally gets

$$\frac{1}{k} \mathrm{H}_\omega \left( \gamma, \Theta^n \circ \gamma, \ldots, \Theta^{n(k-1)} \circ \gamma \right) = \frac{1}{k} \mathrm{H}_\omega \left( \Gamma^{(0)}, \Gamma^{(1)}, \ldots, \Gamma^{(k-1)} \right)$$
$$\leq n \frac{1}{kn} \mathrm{H}_\omega \left( \gamma, \Theta \circ \gamma, \ldots, \Theta^{kn-1} \circ \gamma \right) ,$$

whence the result follows by taking the limit $k \to +\infty$. □

As much as for the *KS* entropy, one needs a means to avoid computing the supremum in (8.25). The structure of *AF* or *UHF* $C^*$ algebras or hyperfinite von Neumann algebras resembles that of classical dynamical systems admitting a generating partition (see Definition 2.3.5) Indeed, by using the continuity properties of the *n-CPU* entropies discussed in Proposition 8.1.2, one can prove a non-commutative counterpart to the Corollary 3.1.1 of the Kolmogorov-Sinai theorem 3.1.1.

**Proposition 8.1.6.** *[88] Let $(\mathcal{A}, \Theta, \omega)$ be a $C^*$ quantum dynamical triple which admits a sequence of CPU maps $\tau_j : \boldsymbol{M}_j \mapsto \mathcal{A}$ and $\sigma_j : \mathcal{A} \mapsto \boldsymbol{M}_j$ from finite-dimensional $C^*$ algebras with identity into $\mathcal{A}$ and back such that $\lim_{j\to+\infty} \|\tau_j \circ \sigma_j[A] - A\| = 0$ for all $A \in \mathcal{A}$. Then*

$$\mathrm{h}^{\mathrm{CNT}}_\omega (\Theta) = \lim_{j\to+\infty} \mathrm{h}^{\mathrm{CNT}}_\omega (\Theta, \tau_j) \ .$$

**Proof:** Let $\gamma : \boldsymbol{M} \mapsto \mathcal{A}$ be any *CPU* map from a finite-dimensional $C^*$ algebra $\boldsymbol{M}$ into $\mathcal{A}$; set $\gamma_j := \tau_j \circ \sigma_j \circ \gamma$. Then, $\gamma_j(M) \to \gamma(M)$ in norm for all $M \in \boldsymbol{M}$ whence $\|\gamma_j - \gamma\| \to 0$ when $j \to +\infty$ for $\boldsymbol{M}$ is finite-dimensional. The same is true for the *CPU* maps $\Theta^k \circ \gamma_j$ and $\Theta^k \circ \gamma$, $k \geq 0$. Since $\|\Theta^k \circ (\gamma_j - \gamma)\|_\omega \leq \|\Theta^k \circ (\gamma_j - \gamma)\|$, Proposition 8.1.2 yields

$$\frac{1}{n} \left| \mathrm{H}_\omega \left( \gamma_j, \Theta \circ \gamma_j, \ldots, \Theta^{n-1} \circ \gamma_j \right) - \mathrm{H}_\omega \left( \gamma, \Theta \circ \gamma, \ldots, \Theta^{n-1} \circ \gamma \right) \right| \leq \varepsilon$$

for any $\varepsilon > 0$ and $j$ sufficiently large, whence

$$\lim_{j\to+\infty} \mathrm{h}^{\mathrm{CNT}}_\omega (\Theta, \gamma_j) = \mathrm{h}^{\mathrm{CNT}}_\omega (\Theta, \gamma) \ .$$

Now, using the monotonicity property (8.15), it turns out that

$$\mathrm{h}^{\mathrm{CNT}}_\omega (\Theta, \gamma) = \liminf_{j\to+\infty} \mathrm{h}^{\mathrm{CNT}}_\omega (\Theta, \gamma_j) \leq \liminf_{j\to+\infty} \mathrm{h}^{\mathrm{CNT}}_\omega (\Theta, \tau_j)$$
$$\leq \limsup_{j\to+\infty} \mathrm{h}^{\mathrm{CNT}}_\omega (\Theta, \tau_j) \leq \mathrm{h}^{\mathrm{CNT}}_\omega (\Theta) \ .$$

The result thus follows by taking the supremum over $\gamma$. □

In case $\mathcal{A}$ is an *AF* or a *UHF* $C^*$ algebra, namely the norm completion of an increasing sequence of finite-dimensional $C^*$ subalgebras $\boldsymbol{M}_{n_i} \subset \mathcal{A}$ or matrix algebras $M_{n_i}(\mathbb{C})$ (see Remark 7.1.1), one chooses as *CPU* maps $\sigma_j$ the corresponding conditional expectations and, as the *CPU* maps $\tau_j$, the natural embeddings $\iota_{\boldsymbol{M}_{n_j}} : \boldsymbol{M}_{n_j} \mapsto \mathcal{A}$ [88, 89, 231, 232].

A similar result as in Proposition 8.1.6 holds for von Neumann quantum dynamical systems with $\mathcal{A}$ a hyperfinite von Neumann algebra (for the proof see [88, 222]).

**Proposition 8.1.7.** *Let* $(\mathcal{A}, \Theta, \omega)$ *be a von Neumann dynamical triple, with* $\mathcal{A}$ *hyperfinite and generated by an increasing sequence of finite-dimensional von Neumann subalgebras* $\{\boldsymbol{M}_k\}_{k\in\mathbb{N}}$*; then,*

$$\mathrm{h}^{\mathrm{CNT}}_{\omega}(\Theta) = \lim_{k\to+\infty} \mathrm{h}^{\mathrm{CNT}}_{\omega}(\Theta, \boldsymbol{M}_k) \ .$$

**Remark 8.1.2.** When a quantum dynamical system under has the algebraic structure as in the above proposition, then the continuity properties of the *CNT* entropy allows to turn the lower bound in (8.27) into an equality [88], namely

$$\mathrm{h}^{\mathrm{CNT}}_{\omega}(\Theta^n) = |n|\, \mathrm{h}^{\mathrm{CNT}}_{\omega}(\Theta) \qquad \forall n \in \mathbb{Z} \ .$$

Moreover, this result can be extended to a one-parameter group of automorphisms $\{\Theta_t\}_{t\in\mathbb{R}}$, that is [214, 222]

$$\mathrm{h}^{\mathrm{CNT}}_{\omega}(\Theta_t) = |t|\, \mathrm{h}^{\mathrm{CNT}}_{\omega}(\Theta) \qquad \forall t \in \mathbb{R} \ ,$$

where $\Theta := \Theta_{t=1}$.

### 8.1.1 *CNT* Entropy: Quasi-Local Algebras

As seen in Section 7.1, in quantum statistical mechanics one often considers quasi-local algebras $\mathcal{A}$ which are generated (inductive limit) by local algebras $\mathcal{A}_V$, indexed by finite volumes $V$, that are not finite dimensional. For instance, this is the case with Bosons in $\mathbb{R}^3$ or with a lattice $\mathbb{Z}^3$ with infinite dimensional Hilbert spaces at its sites; in such a setting one cannot resort to either of the preceding two propositions to compute the *CNT* entropy (8.25).

Nevertheless, a quantum Kolmogorov-Sinai-like theorem holds under the following physically plausible assumptions [232]; we shall consider dynamical triples $(\mathcal{A}, \Theta, \omega)$ consisting of

1. a quasi-local $C^*$-algebra $\mathcal{A}$ which is the norm completion of $\bigcup_V \mathcal{A}_V$ where the local algebras $\mathcal{A}_V$ associated with finite volumes $V \subset \mathbb{R}^3$ share a same identity and are isomorphic to the von Neumann algebras $\mathbb{B}(\mathbb{H}_V)$ [4];

[4] In the following we shall restrict to $\mathbb{R}^3$ for simplicity; the result holds in general for $\mathbb{R}^d$ and $\mathbb{Z}^d$, $d \geq 1$ [232].

2. if $V \subset V'$, set $V'' := V' \setminus V$, then $\mathbb{H}_{V'} = \mathbb{H}_V \otimes \mathbb{H}_{V''}$, $\mathcal{A}_{V'} = \mathcal{A}_V \otimes \mathcal{A}_{V''}$;
3. the $\Theta$-invariant state $\omega$ is locally normal, that is $\omega \upharpoonright \mathcal{A}_V$ is a density matrix $\rho_V \in \mathbb{B}_1^+(\mathbb{H}_V)$.

Let $\iota_V$ denote the embedding of $\mathcal{A}_V$ into $\mathcal{A}$; as shown in [231, 232], using the second assumption one can construct a family of *CPU* conditional expectations $\sigma_V : \mathcal{A} \mapsto \mathcal{A}_V$ such that $\|\iota_V \circ \sigma_V[A] - A\| \to 0$ for all $A \in \mathcal{A}$ when $V \uparrow \mathbb{R}^3$. Consider a *CPU* map $\gamma : \boldsymbol{M} \mapsto \mathcal{A}$ where $\boldsymbol{M}$ is a finite-dimensional unital $C^*$ algebra and set $\gamma_V := \iota_V \circ \sigma_V \circ \gamma$. Now, $\lim_{V\uparrow\mathbb{R}^3} \|\gamma_V - \gamma\| = 0$ for $\boldsymbol{M}$ is finite dimensional whence Proposition 8.1.6 yields

$$\lim_{V\uparrow\mathbb{R}^3} \mathrm{h}_\omega^{\mathrm{CNT}}(\Theta, \gamma_V) = \mathrm{h}_\omega^{\mathrm{CNT}}(\Theta, \gamma) .$$

From (8.25) and the previous equality one derives

$$\begin{aligned} \mathrm{h}_\omega^{\mathrm{CNT}}(\Theta) &= \sup_\gamma \lim_{V\uparrow\mathbb{R}^3} \mathrm{h}_\omega^{\mathrm{CNT}}(\Theta, \gamma_V) \leq \limsup_{V\uparrow\mathbb{R}^3} \sup_\gamma \mathrm{h}_\omega^{\mathrm{CNT}}(\Theta, \gamma_V) \\ &\leq \limsup_{V\uparrow\mathbb{R}^3} \sup_{\lambda_V : \boldsymbol{M} \mapsto \mathcal{A}_V} \mathrm{h}_\omega^{\mathrm{CNT}}(\Theta, \iota_V \circ \lambda_V) \leq \mathrm{h}_\omega^{\mathrm{CNT}}(\Theta) , \end{aligned}$$

where the second inequality holds for not all *CPU* maps $\lambda_V : \boldsymbol{M} \mapsto \mathcal{A}_V$ are of the form of $\gamma_V$. Thus,

$$\mathrm{h}_\omega^{\mathrm{CNT}}(\Theta) = \lim_{V\uparrow\mathbb{R}^3} \sup_{\gamma_V : \boldsymbol{M} \mapsto \mathcal{A}_V} \mathrm{h}_\omega^{\mathrm{CNT}}(\Theta, \gamma_V) . \tag{8.28}$$

Fix a volume $V \subset \mathbb{R}^3$ with local density matrix $\rho_V = \sum_{i=1}^{+\infty} r_V^i |r_V^i\rangle\langle r_V^i|$ where the eigenvalues $r_V^i$ are repeated according to their multiplicities and decreasingly ordered. Let $P_V^{(k)} := \sum_{i=1}^k |r_V^i\rangle\langle r_V^i|$, $Q_V^{(k)} := \mathbb{1} - P_V^{(k)}$ and

$$\mathcal{A}_V^{(k)} := P_V^{(k)} \mathcal{A}_V P_V^{(k)} \oplus \mathbb{C} Q_V^{(k)} .$$

The latter is a finite von Neumann subalgebra of $\mathcal{A}_V$; consider the map

$$\sigma_V^{(k)}[A] := P_V^{(k)} A P_V^{(k)} + \frac{\omega(Q_V^{(k)} A Q_V^{(k)})}{\omega(Q_V^{(k)})} Q_V^{(k)} \qquad \forall A \in \mathcal{A}_V .$$

It linearly maps $\mathcal{A}_V$ into $\mathcal{A}_V^{(k)}$, is unital and positive; further, if $A \in \mathcal{A}_V$ and $\mathcal{A}_V^{(k)} \ni B = P_V^{(k)} Z P_V^{(k)} + c_B Q_V^{(k)}$, with $c_B \in \mathbb{C}$ and $Z \in \mathcal{A}_V$,

$$\sigma_V^{(k)}[A B] = P_V^{(k)} A P_V^{(k)} Z P_V^{(k)} + c_B \frac{\omega(Q_V^{(k)} A Q^{(k)})}{\omega(Q_V^{(k)})} Q_V^{(k)} = \sigma_V^{(k)}[A] B .$$

Therefore, according to Definition 5.2.3, $\sigma_V^{(k)} : \mathcal{A}_V \mapsto \mathcal{A}_V^{(k)}$ is a conditional expectation; then, for any *CPU* map $\gamma_V : \boldsymbol{M} \mapsto \mathcal{A}_V$ set $\tau_V := \iota_V \circ \gamma_V$ and

$\tau^{(k)} := \iota_V \circ \iota_V^{(k)} \circ \sigma_V^{(k)} \circ \gamma$, where $\iota_V^{(k)}$ is the embedding of $\mathcal{A}_V^{(k)}$ into $\mathcal{A}_V$. We now show that, for any $\gamma_V : \boldsymbol{M} \mapsto \mathcal{A}_V$,

$$\mathrm{h}_\omega^{\mathrm{CNT}}(\Theta, \tau_V) = \lim_{k\to+\infty} \mathrm{h}_\omega^{\mathrm{CNT}}\left(\Theta, \tau_V^{(k)}\right) . \tag{8.29}$$

This lead to the result that, in order to compute the *CNT* entropy, one must essentially compute the *CNT* entropy rates of the finite-dimensional subalgebras projected out by the spectral projections of local states.

**Theorem 8.1.1.** *[232] Under the assumptions* $1-3$ *on* $(\mathcal{A}, \Theta, \omega)$,

$$\mathrm{h}_\omega^{\mathrm{CNT}}(\Theta) = \lim_{V\uparrow\mathbb{R}^3} \lim_{k\to+\infty} \mathrm{h}_\omega^{\mathrm{CNT}}\left(\Theta, \mathcal{A}^{(k)}\right) .$$

**Proof:** Writing $\mathrm{h}_\omega^{\mathrm{CNT}}\left(\Theta, \mathcal{A}_V^{(k)}\right) = \mathrm{h}_\omega^{\mathrm{CNT}}\left(\Theta, \iota_V \circ \iota_V^{(k)}\right)$ and using (8.29) and (8.15) one gets

$$\begin{aligned}
\lim_{k\to+\infty} \mathrm{h}_\omega^{\mathrm{CNT}}\left(\Theta, \mathcal{A}_V^{(k)}\right) &\le \sup_{\gamma_V:\boldsymbol{M}\mapsto\mathcal{A}_V} \mathrm{h}_\omega^{\mathrm{CNT}}\Bigg(\Theta, \underbrace{\iota_V \circ \gamma_V}_{\tau_V}\Bigg) \\
&= \sup_{\gamma_V:\boldsymbol{M}\mapsto\mathcal{A}_V} \lim_{k\to+\infty} \mathrm{h}_\omega^{\mathrm{CNT}}\Bigg(\Theta, \underbrace{\iota_V \circ \iota_V^{(k)} \circ \sigma_V^{(k)} \circ \gamma_V}_{\tau_V^{(k)}}\Bigg) \\
&\le \lim_{n\to+\infty} \sup_{\gamma_V:\boldsymbol{M}\mapsto\mathcal{A}_V} \mathrm{h}_\omega^{\mathrm{CNT}}\left(\Theta, \tau_V^{(n)}\right) \\
&\le \lim_{k\to+\infty} \mathrm{h}_\omega^{\mathrm{CNT}}\left(\Theta, \iota_V \circ \iota_V^{(k)}\right) = \lim_{k\to+\infty} \mathrm{h}_\omega^{\mathrm{CNT}}\left(\Theta, \mathcal{A}_V^{(k)}\right) .
\end{aligned}$$

Therefore, the result follows from (8.28) and the above estimates which yield

$$\sup_{\gamma_V:\boldsymbol{M}\mapsto\mathcal{A}_V} \mathrm{h}_\omega^{\mathrm{CNT}}(\Theta, \iota_V \circ \gamma_V) = \lim_{k\to+\infty} \mathrm{h}_\omega^{\mathrm{CNT}}\left(\Theta, \mathcal{A}_V^{(k)}\right) .$$

□

**Proof of (8.29)** We argue as in the proof of Proposition 8.1.2; we thus set $\gamma_j := \Theta^j \circ \tau_V$, $\gamma_j' := \Theta^j \circ \tau_V^{(k)}$ and estimate the norms

$$\left\| \frac{\omega(X_i'\Theta^j \circ \tau_V[M])}{\omega(X_i')} - \frac{\omega(X_i'\Theta^j \circ \tau_V^{(k)}[M])}{\omega(X_i')} \right\| , \tag{$*$}$$

where a short hand notation for (8.5) has been used and the $X_i'$ are positive elements of the commutant $\pi_\omega(\mathcal{A})'$ such that $\sum_i X_i' = 1\!\!1$.

By writing $\mathcal{A}_V \ni X = (P_V^{(n)} + Q_V^{(n)}) X (P_V^{(n)} + Q_V^{(n)})$, and setting $\nu_i(M) := \omega(X_i' \Theta^j \circ \tau_V[M])$ and $\nu_i^{(k)}(M) := \omega(X_i' \Theta^j \circ \tau_V^{(k)}[M])$, one finds

$$\begin{aligned}\nu_i(M) - \nu_i^{(k)}(M) &= \underbrace{\omega(\Theta^{-j}[X_i'] Q_V^{(k)}\, \gamma_V[M]\, Q_V^{(k)})}_{A} \\ &+ \underbrace{\omega(\Theta^{-j}[X_i'] P_V^{(k)}\, \gamma_V[M]\, Q_V^{(k)})}_{B} + \underbrace{\omega(\Theta^{-j}[X_i'] Q_V^{(k)}\, \gamma_V[M]\, P_V^{(k)})}_{C} \\ &- \underbrace{\omega(\Theta^{-j}[X_i'] Q_V^{(k)})\, \frac{\omega(Q_V^{(k)} \gamma_V[M] Q_V^{(k)})}{\omega(Q_V^{(k)})}}_{D} .\end{aligned}$$

Since $0 \le \Theta^{-j}[X_i'] =: Z \in \pi_\omega(\mathcal{A})'$ commutes with the projections $Y = P_V^{(k)}, Q_V^{(k)}$, one can write $ZY = \sqrt{Z} Y \sqrt{Z} = \sqrt{Z} Y \sqrt{Z} Y$; thus, using the Cauchy-Schwartz inequality (5.49), one estimates

$$\begin{aligned}|A|^2 &\le \omega(Z Q_V^{(k)})\, \omega(Z Q_V^{(k)} \gamma_V[M^\dagger] Q_V^{(k)} \gamma_V[M] Q_V^{(k)}) \\ &\le \omega(X_i')\, \omega(Z Q_V^{(k)}) \|\gamma_V\|^2\, \|M\|^2 \\ |B|^2 &\le \omega(Z P_V^{(k)})\, \omega(Z Q_V^{(k)} \gamma_V[M^\dagger] P_V^{(k)} \gamma_V[M] Q_V^{(k)}) \\ &\le \omega(X_i')\, \omega(Z Q_V^{(k)}) \|\gamma_V\|^2\, \|M\|^2 \\ |C|^2 &\le \omega(Z P_V^{(k)})\, \omega(Z Q_V^{(k)} \gamma_V[M] P_V^{(k)} \gamma_V[M^\dagger] Q_V^{(k)}) \\ &\le \omega(X_i')\, \omega(Z P_V^{(k)}) \|\gamma_V\|^2\, \|M\|^2 \\ |D|^2 &\le \omega(X_i')\, \omega(Z Q_V^{(n)})\, \|\gamma_V\|^2\, \|M\|^2 ,\end{aligned}$$

where (5.33) and $Y \le 1\!\!1 \Longrightarrow X^\dagger Y X \le X^\dagger X$ have been repeatedly used; further, in the expression $C$, $Z$ has been transferred to the right side of $\omega(\cdot)$ before applying (5.49).

Since $\sum_i X_i' = 1\!\!1$, these estimates obtain

$$\sum_i \frac{1}{\omega(X_i')} \|\nu_i - \nu_i^{(k)}\|^2 \le 16\, \|\gamma_V\| \sum_i \omega(\Theta^{-j}[X_i'] Q_V^{(k)}) = 16\, \|\gamma_V\|^2 \underbrace{\sum_{j=k+1} r_V^j}_{\le \varepsilon}$$

for and $\varepsilon > 0$ and $k$ sufficiently large. Notice that the summands are $\omega(X_i')$ times the squares of the norms $(*)$; we can now distinguish between the set $I$ of those $i$'s such that the norms $(*) \le \varepsilon^{1/3}$ and the rest $I^c$. Then,

$$\varepsilon \ge \sum_{i \in I^c} \omega(X_i') \left\| \frac{\nu_i - \nu_i^{(k)}}{\omega(X_i')} \right\|^2 \ge \varepsilon^{2/3} \sum_{i \in I^c} \omega(X_i')$$

implies that the total weight of $I^c$ is smaller than $\varepsilon^{1/3}$. As in the proof of Proposition 8.1.2, this can be used to show that, for any $\eta > 0$,

$$\left|\mathrm{H}_\omega\left(\tau_V,\Theta\circ\tau_V,\ldots,\Theta^{n-1}\circ\tau_V\right)-\mathrm{H}_\omega\left(\tau_V^{(k)},\Theta\circ\tau_V^{(k)},\ldots,\Theta^{n-1}\circ\tau_V^{(k)}\right)\right|\leq n\eta$$

for $k$ sufficiently large. □

### 8.1.2 *CNT* Entropy: Stationary Couplings

In this section we reconsider the expressions (8.21) and (8.22) in the following algebraic setting:

- a von Neumann algebra $\mathcal{A}$ with a normal state $\omega$;
- an Abelian von Neumann algebra $\mathcal{B}$ with a normal state $\omega_\mu$ [5];
- the tensor product von Neumann algebra $\mathcal{A}\otimes\mathcal{B}$ equipped with a normal state $\widetilde{\omega}$ such that its marginal states are $\widetilde{\omega}\lceil\mathcal{A}=\omega$ and $\widetilde{\omega}\lceil\mathcal{B}=\omega_\mu$.

Let $\boldsymbol{A}\subseteq\mathcal{A}$ and $\boldsymbol{B}\subseteq\mathcal{B}$ be finite-dimensional $C^*$ subalgebras; as *CPU* maps $\gamma_1$, respectively $\gamma_2$ in (8.21) we shall take the embeddings $\gamma_i=\iota_{\boldsymbol{B}}$, respectively $\gamma_2=\iota_{\boldsymbol{A}}$ of $\boldsymbol{B}$, respectively $\boldsymbol{A}$ into $\mathcal{A}\otimes\mathcal{B}$. We shall focus upon the quantity $\mathrm{H}_{\widetilde{\omega}}(\boldsymbol{B}\,|\,\boldsymbol{A})$: one has the following result [210].

**Proposition 8.1.8.** *Let $B$ be the random variable corresponding to the subalgebra $\boldsymbol{B}$ and $H_\mu(B)$ denote the Shannon entropy corresponding to the von Neumann entropy of the state $\omega_\mu$ restricted to $\boldsymbol{B}$. Then,*

$$\mathrm{H}_{\widetilde{\omega}}(\boldsymbol{B}\,|\,\boldsymbol{A})=H_\mu(B)-S\left(\omega\otimes\omega_\mu\lceil\boldsymbol{A}\otimes\boldsymbol{B}\,,\,\widetilde{\omega}\lceil\boldsymbol{A}\otimes\boldsymbol{B}\right)\ . \tag{8.30}$$

**Proof:** Given the minimal projections $\{\widehat{b}_j\}_{j\in I_B}$, $I_B=\{1,2,\ldots,d\}$ of the finite dimensional Abelian algebra $\boldsymbol{B}\subseteq\mathcal{B}$ and a convex decomposition $\widetilde{\omega}=\sum_{i\in I}\lambda_i\widetilde{\omega}_i$, one can construct a finer decomposition of the form $\widetilde{\omega}=\sum_{i\in I;j\in I_B}\lambda_i\mu_{ij}\widetilde{\omega}_{ij}$, by further decomposing $\widetilde{\omega}_i=\sum_{j\in I_B}\mu_{ij}\widetilde{\omega}_{ij}$, where the states $\widetilde{\omega}_{ij}$ on $\mathcal{A}\otimes\mathcal{B}$ and their weights $\mu_{ij}$ are defined by

$$\widetilde{\omega}_{ij}(a\otimes b):=\frac{\omega_i(a\otimes\widehat{b}_jb)}{\widetilde{\omega}_i(\widehat{b}_j)}\ ,\qquad\mu_{ij}:=\widetilde{\omega}_i(\widehat{b}_j)\qquad(*)$$

for all $a\in\boldsymbol{A}$ and $b\in\boldsymbol{B}$. Since for all $i\in I$

$$\widetilde{\omega}_{ij}(\widehat{b}_k)=\frac{\widetilde{\omega}_i(\widehat{b}_j\widehat{b}_k)}{\widetilde{\omega}_i(\widehat{b}_j)}=\delta_{jk}\ ,$$

the restrictions $\widetilde{\omega}_{ij}\lceil\boldsymbol{B}$ are probability distributions $\Lambda_{ij}=\{\delta_{jk}\}_{k\in I_B}$ with zero Shannon entropy. Then, using (8.22) and (6.23), one computes

[5] According to Section 5.3.2, the state $\omega_\mu$ corresponds to integration with respect to a suitable measure $\mu$ and measure space $\mathcal{X}$.

$$\begin{aligned}\mathrm{H}^{\boldsymbol{B},\boldsymbol{A}}_{\widetilde{\omega}}\left(\{\lambda_i\mu_{ij},\widetilde{\omega}_{ij}\}_{i\in I,j\in I_B}\right) &= H_\mu(B) - S\left(\omega\restriction\boldsymbol{A}\right)\\ &\quad + \sum_{j\in I_B}\lambda_i\mu_{ij}\,S\left(\widetilde{\omega}_{ij}\restriction\boldsymbol{A}\right) \qquad (**)\\ \mathrm{H}^{\boldsymbol{B},\boldsymbol{A}}_{\widetilde{\omega}}\left(\{\lambda_i,\widetilde{\omega}_i\}_{i\in I}\right) &= H_\mu(B) - S\left(\omega\restriction\mathcal{A}\right)\\ &\quad + \sum_{i\in I}\lambda_i\Big(S\left(\widetilde{\omega}_i\restriction\boldsymbol{A}\right) - S\left(\widetilde{\omega}_i\restriction\boldsymbol{B}\right)\Big)\,.\end{aligned}$$

It thus turns out that

$$\mathrm{H}^{\boldsymbol{B},\boldsymbol{A}}_{\widetilde{\omega}}\left(\{\lambda_i\mu_{ij},\widetilde{\omega}_{ij}\}_{i\in I,j\in I_B}\right) - \mathrm{H}^{\boldsymbol{B},\boldsymbol{A}}_{\widetilde{\omega}}\left(\{\lambda_i,\widetilde{\omega}_i\}_{i\in I}\right) = -\sum_{i\in I;j\in I_B}\lambda_i\mu_{ij}\log\mu_{ij}$$

$$-\sum_{i\in I}\lambda_i S\left(\widetilde{\omega}_i\restriction\boldsymbol{A}\right) + \sum_{i\in I;j\in I_B}\lambda_i\mu_{ij}S\left(\widetilde{\omega}_{ij}\restriction\boldsymbol{A}\right) \geq 0\,.$$

Indeed, since $\mu_{ij}\widetilde{\omega}_{ij} \leq \widetilde{\omega}_i$, the monotonicity of $f(x) = \log x$ as an operator function (see Example 5.2.3.9) gives

$$\begin{aligned}\sum_{j\in I_B}\mu_{ij}\widetilde{\omega}_{ij}\restriction\boldsymbol{A}\log\widetilde{\omega}_{ij}\restriction\boldsymbol{A} &= \sum_{j\in I_B}\mu_{ij}\widetilde{\omega}_{ij}\restriction\boldsymbol{A}\Big(\log\Big(\mu_{ij}\,\widetilde{\omega}_{ij}\restriction\boldsymbol{A}\Big) - \log\mu_{ij}\Big)\\ &\leq \sum_{j\in I_B}\mu_{ij}\widetilde{\omega}_{ij}\restriction\boldsymbol{A}\Big(\log\widetilde{\omega}_i\restriction\boldsymbol{A} - \log\mu_{ij}\Big)\\ &= \widetilde{\omega}_i\restriction\boldsymbol{A}\log\widetilde{\omega}_i\restriction\boldsymbol{A} - \sum_{j\in I_B}\mu_{ij}\,\widetilde{\omega}_{ij}\restriction\boldsymbol{A}\,\log\mu_{ij}\,.\end{aligned}$$

Therefore, after multiplying by the weights $\lambda_i$ and summing over $i \in I$, by taking the trace and considering that the marginal state $\omega_{ij}\restriction\boldsymbol{A}$ has trace 1, one finally gets

$$-\sum_{i\in I;j\in I_B}\lambda_i\mu_{ij}S\left(\widetilde{\omega}_{ij}\restriction\boldsymbol{A}\right) \leq -\sum_{i\in I}\lambda_i S\left(\widetilde{\omega}_i\restriction\boldsymbol{A}\right) - \sum_{i\in I;j\in I_B}\lambda_i\mu_{ij}\log\mu_{ij}\,.$$

One thus concludes that in order to compute $\mathrm{H}_{\widetilde{\omega}}\left(\boldsymbol{B}\,|\,\boldsymbol{A}\right)$ one can start with decompositions of $\widetilde{\omega}$ in terms of states of the form $(*)$. Then, consider the decomposition $\widetilde{\omega} = \sum_{j\in I_B}\nu_j\widetilde{\omega}_j$, where

$$\widetilde{\omega}_j(a\otimes b) := \frac{\widetilde{\omega}(a\otimes\widehat{b}_j b)}{\widetilde{\omega}(\widehat{b}_j)}\,,\qquad \nu_j := \widetilde{\omega}(\widehat{b}_j)\quad a\in\boldsymbol{A}\,,\ b\in\boldsymbol{B}\,.$$

Since $S\left(\widetilde{\omega}_j\restriction\boldsymbol{B}\right) = 0$, this decomposition contributes with

$$\mathrm{H}^{\boldsymbol{B},\boldsymbol{A}}_{\widetilde{\omega}}\left(\{\lambda_j,\widetilde{\omega}_j\}_{j\in I_B}\right) = H_\mu(B) - S\left(\omega\restriction\boldsymbol{A}\right) + \sum_{j\in I_B}\nu_j S\left(\widetilde{\omega}_j\restriction\boldsymbol{A}\right) \qquad (***)\,.$$

Further, notice that the decomposition appearing in equation $(**)$ is such that

$$\sum_{i\in I}\lambda_i\mu_{ij}=\nu_j=\omega_\mu(\widehat{b}_j)\ ,\quad \sum_{i\in I}\frac{\lambda_i\mu_{ij}}{\nu_j}\,\widetilde{\omega}_{ij}=\widetilde{\omega}_j\ .$$

Therefore, by the concavity of the von Neumann entropy (see (5.156))

$$\sum_{j\in I_B}\lambda_i\mu_{ij}\,S\left(\widetilde{\omega}_{ij}\restriction \boldsymbol{A}\right)\leq\sum_{j\in I_B}\nu_j\,S\left(\widetilde{\omega}_j\restriction \boldsymbol{A}\right)\ ,$$

whence decompositions of the form $(***)$ are optimal. The proof is finally completed by calculating

$$\begin{aligned}
&S\left(\omega\otimes\omega_\mu\restriction \boldsymbol{A}\otimes\boldsymbol{B}\,,\,\widetilde{\omega}\restriction \boldsymbol{A}\otimes\boldsymbol{B}\right)=\\
&\mathrm{Tr}\Big(\omega\restriction \boldsymbol{A}\otimes\omega_\mu\restriction \boldsymbol{B}\Big(\log\omega\restriction \boldsymbol{A}\otimes\omega_\mu\restriction \boldsymbol{B}\ -\ \log\widetilde{\omega}\restriction \boldsymbol{A}\otimes\boldsymbol{B}\Big)\Big)=\\
&\quad=-H_\mu(B)-S\left(\omega\restriction \boldsymbol{A}\right)+\sum_{j\in I_B}\nu_j\,\mathrm{Tr}\Big(\widetilde{\omega}_j\restriction \boldsymbol{A}\log\Big(\nu_j\,\widetilde{\omega}_j\restriction \boldsymbol{A}\Big)\Big)=\\
&\quad=\sum_{j\in I_B}\nu_j\,S\left(\widetilde{\omega}_j\restriction \boldsymbol{A}\right)\ -\ S\left(\omega\restriction \boldsymbol{A}\right)\ .
\end{aligned}$$

□

The previous considerations are useful in a different approach to the *CNT* entropy which was developed in [264] (see also [222]). We shall refer to the formulation used in [13]

**Definition 8.1.3.** *Let $(\mathcal{A},\Theta,\omega)$ be a dynamical triple with $\mathcal{A}$ a hyperfinite von Neumann algebra and $\omega$ a normal $\Theta$-invariant state; a* **stationary coupling** *to a commutative dynamical triple $(\mathcal{B},\theta,\omega_\mu)$ where $\mathcal{B}$ is an Abelian von Neumann algebra, is any triplet of the form $(\mathcal{A}\otimes\mathcal{B},\Theta\otimes\theta,\widetilde{\omega})$ where $\widetilde{\omega}$ is a $\Theta\otimes\theta$-invariant state such that $\widetilde{\omega}\restriction\mathcal{A}=\omega$ and $\widetilde{\omega}\restriction\mathcal{B}=\omega_\mu$.*

The quantity $\mathrm{H}_{\widetilde{\omega}}\left(\boldsymbol{B}\,|\,\boldsymbol{A}\right)$ and its expression (8.30) can be generalized as follows [210]. For any finite dimensional subalgebra $\boldsymbol{B}\subset\mathcal{B}$ let

$$\begin{aligned}
\mathrm{H}_{\widetilde{\omega}}\left(\boldsymbol{B}\,|\,\mathcal{A}\right)&:=\sup_{\widetilde{\omega}=\lambda_i\widetilde{\omega}_i}\sum_i\lambda_i\Big(S\left(\widetilde{\omega}_i\restriction \boldsymbol{B}\,,\,\widetilde{\omega}\restriction \boldsymbol{B}\right)\ -\ S\left(\widetilde{\omega}_i\restriction\mathcal{A}\,,\,\widetilde{\omega}\restriction\mathcal{A}\right)\Big) &(8.31)\\
&=\ S\left(\omega_\mu\restriction \boldsymbol{B}\right)\ -\ S\left(\widetilde{\omega}\restriction\mathcal{A}\otimes\boldsymbol{B}\,,\,\omega\otimes\omega_\mu\restriction\mathcal{A}\otimes\boldsymbol{B}\right)\ . &(8.32)
\end{aligned}$$

It then turns out [264, 222, 210] that

$$\mathrm{h}^{\mathrm{CNT}}_{\omega}\left(\Theta\right)=\sup_{\boldsymbol{B},\mathcal{B},\theta,\widetilde{\omega}}\Big\{\mathrm{h}^{\mathrm{KS}}_{\omega_\mu}\left(\theta,\boldsymbol{B}\right)\ -\ \mathrm{H}_{\widetilde{\omega}}\left(\boldsymbol{B}\,|\,\mathcal{A}\right)\Big\}\ . \tag{8.33}$$

where the supremum is computed over all possible stationary couplings and all finite-dimensional subalgebras $\boldsymbol{B}\subset\mathcal{B}$. Notice also that in the expression of the *KS* entropy, in according with Section 5.3.2, we have kept the algebraic notation whereby $\boldsymbol{B}$ stands for a partition of a phase-space $\mathcal{X}$, the automorphism $\theta$ for a measurable, invertible dynamical map $T : \mathcal{X} \mapsto \mathcal{X}$ and the state $\omega_\mu$ for a $T$-invariant measure $\mu$.

**Remark 8.1.3.** The relative entropy as it has been used so far has always involved density matrices or restrictions of states to finite dimensional subalgebra, whereas in (8.31), because of the presence of the generic von Neumann algebra $\mathcal{A}$, it apparently works in a more general context. It turns out that the expression (6.23) for the relative entropy has a generalization to any unital $C^*$ algebra $\mathcal{A}$ and to generic positive, linear functionals (not even normalized) $\omega_{1,2}$ on it [88, 222, 300]:

$$S(\omega_1, \omega_2) = \sup \int_0^{+\infty} \frac{\mathrm{d}t}{t}\left[\frac{\omega_1(\mathbb{1})}{1+t} - \omega_1(y(t)^\dagger\, y(t)) - \frac{1}{t}\,\omega_2(x(t)\, x(t)^\dagger)\right] ,$$

where $y(t) = \mathbb{1} - x(t)$ and $t \mapsto x(t) \in \mathcal{A}$ is any step function with values in $\mathcal{A}$ vanishing in a neighborhood of $t = 0$.

### 8.1.3 *CNT* entropy: Applications

We start the presentation of various concrete applications of the *CNT* entropy by showing that in a commutative context it reduces to the *KS* entropy.

Consider a classical dynamical system $(\mathcal{X}, T, \mu)$ that possesses a generating partition $\mathcal{P} = \{P_i\}_{i=1}^p$ (see Definition 2.3.5) and its corresponding von Neumann triplet $(\mathcal{M} := \mathbb{L}^\infty_\mu(\mathcal{X}), \Theta_T, \omega_\mu)$ (compare Definition 2.2.4). In this framework, the partition $\mathcal{P}$ is identified with the finite-dimensional subalgebra $\boldsymbol{M}_{\mathcal{P}}$ generated by the characteristic functions $\chi_{P_i}$ of the atoms $P_i$ of $\mathcal{P}$. Furthermore, the partitions $\mathcal{P}^k_{-k} := \bigvee_{j=-k}^k T^{-j}(\mathcal{P})$ (which generate the $\Sigma$-algebra of $\mathcal{X}$ when $k \to +\infty$) correspond to the Abelian finite-dimensional subalgebras $\boldsymbol{M}_k := \bigvee_{j=-k}^k \Theta_T^j(\boldsymbol{M}_{\mathcal{P}})$ generated by the characteristic functions of the atoms of $\mathcal{P}^k_{-k}$ (these subalgebras generate $\mathcal{M}$). We can thus apply the argument of Example 8.1.1.2 to deduce that

$$\begin{aligned}\mathrm{H}_{\omega_\mu}\left(\boldsymbol{M}_k, \Theta_T(\boldsymbol{M}_k, \dots, \Theta_T^{n-1}(\boldsymbol{M}_k)\right) &= S\left(\omega_\mu \lceil \boldsymbol{M}_k^{(n)}\right) = S\left(\omega_\mu \lceil \boldsymbol{M}^{(n+2k-1)}\right) \\ &= H_\mu(\mathcal{P}^{(n+2k-1)}) ,\end{aligned}$$

where we used that $\omega_\mu$ is $\Theta_T$-invariant and that

$$\boldsymbol{M}_k^{(n)} = \bigvee_{\ell=0}^{n-1} \Theta_T^j(\boldsymbol{M}_k) = \bigvee_{\ell=-k}^{n+k-1} \Theta_T^j(\boldsymbol{M}) = \Theta_T^{-k}\Big(\bigvee_{j=0}^{n+2k-1} \Theta_T^j(\boldsymbol{M})\Big) ,$$

together with (3.1). Thus, from Theorem 3.1.1,

$$\mathrm{h}^{\mathrm{CNT}}_{\omega_\mu}(\Theta, \boldsymbol{M}_k) = \mathrm{h}^{\mathrm{KS}}_\mu(T, \mathcal{P}) = \mathrm{h}^{\mathrm{KS}}_\mu(T) .$$

Then, $\mathrm{h}^{\mathrm{CNT}}_\omega(\Theta_T) = \mathrm{h}^{\mathrm{KS}}_\mu(T)$ follows from Proposition 8.1.7.

### *CNT* entropy: Finite Quantum Systems

For finite-dimensional quantum dynamical systems, the $C^*$ algebra $\mathcal{A}$ is a matrix algebra $M_d(\mathbb{C})$ and the states density matrices $\rho \in M_d(\mathbb{C})$ with von Neumann entropy always bounded from above by $\log d$. All these systems cannot support a non-zero *CNT* entropy rate, in agreement with the fact that their dynamics, given by a unitary $U \in M_d(\mathbb{C})$, is quasi-periodic and shows the behavior discussed in Remark 7.1.7, at the most. We shall prove this by considering the slightly more general scenario studied in [39].

**Proposition 8.1.9.** *Let $(\mathcal{A}, \Theta_\sigma, \omega)$ be a quantum dynamical system with $\mathcal{A} = \mathbb{B}(\mathbb{H})$, $\omega$ corresponding to a density matrix $\rho \in \mathbb{B}_1^+(\mathbb{H})$ with finite von Neumann entropy $S(\rho)$ and invariant under the automorphism $\Theta$ such that*

$$\mathbb{B}(\mathbb{H}) \ni X \mapsto \Theta[X] = \mathrm{e}^{iH} X \, \mathrm{e}^{-iH} ,$$

*where the Hamiltonian $H$ has a discrete spectrum. Then, $\mathrm{h}_\omega^{\mathrm{CNT}}(\Theta) = 0$.*

**Proof:** Let $P^{(n)}$ be the projector onto the subspace of $\mathbb{H}$ spanned by the eigenvectors relative to the first $n$ decreasingly ordered eigenvalues of $H$ and $Q^{(n)} := \mathbb{1} - P^{(n)}$. Then $\mathcal{A}$ is generated as a von Neumann algebra by the increasing sequence of subalgebras $\mathcal{A}^{(n)} := P^{(n)} \mathcal{A} P^{(n)} \oplus \mathbb{C} \, Q^{(n)}$. These subalgebras are $\Theta$-invariant; also, they diagonalize $\rho$ for it commutes with $H$ since $\omega \circ \Theta = \omega$. Then, from Proposition 8.1.7, (8.12) and (8.10) it follows that

$$\begin{aligned} \mathrm{h}_\omega^{\mathrm{CNT}}(\Theta) &= \lim_{n\to+\infty} \mathrm{h}_\omega^{\mathrm{CNT}}\left(\Theta, \mathcal{A}^{(n)}\right) = \lim_{n\to+\infty} \frac{1}{n} \mathrm{H}_\omega\left(\mathcal{A}^{(n)}, \mathcal{A}^{(n)}, \ldots, \mathcal{A}^{(n)}\right) \\ &= \lim_{n\to+\infty} \frac{1}{n} \mathrm{H}_\omega\left(\mathcal{A}^{(n)}\right) \leq \lim_{n\to+\infty} \frac{1}{n} S\left(\rho \restriction \mathcal{A}^{(n)}\right) = 0 . \end{aligned}$$

□

### *CNT* Entropy: Quantum Spin Chains

The algebraic structure of a quantum spin chain $(\mathcal{A}_\mathbb{Z}, \Theta_\sigma, \omega)$ is such that we can apply Proposition 8.1.6. Let then $\{\mathcal{A}_{[-\ell,\ell]}\}_{\ell\in\mathbb{N}}$ be an increasing sequence of finite-dimensional local subalgebras that generate $\mathcal{A}_\mathbb{Z}$, then

$$\mathrm{h}_\omega^{\mathrm{CNT}}(\Theta_\sigma) = \lim_{\ell\to\infty} \mathrm{h}_\omega^{\mathrm{CNT}}\left(\Theta_\sigma, \mathcal{A}_{[-\ell,\ell]}\right) = \lim_{\ell\to\infty} \mathrm{h}_\omega^{\mathrm{CNT}}\left(\Theta_\sigma, \mathcal{A}_{[1,\ell]}\right) ,$$

where the second equality follows from the translation invariance of $\omega$ and the property (8.13). Further, since $\Theta_\sigma^j[\mathcal{A}_{[1,\ell]}] = \mathcal{A}_{[1+j,\ell+j]} \subset \mathcal{A}_{[1,\ell+n-1]}$, for $0 \leq j \leq n-1$, using (8.15), (8.12) and (8.10) one derives

$$\begin{aligned} \mathrm{H}_\omega\left(\mathcal{A}_{[1,\ell]}, \Theta_\sigma(\mathcal{A}_{[1,\ell]}), \ldots, \Theta_\sigma^{n-1}(\mathcal{A}_{[1,\ell]})\right) &\leq \mathrm{H}_\omega\left(\mathcal{A}_{[1,\ell+n-1]}, \ldots, \mathcal{A}_{[1,\ell+n-1]}\right) \\ &= \mathrm{H}_\omega\left(\mathcal{A}_{[1,\ell+n-1]}\right) \leq S\left(\rho_{[1,\ell+n-1]}\right) , \end{aligned}$$

where $\rho_{[1,\ell+n-1]}$ is the local density matrix corresponding to the state $\omega$ restricted to the local subalgebra $\mathcal{A}_{[1,\ell+n-1]}$. By using (8.24) and Example 7.2.1.1 one thus concludes with

**Proposition 8.1.10.** $\mathrm{h}^{\mathrm{CNT}}_{\omega}(\Theta_\sigma) \leq s(\omega)$ *for any* $(\mathcal{A}_{\mathbb{Z}}, \Theta_\sigma, \omega)$.

In order to show whether and when $\mathrm{h}^{\mathrm{CNT}}_{\omega}(\Theta_\sigma) \geq s(\omega)$, we use the following strategy. Consider a local subalgebra $\mathcal{A}_{[1,\ell]}$ with fixed $\ell \geq 1$ and set $n = k\ell + p$, $0 \leq p < \ell$ in (8.24); using Definition 8.1.2 and property (8.16) we get the following lower bound:

$$\begin{aligned}
\mathrm{h}^{\mathrm{CNT}}_{\omega}(\Theta_\sigma) &\geq \lim_{n\to\infty} \mathrm{h}^{\mathrm{CNT}}_{\omega}\left(\Theta_\sigma, \mathcal{A}_{[1,\ell]}\right) \\
&\geq \lim_{n\to\infty} \frac{1}{k\ell+p} \mathrm{H}_\omega\left(\mathcal{A}_{[1,\ell]}, \Theta_\sigma(\mathcal{A}_{[1,\ell]}), \dots, \Theta^{k\ell+p-1}(\mathcal{A}_{[1,\ell]})\right) \\
&\geq \lim_{k\to\infty} \frac{1}{k\ell} \mathrm{H}_\omega\left(\mathcal{A}_{[1,\ell]}, \Theta^{\ell}_\sigma(\mathcal{A}_{[1,\ell]}), \dots, \Theta^{(k-1)\ell}(\mathcal{A}_{[1,\ell]})\right) \\
&= \lim_{k\to\infty} \frac{1}{k\ell} \mathrm{H}_\omega\left(\mathcal{A}_{[1,\ell]}, \mathcal{A}_{[\ell+1,2\ell]}, \dots, \mathcal{A}_{[(k-1)\ell+1,k\ell]}\right) \\
&\geq \lim_{k\to\infty} \frac{1}{k\ell} H^{\{\mathcal{A}_{[j\ell+1,(j+1)\ell]}\}_{j=0}^{k-1}}_{\omega}\left(\{\lambda_{\boldsymbol{i}^{(k)}}, \omega_{\boldsymbol{i}^{(k)}}\}\right),
\end{aligned} \tag{8.34}$$

where we used (8.7) with $\omega = \sum_{\boldsymbol{i}^{(k)}} \lambda_{\boldsymbol{i}^{(k)}} \omega_{\boldsymbol{i}^{(k)}}$ any chosen decomposition adapted to the $k$ commuting local subalgebras $\mathcal{A}_{[j\ell+1,(j+1)\ell]}$.

### *CNT* Entropy: *FCS* States

If a quantum spin chain is endowed with a finitely correlated state $\omega$ as defined in Section 7.1.5, then its *CNT* entropy coincides with the entropy density $s(\omega)$ (see Section 7.2) [133].

**Proposition 8.1.11.** *Let* $(\mathcal{A}_{\mathbb{Z}}, \Theta_\sigma, \omega)$ *be a quantum spin chain with a FCS* $\omega$*, then* $\mathrm{h}^{\mathrm{CNT}}_{\omega}(\Theta_\sigma) = s(\omega)$.

**Proof:** Because of Proposition 8.1.10, the result follows if we show that $\mathrm{h}^{\mathrm{CNT}}_{\omega}(\Theta_\sigma) \geq s(\omega)$; for this we use the lower bound (8.34) and Remark 7.1.15. Therefore, we fix a local subalgebra $\mathcal{A}_{[1,\ell]}$; since altogether the arguments of the $k$-subalgebra entropy in (8.34) generate the local subalgebra $\mathcal{A}_{[1,k\ell]}$ we need consider the local state $\rho_{[1,k\ell]}$. We start from a decomposition as in (7.101) with $n = k\ell$ and regroup the indices $\boldsymbol{j}^{(k\ell)}$ as follows

$$\boldsymbol{j}^{(k\ell)} = \underbrace{j_1 j_2 \cdots j_\ell}_{i_1} \underbrace{j_{\ell+1} j_{\ell+2} \cdots j_{2\ell}}_{i_2} \cdots \underbrace{j_{(k-1)\ell+1} j_{(k-1)\ell+2} \cdots j_{k\ell}}_{i_k} .$$

Notice that the index $j$ of the Kraus operators in the *CPU* map $\mathbb{E}$ defining the *FCS* $\omega$ runs over a finite set $J$, whence $\boldsymbol{j}^{(k\ell)} \in I_J^{k\ell}$, while each $i_p$ in the regrouped index $\boldsymbol{i}^{(k)} = i_1 i_2 \cdots i_k$ belongs to the index set $I_J^\ell$; thus $\boldsymbol{i}^{(k)} \in I^k_{I^\ell_J}$. We have seen in Example 7.1.15 that the weights $p(\boldsymbol{j}^{k\ell})$ assigned to the indices $\boldsymbol{j}^{(k\ell)}$ give rise to a compatible family of local probability distributions $\pi^{(k\ell)}$ and that these define a global shift-invariant state $\omega_\pi$ over the classical spin chain $\mathfrak{D}_J^{\otimes\infty}$. We then construct the decomposition $\omega = \sum_{\boldsymbol{i}^{(k)}} \lambda_{\boldsymbol{i}^{(k)}} \omega_{\boldsymbol{i}^{(k)}}$, where $\lambda_{\boldsymbol{i}^{(k)}} := p(\boldsymbol{i}^{(k)})$ and $\omega_{\boldsymbol{i}^{(k)}} := \rho^{\boldsymbol{i}^{(k)}}_{[1,k\ell]}$ and use it to compute

$$H_\omega^{\{\mathcal{A}_{[j\ell+1,(j+1)\ell]}\}_{j=0}^{k-1}}\left(\{\lambda_{\boldsymbol{i}^{(k)}}, \omega_{\boldsymbol{i}^{(k)}}\}\right) = \sum_{\boldsymbol{i}^{(k)} \in I^k_{I^\ell_J}} \eta(p(\boldsymbol{i}^{(k)}) - \sum_{j=1}^{k} \sum_{i_j \in I^\ell_J} \eta(p^j(i_j)) +$$

$$+ \sum_{j=1}^{k} S\left(\omega \restriction \mathcal{A}_{[(j-1)\ell+1, j\ell]}\right) - \sum_{j=1}^{k} \sum_{i_j \in I^\ell_J} p^j(i_j) S\left(\omega^j_{i_j} \restriction \mathcal{A}_{[(j-1)\ell+1, j\ell]}\right) ,$$

where, with the notation of Example 7.1.15,

$$p^j(i_j) := \sum_{\substack{\boldsymbol{i}^{(k)} \in I^k_{I^\ell_J} \\ i_j\, fixed}} p(\boldsymbol{i}^{(k)}) , \quad \omega^j_{i_j} := \sum_{\substack{\boldsymbol{i}^{(k)} \in I^k_{I^\ell_J} \\ i_j\, fixed}} \frac{p(\boldsymbol{i}^{(k)})}{p^j(i_j)} \rho^{\boldsymbol{i}^{(k)}}_{[1,k\ell]} = \rho^{i_j}_{[1,\ell]} .$$

From translation invariance of *FCS* it follows that $\omega \restriction \mathcal{A}_{[(j-1)\ell+1, j\ell]} = \rho_{[1,\ell]}$, $\omega^j_{i_j} \restriction \mathcal{A}_{[(j-1)\ell+1, j\ell]} = \rho^{\boldsymbol{j}^{(\ell)}}_{[1,\ell]}$ and $p^j(i_j) = p(\boldsymbol{j}^{(\ell)})$ for some $\boldsymbol{j}^{(\ell)} \in I^\ell_J$. Therefore, (8.34) reads

$$\mathrm{h}^{\mathrm{CNT}}_\omega(\Theta_\sigma) \geq \lim_{k\to\infty} \left\{ \frac{1}{k\ell} \sum_{\boldsymbol{i}^{(k)} \in I^k_{I^\ell_J}} \eta(p(\boldsymbol{i}^{(k)})) \right\} - \frac{1}{\ell} \sum_{\boldsymbol{j}^{(\ell)} \in I^\ell_J} \eta(p(\boldsymbol{j}^{(\ell)}))$$

$$+ \frac{1}{\ell} S\left(\rho_{[1,\ell]}\right) - \frac{1}{\ell} \sum_{\boldsymbol{j}^{(\ell)} \in I^\ell_J} p(\boldsymbol{j}^{(\ell)}) S\left(\rho^{\boldsymbol{j}^{(\ell)}}_{[1,\ell]}\right) .$$

In the limit $\ell \to \infty$, the second term in the first line gives the Shannon entropy rate of the classical spin chain $(\mathfrak{D}_J^{\otimes\infty}, \Theta_\sigma, \omega_\pi)$ as well as the limit $k \to \infty$ in the first term, while the first contribution in the second line gives the von Neumann entropy density of $(\mathcal{A}_\mathbb{Z}, \Theta_\sigma, \omega)$. Thus,

$$\mathrm{h}^{\mathrm{CNT}}_\omega(\Theta_\sigma) \geq s(\omega) - \lim_{\ell\to\infty} \frac{1}{\ell} \sum_{\boldsymbol{j}^{(\ell)} \in I^\ell_J} p(\boldsymbol{j}^{(\ell)}) S\left(\rho^{\boldsymbol{j}^{(\ell)}}_{[1,\ell]}\right) .$$

The proof is then completed by using that, as discussed in Example (7.1.15), $S\left(\rho^{\boldsymbol{j}^{(\ell)}}\right) \leq 2\log\ell$. □

### *CNT* Entropy: Price-Powers Shifts

The non-commutative shifts discussed in Section 7.1.5 offer an interesting variety of behaviors of the *CNT* entropy [13].

By construction the quasi-local algebra $\mathcal{A}_g$ is generated by the Abelian algebra $\boldsymbol{A}_1$ consisting of the orthogonal projections $\frac{\mathbb{1} \pm e_1}{2}$ and by its images $\boldsymbol{A}_n := \Theta_\sigma^n(\boldsymbol{A}_1)$. These are also Abelian subalgebras, but in general they do not commute with each other; moreover, denoting by $\boldsymbol{M}_k$ the subalgebra generated by $\boldsymbol{A}_\ell$, $\ell = 1, 2, \ldots, k$, these generate the von Neumann algebra $\mathcal{M}_g$ in Example 7.1.17.1. Therefore, one can compute $\mathrm{h}_\omega^{\mathrm{CNT}}(\Theta_T)$ by means of Proposition 8.1.7. Notice that, because of (7.122), the $\boldsymbol{M}_k$ can be represented as subalgebras of the spin algebras $M_2(\mathbb{C})^{\otimes k}$; this fact allows to derive a bitstream-independent upper bound to the *CNT* entropy. Indeed, by using the properties in Proposition 8.1.3, one estimates

$$\begin{aligned}
&\mathrm{H}_\omega\left(\boldsymbol{M}_k, \Theta_\sigma(\boldsymbol{M}_k), \ldots, \Theta^{n-1}(\boldsymbol{M}_k)\right) \leq \mathrm{H}_\omega\left(\boldsymbol{M}_{n+k-1}\right) \\
&\quad \leq \mathrm{H}_\omega\left(M_2(\mathbb{C})^{\otimes(n+k-1)}\right) = (n+k-1)\log 2 \quad \text{whence} \\
&\mathrm{h}_\omega^{\mathrm{CNT}}(\Theta_\sigma, \boldsymbol{M}_k) \leq \log 2 \Longrightarrow \mathrm{h}_\omega^{\mathrm{CNT}}(\Theta_\sigma) \leq \log 2 \ .
\end{aligned}$$

We discuss a few particular cases, a thorough analysis of the dependence of the *CNT* entropy on the bitstream being provided by [222].

1. For a bitstream $g \equiv 0$, $\left(\mathcal{M}_g, \Theta_\sigma, \omega\right)$ amounts to a classical Bernoulli shift and
$$\mathrm{h}_{\mathcal{M}_g}^{\mathrm{CNT}}(\Theta_\sigma) = \log 2 \ . \tag{8.35}$$
2. If $g(n) = 1$ for all $n \geq 1$, then $\mathcal{M}_g$ describes a Fermi system on a one-dimensional lattice at infinite temperature (see Example 7.1.17.3), where pairs $e_{2i}, e_{2i+1}$ give rise to annihilation and creation operators $a_i, a_i^\dagger$ fulfilling the *CAR*. Since $n$ such operators generate an algebra isomorphic to $M_{2^n}(\mathbb{C})$, an argument similar to the one that gave the universal upper $\log 2$ yields
$$\begin{aligned}
&\mathrm{H}_\omega\left(\boldsymbol{M}_k, \Theta_\sigma(\boldsymbol{M}_k), \ldots, \Theta^{n-1}(\boldsymbol{M}_k)\right) \leq \mathrm{H}_\omega\left(\boldsymbol{M}_{n+k-1}\right) \\
&\quad \leq \mathrm{H}_\omega\left(M_2(\mathbb{C})^{\otimes[(n+k-1)/2]}\right) = [\frac{n+k-1}{2}] \log 2 \quad \text{whence} \\
&\mathrm{h}_\omega^{\mathrm{CNT}}(\Theta_\sigma, \boldsymbol{M}_k) \leq \frac{\log 2}{2} \Longrightarrow \mathrm{h}_\omega^{\mathrm{CNT}}(\Theta_\sigma) \leq \frac{\log 2}{2} \ ,
\end{aligned}$$
where we have set $[j/2] = j/2$ for $j$ even and $[j/2] = (j+1)/2$ for $j$ odd. On the other hand, (7.121) implies
$$\begin{aligned}
\left[e_{2j-1}e_{2j}\,,\, e_{2k-1}e_{2k}\right] &= \left(e_{2j-1}e_{2j}\right)\left(e_{2k-1}e_{2k}\right) \times \\
&\times \left(1 - (-1)^{g(|2(j-k)+1|)+g(|2(j-k)-1|)}\right) = 0 \ ,
\end{aligned}$$

for all $j,k \geq 1$; therefore, operators of the form $e_{2j-1}e_{2j}$ commute. Let $\boldsymbol{A}$ denote the Abelian algebra generated by $e_1 e_2$ (which is isomorphic to the diagonal matrix algebra $D_2(\mathbb{C})$); then, the Abelian algebras $\{\Theta_\sigma^{2j}(\boldsymbol{A})\}_{j=0}^{n-1}$ commute and generate an Abelian algebra isomorphic to the diagonal matrix algebra $D_{2^n}(\mathbb{C})$. Then, using Example 8.1.1.1, one gets

$$\mathrm{H}_\omega\left(\boldsymbol{A}, \Theta_\sigma^2(\boldsymbol{A}), \ldots, \Theta_\sigma^{2(n-1)}(\boldsymbol{A})\right) = n \log 2 \ .$$

Finally, (8.27) yields

$$\mathrm{h}_\omega^{\mathrm{CNT}}(\Theta_\sigma) \geq \mathrm{h}_\omega^{\mathrm{CNT}}(\Theta, \boldsymbol{A}) \geq \frac{1}{2}\mathrm{h}_\omega^{\mathrm{CNT}}(\Theta_\sigma^2, \boldsymbol{A}) = \frac{\log 2}{2} \ ,$$

whence

$$\mathrm{h}_\omega^{\mathrm{CNT}}(\Theta_\sigma) = \frac{\log 2}{2} \ . \tag{8.36}$$

3. In the case of an asymptotically highly anti-commutative Price-Powers shift [220], for any $W_{\boldsymbol{i}} \in \mathcal{A}_g$ there exists an infinite set $I(\boldsymbol{i})$ of integers such that

$$\left\{\Theta_\sigma^n[W_{\boldsymbol{i}}^\dagger]\,,\,\Theta_\sigma^m[W_{\boldsymbol{i}}]\right\} = W_{\boldsymbol{i}+n}^\dagger W_{\boldsymbol{i}+m} + W_{\boldsymbol{i}+m} W_{\boldsymbol{i}+n}^\dagger = 0 \ ,$$

for all $n,m \in I(\boldsymbol{i})$. We shall show that

$$\mathrm{h}_\omega^{\mathrm{CNT}}(\Theta_\sigma) = 0 \ . \tag{8.37}$$

In order to do that, we shall consider a stationary coupling of $(\mathcal{M}_g, \Theta_\sigma, \omega)$ to commutative dynamical triple $(\mathcal{B}, \theta, \omega_\mu)$ (see Definition 8.1.3 and the preceding discussion), namely a triplet of the form $(\mathcal{A}\otimes\mathcal{B}, \Theta\otimes\theta, \widetilde{\omega})$ where $\widetilde{\omega}$ is a $\Theta_\sigma\otimes\theta$-invariant state such that $\widetilde{\omega}\upharpoonright\mathcal{A} = \omega$ and $\widetilde{\omega}\upharpoonright\mathcal{B} = \omega_\mu$. Let $p \in \mathcal{B}$ be any projection; then,

$$\left\{W_{\boldsymbol{i}+n}^\dagger \otimes \theta^n[p]\,,\, W_{\boldsymbol{i}} \otimes \theta^m[p]\right\} = \left\{W_{\boldsymbol{i}+n}^\dagger\,,\, W_{\boldsymbol{i}}\right\} \otimes \theta^n[p]\theta^m[p] = 0$$

for all $n,m \in I(\boldsymbol{i})$. As done in Example 7.1.17.5, by setting

$$X := \frac{1}{N} \sum_{i=1:n_i \in I(\boldsymbol{i})}^{N} W_{\boldsymbol{i}} \otimes p$$

for an arbitrary $N \in \mathbb{N}$, one estimates

$$\left|\widetilde{\omega}(W_{\boldsymbol{i}} \otimes p)\right| = \left|\widetilde{\omega}(X)\right| \leq \sqrt{\frac{1}{N}} \ .$$

Since $N$ is arbitrary, we deduce that $\widetilde{\omega}(W_{\boldsymbol{i}} \otimes p) = 0$ for all $W_{\boldsymbol{i}} \in \mathcal{M}_g$ and all $p \in \mathcal{B}$ which means that the global state factorizes: $\widetilde{\omega} = \omega \otimes \omega_\mu$. This fact in turn implies that the relative entropy contributions in (8.32) vanish so that $\mathrm{H}_{\widetilde{\omega}}(\boldsymbol{B} \mid \mathcal{A}) = S(\omega_\mu \upharpoonright \boldsymbol{B})$ for all finite-dimensional subalgebras $\boldsymbol{B} \subset \mathcal{B}$. Since $\mathrm{h}_{\omega_\mu}^{\mathrm{KS}}(\theta, \omega_\mu) \leq S(\omega_\mu \upharpoonright \boldsymbol{B})$, the result follows from (8.33).

### *CNT* Entropy: Quasi-Free Bosons and Fermions

For quasi-local algebras of Bosons and Fermions in translation-invariant quasi-free states as those considered in Example 7.2.1.3, one can consider the discrete space-translation group $\boldsymbol{\Theta} := \{\Theta_{\boldsymbol{n}}\}_{\boldsymbol{n}\in\mathbb{Z}^3}$ (see Example 7.1.2.1) and enlarge the scope of Definition (8.1.2) to cover the fact that there are now three directions along which the $n$-*CPU* entropy (8.3) can increase.

Given a *CPU* map $\gamma : \boldsymbol{M} \mapsto \mathcal{A}^{B,F}$ from a finite-dimensional unital $C^*$ algebra into the Fermi, respectively Bose quasi-local algebra, a natural way to proceed [233] is to consider, for each $\boldsymbol{k} = (k_1, k_2, k_3) \in \mathbb{N}^3$, the parallelepipeds

$$B(\boldsymbol{k}) := \left\{\boldsymbol{n} \in \mathbb{N}^3 \,:\, 0 \le n_i \le k_i \;,\; i = 1,2,3\right\} ,$$

*CPU* maps of the form $\Theta_{\boldsymbol{n}} \circ \gamma$ and then to replace (8.24) by

$$\mathrm{h}^{\mathrm{CNT}}_{\omega_A}(\boldsymbol{\Theta}, \gamma) := \lim_{k_1,k_2,k_3 \to +\infty} \frac{1}{k_1 k_2 k_3} \mathrm{H}_{\omega_A}\left(\{\Theta_{\boldsymbol{n}} \circ \gamma\}_{\boldsymbol{n}\in B(\boldsymbol{k})}\right) . \qquad (8.38)$$

while keeping the definition of (8.25) for the *CNT* entropy of $\boldsymbol{\Theta}$. Notice that the limit in the right hand side of (8.38) exists because of the subadditivity property (8.14) and the assumed translation-invariance of the quasi-free state $\omega_A$ together with property (8.13).

With the same technical assumptions ensuring the result in Example 7.2.1.3, by means of (8.1.1) it can be showed that the *CNT* entropy of the space-translations coincides with the mean entropy [233]:

$$\mathrm{h}^{\mathrm{CNT}}_{\omega_A}(\boldsymbol{\Theta}) = \frac{1}{(2\pi)^3} \int_{\mathbb{R}^3} \mathrm{d}\boldsymbol{k} \,\left(\eta(\widehat{K}_A(\boldsymbol{k})) \,+\, \eta(1 - \widehat{K}_A(\boldsymbol{k}))\right) \;\text{(Fermions)}$$

$$\mathrm{h}^{\mathrm{CNT}}_{\omega_A}(\boldsymbol{\Theta}) = \frac{1}{(2\pi)^3} \int_{\mathbb{R}^3} \mathrm{d}\boldsymbol{k} \,\left(\eta(\widehat{K}_A(\boldsymbol{k})) \,-\, \eta(1 + \widehat{K}_A(\boldsymbol{k}))\right) \quad \text{(Bosons)} .$$

**Remarks 8.1.4.**

1. Quasi-free automorphisms in the Fermionic case have been considered in [214, 289]; the following result holds [223]: let $f, g \in \mathbb{L}^2_{[0,2\pi]}(\mathrm{d}p)$ be single particle wave-functions for a Fermi system on a lattice ($[0, 2\pi]$ being the momentum space). Consider

$$\Theta_U(a^{\#}(f)) = a^{\#}(Uf) \;, \quad (Uf)(p) = \mathrm{e}^{i\,\omega(p)} f(p) \;;$$

it defines a quasi-free automorphism over the *CAR* algebra with single particle energy $\omega(p)$ assumed to be a real absolutely continuous function of the momentum variable $p$. Further, let

$$\omega(a(f)a^{\dagger}(g)) = \int_0^{2\pi} \mathrm{d}p\; \rho(p)\, f^*(p)\, g(p)$$

define a quasi-free $\Theta_U$-invariant state over the system, with $0 \leq \rho(p) \leq 1$ a measurable one-particle distribution over $[0, 2\pi]$. Then,

$$\mathrm{h}^{\mathrm{CNT}}_{\omega}(\Theta_U) = \int_0^{2\pi} \mathrm{d}p \, |\omega'(p)| \, \big(\eta(\rho(p)) + \eta(1-\rho(p))\big) \; ,$$

where $\omega'(p) := \mathrm{d}\omega(p)/\mathrm{d}p$ is the *group velocity*. The physical interpretation is suggestive [214]: for a quasi-free automorphism the dynamical entropy production as described by the *CNT* entropy amounts to a flux of single particle Fermionic entropy governed by the group velocity.

2. While in one-dimensional quantum dynamical systems the $1/n$ factor controls the asymptotic increase of the $n$-*CPU* map entropies, this is no longer true in higher dimension. An instance of this fact is the previous result where one divides by volumes in order to avoid divergences due to the freedom to move in more than one direction. In general, that is in the case of the time-evolution in dimension $\geq 2$, it turns out that $\mathrm{h}^{\mathrm{CNT}}_{\omega}(\Theta) = +\infty$; this problem arises already on the classical level and a possible way out is to consider space and time translation together [153, 39].

### 8.1.4 Entropic Quantum $K$-systems

In Section 3.1.1 it was proved that the algebraic structure of classical Kolmogorov systems introduced in Section 2.3.1 can be characterized by means of the dynamical entropy rate. In particular, from the proof of Theorem 3.1.3 it emerges that the equivalence between the existence of a $K$-partition, namely property (1) in the theorem, and the entropic properties (3) – (6) hinges upon property (2) that is the triviality of the tail of all finite-dimensional partitions.

Algebraic quantum $K$-systems have been introduced in Section 7.1.4 as generalizations of classical $K$-systems; in this section, we present an entropic characterization of non-commutative $K$-systems that partially mimics that given in Theorem 3.1.3. This gives rise to a class of quantum dynamical systems with particular clustering properties, but in general not $K$-systems from the algebraic point of view.

We start by considering the relations (3) – (6) in the above mentioned theorem and study how they are affected if one substitutes the $n$-subalgebra entropies for the Shannon entropies. For sake of simplicity, we shall restrict to the case of *AF* algebras $\mathcal{A}$ (see Remark7.1.1) so that we can consider finite-dimensional subalgebras $\boldsymbol{A} \subset \mathcal{A}$ as arguments of $n$-subalgebra entropies, namely, we take the natural embeddings $\iota_{\boldsymbol{A}} : \boldsymbol{A} \mapsto \mathcal{A}$ as *CPU* maps $\gamma$. Also, we shall restrict to faithful states $\omega$ so that the only subalgebra with 0 entropy with respect to $\omega$ is the trivial one (see Lemma 6.3.1).

**Theorem 8.1.2.** *Given a quantum dynamical triple $(\mathcal{A}, \Theta, \omega)$ with $\mathcal{A}$ an AF $C^*$ algebra and $\omega$ a faithful state, let $\{c\mathbb{1}\} \subset \mathcal{A}$ denote the trivial subalgebra and consider the following statements*

1. *the CNT entropy is strictly positive, namely for all non-trivial finite-dimensional subalgebras $\mathcal{A} \supset \boldsymbol{A} \neq \{c\mathbb{1}\}$,*
$$\mathrm{h}^{\mathrm{CNT}}_{\omega}(\Theta, \boldsymbol{A}) > 0 \ ; \tag{8.39}$$
2. *for all finite-dimensional subalgebras $\mathcal{A} \supset \boldsymbol{A} \neq \{c\mathbb{1}\}$,*
$$\lim_{n\to+\infty} \mathrm{h}^{\mathrm{CNT}}_{\omega}(\Theta^n, \boldsymbol{A}) = \mathrm{H}_{\omega}(\boldsymbol{A}) \ ; \tag{8.40}$$
3. *for all finite-dimensional subalgebras $\mathcal{A} \supset \boldsymbol{A} \neq \{c\mathbb{1}\}$, $\mathcal{A} \supset \boldsymbol{B}$ and all sequences $\{j_k\}_{k\in\mathbb{N}}$ of positive integers,*
$$\lim_{n\to+\infty} \liminf_{k\to+\infty} \Big[\mathrm{H}_{\omega}\left(\boldsymbol{B}, \Theta^{n+j_1}(\boldsymbol{A}), \ldots, \Theta^{n+j_k}(\boldsymbol{A})\right) \\ -\mathrm{H}_{\omega}\left(\boldsymbol{B}, \Theta^{n+j_1}(\boldsymbol{A}), \ldots, \Theta^{n+j_k}(\boldsymbol{A})\right)\Big] = \mathrm{H}_{\omega}(\boldsymbol{B}) \ ; \tag{8.41}$$
4. *for all finite-dimensional subalgebras $\mathcal{A} \supset \boldsymbol{A} \neq \{c\mathbb{1}\}$ and all sequences $\{j_k\}_{k\in\mathbb{N}}$ of positive integers*
$$\lim_{n\to+\infty} \liminf_{k\to+\infty} \Big[\mathrm{H}_{\omega}\left(\boldsymbol{B}, \Theta^{n+j_1}(\boldsymbol{A}), \ldots, \Theta^{n+j_k}(\boldsymbol{A})\right) \\ -\mathrm{H}_{\omega}\left(\boldsymbol{B}, \Theta^{n+j_1}(\boldsymbol{A}), \ldots, \Theta^{n+j_k}(\boldsymbol{A})\right)\Big] = 0 \Longrightarrow \boldsymbol{B} = \{c\mathbb{1}\} \ . \tag{8.42}$$

*They stand in the following relations*

$$\begin{array}{ccc} (8.40) & \Longrightarrow & (8.39) \\ \Uparrow & & \Uparrow \\ (8.41) & \Longrightarrow & (8.42) \end{array} .$$

**Proof:** (8.40)$\Longrightarrow$ (8.39) Because of the second property in Lemma 6.3.1, one can choose $0 < \varepsilon < \mathrm{H}_{\omega}(\boldsymbol{A})$ and $n$ such that, using the lower bound in (8.27),

$$\mathrm{h}^{\mathrm{CNT}}_{\omega}(\Theta, \boldsymbol{A}) \geq \frac{1}{n}\mathrm{h}^{\mathrm{CNT}}_{\omega}(\Theta^n, \boldsymbol{A}) \geq \frac{\mathrm{H}_{\omega}(\boldsymbol{A}) - \varepsilon}{n} > 0 \ .$$

(8.41)$\Longrightarrow$ (8.40) Consider the sequence $\{j_k = (k-1)n\}_{k\geq 1}$, from the assumption it follows that for any $\varepsilon > 0$ there exists $n_0 \in \mathbb{N}$ such that, for all $n \geq n_0$,

$$\liminf_{k\to+\infty} \Big[\mathrm{H}_{\omega}\left(\boldsymbol{A}, \Theta^n(\boldsymbol{A}), \Theta^{n+n}(\boldsymbol{A}), \ldots, \Theta^{nk}(\boldsymbol{A})\right) - \\ - \mathrm{H}_{\omega}\left(\Theta^n(\boldsymbol{A}), \Theta^{2n}(\boldsymbol{A}), \Theta^{3n}(\boldsymbol{A}), \ldots, \Theta^{nk}(\boldsymbol{A})\right)\Big] = \\ = \liminf_{k\to+\infty} \Big[\mathrm{H}_{\omega}\left(\boldsymbol{A}, \Theta^n(\boldsymbol{A}), \ldots, \Theta^{nk}(\boldsymbol{A})\right) - \mathrm{H}_{\omega}\Big(\boldsymbol{A}, \Theta^n(\boldsymbol{A}), \ldots, \Theta^{n(k-1)}(\boldsymbol{A})\Big)\Big] \\ \geq \mathrm{H}_{\omega}(\boldsymbol{A}) - \varepsilon \ ,$$

where property (8.13) has been used in the first equality. Further, since $\liminf_{k\to+\infty} = \sup_{p\geq 0}\inf_{k\geq p}$ it follows that there exists $p_0 \in \mathbb{N}$ such that, for all $p \geq p_0$,

$$\begin{aligned}\Delta_p &:= \mathrm{H}_\omega\left(\boldsymbol{A}, \Theta^n(\boldsymbol{A}), \ldots, \Theta^{np}(\boldsymbol{A})\right) - \mathrm{H}_\omega\left(\boldsymbol{A}, \Theta^n(\boldsymbol{A}), \ldots, \Theta^{n(p-1)}(\boldsymbol{A})\right) \\ &\geq \liminf_{k\to+\infty}\Big[\mathrm{H}_\omega\left(\boldsymbol{A}, \Theta^n(\boldsymbol{A}), \ldots, \Theta^{nk}(\boldsymbol{A})\right) \\ &\qquad - \mathrm{H}_\omega\left(\boldsymbol{A}, \Theta^n(\boldsymbol{A}), \ldots, \Theta^{n(k-1)}(\boldsymbol{A})\right)\Big] - \varepsilon \geq \mathrm{H}_\omega(\boldsymbol{A}) - 2\varepsilon\ .\end{aligned}$$

Choosing $p > p_0$, one thus estimates

$$\begin{aligned}\frac{1}{p}\mathrm{H}_\omega\left(\boldsymbol{A}, \Theta^n(\boldsymbol{A}), \ldots, \Theta^{n(p-1)}(\boldsymbol{A})\right) &= \frac{1}{p}\sum_{j=1}^{p-1}\Delta_j + \frac{\mathrm{H}_\omega(\boldsymbol{A})}{p} \\ &\geq \frac{\mathrm{H}_\omega(\boldsymbol{A})}{p} + \frac{1}{p}\sum_{j=1}^{p_0-1}\Delta_j + \frac{p-p_0}{p}\left(\mathrm{H}_\omega(\boldsymbol{A}) - 2\varepsilon\right)\ .\end{aligned}$$

Since the left hand side of the first inequality is always smaller than $\mathrm{H}_\omega(\boldsymbol{A})$ (see (8.26)), the result follows from the arbitrariness of $\varepsilon > 0$ by letting $p \to +\infty$.

(8.41)$\Longrightarrow$ (8.42) This follows from property 2 in Lemma 6.3.1.

(8.42)$\Longrightarrow$ (8.39) When $\boldsymbol{A} \neq \{c\mathbb{1}\}$, then, the same argument used to show that (8.41)$\Longrightarrow$ (8.40) implies that, for any $\varepsilon$ and sufficiently large $n$, $\mathrm{h}^{\mathrm{CNT}}_\omega(\Theta^n, \boldsymbol{A}) \geq \varepsilon$. Thus, the lower bound in (8.27) yields

$$\mathrm{h}^{\mathrm{CNT}}_\omega(\Theta, \boldsymbol{A}) \geq \frac{1}{n}\mathrm{h}^{\mathrm{CNT}}_\omega(\Theta^n, \boldsymbol{A}) \geq \frac{\varepsilon}{n} > 0\ .$$

□

While in a commutative context the above relations are equivalent, there are no proofs so far that they are such also for quantum dynamical systems. Also, the classical versions of relations (8.39)– (8.42) are equivalent to $K$-mixing which is the strongest possible way of clustering. If one wants to relate the behavior of the *CNT* entropy to the mixing properties of the dynamics, among the possible choices, (8.40) appears the more appropriate. Indeed, one knows that $\mathrm{h}^{\mathrm{CNT}}_\omega(\Theta^n, \boldsymbol{A}) \leq \mathrm{H}_\omega(\boldsymbol{A})$; this is due to the fact that the dynamics usually create correlations between past and future. Therefore, if, asymptotically, the equality holds as in (8.40) this means that for large intervals between successive events the system is affected by memory loss [216].

**Definition 8.1.4 (Entropic Quantum $K$-Systems).** *[215] A quantum dynamical triple $(\mathcal{A}, \Theta, \omega)$ is called an* entropic $K$-system *if, for any CPU map $\gamma : \boldsymbol{M} \mapsto \mathcal{A}$ from a finite-dimensional algebra $\boldsymbol{M}$ into $\mathcal{A}$, it holds that*

$$\lim_{t\to+\infty}\mathrm{h}^{\mathrm{CNT}}_\omega(\Theta^t, \gamma) = \mathrm{H}_\omega(\gamma)\ .$$

This choice is also convenient because the behavior of $\mathrm{h}^{\mathrm{CNT}}_{\omega}(\Theta^n, \boldsymbol{M})$ is often more informative on the properties of the dynamics than $\mathrm{h}^{\mathrm{CNT}}_{\omega}(\Theta)$. For instance, if $(\mathcal{A}, \Theta, \omega)$ is an entropic $K$-system, then

$$\lim_{t\to+\infty} \mathrm{H}_\omega\left(\boldsymbol{M}, \Theta^t(\boldsymbol{M}), \ldots, \Theta^{t(k-1)}(\boldsymbol{M})\right) = k\,\mathrm{H}_\omega(\boldsymbol{M}) \ , \tag{8.43}$$

for all $k \in \mathbb{N}$ and for all finite-dimensional subalgebras $\boldsymbol{M} \subseteq \mathcal{A}$. Indeed, from (8.23) it follows that

$$\mathrm{h}^{\mathrm{CNT}}_{\omega}(\Theta^t, \boldsymbol{M}) \le \frac{1}{k}\mathrm{H}_\omega\left(\boldsymbol{M}, \Theta^t(\boldsymbol{M}), \ldots, \Theta^{t(k-1)}(\boldsymbol{M})\right) \le \mathrm{H}_\omega(\boldsymbol{M}) \ ,$$

whence the result follows by taking the limit $\lim_{t\to+\infty}$.

The asymptotic behavior (8.43) is an effective expression of the memory-loss properties of the dynamics of entropic $K$-systems. Comparing the contributions in (8.3) to the $n$-*CPU* entropies, one sees that, for large $t$, the optimal decompositions $\omega = \sum_{\boldsymbol{i}^{(k)}} \lambda^{(k)}_{\boldsymbol{i}^{(k)},t}\omega_{\boldsymbol{i}^{(k)},t}$ for $\mathrm{H}_\omega\left(\boldsymbol{M}, \Theta^t(\boldsymbol{M}), \ldots, \Theta^{t(k-1)}(\boldsymbol{M})\right)$ must be such that

$$\lim_{t\to+\infty}\left(H(\Lambda^{(k)}_t) - \sum_{j=0}^{k-1} H(\Lambda_{j,t})\right) = 0 \tag{8.44}$$

$$\lim_{t\to+\infty} \sum_{i_j\in I_j} \lambda^j_{i_j,t}\, S\left(\omega^j_{i_j,t}\restriction\Theta^{jt}(\boldsymbol{M})\,,\, \omega\restriction\Theta^{jt}(\boldsymbol{M})\right) = \mathrm{H}_\omega(\boldsymbol{M})\,, \tag{8.45}$$

where $\Lambda^{(k)}_t := \{\lambda^{(k)}_{\boldsymbol{i}^{(k)},t}\}_{\boldsymbol{i}^{(k)}}$, $\Lambda_{j,t} := \{\lambda^j_{i_j,t}\}_{i_j}$ and

$$\lambda^j_{i_j,t} := \sum_{\boldsymbol{i}^{(k)}, i_j\ fixed} \lambda^{(k)}_{\boldsymbol{i}^{(k)},t}\ , \quad \omega^j_{i_j,t} := \sum_{\boldsymbol{i}^{(k)}, i_j\ fixed} \frac{\lambda^{(k)}_{\boldsymbol{i}^{(k)},t}\,\omega_{\boldsymbol{i}^{(k)},t}}{\lambda^j_{i_j,t}} \ .$$

In fact, the difference in (8.44) can at most vanish, but never be positive, while each of the summands in (8.45) is bounded by $\mathrm{H}_\omega(\boldsymbol{M})$. This also means that the optimal decompositions for large $n$ must be such that the corresponding sub-decompositions are close to be optimal for $\mathrm{H}_\omega(\boldsymbol{M})$.

**Example 8.1.3.** Let $(\mathcal{A}_\theta, \Theta_{\mathbb{A}}, \omega)$ denote a quantized hyperbolic automorphism of the torus, where $\mathcal{A}$ is the $C^*$ algebra generated by the Weyl-operators $W_\theta(f)$, with $\theta =< \alpha^2 s >$, $s \in \mathbb{N}$; these quantum dynamical systems are norm asymptotic Abelian (see Example 7.1.12). By using the exponential decay (7.81) of their commutators, we shall show that they are entropic $K$-systems [209].

Let $\gamma : \boldsymbol{M} \mapsto \mathcal{A}_\theta$ be a *CPU* map from a finite dimensional algebra $\boldsymbol{M}$ into $\mathcal{A}$. Fix $\varepsilon > 0$ and choose $k$ such that

$$\mathrm{H}_\omega(\gamma) \geq \mathrm{h}^{\mathrm{CNT}}_\omega\left(\Theta^t_{\mathbb{A}}, \gamma\right) \geq \frac{1}{k}\mathrm{H}_\omega\left(\gamma, \Theta^t_{\mathbb{A}} \circ \gamma, \dots, \Theta^t_{\mathbb{A}} \circ \gamma\right) - \varepsilon$$

$$\geq \frac{1}{k}\Big(H(\Lambda^{(k)}_t) - \sum_{j=0}^{k-1} H(\Lambda_{j,t})\Big) \tag{8.46}$$

$$+ \frac{1}{k}\sum_{j=0}^{k-1}\sum_{i_j} \lambda^j_{i_j,t} S\left(\omega^j_{i_j,t} \circ \Theta^{jt} \circ \gamma \circ \ , \omega \circ \gamma \circ \ \right). \tag{8.47}$$

Notice that $\omega$ is the tracial state; thus, its modular operator is trivial whence its decompositions can be chosen of the form (see (8.6))

$$\omega = \sum_j \lambda_j \omega_j \ , \quad \omega_j(A) = \frac{\omega(X_j\, A)}{\omega(X_j)}$$

for $\mathcal{A} \ni X_j \geq 0$ such that $\sum_j X_j = \mathbb{1}$. Because of the norm-density of the Weyl operators within $\mathcal{A}_\theta$, given $\varepsilon > 0$, we can reach

$$\mathrm{H}_\omega(\gamma) \leq \sum_{i=1}^p \lambda_i S\left(\omega_i \circ \gamma \,,\, \omega \circ \gamma\right) + \varepsilon \tag{8.48}$$

by means of a decomposition $\omega = \sum_{i=1}^p \lambda_i \omega_i$, where

$$X_j := W_\theta(f_i) W_\theta(f_i) \ , \quad \sum_{i=1}^p W_\theta(f_i) W_\theta(f_i) = \mathbb{1} \ ,$$

with functions such that $f_i^*(-\boldsymbol{n}) = f_i(\boldsymbol{n})$. Moreover, we can arrange them in such a way that $f_j(\boldsymbol{n}) = 0$ for all $1 \leq j \leq p-1$ if $\|\boldsymbol{n}\| > K$, while $\|W_0(f_i)\| \leq \varepsilon$.

As a trial decomposition for $\mathrm{H}_\omega\left(\gamma, \Theta^t_{\mathbb{A}} \circ \gamma, \dots, \Theta^{t(k-1)}_{\mathbb{A}}\right)$, consider the positive operators

$$X^{(k)}_{\boldsymbol{i}^{(k)},t} := \left(Y^{(k)}_{\boldsymbol{i}^{(k)},t}\right)^\dagger Y^{(k)}_{\boldsymbol{i}^{(k)},t} \quad \text{where}$$
$$Y^{(k)}_{\boldsymbol{i}^{(k)},t} := W_\theta(f_{i_0})\, \Theta^t_{\mathbb{A}}[W_\theta(f_{i_1})] \cdots \Theta^{t(k-1)}_{\mathbb{A}}[W_\theta(f_{i_{k-1}})] \ ,$$

where $\boldsymbol{i}^{(k)} \in \Omega^{(k)}_p$. They satisfy $\sum_{\boldsymbol{i}^{(k)}} X^{(k)}_{\boldsymbol{i}^{(k)},t} = \mathbb{1}$; further,

$$X^j_{i_j,t} := \sum_{\boldsymbol{i}^{(k)},i_j\ fixed} X^{(k)}_{\boldsymbol{i}^{(k)},t}$$

$$= \Theta^{jt}_{\mathbb{A}}\Big(\sum_{i_{j+1}\cdots i_{k-1}} \left(Y_{[i_{j+1},i_{k+1}]}\right)^\dagger X_{i_j}\, Y_{[i_{j+1},i_{k-1}]}\Big) \tag{8.49}$$

$$Y_{[i_{j+1},i_{k-1}]} := \Theta^t_{\mathbb{A}}[W_\theta(f_{i_{j+1}})] \cdots \Theta^{(k-j-1)t}_{\mathbb{A}}[W_\theta(f_{i_{k-1}})] \ . \tag{8.50}$$

Using the tracial properties, the coefficients of the convex decompositions $\omega = \sum_{i_j} \lambda^j_{i_j,t}\omega^j_{i_j,t}$ are

$$\lambda^j_{i_j,t} = \omega(X^j_{i_j,t}) = \omega(X_{i_j}) = \lambda_{i_j} \ . \tag{8.51}$$

Therefore, in (8.46), $H(\Lambda_{j,t}) = H(\Lambda)$ for all $0 \leq j \leq k-1$, where, in terms of $\eta(x) = -x\log x$, $H(\Lambda) = \sum_{i=1}^{p} \eta(\lambda_i)$. Therefore,

$$\frac{1}{k}\Big(H(\Lambda^{(k)}_t) - \sum_{j=0}^{k-1} H(\Lambda_{j,t})\Big) = \frac{1}{k}H(\Lambda^{(k)}_t) - H(\Lambda) \tag{8.52}$$

In order to control the probability distribution $\Lambda^{(k)}_t$ consisting of the coefficients $\lambda^{(k)}_{\boldsymbol{i}^{(k)},t}$, we expand

$$W_\theta(f_{i_r}) = \sum_{\boldsymbol{n}_r} f_{i_r}(\boldsymbol{n}_r)W_\theta(\boldsymbol{n}_r) \ , \quad \boldsymbol{n}_r \in \mathrm{Supp}(f_{i_r}) \ .$$

By means of the Weyl relations (7.29), they thus read

$$\lambda^{(k)}_{\boldsymbol{i}^{(k)},t} = \omega(X^{(k)}_{\boldsymbol{i}^{(k)},t}) = \sum_{\substack{\boldsymbol{n}_0\cdots\boldsymbol{n}_{k-1}\\ \boldsymbol{m}_0\cdots\boldsymbol{m}_{k-1}}} \Big(\prod_{r=0}^{k-1} f_{i_r}(\boldsymbol{n}_r)f^*_{i_r}(\boldsymbol{m}_r)\Big)\, \mathrm{e}^{2\pi i\theta\beta(\{\boldsymbol{n}_r\},\{\boldsymbol{m}_r\})} \times$$
$$\times\, \omega\Big(W_\theta\Big(\sum_{j=0}^{k-1} \mathbb{B}^{jt}(\boldsymbol{n}_j - \boldsymbol{m}_j)\Big)\Big) \quad \text{where}$$

$$\beta(\{\boldsymbol{n}_r\},\{\boldsymbol{m}_r\}) = \sum_{a=1}^{k-1}\sum_{b=0}^{a-1}\Big(\sigma(\mathbb{B}^{nb}\boldsymbol{n}_b, \mathbb{B}^{na}\boldsymbol{n}_a) - \sigma(\mathbb{B}^{nb}\boldsymbol{m}_b, \mathbb{B}^{na}\boldsymbol{m}_a)\Big)$$

$$\omega\Big(W_\theta\Big(\sum_{j=0}^{k-1} \mathbb{B}^{jt}(\boldsymbol{n}_j - \boldsymbol{m}_j)\Big)\Big) = \delta_{\sum_{j=0}^{k-1}\mathbb{B}^{jt}(\boldsymbol{n}_j-\boldsymbol{m}_j),\boldsymbol{0}} \ ,$$

where $\mathbb{B} = \mathbb{A}^T$ is the transpose of the dynamical matrix $\mathbb{A}$. By expanding $|\,\boldsymbol{n}_j - \boldsymbol{m}_j\,\rangle = \gamma_j|\,a_+\,\rangle + \delta_j|\,a_-\,\rangle$ along the eigenvectors of $\mathbb{B}$, one gets

$$\Big(\sum_{j=0}^{k-1}\gamma_j\alpha^{jt}\Big)|\,a_+\,\rangle + \Big(\sum_{j=0}^{k-1}\gamma_j\alpha^{-jt}\Big)|\,a_-\,\rangle = 0 \ ,$$

where $\alpha > 1$ ad $\alpha^{-1}$ are the eigenvalues of $\mathbb{B}$, whence

$$\gamma_{k-1} + \sum_{j=0}^{k-2}\gamma_j\alpha^{-(k-1-j)t} = 0 = \delta_0 + \sum_{j=1}^{k-1}\gamma_j\alpha^{-jt} \ .$$

Suppose $1 \leq i_j \leq p-1$ for all $i_j \in \boldsymbol{i}^{(k)}$, then the $f_{i_j}$ have compact support and, for $t$ sufficiently large, the above equalities imply $\gamma_{k-1} = 0 = \delta_0$. Iterating this argument, one gets $\boldsymbol{n}_j = \boldsymbol{m}_j$ for $0 \leq j \leq k-1$, whence

$$\lambda^{(k)}_{\boldsymbol{i}^{(k)},t} = \sum_{\boldsymbol{n}_0 \ldots \boldsymbol{n}_{k-1}} \prod_{r=0}^{k-1} |f_{i_r}(\boldsymbol{n}_r)|^2 = \prod_{j=0}^{k-1} \omega(X_{i_j}) .$$

Then, $\widetilde{\sum}_{\boldsymbol{i}^{(k)}} \eta(\lambda^{(k)}_{\boldsymbol{i}^{(k)},t}) = k \sum_{i=1}^{p-1} \eta(\lambda_i)$, where $\widetilde{\sum}$ denotes the sum over $i_j \neq p$ for all $0 \leq j \leq k-1$. Therefore,

$$\frac{1}{k} H(\Lambda^{(k)}_t) = \frac{1}{k} \sum_{\boldsymbol{i}^{(k)}} \eta(\lambda^{(k)}_{\boldsymbol{i}^{(k)},t}) \geq \frac{1}{k} \widetilde{\sum}_{\boldsymbol{i}^{(k)}} \eta(\lambda^{(k)}_{\boldsymbol{i}^{(k)},t}) = H(\Lambda) - \eta(\lambda_p) .$$

Furthermore, from the assumptions, $\lambda_p = \omega(X_p) \leq \varepsilon$; thus (8.52) can be estimated as follows

$$\frac{1}{k}\Big( H(\Lambda^{(k)}_t) - \sum_{j=0}^{k-1} H(\Lambda_{j,t}) \Big) \geq \varepsilon \log \varepsilon . \tag{8.53}$$

In order to lowerbound (8.47), we first rewrite it by means of (8.51) as

$$\begin{aligned} \frac{1}{k} \sum_{j=0}^{k-1} \sum_{i_j} \lambda^j_{i_j,t} S\left(\omega^j_{i_j,t} \circ \Theta^{jt} \circ \gamma \,,\, \omega \circ \gamma\right) &= S(\omega \circ \gamma) - \sum_{i=1}^{p} \lambda_i S(\omega_i \circ \gamma) \\ &+ \frac{1}{k} \sum_{j=0}^{k-1} \sum_{i_j} \lambda_{i_j} \Big( S(\omega_{i_j} \circ \gamma) - S\left(\omega^j_{i_j,t} \circ \Theta^{jt} \circ \gamma\right) \Big) . \end{aligned}$$

Secondly, by means of (8.48), we lowerbound it by

$$\mathrm{H}_\omega(\gamma) - \varepsilon + \frac{1}{k} \sum_{j=0}^{k-1} \sum_{i_j} \lambda_{i_j} \Big( S(\omega_{i_j} \circ \gamma) - S\left(\omega^j_{i_j,t} \circ \Theta^{jt} \circ \gamma\right) \Big) . \tag{8.54}$$

Thirdly, we consider the expectations

$$\begin{aligned} \omega\Big(X^j_{i_j,t} \Theta^{jt}[W_\theta(g)]\Big) &= \sum_{i_{j+1} \ldots i_{k-1}} \omega\Big( \big(Y_{[i_{j+1}, i_{k+1}]}\big)^\dagger \, X_{i_j} \, Y_{[i_{j+1}, i_{k-1}]} W_\theta(g) \Big) \\ &= \omega(X_{i_j} W_\theta(g)) + \sum_{i_{j+1} \ldots i_{k-1}} \omega\Big( \big(Y_{[i_{j+1}, i_{k+1}]}\big)^\dagger \, X_{i_j} \left[ Y_{[i_{j+1}, i_{k-1}]}, W_\theta(g) \right] \Big) , \end{aligned}$$

and expand the commutator as

$$\left[Y_{[i_{j+1},i_{k-1}]}, W_\theta(g)\right] = \sum_{a=1}^{k-j-1} \Theta^t_{\mathbb{A}}[W_\theta(f_{i_{j+1}})] \cdots$$
$$\cdots \Theta^{(a-1)t}_{\mathbb{A}}[W_\theta(f_{i_{j+a-1}})] \left[\Theta^{at}_{\mathbb{A}}[W_\theta(f_{i_{j+a}})]\,,\, W_\theta(g)\right] \Theta^{(a+1)t}_{\mathbb{A}}[W_\theta(f_{i_{j+a+1}})] \cdots$$
$$\cdots \Theta^{t(k-j-1)}_{\mathbb{A}}[W_\theta(f_{i_{k-1}}) \,.$$

Since $W_\theta(f_i)W_\theta(f_i) \leq \sum_{j=1}^p W_\theta(f_j)W_\theta(f_j) = 1\!\!1$, then $\|W_\theta(f_i)\| \leq 1$; therefore, using (7.81), one can estimate

$$\left\|\left[Y_{[i_{j+1},i_{k-1}]}, W_\theta(g)\right]\right\| \leq \sum_{a=1}^{k-j-1} \left\|\left[\Theta^{at}_{\mathbb{A}}[W_\theta(f_{i_{j+a}})]\,,\, W_\theta(g)\right]\right\|$$
$$\leq \sum_{a=1}^{k-j-1} \alpha^{-at} \sum_{\boldsymbol{n}_a,\boldsymbol{m}_a} |f_{i_{j+a}}(\boldsymbol{n}_a)|\, |g(\boldsymbol{m}_a)|\, C_{\boldsymbol{n}_a,\boldsymbol{m}_a} \,.$$

Consequently, if all functions have compact support, the commutator goes to 0 exponentially fast with $n$; now, the function $f_p$ not necessarily with compact support is in any case such that $\|W_\theta(f_p)\| \leq \varepsilon$ and any element in $\gamma(\boldsymbol{M})$ can be approximated in norm within $\varepsilon$ by suitable $W_\theta(g)$ where $g$ has compact support. Therefore, one can adjust $n$ so that $\left\|\omega^j_{i_j,t}\circ\gamma - \omega_{i_j}\circ\gamma\right\| \leq \varepsilon$, whence (8.54) and the Fannes inequality (5.157) yield

$$\frac{1}{k}\sum_{j=0}^{k-1}\sum_{i_j} \lambda^j_{i_j,t} S\left(\omega^j_{i_j,t}\circ\Theta^{jt}\circ\gamma\,,\,\omega\circ\gamma\right) \geq \mathrm{H}_\omega(\gamma) - \varepsilon + h(\varepsilon)$$

where $h(\varepsilon) \to 0$ when $\varepsilon \to 0$. Together with (8.46), (8.47) and (8.53), this last estimate proves the result; indeed, for all $\gamma : \boldsymbol{M} \mapsto \mathcal{A}_\theta$ and $\varepsilon > 0$, one can choose $t$ large enough so that

$$\mathrm{H}_\omega(\gamma) \geq \mathrm{h}^{\mathrm{CNT}}_\omega\left(\Theta^t_{\mathbb{A}}, \gamma\right) \geq \mathrm{H}_\omega(\gamma) + \varepsilon\log\varepsilon - 2\varepsilon + h(\varepsilon)\,.$$

The previous result puts into evidence the role of asymptotic commutativity in establishing the existence of a memory loss mechanism. One wonders whether the vice versa is also true, namely whether asymptotic memory loss implies asymptotic Abelianess and to which degree. The following result whose proof can be found in [42, 222] gives a partial answers to this question.

**Proposition 8.1.12.** *Let $(\mathcal{A}, \Theta, \omega)$ be a quantum dynamical triple with $\mathcal{A}$ a hyperfinite von Neumann algebra of type $II_1$ equipped with the tracial state $\omega$. Then, this dynamical system is strongly asymptotically Abelian.*

The following corollary regards quantized hyperbolic automorphisms of the torus with rational deformation parameter $\theta = p/q$ which are algebraic quantum $K$-systems and expresses the generic inequivalence of this notion and the one of entropic quantum $K$-system.

**Corollary 8.1.1.** *The quantized hyperbolic automorphisms of the torus with rational deformation parameter $\theta = p/q$ cannot be entropic $K$-systems.*

**Proof:** As seen in Example (7.1.13), these quantum dynamical systems cannot be strongly asymptotically Abelian. □

## 8.2 *AFL* Entropy: *OPUs*

The quantum dynamical entropy developed by R. Alicki and M. Fannes [9] is based on an earlier approach of Lindblad [194] to the non-commutative generalization of the *KS* entropy and considers the description of $C^*$ quantum dynamical systems $(\mathcal{A}, \Theta, \omega)$ by means of quantum symbolic models. In analogy with classical symbolic models (see Section 2.2), the time-evolution $\Theta$ is coarsely reconstructed by means of a shift automorphism $\Theta_\sigma$ on a quantum spin half-chain $\mathcal{A}_\mathcal{X}$ (see Section 7.1.5) equipped with a particular (unlike for classical dynamical systems, in general not translation-invariant) state $\omega_\mathcal{X}$. We shall denote these quantum symbolic models by quantum dynamical triples $(\mathcal{A}_\mathcal{X}, \Theta_\sigma, \omega_\mathcal{X})$, where the subindex $\mathcal{X}$ denotes the fact that they are constructed by means of *operational partitions of unity* (*OPU*s) in a way that can be physically interpreted as corresponding to repeated measurements performed on the system $(\mathcal{A}, \Theta, \omega)$.

**Definition 8.2.1 (Operational Partitions of Unity).** *An operational partition of unity in $\mathcal{A}$ is any finite collection of operators $\mathcal{Z} = \{Z_i\}_{i=1}^{|\mathcal{Z}|}$, $Z_i \in \mathcal{A}$, such that*

$$\sum_{i=1}^{|\mathcal{Z}|} Z_i^\dagger Z_i = 1\!\!1 , \tag{8.55}$$

*where $|\mathcal{Z}|$ is the cardinality of $\mathcal{Z}$.*

*OPU*s correspond to the *POVM* measurements typical of quantum information (see Definition 5.6.1); as already observed, they are the most general algebraic extension of the notion of classical partitions to quantum systems. Furthermore, we shall see that *OPU*s can profitably be used instead of partitions even in a classical context.

Given two *OPU*s $\mathcal{Z}_1 = \{Z_{1i}\}_{i=1}^{|\mathcal{Z}_1|}$ and $\mathcal{Z}_2 = \{Z_{2j}\}_{j=1}^{|\mathcal{Z}_2|}$, the algebraic extension of the notion of refinement of two partitions (see Section 2.2) is as follows

$$\mathcal{Z}_1 \circ \mathcal{Z}_2 := \{Z_{1i} Z_{2j}\}_{i,j=1}^{|\mathcal{Z}_1|\,|\mathcal{Z}_2|} . \tag{8.56}$$

What one gets by this definition is a finer *OPU*; indeed,

$$\sum_{i,j=1}^{|\mathcal{Z}_1|\,|\mathcal{Z}_2|} Z_{2j}^\dagger\, Z_{1i}^\dagger\, Z_{1i}\, Z_{2j} = \sum_{j=1}^{|\mathcal{Z}_2|} Z_{2j}^\dagger\, Z_{2j} = \mathbb{1}\ .$$

Moreover, consider a measure-theoretic triple $(\mathcal{X}, T, \mu)$ and the corresponding von Neumann algebraic commutative triple $(\mathbb{L}^\infty_\mu(\mathcal{X}), \Theta_T, \omega_\mu)$. One can associate to two finite, measurable partitions $\mathcal{P} = \{P_i\}_{i=1}^{|\mathcal{P}|}$ and $\mathcal{Q} = \{Q_j\}_{j=1}^{|\mathcal{Q}|}$ of the measure space $\mathcal{X}$ the *OPUs* $\mathcal{Z}_P$ and $\mathcal{Z}_Q$ from $\mathbb{L}^\infty_\mu(\mathcal{X})$ consisting of the characteristic functions $\chi_{P_i}$ and $\chi_{Q_j}$ of their atoms. It then turns out that $\mathcal{Z}_P \circ \mathcal{Z}_Q = \{\chi_{P_i \cap Q_j}\}_{i,j=1}^{|\mathcal{P}|,|\mathcal{Q}|}$ corresponds to the refined partition $\mathcal{P} \vee \mathcal{Q}$.

**Definition 8.2.2.** *Given an OPU* $\mathcal{Z} = \{Z_i\}_{i=1}^{|\mathcal{Z}|} \subset \mathcal{A}$, *its time-evolution at time* $t = k \in \mathbb{Z}$ *under the dynamics* $\Theta$ *is defined as*

$$\mathcal{Z}^k := \Theta^k(\mathcal{Z}) = \{\Theta^k(Z_i)\}_{i=1}^{|\mathcal{Z}|}\ . \tag{8.57}$$

*Further,* $\mathcal{Z}^{(n)}$ *will denote the OPU*

$$\mathcal{Z}^{(n)} := \mathcal{Z} \circ \Theta(\mathcal{Z}) \circ \cdots \Theta^{n-1}(\mathcal{Z}) = \{Z_{\boldsymbol{i}^{(n)}}\}_{\boldsymbol{i}^{(n)} \in \Omega^{(n)}_{|\mathcal{Z}|}} \tag{8.58}$$

$$\text{where } Z_{\boldsymbol{i}^{(n)}} = \Theta^{n-1}(Z_{i_{n-1}}) \cdots \Theta(Z_{i_1}) Z_{i_0}\ , \tag{8.59}$$

*and* $\Omega^{(n)}_{|\mathcal{Z}|} \ni \boldsymbol{i}^{(n)} := i_0 i_1 \cdots i_{n-1}$ *with* $i_j \in \{1, 2, \ldots, |\mathcal{Z}|\}$.

Note that $\mathcal{Z}^k$ is an *OPU* follows since $\Theta$ is an automorphism of $\mathcal{A}$:

$$\sum_{i=1}^{|\mathcal{Z}|} \Theta(Z_i)^\dagger \Theta(Z_i) = \Theta\Big(\sum_{i=1}^{|\mathcal{Z}|} Z_i^\dagger Z_i\Big) = \Theta(\mathbb{1}) = \mathbb{1}\ .$$

Then, $\mathcal{Z}^{(n)}$ is also an *OPU* .

### 8.2.1 Quantum Symbolic Models and *AFL* Entropy

Given an *OPU* $\mathcal{Z} = \{Z_j\}_{j=1}^{|\mathcal{Z}|}$, let $\{|\, z_i\,\rangle\}_{i=1}^{|\mathcal{Z}|}$ denote a fixed orthonormal basis in the finite-dimensional Hilbert space $\mathbb{C}^{|\mathcal{Z}|}$. The $|\mathcal{Z}| \times |\mathcal{Z}|$ matrix

$$M_{|\mathcal{Z}|}(\mathbb{C}) \ni \rho[\mathcal{Z}] := \sum_{i,j=1}^{|\mathcal{Z}|} |\, z_i\,\rangle\langle\, z_j\,|\ \omega(Z_j^\dagger Z_i) \tag{8.60}$$

is a density matrix. Indeed, normalization comes from Definition 8.2.1, while positivity is ensured by the fact that

$$\langle \psi \mid \rho[\mathcal{Z}] \mid \psi \rangle = \sum_{i,j=1}^{|\mathcal{Z}|} \psi_i^* \psi_j \; \omega(Z_j^\dagger Z_i) = \omega(Z_\psi^\dagger Z_\psi) \geq 0 \,,$$

where $\mathbb{C}^{|\mathcal{Z}|} \ni |\psi\rangle = \sum_{i=1}^{|\mathcal{Z}|} \psi_i | z_i \rangle$ and $Z_\psi := \sum_{i=1}^{|\mathcal{Z}|} \psi_i^* Z_i$.
Furthermore, from $\omega \circ \Theta = \omega$, it follows that

$$\rho[\mathcal{Z}^t] = \rho[\mathcal{Z}] \qquad \forall\, t \in \mathbb{Z} \,. \tag{8.61}$$

Consider the time-refined *OPU* $\mathcal{Z}^{(n)}$; the corresponding density matrix is of the form

$$M_{|\mathcal{Z}|}(\mathbb{C})^{\otimes n} \ni \rho[\mathcal{Z}^{(n)}] = \sum_{\boldsymbol{i}^{(n)}, \boldsymbol{j}^{(n)} \in \Omega_{|\mathcal{Z}|}^{(n)}} | z_{\boldsymbol{i}^{(n)}} \rangle\langle z_{\boldsymbol{j}^{(n)}} | \, \omega\Big(Z_{\boldsymbol{j}^{(n)}}^\dagger Z_{\boldsymbol{i}^{(n)}}\Big) \,, \tag{8.62}$$

where

$$| z_{\boldsymbol{i}^{(n)}} \rangle := | z_{i_1} \rangle \otimes | z_{i_2} \rangle \otimes \cdots | z_{i_n} \rangle \,.$$

At each iteration of the dynamics $\Theta$, one component is added to the *OPU* $\mathcal{Z}^{(n)}$ and one factor to the corresponding algebraic tensor product $M_{|\mathcal{Z}|}(\mathbb{C})^{\otimes n}$. Therefore, to any given *OPU* $\mathcal{Z} \subset \mathcal{A}$ there remains associated a quantum spin half-chain $\mathcal{A}_{\mathcal{Z}}$ (see Section 7.1.5), with a $|\mathcal{Z}|$-dimensional spin at each site and a family of density matrices $\rho\left[\mathcal{Z}^{(n)}\right]$, $n \in \mathbb{N}$. Since $\Theta$ is an automorphism, applying Definition 8.2.1, it turns out that these density matrices form a compatible family in the sense of (7.85), namely

$$\begin{aligned}
\mathrm{Tr}_{\{n+1\}}\Big(\rho[\mathcal{Z}^{(n+1)}]\Big) &= \mathrm{Tr}_{\{n+1\}}\Big(\sum_{\substack{\boldsymbol{i}^{(n+1)}\\ \boldsymbol{j}^{(n+1)}}} | z_{\boldsymbol{i}^{(n+1)}} \rangle\langle z_{\boldsymbol{j}^{(n+1)}} | \, \omega\Big(Z_{\boldsymbol{j}^{(n+1)}}^\dagger Z_{\boldsymbol{i}^{(n+1)}}\Big)\Big) \\
&= \sum_{\substack{\boldsymbol{i}^{(n+1)}\\ \boldsymbol{j}^{(n+1)}}} | z_{\boldsymbol{i}^{(n)}} \rangle\langle z_{\boldsymbol{j}^{(n)}} | \, \langle z_{j_{n+1}} | z_{i_{n+1}} \rangle \, \omega\Big(Z_{\boldsymbol{j}^{(n+1)}}^\dagger Z_{\boldsymbol{i}^{(n+1)}}\Big) \\
&= \sum_{\boldsymbol{i}^{(n)}, \boldsymbol{j}^{(n)}} | z_{\boldsymbol{i}^{(n)}} \rangle\langle z_{\boldsymbol{j}^{(n)}} | \sum_{i=1}^{|\mathcal{Z}|} \omega\Big(Z_{\boldsymbol{j}^{(n)}}^\dagger \Theta^n(Z_i^\dagger Z_i) Z_{\boldsymbol{i}^{(n)}}\Big) = \rho[\mathcal{Z}^{(n)}] \,.
\end{aligned}$$

Thus the family $\rho[\mathcal{Z}^{(n)}]$, $n \in \mathbb{N}$, provides a state $\omega_{\mathcal{Z}}$ over $\mathcal{A}_{\mathcal{Z}}$.

The dynamics $\Theta$ on $\mathcal{A}$ corresponds to moving right along $\mathcal{A}_{\mathcal{Z}}$ with the shift automorphism $\Theta_\sigma$; however, unlike the states of quantum spin chains (see Definition 7.1.11) which are $\Theta$-invariant, the compatible family $\{\rho[\mathcal{Z}^{(n)}]\}_{n \in \mathbb{N}}$, need not satisfy condition (7.86); namely, in general, $\omega_{\mathcal{X}} \circ \Theta_\sigma \neq \omega_{\mathcal{X}}$. For instance, in general,

$$\mathrm{Tr}_{\{1\}} \rho[\mathcal{Z}^{(2)}] = \sum_{i_1, j_1 = 1}^{|\mathcal{Z}|} | z_{i_1} \rangle\langle z_{j_1} | \sum_{k=1}^{|\mathcal{Z}|} \omega\left(Z_k^\dagger \Theta(Z_{j_1}^\dagger Z_{i_1}) Z_k\right) \neq \rho[\mathcal{Z}] \,.$$

In the classical setting the *KS*-entropy is the maximal Shannon entropy per symbol over all symbolic models built upon finite measurable partitions. The *AFL* construction defines the quantum dynamical entropy of $(\mathcal{A}, \Theta, \omega)$ as the largest mean von Neumann entropy over all its symbolic models $(\mathcal{A}_{\mathcal{Z}}, \Theta_\sigma, \omega_{\mathcal{Z}})$ constructed from *OPU*s $\mathcal{Z}$ chosen from a selected $\Theta$-invariant subalgebra $\mathcal{A}_0 \subseteq \mathcal{A}$. Because of the lack of translation invariance, it is not guaranteed that the mean von Neumann entropy of $(\mathcal{A}_{\mathcal{Z}}, \Theta_\sigma, \omega_{\mathcal{Z}})$ exists as a limit.

**Definition 8.2.3 (*AFL*-Entropy).** *Let $\mathcal{A}_0 \subseteq \mathcal{A}$ be a $\Theta$-invariant subalgebra and let $\mathcal{Z} \subset \mathcal{A}_0$ be an OPU ; set*

$$\mathrm{h}_\omega^{\mathrm{AFL}}(\Theta, \mathcal{Z}) := \limsup_{n\to\infty} \frac{1}{n} S\left(\rho[\mathcal{Z}^{(n)}]\right) , \tag{8.63}$$

*where $S\left(\rho[\mathcal{Z}^{(n)}]\right)$ is the von Neumann entropy of the density matrix associated with the OPUs $\mathcal{Z}^{(n)}$. The AFL -entropy of $(\mathcal{A}, \Theta, \omega)$ is then defined as*

$$\mathrm{h}_\omega^{\mathrm{AFL}}(\Theta) := \sup_{\mathcal{Z}\in\mathcal{A}_0} \mathrm{h}_\omega^{\mathrm{AFL}}(\Theta, \mathcal{Z}) . \tag{8.64}$$

*When needed, we shall explicitly refer to the dependence on $\mathcal{A}_0$ by writing $\mathrm{h}_\omega^{\mathrm{AFL}}(\Theta, \mathcal{A}_0)$.*

As for the *CNT* entropy (compare (8.27)), when one considers powers $\Theta^q$, $q \geq 0$, of the automorphism $\Theta$, one has the following bound.

**Proposition 8.2.1.** *For all $\mathbb{N} \ni q \geq 1$, it holds that $\frac{1}{q}\mathrm{h}_\omega^{\mathrm{AFL}}(\Theta^q) \geq \mathrm{h}_\omega^{\mathrm{AFL}}(\Theta)$.*

**Proof:** For any given *OPU* $\mathcal{Z} = \{Z_i\}_{i=1}^{|\mathcal{Z}|}$, set

$$\mathcal{Z}^{(q,n)} := \Theta^{q(n-1)}[\mathcal{Z}] \circ \Theta^{q(n-2)}[\mathcal{Z}] \cdots \Theta^q[\mathcal{Z}] \circ \mathcal{Z} .$$

Given the *OPU* $\mathcal{Z}^{(q)}$, $q \geq 1$, one verifies that $(\mathcal{Z}^{(q)})^{(q,n)} = \mathcal{Z}^{(qn)}$. Therefore, writing $n = kp + q$ with $0 \leq q \leq p$, by means of (5.161) one gets

$$\begin{aligned}
\mathrm{h}_\omega^{\mathrm{AFL}}(\Theta, \mathcal{Z}) &= \limsup_{n\to\infty} \frac{1}{n} S\left(\rho[\mathcal{Z}^{(n)}]\right) = \limsup_{k\to\infty} \frac{1}{k\,q+p} S\left(\rho[\mathcal{Z}^{(k\,q+p)}]\right) \\
&\leq \frac{1}{q} \limsup_{k\to\infty} \frac{1}{k} S\left(\rho[\mathcal{Z}^{(k\,q)}]\right) = \frac{1}{q} \limsup_{k\to\infty} \frac{1}{k} S\left(\rho[\mathcal{Z}^{(q,k)}]\right) \\
&= \frac{1}{q} \mathrm{h}_\omega^{\mathrm{AFL}}\left(\Theta^q, \mathcal{Z}^{(q)}\right) .
\end{aligned}$$

In fact, since the states $\rho[\mathcal{Z}^{(n)}]$ are density matrices on a spin-algebra $M_{|\mathcal{Z}|}(\mathbb{C})^{\otimes n} = \bigotimes_{j=0}^{n-1}(M_{|\mathcal{Z}|}(\mathbb{C}))_j$, one derives the bound

$$\left|S\left(\rho[\mathcal{Z}^{(k\,q+p)}\right) - S\left(\rho[\mathcal{Z}^{(k\,q)}]\right)\right| \leq S\left(\rho_{[k\,q,k\,q+p]}\right) \leq q\,\log|\mathcal{Z}|\ ,$$

where $\rho_{[k\,q,k\,q+p]}$ is the marginal density matrix on $\bigotimes_{j=k\,q}^{k\,q+p}(M_{|\mathcal{Z}|}(\mathbb{C}))_j$. Since *OPUs* of the form $\mathcal{Z}^{(q)}$ are a subclass of all possible *OPUs*, one concludes

$$\mathrm{h}_\omega^{\mathrm{AFL}}(\Theta) = \sup_{\mathcal{Z}\in\mathcal{A}_0} \mathrm{h}_\omega^{\mathrm{AFL}}(\Theta,\mathcal{Z}) \leq \frac{1}{q}\mathrm{h}_\omega^{\mathrm{AFL}}(\Theta^q)\ .$$

□

**Remarks 8.2.1.**

1. Suppose the dynamics is trivial, $\Theta = \mathrm{id}_{\mathcal{A}}$, namely $\Theta[A] = A$ for all $A \in \mathcal{A}$; from the previous result it follows that, if $\mathrm{h}_\omega^{\mathrm{AFL}}(\mathrm{id}_{\mathcal{A}}) > 0$, then it is infinite for one can choose an arbitrarily large $q$ and $\mathrm{id}_{\mathcal{A}}^q = \mathrm{id}_{\mathcal{A}}$. This effect is clearly due to the perturbing action of the *OPUs* which themselves act as a source of entropy. Therefore, the $\Theta$-invariant subalgebra $\mathcal{A}_0$ from where the *OPUs* are taken has to be chosen in such a way to minimize these perturbing effects.
2. The request that *OPUs* consist of elements from a selected $\Theta$-invariant subalgebra $\mathcal{A}_0 \subset \mathcal{A}$ usually comes from physical considerations. Indeed, *OPUs* as *POVMs* should correspond to physically realizable measurement processes which are always strictly local, namely they should consist of operators from local subalgebras of $\mathcal{A}$. The obvious choice for $\mathcal{A}_0$ is thus the $*$-algebra containing all strictly local $C^*$ algebras.
3. Unlike for the *CNT* entropy (see Remark 8.1.2), it is not known whether an equality of the form $\mathrm{h}_\omega^{\mathrm{AFL}}(\Theta^q) = q\,\mathrm{h}_\omega^{\mathrm{AFL}}(\Theta)$ holds. Indeed, the key ingredient in the proof of this equality for the *CNT* entropy is its strong continuity which is not usable in the case of the *AFL* entropy. Continuity is also important to check on its dependence on the *OPUs* : for results in this direction see [10, 133].

### 8.2.2 *AFL* Entropy: Interpretation

Like the *KS* entropy, one can interpret the *AFL* entropy as the asymptotic rate of information provided by repeated, coarse-grained observations of the time-evolution; the difference from the classical setting is in that a coupling to an external ancillary system is required. This can be seen by going to the *GNS* construction $\left(\mathbb{H}_\omega, \pi_\omega, \Omega_\omega\right)$ based on the $\Theta$-invariant state $\omega$.

Consider an *OPU* $\mathcal{Z} = \{Z_i\}_{i=1}^{|\mathcal{Z}|}$, the pure state projection onto

$$\mathbb{H}_\omega \otimes \mathbb{C}^{|\mathcal{Z}|} \ni |\,\Psi_{\mathcal{Z}}^{\omega}\,\rangle := \sum_{i=1}^{|\mathcal{Z}|} \pi_\omega(Z_i)|\,\Omega_\omega\,\rangle \otimes |\,z_i\,\rangle$$

has the following marginal density matrices (see (8.62))

$$\begin{aligned}\rho_I &= \mathrm{Tr}_{\mathbb{C}^{|\mathcal{Z}|}} |\Psi^\omega_{\mathcal{Z}}\rangle\langle\Psi^\omega_{\mathcal{Z}}| = \sum_{i=1}^{|\mathcal{Z}|} \pi_\omega(Z_i)|\,\Omega_\omega\,\rangle\langle\,\Omega_\omega\,|\pi_\omega(Z_i^\dagger) \\ &=: \mathbb{F}^\omega_{\mathcal{Z}}[|\,\Omega_\omega\,\rangle\langle\,\Omega_\omega\,|] , \qquad (8.65)\\ \rho_{II} &:= \mathrm{Tr}_{\mathbb{H}_\omega}\Big(|\Psi^\omega_{\mathcal{Z}}\rangle\langle\Psi^\omega_{\mathcal{Z}}|\Big) = \sum_{i,j=1}^{|\mathcal{Z}|} \langle\,\Omega_\omega\,|\,\pi_\omega(Z_j^\dagger Z_i)\,|\Omega_\omega\,\rangle\,|\,z_i\,\rangle\langle\,z_j\,| = \rho[\mathcal{Z}] .\end{aligned}$$

The first marginal state is a mixed state on $\mathbb{H}_\omega$ resulting from a *POVM* measurement (see Definition 5.6.1) on the *GNS* state $|\,\Omega_\omega\,\rangle\langle\,\Omega_\omega\,|$. This effect corresponds to the action of a map which, in the *GNS* representation, is the dual of the following *CPU* map on $\mathcal{A}$:

$$\mathcal{A} \ni A \mapsto \mathbb{E}_{\mathcal{Z}}[A] = \sum_{i=1}^{|\mathcal{Z}|} Z_i^\dagger\, A\, Z_i .$$

Since $|\,\Psi^\omega_{\mathcal{Z}}\,\rangle\langle\,\Psi^\omega_{\mathcal{Z}}\,|$ is a pure state, Proposition 5.5.5 ensures that $\rho[\mathcal{Z}]$ and $\mathbb{F}^\omega_{\mathcal{Z}}[|\,\Omega_\omega\,\rangle\langle\,\Omega_\omega\,|]$ have the same von Neumann entropy. The same argument applies to the case of the refined *OPU* $\mathcal{Z}^{(n)}$: the von Neumann entropy $S\left(\rho[\mathcal{Z}^{(n)}]\right)$ equals that of

$$\mathbb{F}^\omega_{\mathcal{Z}^{(n)}}[|\,\Omega_\omega\,\rangle\langle\,\Omega_\omega\,|] = \sum_{\boldsymbol{i}^{(n)}\in\Omega^{(n)}_{|\mathcal{Z}|}} \pi_\omega(Z_{\boldsymbol{i}^{(n)}})|\,\Omega_\omega\,\rangle\langle\,\Omega_\omega\,|\pi_\omega(Z^\dagger_{\boldsymbol{i}^{(n)}}) , \qquad (*)$$

with $Z_{\boldsymbol{i}^{(n)}}$ as in Definition 8.2.2. Using the *GNS* implementation of the dynamics, $\pi_\omega(\Theta(X)) = U_\omega^\dagger \pi_\omega(X) U_\omega$, and the fact that $U_\omega|\,\Omega_\omega\,\rangle = |\,\Omega_\omega\,\rangle$, one rewrites

$$\begin{aligned}\pi_\omega\left(Z_{\boldsymbol{i}^{(n)}}\right)|\,\Omega_\omega\,\rangle &= U_\omega^{n-1}\pi_\omega(Z_{i_{n-1}})U_\omega^\dagger)^{n-1})\cdots U_\omega\pi_\omega(Z_{i_1})U_\omega^\dagger\pi_\omega(Z_{i_0})|\,\Omega_\omega\,\rangle \\ &= U_\omega^n\left(U_\omega^\dagger\pi_\omega(Z_{i_{n-1}})U_\omega^\dagger\pi_\omega(Z_{i_{n-2}})\cdots U_\omega^\dagger\pi_\omega(Z_{i_0}\right)|\,\Omega_\omega\,\rangle ,\end{aligned}$$

whence, setting $\mathbb{U}_\omega[A] := U_\omega\, A\, U_\omega^\dagger$, $A \in \mathcal{A}$, $(*)$ can be recast as

$$\mathbb{F}^\omega_{\mathcal{Z}^{(n)}}[|\,\Omega_\omega\,\rangle\langle\,\Omega_\omega\,|] = \mathbb{U}_\omega^n \circ \left(\mathbb{U}_\omega^\dagger \circ \mathbb{F}^\omega_{\mathcal{Z}}\right)^n[|\,\Omega_\omega\,\rangle\langle\,\Omega_\omega\,|] . \qquad (8.66)$$

It thus follows that

$$S\left(\mathbb{F}^\omega_{\mathcal{Z}^{(n)}}[|\,\Omega_\omega\,\rangle\langle\,\Omega_\omega\,|]\right) = S\left(\rho[\mathcal{Z}^{(n)}]\right) . \qquad (8.67)$$

Therefore, the *AFL* entropy can be regarded as the largest entropy production provided by *POVM* measurements based on a selected class of *OPU*s and performed at each tick of time on the evolving system coupled to a purifying GNS ancilla.

Because of this fact, while the *CNT* entropy which corresponds to the maximal compression rate of ergodic quantum sources, the *AFL* entropy appears to be related to the classical capacity of quantum channels. We shall show this after providing examples of quantum dynamical systems where the *AFL* entropy can be explicitly computed.

### 8.2.3 *AFL*-Entropy: Applications

As for the *CNT* entropy, we first ascertain whether the *AFL* entropy reduces to the *KS* entropy when the algebraic dynamical triple $(\mathcal{A},\Theta,\omega)$ describes a classical dynamical system $(\mathcal{X},T,\mu)$.

As already remarked, classical partitions $\mathcal{P}=\{P_i\}_{i=1}^{|\mathcal{P}|}$ of $\mathcal{X}$, are associated with *OPUs* $\mathcal{Z}_P=\{\chi_{P_i}\}_{i=1}^{|\mathcal{P}|}$, then

$$\mathcal{Z}_P^{(n)}=\mathcal{Z}_P^{n-1}\circ\mathcal{Z}_P^{n-2}\circ\cdots\mathcal{Z}_P=\left\{\Theta^{n-1}(\chi_{P_{i_{n-1}}})\Theta^{n-2}(\chi_{P_{i_{n-2}}})\cdots\chi_{P_{i_0}}\right\}$$
$$=\left\{\chi_{T^{-n+1}(P_{i_{n-1}})\cap T^{-n+2}(P_{i_{n-2}})\cap\cdots P_{i_0}}\right\}=\left\{\chi_{P^{(n)}_{\boldsymbol{i}^{(n)}}}\right\},$$

so that $\mathcal{Z}^{(n)}$ is the *OPU* associated with the refined partition $\mathcal{P}^{(n)}$ [6] and

$$\rho[\mathcal{Z}^{(n)}]=\sum_{\boldsymbol{i}^{(n)},\boldsymbol{j}^{(n)}\in\Omega^{(n)}_{|\mathcal{Z}_P|}}|\,z_{\boldsymbol{i}^{(n)}}\,\rangle\langle\,z_{\boldsymbol{j}^{(n)}}\,|\;\mu\left(P_{\boldsymbol{j}^{(n)}}P_{\boldsymbol{i}^{(n)}}\right)$$
$$=\sum_{\boldsymbol{i}^{(n)}\in\Omega^{(n)}_{|\mathcal{Z}_P|}}|\,z_{\boldsymbol{i}^{(n)}}\,\rangle\langle\,z_{\boldsymbol{i}^{(n)}}\,|\;\mu\left(P_{\boldsymbol{i}^{(n)}}\right)$$

is diagonal with eigenvalues $\mu\left(P_{\boldsymbol{i}^{(n)}}\right)$ so that (see (3.1))

$$S\left(\rho[\mathcal{Z}_P^{(n)}]\right)=-\sum_{\boldsymbol{i}^{(n)}\in\Omega^{(n)}_{|\mathcal{Z}_P|}}\mu(P_{\boldsymbol{i}^{(n)}})\log\mu(P_{\boldsymbol{i}^{(n)}})=H_\mu(\mathcal{P}^{(n)})$$

$$\limsup_{n\to+\infty}\frac{1}{n}S\left(\rho[\mathcal{Z}_P^{(n)}]\right)=\mathrm{h}_\mu^{\mathrm{KS}}\left(T,\mathcal{P}\right)\;.\tag{8.68}$$

However, in view of the fact that one is free to choose more general *OPUs* than those arising from classical partitions, Definition (8.2.3) may in general lead to an *AFL* entropy of $(\mathcal{X},T,\mu)$ which is larger than its *KS* entropy. Actually, this is not the case: in order to prove it let us consider a generic *OPU* given by a finite collection $\mathcal{F}:=\{f_i\}_{i=1}^{|\mathcal{F}|}$ of essentially bounded functions such that

$$\sum_{i=1}^{|\mathcal{F}|}|f_i|^2=1\in\mathbb{L}^\infty_\mu(\mathcal{X})\;.$$

---

[6] The notations is that used in (2.40) and (2.39).

The corresponding density matrix (see (8.60)) reads

$$\rho[\mathcal{F}] = \sum_{i,j=1}^{|\mathcal{F}|} \omega_\mu(f_j^* f_i) \,|\, z_i \,\rangle\langle\, z_j \,| = \int_{\mathcal{X}} \mathrm{d}\mu\,(x) \sum_{i,j=1}^{|\mathcal{F}|} f_j^*(x) f_i(x) \,|\, z_i \,\rangle\langle\, z_j \,|$$
$$= \int_{\mathcal{X}} \mathrm{d}\mu\,(x)\, P_{\mathcal{F}}(x) \tag{8.69}$$

with $\{|\, z_i \,\rangle\}_{i=1}^{|\mathcal{F}|}$ an *ONB* in $\mathbb{C}^{|\mathcal{Z}|}$ and

$$M_{|\mathcal{F}|}(\mathbb{C}) \ni P_{\mathcal{F}}(x) = |\, \Psi_{\mathcal{F}}(x) \,\rangle\langle\, \Psi_{\mathcal{F}}(x) \,| \;, \quad |\, \Psi_F(x) \,\rangle := \sum_{i=1}^{|\mathcal{F}|} f_i(x) \,|\, z_i \,\rangle \;. \tag{8.70}$$

Notice that, because $\mathcal{F}$ is an *OPU*, the $P_F(x)$ are projections onto normalized vector states. If an *OPU* results from the refinement of other *OPUs* , then the associated density matrix is a continuous convex combination of tensor products of projections of the form (8.70). Concretely, if $\mathcal{F} = \mathcal{F}_1 \circ \mathcal{F}_2 \circ \cdots \mathcal{F}_n$,

$$\bigotimes_{j=1}^{n} M_{|\mathcal{F}_j|}(\mathbb{C}) \ni \rho[\mathcal{F}] = \int_{\mathcal{X}} \mathrm{d}\mu\,(x)\, P_{\mathcal{F}_1}(x) \otimes P_{\mathcal{F}_2}(x) \otimes \cdots P_{\mathcal{F}_n}(x) \;. \tag{8.71}$$

Using this expression it is possible to prove that, without restrictions on the *OPUs* taken from $\mathcal{M} := \mathbb{L}^\infty_\mu(\mathcal{X})$, the *AFL* entropy of $(\mathcal{M}, \Theta_T, \omega_\mu)$ coincides with the *KS* entropy of $(\mathcal{X}, T, \mu)$.

**Proposition 8.2.2.** $\mathrm{h}^{\mathrm{AFL}}_{\omega_\mu}(\Theta_T, \mathcal{M}) = \mathrm{h}^{\mathrm{KS}}_\mu(T)$ *(see Definition 8.2.3).*

**Proof:** Let $\mathcal{P} = \{P_i\}_{i=1}^d$ be any finite measurable partition of $\mathcal{X}$ with $\mathcal{P}^{(n)} = \{P^{(n)}_{\boldsymbol{i}^{(n)}}\}_{\boldsymbol{i}^{(n)} \in \Omega_d^{(n)}}$ its dynamical refinement up to time $t = n-1$ and let $\mathcal{F}$ be any other *OPU* from $\mathcal{M}$. Given $\mathcal{F}^{(n)}$ as in (8.58), let

$$P_{\mathcal{F}^{(n)}}(x) := P_{\mathcal{F}}(x) \otimes P_{\mathcal{F}^1}(x) \otimes \cdots P_{\mathcal{F}^{n-1}}(x) \;. \tag{$*$}$$

Since the atoms of $\mathcal{P}^{(n)}$ are disjoint, one can decompose $\rho[\mathcal{F}^{(n)}]$ into a convex sum of other density matrices in $M_{|\mathcal{F}|}(\mathbb{C})^{\otimes n}$:

$$\rho[\mathcal{F}^{(n)}] = \int_{\mathcal{X}} \mathrm{d}\mu\,(x)\, P_{\mathcal{F}^{(n)}}(x) = \sum_{\boldsymbol{i}^{(n)} \in \Omega_d^{(n)}} \mu(P^{(n)}_{\boldsymbol{i}^{(n)}})\, \rho_{\boldsymbol{i}^{(n)}}$$
$$\rho_{\boldsymbol{i}^{(n)}} := \frac{1}{\mu(P^{(n)}_{\boldsymbol{i}^{(n)}})} \int_{P^{(n)}_{\boldsymbol{i}^{(n)}}} \mathrm{d}\mu\,(x)\, P_{\mathcal{F}^{(n)}}(x) \;.$$

Thus, the concavity of the von Neumann entropy (5.156) and the triangle inequality (5.161) implies

$$
\begin{aligned}
S\left(\rho[\mathcal{F}^{(n)}]\right) &\leq -\sum_{\boldsymbol{i}^{(n)}\in\Omega_d^{(n)}} \mu(P^{(n)}_{\boldsymbol{i}^{(n)}}) \log \mu(P^{(n)}_{\boldsymbol{i}^{(n)}}) + \sum_{\boldsymbol{i}^{(n)}\in\Omega_d^{(n)}} \mu(P^{(n)}_{\boldsymbol{i}^{(n)}})\, S\left(\rho_{\boldsymbol{i}^{(n)}}\right) \\
&= H_\mu(\mathcal{P}^{(n)}) + \sum_{\boldsymbol{i}^{(n)}\in\Omega_d^{(n)}} \mu(P^{(n)}_{\boldsymbol{i}^{(n)}})\, S\left(\rho_{\boldsymbol{i}^{(n)}}\right) \\
&\leq H_\mu(\mathcal{P}^{(n)}) + \sum_{\boldsymbol{i}^{(n)}\in\Omega_d^{(n)}} \sum_{j=0}^{n-1} \mu(P^{(n)}_{\boldsymbol{i}^{(n)}})\, S\left(\rho_{i_j}\right) , \qquad (**)
\end{aligned}
$$

where, from $(*)$, $\rho_k := \dfrac{1}{\mu(P^{(n)}_{\boldsymbol{i}^{(n)}})} \displaystyle\int_{P^{(n)}_{\boldsymbol{i}^{(n)}}} \mathrm{d}\mu\,(x)\, P_{\mathcal{F}^k}(x)$.

Let $Q_k \in M_{|\mathcal{F}|}(\mathbb{C})$ be a projection; since $S\left(Q_k\right) = 0$, the Fannes inequality (5.157) implies

$$
S\left(\rho_k\right) \leq \|\rho_k - Q_k\|_1 \log|\mathcal{F}| + \eta(\|\rho_k - Q_k\|_1) \ .
$$

Notice that each $\rho_k \in \mathbb{C}^{|\mathcal{F}|}$ is a continuous convex combination of pure state projections $P_{\mathcal{F}^{(n)}}(x)$; the partition $\mathcal{P}$ is arbitrary and can always be chosen in such a way that each $\rho_k$ stays sufficiently close to a projection $\mathcal{Q}_k$ so that the right hand side of the previous inequality can be upperbounded by a quantity independent of $n$ and $\boldsymbol{i}^{(n)}$. Consequently, dividing both sides of $(**)$ by $n$ and taking the lim sup obtains

$$
\mathrm{h}^{\mathrm{AFL}}_{\omega_\mu}(\Theta_T, \mathcal{F}) \leq \limsup_{n\to+\infty} \frac{1}{n} H_\mu(\mathcal{P}^{(n)}) = \mathrm{h}^{\mathrm{KS}}_\mu(T, \mathcal{P}) \leq \mathrm{h}^{\mathrm{KS}}_\mu(T) \ .
$$

On the other hand, from (8.68) one gets

$$
\mathrm{h}^{\mathrm{AFL}}_{\omega_\mu}(\Theta_T, \mathcal{M}) \geq \sup_{\mathcal{Z}_P\in\Pi} \mathrm{h}^{\mathrm{AFL}}_{\omega_\mu}(\Theta_T, \mathcal{Z}_P) = \mathrm{h}^{\mathrm{KS}}_\mu(T) \ ,
$$

where $\Pi$ is the $*$-subalgebra of $\mathcal{M}$ containing the *OPUs* $\mathcal{Z}_P$ arising from all possible measurable partitions of $\mathcal{X}$. □

Evidently, one would like to reach the *KS* entropy by computing the *AFL* entropy on a smaller set than the whole of $\mathcal{M} = \mathbb{L}^\infty_\mu(\mathcal{X})$. The search for a suitable $*$-subalgebra $\mathcal{M}_0 \subset \mathcal{M}$ starts with the introduction [10] of an *entropic distance* between two *OPUs* $\mathcal{F}_{1,2} \subset \mathcal{M}$ defined by

$$
\Delta[\mathcal{F}_1|\mathcal{F}_2] := S\left(\rho[\mathcal{F}_1 \circ \mathcal{F}_2]\right) - S\left(\rho[\mathcal{F}_2]\right) \ . \qquad (8.72)
$$

Some useful properties of the entropic distance can be extracted by inspecting more closely the consequences of (8.71). Indeed, it turns out that, in a commutative context, the entropy of a composite *OPU* is invariant under permutations of the constituent *OPUs* :

$$
S\left(\rho[\mathcal{F}_1 \circ \mathcal{F}_2 \circ \mathcal{F}_3]\right) = S\left(\rho[\mathcal{F}_2 \circ \mathcal{F}_1 \circ \mathcal{F}_3]\right) \ , \qquad (8.73)
$$

for all *OPUs* $\mathcal{F}_{1,2,3} \subset \mathcal{M}$. In fact, because of their tensor product form, the density matrices $\rho[\mathcal{F}_1 \circ \mathcal{F}_2 \circ \mathcal{F}_3]$ and $\rho[\mathcal{F}_2 \circ \mathcal{F}_1 \circ \mathcal{F}_3]$ are unitarily equivalent. Also, (5.161) yields

$$S\left(\rho[\mathcal{F}_1 \circ \mathcal{F}_2]\right) \leq S\left(\rho[\mathcal{F}_1]\right) + S\left(\rho[\mathcal{F}_2]\right) , \tag{8.74}$$

for all *OPUs* $\mathcal{F}_{1,2} \subset \mathcal{M}$. Indeed, by partial tracing $\rho[\mathcal{F}_1 \circ \mathcal{F}_2]$ over $\mathbb{C}^{|\mathcal{F}_1|}$, respectively $\mathbb{C}^{|\mathcal{F}_2|}$, one gets

$$\begin{aligned}
\mathrm{Tr}_2(\rho[\mathcal{F}_1 \circ \mathcal{F}_2]) &= \int_{\mathcal{X}} \mathrm{d}\mu\,(x)\, P_{\mathcal{F}_1}(x)\, \mathrm{Tr}(P_{\mathcal{F}_2}) = \rho[\mathcal{F}_1] \\
\mathrm{Tr}_1(\rho[\mathcal{F}_1 \circ \mathcal{F}_2]) &= \int_{\mathcal{X}} \mathrm{d}\mu\,(x)\, \mathrm{Tr}(P_{\mathcal{F}_1}(x))\, P_{\mathcal{F}_2}) = \rho[\mathcal{F}_2] \ .
\end{aligned}$$

A more interesting property is the following one: for all *OPUs* $\mathcal{F}_{1,2} \subset \mathcal{M}$,

$$S\left(\rho[\mathcal{F}_1 \circ \mathcal{F}_2]\right) \geq S\left(\rho[\mathcal{F}_1]\right) \ . \tag{8.75}$$

To prove this, consider the case in which the integration measure in (8.71) is a discrete probability distribution $\mu = \{p_j\}_{j=1}^d$, that is $\rho[\mathcal{F}_1 \circ \mathcal{F}_2] = \sum_{j=1}^d p_j\, P_{\mathcal{F}_1}(j) \otimes P_{\mathcal{F}_2}(j)$. Then, construct the density matrix

$$\rho_{123} := \sum_{j=1}^d p_j\, |\, j\,\rangle\langle\, j\,| \otimes P_{\mathcal{F}_1}(j) \otimes P_{\mathcal{F}_2}(j) \ ,$$

where $\{|\, j\,\rangle\}_{j=1}^d$ is an *ONB* in $\mathbb{C}^d$, where the labels denote the factors from left to right. Notice that

$$\begin{aligned}
\rho_2 := \mathrm{Tr}_{13}(\rho_{123}) &= \sum_{j=1}^d p_j\, P_{\mathcal{F}_1}(j) = \rho[\mathcal{F}_1] \\
\rho_{12} := \mathrm{Tr}_3(\rho_{123}) &= \sum_{j=1}^d p_j\, |\, j\,\rangle\langle\, j\,| \otimes P_{\mathcal{F}_1}(j) \\
\rho_{23} := \mathrm{Tr}_1(\rho_{123}) &= \sum_{j=1}^d p_j\, P_{\mathcal{F}_1}(j) \otimes P_{\mathcal{F}_1}(j) = \rho[\mathcal{F}_1 \circ \mathcal{F}_2] \ .
\end{aligned}$$

Then, strong subadditivity (5.162) yields (8.74): indeed, because of Remark 5.5.5 and of the fact that $P_{\mathcal{F}_i}(j)$ projects onto a normalized vector in $\mathbb{C}^{|\mathcal{F}_i|}$, it turns out that

$$S\left(\rho_{123}\right) = S\left(\rho_{12}\right) = -\sum_j p_j \log p_j \ .$$

Finally, probability distributions given by regular Borel measures $\mu$ can be approximated by discrete ones; thus, the inequality (8.75) can be extended to generic $\mu$ [10] by means of the continuity of the von Neumann entropy (5.157) (notice that all the density matrix considered act on a Hilbert space of fixed finite dimension).

**Lemma 8.2.1.** *Given OPUs* $\mathcal{F}_{1,2,3} \subset \mathcal{M}$, *the entropic distance satisfies*

$$\Delta[\mathcal{F}_1|\mathcal{F}_2] \geq 0 \tag{8.76}$$

$$\Delta[\Theta_T[\mathcal{F}_1]|\Theta_T[\mathcal{F}_2]] = \Delta[\mathcal{F}_1|\mathcal{F}_2] \tag{8.77}$$

$$\Delta[\mathcal{F}_1 \circ \mathcal{F}_2|\mathcal{F}_3] \leq \Delta[\mathcal{F}_1|\mathcal{F}_3] + \Delta[\mathcal{F}_2|\mathcal{F}_3| \tag{8.78}$$

$$\Delta[\mathcal{F}_1|\mathcal{F}_2 \circ \mathcal{F}_3] \leq \Delta[\mathcal{F}_1|\mathcal{F}_2] \tag{8.79}$$

$$\Delta[\mathcal{F}_1^{(n)}|\mathcal{F}_2^{(n)}] \leq n\Delta[\mathcal{F}_1|\mathcal{F}_2] \ . \tag{8.80}$$

**Proof:** Positivity is a consequence of (8.72) and (8.75) while time-invariance comes from (8.61). Subadditivity in the first argument can be derived as follows. By using (8.73) one gets

$$\begin{aligned}\Delta[\mathcal{F}_1 \circ \mathcal{F}_2|\mathcal{F}_3] &= S\left(\rho[\mathcal{F}_1 \circ \mathcal{F}_2 \circ \mathcal{F}_3]\right) - S\left(\rho[\mathcal{F}_1 \circ \mathcal{F}_2]\right) \\ &= S\left(\rho[\mathcal{F}_1 \circ \mathcal{F}_3 \circ \mathcal{F}_2]\right) - S\left(\rho[\mathcal{F}_1 \circ \mathcal{F}_2]\right) \ .\end{aligned}$$

Setting $\rho_{123} = \rho[\mathcal{F}_1 \circ \mathcal{F}_3 \circ \mathcal{F}_2]$, it turns out that $\rho_2 = \mathrm{Tr}_{13}(\rho_{123}) = \rho[\mathcal{F}_3]$, while $\rho_{12} = \mathrm{Tr}_3(\rho_{123}) = \rho[\mathcal{F}_1 \circ \mathcal{F}_3]$ and $\rho_{23} = \mathrm{Tr}_1(\rho_{123}) = \rho[\mathcal{F}_2 \circ \mathcal{F}_3]$. Then, strong subadditivity (5.162) yields

$$S\left(\rho[\mathcal{F}_1 \circ \mathcal{F}_3 \circ \mathcal{F}_2]\right) + S\left(\rho[\mathcal{F}_3]\right) \leq S\left(\rho[\mathcal{F}_1 \circ \mathcal{F}_3]\right) + S\left(\rho[\mathcal{F}_2 \circ \mathcal{F}_3]\right) \ ,$$

whence

$$\begin{aligned}\Delta[\mathcal{F}_1 \circ \mathcal{F}_2|\mathcal{F}_3] &\leq S\left(\rho[\mathcal{F}_1 \circ \mathcal{F}_3]\right) + S\left(\rho[\mathcal{F}_2 \circ \mathcal{F}_3]\right) - 2S\left(\rho[\mathcal{F}_3]\right) \\ &= \Delta[\mathcal{F}_1|\mathcal{F}_2] + \Delta[\mathcal{F}_2|\mathcal{F}_3] \ .\end{aligned}$$

Further, using (8.78) and (8.73),

$$\begin{aligned}\Delta[\mathcal{F}_1|\mathcal{F}_2 \circ \mathcal{F}_3] &= S\left(\rho[\mathcal{F}_1 \circ \mathcal{F}_2 \circ \mathcal{F}_3]\right) - S\left(\rho[\mathcal{F}_2 \circ \mathcal{F}_3]\right) \\ &= \Delta[\mathcal{F}_1 \circ \mathcal{F}_3|\mathcal{F}_2] - \Delta[\mathcal{F}_3|\mathcal{F}_2] \\ &\leq \Delta[\mathcal{F}_1|\mathcal{F}_2] + \Delta[\mathcal{F}_3|\mathcal{F}_2] - \Delta[\mathcal{F}_3|\mathcal{F}_2] = \Delta[\mathcal{F}_1|\mathcal{F}_2] \ .\end{aligned}$$

Finally, if in (8.78) one puts $\mathcal{F}_1^{(n)}$ (see (8.58)) in the place of $\mathcal{F}_1 \circ \mathcal{F}_2$ and $\mathcal{F}_2^{(n)}$ in the place of $\mathcal{F}_3$, then using of (8.79), (8.73) and (8.77) one gets

$$\begin{aligned}\Delta[\mathcal{F}_1^{(n)}|\mathcal{F}_2^{(n)}] &\leq \sum_{k=0}^{n-1} \Delta[\mathcal{F}^k|\mathcal{F}_2^{(n)}] \leq \sum_{k=0}^{n-1} \Delta[\mathcal{F}^k|\mathcal{F}_2 k] \\ &= n\Delta[\mathcal{F}_1|\mathcal{F}_2] \ .\end{aligned}$$

□

**Definition 8.2.4.** *[10] A $*$-subalgebra with identity $\mathcal{M}_0 \subset \mathcal{M}$ is entropy-dense in $\mathcal{M} = \mathbb{L}^\infty_\mu(\mathcal{X})$ if for any finite, measurable partition $\mathcal{P}$ of $\mathcal{X}$ and any $\varepsilon > 0$ there exists an OPU $\mathcal{F} \subset \mathcal{M}_0$ such that $\Delta[\mathcal{Z}_P|\mathcal{F}] \leq \varepsilon$, where $\mathcal{Z}_P \in \mathcal{M}$ denotes the OPU corresponding to $\mathcal{P}$.*

**Theorem 8.2.1.** *Let $(\mathbb{L}^\infty_\mu(\mathcal{X}), \Theta_T, \omega_\mu)$ be the algebraic triple corresponding to a classical dynamical system $(\mathcal{X}, T, \mu)$. Let $\mathcal{M}_0 \subset \mathcal{M} = \mathbb{L}^\infty_\mu(\mathcal{X})$ be a $\Theta$-invariant entropy-dense $*$-subalgebra of $\mathcal{M}$ with identity, then*

$$\mathrm{h}^{\mathrm{AFL}}_{\omega_\mu}(\Theta_T, \mathcal{M}_0) = \mathrm{h}^{\mathrm{KS}}_\mu(T) \ . \tag{8.81}$$

**Proof:** Since $\mathcal{M}_0 \subset \mathcal{M}$, Proposition 8.2.2 gives $\mathrm{h}^{\mathrm{KS}}_\mu(T) \geq \mathrm{h}^{\mathrm{AFL}}_{\omega_\mu}(\Theta_T, \mathcal{M}_0)$. Let $\mathcal{P}$ be any finite, measurable partition of $\mathcal{X}$, $\mathcal{P}^{(n)}$ its dynamical refinement up to time $t = n-1$ and $\mathcal{Z}_P$, $\mathcal{Z}_P^{(n)}$ the corresponding *OPU*s in $\mathcal{M}$. Fix $\varepsilon > 0$ and choose $\mathcal{F}$ to be an *OPU* in the entropy-dense $\mathcal{M}_0$ such that $\Delta[\mathcal{Z}_P|\mathcal{F}] \leq \varepsilon$; then, using (8.75), (8.73), (8.72) and (8.80) one gets

$$\begin{aligned} S\left(\rho[\mathcal{Z}_P^{(n)}]\right) &\leq S\left(\rho[\mathcal{Z}_P^{(n)} \circ \mathcal{F}^{(n)}]\right) = S\left(\rho[\mathcal{F}^{(n)} \circ \mathcal{Z}_P^{(n)}]\right) \\ &= S\left(\rho[\mathcal{F}^{(n)}]\right) + \Delta[\mathcal{Z}_P^{(n)}|\mathcal{F}^{(n)}] \leq S\left(\rho[\mathcal{F}^{(n)}]\right) + n\varepsilon \ . \end{aligned}$$

By dividing by $n$ and taking the lim sup, (8.68) obtains

$$\mathrm{h}^{\mathrm{AFL}}_{\omega_\mu}(\Theta_T, \mathcal{Z}_P) = \mathrm{h}^{\mathrm{KS}}_\mu(T, \mathcal{P}) \leq \mathrm{h}^{\mathrm{AFL}}_{\omega_\mu}(\Theta_T, \mathcal{F}) + \varepsilon \leq \mathrm{h}^{\mathrm{AFL}}_{\omega_\mu}(\Theta_T, \mathcal{M}_0) + \varepsilon$$

for all $\varepsilon > 0$. Therefore, $\mathrm{h}^{\mathrm{KS}}_\mu(T, \mathcal{P}) \leq \mathrm{h}^{\mathrm{AFL}}_{\omega_\mu}(\Theta_T, \mathcal{M}_0)$ for all finite, measurable partitions of $\mathcal{X}$ whence $\mathrm{h}^{\mathrm{KS}}_\mu(T) \leq \mathrm{h}^{\mathrm{AFL}}_{\omega_\mu}(\Theta_T, \mathcal{M}_0)$. □

**Example 8.2.1.** Consider the hyperbolic automorphisms of the torus $\mathbb{T}^2$ studied in Example 2.1.3. To the measure-theoretic triple $(\mathbb{T}^2, T_A, \mathrm{d}\boldsymbol{r})$ one associates the algebraic dynamical triple $(\mathcal{M}, \Theta_A, \omega)$ where $\mathcal{M} := \mathbb{L}^\infty_{\mathrm{d}\boldsymbol{r}}(\mathbb{T}^2)$, $\Theta_A := \Theta_{T_A}$ and $\omega$ is the state obtained by integration with respect to $\mathrm{d}\boldsymbol{r}$. We now show that the $*$-subalgebra $\mathcal{M}_0 \subset \mathcal{M}$ linearly spanned by the exponential functions $e_{\boldsymbol{n}}(\boldsymbol{r}) = \exp(2\pi i \boldsymbol{n} \cdot \boldsymbol{r})$ is entropy-dense in $\mathcal{M}$.

Given a fixed $N \in \mathbb{N}$, the following collection of exponential functions

$$\mathcal{F}_N = \left\{\frac{e_{\boldsymbol{n}}}{\sqrt{M}}\right\}_{\boldsymbol{n} \in I_N} \ , \quad I_N := \left\{\boldsymbol{n} = (n_1, n_2) \ : \ -N \leq n_i \leq N\right\} ,$$

where $M := (2N+1)^2$, is an *OPU*; indeed,

$$\sum_{\boldsymbol{n} \in I_N} \left|\frac{e_{\boldsymbol{n}}}{\sqrt{M}}\right|^2 = \sum_{\boldsymbol{n} \in I_N} \frac{1}{M} = \mathbb{1} \ ,$$

where $1\!\!1$ is the identity function on $\mathbb{T}^2$. Notice that the functions $e_{\boldsymbol{n}}$ are orthogonal, namely $\omega(e^*_{\boldsymbol{n}}\, e_{\boldsymbol{m}}) = \delta_{\boldsymbol{n},\boldsymbol{m}}$; thus, (8.60) reads

$$\rho[\mathcal{F}_N] = \frac{1}{M} \sum_{\boldsymbol{n}\in I_N} |\, z_{\boldsymbol{n}}\,\rangle\langle\, z_{\boldsymbol{n}}\,| \ ,$$

where $\{|\, z_{\boldsymbol{n}}\,\rangle\}_{\boldsymbol{n}\in I_N}$ is an *ONB* in $\mathbb{C}^M$. Then, $S\left(\rho[\mathcal{F}_N]\right) = \log M$ which is also the von Neumann entropy of the state in (8.65),

$$\mathbb{F}^{\omega}_{\mathcal{F}_N}[|\, \Omega_\omega\,\rangle\langle\, \Omega_\omega\,|] = \frac{1}{M} \sum_{\boldsymbol{n}\in I_N} |\, e_{\boldsymbol{n}}\,\rangle\langle\, e_{\boldsymbol{n}}\,| =: \frac{1}{M} P_N \ ,$$

where $P_N$ is the orthogonal projection onto the $M$-dimensional subspace spanned by the $M$ orthogonal vectors $|\, e_{\boldsymbol{n}}\,\rangle$. Also, we have chosen the *GNS* representation where $|\, \Omega_\omega\,\rangle$ is the identity function on $\mathbb{T}^2$ and the action of $\pi_\omega(f)$ on vectors of $\mathbb{L}^2_{\mathrm{d}\boldsymbol{r}}(\mathbb{T}^2)$ is the multiplication by $f \in \mathcal{M}$: $\langle\, \boldsymbol{r}\,|\, \pi_\omega(f)\,|\psi\,\rangle = f(\boldsymbol{r})\psi(\boldsymbol{r})$ (see Example 5.3.2.2).

Consider now a finite, measurable partition $\mathcal{P} = \{P_j\}_{j=1}^m$ of $\mathbb{T}^2$ and the corresponding *OPU* $\mathcal{Z}_P$; we want to estimate

$$\Delta[\mathcal{Z}_P|\mathcal{F}_N] = S\left(\rho[\mathcal{Z}_P \circ \mathcal{F}_N]\right) - S\left(\rho[\mathcal{F}_N]\right) = S\left(\rho[\mathcal{Z}_P \circ \mathcal{F}_N]\right) - \log M \ .$$

The von Neumann entropy of $\rho[\mathcal{Z}_P \circ \mathcal{F}_N]$ is the same as the von Neumann entropy of (see (8.67))

$$\begin{aligned}
\sigma_N := \mathbb{F}^{\omega}_{\mathcal{Z}_P\circ\mathcal{F}_N}[|\, \Omega_\omega\,\rangle\langle\, \Omega_\omega\,|] &= \frac{1}{M}\sum_{j=1}^m \sum_{\boldsymbol{n}\in I_N} \pi_\omega(\chi_{P_j} e_{\boldsymbol{n}})|\, \Omega_\omega\,\rangle\langle\, \Omega_\omega\,|\pi_\omega(\chi_{P_j} e^*_{\boldsymbol{n}}) \\
&= \frac{1}{M}\sum_{j=1}^m Q_j\, P_N\, Q_j \ , \quad \text{where} \quad \langle\, \boldsymbol{r}\,|\, Q_j\,|\psi\,\rangle = \chi_{P_j}(\boldsymbol{r})\psi(\boldsymbol{r}) \\
&= \frac{1}{M}\sum_{j=1}^m \mathrm{Tr}(Q_j\, P_N\, Q_j)\, \sigma_j \ , \quad \text{where} \quad \sigma_j := \frac{Q_j\, P_N\, Q_j}{\mathrm{Tr}(Q_j\, P_N\, Q_j)} \ .
\end{aligned}$$

To compute $\mathrm{Tr}(Q_j\, P_N\, Q_j)$ we use Example 5.2.3.8; since $Q_j\, P_N\, Q_j = X^\dagger\, X$, $X := P_N Q_j$, its spectrum is the same as that of $X\, X^\dagger = P_N\, Q_j\, P_N$ and thus their traces are the same. Since

$$\begin{aligned}
\mathrm{Tr}(P_N\, Q_j\, P_N) &= \sum_{\boldsymbol{n}\in I_N} \langle\, e_{\boldsymbol{n}}\,|\, Q_j\,|e_{\boldsymbol{n}}\,\rangle \\
&= \sum_{\boldsymbol{n}\in I_N} \int_{\mathbb{T}^2} \mathrm{d}\boldsymbol{r}\ |e_{\boldsymbol{n}}(\boldsymbol{r})|^2\, \chi_{P_j}(x) = M\mu(P_j) \ ,
\end{aligned}$$

it follows that $\sigma_N = \sum_{j=1}^m \mu(P_j)\, \sigma_j$. Furthermore, as the $\sigma_j$ are density matrices with orthogonal ranges, Remark 5.5.5 yields

$$\begin{aligned} S(\sigma_N) &= -\sum_{j=1}^m \eta(\mu(P_j)) + \sum_{j=1}^m \mu(P_j)\, S(\sigma_j) \\ &= -\sum_{j=1}^m \eta(\mu(P_j)) + \sum_{j=1}^m \mu(P_j) \left( \frac{\mathrm{Tr}(\eta(Q_j\, P_N\, Q_j))}{M\mu(P_j)} + \log(M\mu(P_j)) \right) \\ &= \frac{1}{M}\sum_{j=1}^m \mathrm{Tr}(\eta(Q_j\, P_N\, Q_j)) + \log M \qquad \text{whence} \end{aligned}$$

$$\Delta[\mathcal{P}|\mathcal{F}_N] = \frac{1}{M}\sum_{j=1}^m \mathrm{Tr}(\eta(Q_j\, P_N\, Q_j))\ , \qquad \eta(x) = -x\log x\ .$$

We now conclude tgeh proof by showing that, for $N$ large enough, the right hand side of the last inequality can be made negligibly small. As already seen, $Q_j\, P_N\, Q_j$ and $P_N\, Q_j\, P_N$ have the same spectrum; therefore,

$$\mathrm{Tr}(\eta(Q_j\, P_N\, Q_j)) = \mathrm{Tr}(\eta(P_N\, Q_j\, P_N))\ .$$

Further, $\eta(x) \geq x(1-x)$ for all $0 \leq x \leq 1$ and $\eta(x) - x(1-x)$ is bounded; thus, for all $\varepsilon > 0$, there exists $C(\varepsilon) > 0$ such that [10] $\eta(x) \leq \varepsilon + C(\varepsilon)\, x(1-x)$. Applying this inequality to the eigenvalues $\pi_{jk}$ of $P_N\, Q_j\, P_N$ and observing that $0 \leq \pi_{jk} \leq 1$ for $P_N\, Q_j\, P_N \leq 1\!\!1$, one gets

$$\begin{aligned} \frac{1}{M}\mathrm{Tr}(\eta(P_N\, Q_j\, P_N)) &= \frac{1}{M}\sum_k \eta(\pi_{jk}) \leq \frac{1}{M}\sum_k \Big( \varepsilon + C(\varepsilon)\, \pi_{jk}(1-\pi_{jk}) \Big) \\ &\leq \varepsilon + C(\varepsilon)\frac{1}{M}\Big( \mathrm{Tr}(P_N\, Q_j\, P_N) - \mathrm{Tr}((P_N\, Q_j\, P_N)^2) \Big) \\ &= \varepsilon + C(\varepsilon)\left( \mu(P_j) - \frac{1}{M}\sum_{n\in I_N} \langle\, e_n \,|\, Q_j\, P_N\, Q_j \,|e_n\,\rangle \right)\ . \end{aligned}$$

In the second inequality it has been used that the range of $P_N$ has dimension $M$. Since the exponential functions $|\, e_n\,\rangle$ form an *ONB* in $\mathbb{L}^2_{\mathrm{d}r}(\mathbb{T}^2)$, by increasing $N$ (and thus $M$) one makes $P_N \to 1\!\!1$ so that

$$\frac{1}{M}\sum_{n\in I_N} \langle\, e_n \,|\, Q_j\, P_N\, Q_j \,|e_n\,\rangle = \mu(P_j) - \frac{1}{M}\sum_{n\in I_N} \langle\, e_n \,|\, Q_j(1\!\!1 - P_N)Q_j \,|e_n\,\rangle$$

tends to $\mu(P_j)$ when $N \to \infty$.

### *AFL* Entropy: Finite Quantum Systems

Like the *CNT* entropy, also the *AFL* entropy vanishes for finite-level quantum systems. In order to show this we start by deriving a useful bound [10] on the von Neumann entropy $S(\rho[\mathcal{Z}])$ of a given *OPU* $\mathcal{Z} = \{Z_i\}_{i=1}^{|\mathcal{Z}|} \subset \mathbb{B}(\mathbb{H})$, when $\mathbb{B}(\mathbb{H})$ is equipped with a state represented by a density matrix $\rho$.

Let $0 \leq r_j \leq 1$ and $|\, r_j\,\rangle$ be the eigenvalues and eigenvectors of $\rho$ and $\{|\, z_j\,\rangle\}_{j=1}^{|\mathcal{Z}|}$ an *ONB* in $\mathbb{C}^{|\mathcal{Z}|}$; because of Definition 8.2.1, the vectors

$$|\, \Psi_j^{\mathcal{Z}}\,\rangle := \sum_{k=1}^{|\mathcal{Z}|} Z_k |\, r_j\,\rangle \otimes |\, z_k\,\rangle$$

are orthogonal, indeed (8.55) yields

$$\langle\, \Psi_j^{\mathcal{Z}}\,|\, \Psi_\ell^{\mathcal{Z}}\,\rangle = \sum_{k=1}^{|\mathcal{Z}|} \langle\, r_j\,|\, Z_k^\dagger Z_k\, |r_\ell\,\rangle = \langle\, r_j\,|\, r_\ell\,\rangle = \delta_{j\ell}\ .$$

Set $\rho_{\mathcal{Z}} := \sum_j r_j |\, \Psi_j^{\mathcal{Z}}\,\rangle\langle\, \Psi_j^{\mathcal{Z}}\,|$; then, $S\,(\rho_{\mathcal{Z}}) = S\,(\rho)$ and

$$\mathbb{B}_1(\mathbb{H}) \ni \rho_I := \mathrm{Tr}_{II}(\rho_{\mathcal{Z}}) = \sum_j \sum_{k=1}^{|\mathcal{Z}|} r_j\, Z_k\,|\, r_j\,\rangle\langle\, r_j\,|\, Z_k^\dagger =: \mathbb{F}_{\mathcal{Z}}[\rho]$$

$$\mathbb{B}_1(\mathbb{C}^{|\mathcal{Z}|}) \ni \rho_{II} := \mathrm{Tr}_I(\rho_{\mathcal{Z}}) = \sum_{j,k=1}^{|\mathcal{Z}|} \mathrm{Tr}(\rho\, Z_k^\dagger Z_j)\ |\, j\,\rangle\langle\, k\,| = \rho[\mathcal{Z}]\ .$$

Applying subadditivity 5.161 to these marginal density matrices one obtains the following upper bound to $S\,(\rho[\mathcal{Z}])$.

**Proposition 8.2.3.** *Let* $\rho \in \mathbb{B}_1(\mathbb{H})$ *be a state on* $\mathcal{A} = \mathbb{B}(\mathbb{H})$ *and* $\mathcal{Z} = \{Z_i\}_{i=1}^{|\mathcal{Z}|} \subset \mathbb{B}(\mathbb{H})$ *any fixed OPU ; then*

$$S\,(\rho[\mathcal{Z}]) \leq S\,(\rho) + S\,(\mathbb{F}_{\mathcal{Z}}[\rho])\ . \tag{8.82}$$

If $\mathbb{B}(\mathbb{H}) = M_d(\mathbb{C})$, then the von Neumann entropy of both $\rho$ and $\mathbb{F}_{\mathcal{Z}}[\rho]$ are upperbounded by $\log d$ independently of $\mathcal{Z}$. Since in the case of a finite-level system the dynamics is implemented by a unitary operator which belongs to $M_d(\mathbb{C})$, all *OPUs* from $M_d(\mathbb{C})$ are such that also the refined *OPUs* $\mathcal{Z}^{(n)}$ up to discrete time $t = n - 1$ also belong to $M_d(\mathbb{C})$. Then, the following results holds.

**Proposition 8.2.4.** *Let* $(\mathcal{A}, \Theta_\sigma, \omega)$ *be a finite-level quantum system, where* $\mathcal{A} = M_d(\mathbb{C})$, $\omega$ *corresponds to a density matrix* $\rho \in \mathbb{B}_1(\mathbb{C}^d)$ *and* $\Theta$ *is implemented by a unitary* $U \in M_d(\mathbb{C})$. *Then, for all OPUs* $\mathcal{Z} \subset M_d(\mathbb{C})$,

$$\mathrm{h}_\omega^{\mathrm{AFL}}\,(\Theta, \mathcal{X}) = 0\ , \qquad \mathrm{h}_\omega^{\mathrm{AFL}}\,(\Theta) = 0\ .$$

**Remark 8.2.2.** The fact that for finite-level systems the *AF* entropy vanishes is neither a surprise nor is it the end of the story. Indeed, in quantum chaotic phenomena [75, 322], namely when studying the behavior for $\hbar \to 0$ of quantum systems with a chaotic classical limit, the associated classical instability manifests itself in the presence of a logarithmic time-scale as in Remark 2.1.3.4. Roughly speaking, the explanation of this fact stems from the fact that, in the semi-classical approximation, quantizing means operating a coarse graining of the phase-space into atoms of size $2\pi\hbar$: this forbids the existence of a *bona fide* Lyapounov exponent, but makes its classical existence felt up to times that scale as $-\log(\hbar/S)$ (where $\hbar$ is normalized to a reference classical action $S$). In some models, as for instance the quantized finite Arnold cat map in Example 5.6.1.3 and the *kicked top* in [184], the classical limit can be mimicked by the dimension of the underlying Hilbert space $N \to +\infty$. The *AFL* construction, notably the entropy of a time-evolving *OPU* , has been applied to such cases and proved to increase linearly with the number of timesteps $T$ up to $T \simeq \log N$ [10, 12, 25]. Interestingly, the *AFL* entropy has also been applied to study the emergence of chaos in the continuous limit of discretized classical dynamical systems [24, 26], where the suppression of instability also finds its root in a finite coarse graining of the phase space.

### *AFL* Entropy: Quantum Spin Chains

Interestingly, the *AFL* entropy of quantum sources differs from the *CNT* entropy by a correction term which increases with the dimension of single site algebras. According to Remark 8.2.1, the *OPU*s will be taken from strictly local subalgebras of the quasi-local source algebra $\mathcal{A}$.

**Proposition 8.2.5.** *Let $(\mathcal{A}_{\mathbb{Z}}, \omega)$ be a quantum spin chain with single site matrix algebras $M_d(\mathbb{C})$. Relative to OPUs from any local subalgebra $\mathcal{A}_{[p,q]}$, $\mathcal{X} \subset \mathcal{A}_{[p,q]} \subseteq \mathcal{A}_0 := \mathcal{A}_{\mathbb{Z}}^{loc}$, the AFL entropy is given by*

$$\mathrm{h}_{\omega}^{\mathrm{AFL}}(\Theta_\sigma) = s(\omega) + \log d \ ,$$

*where the dynamics is the shift $\Theta_\sigma$ over $\mathcal{A}_{\mathbb{Z}}$, and the translation-invariant state $\omega \circ \Theta_\sigma = \omega$ has mean von Neumann entropy $s(\omega)$.*

**Proof:** Because of translation-invariance of $\omega$, it is no restriction to take $\mathcal{X} = \{X_i\}_{i=1}^{p}$, $X_i \in \mathcal{A}_{[0,\ell]}$. It follows that the dynamical refinements $\mathcal{X}^{(k)}$ are localized within $[0, \ell + k - 1]$. With $\rho = \omega \lceil \mathcal{A}_{[0,\ell+k-1]}$ and $\mathbb{F}_{\mathcal{X}^{(k)}}[\rho]$, both density matrices in $M_d(\mathbb{C})^{\otimes (\ell+k)}$, (8.82) yields

$$\mathrm{h}_{\omega}^{\mathrm{AFL}}(\Theta_\sigma, \mathcal{X}) \leq \limsup_{k\to\infty} \frac{S(\omega \lceil \mathcal{A}_{[0,\ell+k-1]})}{k} + \log d = s(\omega) + \log d \ ,$$

whence $\mathrm{h}^{\mathrm{AFL}}_{\omega}(\Theta_\sigma) \leq s(\omega) + \log d$. The upper bound is reached by an *OPU* consisting $\mathcal{X}$ made of matrix units $e^0_{p_0 q_0} := |\, p_0 \,\rangle\langle\, q_0 \,|$ from the algebra $M_n(\mathbb{C})$ at site 0, where $\{|\, i \,\rangle\}_{i=1}^d$ is an *ONB* in $\mathbb{C}^n$ [133].

Explicitly, $\mathcal{X} = \left\{ \frac{1}{d^{1/2}} e^0_{p_0 q_0} \right\}_{(q_0, p_0)}$ and

$$\mathcal{X}^{(k)} = \left\{ \frac{e_{\boldsymbol{p}^{(k)}\boldsymbol{q}^{(k)}}}{\sqrt{d^k}} \right\}_{(\boldsymbol{p}^{(k)}, \boldsymbol{q}^{(k)})} , \quad e_{\boldsymbol{p}^{(k)}\boldsymbol{q}^{(k)}} := \bigotimes_{i=0}^{k-1} e^i_{p_i q_i} .$$

According to (8.60), the matrix elements of the $(d^k \times d^k) \times (d^k \times d^k)$ density matrix $\rho[\mathcal{X}^{(k)}]$ are thus given by

$$\begin{aligned} \rho[\mathcal{X}^{(k)}]_{\boldsymbol{pq};\boldsymbol{rs}} &= \frac{1}{d^k} \omega \left( e^{(k)}_{\boldsymbol{sr}} e^{(k)}_{\boldsymbol{pq}} \right) = \frac{1}{d^k} \omega \left( \bigotimes_{i=0}^{k-1} e^i_{s_i r_i} e^i_{p_i q_i} \right) \\ &= \frac{1}{d^k} \prod_{i=0}^{k-1} \delta_{r_i p_i}\, \omega \left( \bigotimes_{i=0}^{k-1} e^i_{s_i q_i} \right) . \end{aligned}$$

The expectations on the right hand side of the last equality define the local state $\rho_{[0,k-1]} := \omega \restriction \mathcal{A}_{[0,k-1]}$ so that

$$\rho[\mathcal{X}^{(k)}] = \frac{1}{d^k} \mathbb{1}_{\mathbb{C}^{d^k}} \otimes \rho_{[0,k-1]} , \Longrightarrow S\left( \rho[\mathcal{X}^{(k)}] \right) = S(\rho_{[0,k-1]}) + k \log d$$

and $\mathrm{h}^{\mathrm{AFL}}_{\omega}(\Theta_\sigma) \geq \mathrm{h}^{\mathrm{AFL}}_{\omega}(\Theta_\sigma)\, \mathcal{X} = s(\omega) + \log d$. □

## *AFL* Entropy: Price-Powers Shifts

Price-Powers shifts (see Definition 7.1.14) provide non-commutative contexts whereby the differences between *CNT* and *AFL* entropies can be better appreciated [13]: indeed, it turns out that, while the former depends on the *bit-stream* $g$, the latter does not.

**Proposition 8.2.6.** *Let the triplet* $(\mathcal{U}_g, \Theta_\sigma, \omega)$ *represent a Price-Powers shift with bitstream* $g$*; then, relative to local OPUs ,* $\mathrm{h}^{\mathrm{AFL}}_{\omega}(\Theta_\sigma) = 1$ *independently of* $g$*.*

**Proof:** As for quantum spin chains, *OPU*s will be taken from local subalgebras, $\mathcal{X} \subset \mathcal{A}_0 = \mathcal{U}^{loc}_g$; by translation-invariance of the state $\omega$, we can always suppose $\mathcal{X} = \{X_i\}_{1=1}^d \subset \mathcal{U}_{[0,\ell]}$, so that $\mathcal{X}^{(k)} \subset \mathcal{U}_{[0,\ell+k-1]}$. The latter local subalgebra is not isomorphic to a full-matrix algebra, rather to an orthogonal sum of $m$ $\nu_j \times \nu_j$ full-matrix algebras: $\mathcal{U}_{[0,\ell+k-1]} = \bigoplus_{j=1}^m M_{\nu_j}(\mathbb{C})$.

As a linear space $\mathcal{U}_{[0,\ell+k-1]}$ is $2^{\ell+k}$ dimensional (this is the number of independent $W_i$ that generate it), while each of the contributing $M_{\nu_j}(\mathbb{C})$ is a $\nu_j^2$-dimensional linear space, whence the constraint $2^{\ell+k} = \sum_{j=1}^m \nu_j^2$. From the splitting of $\mathcal{U}_{[0,\ell]}$, the elements $X_{\boldsymbol{i}^{(k)}}$, $\boldsymbol{i}^{(k)} = i_0 i_1 \cdots i_{n-1} \in \Omega_d^{(n)}$, of $\mathcal{X}^{(k)}$ can be decomposed as $X_{\boldsymbol{i}^{(k)}} = \bigoplus_{j=1}^m X^j_{\boldsymbol{i}^{(k)}}$, and

$$\rho_{[0,\ell+k-1]} = \bigoplus_{j=1}^m \delta_j \, \tau_j \ ,$$

where $\tau_j = \frac{1}{\nu_j} \mathbb{1}_{\nu_j}$ are tracial states on $M_{\nu_j}(\mathbb{C})$, while $0 < \delta_j$, $\sum_{j=1}^m \delta_j = 1$ account for the various multiplicities. It follows that

$$\rho[\mathcal{X}^{(k)}] = \sum_{j=1}^m \delta_j \rho_j^{(k)} \ , \quad (\rho_j^{(k)})_{\boldsymbol{q}^{(k)}\boldsymbol{p}^{(k)}} := \tau_j\left((X^j_{\boldsymbol{p}^{(k)}})^\dagger X^j_{\boldsymbol{q}^{(k)}}\right) \ .$$

Then, (8.82), (5.156), (5.155) and concavity of $\log x$ yield

$$\begin{aligned} S(\rho[\mathcal{X}^{(k)}]) &\leq \sum_{j=1}^m \delta_j \, S\left(\rho_j^{(k)}\right) + \log m \leq \sum_{j=1}^m \delta_j \, \log \frac{\nu_j^2}{\delta_j} \\ &\leq \log(\sum_{j=1}^m \nu_j^2) = \ell + k \ , \end{aligned}$$

whence $\mathrm{h}^{\mathrm{AFL}}_\omega(\Theta_\sigma) \leq 1$. The bound is attained at the *OPU* consisting of orthogonal projectors at site $j = 0$,

$$\mathcal{X} = \{p_1^0, p_2^0\} \ , \quad p_i^0 := \frac{\mathbb{1} + (-)^i \, e_0}{2} \ .$$

In fact, $\mathcal{X}^{(k)} = \left\{p_{\boldsymbol{i}^{(k)}} := \prod_{j=k-1}^0 p^j_{i_j}\right\}_{\boldsymbol{i}^{(k)} \in \Omega_2^{(k)}}$ and

$$\rho[\mathcal{X}^{(k)}]_{\boldsymbol{i}^{(k)}\boldsymbol{j}^{(k)}} = \omega(p_{\boldsymbol{j}^{(k)}} p_{\boldsymbol{i}^{k)}}) = 2^{-k} \prod_{\ell=0}^{k-1} \delta_{i_\ell j_\ell} = 2^{-k} \delta_{\boldsymbol{i}^{(k)}\boldsymbol{j}^{(k)}} \ .$$

The last equality follows by using (7.119) and (7.125); $\omega$ is tracial and the $p^i_j$ orthogonal for fixed $i$, thus

$$\begin{aligned} \omega(p_{\boldsymbol{j}^{(k)}} p_{\boldsymbol{i}^{(k)}}) &= \omega\left(\prod_{r=0}^{k-1} p^r_{j_r} \prod_{s=1}^{k-1} p^s_{i_s}\right) \\ &= \delta_{i_0 j_0} \delta_{i_{k-1} j_{k-1}} \omega\left(p^0_{i_0} \cdots p^{k-2}_{j_{k-2}} p^{k-1}_{i_{k-1}} p^{k-2}_{i_{k-2}} \cdots p^1_{i_1}\right) \\ &= \delta_{i_0 j_0} \delta_{i_{k-1} j_{k-1}} \delta_{i_{k-2} j_{k-2}} \omega\left(p^0_{i_0} \cdots p^{k-2}_{j_{k-2}} p^{k-1}_{i_{k-1}} p^{k-3}_{i_{k-3}} \cdots p^1_{i_1}\right) + \\ &+ \delta_{i_0 j_0} \delta_{i_{k-1} j_{k-1}} \omega\left(\cdots p^{k-3}_{j_{k-3}} p^{k-2}_{j_{k-2}} \left[p^{k-1}_{i_{k-1}}, p^{k-2}_{i_{k-2}}\right] p^{k-3}_{i_{k-3}} \cdots\right) \ . \qquad (*) \end{aligned}$$

Using (7.119) one calculates

$$\left[p^{k-1}_{i_{k-1}}, p^{k-2}_{i_{k-2}}\right] = \frac{1}{4}(-)^{i_{k-2}+j_{k-1}}\left(1-(-)^{g(1)}\right) e_{k-1}e_{k-2} \ .$$

Notice that either this commutator vanishes because $g(1) = 1$ or the operator $p^{k-2}_{j_{k-2}}\left[p^{k-1}_{i_{k-1}}, p^{k-2}_{i_{k-2}}\right]$ (it belongs to the subalgebra $\mathcal{U}_{[k-2,k-1]}$) cannot be turned into an identity by means of (7.117) as all the other $p$ come from different sites. Thus, $(*)$ vanishes. Iteration of this argument yields

$$\omega(p^{(k)}_{\boldsymbol{j}}{}^{*}p^{(k)}_{\boldsymbol{i}}) = \prod_{\ell=0}^{k-1}\delta_{i_\ell j_\ell}\,\omega\left(\prod_{r=0}^{k-1} p_{r j_r}\right) = 2^{-k}\delta_{\boldsymbol{i}\boldsymbol{j}} \ .$$

The density matrix $\rho[\mathcal{X}^{(k)}]$ is thus diagonal with eigenvalues $2^{-k}$ whence $S(\rho[\mathcal{X}^{(k)}]) = k\log 2$ and $\mathrm{h}^{\mathrm{AFL}}_{\omega}(\Theta_\sigma) \geq 1$. □

**Remark 8.2.3.** While the *AFL* entropy is always $\log 2$ for all bitstreams, instead the *CNT* entropy varies from 0 to $\log 2$ (see (8.35)– (8.37)). Since the bitstream fixes the degree of departure from commutativity, this fact indicates the *CNT* entropy is sensitive to the dynamics, but also to the algebraic structure of quantum dynamical systems, in particular to whether they are asymptotically Abelian. On the contrary, the *AFL* entropy accounts for the effects of the dynamics not directly, rather through a particular family, in general not translation invariant, of local density matrices over a quantum spin chain. As such, it is more sensitive to the properties of the state $\omega$ and in some cases strongly depends on the *OPUs* that are used to construct the local density matrices. The effects of the *OPUs* are at the root of the fact that the *AFL* entropy of a spin chain is the entropy density augmented by the logarithm of the dimension of the spin algebras, $\mathrm{h}^{\mathrm{AFL}}_{\omega}(\Theta_\sigma) = s(\omega) + \log d$, whereas the *CNT* entropy equals the entropy density $\mathrm{h}^{\mathrm{CNT}}_{\omega}(\Theta_\sigma) = s(\omega)$. If freely chosen and not carefully selected from a suitable $\Theta$-invariant $\mathcal{A}_0$ in such a way that the perturbations are kept to a minimum, it may happen that even dynamical systems without dynamics may have non-zero *AFL* entropy. An abstract though revealing example is that of a so-called *Cuntz algebra* [10], namely the $C^*$ algebra $\mathcal{A}$ generated by the identity $\mathbb{1}$ and by linear combinations of products $W_{\boldsymbol{i}} := S_{i_1}S_{i_2}\cdots S_{i_n}$ of two isometries $S_i$, $i = 0, 1$, such that

$$S_0^\dagger S_0 = S_1^\dagger S_1 = \mathbb{1}\ , \qquad S_0 S_0^\dagger + S_1 S_1^\dagger = \mathbb{1}\ .$$

It turns out that

$$S_0^\dagger\, S_1 = S_0^\dagger(\, S_0 S_0^\dagger \ + \ S_1 S_1^\dagger) S_1 = S_0^\dagger\, S_1 \ + \ S_0^\dagger\, S_1 \Longrightarrow S_0^\dagger S_1 = 0\ .$$

Let the Cuntz algebra $\mathcal{A}$ be equipped with the tracial state $\omega(W_{\boldsymbol{i}}) = 0$ unless $W_{\boldsymbol{i}} = \mathbb{1}$ in which case $\omega(\mathbb{1}) = 1$ and take the dynamics as trivial $\Theta = \mathrm{id}_{\mathcal{A}}$

namely, $\Theta[W_i] = W_i$ for all $W_i$. While $\mathrm{h}^{\mathrm{CNT}}_{\omega}(\mathrm{id}_{\mathcal{A}}) = 0$, the *AFL* entropy of the *OPU* $\mathcal{X} = \{S_i/\sqrt{2}\}_{i=0}^{1}$ diverges. Indeed, the elements of the partition $\mathcal{X}^{(k)}$ are of the form $X_{\boldsymbol{i}^{(k)}} := 2^{-k/2}\, S_{i_k} S_{i_{k-1}} \cdots S_{i_1}$; thus,

$$X^{\dagger}_{\boldsymbol{i}^{(k)}}\, X_{\boldsymbol{j}^{(k)}} = 2^{-k} S^{\dagger}_{i_1} \cdots S^{\dagger}_{i_{k-1}} S^{\dagger}_{i_k}\, S_{j_k} S_{j_{k-1}} \cdots S_{j_1} = 2^{-k}\, \delta_{\boldsymbol{i}^{(k)},\boldsymbol{j}^{(k)}} \quad \text{whence}$$
$$S\left(\rho[\mathcal{X}^{(k)}]\right) = k \log 2 \Longrightarrow \mathrm{h}^{\mathrm{AFL}}_{\omega}(\mathrm{id}_{\mathcal{A}}, \mathcal{X}) = \log 2 \ .$$

Therefore, using Remark 8.2.1.1, one deduces that, if the *OPU*s are freely taken from $\mathcal{A}$, then $\mathrm{h}^{\mathrm{AFL}}_{\omega}(\mathrm{id}_{\mathcal{A}}) = +\infty$.

### *AFL* Entropy: Arnold Cat Maps

We now consider the infinite dimensional quantized hyperbolic automorphisms of the torus in Example 7.1.6, namely the triplets $(\mathcal{M}_\theta, \Theta_{\mathbb{A}}, \omega)$, where $\mathcal{M}_\theta$ is the von Neumann algebra generated by the Weyl operators (7.30), equipped with the automorphism (7.34) and the $\Theta_{\mathbb{A}}$-invariant tracial state (7.34).

We shall show that, independently of the deformation parameter, when the *OPU*s are taken from the $\Theta_{\mathbb{A}}$-invariant $*$ subalgebra $\mathcal{A}_0$ generated the by Weyl operators $W_\theta(f)$ where the $f$ have compact support, the *AFL* entropy of $(\mathcal{M}_\theta, \Theta_{\mathbb{A}}, \omega)$ coincides with the *KS* entropy [15] (see Proposition 3.1.1).

**Proposition 8.2.7.** $\mathrm{h}^{\mathrm{AFL}}_{\omega}(\Theta_{\mathbb{A}}) = \log \alpha$ *for all* $(\mathcal{M}_\theta, \Theta_{\mathbb{A}}, \omega)$*, where* $\alpha > 1$ *is the largest eigenvalue of* $\mathbb{A}$ *(see Example 2.1.3) and the OPUs are taken from the* $\Theta_{\mathbb{A}}$*-invariant subalgebra* $\mathcal{A}_0 \subset \mathcal{M}_\theta$ *which is generated by the Weyl operators* $W_\theta(f)$ *with compact Supp*$(f)$*.*

**Remark 8.2.4.** Since the *AFL* entropy does not depend on $\theta$ and because, for $\theta = 0$, the quantum dynamical system $(\mathcal{M}_\theta, \Theta_{\mathbb{A}}, \omega)$ reduces to the classical hyperbolic automorphisms of the torus (see Example 8.2.1), the proof which follows is another way to compute $\mathrm{h}^{\mathrm{KS}}_{\mu}(T_{\mathbb{A}})$ and thus the Lyapounov exponents.

The *OPU*s that will be repeatedly used in the following have the form

$$\mathcal{Z} = \left\{ Z_i := \frac{\mathrm{e}^{i\beta_i}}{\sqrt{p}}\, W_\theta(\boldsymbol{n}_i) \right\}_{i=1}^{p} \ , \qquad \boldsymbol{n}_i \in \mathbb{Z}^2 \ . \tag{8.83}$$

Since $\omega(Z_j^{\dagger}\, Z_i) = \dfrac{\delta_{\boldsymbol{n}_i \boldsymbol{n}_j}}{p} \mathrm{e}^{i(\beta_i - \beta_j)}$, from (8.60) one computes

$$\rho[\mathcal{Z}] = \sum_{i,j=1}^{p} |\, z_i \,\rangle\langle\, z_j \,|\; \omega(Z_j^{\dagger} Z_i) = \sum_{i,j\,:\,\boldsymbol{n}_i = \boldsymbol{n}_j} \frac{\mathrm{e}^{i(\beta_i - \beta_j)}}{p} \,|\, z_i \,\rangle\langle\, z_j \,| \ .$$

The index set $\{1, 2, \ldots p\}$ can thus be divided into disjoint equivalence classes

$$\{1,2,\ldots,p\} = \bigcup_i [i] \ , \quad [i] := \{1 \leq a \leq p \,:\, \boldsymbol{n}_a = \boldsymbol{n}_i\} \ .$$

If $\#[i]$ denotes the cardinality of $[i]$, one can then write

$$\rho[\mathcal{Z}] := \sum_{[i]} \frac{1}{p} \sum_{a,b \in [i]} \mathrm{e}^{i(\beta_i - \beta_j)} \, | \, z_i \,\rangle\langle\, z_j \, | = \sum_{[i]} \frac{\#[i]}{k} | \, [i] \,\rangle\langle\, [i] \, | \tag{8.84}$$

where the vectors $| \, [i] \,\rangle := \frac{1}{\#[i]} \sum_{a \in [i]} \mathrm{e}^{i\beta_i} | \, z_a \,\rangle$, are orthogonal. Thus,

$$S\left(\rho[\mathcal{Z}]\right) = -\sum_{[i]} \frac{\#[i]}{p} \log \frac{\#[i]}{p} = \sum_{[i]} \eta\left(\frac{\#[i]}{p}\right) \ . \tag{8.85}$$

where (2.84) has been used. Consider now two *OPUs* of the form (8.83),

$$\mathcal{Z}_1 = \left\{ \mathrm{e}^{i\beta_i^{(1)}} \frac{W_\theta(\boldsymbol{n}_i^{(1)})}{\sqrt{p_1}} \right\}_{i=1}^{p_1} \ , \quad \mathcal{Z}_2 = \left\{ \mathrm{e}^{i\beta_i^{(2)}} \frac{W_\theta(\boldsymbol{n}_i^{(2)})}{\sqrt{p_2}} \right\}_{i=1}^{p_2} \ .$$

According to (8.56) and (7.29), their refinement is an *OPU* of the same form

$$\mathcal{Z}_1 \circ \mathcal{Z}_2 = \left\{ \frac{\mathrm{e}^{i\beta(i,j)}}{\sqrt{p_1 p_2}} W_\theta(\boldsymbol{n}_i^{(1)} + \boldsymbol{n}_j^{(2)}) \right\}_{i,j=1}^{p_1,p_2} \ . \tag{8.86}$$

One now introduces the equivalent classes

$$[i,j] := \left\{ (a,b) \,:\, \boldsymbol{n}_a^{(1)} + \boldsymbol{n}_b^{(2)} = \boldsymbol{n}_i^{(1)} + \boldsymbol{n}_j^{(2)} \ , \ 1 \leq a \leq p_1 \ .1 \leq b \leq p_2 \right\} \ ,$$

Notice that if $x \in [a]_1$ is such that there exist $y \in [b]_2$ such that $(x,y) \in [i,j]$, then this is true for all pairs $(u,v)$ with $u \in [a]_1$ and $v \in [b]_2$. Therefore, one can write

$$\#[i,j] = \sum_{[a]_1 \,:\, \exists b \ s.t. \ [a,b]=[i,j]} \#[a]_1 \, \#[b]_2 \ .$$

**Lemma 8.2.2.** *The following two properties hold:*

$$S\left(\rho[\mathcal{Z}_1 \circ \mathcal{Z}_2]\right) = S\left(\rho[\mathcal{Z}_2 \circ \mathcal{Z}_1]\right) \tag{8.87}$$

$$S\left(\rho[\mathcal{Z}_1 \circ \mathcal{Z}_2]\right) \geq S\left(\rho[\mathcal{Z}_2]\right) \ . \tag{8.88}$$

**Proof:** The first equality follows from (8.85) applied to (8.86) which gives

$$S\left(\rho[\mathcal{Z}_1 \circ \mathcal{Z}_2]\right) = \sum_{[i,j]} \eta\left(\frac{\#[i,j]}{p_1 p_2}\right) \ .$$

The second one can be derived as follows: set

$$N(i,j) := \sum_{[a]_1\,:\,\exists b\; s.t.\; [a,b]=[i,j]} \frac{\#[a]_1}{p_1}\ ,$$

and use that $\eta(xy) = x\,\eta(y) + y\,\eta(x) \geq x\,\eta(y)$ to estimate

$$\begin{aligned}
S\left(\rho[\mathcal{Z}_1 \circ \mathcal{Z}_2]\right) &= \sum_{[i,j]} \eta\Bigg(N(i,j) \Bigg( \sum_{[a]_1\,:\,\exists b\; s.t.\; [a,b]=[i,j]} \frac{1}{N(i,j)} \frac{\#[a]_1}{p_1} \frac{\#[b]_2}{p_2}\Bigg)\Bigg) \\
&\geq \sum_{[i,j]} N(i,j)\, \eta \Bigg( \sum_{[a]_1\,:\,\exists b\; \text{s.t.}\; [a,b]=[i,j]} \frac{\#[a]_1}{N(i,j)\, p_1} \frac{\#[b]_2}{p_2}\ .\Bigg)
\end{aligned}$$

Since

$$\sum_{[a]_1\,:\,\exists b\; s.t.\; [a,b]=[i,j]} \frac{1}{N(i,j)} \frac{\#[a]_1}{p_1} = 1\ ,$$

the concavity of $\eta(x)$ yields

$$\begin{aligned}
S\left(\rho[\mathcal{Z}_1 \circ \mathcal{Z}_2]\right) &\geq \sum_{[i,j]} \sum_{[a]_1\,:\,\exists b\; s.t.\; [a,b]=[i,j]} \frac{\#[a]_1}{p_1}\, \eta\Big(\frac{\#[b]_2}{p_2}\Big) = \sum_{[b]_2} \eta\Big(\frac{\#[b]_2}{p_2}\Big) \\
&= S\left(\rho[\mathcal{Z}_2]\right)\ .
\end{aligned}$$

In fact, by summing over all $[i,j]$, one sums over all $[b]_2$ and $[a]_1$, the latter being as many as $p_1/\#[a]$. □

We now concentrate on a special *OPU* , $\mathcal{Z} = \left\{\frac{1}{\sqrt{q+1}}\, W_\theta(\boldsymbol{n}_j)\right\}_{j=0}^{q}$, where, for a fixed $q \in \mathbb{N}$, we choose $\boldsymbol{n}_j := ([\alpha^j] - 1)\boldsymbol{n}$, $0 \leq j \leq q$, with $\mathbb{Z}^2 \ni \boldsymbol{n} \neq 0$. Also, $[\alpha^j] < \alpha^j$ denotes the integer part of the $j$-th power of the eigenvalue $\alpha > 1$. The refined *OPU*

$$\mathcal{Z}^{q,\ell} := \Theta_{\mathbb{A}}^{q(\ell-1)}[\mathcal{Z}] \circ \Theta_{\mathbb{A}}^{q(\ell-2)}[\mathcal{Z}] \circ \cdots \Theta_{\mathbb{A}}^{q}[\mathcal{Z}] \circ \mathcal{Z}$$

has $|\mathcal{Z}^{q,\ell)}| = [\alpha^q]^\ell$ elements of the form

$$\mathrm{e}^{i\beta(\boldsymbol{j}^{(n)})} W_\theta(\boldsymbol{n}(\boldsymbol{j}^{(\ell)}))\ , \quad \boldsymbol{n}(\boldsymbol{j}^{(\ell)}) := \sum_{k=0}^{\ell-1} ([\alpha^{j_k}] - 1)(\mathbb{A}^T)^k \boldsymbol{n}\ ,$$

where $\boldsymbol{j}^{(\ell)} = j_0 j_1 \cdots j_{\ell-1}$ with $j_\ell \in I(q) := \{0, 1, \ldots, [\alpha^{q-1}] - 1$. In order to evaluate $S\left(\rho[\mathcal{Z}^{q,\ell}]\right)$, we have to investigate the equivalence classes determined by relations of the form $\boldsymbol{n}(\boldsymbol{r}^{(\ell)}) = \boldsymbol{n}(\boldsymbol{s}^{(\ell)})$, $\boldsymbol{r}^{(\ell)}, \boldsymbol{s}^{(\ell)} \in I^\ell$. By expanding $\boldsymbol{n}$ along the (linearly independent) eigenvectors $|\pm\rangle$ of $\mathbb{A}^T$ relative to the eigenvalues $\alpha^{\pm 1}$, $|\,\boldsymbol{n}\,\rangle = \gamma|+\rangle + \delta|-\rangle$, one gets

$$\sum_{k=0}^{\ell-1}\lambda^{qk}([\lambda^{r_k}]-[\lambda^{s_k}])=\lambda^{q(\ell-1)}\Big([\lambda^{r_k}]-[\lambda^{s_j}]+$$

$$+\sum_{k=0}^{\ell-2}\lambda^{-q\,(\ell-k-1)}([\lambda^{r_k}]-[\lambda^{s_k}])\Big)=0\,.$$

By choosing $q$ large enough, such an equality can only be true if $r_{\ell-1}=k_{\ell-1}$; iterating this argument yields

$$\boldsymbol{n}(\boldsymbol{r}^{(\ell)})=\boldsymbol{n}(\boldsymbol{s}^{(\ell)})\Longleftrightarrow\boldsymbol{r}^{(\ell)}=\boldsymbol{s}^{(\ell)}\,.$$

This implies that the equivalence classes contain one element only, $[\boldsymbol{r}^{(\ell)}]=\boldsymbol{r}^{\ell)}$, so that (8.85) obtains

$$\frac{1}{\ell}S\left(\rho[\mathcal{Z}^{q,\ell}]\right)=\frac{1}{\ell}\log|\mathcal{Z}^{(q,\ell)}|=\log[\alpha^q]\,. \tag{8.89}$$

**Lemma 8.2.3.** $\mathrm{h}^{\mathrm{AFL}}_\omega(\Theta_{\mathbb{A}})\geq\log\alpha$.

**Proof:** Given the *OPU* of above, consider the refined *OPU*

$$\mathcal{Z}^{(q(\ell-1)+1}=\Theta_{\mathbb{A}}^{q(\ell-1)}[\mathcal{Z}]\circ\Theta_{\mathbb{A}}^{q(\ell-1)-1}[\mathcal{Z}]\circ\cdots\Theta_{\mathbb{A}}[\mathcal{Z}]\circ\mathcal{Z}\,.$$

As already remarked, by refining any pairs of the constituent *OPUs* one gets an *OPU* of the form (8.83); thus, by repeatedly applying (8.87) and (8.88), one gets

$$S\left(\rho[\mathcal{Z}^{(q(\ell-1)+1}]\right)\geq S\left(\rho[\mathcal{Z}^{(q,\ell)}]\right)\,.$$

One finally estimates

$$\begin{aligned}\mathrm{h}^{\mathrm{AFL}}_\omega(\Theta_{\mathbb{A}})&\geq\limsup_{\ell\to+\infty}\frac{1}{q(\ell-1)+1}S\left(\rho[\mathcal{Z}^{q(\ell-1)+1}]\right)\\&\geq\frac{1}{q}\limsup_{\ell\to+\infty}\frac{1}{\ell}S\left(\rho[\mathcal{Z}^{q,\ell)}]\right)\geq\frac{1}{q}\log[\alpha^q]\,,\end{aligned}$$

and the result follows by choosing $q$ arbitrarily large. □

In order to reverse the inequality in the previous lemma, we shall consider *OPUs* of the form $\mathcal{Z}=\{W(f_i)\}_{i=1}^p$; notice that

$$\sum_{i=1}^p W_\theta^\dagger(f_i)W_\theta(f_i)=\mathbb{1}\Longrightarrow\sum_{i=1}^p\|f_i\|^2=1$$

as turns out by computing the expectations with respect to $\omega$ of both sides of the operatorial equation and by using (7.35). Let the support of $\mathcal{Z}$ be defined as the union of the supports of the constituent $f_i$, $\mathrm{Supp}(\mathcal{Z}) := \bigcup_{i=1}^{p} \mathrm{Supp}(f_i)$, where $\mathrm{Supp}(f)$ is the set of $\boldsymbol{n} \in \mathbb{Z}^2$ where $f(\boldsymbol{n}) \neq 0$ (see (7.31)). The compactness assumption means that, given $\mathcal{Z}$ there exist finite real constants $g$, $d$ such that $\mathrm{Supp}(\mathcal{Z}) \subseteq R_{g,d}$, where

$$R_{gd} := \left\{\boldsymbol{n} \in \mathbb{Z}^2 \,:\, |\,\boldsymbol{n}\,\rangle = \gamma |\, a_+ \,\rangle + \delta |\, a_- \,\rangle \,,\, |\gamma| \leq g \,,\, |\delta| \leq d\right\} \,,$$

with $|\, a_\pm \,\rangle$ the eigenvectors of $\mathbb{A}$.

As seen in Example 7.1.6, the vectors $\pi_\omega(W_\theta(f))|\,\Omega_\omega\,\rangle$ in the *GNS* representation, amount to the $\ell^2(\mathbb{Z}^2)$ vectors $|\, f_i \,\rangle = \{f(\boldsymbol{n})\}_{\boldsymbol{n}\in\mathbb{Z}^2}$; thus, (8.65) gives

$$S\left(\rho[\mathcal{Z}]\right) = S\left(\sum_{i=1}^{p} |\, f_i \,\rangle\langle\, f_i \,|\right) \leq \log \#(\mathrm{Supp}(\mathcal{Z}))\,, \tag{8.90}$$

where $\#(\mathrm{Supp}(\mathcal{Z}))$ denotes the cardinality of the support.

Since the refined *OPU* $\mathcal{Z}^{(\ell)}$ has elements

$$\Theta_{\mathbb{A}}^{\ell-1}[W_\theta(f_{i_{\ell-1}})]\,\Theta_{\mathbb{A}}^{\ell-2}[W_\theta(f_{i_{\ell-2}})]\,\cdots\,\Theta_{\mathbb{A}}[W_\theta(f_{i_1})]\,W_\theta(f_{i_0})\,,$$

and each $\Theta_{\mathbb{A}}^{j}[W_\theta(f_{i_j})]$ is supported by vectors of the form

$$\mathbb{A}^j |\,\boldsymbol{n}_j\,\rangle = \gamma_j \alpha^j |\, a_+ \,\rangle + \delta_j \alpha^{-j} |\, a_- \,\rangle\,,$$

with $\boldsymbol{n}_j \in \mathrm{Supp}(f_i)$, it turns out that $\mathrm{Supp}(\mathcal{Z}^{(\ell)})$ consists of vectors

$$|\,\boldsymbol{n}(\ell)\,\rangle = \Big(\sum_{j=0}^{\ell-1} \alpha^j \gamma_j\Big)|\, a_+ \,\rangle + \Big(\sum_{j=0}^{\ell-1} \alpha^{-j} \delta_j\Big)|\, a_- \,\rangle$$

$$\left|\sum_{j=0}^{\ell-1} \alpha^j \gamma_j\right| \leq g\,\frac{\alpha^n - 1}{\alpha - 1}\,, \quad \left|\sum_{j=0}^{\ell-1} \alpha^{-j} \gamma_j\right| \leq d\,\frac{\alpha}{\alpha - 1}\,,$$

whence $\#(\mathcal{Z}^{(n)}) = O(\alpha^n)$ and, from (8.90),

$$\limsup_{n\to+\infty} \frac{1}{n} S\left(\rho[\mathcal{Z}^{(n)}]\right) \leq \log \alpha$$

for all *OPU*s from the chosen $\mathcal{A}_0$.

**Lemma 8.2.4.** $\mathrm{h}_\omega^{\mathrm{AFL}}(\Theta_A) = \sup_{\mathcal{Z}\in\mathcal{A}_0} \mathrm{h}_\omega^{\mathrm{AFL}}(\Theta_{\mathbb{A}})\,\mathcal{Z} \leq \log \alpha$.

Finally, Lemma 8.2.3 and Lemma 8.2.4 together prove Proposition 8.2.7

### 8.2.4 *AFL* Entropy and Quantum Channel Capacities

In Section 7.3.2 we have discussed the encoding of strings $\boldsymbol{i}^{(n)}$ of symbols from an alphabet $I_A$ emitted by a classical source by means of quantum code-words $\rho(\boldsymbol{i}^{(n)})$ taken from a statistical ensemble with weights $p(\boldsymbol{i}^{(n)})$. In [8] a different encoding protocol is proposed; it uses a quantum dynamical system $(\mathcal{A}, \Theta, \omega)$ and the *CPU* maps $\mathcal{M}(\mathcal{A}_0) \ni \mathbb{E} : \mathcal{A} \mapsto \mathcal{A}$. The idea is to encode strings $\boldsymbol{i}^{(n)} = i_1 i_2 \cdots i_n$ by perturbing the state $\omega$ with *CPU* maps $\mathbb{E}_{i_j}$ at each stroke of time $t = j$, $1 \leq j \leq n$.

**Remark 8.2.5.** As $\mathcal{A}$ is in general a quasi-local algebra, in order that the encoding protocols be physically implementable, the *CPU* maps are chosen to consist of finitely many Kraus operators taken from the union $\mathcal{A}_0$ of all strictly local subalgebras,

$$\mathcal{A} \ni A \mapsto \mathbb{E}_{i_j}[A] = \sum_{k \in I(i_j)} X_{i_j k}^\dagger \, A \, X_{i_j k} \ , \qquad X_{i_j k} \in \mathcal{A}_0 \ ,$$

where $\sum_{k \in I(i_j)} X_{i_j k}^\dagger X_{i_j k} = \mathbb{1}$ and the index set $I(i_j)$ is of finite cardinality. The *CPU* maps are further distinguished in $\mathbb{E} \in \mathcal{M}_b(\mathcal{A}_0)$ when they are bistochastic, see Example 6.3.3.1, in which case they are entropy increasing, and $\mathbb{E} \in \mathcal{M}_u(\mathcal{A}_0)$ when the $X_{i_j k}$ are unitary: $\mathcal{M}_u(\mathcal{A}) \subset \mathcal{M}_b(\mathcal{A}_0) \subset \mathcal{M}(\mathcal{A}_0)$.

In order to proceed with the explicit encoding, it is convenient to pass to the *GNS* triple $(\mathbb{H}_\omega, U_\omega, \Omega_\omega)$; set $\widehat{X}_{i_j k} := \pi_\omega(X_{i_j k})$ and denote by

$$\widehat{\mathbb{E}}_{i_j}[\widehat{B}] = \sum_{k \in I(i_j)} \widehat{X}_{i_j k}^\dagger \, \widehat{B} \, \widehat{X}_{i_j k} \ , \qquad B \in \mathbb{B}(\mathbb{H}_\omega) \ , \tag{8.91}$$

the *GNS* representation of the *CPU* maps $\mathbb{E}_{i_j}$ as *CPU* maps on $\mathbb{B}(\mathbb{H}_\omega)$. Moreover, let $\widehat{\mathbb{F}}_{i_j}$ be their dual maps acting on $\mathbb{B}_1(\mathbb{H}_\omega)$,

$$\mathbb{B}_1(\mathbb{H}_\omega) \ni \widehat{\rho} \mapsto \widehat{\mathbb{F}}_{i_j}[\widehat{\rho}] = \sum_{k \in I(i_j)} \widehat{X}_{i_j k} \, \widehat{\rho} \, \widehat{X}_{i_j k}^\dagger \ , \tag{8.92}$$

and let $\mathbb{U}_\omega^{-1}[\widehat{\rho}] := U_\omega \widehat{\rho} U_\omega^\dagger$ denote the Schrödinger time-evolution in the *GNS* representation. Then, the encoding procedure proposed in [8] is to assign to a string $\boldsymbol{i}^{(n)} \in I_A^n$ a density matrix $\widehat{\rho}(\boldsymbol{i}^{(n)})$ according to the following scheme:

$$\begin{aligned} \boldsymbol{i}^{(n)} \mapsto \mathcal{E}^{(n)}(\boldsymbol{i}^{(n)}) =: \widehat{\rho}(\boldsymbol{i}^{(n)}) &= \left( \prod_{j=n}^{1} \mathbb{U}_\omega^{-1} \circ \widehat{\mathbb{F}}_{i_j} \right) [| \, \Omega_\omega \rangle\langle \, \Omega_\omega \, |] \\ &= (\mathbb{U}_\omega^{-1} \circ \widehat{\mathbb{F}}_{i_n}) \circ (\mathbb{U}_\omega^{-1} \circ \widehat{\mathbb{F}}_{i_{n-1}}) \circ \cdots (\mathbb{U}_\omega^{-1} \circ \widehat{\mathbb{F}}_{i_1})[| \, \Omega_\omega \rangle\langle \, \Omega_\omega \, |] \ , \end{aligned} \tag{8.93}$$

The states $\widehat{\rho}(\boldsymbol{i}^{(n)})$ are the *GNS* representations of perturbed states obtained from the given $\Theta$-invariant state $\omega$ as follows:

$$\omega_{\boldsymbol{i}^{(n)}} := \omega \circ \left( \prod_{j=1}^{n} \mathbb{E}_{i_j} \circ \Theta \right) = \omega \circ (\mathbb{E}_{i_1} \circ \Theta) \circ (\mathbb{E}_{i_2} \circ \Theta) \circ \cdots (\mathbb{E}_{i_n} \circ \Theta) \ . \quad (8.94)$$

**Example 8.2.2.** Let the encoding (8.93) be based on a Bernoulli quantum spin chain $(\mathcal{A}_{\mathbb{Z}}, \Theta_\sigma, \omega)$, where $\mathcal{A}_{\mathbb{Z}}$ consists of single site algebras $M_d(\mathbb{C})$, $\Theta_\sigma$ is the right shift and $\omega$ the product state

$$\omega(A) = \mathrm{Tr}(\underbrace{\rho \otimes \rho \otimes \cdots \rho}_{p-q+1 \ times} A) \ , \qquad A \in \mathcal{A}_{[p,q]} \ .$$

Since the Kraus operators from the various *CPU* maps in (8.93) are finitely many and belong to local subalgebras of $\mathcal{A}_{\mathbb{Z}}$, there exists an $\ell \in \mathbb{N}$ such that $X_{i_j k} \in \mathcal{A}_{[-\ell,\ell]}$ for all $i_j \in I_A$ and $k \in I(i_j)$. With respect to (8.94), each $\Theta = \Theta_\sigma$ shifts the Kraus operators of the *CPU* map to the right by one site; therefore, the $X_{i_j k}$ of $\mathbb{E}_{i_j}$, $2 \leq j \leq n$, will be shifted to the right by $j-1$ sites. Therefore, the perturbed states $\omega_{\boldsymbol{i}^{(n)}}$ have the form

$$\omega_{\boldsymbol{i}^{(n)}} = \omega \circ \mathbb{E}^{(0)}_{i_1} \circ \mathbb{E}^{(1)}_{i_2} \circ \cdots \mathbb{E}^{(n-1)}_{i_n} \circ \Theta^{n-1} \ ,$$

where the Kraus operators of $\mathbb{E}^{(j-1)}_{i_j}$, $\Theta^{j-1}_\sigma(X_{i_j k})$, belong to $\mathcal{A}_{[-\ell+j-1\,,\,\ell+j-1]}$. It thus turns out that the state $\omega_{\boldsymbol{i}^{(n)}}$ in (8.94) amounts to a density matrix $\rho^{\mathrm{loc}}_{\boldsymbol{i}^{(n)}} \in \mathcal{A}_{[\ell\,,\,\ell-n-1]}$ tensorized with $\rho$ over the sites $k \notin [\ell\,,\,\ell-n-1]$. In the *GNS* representation, the corresponding $\widehat{\rho}(\boldsymbol{i}^{(n)})$ in (8.93) acts as a density matrix $\widehat{\rho}^{\mathrm{loc}}_{\boldsymbol{i}^{(n)}}$ on $\pi_\omega(\mathcal{A}_{[-\ell,\ell+n-1]}) \otimes \pi_\omega(\mathcal{A}_{[-\ell,\ell+n-1]})$.

**Example 8.2.3.** Given a quantum spin chain as encoder, single-site encodings turn out to be particularly useful; that is, we will use *CPU* maps consisting of Kraus operators $X_{ik}$ belonging to single site algebras $M_d(\mathbb{C})$. The following *CPU* maps are three interesting possibilities.

1. Let $\{|\,i\,\rangle\}_{i=1}^d$ denote a fixed *ONB* in $\mathbb{C}^d$; then, consider the purifying maps

$$M_d(\mathbb{C}) \ni \rho \mapsto \mathbb{F}_i[\rho] := \mathrm{Tr}(\rho)\,|\,i\,\rangle\langle\,i\,| \ , \quad i = 1, 2, \ldots, d \ .$$

These are the dual maps of the *CPU* maps

$$M_d(\mathbb{C}) \ni A \mapsto \mathbb{E}_i[B] := \sum_{k=1}^{d} |\,k\,\rangle\langle\,i\,|\,A\,|\,i\,\rangle\langle\,k\,| = \langle\,i\,|\,A\,|i\,\rangle\,\mathbb{1} \ .$$

The encoding is performed by choosing $X_{ik} = |\,i\,\rangle\langle\,k\,|$, $i,k = 1,2,\ldots,d$, whence if $Y \in \mathcal{A}$ is such that $\Theta_\sigma^{n-1}(Y) \in \mathcal{A}^{(n)} = \mathcal{A}_{[0,n-1]}$, then $\omega_{\boldsymbol{i}^{(n)}}(Y) = \langle\,\boldsymbol{i}^{(n)}\,|\,\Theta_\sigma^{n-1}(Y)\,|\boldsymbol{i}^{(n)}\,\rangle$, where $|\,\boldsymbol{i}^{(n)}\,\rangle := |\,i_1\,\rangle \otimes |\,i_2\,\rangle \otimes \cdots |\,i_n\,\rangle$. As a consequence, this particular encoding corresponds to a perturbation of $\omega$ such that $\rho^{\mathrm{loc}}_{\boldsymbol{i}^{(n)}} = |\,\boldsymbol{i}^{(n)}\,\rangle\langle\,\boldsymbol{i}^{(n)}\,| \in \mathcal{A}^{(n)}$.

2. Consider the discrete Weyl operators in Example 5.4.2 with $N = d$ and perform the encoding corresponding to the *CPU* maps

$$M_d(\mathbb{C}) \ni A \mapsto \mathbb{E}_{\boldsymbol{n}}[A] = W_d(\boldsymbol{n})^\dagger\, A\, W_d(\boldsymbol{n})\ ,$$

with $\boldsymbol{n} = (n_1, n_2)$, $n_i = 0,1,\ldots d-1$. The Kraus operators involved are thus of the form $X_{ik} = W_d(\boldsymbol{n}_i)$ where $i = 1,2,\ldots,d^2$ enumerates the $d^2$ pairs $\boldsymbol{n}$ and $k = 1$ for all $i$. Then, choosing $Y \in \mathcal{A}$ as in the previous case, the perturbed states turn out to be

$$\omega_{\boldsymbol{i}^{(n)}}(Y) = \mathrm{Tr}\left(\bigotimes_{j=1}^{n} W_d(\boldsymbol{n}_{i_j})\,\rho\, W_d^\dagger(\boldsymbol{n}_{i_j})\,\Theta_\sigma^{n-1}(Y)\right)\ ,$$

corresponding to $\mathcal{A}^{(n)} \ni \rho^{\mathrm{loc}}_{\boldsymbol{i}^{(n)}} = \bigotimes_{j=1}^{n} W_d(\boldsymbol{n}_{i_j})\,\rho\, W_d^\dagger(\boldsymbol{n}_{i_j})$.

3. Let $\rho = \sum_{i=1}^{d} r_i |\,r_i\,\rangle\langle\,r_i\,|$ be the spectral representation of the single site density matrix and $|\,\sqrt{\rho}\,\rangle = \sum_{i=1}^{d} \sqrt{r_j}\,|\,r_j\,\rangle \otimes |\,r_j\,\rangle$ its purification. Consider the *GNS* representation where $|\,\Omega_\omega\,\rangle = (|\,\sqrt{\rho}\,\rangle\langle\,\sqrt{\rho}\,|)^{\otimes\infty}$; then, the encoding in the previous point yields a perturbed state (8.93) that amounts to a local density matrix of the form

$$\pi_\omega(\mathcal{A}^{(n)}) \otimes \pi_\omega(\mathcal{A}^{(n)})' \ni \widehat{\rho}^{\mathrm{loc}}_{\boldsymbol{i}^{(n)}} = \bigotimes_{j=1}^{n} W_d(\boldsymbol{n}_{i_j}) \otimes \mathbb{1}_d\,|\,\sqrt{\rho}\,\rangle\langle\,\sqrt{\rho}\,|\,W_d^\dagger(\boldsymbol{n}_{i_j}) \otimes \mathbb{1}_d\ .$$

As in Section 6.3.1, the classical source $A$ emitting symbols $\boldsymbol{i}^{(n)} \in I_A^n$ with probabilities $p(\boldsymbol{i}^{(n)})$ is described as a stochastic variable $A^{(n)}$ with probability distribution $\pi^{(n)} = \{p(\boldsymbol{i}^{(n)})\}_{\boldsymbol{i}^{(n)} \in I_A^n}$. The encoding (8.93) provides a statistical mixture described by the density matrix

$$\mathbb{B}_1(\mathbb{H}_\omega) \ni \widehat{\rho}_{\mathcal{E}^{(n)}} = \sum_{\boldsymbol{i}^{(n)} \in I_A^n} p(\boldsymbol{i}^{(n)})\,\widehat{\rho}(\boldsymbol{i}^{(n)})\ ,$$

and any decoding *POVM* $\widehat{\mathcal{B}}^{(n)} = \{B_j^{(n)}\}_{j \in I_B}$ by means of operators in $\mathbb{B}(\mathbb{H}_\omega)$ defines another stochastic variable $\widehat{B}^{(n)}$. The mutual information (6.32) of $A^{(n)}$ and $\widehat{B}^{(n)}$ is bounded by the Holevo $\chi$ quantity (6.33),

$$I(A^{(n)}; \widehat{B}^{(n)}) \leq S\left(\widehat{\rho}_{\mathcal{E}^{(n)}}\right) - \sum_{\boldsymbol{i}^{(n)} \in I_A^n} p(\boldsymbol{i}^{(n)})\, S\left(\widehat{\rho}(\boldsymbol{i}^{(n)})\right)\ , \tag{8.95}$$

and depends on the source probability $\pi_{A^{(n)}}$, on the *CPU* maps implementing the encoding $\mathcal{E}^{(n)}$ and on the *POVM* $\widehat{\mathcal{B}}^{(n)}$.

If the encoding (8.93) is based on Bernoulli quantum spin chains as in Example 8.2.2, then

$$\widehat{\rho}_{\mathcal{E}^{(n)}} = \sum_{\boldsymbol{i}^{(n)}\in I_A^n} p(\boldsymbol{i}^{(n)})\,\widehat{\rho}(\boldsymbol{i}^{(n)}) = \mathbb{F}^{\omega}_{\mathcal{Y}^{(n)}}[|\,\Omega\,\rangle\langle\,\Omega\,|] \ , \tag{8.96}$$

where, using the argument which led to (8.66), $\mathcal{Y}^{(n)}$ is a localized *POVM* whose elements are operators of the form

$$\Theta_\sigma^{n-1}(X_{i_{n-1}k_{n-1}})\Theta^{n-2}(X_{i_{n-2}k_{n-2}})\cdots X_{i_0k_0} \in \mathcal{A}_{[-\ell,\ell+n-1]} \ ,$$

with $X_{i_jk} \in \mathcal{A}_{[-\ell,\ell]}$. Then, from (8.95), (8.67) and Proposition 8.2.3

$$\begin{aligned} I(A^{(n)};\widehat{B}^{(n)}) &\leq S\left(\widehat{\rho}_{\mathcal{E}^{(n)}}\right) = S\left(\mathbb{F}^{\omega}_{\mathcal{Y}^{(n)}}[|\,\Omega\,\rangle\langle\,\Omega\,|]\right) = S\left(\rho^{(n)}[\mathcal{Y}]\right) \\ &\leq (n-2\ell)(S\left(\rho\right)+\log\, d) \ . \end{aligned} \tag{8.97}$$

Indeed, $\rho^{(n)}[\mathcal{Y}]$ results from the tensor product density matrix $\rho^{\otimes\,(n-2\ell)}$ on the algebra $M_d(\mathbb{C})^{\otimes\,(n-2\ell)}$.

Furthermore, if the decoding is operated by means of local *POVMs* $\mathcal{B}$ consisting of operators $B_i \in \mathcal{A}_{[p,q]}$, then in (8.95) one can substitute the density matrices $\widehat{\rho}(\boldsymbol{i}^{(n)})$ and $\widehat{\rho}_{\mathcal{E}^{(n)}}$ in the *GNS* representation with local density matrices $\rho^{\rm loc}(\boldsymbol{i}^{(n)})$ and $\rho^{loc}_{\mathcal{E}^{(n)}} = \sum_{\boldsymbol{i}^{(n)}} p(\boldsymbol{i}^{(n)})\,\rho^{\rm loc}(\boldsymbol{i}^{(n)})$. The corresponding Holevo's bound reads

$$I(A^{(n)};B^{(n)}) \leq S\left(\rho^{\rm loc}_{\mathcal{E}^{(n)}}\right) - \sum_{\boldsymbol{i}^{(n)}\in I_A^n} p(\boldsymbol{i}^{(n)})S\left(\rho^{\rm loc}(\boldsymbol{i}^{(n)})\right) \ . \tag{8.98}$$

This bound and the fact that the various states are matrices in $M_d(\mathbb{C})^{\otimes\,(n-2\ell)}$ imply that, for encodings $\mathcal{B}$ by means of generic *POVMs* in $\mathcal{A}$,

$$I(A^{(n)};B^{(n)}) \leq (n-2\ell)\log\, d \ , \tag{8.99}$$

while, for *POVMs* $\mathcal{B}$ consisting of bistochastic maps

$$I(A^{(n)};B^{(n)}) \leq (n-2\ell)(\log\, d - S(\rho)) \ , \tag{8.100}$$

for the encodings are entropy increasing so that $S(\rho^{\rm loc}(\boldsymbol{i}^{(n)}) \geq S\left(\rho^{\otimes\,(n-2\ell)}\right)$.

**Example 8.2.4.** With reference to the three encodings in Example 8.2.3, the Holevo $\chi$ quantity, denoted by $\chi_{1,2,3}$ for sake of simplicity, depend only on the structure of the perturbed states restricted to the first $n$ sites of the quantum spin chain $\mathcal{A}$. By choosing uniform Bernoulli probability distributions $p(\boldsymbol{i}^{(n)}) = \prod_{j=1}^n p_{i_j}$ over the indices $\boldsymbol{i}^{(n)}$, the product structure of the perturbed states yields:

1. Let $\pi = \{p_i = 1/d\}_{i=1}^d$; since $\rho^{\rm loc}_{\boldsymbol{i}^{(n)}} = |\,\boldsymbol{i}^{(n)}\,\rangle\langle\,\boldsymbol{i}^{(n)}\,|$, $S(\rho^{\rm loc}_{\boldsymbol{i}^{(n)}}) = 0$ and
$$\sum_{\boldsymbol{i}^{(n)}} p(\boldsymbol{i}^{(n)})\,\rho^{\rm loc}_{\boldsymbol{i}^{(n)}} = \left(\frac{\mathbb{1}_d}{d}\right)^{\otimes n} \Longrightarrow \chi_1 = n\log\,d\ .$$

2. Let $\pi = \{p_i = 1/d^2\}_{i=1}^{d^2}$; since $\rho^{\rm loc}_{\boldsymbol{i}^{(n)}} = \bigotimes_{j=1}^n W_d(\boldsymbol{n}_{i_j})\,\rho\,W_d^\dagger(\boldsymbol{n}_{i_j})$, additivity of the von Neumann entropy (5.160) and unitarity of the Weyl operators imply $S\left(\rho^{\rm loc}_{\boldsymbol{i}^{(n)}}\right) = nS\left(\rho\right)$. Further, from (5.30) it follows that $\sum_{i=1}^{d^2} W_d(\boldsymbol{n}_i)\,\rho\,W_d^\dagger(\boldsymbol{n}_i) = d\,\mathbb{1}_d$. Thus
$$\sum_{\boldsymbol{i}^{(n)}} p(\boldsymbol{i}^{(n)})\,\rho^{\rm loc}_{\boldsymbol{i}^{(n)}} = \left(\frac{\mathbb{1}_d}{d}\right)^{\otimes n} \Longrightarrow \chi_2 = n(\log\,d - S(\rho))\ .$$

3. Since $\widehat{\rho}^{\rm loc}_{\boldsymbol{i}^{(n)}} = \bigotimes_{j=1}^n W_d(\boldsymbol{n}_{i_j}) \otimes \mathbb{1}_d\,|\,\sqrt{\rho}\,\rangle\langle\,\sqrt{\rho}\,|\,W_d^\dagger(\boldsymbol{n}_{i_j}) \otimes \mathbb{1}_d$ are pure, $S(\widehat{\rho}^{\rm loc}_{\boldsymbol{i}^{(n)}}) = 0$. Also, using again (5.30), it follows that
$$\frac{1}{d}\sum_{i=1}^{d^2} W_d(\boldsymbol{n}_i)\otimes\mathbb{1}_d\,|\,\sqrt{\rho}\,\rangle\langle\,\sqrt{\rho}\,|\,W_d^\dagger(\boldsymbol{n}_i)\otimes\mathbb{1}_d = \mathbb{1}\otimes{\rm Tr}_I(|\,\sqrt{\rho}\,\rangle\langle\,\sqrt{\rho}\,|) = \mathbb{1}\otimes\rho\,,$$
where ${\rm Tr}_I$ denotes partial trace over the first factor. Therefore, choosing the probability $\pi$ as in the previous point,
$$\sum_{\boldsymbol{i}^{(n)}} p(\boldsymbol{i}^{(n)})\,\widehat{\rho}^{\rm loc}_{\boldsymbol{i}^{(n)}} = \left(\frac{\mathbb{1}_d}{d}\otimes\rho\right)^{n} \Longrightarrow \chi_3 = n(\log\,d + S(\rho))\ .$$

According to Section 3.2.2, the classical capacity of the channel resulting from the considered encodings is:
$$C := \sup_{\pi^{(n)}\,,\ \mathcal{E}^{(n)}\,,\ \mathcal{B}^{(n)}}\ \limsup_{n\to\infty} \frac{1}{n} I(A^{(n)};B^{(n)})\ , \tag{8.101}$$
where the supremum is computed varying probabilities, encoding and decoding protocols. The following possibilities are envisaged:

1. entanglement assisted capacity, $C_{\rm ent}$, when $\widehat{\mathcal{B}} = \{\widehat{B}_i\}_{i\in I_B} \subset \mathbb{B}(\mathbb{H}_\omega)$ consists of bounded operators on the *GNS* Hilbert space;
2. ordinary capacities, $C \geq C_b \geq C_u$, when $\mathcal{B} = \{B_i\}_{i\in I_B} \subset \mathcal{A}$ and the encoding $\mathcal{E}^{(n)}$ is performed with any localized *CPU* map ($C$) with localized bistochastic *CPU* maps ($C_b$) and with localized *CPU* maps consisting of unitary Kraus operators ($C_u$);
3. Bernoulli capacities, $C^0_{\rm ent} \geq C^0 \geq C^0_b \geq C^0_u$, when the supremum is taken over input probabilities that factorize $p(\boldsymbol{i}^{(n)}) = \prod_{i=1}^n p_{i_j}$.

**Remark 8.2.6.** The entanglement in the capacity $C_{\rm ent}$ is due to the *GNS* vector $|\,\Omega_\omega\,\rangle$ being entangled over the algebra $\pi_\omega(\mathcal{A})\otimes\pi_\omega(\mathcal{A})'$ and the considered *POVMs* $\widehat{\mathcal{B}}$ consist of generic operators in $\mathbb{B}(\mathbb{H}_\omega)$. The entanglement of $|\,\Omega_\omega\,\rangle$ is most simply seen in the case of a Bernoulli quantum spin chain as the one discussed in Example 8.2.3; there, the *GNS* construction amounts to purifying the single site density matrix $\rho$ into a vector $|\,\sqrt{\rho}\,\rangle\in\mathbb{C}^d\otimes\mathbb{C}^d$ which entangles $M_d(\mathbb{C})\otimes\mathbb{1}$ with $\mathbb{1}\otimes M_d(\mathbb{C})$ at each single site.

The entanglement assisted capacity of Bernoulli quantum sources is bounded by the *AFL* entropy, for all triplets $(\mathcal{A},\Theta,\omega)$, while the capacity equals the *AFL* entropy in the case of Bernoulli quantum spin chains.

**Proposition 8.2.8.** *The entanglement assisted capacity relative to Bernoulli classical sources encoded by using quantum dynamical systems $(\mathcal{A},\Theta,\omega)$, is bounded by the AFL entropy: $C^0_{\rm ent}\leq {\rm h}^{\rm AFL}_\omega(\Theta,\mathcal{A}_0)$.*

*Moreover, the capacities of encodings by Bernoulli quantum spin chains $(\mathcal{A},\Theta_\sigma,\omega_\rho)$ can be explicitly computed:*

$$C^0_u = C^0_b = C_u = C_b = \log d - S(\rho) \tag{8.102}$$
$$C^0 = C = \log d \tag{8.103}$$
$$C^0_{\rm ent} = C_{\rm ent} = S(\rho) + \log d\ . \tag{8.104}$$

**Proof:** As regards the first part of the proposition, the source probabilities $p(\boldsymbol{i}^{(n)})=\prod_{j=1}^n p_{i_j}$ by assumption; therefore, (8.93) and (8.66) yield

$$\begin{aligned}\widehat{\rho}_{\mathcal{E}^{(n)}} &:= \sum_{\boldsymbol{i}^{(n)}\in I^n_A} p(\boldsymbol{i}^{(n)})\widehat{\rho}(\boldsymbol{i}^{(n)}) = \left(\prod_{j=n}^{1}\mathbb{U}_\omega^{-1}\circ\sum_{i_j\in I_A} p_{i_j}\widehat{\mathbb{F}}_{i_j}\right)[|\,\Omega_\omega\,\rangle\langle\,\Omega_\omega\,|]\\ &= \mathbb{U}^{-n}_\omega\circ\widehat{\mathbb{F}}^\omega_{\mathcal{X}^{(n)}}[|\,\Omega_\omega\,\rangle\langle\,\Omega_\omega\,|]\ ,\quad \mathcal{X}:=\{\sqrt{p_i}X_{ik}\}_{\substack{i\in I_a\\ k\in I(i)}}\ .\end{aligned}$$

Then, from (8.67) and (8.95) $I(A^{(n)};\widehat{B}^{(n)})\leq S\Big(\widehat{\rho}_{\mathcal{E}^{(n)}}\Big)=S\left(\rho^{(n)}[\mathcal{X}]\right)$, whence the result follows from Definition 8.63.

Concerning the second part of the proposition, for single site encodings of Bernoulli classical sources by means of Bernoulli quantum spin chains, we can use the result in Theorem 7.3.4. It ensures that one can always find a suitable decoding *POVM* such that the asymptotic amount of transmitted information per symbol, namely the argument of the supremum in (8.101) equals the corresponding Holevo $\chi$ quantity. Consider now the first case in Example 8.2.4, $\chi_1/n=\log d$ and (8.99) imply that $\log d\leq C^0\leq C\leq\log d$, whence (8.102). In the second case $\chi_2/n=\log d-S(\rho)$, this and (8.99) imply

$$\log d - S(\rho)\leq C^0_b\leq C_b\leq \log d - S(\rho)\ .$$

Thus, (8.103) results from the fact that $C_u^0 \leq C_b^0$ and $C_u \leq C_b$. Finally, by means of the same argument, (8.104) follows from the third case in the quoted example and from (8.97):

$$\log d + S(\rho) = \chi_3 \leq C_E^0 \leq C_E \leq \log d + S(\rho) \ .$$

□

## Bibliographical Notes

The recent book [222] represents a most up to date and complete approach to *CNT* entropy and its applications to $C^*$ and von Neumann dynamical systems of physical and mathematical origin. It also presents in full detail the approach to the *CNT* entropy developed in [264] and the construction of dynamical and topological entropies due to Voiculescu [311]. Another formulation of a quantum topological entropy, namely of a quantum dynamical entropy independent of the given invariant state, can be found in [154].

Older books also dealing with the *CNT* entropy are [22, 226], while a detailed presentation of the *AFL* entropy ad its applications is in [10]. The relations between the *CNT* entropy and the *AFL* entropy are reviewed in [155].

A different proposal of quantum dynamical entropy based on coherent states and suitable to applications to quantum chaos and the semi-classical limit is in [282, 281, 74]. Possible applications of quantum dynamical entropies to chaotic phenomena in quantum spin chains are discussed in [243].

# 9 Quantum Algorithmic Complexities

As already emphasized in Chapter 4 information is physical and the limits to information processing tasks are ultimately set by the underlying physical laws. For instance, the standing models of computation are based on the physics of deterministic and/or stochastic classical processes; instead, *quantum computation theory* [66, 128, 224, 165, 71] studies the new possibilities offered by a model of computation based on quantum mechanical laws. The birth of such a theory finds a technological motivation in the high pace at which chip miniaturization proceeds. Indeed, information processing at the atomic level, namely at a scale where the physical laws are those of quantum mechanics, might soon become a concrete practical issue [66]. On the other hand, a strong theoretical impulse to the development of quantum computation theory came from Feynman's suggestion [116, 139] that quantum computers might provide a more efficient description of quantum systems than classical (probabilistic) computers and, above all, from the discovery of quantum algorithms with more efficient performances with respect to what is classical achievable [224].

A first theoretical step in this direction was the extension of the notions of *TM* and of *UTM* to those of *quantum Turing machines* (*QTM*s ) [99] and to *universal QTMs* (*UQTM*s ) [56]: very roughly speaking, these latter are computing devices that work as classical *TM*s and *UTM*s , the only difference being that their configurations behave as vector states of a suitable Hilbert space. Namely, given any set of possible configurations, their linear superpositions are also possible configurations.

Once the existence of *UQTM*s is foreseen, a very natural theoretical step is to try to formulate quantum versions of the concepts introduced in Chapter 4; in particular, by extending algorithmic complexity theory to the quantum setting, one may try to set up a theory of randomness of individual quantum states [120] and of quantum processes.

In the following, we shall consider some proposals that have recently been put forward concerning different ways in which one might approach the algorithmic complexity of quantum states. All proposals start from the basic intuitive idea that complexity should characterize properties of systems that are difficult to describe; they can roughly be summarized as follows:

F. Benatti, *Dynamics, Information and Complexity in Quantum Systems*,
Theoretical and Mathematical Physics, DOI 10.1007/978-1-4020-9306-7_9,

1. one may attempt to describe quantum states by means of other quantum states that are processed by *UQTMs* [57]: the corresponding complexity will be referred to as *qubit quantum complexity* and denoted by $QC_q$;
2. one may decide to describe quantum states by classical [309] programs run by *UQTMs* : the corresponding complexity will be denoted by $QC_c$ and referred to as *bit quantum complexity*;
3. one may choose to relate the complexity of a *qubit* string to the complexity of the (classical) description of the quantum circuits that construct the *qubit* string [205, 206]. The corresponding complexity will be denoted by $QC_{net}$ and referred to as *circuit quantum complexity*;
4. one may extend the notion of universal probability (see Section 4) and define a *quantum universal semi-density matrix* [119]. There then arise two possible definitions of quantum complexity, denoted by $QC^{\pm}_{u.p.}$ that do not refer either to *QTMs* or to circuits.

We shall mainly concentrate on the *qubit* quantum complexity $QC_q$: it allows for a quantum generalization of Brudno's theorem that will be presented in detail. On general grounds, one should not expect the above proposals to yield equivalent notions; very likely, each one of them will be sensitive to different specific quantum properties, as we have seen to be the case with the quantum extensions of the *KS* dynamical entropy. Unlike in the classical domain (see Remark 4.3.1.4) where chaoticness and typicalness appear to be equivalent characterization of random *bit* sequences, *qubit* sequences are likely to be random in different inequivalent ways.

## 9.1 Effective Quantum Descriptions

The notions of *qubit* and *bit* quantum complexity are based on the use of *QTMs*. In the following, we will not consider what *quantum computers* might do that classical computers do not, nor will we address their practical implementation (see for instance [128, 224]). We shall simply assume that such devices exist and proceed to define:

1. the targets of the algorithmic descriptions processed by *QTMs* ;
2. which kinds of algorithms are processed by *QTMs* ;
3. how these algorithms are processed by *QTMs* ;
4. which are the outputs of these processes.

1. In the quantum setting, the targets of the effective descriptions will be *qubit* strings; since one is always interested in targets of increasing length, a convenient mathematical framework is provided by quantum spin chains (see Section 7.1.5), namely by algebraic triples of the form $(\mathcal{A}_{\mathbb{Z}}, \Theta_\sigma, \omega)$ that have been introduced in Definition (7.1.11), with $2 \times 2$ matrix algebras at each site. As already noted in Remarks 7.3.1, in going from *bit* to *qubit* strings there are similarities, but also differences. In particular, there is a

larger variety of *qubit* strings. Therefore, by *qubit* strings it will be meant any local density matrix corresponding to generic mixed and entangled states on local subalgebras $\mathcal{A}^{(n)} = (M_2(\mathbb{C}))^{\otimes n}$.

2. The inputs to *QTMs* will be generic *qubit* strings, loosely referred to as quantum programs; a subclass of these are the classical programs or *bit* strings that *QTMs*, as extensions of classical *TMs*, must also be able to process.

3. While classical *TMs* ultimately amount to specific transition functions between their configurations, *QTMs* are defined by transition amplitudes between their configurations which form a Hilbert space. Any *QTM* will thus identify a specific quantum computation that is a specific unitary operators acting on the Hilbert space of its configurations.

4. Finally, the outputs of a quantum computation operated by a *QTM* will be a *qubit* string read out by a measurement process.

Within the framework just outlined, the first two generalizations of classical algorithmic complexity previously mentioned are based on *qubit* strings effectively described by *bit* strings in the first case and by *qubit* strings in the second one.

### 9.1.1 Effective Descriptions by *qubit* Strings

Given the quasi-local structure of quantum sources as $C^*$ algebras generated by local *n-qubit* sub-algebras $\mathcal{A}^{(k)} = M_2(\mathbb{C})^{\otimes k}$, let us denote by $\mathbb{H}_k := (\mathbb{C}^2)^{\otimes k}$ the Hilbert space of $k$ *qubits* ($k \in \mathbb{N}_0$) and fix in each single *qubit* Hilbert space $\mathbb{C}^2$ a *computational basis* $|\,0\,\rangle, |\,1\,\rangle$. In order to be as general as possible, superpositions of *qubit* states of different lengths $k$ are allowed: they correspond to vectors in the Fock-like Hilbert space $\mathbb{H}_F := \bigoplus_{k=0}^{\infty} \mathbb{H}_k$. More in general, *qubit* strings will be represented by density matrices $\rho \in \mathbb{B}_1^+(\mathbb{H}_F)$ acting on $\mathbb{H}_F$.

**Example 9.1.1.** Any *bit* string $\boldsymbol{i} \in \{0,1\}^*$ identifies a computational basis vector in $\mathbb{H}_F$: the empty string $\lambda$ corresponds to the vacuum $|\,\Omega_F\,\rangle$, the 1-*qubit* subspace $\mathbb{H}_1$ is spanned by $|\,0\,\rangle$, $|\,1\,\rangle$, while the *k-qubit* subspace $\mathbb{H}_k$ is generated by the vectors corresponding to the *bit* strings of length $k$, $\boldsymbol{i}^{(k)} \in \Omega_2^{(k)}$, namely by $|\,\boldsymbol{i}^{(k)}\,\rangle = |\,i_1 i_2 \cdots i_k\,\rangle$, $i_j = 0, 1$. Generic *qubit* strings amount to density matrices in $\mathbb{B}_1^+(\mathbb{H}_{\leq n})$ acting on $\mathbb{H}_{\leq n} := \bigoplus_{k=0}^{n} \mathbb{H}_k$, its dimension being $\sum_{k=0}^{n} 2^k = 2^{n+1} - 1$.

In the commutative setting, the length of a *bit* string is simply the number of bits it consists of; in the quantum setting, the number of *qubits* involved fixes the Hilbert space dimension. Therefore, the following definition naturally extends the notion of length of a program.

**Definition 9.1.1.** *The length, $\ell(\rho)$, of a qubit string $\rho \in \mathbb{B}_1^+(\mathbb{H}_F)$ is*

$$\ell(\rho) := \min\{n \in \mathbb{N}_0 \,|\, \rho \in \mathbb{B}_1^+(\mathbb{H}_{\leq n})\} , \tag{9.1}$$

*setting $\ell(\rho) = \infty$ if this set is empty.*

As we shall see, *QTMs* act on and construct superpositions of vector *qubit* strings and, more in general, convex combinations of projection onto vector *qubit* strings of different lengths. Moreover, like their classical counterparts, *QTMs* comprise different parts as a read/write head, a control unit and one or more tapes all of them capable of being in states that are either Hilbert space vectors or density matrices acting on them. Therefore, the *QTMs* configurations too are generically described by density matrices acting on appropriate Hilbert spaces. Notice that mixed states are quite typical in such a context for they naturally appear when one is interested in the state of the read/write head, say, and therefore traces over the Hilbert spaces corresponding to the other *QTMs* components.

As observed in Remark 7.3.1.3, unlike in the classical situation where there are countably many *bit* strings, there are uncountably many *qubit* strings that can be arbitrarily close to one another. In order to quantify how close two *qubit* strings $\rho, \sigma \in \mathbb{B}_1^+(\mathbb{H}_F)$ actually are, it is convenient to use the trace-distance introduced in Definition 6.3.4, $D(\rho, \sigma)$.

### 9.1.2 Quantum Turing Machines

Any model of computation is based on the physics of the processors performing the computations; both deterministic and probabilistic Turing machines (see Example 4.1.4) work according to the laws of classical physics. It was Feynman [116] who was the first one to argue that quantum processes, to be efficiently simulated, require quantum computers. Indeed, quantum mechanics allow superpositions of states; in the case of *TMs*, the natural classical states are their configurations $c := \Big((\sigma_i)_{i\in\mathbb{Z}}, q, k\Big)$ (see Definition 4.1.1). The main feature of quantum computing machines is the possibility of producing and acting on linear superpositions of classical configurations, thus of performing in one single step of a computation what, classically, would only be achieved by an enormous number of *TMs* working in parallel (this is *quantum parallelism*, a phenomenon briefly sketched in Example 6.1.1).

The Hilbert space spanned by the classical configurations $|\,c\,\rangle$ provides (vector) states $|\,\Psi\,\rangle = \sum_{c\in C} \Psi(c)|\,c\,\rangle$ of the *QTM*, with Fourier coefficients $\Psi(c)$ that represent the *complex amplitudes* associated to the computational steps $c$. As in the case of *PTMs*, a quantum computation corresponds to a level-tree with an initial configuration branching into others, the main difference being that the edges leading from one level to the next do not carry branching probabilities, rather branching amplitudes that give rise to interference effects.

**Example 9.1.2.** With reference to Example 4.1.4 [128], consider the following branching tree that resembles the scheme of a Mach-Zender interferometer (see Figure 5.5). It describes a computational process starting off with an initial configuration $c_0$ that branches into two different configurations $c_{11}$ and $c_{12}$ at level 1 with amplitudes $a_{01} := a(c_0 c_{11})$ and $a_{02} := a(c_0 c_{12}) = 2^{-1/2}$, so that $|a_{01}|^2 + |a_{02}|^2 = 1$. This first computational step is then followed by a second one with two configurations at level 1 branching as follows: $c_{11}$ into $c_{21}$ and $c_{22}$ with amplitudes $a_{11} := a(c_{11}c_{21}) = 1/\sqrt{2}$ and $a_{12} := a(c_{11}c_{22}) = 1/\sqrt{2}$, while $c_{12}$ into $c_{23}$ and $c_{24}$ with amplitudes $a_{23} := a(c_{12}c_{23}) = -1/\sqrt{2}$ and $a_{24} := a(c_{12}c_{24}) = 1/\sqrt{2}$.

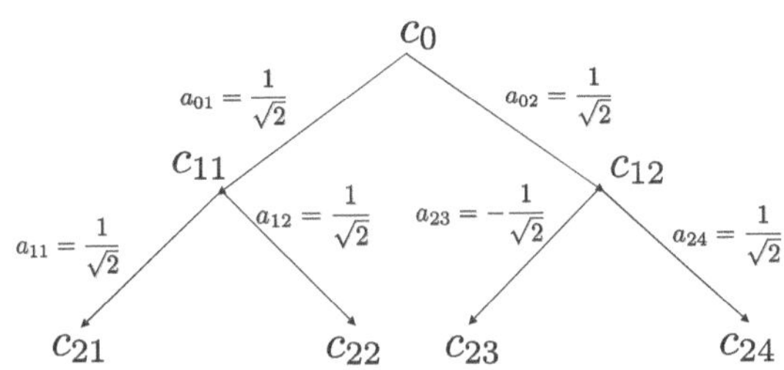


**Fig. 9.1.** Quantum Turing Machines: Level Tree

Thus, the overall amplitudes for the 4 configurations at step 2 are

$$a(c_{21}) := a_{01}\, a_{11} = \frac{1}{2}\;, \quad a(c_{22}) := a_{01}a_{12} = \frac{1}{2}\;,$$
$$a(c_{23}) := a_{02}a_{23} = -\frac{1}{2}\;, \quad a(c_{24}) := a_{02}a_{24} = \frac{1}{2}\;.$$

The most important difference with respect to classical *PTMs* is now apparent; indeed, consider the case of equal configurations $c_{22}$ and $c_{23}$, say $c_{22} = c_{23} = c^*$. Then, the amplitude for $c^*$ is the sum of the amplitudes for $c_{22}$ and $c_{23}$, $a(c^*) = 0$, whence $p(c^*) = |a(c^*)|^2 = 0$. The corresponding destructive interference eliminates the configuration $c^*$ from the computation. On the other hand, assume $c_{21} = c_{24} = c_*$; these two configurations constructively interfere at level 2 so that $a(c_*) = 1$, whence $c_*$ appears among the computational steps with probability $p(c_*) = 1$.

The notion of *QTM* as a computing device working according to quantum mechanics was first proposed by Deutsch[99]. A full and detailed analysis can be found in [56] and [3] and further developments in connection with the notion of universality in [208]. In the following, we shall assume the existence of such machines and provide a schematic presentation of how they perform their tasks. *QTMs* work analogously to classical *TMs*, that is they consist of

1. An internal control unit $\mathbf{C}$ with associated Hilbert space $\mathbb{H}_{\mathbf{C}}$ linearly spanned by the classical control states $q_i$, $i = 1, 2, \ldots, |Q|$, the typical control vector being
$$|\Psi_{\mathbf{C}}\rangle = \sum_{i=1}^{|Q|} c(i)\,|q_i\rangle\ , \quad \sum_{i=1}^{|Q|} |c(i)|^2 = 1\ .$$
We shall distinguish special initial and final control states $q_0$ and $q_f$, respectively.
2. An input/output tape, whose vector states are of the form
$$|\Psi_{\mathbf{T}}\rangle = \sum_{\boldsymbol{\sigma}\in\Sigma^{\mathbb{Z}}} t(\boldsymbol{\sigma})\,|\boldsymbol{\sigma}\rangle\ ,$$
where $\boldsymbol{\sigma} \in \Sigma^{\mathbb{Z}}$ denotes any sequence consisting of infinitely many blanks and only finitely many symbols from the alphabet $\widetilde{\Sigma} = \{0, 1\}$ (see Section 4.1.1). The basis states $|\boldsymbol{\sigma}\rangle$ correspond to classical tape-configurations and span a (separable) tape Hilbert space $\mathbb{H}_{\mathbf{T}}$.
3. A read/write head $\mathbf{H}$ that can position itself on the tape cells labeled by the integers $k \in \mathbb{Z}$. The head Hilbert space $\mathbb{H}_{\mathbf{H}}$ is formed by square-summable sequences and the typical head vector state is
$$|\Psi_{\mathbf{C}}\rangle = \sum_{k\in\mathbb{Z}} h(k)\,|k\rangle\ , \quad \sum_{k\in\mathbb{Z}} |h(k)|^2 = 1\ .$$

A *QTM* $\mathfrak{U}$ will then be described by means of a Hilbert space of the form $\mathbb{H}_{\mathfrak{U}} = \mathbb{H}_{\mathbf{T}} \otimes \mathbb{H}_{\mathbf{C}} \otimes \mathbb{H}_{\mathbf{H}}$ with the configuration basis vectors $|\boldsymbol{\sigma}, q, k\rangle$ providing a distinguished orthonormal basis.

The time-evolution of standard quantum mechanical systems, that is isolated from their environment, is linear and reversible; as any step of a quantum computation corresponds, in absence of external noise, to a physical quantum process, it must be described by means of a unitary operator $U_{\mathfrak{U}} : \mathbb{H}_{\mathfrak{U}} \mapsto \mathbb{H}_{\mathfrak{U}}$.

**Remarks 9.1.1.**

1. The probabilistic transition functions (4.4) are replaced by a *quantum transition function* which assigns amplitudes (**not** probabilities) to the transitions $(q, \sigma) \mapsto (q', \sigma', d)$:
$$(q, \sigma; q', \sigma', d) \mapsto \delta(q, \sigma; q', \sigma', d) \in \widetilde{\mathbb{C}}^{[0,1]}\ , \tag{9.2}$$
where $\tilde{\mathbb{C}}$ denotes the set of complex numbers $\alpha \in \mathbb{C}$, $0 \leq |\alpha| \leq 1$, such that there is a deterministic algorithm that computes the real and imaginary parts of $\alpha$ to within any fixed precision $2^{-n}$ in time polynomial in $n$.

Consider the linear operator $U_{\mathfrak{U}}$ on $\mathbb{H}_{\mathfrak{U}}$ whose matrix elements with respect to the configuration basis vectors are defined by [229]

$$\langle q', \boldsymbol{\sigma}', k' | U_{\mathfrak{U}} | q, \boldsymbol{\sigma}, k\rangle = \begin{cases} \delta(q, \sigma_k; q', \sigma'_k, -1) & \text{if} \quad k' = k - 1 \\ \delta(q, \sigma_k; q', \sigma'_k, +1) & \text{if} \quad k' = k + 1 \,, \end{cases} \tag{9.3}$$

where $d = \pm 1$ identify a head's movement to the left $d = -1$), respectively to the right ($d = +1$), and the tape (classical) configurations $\boldsymbol{\sigma}$ and $\boldsymbol{\sigma}'$ are such that their symbols $\boldsymbol{\sigma}_j = \boldsymbol{\sigma}'_j$ for all $j \neq k$.

In this way the quantum transition function $\delta$ identifies the possible transitions operated by the linear operator $U_{\mathfrak{U}}$; namely, $\mathbb{U}_{\mathfrak{U}}$ operates a transition

$$\left.\begin{array}{c} \text{tape conf. : } \boldsymbol{\sigma} \\ \text{cell with head on it : } k \\ \text{symbol in cell } k : \sigma_k \end{array}\right\} \longmapsto \left\{\begin{array}{c} \text{tape conf. } \boldsymbol{\sigma}' \\ \text{cell with head on it: } k + d \\ \text{symbol in cell } k : \sigma'_k \end{array}\right.$$

if and only if $\delta(q, \sigma_k, q', \sigma'_k, d) \neq 0$.

Using the orthogonality of the configuration vectors, one explicitly computes the action of $U_{\mathfrak{U}}$ as

$$U_{\mathfrak{U}} | \, q, \boldsymbol{\sigma}, k \, \rangle = \sum_{q', \boldsymbol{\sigma}'_k, d} \delta(q, \sigma_k; q', \sigma'_k, k + d) \, | \, q', \boldsymbol{\sigma}'_k, j + d \, \rangle \,, \tag{9.4}$$

where $\boldsymbol{\sigma}'_k$ denotes the tape-configuration with all symbols equal to those of $\boldsymbol{\sigma}$, but for the $k$-th one. In [229] necessary and sufficient conditions are given on the quantum transition function $\delta$ so that $U_{\mathfrak{U}}$ acts unitarily on $\mathbb{H}_{\mathfrak{U}}$ and thus appropriately describes a quantum computation as a unitary discrete-time quantum evolution.

2. A possible model of a *QTM* is obtained via a quantum circuit consisting of unitary gates (see Example 5.5.9), a so-called *circuit model*. A quantum computation on, say, $N$ *qubits* thus amounts to a unitary operator $U$ acting on a $2^N$ dimensional Hilbert space $\mathbb{H}_N$. It requires a certain number of gates to be implemented; if one had at disposal all 1-*qubit* unitary gates plus the *CNOT* 2-*qubit* gate, then any $U$ would be exactly implementable [165]. In particular, the action of $U : \mathbb{H}_N \mapsto \mathbb{H}_N$ on a given state $| \, \psi \, \rangle$ requires $O(2^N)$ of these gates to be implemented [206]. More constructively, one seeks finite sets of gates $\mathcal{G}$ that would provide a so-called *complete gate basis* in the sense that the action of any 1-*qubit* gate can be mimicked by gates from $\mathcal{G}$ up to an arbitrary precision: one such set consists [165] of the *CNOT* gate, the Hadamard gate and the 1-*qubit* gate

$$T := \begin{pmatrix} \mathrm{e}^{-i\pi/8} & 0 \\ 0 & \mathrm{e}^{i\pi/8} \end{pmatrix} .$$

Consider a generic unitary action $U : \mathbb{H}_N \mapsto \mathbb{H}_N$ of a quantum circuit consisting of $m$ 1-*qubit* and *CNOT* gates; a result known as *Solovay-Kitaev theorem* [170, 306, 165, 206] states that that $U$ can be reproduced up to any $\varepsilon > 0$ by $O(m \log^c m/\varepsilon)$ gates from $\mathcal{G}$ with $c \in [1,2]$. Then, on a given $|\psi\rangle \in \mathbb{H}_N$, $\|(U-V)|\psi\rangle\| \leq \varepsilon$ where $V : \mathbb{H}_N \mapsto \mathbb{H}_N$ is a unitary operator corresponding to a quantum circuit consisting of $N(U,\varepsilon)$ gates from $\mathcal{G}$, where [206],

$$N(U,\varepsilon) = O\left(2^N \log^c\left(\frac{2^N}{\varepsilon}\right)\right) .$$

3. Another model of quantum computation is based on the possibility of implementing unitary operations on *qubits* by means of the mechanism outlined in Example 6.1.5. This latter is at the root of the so-called *one-way quantum computation* [249, 323], whereby quantum gates, that is unitary transformations, on *qubits* are implemented by performing measurements, that is irreversible operations, on some other *qubits*, all of them prepared in certain entangled multipartite states called *cluster states.*

**Definition 9.1.2 (*QTMs*: Starting and Evolution Conventions).** *Given a UQTM $\mathfrak{U}$ and an input qubit string $\sigma \in \mathbb{B}_1^+(\mathbb{H}_F)$, we shall identify it with the initial state of a quantum computation by $\mathfrak{U}$ again denoted by $\sigma$. It corresponds to a density matrix acting on $\mathbb{H}_{\mathfrak{U}}$ with $\sigma$ written on the input track over the cells indexed by $[0, l(\sigma)-1]$, and blank states # on the remaining cells of the input track and on the whole output track, while the control is in the distinguished initial state $q_0$ and the head is in the state corresponding to its being positioned upon the 0 cell. The state $\mathfrak{U}^t(\sigma)$ of $\mathfrak{U}$ on input $\sigma$ after $t \in \mathbb{N}_0$ computational steps will be given by $\mathfrak{U}^t(\sigma) := U_{\mathfrak{U}}^t \sigma \left(U_{\mathfrak{U}}^t\right)^\dagger$.*

In the rest of this section we shall deal with the halting conditions for *QTMs* and with showing that their actions amount to definite *quantum operations*, that is to trace-preserving completely positive maps on $\mathbb{B}_1^+(\mathbb{H}_F)$. For this observe that, in accordance to the previous definition, the state of the control after $t$ steps is given by partial trace over all the other parts of the machine, that is over the head and tape Hilbert spaces, $\mathbb{H}_{\mathbf{C}}$ and $\mathbb{H}_{\mathbf{T}}$, respectively, $\mathfrak{U}_{\mathbf{C}}^t(\sigma) := \mathrm{Tr}_{\mathbf{H},\mathbf{T}}\left(\mathfrak{U}^t(\sigma)\right)$.

**Definition 9.1.3 (*QTMs*: Halting Convention).** *A QTM $\mathfrak{U}$ halts at time $t \in \mathbb{N}_0$, that is after $t$ computational steps, on input $\sigma \in \mathbb{B}_1^+(\mathbb{H}_F)$, iff*

$$\langle q_f \,|\, \mathfrak{U}_{\mathbf{C}}^t(\sigma)\,| q_f\rangle = 1 \qquad \text{and} \quad \langle q_f \,|\, \mathfrak{U}_{\mathbf{C}}^{t'}(\sigma)\,| q_f\rangle = 0 \quad \text{for every } t' < t \;, \tag{9.5}$$

*where $q_f$ is a special control state.*

**Remark 9.1.2.** The above halting convention expresses the possibility of checking whether a *QTM* halts on a certain input by measuring the orthogonal projection $|\, q_f \,\rangle\langle\, q_f \,|$: if $\mathfrak{U}$ has halted on input $\sigma$ then, by measuring $|\, q_f \,\rangle\langle\, q_f \,|$, one ascertains this fact with certainty. On the other hand, if $\mathfrak{U}$ has not yet halted then measuring $|\, q_f \,\rangle\langle\, q_f \,|$ has no effect on the still going on computation. In general, for a generic input $\sigma = |\, \psi \,\rangle\langle\, \psi \,| \in \mathbb{B}_1^+(\mathbb{H}_F)$, $0 < \langle\, q_f \,|\mathfrak{U}^t_{\mathbf{C}}(|\, \psi \,\rangle\langle\, \psi \,|)\, |q_f \,\rangle < 1$; in such a case the vector $|\, \psi \,\rangle$ will be called non-halting, otherwise *t-halting*. Let $\tilde{\mathbb{H}}(t) \subset \mathbb{H}_F$ denote the set of vector inputs with equal halting time $t$: their linear combinations are also inputs such that $\mathfrak{U}$ halts on them at time $t$. Therefore, $\tilde{\mathbb{H}}(t)$ is a linear subspace of $\mathbb{H}_F$; what is more important, if $t \neq t'$, the corresponding subspaces $\mathbb{H}(t)$ and $\mathbb{H}(t')$ are mutually orthogonal. Indeed, were this not true, non-orthogonal vectors could be perfectly distinguished by means of their different halting times. It follows that the subset $\mathbb{B}_1^+(\mathbb{H}_F)$ on which $\mathfrak{U}$ halts is the union $\bigcup_{t\in\mathcal{N}} \mathbb{B}_1^+(\mathbb{H}(t))$.

It proves convenient to consider a special class of *QTMs* with the property that their tape $\mathbf{T}$ consists of two different tracks, an *input track* $\mathbf{I}$ and an *output track* $\mathbf{O}$. This can be achieved by having an alphabet which is a Cartesian product of two alphabets, in our case $\Sigma = \{0, 1, \#\} \times \{0, 1, \#\}$. Then, the tape Hilbert space $\mathbb{H}_{\mathbf{T}}$ can be written as $\mathbb{H}_{\mathbf{T}} = \mathbb{H}_{\mathbf{I}} \otimes \mathbb{H}_{\mathbf{O}}$.

**Definition 9.1.4 (Quantum Turing Machines).**
*A map $\boldsymbol{U} : \mathbb{B}_1^+(\mathbb{H}_F) \to \mathbb{B}_1^+(\mathbb{H}_F)$ will be called a QTM, if there is a two-track QTM $\mathfrak{U}$ with the following properties [56]:*

1. *the alphabet consists of $\Sigma = \{0, 1, \#\} \times \{0, 1, \#\}$;*
2. *the corresponding time evolution operator $U_{\mathfrak{U}}$ is unitary,;*
3. *if $\mathfrak{U}$ halts on input $\sigma$ with a variable-length qubit string $\rho \in \mathbb{B}_1^+(\mathbb{H}_F)$ on the output track starting in cell 0 such that the $i$–th cell is empty for every $i \notin [0, \ell(\rho) - 1]$, then $\boldsymbol{U}(\sigma) = \rho$; otherwise, $\boldsymbol{U}(\sigma)$ is undefined.*

In general, different inputs $\sigma$ have different halting times $t(\sigma)$ and the corresponding outputs result from different unitary transformations $U_{\mathfrak{U}}^{t(\sigma)}$. However, notice that the subset of $\mathbb{B}_1^+(\mathbb{H}_F)$ on which $\boldsymbol{U}$ is defined is of the form $\bigcup_{t\in\mathrm{N}} \mathbb{B}_1^+(\mathbb{H}(t))$. Therefore, by introducing an internal clock that keeps track of the halting times, the action of $\boldsymbol{U}$ restricted to this subset amounts to a well-defined quantum operation, that is to a completely positive map $\mathbb{U} : \mathbb{B}_1^+(\mathbb{H}_F) \to \mathbb{B}_1^+(\mathbb{H}_F)$.

**Lemma 9.1.1 (*QTMs* as Quantum Operations).**
*For every QTM $\mathfrak{U}$ there is a quantum operation $\mathbb{U} : \mathbb{B}_1^+(\mathbb{H}_F) \to \mathbb{B}_1^+(\mathbb{H}_F)$, such that $\mathfrak{U}(\sigma) = \mathbb{U}[\sigma]$ for every $\sigma \in \bigcup_{t\in\mathrm{N}} \mathbb{B}_1^+(\mathbb{H}(t))$.*

**Proof:** Let $\mathcal{B}_t$ be an orthonormal basis of $\mathbb{H}(t)$, $t \in \mathbb{N}$, and $\mathcal{B}_\perp$ an orthonormal basis in the orthogonal complement of $\bigoplus_{t\in\mathbb{N}} \mathbb{H}(t)$ within $\mathbb{H}_F$. Let an ancilla Hilbert space $\mathbb{H}_{\mathbf{A}} := \ell^2(\mathbb{N}_0)$ be added to the *QTM*, and define a linear operator $V_{\mathfrak{U}} : \mathbb{H}_F \to \mathbb{H}_{\mathfrak{U}} \otimes \mathbb{H}_{\mathbf{A}}$ by specifying its action on the orthonormal basis vectors $\cup_{t\in\mathbb{N}}\{\mathcal{B}_t\} \cup \mathcal{B}_\perp$:

$$V_{\mathfrak{U}} |\, b\,\rangle := \begin{cases} U^t_{\mathfrak{U}} |\, b\,\rangle \otimes |\, t\,\rangle \ \text{ if } |\, b\,\rangle \in \mathcal{B}_t , \\ \quad |\, b\,\rangle \otimes |\, 0\,\rangle \ \text{if} |\, b\,\rangle \in \mathcal{B}_\perp \ . \end{cases}$$

The ancilla acts as a sort of internal clock which registers the halting times of the components of a vectors belonging to the halting subspaces and assigns time 0 to the non-halting components. With $\mathcal{B}_t = \{|\, b^t_{j_t}\,\rangle\}$, $\mathcal{B}^\perp = \{|\, b^\perp_j\,\rangle\}$:

$$\mathbb{H}_{\mathfrak{U}} \otimes \mathbb{H}_{\mathbf{A}} \ni |\,\Psi\,\rangle = \sum_{t=0}^{\infty} \sum_{j_t} C^t_\Psi(j_t)\,|\, b^t_{j_t}\,\rangle + \sum_j C^\perp_\Psi(j)\,|\, b^\perp_j\,\rangle \ ,$$

$$V_{\mathfrak{U}} |\,\Psi\,\rangle = \sum_{t=0}^{\infty} \sum_{j_t} C^t_\Psi(j_t)\, U^t_{\mathfrak{U}} |\, b^t_{j_t}\,\rangle \otimes |\, t\,\rangle + \sum_j C^\perp_\Psi(j)\,|\, b^\perp_j\,\rangle \otimes |\, 0\,\rangle \ .$$

From orthogonality, it turns out that the map $V_{\mathfrak{U}}$ is a partial isometry:

$$\langle\,\Psi\,|\, V^\dagger_{\mathfrak{U}} V_{\mathfrak{U}}\,|\Phi\,\rangle = \langle\,\Psi\,|\,\Phi\,\rangle \ , \qquad \Psi, \Phi \in \mathbb{H}_{\mathfrak{U}} \otimes \mathbb{H}_{\mathbf{A}} \ .$$

Thus, the map $\sigma \mapsto V_{\mathfrak{U}} \sigma V^\dagger_{\mathfrak{U}}$ is trace-preserving and completely positive (see Section 5.2.2). Further, by partial tracing over the Hilbert spaces of the head, of the control unit, of the input tape and of the internal clock Hilbert spaces, one obtains the quantum operation $\mathbb{U}[\sigma] := \mathrm{Tr}_{\mathbf{C},\mathbf{H},\mathbf{I},\mathbf{A}}(V_{\mathfrak{U}} \sigma V^\dagger_{\mathfrak{U}})$. □

We have seen in Chapter 4 that the definition of algorithmic complexity rests on a solid ground because the length of the shortest effective description of a *bit* string is essentially independent of the computer that computes it once this is chosen from the class of universal Turing machines. Clearly, any definition of quantum complexity based on using *QTMs* will also need the existence of universal *QTMs* in order to be essentially machine-independent. In [56], a *UQTM* $\mathfrak{U}$ was constructed that works as follows: for any *QTM* $\mathfrak{A}$ there exists a classical description (*bit* string) $\boldsymbol{i}_{\mathfrak{A}}$ of $\mathfrak{A}$ such that

$$D\left(\mathfrak{U}(\boldsymbol{i}_{\mathfrak{A}}, T, |\,\psi\,\rangle\langle\,\psi\,|, \delta)\,,\, \mathfrak{A}^T(|\,\psi\,\rangle\langle\,\psi\,|)\right) \leq \delta \ ,$$

for all inputs $|\,\psi\,\rangle\langle\,\psi\,| \in \mathbb{B}^+_1(\mathbb{H}_F)$, computational steps $T$ and $\delta > 0$ with $D(\cdot,\cdot)$ the trace-distance introduced in Definition 6.3.4.

According to this definition $\mathfrak{U}$ is universal in that it simulates any other *QTM* up to an arbitrary accuracy for a given number of steps; notice that this latter piece of information must be part of the input. This means that, if $\mathfrak{A}$ halts on a certain input, $\mathfrak{U}$ is able to approximate the output of $\mathfrak{A}$ only if provided with the halting time. While such a definition works perfectly well for the aims of [56] which are directed to see the impact of *QTMs* on computational complexity (see Remark 4.1.2), it is on the other hand not appropriate for an approach to quantum algorithmic complexity simply because the halting times are likely to be enormous and in any case cannot be given beforehand.

A useful definition of a *UQTM* $\mathfrak{U}$ for algorithmic purposes must then be independent from the halting time of the simulated *QTM* $\mathfrak{A}$. The main problem is that, as the simulation is only approximate, such is in particular the simulation of the control state of $\mathfrak{A}$ whence, when $\mathfrak{A}$ halts, $\mathfrak{U}$ will in general do it only with a certain probability thus violating the halting convention in Definition 9.1.3. In [208] it is showed how such a problem can be circumvented and how one can arrive at the following operative definitions of *UQTM* which is fully consistent from the point of view of quantum algorithmic complexity.

**Theorem 9.1.1 (Strongly *UQTMs*).** *[208] There is a QTM $\mathfrak{U}$ such that for every QTM $\mathfrak{A}$ and every qubit string $\sigma$ for which $\mathfrak{A}(\sigma)$ is defined, there is a qubit string $\sigma_{\mathfrak{A}}$ such that*

$$D\left(\mathfrak{U}(\delta,\sigma_{\mathfrak{A}})\,,\,\mathfrak{A}(\sigma)\right)\leq\delta\qquad\forall\delta\in\mathbb{Q}^{+}\ ,$$

*where (see Definition 9.1.1) $\ell(\sigma_{\mathfrak{A}})\leq\ell(\sigma)+C_M$, $D(\cdot\,\cdot)$ is the trace-distance and $C_{\mathfrak{A}}\in\mathbb{N}$ is a constant depending only on $\mathfrak{A}$.*

In the following theorem, both the universal simulator, $\mathfrak{U}$, and the *QTM* to be simulated, $\mathfrak{A}$, are provided with a quantum input and a classical input fixing the accuracy $\delta$ of the approximation.

**Theorem 9.1.2 (Parameter Strongly *UQTMs*).** *[208] There is a UQTM $\mathfrak{U}$ with the properties of the previous theorem such that for every QTM $\mathfrak{A}$ and every qubit string $\sigma$, there is a qubit string $\sigma_{\mathfrak{A}}$ such that, if $\mathfrak{A}(2k,\sigma)$ is defined, then*

$$D\left(\mathfrak{U}(k,\sigma_{\mathfrak{A}})\,,\,\mathfrak{A}(2k,\sigma)\right)\leq\frac{1}{2k}\qquad\forall k\in\mathbb{N}\ ,$$

*and $\ell(\sigma_{\mathfrak{A}})\leq\ell(\sigma)+C_{\mathfrak{A}}$, $C_{\mathfrak{A}}\in\mathbb{N}$ depending only on $\mathfrak{A}$.*

This result is not just a corollary of the preceding theorem: indeed, according to Theorem 9.1.1, the input, $\sigma_{\mathfrak{A}}$, may in general depend on $k$.

**Remark 9.1.3.** A *UQTM* is able to apply a unitary transformation $U$ on some segment of its tape within an accuracy of $\delta$, if it is supplied with a complex matrix $\tilde{U}$ as input such that

$$\|U - \tilde{U}\| \leq \frac{\delta}{2(10\sqrt{d})^d} ,$$

$d$ being the size of the matrix. The machine cannot apply $U$ exactly; in fact, it only knows an approximation $\tilde{U}$. It also cannot apply $\tilde{U}$ directly, for $\tilde{U}$ is only approximately unitary, and the machine can only work unitarily. Instead, it will effectively apply another unitary transformation $V$ which is close to $\tilde{U}$ and thus close to $U$, such that $\|V - U\| < \delta$. Let $|\psi\rangle := U|\psi_0\rangle$ be the output that one wants to have from $\mathfrak{U}$ and let $|\phi\rangle := V|\psi_0\rangle$ be the approximation that is really computed by the machine. Then, both the norm and trace-distance are small: $\||\phi\rangle - |\psi\rangle\| < \delta$, $D\Big(|\phi\rangle\langle\phi|, |\psi\rangle\langle\psi|\Big) < \delta$.

## 9.2 *qubit* Quantum Complexity

As already remarked, unlike *bit* strings, *qubit* strings, are uncountably many and cannot be expected to be exactly reproducible by a *QTM* . It rather makes sense to try to approximate a target *qubit* string $\rho$ by a *qubit* string $\tilde{\rho}$ within a trace-distance $0 \leq D(\rho, \tilde{\rho}) \ll 1$ ($\tilde{\rho} \approx \rho$). According to the previous section, $\tilde{\rho}$ will be the output of a *QTM* $\mathfrak{U}$ that executes a quantum program $\sigma \in \mathbb{B}_1^+(\mathbb{H}_F)$: $\tilde{\rho} := \mathbb{U}[\sigma] \approx \rho$.

**Remark 9.2.1.** In view of the definition of classical algorithmic complexity, one is particularly interested to seek whether the length of $\sigma$ (see (9.1)) can be made shorter than that of $\rho$ itself: $\ell(\sigma) < \ell(\rho)$. The minimum possible length $\ell(\sigma)$ for reproducing $\rho$ will get us close to the notion of *qubit* quantum complexity $\mathrm{QC_q}$. There are at least two natural possible definitions. The first one is to demand only optimal (in the sense of minimal length) approximate reproductions of $\rho$ within some trace distance $\delta$. The second one is based on the notion of an approximation scheme. In order to define the latter, the chosen *QTM* has to be supplied with two inputs, the *qubit* string and a parameter.

**Definition 9.2.1.** *Let $k \in \mathbb{N}$ and $\sigma \in \mathbb{B}_1^+(\mathbb{H}_F)$. Let $\beta(k)$ denote the string that consists of the at most $\lfloor \log_2 k \rfloor$ bits of the binary expansion of $k$, each repeated twice and ends with 01. Let $|\,\beta(k)\,\rangle\langle\,\beta(k)\,|$ be the corresponding projector in the computational basis. The map $(k, \sigma) \mapsto C(k, \sigma) := |\,\beta(k)\,\rangle\langle\,\beta(k)\,| \otimes \sigma$ defines an encoding $C : \mathbb{N} \times \mathbb{B}_1^+(\mathbb{H}_F) \to \mathbb{B}_1^+(\mathbb{H}_F)$ of a the pair $(k, \sigma)$ into a single qubit string $C(k, \sigma)$. Note that*

$$\ell(C(k, \sigma)) = 2\lfloor \log k \rfloor + 2 + \ell(\sigma) . \tag{9.6}$$

*We shall denote by $\mathfrak{U}(k, \sigma)$ the result of the action of a QTM $\mathfrak{U}$ on $C(k, \sigma)$.*

The above encoding has the typical self-delimiting form that we have already met in Example 4.1.5. In this way, the *QTM* $\mathfrak{U}$ is able to detach in $\mathcal{C}(k,\sigma)$ the information about $k$ from that about $\sigma$.

**Definition 9.2.2** (*qubit* **Quantum Complexity).**
*Let $\mathfrak{U}$ be a QTM and $\rho \in \mathbb{B}_1^+(\mathbb{H}_F)$ a qubit string. For every $\delta \geq 0$, the finite-accuracy quantum complexity $\mathrm{QC}_{\mathfrak{U}}^{\delta}(\rho)$ is defined as the minimal length $\ell(\sigma)$ of any quantum program $\sigma \in \mathbb{B}_1^+(\mathbb{H}_F)$ such that the corresponding output $\mathfrak{U}(\sigma)$ has a trace-distance from $\rho$ smaller than $\delta$,*

$$\mathrm{QC}_{\mathfrak{U}}^{\delta}(\rho) := \min\Big\{\ell(\sigma)\ :\ D\left(\rho, \mathfrak{U}(\sigma)\right) \leq \delta\Big\}\ . \tag{9.7}$$

*Similarly, an approximation-scheme quantum complexity $\mathrm{QC}_{\mathfrak{U}}$ is defined as the minimal length $\ell(\sigma)$ of any density operator $\sigma \in \mathbb{B}_1^+(\mathbb{H}_F)$, such that when processed by $\mathfrak{U}$ together with any integer $k$, the output $\mathfrak{U}(k,\sigma)$ has trace-distance from $\rho$ smaller than $1/k$, for all $k$:*

$$\mathrm{QC}_{\mathfrak{U}}(\rho) := \min\Big\{\ell(\sigma)\ :\ D\left(\rho\,,\,\mathfrak{U}(k,\sigma)\right) \leq \frac{1}{k}\ \textit{for every}\ k \in \mathbb{N}\Big\}\ . \tag{9.8}$$

We now show that theorems 9.1.1 and 9.1.2 allow one to prove the independence (up to an additive constant) of the above definitions from the chosen *QTM* $\mathfrak{U}$ if this is universal as specified in those theorems. Accordingly, we will fix an arbitrary *UQTM* and, like in the classical case, drop reference to it and set

$$\mathrm{QC}_q\left(\rho\right) := QC_{\mathfrak{U}}(\rho)\ ,\quad \mathrm{QC}_q^{\delta}\left(\rho\right) := \mathrm{QC}_{\mathfrak{U}}^{\delta}(\rho)\ . \tag{9.9}$$

**Theorem 9.2.1.** *There is a QTM $\mathfrak{U}$ such that for every QTM $\mathfrak{A}$ there exist constants $C_{\mathfrak{A}} \geq 0$ and $C_{\mathfrak{A},\delta,\Delta}$ such that for every qubit string $\rho \in \mathbb{B}_1^+(\mathbb{H}_F)$ and $0 \leq \delta < \Delta$, it holds that*

$$\mathrm{QC}_{\mathfrak{U}}(\rho) \leq \mathrm{QC}_{\mathfrak{A}}(\rho) + C_{\mathfrak{A}}\ ,\quad \mathrm{QC}_{\mathfrak{U}}^{\Delta}(\rho) \leq \mathrm{QC}_{\mathfrak{A}}^{\delta}(\rho) + C_{\mathfrak{A},\delta,\Delta}\ .$$

**Proof:** Let $\ell = \mathrm{QC}_{\mathfrak{A}}^{\delta}(\rho)$, then there exists $\sigma$ such that, according to (9.7), $\ell = \ell(\sigma)$ and $D(\mathfrak{A}[\sigma],\rho) \leq \delta$. On the other hand, Theorem 9.1.1 implies that there exists a *QTM* $\mathfrak{U}$ and a density matrix $\sigma_{\mathfrak{A}}$ such that

$$D\Big(\mathfrak{U}(\Delta-\delta,\sigma_{\mathfrak{A}})\,,\,\mathfrak{A}(\sigma)\Big) \leq \Delta-\delta$$

whence, by the triangle inequality,

$$D\Big(\mathfrak{U}(\Delta-\delta,\sigma_{\mathfrak{A}})\,,\,\rho\Big) \leq \Delta\ .$$

Moreover, $\ell(\sigma_{\mathfrak{A}}) \leq \ell(\sigma) + C_{\mathfrak{A}} = \mathrm{QC}^{\delta}_{\mathfrak{A}}(\rho) + C_{\mathfrak{A}}$; thus, with $\mathcal{C}(\Delta - \delta, \sigma_{\mathfrak{A}})$ as in Definition 9.2.1, using (9.6) it follows that

$$\ell\Big(\mathcal{C}(\Delta - \delta, \sigma_{\mathfrak{A}})\Big) \leq \ell(\sigma_{\mathfrak{A}}) + C_{\delta,\Delta} \leq \mathrm{QC}^{\delta}_{\mathfrak{A}}(\rho) + C_{\mathfrak{A},\delta,\Delta} \cdot$$

whence $\mathrm{QC}^{\Delta}_{\mathfrak{U}}(\rho) \leq \mathrm{QC}^{\delta}_{\mathfrak{A}}(\rho) + C_{\mathfrak{A},\delta,\Delta}$.

If $\ell = \mathrm{QC}_{\mathfrak{A}}(\rho)$, then there exists a *qubit* string $\sigma$ such that $\ell = \ell(\sigma)$ and $D(\mathfrak{A}(k,\sigma),\rho) \leq 1/k$ for all $k \in \mathbb{N}$. On the other hand, Theorem 9.1.2 says that there exists a *QTM* $\mathfrak{U}$ and a density matrix $\sigma_{\mathfrak{A}}$ such that $D\Big(\mathfrak{U}(k,\sigma_{\mathfrak{A}})\,,\,\mathfrak{A}(2k,\sigma)\Big) \leq \frac{1}{2k}$. It follows that

$$\begin{aligned} D\Big(\mathfrak{U}(k,\sigma_{\mathfrak{A}})\,,\,\rho\Big) &\leq D\Big(\mathfrak{U}(k,\sigma_{\mathfrak{A}})\,,\,\mathfrak{A}(2k,\sigma)\Big) + D\Big(\mathfrak{A}(2k,\sigma)\,,\,\rho\Big) \\ &\leq \frac{1}{2k} + \frac{1}{2k} = \frac{1}{k} \,. \end{aligned}$$

Together with the fact that $\ell(\sigma_{\mathfrak{A}}) \leq \ell(\sigma) + C_{\mathfrak{A}} \leq \mathrm{QC}_{\mathfrak{A}}(\rho) + C_{\mathfrak{A}}$, this implies $\mathrm{QC}_{\mathfrak{U}}(\rho) \leq \mathrm{QC}_{\mathfrak{A}}(\rho) + C_{\mathfrak{A}}$. □

**Remarks 9.2.2.**

1. Definition 9.2.2 is essentially equivalent to that in [57], the only technical difference being the use of the trace distance rather than the fidelity.
2. The *same qubit* program $\sigma$ is accompanied by a classical specification of an integer $k$, which tells the program to what accuracy the computation of the output state must be accomplished. Notice that in (9.8) the minimal length has to be sought among those $\sigma$ such that anyone of them yields an approximation of $\rho$ within $1/k$ for all $k$: this is an effective procedure.
3. The exact choice of the accuracy $1/k$ is not important; choosing any computable function that tends to zero for $k \to \infty$ will get an equivalent definition (in the sense of being equal up to some constant). The same is true for the choice of the encoding $\mathcal{C}$: as long as $k$ and $\sigma$ can both be computably decoded from $\mathcal{C}(k,\sigma)$ and as long as there is no way to extract additional information on the desired output $\rho$ from the $k$-description part of $\mathcal{C}(k,\sigma)$, the results will be equivalent up to a suitable constant.

**Examples 9.2.1.**

1. If $\mathfrak{U}$ is a *UQTM*, a noiseless transmission channel (implementing the identity transformation) between the input and output tracks can always be realized: this corresponds to classical literal transcription, so that automatically $\mathrm{QC}^{\delta}_{\mathfrak{U}}(\rho) \leq \ell(\rho) + c_U$ for some constant $c_U$. Of course, the key point in classical as well as in quantum algorithmic complexity is that there sometimes exist much shorter *qubit* programs than just literal transcription.

2. The finite accuracy and approximation scheme $\mathrm{QC_q}$ are related to each other by the following inequality: for every *QTM* $\mathfrak{U}$ and every $k \in \mathcal{N}$,

$$\mathrm{QC}_q^{1/k}(\rho) \le \mathrm{QC}_q(\rho) + 2\lfloor \log k \rfloor + 2, \qquad \forall \rho \in \mathbb{B}_1^+(\mathbb{H}_F) .$$

Indeed, if $\mathrm{QC}_{\mathfrak{U}}(\rho) = \ell$, there is $\sigma \in \mathbb{B}_1^+(\mathbb{H}_F)$ with $\ell(\sigma) = \ell$, such that $D(\mathbb{U}[k,\sigma],\rho) \le 1/k$ for every $k \in \mathbb{N}$. Then $\sigma' := \mathcal{C}(k,\sigma)$, where $\mathcal{C}$ is the encoding in Definition 9.2.1, is such that $D(\mathbb{U}[\sigma'],\rho) \le 1/k$ and

$$\mathrm{QC}_q^{1/k}(\rho) \le \ell(\sigma') \le 2\lfloor \log k \rfloor + 2 + \ell = 2\lfloor \log k \rfloor + 2 + \mathrm{QC}_q(\rho) ,$$

where the second equality follows from (9.6).

### 9.2.1 Quantum Brudno's Theorem

In this section, we prove a quantum version of Brudno's theorem (Theorem 4.2.1), by means of which we shall connect the quantum entropy rate $s$ of an ergodic quantum spin chain to the *qubit* complexities $\mathrm{QC}_q(\rho)$ and $\mathrm{QC}_q^\delta(\rho)$ of *qubit* strings that are pure states $\rho = |\psi\rangle\langle\psi|$ of the chain. It will be showed that there are sequences of typical subspaces of $(\mathbb{C}^2)^{\otimes n}$, such that the complexity rates $\frac{1}{n}\mathrm{QC}_q(q)$ and $\frac{1}{n}\mathrm{QC}_q^\delta(q)$ of any one-dimensional projector $q$ onto a state belonging to these subspaces can be made arbitrarily close to the entropy rate by choosing $n$ large enough. Moreover, there are no such sequences with a smaller expected complexity rate.

**Theorem 9.2.2 (Quantum Brudno's Theorem).**
*Let $(\mathcal{A},\omega)$ be an ergodic quantum source with entropy rate $s$. For every $\delta > 0$, there exists a sequence of $\omega$-typical projectors $q_n(\delta) \in \mathcal{A}^{(n)}$, $n \in \mathbb{N}$, i.e. $\lim_{n\to\infty} \mathrm{Tr}(\rho^{(n)} q_n(\delta)) = 1$, such that for every one-dimensional projector $q \le q_n(\delta)$ and $n$ large enough*

$$\frac{1}{n}\mathrm{QC}_q(q) \in (s-\delta, s+\delta) , \tag{9.10}$$

$$\frac{1}{n}\mathrm{QC}_q^\delta(q) \in (s-\delta(2+\delta)s, s+\delta) . \tag{9.11}$$

*Moreover, $s$ is the optimal expected asymptotic complexity rate, in the sense that every sequence of projectors $q_n \in \mathcal{A}^{(n)}$, $n \in \mathbb{N}$, that for large $n$ may be represented as a sum of mutually orthogonal one-dimensional projectors that all violate the lower bounds in (9.10) and (9.11) for some $\delta > 0$, has an asymptotically vanishing expectation value with respect to $\omega$.*

As for the proof of Brudno's theorem 9.2.2, we first prove upper and then lower bounds [40].

### Lower Bounds

In the classical case, it has been showed that there cannot be more than $2^{c+1} - 1$ different programs of length $\ell \leq c$ and this fact has been used to prove the lower bound to complexity in Brudno's Theorem.

A similar result holds for *QTMs* , too. In order to show this, one can adapt an argument due to [57] which states that there cannot be more than $2^{\ell+1} - 1$ mutually orthogonal one-dimensional projectors $p$ with quantum complexity $\mathrm{QC}_q(p) \leq \ell$. The proof is based on the Holevo's $\chi$-quantity (see Proposition 6.3.3); we shall use it to provide an explicit upper bound on the maximal number of orthogonal one-dimensional projectors that can be approximated within trace-distance $\delta$ by the action of completely positive maps $\mathbb{E}$ on density matrices $\sigma$ of length $\ell(\sigma) \leq c$.

**Lemma 9.2.1 (Quantum Counting Argument).**
*Let $0 < \delta < 1/e$, $c \in \mathbb{N}$ such that $c \geq \frac{1}{\delta}\left(4 + 2\log\frac{1}{\delta}\right)$, $\mathcal{K}$ a linear subspace of an arbitrary Hilbert space $\mathbb{K}$, and $\mathbb{E} : \mathbb{B}_1^+(\mathbb{H}_F) \to \mathbb{B}_1^+(\mathbb{K})$ a quantum operation. Let $N_c^\delta$ be a maximum cardinality subset of orthonormal vectors from the set $A_c^\delta(\mathbb{E}, \mathcal{K})$ of all normalized vectors in $\mathcal{K}$ which are reproduced within $\delta$ by the operation $\mathbb{E}$ on some input of length $\leq c$:*

$$A_c^\delta(\mathbb{E}, \mathcal{K}) := \left\{ |\,\phi\,\rangle \in \mathcal{K} : \ \exists \sigma_\phi \in \mathbb{B}_1^+(\mathbb{H}_{\leq c}), D\left(\mathbb{E}[\sigma_\phi], |\,\phi\,\rangle\langle\,\phi\,|\right) \leq \delta \right\} ,$$

*Then,* $\log_2 |N_c^\delta| < c + 1 + \dfrac{2+\delta}{1-2\delta}\delta c.$

**Proof:** Let $\phi_j \in A_c^\delta(\mathbb{E}, \mathcal{K})$, $j = 1, \ldots, N$, a set of orthonormal vectors and $\mathcal{V}$ denote the Abelian subalgebra of $\mathcal{B}(\mathbb{K})$ generated by the corresponding projectors $P_j := |\phi_j\rangle\langle\phi_j|$ and $P_{N+1} := \mathbf{1}_\mathbb{K} - \sum_{i=1}^N P_i$. By the definition of $A_c^\delta(\mathbb{E}, \mathcal{K})$, for every $1 \leq i \leq N$, there are density matrices $\sigma_i$ acting on $\mathbb{H}_{\leq c}$ with $D(\mathbb{E}[\sigma_i], P_i) \leq \delta$.

Let $\sigma := \dfrac{1}{N}\sum_{i=1}^N \sigma_i$; it also acts on $\mathbb{H}_{\leq c}$ and $\dim \mathbb{H}_{\leq c} = 2^{c+1} - 1$, whence (6.33) yields $\chi(\mathcal{E}_\sigma) < c+1$, where $\mathcal{E}_\sigma := \{\sigma, \sigma_i/N\}$. Then, consider the completely positive map $\mathbb{E}_\mathcal{V} : \mathbb{B}_1^+(\mathbb{K}) \to \mathbb{B}_1^+(\mathbb{K})$, $\rho \mapsto \mathbb{E}_\mathcal{V}[\rho] := \sum_{i=1}^{N+1} P_i \rho P_i$. Applying twice the monotonicity of the relative entropy under completely positive maps,

$$\begin{aligned}\frac{1}{N}\sum_{i=1}^N S\left(\mathbb{E}_\mathcal{V} \circ \mathbb{E}[\sigma_i], \mathbb{E}_\mathcal{V} \circ \mathbb{E}[\sigma]\right) &\leq \frac{1}{N}\sum_{i=1}^N S\left(\mathbb{E}[\sigma_i], \mathbb{E}[\sigma]\right) \\ &\leq \chi(\mathcal{E}_\sigma) .\end{aligned}$$

For every $i \in \{1, \ldots, N\}$, the density matrix $\mathbb{E}_\mathcal{V} \circ \mathbb{E}[\sigma_i]$ is close to the corresponding one-dimensional projector $\mathbb{E}_\mathcal{V}[P_i] = P_i$. Indeed, (6.68) yields

$$D(\mathbb{E}_{\mathcal{V}} \circ \mathbb{E}[\sigma_i], \mathbb{E}_{\mathcal{V}}[P_i]) \leq D(\mathbb{E}[\sigma_i], P_i) \leq \delta \ .$$

Let $\Delta := \frac{1}{N}\sum_{i=1}^{N} P_i$. The trace-distance is jointly convex (see (6.69)), thus

$$D(\mathbb{E}_V \circ \mathbb{E}[\sigma], \Delta) \leq \frac{1}{N}\sum_{i=1}^{N} D(\mathbb{E}_{\mathcal{V}} \circ \mathbb{E}[\sigma_i]), P_i) \leq \delta \ .$$

Since $\delta < \frac{1}{e}$, Fannes inequality (5.157) gives

$$S\left(\mathbb{E}_{\mathcal{V}} \circ \mathbb{E}[\sigma_i]\right) = \left|S\left(\mathbb{E}_{\mathcal{V}} \circ \mathbb{E}[\sigma_i]\right) - S(P_i)\right| \leq \delta \log_2(N+1) + \eta(\delta)$$

$$\Big|S\left(\mathbb{E}_{\mathcal{V}} \circ \mathbb{E}[\sigma]\right) - S(\Delta)\Big| \leq \delta \log(N+1) + \eta(\delta) \ ,$$

where $\eta(\delta) := -\delta \log_2 \delta$. Combining the previous estimates yields

$$c + 1 > \chi(\mathcal{E}_\sigma) \geq (1 - 2\delta) \log_2 N - 2\delta - 2\eta(\delta) \ .$$

If $\log_2 N \geq c+1+\frac{2+\delta}{1-2\delta}\delta c$, then $c+1 > c+1+\delta(c\delta-4)+2\delta\log\delta$, whence $c < \frac{2}{\delta}\left(2 + \log\frac{1}{\delta}\right)$. Therefore, the maximum number $|N_c^\delta|$ of orthonormal vectors in $A_c^\delta(\mathcal{E}, \mathcal{K})$ must fulfil $\log_2 |N_c^\delta| < c + 1 + \dfrac{2+\delta}{1-2\delta}\delta c$. □

The second step uses the previous lemma together with Proposition 7.3.1 about the minimum dimension of the typical subspaces. Notice that the limit (7.153) is valid for all $0 < \varepsilon < 1$. By means of this property, one proves the lower bound for the finite-accuracy complexity $\mathrm{QC}_q^\delta(\rho)$, and then use Example 9.2.1.2 to extend it to $\mathrm{QC}_q(\rho)$.

**Corollary 9.2.1 (Lower Bound for $\frac{1}{n}\mathrm{QC}_q^\delta(\rho)$).**

*Let $(\mathcal{A}_{\mathbb{Z}}, \omega)$ be an ergodic quantum source with entropy rate $s$. Further, let $0 < \delta < 1/e$, and let $(p_n)_{n\in\mathbb{N}}$ be a sequence of typical projectors, according to Definition 7.3.2. Then, there is another sequence of typical projectors $\widetilde{p}_n(\delta) \leq p_n$, such that for $n$ large enough*

$$\frac{1}{n}\mathrm{QC}_q^\delta(\widetilde{p}) > s - \delta(2+\delta)s$$

*is true for every one-dimensional projector $p \leq \widetilde{p}_n(\delta)$.*

**Proof:** The case $s = 0$ is trivial, so let $s > 0$. Fix $n \in \mathbb{N}$, $0 < \delta < 1/e$ and consider the set

$$\tilde{A}_n(\delta) := \Big\{p \leq p_n \ : \ p = |\,\psi\,\rangle\langle\,\psi\,| \,,\ \mathrm{QC}_q^\delta(p) \leq ns(1 - \delta(2+\delta))\Big\} \ .$$

From the definition of $\mathrm{QC}_q^\delta(p)$, for any of such $p$'s there exists a density matrix $\sigma_p$ with $\ell(\sigma_p) \leq ns(1-\delta(2+\delta))$ such that $D(\mathfrak{U}(\sigma_p), p) \leq \delta$, where, as explained in Lemma 9.1.1, $\mathfrak{U}(\sigma_p)$ is the result of the quantum operation $\mathbb{U} : \mathbb{B}_1^+(\mathbb{H}_F) \to \mathbb{B}_1^+(\mathbb{H}_F)$ associated with the *UQTM* $\mathfrak{U}$ that has been fixed as explained before Theorem 9.1.1. Then, using the notation of Lemma 9.2.1, $\tilde{A}_n(\delta) \subset A^{\delta}_{\lceil ns(1-\delta(2+\delta))\rceil}(\mathbb{U}, \mathbb{K}_n)$, where $\mathbb{K}_n$ is the typical subspace supporting $p_n$. Let $p_n(\delta) \leq p_n$ be the sum of any maximal number of mutually orthogonal projectors from $A^{\delta}_{\lceil ns(1-\delta(2+\delta))\rceil}(\mathbb{U}, \mathbb{K}_n)$. If $n$ is such that

$$ns(1-\delta(2+\delta)) \geq \frac{1}{\delta}\left(4 + 2\log_2 \frac{1}{\delta}\right) ,$$

Lemma 9.2.1 implies that

$$\log_2 \operatorname{Tr} p_n(\delta) < \lceil ns(1-\delta(2+\delta))\rceil + 1 + \frac{2+\delta}{1-2\delta}\delta\lceil ns(1-\delta(2+\delta))\rceil \ . \tag{9.12}$$

Therefore, no one-dimensional projectors $p \leq p_n(\delta)^{\perp} := p_n - p_n(\delta)$ exist such that $p \in A^{\delta}_{\lceil ns(1-\delta(2+\delta))\rceil}(\mathbb{U}, \mathbb{K}_n)$. Namely, one-dimensional projectors $p \leq p_n(\delta)^{\perp}$ must satisfy

$$\frac{1}{n}\mathrm{QC}^{\delta}_q(p) > s - \delta(2+\delta)s \ .$$

Since inequality (9.12)) is valid for every $n \in \mathbb{N}$ large enough,

$$\limsup_{n\to\infty} \frac{1}{n} \log_2 \operatorname{Tr}_n p_n(\delta) \leq s - 2\delta^3 s - \frac{5\delta^4 s}{1-2\delta} < s. \tag{9.13}$$

From Proposition 7.3.1 $\lim_{n\to\infty} \operatorname{Tr}(\rho^{(n)} p_n(\delta)) = 0$, whence $\widetilde{p}_n(\delta) := p_n(\delta)^{\perp}$ provide the required sequence of typical projectors. □

**Corollary 9.2.2 (Lower Bound for $\frac{1}{n}\mathrm{QC}_q(\rho)$).**
*Let $(\mathcal{A}_{\mathbb{Z}}, \omega)$ be an ergodic quantum source with entropy rate $s$. Let $(p_n)_{n\in\mathbb{N}}$ with $p_n \in \mathcal{A}^{(n)}$ be an arbitrary sequence of typical projectors. Then, for every $0 < \delta < 1/e$, there is a sequence of typical projectors $\widetilde{p}_n(\delta) \leq p_n$ such that, for $n$ large enough, $\frac{1}{n}\mathrm{QC}_q(p) > s - \delta$ is satisfied for every one-dimensional projector $p \leq \widetilde{p}_n(\delta)$.*

**Proof:** From Corollary 9.2.1, for every $k \in \mathbb{N}$, there exists a sequence of typical projectors $p_n(\frac{1}{k}) \leq p_n$, such that, if $n$ is large enough,

$$\frac{1}{n}\mathrm{QC}^{1/k}_q(p) > s - \frac{1}{k}(2 + \frac{1}{k})s$$

for every one-dimensional projector $p \leq p_n(1/k)$. Then

$$\frac{1}{n}\mathrm{QC}_q(p) \geq \frac{1}{n}\mathrm{QC}_q^{1/k}(p) - \frac{2+2\lfloor\log_2 k\rfloor}{n}$$
$$> s - \frac{1}{k}\left(2+\frac{1}{k}\right)s - \frac{2(2+\log_2 k)}{n} ,$$

where the first estimate is by Example 9.2.1.2 and the second one is true for one-dimensional projectors $p \leq p_n(\frac{1}{k})$ and $n \in \mathbb{N}$ large enough. Fix a large $k$ satisfying $\frac{1}{k}(2+\frac{1}{k})s \leq \frac{\delta}{2}$. The result follows by setting $\widetilde{p}_n(\delta) = p_n(\frac{1}{k})$ with $k$, and $n$ such that

$$\frac{1}{k}(2+\frac{1}{k})s \leq \frac{\delta}{2} , \quad \frac{2(2+\log_2 k)}{n} \leq \frac{\delta}{2} .$$

□

## Upper Bounds

The lower bound shows that, for large $n$, with high probability the *qubit* complexity of pure states of a quantum spin chain is bounded from below by a quantity which is close to the entropy rate of the chain. Similar upper bounds also hold from which Theorem 9.2.2 follows.

**Proposition 9.2.1 (Upper Bound).**
*Let $(\mathcal{A}_{\mathbb{Z}}, \omega)$ be an ergodic quantum source with entropy rate $s$. Then, for every $0 < \delta < 1/e$, there is a sequence of typical projectors $p_n(\delta) \in \mathcal{A}^{(n)}$ such that for every one-dimensional projector $p \leq p_n(\delta)$ and $n$ large enough*

$$\frac{1}{n}\mathrm{QC}_q(p) < s + \delta \qquad \textit{and} \qquad \frac{1}{n}\mathrm{QC}_q^{\delta}(p) < s + \delta .$$

The proof of this statement is obtained by explicitly providing, for any minimal projector $p \leq p_n(\delta) \in \mathcal{A}^{(n)}$, a *qubit* string $\widetilde{p}$ of length $\ell(\widetilde{p}) \simeq n(s+\delta)$, that computes $p$ with arbitrary accuracy. Such a *qubit* string is constructed by means of universal quantum typical subspaces introduced (see Definition 7.3.3); its length is in general not minimal and only upperbounds the quantum complexities $\mathrm{QC}_q^{\delta}(p)$ and $\mathrm{QC}_q(p)$. However, it is shorter than the literal transcription of $p$ (see Example 9.2.1.2): recall that the latter corresponds to a *qubit* string $\widetilde{p}$ comprising $p$ itself plus the instructions to the *UQTM* $\mathfrak{U}$ to copy $p$, whence its length $\ell(\widetilde{p}) \simeq n > n(s+\delta)$ for $n$ large enough and $\delta$ sufficiently small.

Let $0 < \varepsilon < \delta/2$ be an arbitrary real number such that $r := s + \varepsilon$ is rational, and let $\{\widetilde{p}_n := Q_{s,\varepsilon}^{(n)}\}_{n\in\mathbb{N}}$ be the universal projector sequence of Theorem 7.3.3, which is *independent* of the given state $\omega$ as long as $s(\omega) \leq s$.

Though the dimension of the subspace supporting $\widetilde{p}_n$ is $\simeq 2^{nr}$, generic one dimensional projections $\widetilde{q} \leq \widetilde{p}_n$ are not *qubit* strings as defined in

Section 9.1.1 and their lengths need not be $\ell(\widetilde{q}) \simeq nr$. However, because of (7.154), if $n$ is large enough then there exists some unitary transformation $U^\dagger$ that transforms the projector $\widetilde{p}_n$ into a projector belonging to the state-space $\mathbb{B}_1^+(\mathbb{H}_{\lceil nr \rceil})$, where $\lceil nr \rceil$ is the smallest integer larger then $nr$. It follows that every one-dimensional projector $\widetilde{p} \leq \widetilde{p}_n$ can be transformed into a *qubit* string $p := U^\dagger \widetilde{p} U$ of length $\ell(p) = \lceil nr \rceil$.

According to Remark 9.1.3, $p$ can be presented to the *UQTM* $\mathfrak{U}$ together with some classical instructions including a subprogram for the computation of the necessary unitary rotation $U$. This *UQTM* starts by computing a classical description of the transformation $U$, and subsequently applies $U$ to $p$, recovering the original projector $\widetilde{p} = U\,p\,U^\dagger$ on the output tape.

Apart from technical details, the main point in the proof is the following: since the unitary operator $U$ depends on $\omega$ only through the entropy rate $s$, the subprogram that computes $U$ does not have to be supplied with additional information on $\omega$ and its restriction to $\mathcal{A}^{(n)}$. Therefore, the additional instruction for the implementation of $U$ will contribute with a number of extra *qubits* which is independent of the universal projection index $n$.

The quantum decompression algorithm $\mathfrak{D}$ will formally amount to a mapping ($r$ is rational)

$$\mathfrak{D} : \mathbb{N} \times \mathbb{N} \times \mathbb{Q} \times \mathbb{H}_F \to \mathbb{H}_F \ , \ (k, n, r, \widetilde{p}) \mapsto p = \mathfrak{D}(k, n, r, \widetilde{p}) \ .$$

**Remark 9.2.3.** The decompression algorithm $\mathfrak{D}$ is due to be *short* in the sense of being "short in description", not short (fast) in running time or resource consumption. Indeed, the algorithm $\mathfrak{D}$ is in general slow and memory consuming; however, this does not matter. In fact, algorithmic complexity only cares about the length of the programs and not either in how fast they are computed or in how much resources they consume.

In the following steps, $\mathfrak{D}$ will deal with rational numbers, square roots of rational numbers, *bit*-approximations (up to some specified accuracy) of real numbers and vectors and matrices containing such numbers. Classical *TMs* as well *QTMs* can of course deal with all such objects. For example, rational numbers can be stored as lists of two integers (containing numerator and denominator), square roots can be stored as such lists supplemented with an additional bit to denote the square root operation, and, also, binary-digit-approximations can be stored as binary strings. Vectors and matrices are arrays containing those objects. They will be presented to the *UQTM* $\mathfrak{U}$ as vectors of the computational basis and operations on them, like addition or multiplication, will as easily be implemented as by classical computers.

The instructions defining the quantum algorithm $\mathfrak{D}$ are as follows.

1. Read $n, r$; find $\ell \in \mathbb{N}$ such that $\ell \cdot 2^{3\ell} \leq n < 2 \cdot \ell \cdot 2^{3 \cdot 2\ell}$ with $\ell$ a power of two (there is only one such $\ell$). Compute $\widetilde{n} := \lfloor \frac{n}{\ell} \rfloor$. Compute $R := r\ell$.
2. Compute a list of codewords $\Omega_{\ell,R}^{(\widetilde{n})}$, belonging to a classical universal block code sequence of rate $R$. The construction of an appropriate algorithm can be found for instance in [168].

Since $\Omega_{l,R}^{(\tilde{n})} \subset (\{0,1\}^l)^{\tilde{n}}$, $\Omega_{l,R}^{(\tilde{n})} = \{\omega_1, \omega_2, \ldots, \omega_M\}$ can be stored as a list of binary strings. Every string has length $\ell(\omega_i) = \tilde{n}l$ and the exact value of the cardinality $M \approx 2^{\tilde{n}R}$ depends on the choice of $\Omega_{\ell,R}^{(\tilde{n})}$.

3. Compute a basis $\{A_{\{i_1,\ldots,i_{\tilde{n}}\}}\}$ of the symmetric subspace

$$SYM^{\tilde{n}}(\mathcal{A}^{(\ell)}) := \mathrm{span}\{A^{\otimes\tilde{n}} \,:\, A \in \mathcal{A}^{(\ell)}\} .$$

Namely, for every $\tilde{n}$-tuple $\{i_1, \ldots, i_{\tilde{n}}\}$, where $i_k \in \{1, \ldots, 2^{2\ell}\}$, there is one basis element $A_{\{i_1,\ldots,i_{\tilde{n}}\}} \in \mathcal{A}^{(\tilde{n}\ell)}$, given by

$$A_{\{i_1,\ldots,i_{\tilde{n}}\}} = \sum_\sigma e^{(\ell,\tilde{n})}_{\sigma(i_1,\ldots,i_{\tilde{n}})} , \tag{9.14}$$

where the summation runs over all $\tilde{n}$-permutations $\sigma$, and

$$e^{(\ell,\tilde{n})}_{i_1,\ldots,i_{\tilde{n}}} := e^{(\ell)}_{i_1} \otimes e^{(\ell)}_{i_2} \otimes \ldots \otimes e^{(\ell)}_{i_{\tilde{n}}} ,$$

with $\left\{e_k^{(\ell)}\right\}_{k=1}^{2^{2\ell}}$ a system of matrix units in $\mathcal{A}^{(\ell)}$. In the computational basis, all entries of such matrices are zero, except for one entry which is one. There is a number of $d = \binom{\tilde{n}+2^{2l}-1}{2^{2\ell}-1} = \dim(SYM^{\tilde{n}}(\mathcal{A}^{(\ell)}))$ different matrices $A_{\{i_1,\ldots,i_{\tilde{n}}\}}$ which we can label by $\{A_k\}_{k=1}^d$. It follows from (9.14) that these matrices have integer entries and can thus be stored as lists of $2^{\tilde{n}\ell} \times 2^{\tilde{n}\ell}$-tables of integers without any need of approximations.

4. For every $i \in \{1, \ldots, M\}$ and $k \in \{1, \ldots, d\}$, let $|u_{k,i}\rangle := A_k|\omega_i\rangle$, where $|\omega_i\rangle$ denotes the computational basis vector which is a tensor product of $|0\rangle$'s and $|1\rangle$'s according to the bits of the string $\omega_i$. Compute the vectors $|u_{k,i}\rangle$ one after the other. For every vector that has been computed, check if it can be written as a linear combination of already computed vectors. (The corresponding system of linear equations can be solved exactly, since every vector is given as an array of integers.) If yes, then discard the new vector $|u_{k,i}\rangle$, otherwise store it and give it a number. This way, a set of vectors $\{|u_k\rangle\}_{k=1}^D$ is computed. These vectors linearly span the support of the projector $W_{\ell,R}^{(l\tilde{n})}$ given in (7.157).

5. Denote by $\{|\phi_i\rangle\}_{i=1}^{2^{n-\tilde{n}\ell}}$ the computational basis vectors of $\mathbb{H}_{n-\tilde{n}\ell}$. If $n = \ell\, 2^{3\cdot l}$, then let $\tilde{D} := D$, and let $|x_k\rangle := |u_k\rangle$. Otherwise, compute $|u_k\rangle \otimes |\phi_i\rangle$ for every $k \in \{1, \ldots, D\}$ and $i \in \{1, \ldots, 2^{n-\tilde{n}\ell}\}$. The resulting set of vectors $\{|x_k\rangle\}_{k=1}^{\tilde{D}}$ has cardinality $\tilde{D} := D \cdot 2^{n-\tilde{n}\ell}$. In both cases, the resulting vectors $|x_k\rangle \in \mathbb{H}_n$ span the support of the projector $Q_{s,\varepsilon}^{(n)} = p_n$.

6. The set $\{|x_k\rangle\}_{k=1}^{\tilde{D}}$ is completed to linearly span the whole space $\mathbb{H}_n$. This will be accomplished as follows. Consider the sequence of vectors

$$\left\{|\,\tilde{x}_j\,\rangle\right\}_{j=1}^{\tilde{D}+2^n} := \left\{|\,x_j\,\rangle\right\}_{j=1}^{\tilde{D}} \cup \left\{|\,\Phi_j\,\rangle\right\}_{j=1}^{2^n} ,$$

where $\{\Phi_k\}_{k=1}^{2^n}$ denotes the computational basis vectors of $\mathbb{H}_n$. Find the smallest $i$ such that $|\tilde{x}_i\rangle$ can be written as a linear combination of $\left\{|\,\tilde{x}_j\,\rangle\right\}_{j=1}^{i-1}$, and discard it (this can still be decided exactly, since all the vectors are given as tables of integers). Repeat this step $\tilde{D}$ times until there remain only $2^n$ linearly independent vectors, namely all the $|x_j\rangle$ and $2^n - \tilde{D}$ of the $|\Phi_j\rangle$.

7. Finally, apply the Gram-Schmidt orthonormalization procedure to the resulting vectors, to get an orthonormal basis $\left\{|y_k\rangle\right\}_{k=1}^{2^n}$ of $\mathbb{H}_n$, such that the first $\tilde{D}$ vectors are a basis for the support of $Q_{s,\varepsilon}^{(n)} = p_n$. Since every vector $|x_j\rangle$ and $|\Phi_j\rangle$ has only integer entries, all the resulting vectors $|y_k\rangle$ will have only entries that are (plus or minus) the square root of some rational number.

Up to this point, the previous steps did not involve any kind of numerical approximation. Instead, the next ones will compute an approximate description of the desired unitary decompression map $U$ and apply it to the quantum state $p$. In view of Remark 9.1.3, the task is to calculate the number $N$ of *bits* necessary to guarantee that the output will be within trace-distance $\delta = 1/k$ of $\widetilde{p}$.

8. Read the value of $k$ (which denotes an approximation parameter; the larger $k$, the more accurate the output of the algorithm will be). Due to the considerations above and the calculations below, the necessary number of *bits* $N$ turns out to be $N = 1 + \lceil \log(2k2^n(10\sqrt{2^n})^{2^n}) \rceil$. Compute this number. Then, compute the components of all the vectors $\{|y_k\rangle\}_{k=1}^{2^n}$ up to $N$ *bits* of accuracy. (This involves only calculation of the square root of rational numbers, which can be done to any desired accuracy.) Denote the resulting numerically approximated vectors by $|\tilde{y}_k\rangle$ and write them as columns into an array (a matrix) $\tilde{U} := (\tilde{y}_1, \tilde{y}_2, \ldots, \tilde{y}_{2^n})$. Let $U := (y_1, y_2, \ldots, y_{2^n})$ denote the unitary matrix with the exact vectors $|y_k\rangle$ as columns. Since $N$ binary digits give an accuracy of $2^{-N}$, it follows that
$$\left|\tilde{U}_{i,j} - U_{i,j}\right| < 2^{-N} < \frac{1/k}{2 \cdot 2^n (10\sqrt{2^n})^{2^n}} .$$
If two $2^n \times 2^n$-matrices $U$ and $\tilde{U}$ are $\varepsilon$-close in their entries, they must be $2^n \cdot \varepsilon$-close in norm, too. Whence we get
$$\|\tilde{U} - U\| < \frac{1/k}{2(10\sqrt{2^n})^{2^n}} .$$

So far, every step could have been performed on a classical computer; the intrinsically quantum part starts when one consider the *qubit* string $p$, that is the input quantum program.

9. Compute $\lceil nr \rceil$, which gives the length $\ell(\widetilde{p})$. Afterwards, move $\widetilde{p}$ to some free space on the input tape, and append zeroes, i.e. create the state
$$p' \equiv |\psi_0\rangle\langle\psi_0| := (|0\rangle\langle 0|)^{\otimes(n-\ell(\widetilde{p}))} \otimes \widetilde{p}$$
on some segment of $n$ cells on the input tape.
10. According to Remark 9.1.3, apply a unitary approximation to the unitary transformation $U$ on the tape segment that contains the state $p'$, move the result onto the output tape and halt.

**Proof of Proposition 9.2.1** The triple $(n, r, \tilde{q})$ can be encoded into a single *qubit* string $\sigma$ (note that the parameter $k$ is not a part of $\sigma$) as follows. First, write both $r$ and $n$ in a self-delimiting way as computational basis vectors $|\beta(r)\rangle$, respectively $|\beta(n)\rangle$ (see Definition9.2.1), of length $2\log_2 r + 2$, respectively $2\log_2 n + 2$,.

Then, consider the projectors $P_n := |\beta(n)\rangle\langle\beta(n)|$, $P_r := |\beta(r)\rangle\langle\beta(r)|$ and attach to them the rotated projector $p = U^\dagger\, \widetilde{p}\, U$, so that the resulting input *qubit* string is $\sigma(p) := P_r \otimes P_n \otimes p$. If $n$ fulfils (7.162), then

$$\ell(\sigma(p)) = 2\lfloor \log_2 n \rfloor + 2 + c + \lceil nr \rceil \ ,$$

where $C(r) \in \mathbb{N}$ is some constant which depends on $C(r)$, but not on $n$.

This *qubit* string is presented to the *UQTM* $\mathfrak{U}$ together with a description of the decompression algorithm $\mathfrak{D}$ of fixed length $C'(r)$ which depends on $r$, but not on $n$. This will give a *qubit* string $\sigma_{\mathfrak{U}}(p)$ of length

$$\begin{aligned}\ell(\sigma_{\mathfrak{U}}(p)) &= 2\lfloor \log_2 n \rfloor + 2 + C(r) + \lceil nr \rceil + C'(r) \\ &\leq 2\log_2 n + n\left(s + \frac{1}{2}\delta\right) + C''(r) \ ,\end{aligned}$$

where $C''(r)$ is again a constant which depends on $r$, but not on $n$. The matrix $U$, whose construction is part of the decompression algorithm $\mathfrak{D}$, rotates (decompresses) a compressed (short) *qubit* string $p$ back into the typical subspace. Conversely, for every one-dimensional projector $\widetilde{p} \leq \widetilde{p}_n$, where $\widetilde{p}_n = Q^{(n)}_{s,\varepsilon}$ was defined in (7.161), let $p \in \mathbb{H}_{\lceil nr \rceil}$ be the projector given by $(|\,0\,\rangle\langle\,0\,|)^{\otimes(n-\lceil nr \rceil)} \otimes p = U^\dagger \widetilde{p} U$. Then, since $\mathfrak{D}$ is such that the trace-distance fulfils $D\left(\mathfrak{U}(\sigma_{\mathfrak{U}}(p), k), \widetilde{p}\right) < \frac{1}{k}$ for every $k \in \mathbb{N}$, it follows that

$$\frac{1}{n}\mathrm{QC}_q\left(\widetilde{p}\right) \leq 2\frac{\log_2 n}{n} + s + \frac{1}{2}\delta + \frac{C''(r)}{n} \ .$$

If $n$ is large enough, then the first inequality in Proposition 9.2.1 follows, while the second inequality is proved by letting $k := \lceil \frac{1}{2\delta} \rceil$. Then, for every one-dimensional projector $\widetilde{p} \leq \widetilde{p}_n$ and $n$ large enough

$$\begin{aligned}\frac{1}{n}\mathrm{QC}^{2\delta}_q\left(\widetilde{p}\right) &\leq \frac{1}{n}\mathrm{QC}^{1/k}_q\left(\widetilde{p}\right) \leq \frac{1}{n}\mathrm{QC}_q\left(\widetilde{p}\right) + \frac{2\lfloor \log_2 k \rfloor + 2}{n} \\ &< s + \delta + \frac{2\log_2 k + 2}{n} < s + 2\delta \ ,\end{aligned} \tag{9.15}$$

where the first inequality follows from the obvious monotonicity property $\delta \geq \varepsilon \Rightarrow \mathrm{QC}^{\delta}_q(\rho) \leq \mathrm{QC}^{\varepsilon}_q(\rho)$, the second one is by Example 9.2.1.2 and the third estimate is due to the first inequality in Proposition 9.2.1. □

**Proof of Theorem 9.2.2** Let $\widetilde{p}_n(\delta)$ be the typical projector sequence given in Proposition 9.2.1, i.e. the complexities $\frac{1}{n}\mathrm{QC}_q\left(\widetilde{p}\right)$ and $\frac{1}{n}\mathrm{QC}^{\delta}_q\left(\widetilde{p}\right)$ of every one-dimensional projector $\widetilde{p} \leq \widetilde{p}_n(\delta)$ are upperbounded by $s + \delta$.

Due to Corollary 9.2.1, there exists another sequence of typical projectors $\pi_n(\delta) \leq \widetilde{p}_n(\delta)$ such that additionally, $\frac{1}{n}\mathrm{QC}_q^\delta(\pi) > s - \delta(2+\delta)s$ is satisfied for all one-dimensional projections $\pi \leq \pi_n(\delta)$.

Also, from Corollary 9.2.2, there is another sequence of typical projectors $\widetilde{\pi}_n(\delta) \leq \pi_n(\delta)$ such that $\frac{1}{n}\mathrm{QC}_q(\pi) > s - \delta$ holds for all one-dimensional projections $\pi \leq \widetilde{\pi}_n(\delta)$.

Further, the optimality of these upper and lower bounds, and thus of $s$ as optimal expected asymptotic complexity rate, follows from applying Lemma 9.2.1 together with Proposition 7.3.1. □

**Remark 9.2.4.** Unlike in Theorem 4.2.1 where the result holds almost everywhere, its quantum generalization given above essentially holds in probability. The major obstruction to a stronger quantum version comes form the difficulty of extending to *qubit* strings what is natural for *bit* strings, namely their concatenation [60].

**Example 9.2.2.** Consider a quantum spin chain $(\mathcal{A}, \omega)$ of Bernoulli type with a state $\omega$ which is the tensor product of tracial states $\rho = \mathbb{1}_2/2$ for each *qubit*; this quantum source is mixing, thus ergodic and its entropy rate is $s = -\mathrm{Tr}\rho\log_2\rho = 1$. Then, the quantum version of Brudno's theorem states that there exists a sequence of subspaces $\mathbb{K}_n \subseteq \mathbb{H}_F$ of high probability, such that for any $\varepsilon > 0$, by taking $n$ sufficiently large,

$$1 - \varepsilon \leq \frac{1}{n}\mathrm{QC}_q(|\Psi\rangle\langle\Psi|) \leq 1 + \varepsilon\ ,$$

for all *qubit* pure state $\Psi \in \mathbb{K}_n$.

## 9.3 *cbit* Quantum Complexity

A different approach to quantum algorithmic complexity is proposed in [309] where as effective descriptions of *n-qubit* strings $|\Psi\rangle \in \mathbb{H}_n$ one chooses *bit* strings corresponding to self-delimiting classical programs $p \in \Omega_2^*$ instead of generic *qubit* strings. These classical programs are presented to a fixed *UQTM* $\mathfrak{U}$ as computational basis vectors $|p\rangle$ which, after being processed by $\mathfrak{U}$, outputs normalized vectors $|\mathfrak{U}(p)\rangle \in \mathbb{H}_n$. Furthermore, the difference between the output $|\mathfrak{U}(p)\rangle$ and the target $|\Psi\rangle$ is taken care of by the scalar product $\langle\Psi|\mathfrak{U}[p]\rangle$.

**Definition 9.3.1 (*bit* Quantum Complexity).** *The bit quantum complexity $\mathrm{QC}_c(\Psi)$ of n-qubit vector states $|\Psi\rangle \in \mathbb{H}_n$ is*

$$\mathrm{QC}_c(\Psi) := \min\Big\{\ell(p) + \Big\lceil -\log_2|\langle\Psi|\mathfrak{U}[p]\rangle|^2\Big\rceil\Big\}\ ,$$

*where $p \in \Omega_2^*$ is any self-delimiting binary program.*

The logarithmic correction acts as a *penalty for bad approximations*: $-\log_2 |\langle \Psi \,|\, \mathfrak{U}[p] \rangle|$ diverges for an effective description of $\Psi$ which yields a vector orthogonal to it, while it vanishes when $|\,\mathfrak{U}(p)\,\rangle \simeq |\,\Psi\,\rangle$. Therefore, the *bit* quantum complexity results from a tradeoff between the length of the classical description and the permitted errors.

**Example 9.3.1.** Let a vector $\Psi \in \mathbb{H}_n$ be called *directly computable* if there exists a self-delimiting program $p \in \Omega_2^*$ such that $|\,\mathfrak{U}(p)\,\rangle = |\,\Psi\,\rangle$. Then, consider an orthonormal basis $\mathcal{B} := \{|\,b_i\,\rangle\}_{i=1}^{2^n}$ in $\mathbb{H}_n$ entirely consisting of directly computable vectors. Let $\mathrm{K}(\mathcal{B})$ denote its classical prefix-complexity achieved by a self-delimiting program $q_{\mathcal{B}}$, $\mathrm{K}(\mathcal{B}) = \ell(q_{\mathcal{B}})$. Let us fix $|\,b_i\,\rangle \in \mathcal{B}$; if $p_i$ is any program such that $|\,\mathfrak{U}(p_i)\,\rangle = |\,b_i\,\rangle$ then no penalty for a bad approximation is to be payed and, with $p_*$ the shortest among such programs,

$$\mathrm{QC}_c\,(b_i) \leq \ell(p_*)\,. \qquad (*)$$

On the other hand, let $\mathrm{QC}_c\,(b_i)$ be attained at $q_* \in \Omega_2^*$, namely

$$\mathrm{QC}_c\,(b_i) = \ell(q_*) + \left\lceil -\log_2 |\langle\, \mathfrak{U}(q_*) \,|\, b_i \,\rangle|^2 \right\rceil .$$

By letting $\mathfrak{U}$ process the binary programs in dovetailed fashion (see Remark 4.1.5), $q_*$ can be used to construct the vector $|\,\mathfrak{U}(q_*)\,\rangle \in \mathbb{H}_n$ whose coefficients $\langle\, b_j \,|\, \mathfrak{U}(q_*)\,\rangle$ in the expansion with respect to the *ONB* $\mathcal{B}$ provide probabilities $|\langle\, b_j \,|\, \mathfrak{U}(q_*)\,\rangle|^2$ that can be used to construct a Shannon-Fano-Elias code word $q(i)$ for $|\,b_i\,\rangle$ (see Example 3.2.3). Therefore, $q_{\mathcal{B}}$, $q_*$ and $q(i)$ can be used to construct a self-delimiting program $q = q_{\mathcal{B}} q_* q(i)$ such that $\mathfrak{U}$ does the following:

- it constructs the directly computable basis $\mathcal{B}$ and the vector $|\,\mathfrak{U}(q_*)\,\rangle$;
- it computes the Shannon-Fano-Elias code for $\mathcal{B}$ with respect to $|\,\mathfrak{U}(q_*)\,\rangle$;
- it outputs the vector with code-word $q(i)$.

Since $|\,\mathfrak{U}(q)\,\rangle = |\,b_i\,\rangle$, from $(*)$ one gets

$$\ell(p_*) \leq \ell(q) \leq \ell(q_*) + \ell(q(i)) + \mathrm{K}(\mathcal{B}) + C = \mathrm{QC}_c\,(b_i) + \mathrm{K}(\mathcal{B}) + C\,, \qquad (**)$$

whence, up to an additive constant,

$$\mathrm{QC}_c\,(\Psi) = \min\Big\{\ell(p)\,:\,|\,\mathfrak{U}(p)\,\rangle = |\,\Psi\,\rangle\Big\}$$

for all $\Psi$ belonging to a directly computable *ONB* .

The preceding example can be used to show that *bit* quantum complexity and classical prefix complexity agree on *bit* strings.

**Proposition 9.3.1.** *For all* $\boldsymbol{i} \in \Omega_2^*$, $\mathrm{QC}_c\,(|\,\boldsymbol{i}\,\rangle) = \mathrm{K}(\boldsymbol{i})$ *up to an additive constant.*

**Proof:** Choosing the computational basis $\{|\,\boldsymbol{i}^{(n)}\,\rangle\}_{\boldsymbol{i}^{(n)}} \in \Omega_2^{(n)}$ as the directly computable *ONB* $\mathcal{B}$ of the previous example, the result follows from $(**)$ because the shortest program that tells $\mathfrak{U}$ how to generate $\mathcal{B}$ is now such that $\ell(q_{\mathcal{B}}) = O(1)$. □

For generic *qubit* strings, a loose upper bound is easily obtained.

**Proposition 9.3.2.** *[309] If $\Psi \in \mathbb{H}_n$ is normalized*

$$\mathrm{QC}_c(\Psi) \leq 2n + C \ ,$$

*where $C$ is a constant independent of $\Psi$.*

**Proof:** Consider the computational basis vectors $|\,\boldsymbol{i}^{(n)}\,\rangle \in \mathbb{H}_n$; by expanding $|\,\Psi^{(n)}\,\rangle = \sum_{\boldsymbol{i}^{(n)} \in \Omega_2^{(n)}} c(\boldsymbol{i}^{(n)})\,|\,\boldsymbol{i}^{(n)}\,\rangle$, there must be at least one $\boldsymbol{i}_*^{(n)}$ such that $|c(\boldsymbol{i}_*^{(n)})|^2 \geq 2^{-n}$. Let $p \in \Omega_2^*$ be a self-delimiting program such that, by literal transcription, $|\,\mathfrak{U}(p)\,\rangle = |\,\boldsymbol{i}_*^{(n)}\,\rangle$. Then, with such a choice of effective description of $|\,\Psi\,\rangle$ one gets the upper bound

$$\mathrm{QC}_c(\Psi) \leq \ell(p) + \left\lceil -\log_2 |\langle\,\mathfrak{U}(p)\,|\,\Psi\,\rangle|^2 \right\rceil \leq 2n + C \ .$$

□

A lower bound to the *bit* quantum complexity of a subset of $\Psi \in \mathbb{H}_n$ can be obtained following an argument developed in [119]. For any $\Psi \in \mathbb{H}_n$ and $\alpha \geq 0$, let us define the subsets

$$\Omega_2^* \supseteq \Pi_\alpha(\Psi) := \left\{p \in \Omega_2^* \ : \ -\log_2 |\langle\,\mathfrak{U}(p)\,|\,\Psi\,\rangle|^2 < \alpha\right\}$$

and the quantities $\mathrm{QC}_\alpha(\Psi) := \min\{\ell(p) \ : \ p \in \Pi_\alpha(\Psi)\}$. If $\alpha \geq \beta$, then $\Pi_\beta(\Psi) \subseteq \Pi_\alpha(\Psi)$, whence

$$\alpha \geq \beta \Longrightarrow \mathrm{QC}_\beta(\Psi) \geq \mathrm{QC}_\alpha(\Psi) \geq \mathrm{QC}_\infty(\Psi) \ .$$

Notice that $\mathrm{QC}_\infty(\Psi)$ is the length of the shortest classical programs $p$ such that $|\,\mathfrak{U}[p]\,\rangle$ is non-orthogonal to $|\,\Psi\,\rangle$, $|\langle\,\mathfrak{U}(p)\,|\,\Psi\,\rangle| > 0$. Therefore, if $\mathrm{QC}_c(\Psi)$ is attained at $q$, that is if

$$\mathrm{QC}_c(\Psi) = \ell(q) + \underbrace{\left\lceil -\log_2 |\langle\,\mathfrak{U}(q)\,|\,\Psi\,\rangle|^2 \right\rceil}_{\beta} \ ,$$

then, $\ell(q) \geq QC_\beta(\Psi) \geq QC_\alpha(\Psi)$ for all $\alpha \geq \beta$ and

$$\mathrm{QC}_c(\Psi) \geq \mathrm{QC}_\infty(\Psi) + \beta \ . \tag{9.16}$$

The following Lemma shows that there are vectors $\Psi \in \mathbb{H}_n$ for which $\mathrm{QC}_\infty(\Psi)$ cannot be small.

**Lemma 9.3.1.** *Let $\mathbb{K}(d) \subseteq \mathbb{H}_n$ be a d-dimensional subspace; then, for all $0 \leq a \leq \log_2 d$ there exists a subspace $\mathbb{K}_a \subseteq \mathbb{K}(d)$ of dimension $d_a \geq d - 2^a$ such that $\mathrm{QC}_\infty(\Psi) \geq a$ for all $\Psi \in \mathbb{K}_a$.*

**Proof:** According to (4.6), there are less than $2^a$ programs $p \in \Omega_2^*$ with $\ell(p) < a$; it follows that the subspace $\mathbb{H}(a)$ linearly spanned by the corresponding vectors $|\,\mathfrak{U}[p]\,\rangle$ has dimension $\leq 2^a$. Let $\mathbb{K}(d) \subseteq \mathbb{H}_n$ be any subspace of dimension $d \geq 2^a$ and choose $\mathbb{K}_a \subseteq \mathbb{K}(d)$ orthogonal to $\mathbb{H}(a)$ and thus of dimension $d_a \geq d - 2^a$. Now, $|\,\Psi\,\rangle \in \mathbb{K}_a$ satisfies $\mathrm{QC}_\infty(\Psi) \geq a$, unless there is a program $p$ with $\ell(p) < a$ with $\langle\,\mathfrak{U}(p)\,|\,\Psi\,\rangle \neq 0$; this is impossible since, by construction, $|\,\Psi\,\rangle$ is orthogonal to the linear span of $|\,\mathfrak{U}[p]\,\rangle$ with $\ell(p) < a$. □

Unlike for the *qubit* quantum complexity $\mathrm{QC_q}$ where the corresponding complexity rate could be controlled by means of high probability subspaces, in the case of the *bit* quantum complexity $\mathrm{QC_c}$, one has to argue in terms of volumes of vectors. Indeed, one can estimate how many unit vectors $|\,\Psi\,\rangle \in \mathbb{K}_a$ satisfy $\mathrm{QC}_\alpha(\Psi) < r$. This will be done by representing $\Psi$ as a point $\boldsymbol{u} \in \mathbb{R}^{2d_a}$ on the unit sphere $S_{2d_a}$ whose coordinates are the real and imaginary parts of the Fourier coefficients of the expansion of $|\,\Psi\,\rangle$ with respect to a chosen *ONB* in the subspace $\mathbb{K}_a$.

Let $S_{2d_a}(\theta)$ denote the area of the sector of $S_{2d_a}$ consisting of unit vectors $\boldsymbol{u} \in \mathbb{R}^{2d_a}$ which have scalar product $1 \geq \boldsymbol{u} \cdot \boldsymbol{e} \geq \cos\theta$ with respect to a fixed vector $\boldsymbol{e}$; it is expressed by

$$S_{2d_a}(\theta) = \int_0^\theta \mathrm{d}\phi \, A_{2d_a-1}(\sin\phi) \ ,$$

where

$$A_n(t) = t^{n-1} \frac{2\,\pi^{n/2}}{\Gamma(n/2)} \ ,$$

with $\Gamma(z)$ the Euler Gamma function, is the area of the unit sphere $S_n$ in $\mathbb{R}^n$ of radius $t$ (notice that area of the unit sphere and area of the sector of angle $\theta$ are related by $A_n(1) = S_n(\pi)$). The sector area can be bounded from above as follows:

$$S_{2d_a}(\theta) = \frac{2\pi^{d_a-1/2}}{\Gamma(d_a - 1/2)} \int_0^\theta \mathrm{d}\phi \, \sin^{2d_a-1}\phi \leq \frac{2\pi^{d_a-1/2}}{\Gamma(d_a - 1/2)} \sin^{2(d_a-1)}\theta \ . \tag{9.17}$$

Let $\widetilde{S}_{2d_a}(\theta)$ denote the sector area $S_{2d_a}(\theta)$ normalized to that of the unit sphere, $A_{2d_a}(1)$; then,

$$\widetilde{S}_{2d_a}(\theta) := \frac{S_{2d_a}(\theta)}{A_{2d_a}(1)} \leq \frac{2\pi^{d_a-1/2}}{\Gamma(d_a - 1/2)} \frac{\Gamma(d_a)}{2\pi^{d_a}} \sin^{2(d_a-1)}\theta \tag{9.18}$$

$$< d_a \, \mathrm{e}^{-(d_a-1)(\theta-\pi/2)^2} \ , \tag{9.19}$$

where the last inequality comes from expanding $f(\theta) := \log\sin\theta$ around $\pi/2$,

$$f(\theta) = -\frac{1}{2}(\theta-\pi/2)^2 + \frac{1}{6}f'''(\overline{\theta})\,(\theta-\pi/2)^3 \le -\frac{1}{2}(\theta-\pi/2)^2\,, \qquad \theta \le \overline{\theta} \le \pi/2\,,$$

and from the fact that $f'''(\overline{\theta}) \ge 0$. We now use (9.19) to estimate the relative volume, $F_a(p,\alpha)$, of the subset

$$\mathcal{F}_a(p,\alpha) := \Big\{ |\,\psi\,\rangle \in \mathbb{K}_a \;:\; -\log_2 |\langle\, \mathfrak{U}(p)\,|\,\psi\,\rangle|^2 < \alpha \Big\}$$

consisting of vectors with penalty smaller than $\alpha$ with respect to a given output $|\,\mathfrak{U}[p]\,\rangle$. Since $2^{-\alpha/2} < |\langle\,\mathfrak{U}[p]\,|\,\psi\,\rangle| = \cos\theta = \sin(\pi/2-\theta) \le \pi/2-\theta$,

$$F_a(p,\alpha)) < d_a\, \mathrm{e}^{-(d_a-1)2^{-\alpha}}\ .$$

From this inequality we further deduce

**Lemma 9.3.2.** *The relative volume, $F^r_a(\alpha)$, of the set*

$$\mathcal{F}^r_a(\alpha) := \Big\{ |\,\psi\,\rangle \in \mathbb{K}_a \;:\; \mathrm{QC}_\alpha(\psi) < r \Big\}$$

*has relative volume $F^r_a(\alpha)$ such that*

$$F^r_a(\alpha) < d_a\, 2^r\, \mathrm{e}^{-(d_a-1)2^{-\alpha}}\ .$$

**Proof:** If $|\,\psi\,\rangle \in \mathbb{K}_a$ is such that $\mathrm{QC}_\alpha(\psi) < r$, then $-\log_2 |\langle\,\psi\,|\,\mathfrak{U}[p]\,\rangle|^2 < \alpha$ for at least one program $p$ with $\ell(p) < r$; the result then follows since there are $\le 2^r$ such programs. □

The complement $\mathcal{G}^r_a(\alpha)$ of $\mathcal{F}^r_a(\alpha)$ consists of $|\,\psi\,\rangle \in \mathbb{K}_a$ such that either $-\log_2 |\langle\,\psi\,|\,\mathfrak{U}[p]\,\rangle|^2 \ge \alpha$ or $-\log_2 |\langle\,\psi\,|\,\mathfrak{U}[p]\,\rangle||^2 < \alpha$, but $\ell(p) \ge r$. In other words, from (9.16) and Lemma 9.3.1, it turns out that $\mathcal{G}^r_a(\alpha)$ consists of $|\,\psi\,\rangle \in \mathbb{K}_a$ such that

$$\mathrm{QC}_c(\psi) \ge \mathrm{QC}_\infty(\Psi) + \alpha \ge a + \alpha \tag{9.20}$$

or

$$\mathrm{QC}_c(\psi) \ge r - \log_2 |\langle\,\psi\,|\,\mathfrak{U}[p]\,\rangle||^2 \ge r\ . \tag{9.21}$$

Notice that the relative volume $G^r_a(\alpha)$ of $\mathcal{G}^r_a(\alpha)$ is large, $G^r_a(\alpha) \ge 1-\varepsilon$ if the relative volume of $\mathcal{F}^r_a(\alpha)$ is small, $F^r_a(\alpha) \le \varepsilon$. Lemma 9.3.2 can then be used to prove that for a large fraction of vectors $|\,\psi\,\rangle \in \mathbb{K}_a$ one has $\mathrm{QC}_c(\psi) \simeq 2n$ when $n \to \infty$.

**Proposition 9.3.3.** *For any $\varepsilon \ge 0$ and $\mathbb{N} \ni n$ large enough, there exists a subspace $\mathbb{K}_{n-1} \subset \mathbb{H}_n$ of dimension $\ge 2^{n-1}$ containing a subset $\mathcal{G}_{n-1}$ of relative volume $G_{n-1} \ge 1-\varepsilon$ such that, for all $|\,\Psi\,\rangle \in \mathcal{G}_{n-1}$,*

$$2-\varepsilon \le \frac{\mathrm{QC}_c(\Psi)}{n} \le 2+\varepsilon\ .$$

**Proof:** Choose $\mathbb{H}(d) = \mathbb{H}_n$ in Lemma 9.3.1 and $a = n-1$; then, there exists a subspace $\mathbb{K}_{n-1} \subset \mathbb{H}_n$ of dimension $d_{n-1} \geq 2^n - 2^{n-1} = 2^{n-1}$ such that $\mathrm{QC}_\infty(\Psi) \geq n-1$ for all $\Psi \in \mathbb{K}_{n-1}$. Setting $r = 2n$ and $\alpha = n - 1 - 2\log_2 n$ in Lemma 9.3.2 one gets

$$F^{2n}_{n-1}(n - 1 - 2\log_2 n) < \mathrm{e}^{-(1-2^{-n+1})n^2 + (3n-1)\log 2} .$$

Thus, for $n$ sufficiently large, the subset $\mathcal{G}_{n-1} \subseteq \mathbb{K}_{n-1}$ of $\Psi \in \mathbb{K}_{n-1}$ that violate $\mathrm{QC}_{n-1-2\log_2 n}(\Psi) < 2n$ has relative volume $\mathcal{G}_{n-1} \geq 1 - \varepsilon$. The result then follows by applying the lower bound in Proposition 9.3.2 and the upper bounds (9.20) and (9.21). □

**Remark 9.3.1.** Proposition 9.3.3 states that a large fraction of $n$-*qubit* vector states belonging to a subspace of dimension not less than $2^{n-1}$ has a *bit* quantum complexity per symbol close to 2. Notice that this is twice the *qubit* quantum complexity per symbol of all pure states in the high probability subspaces of a Bernoulli quantum source (see Example 9.2.2). However, in the latter case the fact that the complexity rate $\simeq 1$ follows from the specific structure of the state $\omega$ on the quantum spin chain $\mathcal{A}$. Instead, the result of Proposition 9.3.3 does not refer to the considered $n$ *qubits* belonging to a quantum chain and thus to a reference global state $\omega$. Indeed, the weights of the subsets of vectors with *bit* quantum complexity rate $\simeq 2$ are estimated in terms of relative volumes instead of probabilities as in Theorem 9.2.2.

## Circuit Algorithmic Complexity

Any state $|\Psi\rangle \in \mathbb{H}_n$ of $n$ *qubits* can be obtained as the result of an action on a fixed state $|\Phi\rangle \in \mathbb{H}_n$ by a suitable unitary operator $U : \mathbb{H}_n \mapsto \mathbb{H}_n$. From Remark 9.1.1.2 we know that the action of $U$ can be approximated within any $\varepsilon > 0$ by means of a quantum circuit, that is by another unitary operator $V : \mathbb{H}_n \mapsto \mathbb{H}_n$, consisting of $N(U,\varepsilon)$ gates from a complete gate basis $\mathcal{G}$. Furthermore, to leading order in the number of *qubits* and of the accuracy $\varepsilon$, the number of gates scales as $N(U,\varepsilon) = O\Big(2^n \log \frac{1}{\varepsilon}\Big)$.

This fact is essential in the definition of quantum algorithmic complexity proposed in [205, 206, 207] where the focus is not on the effective description of the $n$-*qubit* states, whether quantum or classical, rather on the effective description of the quantum circuits that can be used to effectively construct those quantum states up to a certain fixed accuracy $\varepsilon$.

Given a complete gate basis $\mathcal{G}$, the fixed ready state $|\Phi\rangle$ and the accuracy parameter $\varepsilon > 0$, a same $|\Psi\rangle$ can be reached up to $\varepsilon$ by a certain set $\mathfrak{C}^{\mathcal{G},\varepsilon}_\Psi$ of quantum circuits $V^{\mathcal{G},\varepsilon}_\Psi$ that will be identified with their unitary actions on $|\Phi\rangle$. Let the description of any of these circuits be encoded by a binary string $\boldsymbol{i}_{V^{\mathcal{G},\varepsilon}_\Psi} \in \Omega_2^*$; then

**Definition 9.3.2 (Circuit Quantum Complexity).** *Let $\Omega_\Psi^{\mathcal{G},\varepsilon} \subset \Omega_2^*$ be the subset of strings that encode the circuits in $\mathfrak{C}_\Psi^{\mathcal{G},\varepsilon}$; the* circuit quantum algorithmic complexity *of an n-qubit state $|\Psi\rangle \in \mathbb{H}_n$ is the least classical prefix algorithmic complexity of the strings in $\Omega_\Psi^{\mathcal{G},\varepsilon}$:*

$$\mathrm{QC}_{\mathrm{net}}^{\mathcal{G},\varepsilon}(\Psi) := \min\left\{\mathrm{K}(\boldsymbol{i}_{V_\Psi^{\mathcal{G},\varepsilon}}) \,:\, \boldsymbol{i}_{V_\Psi^{\mathcal{G},\varepsilon}} \in \Omega_\Psi^{\mathcal{G},\varepsilon}\right\} .$$

**Remark 9.3.2.** The dependence of $\mathrm{QC}_{\mathrm{net}}$ on the encoding of the description of the circuits $V_\Psi^{\mathcal{G},\varepsilon} \in \mathfrak{C}_\Psi^{\mathcal{G},\varepsilon}$ can be handled as in classical algorithmic complexity: a change of code is taken care of by a finite additive constant corresponding to a suitable dictionary which is thus independent of the circuit described.

The physical motivation behind such a definition is that, after all, quantum states can be prepared by means of arrays of unitary gates that can be effectively described; then, the idea is to relate the complexity of vector states to the degree of compressibility of the descriptions of the quantum circuits that provide suitable approximations to them.

**Example 9.3.2.** If one want to reproduce a *bit* string $\boldsymbol{i}^{(n)}$ by a quantum circuit, the first step is to associate it to a *qubit* vector state $|\boldsymbol{i}^{(n)}\rangle$ of the so called computational basis (see Section 4.1.1), where

$$|\boldsymbol{i}^{(n)}\rangle = |i_1\rangle \otimes |i_2\rangle \otimes \cdots |i_n\rangle\,, \quad i_j = 0,1\,, \quad \sigma_3|i_j\rangle = (-1)^{i_j}|i_j\rangle\,.$$

This state can then easily be obtained by flipping with $\sigma_1^{i_j}$ the $j$-th *qubit* of $|0\rangle^{\otimes n}$. The corresponding quantum circuit consists of $n$ 1-*qubit* gates, either trivially the identity matrix $\mathbb{1}_2$ or the Pauli matrix $\sigma_1$; therefore, an upper bound to the algorithmic complexity of the classical description of such a quantum circuit is easily seen to scale as $n$, exactly as the Kolmogorov complexity of a generic *bit* string of length $n$ (see Proposition 4.1.1).

Within this approach one usually estimates the complexity by upper bounds that depend on results as the one quoted in Remark 9.1.1.2, whence the circuit complexity of a state $|\Psi\rangle$ of $n$ *qubits* can be estimated as follows:

$$\mathrm{QC}_{\mathrm{net}}^{\mathcal{G},\varepsilon}(\Psi) = O(n^2 2^n \log 1/\varepsilon)\,, \tag{9.22}$$

where $f(n) = O(g(n))$ if there exists $C_{f,g} > 0$ such that $|f(n)| \leq C_{f,g}|g(n)|$.

**Remark 9.3.3.** As already pointed out (see for instance Example 3.2.2), a *bit* string $\boldsymbol{i}^{(n)}$ of length $n$ can be associated with an interval in $[0,1]$ of length $2^{-n}$; this latter can also be interpreted as the volume of the subset $V(\boldsymbol{i}^{(n)})$

of *bit* strings of any length that are prefixed by $\boldsymbol{i}^{(n)}$. Therefore, the upper bound (4.5) to the algorithmic complexity of $\boldsymbol{i}^{(n)}$ scales as $-\log V(\boldsymbol{i}^{(n)}) = n$. In a quantum context, because of the lack of discreteness, given a fixed $n$-*qubit* vector $\Psi$, one can in general only hope to construct it within an error $\varepsilon$; namely, a quantum circuit can be devised that outputs $|\,\psi\,\rangle \in (\mathbb{C}^2)^{\otimes n}$ such that $|\langle\,\psi\,|\,\Psi\,\rangle| \geq 1-\varepsilon$ for some accuracy parameter $0 \leq \varepsilon \leq 1$. Using (9.17), one finds that the logarithm of the volume $V_\varepsilon(\Psi)$ of such a cone scales as $2^n \log \varepsilon$ in agreement with (9.22).

Notice that the upper bound to the *bit* quantum complexity in Proposition 9.3.2 is obtained by choosing $|\,\psi\,\rangle$ such that $|\langle\,\psi\,|\,\Psi\,\rangle| \geq 2^{-n}$; this corresponds to a parameter in (9.22) which scales as $\varepsilon \simeq 1 - 2^{-n}$ and to an upper bound to $\mathrm{QC}^{\mathcal{G},\varepsilon}_{\mathrm{net}}(\Psi)$ which is only polinomially different from the one to $\mathrm{QC}_c(\Psi)$ [207].

By means of the upper bound (9.22), it is possible to put into evidence the difference in circuit complexity between separable and entangled states. The important point is that a product state $|\,\Psi\,\rangle = \bigotimes_{j=1}^{J} |\,\Phi_j\,\rangle \in \mathbb{H}_n$ can be constructed with accuracy $\varepsilon$ by means of $n$ circuits that construct the vectors $|\,\Phi_j\,\rangle$ with accuracy $\varepsilon/n$ [205]. Suppose that the state $|\,\Psi\,\rangle$ shows some entanglement between its constituent *qubits*; namely, $|\,\Psi\,\rangle = \bigotimes_{j=1}^{J} |\,\Phi_j\,\rangle$, where $\sum_{j=1}^{J} n_j = n$ and $|\,\Phi_j\,\rangle \in \mathbb{H}_{n_j}$ are entangled states of $n_j$ *qubits*. Then, one obtains

$$\mathrm{QC}^{\mathcal{G},\varepsilon}_{\mathrm{net}}(\Psi) = O\Big(\sum_{j=1}^{J} n_j^2 2^{n_j} \log\frac{J}{\varepsilon}\Big) .$$

For sufficiently small $\varepsilon$, the sum is upper bounded by

$$\sum_{j=1}^{J} n_j^2 2^{n_j} \log\frac{J}{\varepsilon} \leq n^2 2^n \frac{J}{2^{J-1}} \log\frac{J}{\varepsilon} \leq n^2 2^n \log\frac{1}{\varepsilon} .$$

Therefore, the upper bound is the largest when the state $|\,\Psi\,\rangle$ is completely entangled, that is when $J = 1$ and $n_j = n$; vice versa, in the case of complete separability, namely when $J = n$ and $n_j = 1$, one gets

$$\mathrm{QC}^{\mathcal{G},\varepsilon}_{\mathrm{net}}(\Psi) = O\Big(2n \log\frac{n}{\varepsilon}\Big) ,$$

with polynomial instead of exponential increase with the number of *qubits*.

### Quantum Universal Semi-Density Matrix

Like in Section 9.3, the quantum extension of classical algorithmic complexity proposed in [119] starts from the classical description of quantum states

$|\Psi\rangle \in \mathbb{H}_n$ of $n$ *qubit* systems by means of *bit* strings $\boldsymbol{i} \in \Omega_2^*$. However, these descriptions are not considered as programs of a certain length which has to be minimized, rather as *bit* strings characterized by a given universal probability $P_{\mathfrak{U}}$ as explained in Remark 4.3.2.3.

For instance, if a state $|\Psi\rangle$ can be expanded with respect to the computational basis $\{|\boldsymbol{i}^{(n)}\rangle\}_{\boldsymbol{i}^{(n)}\in\Omega_2^{(n)}}$ by means of coefficients which are exactly computable by a program $\boldsymbol{j} \in \Omega_2^*$ processed by a fixed *UTM* $\mathfrak{U}$, then

$$\boldsymbol{m}(\Psi) := P_{\mathfrak{U}}(\boldsymbol{j})$$

naturally represents the universal probability of this state. Exactly computable states are termed *elementary* as well as linear operators $X$ on $\mathbb{H}_n$ whose matrix elements with respect to the computational basis can be exactly computed; operators which can be approximated from below by an increasing sequence of elementary operators are called lower semi-computable (see Definition 4.1.4). Then, an argument similar to the one in Example 4.1.7.3 leads to the following result [119].

**Theorem 9.3.1.** *A lower semi-computable semi-density matrix* $\boldsymbol{\rho} \in \mathbb{B}_1(\mathbb{H}_n)$, *namely* $\boldsymbol{\rho} \geq 0$ *and* $\boldsymbol{\rho} \leq 1$, *can be effectively constructed such that, for any other semi-computable semi-density matrix* $\sigma \in \mathbb{B}_1(\mathbb{H}_n)$, *there is a constant* $C_\sigma$ *for which* $C_\sigma\,\sigma \leq \boldsymbol{\rho}$. *Moreover,* $\boldsymbol{\rho}$ *can be identified with*

$$\boldsymbol{\rho} = \sum_{|\Psi_{el}\rangle\in\mathbb{H}_n} \boldsymbol{m}(\Psi_{el})\,|\Psi_{el}\rangle\langle\Psi_{el}|\ ,$$

*where the sum runs over all elementary vector states of* $n$ *qubits.*

The operator $\boldsymbol{\rho}$ is a convex combination over elementary projections weighted with their universal probabilities; since the universal probability $P_{\mathfrak{U}}$ is not normalized, neither is $\boldsymbol{\rho}$. Inspired by Remark 4.3.3, it is thus suggestive to introduce an *operatorial complexity*

$$\boldsymbol{\kappa} := -\log_2 \boldsymbol{\rho}\ ,$$

and two possible definitions of algorithmic complexity of a state $|\Psi\rangle$:

$$\mathrm{QC}^-_{u.p.}(\Psi) := -\log_2(\langle\Psi|\boldsymbol{\rho}|\Psi\rangle)\ , \qquad \mathrm{QC}^+_{u.p.}(\Psi) := \langle\Psi|\boldsymbol{\kappa}|\Psi\rangle\ .$$

From the the concavity of the function $f(x) = \log_2 x$ it follows that

$$\mathrm{QC}^-_{u.p.}(\Psi) \leq \mathrm{QC}^+_{u.p.}(\Psi)\ .$$

Indeed (see the proof of Proposition 5.5.3), with $\rho = \sum_i r_i|r_i\rangle\langle r_i|$ the spectral decomposition of $\rho$,

$$\log_2(\langle \Psi \,|\, \rho \,|\Psi \rangle) = \log_2 \left( \sum_i r_i \, |\langle \Psi \,|\, r_i \rangle|^2 \right)$$
$$\geq \sum_i |\langle \Psi \,|\, r_i \rangle|^2 \log_2 r_i = \langle \Psi \,|\, \log_2 \rho \,|\Psi \rangle \ .$$

For *bit* strings $\boldsymbol{i}^{(n)} \in \Omega_2^*$ the two possibilities coincide with the algorithmic complexity $\mathrm{K}(\boldsymbol{i}^{(n)})$. Indeed, consider the computational basis vectors $\boldsymbol{i}^{(n)}$, then $\{\langle \boldsymbol{i}^{(n)} \,|\, \boldsymbol{\rho} \,|\boldsymbol{i}^{(n)} \rangle\}_{\boldsymbol{i}^{(n)} \in \Omega_2^{(n)}}$ is a semi-measure. As $\boldsymbol{m}$ is a universal semi-measure on $\Omega_2^*$ it follows that there exists a constant $C_\rho$ such that

$$C_\rho \langle \boldsymbol{i}^{(n)} \,|\, \boldsymbol{\rho} \,|\boldsymbol{i}^{(n)} \rangle \leq \boldsymbol{m}(\boldsymbol{i}^{(n)}) \ ,$$

whence, from Remark 4.3.3, $\mathrm{QC}^-_{u.p.}(\boldsymbol{i}^{(n)}) \geq \mathrm{K}(\boldsymbol{i}^{(n)}) + O(1)$.

On the other hand, $\rho = \sum_{\boldsymbol{i}^{(n)} \in \Omega_2^*} \boldsymbol{m}(\boldsymbol{i}^{(n)}) \,|\, \boldsymbol{i}^{(n)} \rangle\langle \boldsymbol{i}^{(n)} \,|$ is a lower semi-computable, semi-density matrix. Therefore, the monotonicity of the logarithm as an operator function (see Example 5.2.3.6) yields

$$C_\rho \, \rho \leq \boldsymbol{\rho} \Longrightarrow -\log_2 \rho + \log_2 C_\rho \geq \boldsymbol{\kappa} \ ,$$

whence $\mathrm{QC}^+_{u.p.}(\boldsymbol{i}^{(n)}) \leq \mathrm{K}(\boldsymbol{i}^{(n)}) + O(1)$.

The operatorial complexity $\boldsymbol{\kappa}$ has an interesting similarity with the classical algorithmic complexity in that its mean value with respect to a lower semi-computable density matrix $\rho$ equals its von Neumann entropy up to an additive constant [119] (compare Corollary 4.3.2),

$$\mathrm{Tr}(\rho \, \boldsymbol{\kappa}) = S_2(\rho) \, + \, O(1) \ , \quad S_2(\rho) := -\mathrm{Tr}(\rho \log_2 \rho) \ .$$

Setting $\hat{\boldsymbol{\rho}} := \dfrac{\boldsymbol{\rho}}{\mathrm{Tr}(\boldsymbol{\rho})}$, the positivity of the relative entropy, $S\,(\rho\,,\,\hat{\boldsymbol{\rho}}) \geq 0$, yields

$$S_2(\rho) \ \leq \ \mathrm{Tr}(\rho \, \boldsymbol{\kappa}) \, + \, \log_2 \mathrm{Tr}(\boldsymbol{\rho}) \ .$$

By assumption, there exists a constant $C_\rho$ such that $C_\rho \, \rho \leq \boldsymbol{\rho}$; thus, as before,

$$-\log_2 C_\rho \ - \log_2 \rho \ \geq \ \boldsymbol{\kappa} \Longrightarrow \ S_2(\rho) \ \geq \ \mathrm{Tr}(\rho \boldsymbol{\kappa}) + O(1) \ .$$

**Remark 9.3.4.** The topics addressed in this chapter are relatively recent and still in their infancy so that the relations between the various extensions of classical algorithmic complexity theory to quantum systems are largely to be explored (a discussion of those between Vitanyi's and Gács' proposals can be found in [119]).

Further, beside the previous result and Theorem 9.2.2, the connections between quantum algorithmic complexities and the von Neumann entropy or the von Neumann entropy rate have not yet been clarified. In particular, the randomness of the quantum dynamics, rather than of quantum states have not been tackled yet; namely, a quantum extension of the dynamical version of Brudno's theorem (see Corollary 4.2.1) is still missing.

# References


1. L. Accardi, M. Ohya, N. Watanabe: Rep. Math. Phys. **38**, 457 (1996)
2. G. Adesso, F. Illuminati: J. Phys. A **40**, 7821 (2007)
3. L.M. Adleman, J. Demarrais, M.A. Huang: SIAM J. Comput. **26**, 1524 (1997)
4. I. Affleck, T. Kennedy, E.H. Lieb et al: Commmun. Math. Phys. **115**, 477 (1988)
5. I. Affleck, T. Kennedy, E.H. Lieb et al: Phys. Rev. Lett. **59**, 799 (1987)
6. G. Alber et al. Eds.: *Quantum Information: An Introduction to Basic Theoretical Concepts and Experiments*, Springer Tracts in Mod. Phys. **173**, (Springer-Verlag, Berlin 2001)
7. V.M. Alekseev, M.V Yakobson: Phys. Rep. **75**, 287 (1981)
8. R. Alicki: Phys. Rev. A **66**, 052302 (2002)
9. R. Alicki, M. Fannes: Lett. Math. Phys. **32**, 75 (1994)
10. R. Alicki, M. Fannes: *Quantum Dynamical Systems*, (Oxford University Press, Oxford 2001)
11. R. Alicki, K. Lendi: *Quantum Dynamical Semigroups and Applications*, (Springer, Berlin 2007)
12. R. Alicki, D. Makowiec, W. Mikłaszewski: Phys. Rev. Lett. **77**, 838 (1996)
13. R. Alicki, H. Narnhofer: Lett. Math. Phys. **33**, 241 (1995)
14. R. Alicki: *Invitation to Quantum Dynamical Semigroups*, in: Lect. Notes Phys. **597**, P. Garbaczewski and R. Olkiewicz, Eds., (Springer-Verlag, Berlin 2002)
15. J. Andries, M. Fannes, P.Tuyls et al.: Lett. Math. Phys. **35**, 375 (1995)
16. V.I. Arnold: *Mathematical Methods of Classical Mechanics*, (Sprnger, New York 1978)
17. V.I. Arnold, A. Avez: *Ergodic Problems of Classical Mechanics*, (W.A. Benjamin, New York 1968)
18. R. Badii, A. Politi: *Complcxity*, (Cambridge University Press, Cambridge 1997)
19. H. Barnum, C. Fuchs, R. Jozsa et al: Phys. Rev. **A54**, 4707 (1996)
20. A. Barvinok: *A Course in Convexity*, (Graduate Texts in Mathematics **54**, A.M.S. 2002)
21. A. Bassi, G.C. Ghirardi: Phys. Rep. **379**, 257 (2003)
22. F. Benatti: *Deterministic Chaos in Infinite Quantum Systems*, (Trieste Notes in Physics, Springer Heidelberg 1993)
23. F. Benatti: J. Math. Phys. **37**, 5244 (1996)
24. F. Benatti, V. Cappellini: J. Math. Phys. **46**, 062702 (2005)
25. F. Benatti, V. Cappellini, M. De Cock et al.: Rev. Math. Phys. **15**, 847 (2003)
26. F. Benatti, V. Cappellini, F. Zertuche: J. Phys. A **37**, 105 (2004)
27. F. Benatti, R. Floreanini: Int. J. Phys. B **19**, 3063 (2005)

28. F. Benatti, R. Floreanini: Banach Centre Publications **43** 71 (1998)
29. F. Benatti, R. Floreanini Eds.: *Dissipative Quantum Dynamics*, Lect. Notes Phys. **622**, (Springer-Verlag, Berlin 2003)
30. F. Benatti, R. Floreanini: Nucl. Phys. **B 488**, 335 (1997)
31. F. Benatti, R. Floreanini: Nucl. Phys. **B 511**, 550 (1998)
32. F. Benatti, R. Floreanini: Open Sys. & Inf. Dyn. **13**, 229 (2006)
33. F. Benatti, R. Floreanini, A.M. Liguori: J. Math. Phys. **48**, 052103 (2007)
34. F. Benatti, R. Floreanini, M. Piani: Phys. Rev. A **67**, 042110 (2003)
35. F. Benatti, R. Floreanini, M. Piani: Phys. Lett. A **326**, 187 (2004)
36. F. Benatti, R. Floreanini, M. Piani: Open Sys. and Information Dyn. **11**, 325 (2004)
37. F. Benatti, R. Floreanini, R. Romano: J. Phys. A **35**, L551 (2002)
38. F. Benatti, B. Hiesmayr, H. Narnhofer: Europhys. Lett. **72**, 28 (2005)
39. F. Benatti, T. Hudetz, A. Knauf: Commun. Math. Phys. **198**, 607 (1998)
40. F. Benatti, T. Krüger, M. Müller et al: Commun. Math. Phys. **265**, 437 (2006)
41. F. Benatti, H. Narnhofer: Lett. Math. Phys. **15**, 325 (1988)
42. F. Benatti, H. Narnhofer: Commun. Math. Phys. **136**, 231 (1991)
43. F. Benatti, H. Narnhofer: Phys. Rev. **A63**, 042306 (2001)
44. F. Benatti, H. Narnhofer, G.L.Sewell: Lett. Math. Phys. **21**, 157 (1991)
45. F. Benatti, H. Narnhofer, A. Uhlmann: Rep. Math. Phys. **38**, 123 (1996)
46. F. Benatti, H. Narnhofer, A. Uhlmann: Lett. Math. Phys. **47**, 237 (1999)
47. F. Benatti, H. Narnhofer, A. Uhlmann: J. Math. Phys. **44**, 2402 (2003)
48. G. Benenti, G. Casati, G. Strini: *Principles of Quantum Computation and Information*, (World Scientific, Singapore 2004)
49. I. Bengtsson, K. Życzkowski: *Geometry of Quantum States: an Introduction to Quantum Entanglement*, (University Press, Cambridge 2006)
50. C.H. Bennett: Sci. Amer. **241**.5, 22 (1979)
51. C.H. Bennett: Int. J. Th. Phys. **21**, 905 (1982)
52. C.H. Bennett: Sci. Amer. **257**.5, 88 (1987)
53. C.H. Bennett, D.P. Di Vincenzo, J. Smolin et al.: Phys. Rev. A **54** 3824 (1996)
54. C.H. Bennett, P. Gács, M. Li et al.: Proc. 25th ACM Symp. Theory of Computation, (ACM Press 1993)
55. C.H. Bennett, R. Landauer: Sci. Amer. **253**.1, 38 (1985)
56. E. Bernstein, U. Vazirani: SIAM Journal on Computing **26**, 1411 (1997)
57. A. Berthiaume, W. Van Dam, S. Laplante: J. Comput. System Sci. **63**, 201 (2001)
58. I. Bjelaković, T. Krüger, R. Siegmund-Schultze et al: Invent. Math. **155**, 203 (2004)
59. I. Bjelaković, A. Szkoła: Quant. Inform. Proc. **4**, 49 (2005)
60. I. Bjelaković, T. Krüger, R. Siegmund-Schultze et al: *Chained Typical Subspaces - a Quantum Version of Breiman's Theorem*, `arXiv:quant-ph70301177`
61. P. Billingsley: *Ergodic Theory and Information*, (J. Wiley, New York 1965)
62. G. Boffetta, M. Cencini, M. Falcioni et al.: Phys. Rep. **356**, 367 (2002)
63. D. Bouwmeester et al. Eds.: *The Physics of Quantum Information*, (Springer-Verlag, Berlin 2000)
64. O. Bratteli, D.W. Robinson: *Operator Algebra and Quantum Statistical Mechanics I*, (Springer, New York 1987)

65. O. Bratteli, D.W. Robinson: *Operator Algebra and Quantum Statistical Mechanics II*, (Springer, New York 1981)
66. S.L. Braunstein ed.: *Quantum Computing*, (Wiley-VHC, Weinheim, 1999)
67. H.-P. Breuer: Phys. Rev. Lett. **97**, 080501 (2006)
68. H-P. Breuer, F. Petruccione: *The Theory of Open Quantum Systems*, (Oxford University Press 2002)
69. A.A. Brudno: Trans. Moscow Math. Soc. **2**, 127 (1983)
70. D. Bruss: J. Math. Phys. **43**, 4237 (2002)
71. D. Bruss, G. Leuchs: *Lectures on Quantum Information*, (Wiley-Vch GmbH & Co. KGaA, 2007)
72. J. Budimir, J.L. Skinner: J. Stat. Phys. **49**, 1029 (1987)
73. C.S. Calude: *Information and Randomness. An Algorithmic Perspective*, (Springer, Berlin 2002)
74. V. Cappellini, J. Phys. A **38**, 6893 (2005)
75. G. Casati, B.V. Chirikov: *Quantum chaos between order and disorder*, (Cambridge University Press, Cambridge 1995)
76. N.J. Cerf, G. Leuchs, E.S. Polzik eds.: *Quantum Information with Continuous Variables of Atmos and Light*, (Imperial College Press, World Scientific 2007)
77. G.J. Chaitin, J. Assoc. Comp. Mach. **13**, 547 (1966)
78. G.J. Chaitin: J. of the ACM **22**, 329 (1975)
79. G.J. Chaitin: *Information, Randomness and Incompleteness*, (World Scientific, Singapore, New Jersey, Hong Kong 1987)
80. G.J. Chaitin: *The Limits of Mathematics*, (Springer, Singapore 1998)
81. G.J. Chaitin: *Algorithmic Information Theory*, (Cambridge University Press, Cabridge 1988)
82. M.-D. Choi: Can. J. Math. **24**, 520 (1972)
83. M.-D. Choi: Lin. Alg. Appl. **10**, 285 (1975)
84. D. Chruscinski, A. Kossakowski: Phys. Rev. A **74**, 022308 (2006)
85. D. Chruscinski, A. Kossakowski: Open Sys. Information Dyn. **14**, 275 (2007)
86. D. Chruscinski, A. Kossakowski: Phys. Rev. A **76**, 032308 (2007)
87. C. Cohen-Tannoudji, J. Dupont-Roc, G. Grynberg: *Atom-Photon Interactions*, (Wiley, New York 1988)
88. A. Connes, H. Narnhofer, W. Thirring: Commun. Math. Phys. **112**, 691 (1987)
89. A. Connes, E. Størmer: Acta Math. **134**, 289 (1975)
90. J.B. Conway: *A Course in Operator Theory*, Graduated Studies in Mathematics **21** (American Mathematical Society, Providence, Rhode Island 2000)
91. I.P. Cornfeld, S.V. Fomin, Ya.G. Sinai: *Ergodic Theory*, (Springer, New York 1982)
92. T.M. Cover, J.A. Thomas: *Elements of Information Theory*, (Wiley Series in Telecommunications, John Wiley & Sons, New York 1991)
93. N. Cutland: *Computability*, (Cambridge University Press, Cambridge New York 1988)
94. N. Datta, Y. Suhov: Quantum Information Proc. **1**, 257 (2002)
95. E.B. Davies: Commun. Math. Phys. **39**, 91 (1974)
96. E.B. Davies: *Quantum Theory of Open Systems*, (Academic Press, London 1976)
97. K.R. Davidson: *$C^*$-Algebras by Example*, Fields Institute Monographs **7**, (American Mathematical Society, Providence Rhode Island 1996)

98. M. Degli Esposti, S. Graffi eds: *The Mathematical Aspects of Quantum Maps*, Lec. Notes Phys. **618**, Springer (2003)
99. D. Deutsch: Proc. R. Soc. Lond. **A400**, 97 (1985)
100. L. Diosi: *A Short Course in Quantum Information Theory: an Approach from Theoretical Physics*, (Springer, Berlin 2007)
101. J.L. Doob: *Stochastic Processes*, (John Wiley and Sons, New York 1953)
102. K. Fujikawa: *On the separability criterion for continuous variable systems* arXiv:0710.5039
103. L.M. Duan, G. Giedke, J.I. Cirac et al.: Phys. Rev. Lett. **84**, 2722 (2000)
104. R. Dümcke: Commun. Math. Phys. **97**, 331 (1985)
105. R. Dümcke, H. Spohn: Z. Phys. **B34**, 419 (1979)
106. J.P. Eckmann, D. Ruelle: Rev. Mod. Phys. **57**, 617 (1985)
107. G.G. Emch: Commun. Math. Phys. **49**, 191 (1976)
108. G.G. Emch: *Algebraic Methods in Statistical Mechanics and Quantum Field Theory*, (Wiley-Interscience, New York 1972)
109. D.E. Evans, J.T. Lewis: *Dilation of Irreversible Evolutions in Algebraic Quantum Theory*, Communications of the Dublin Institute for Advanced Studies **A 24**, Dublin 1977
110. M. Fannes: Commun. Math. Phys. **31**, 279 (1973)
111. M. Fannes, B. Haegeman, M. Mosonyi: J. Math. Phys. **44**, 600 (2003)
112. M. Fannes, B. Haegeman, D. Vanpeteghem: J. Phys. A: Math. Gen. **38**, 2103 (2005)
113. M. Fannes, B. Nachtergaele, R.F. Werner: Commun. Math. Phys. **144**, 443 (1992)
114. S-M. Fei, X. Li-Jost, B-Z. Sun: Phys. Lett. A **352**, 321 (2006)
115. A. Ferraro, S. Olivares, M. Paris: *Gaussian States in Continuous Variable Quantum Information*, (Bibliopolis, Napoli 2005)
116. R.P. Feynman: *Feynman Lectures on Computation*, (Addison-Wesley, 1995)
117. P.A. Fillmore: *A User's Guide to Operator Algebras*, (Wiley-Interscience, New York 1996)
118. J. Ford, G. Mantica, G.H. Ristow: Physica D **50**, 493 (991)
119. P. Gács: J. Phys. A: Math. Gen. **34**, 6859 (2001)
120. P. Gács: *Lecture Notes on Descriptional Complexity and Randomness. Technical Report*, Comput. Sci. Dept., Boston Univ., 1988
121. C.W. Gardiner, P. Zoller: *Quantum Noise*, (Springer, Berlin 2004)
122. P. Gaspard, M. Nagaoka: J. Chem. Phys. **111**, 5668 (1999)
123. C.C. Gerry, P.L. Knight: *Introductory Quantum Optics*, (Cambridge University Press, Cambridge 2005)
124. S. Gnutzmann, F. Haake: Z. Phys. B **10**, 263 (1996)
125. V. Gorini, A. Kossakowski: J. Math. Phys. **17**, 1298 (1976)
126. V. Gorini, A. Kossakowski, E.C.G. Sudarshan: J. Math. Phys. **17**, 821 (1976)
127. P. Grünwald, P. Vitanyi: *Shannon Information and Kolmogorov Complexity*, arXiv: cs/0410002
128. J. Gruska: *Quantum Computing*, (Mc Graw Hill, London 1999)
129. M.C. Gutzwiller: *Chaos in Classical and Quantum Mechanics*, (Springer, New York 1990)
130. K.-C. Ha, S.-H. Kye and Y.-S. Park: Phys. Lett. **A313**, 163 (2003)
131. R. Haag, N. Hugenholtz and M. Winnink: Commun. Math. Phys. **5**, 848 (1967)

132. F. Haake: *Statistical Treatment of Open Systems by Generalized Master Equations*, in Springer Tracts in Mod. Phys. **95**, (Springer-Verlag, Berlin 1973)
133. B. Haegeman: *Local aspects of quantum entropy*, (PhD Thesis, Katholieke Universitaet Leuven, Belgium 2004)
134. P.R. Halmos: *A Hilbert space problem book*, (Springer-Verlag, New York 1982)
135. J.H. Hannay, M.V. Berry: Physica D **1**, 267 (1980)
136. P. Hausladen, R. Jozsa, B. Schumacher et al.: Phys. Rev. A **54**, 1869 (1996)
137. P.M. Hayden, M. Horodecki, B.M. Terhal: J. Phys. A **34**, 6891 (2001)
138. K. Hepp: Commun. Math. Phys. **35**, 265 (1974)
139. A.J.P. Hey ed.: *Feynman and Computation*, (Perseus Books, Reading MS, 1999)
140. F. Hiai, D. Petz: Commun. Math. Phys. **143**, 99 (1991)
141. F. Hiai, D. Petz: J. Functional Anal. **125**, 287 (1994)
142. A.S. Holevo: Probl. Inf. Transm. (USSR) **9**, 177 (1973)
143. A.S. Holevo: *Probabilistic and Statistical Aspects of Quantum Theory*, (Amsterdam, North Holland, 1982)
144. A.S. Holevo: IEEE Trans. Information Theory **44**, 269 (1998)
145. A.S. Holevo: *Statistical Structure of Quantum Theory*, (Lect. Notes in Physics Monographs **67**, Springer Verlag, Berlin, Heidelberg 2001)
146. A.S. Holevo: *Coding Theorems for Quantum Channels*, `arXiv: quant-ph/9809023`
147. R.A. Horne, C.R. Johnson: *Matrix Analysis*, (Cambridge University Press, Cambridge 1985)
148. M. Horodecki, P. Horodecki, R. Horodecki: Phys. Lett. A **223**, 1 (1996)
149. P. Horodecki: Phys. Lett. A **232**, 333 (1997)
150. M. Horodecki, P. Horodecki, R. Horodecki: Phys. Rev. Lett. **80**, 5239 (1998)
151. M. Horodecki, P. Horodecki: Phys. Rev. A **59**, 4206 (1999)
152. R. Horodecki, P. Horodecki, M. Horodecki et al.: *Quantum Entanglement*, `arXiv: quant-ph/0702225`
153. T. Hudetz: Lett. Math. Phys. **16**, 151 (1988)
154. T. Hudetz: J. Math. Phys. **35**, 4303 (1994)
155. T. Hudetz: Banach Center Publications **43**, 241 (1998)
156. L.P. Hughston, R. Jozsa, W.K. Wootters, Phys. Lett. A **183**, 14 (1993)
157. J. Jacod: *Probability Essentials*, (Springer Berlin 2003)
158. A. Jamiołkowski: Rep. Math. Phys. **3**, 275 (1972)
159. R. Jozsa, B. Schumacher: J. Mod. Opt. **41**, 2343 (1994)
160. R. Jozsa, M. Horodecki, P. Horodecki et al.: Phys. Rev. Lett. **81**, 1714 (1998)
161. R.V. Kadison, J.R. Ringrose: *Fundamentals of the Theory of Operator Algebras 1: Elementary Theory*, (Academic Press, New York 1983)
162. R.V. Kadison, J.R. Ringrose: *Fundamentals of the Theory of Operator Algebras 2: Advanced Theory*, (Academic Press, New York 1986)
163. A. Kaltchenko, E.H. Yang: Quantum Information and Computation **3**, 359 (2003)
164. A. Katok, B. Hasselblatt, L. Mendoza: *Introduction to the Modern Theory of Dynamical systems*, (Cambridge University Press, Cambridge 1995)
165. P. Kaye, R. Laflamme, M. Mosca: *An Introduction to Quantum Computing*, (Oxford University Press, Oxford UK, 2007)
166. G. Keller: *Wahrsheinlichkeittheorie* (Lecture Notes, Universität Erlangen-Nurnberg 359 (2003)

167. A.Y. Khinchin: *Mathematical Foundations of Statistical Mechanics*, (Dover, New York 1949)
168. J. Kieffer: IEEE Trans. Inform. Theory **24**, 674 (1978)
169. C. King, A. Leśniewski: J. Math. Phys. **39**, 88 (1998)
170. A.Y. Kitaev: Russ. Math. Surv. **52**, 1191 (1997)
171. A.N. Kolmogorov: Dokl. Akad. Nauk SSSR **119**, 861 (1958)
172. A.N. Kolmogorov: Dokl. Akad. Nauk SSSR **124**, 754 (1959)
173. A.N. Kolmogorov: Problems of Information Transmission **1**, 4 (1965)
174. A.N. Kolmogorov: IEEE Trans. Inform. Theory **14**, 662 (1968)
175. B.O. Koopman, J. von Neumann: N.A.S Proc. **18** 255 (1932)
176. L.B. Koralov, Y.G. Sinai: *Theory of Probability and Random Processes* (Springer, Berlin 2007)
177. A. Kossakowski: Bull. Acad. Polon. Sci., Sér. Sci. Math. Astronom. Phys., **20** 1021 (1972)
178. A. Kossakowski: Bull. Acad. Plon. Sci., Sér. Sci. Math. Astronom. Phys., **21** 649 (1973)
179. A. Kossakowski, Open Sys. and Inf. Dyn. **10** 1 (2003)
180. K. Kraus: Ann. Phys., **64** 311 (1971)
181. K. Kraus: *States, Effects, and Operations*, Lec. Notes Phys. **190** (Springer, Berlin 1983)
182. D. Kretschmann, R.F. Werner: 2004 New J. Phys. **6**, 26 (2004)
183. R. Kubo: J. Phys. Soc. Japan **12**, 570 (1957)
184. F. Haake, M. Kuś, R. Scharf: Z. Phys. **B 65**, 381 (1987)
185. B.B. Laird, J. Budimir, J.L. Skinner: J. Chem. Phys. **94**, 4391 (1991)
186. B.B. Laird, J.L. Skinner: J. Chem. Phys. **94**, 4405 (1991)
187. R. Landauer: IBM J. Research **3**, 183 (1961)
188. M. Lebellac: *A Short Introduction to Quantum Information and Quantum Computation*, (Cambridge Univeristy Press, 2006)
189. J. Lebowitz, H. Spohn: Adv. Chem. Phys. **39**, 109 (1978)
190. H.S. Leff, A.F. Rex: *Maxwell's Demon : Entropy, Information, Computing*, (Princeton University Press, Princeton 1990)
191. M. Lewenstein, A. Sanpera, V. Ahufinger et al.: Adv. in Phys. **56**, 243 (2007)
192. E.H. Lieb, M.B. Ruskai: J. Math. Phys. **14**, 1938 (1973)
193. G. Lindblad: Commun. Math. Phys. **48**, 119 (1976)
194. G. Lindblad: *Dynamical Entropy for Quantum Systems*, in: Lec. Notes Math. **1303**, (Springer, Berlin 1988)
195. G. Lindblad: Commun. Math. Phys. **65**, 281 (1979)
196. G. Lindblad: J. Phys. A **26**, 7193 (1993)
197. Y. Makhlin, G. Schön, A. Schnirman: Rev. Mod. Phys. **73**, 357 (2001)
198. S. Mancini, S. Severini: *Electronic Notes in Theoretical Computer Science* **169**, 121 (2007)
199. R. Mane: *Ergodic Theory and Differentiable Dynamics*, (Springer, Berlin 1987)
200. J. Manuceau, F. Rocca, D. Testard: Commun. Math. Phys. **12**, 43 (1969)
201. J. Manuceau, A. Verbeure: Commun. Math. Phys. **8**, 293 (1968)
202. J. Manuceau, A. Verbeure: Commun. Math. Phys. **18**, 319 (1970)
203. P.C. Martin, J. schwinger: Phys. Rev. **115**, 1342 (1959)
204. S. Michalakis, B. Nachtergaele: Phys. Rev. Lett. **97** 140601, (2006)

205. C.E. Mora, H.J. Briegel: *Algorithmic Complexity of Quantum States*, arXiv: quant-ph/0412172
206. C.E. Mora, H.J. Briegel: Phys. Rev. Lett. **95**, 200503 (2005)
207. C.E. Mora, H.J. Briegel, B. Kraus: *Quantum Kolmogorov complexity and its applications*, arXiv: quant-ph/0610109
208. M. Müller: *Strongly Universal Quantum Turing Machines and Invariance of Kolmogorov Complexity*, arXiv: quant-ph/0605030
209. H. Narnhofer: J. Math. Phys. **33**, 1502 (1992)
210. H. Narnhofer: Lett. Math. Phys. **28**, 85 (1993)
211. H. Narnhofer: Phys. Lett. A **310**, 423 (2003)
212. H. Narnhofer, H. Pflug, W. Thirring: *Symmetry in Nature*, (Scuola Nomale Superiore di Pisa **597** 1989)
213. H. Narnhofer, W. Thirring: Fizika **17**, 89 (1985)
214. H. Narnhofer, W. Thirring: Lett. Math. Phys. **14**, 89 (1987)
215. H. Narnhofer, W. Thirring: J. Stat.Phys. **57**, 511 (1989)
216. H. Narnhofer, W. Thirring: Commun. Math. Phys. **125**, 565 (1989)
217. H. Narnhofer, W. Thirring: Lett. Math. Phys. **20**, 231 (1990)
218. H. Narnhofer, W. Thirring: Phys. Rev. Lett. **64**, 1863 (1990)
219. H. Narnhofer, W. Thirring: Int. J. Mod. Phys. **17**, 2937 (1991)
220. H. Narnhofer, W. Thirring: Lett. Math. Phys. **30**, 307 (1994)
221. H. Narnhofer, W. Thirring, W. Wiclicki: J. Stat. Phys. **52**, 1097 (1989)
222. S. Neshveyev, E. Störmer: *Dynamical Entropy in Operator Algebras*, (Springer, Berlin 2006)
223. S. Neshveyev, E. Störmer: Ergod. Th. & Dynam. Sys. **22**, 889 (2002)
224. M.A. Nielsen, I.L. Chuang: *Quantum Information and Quantum Computation*, (Cambridge University Press, Cambridge UK 2000)
225. M.A. Nielsen, D. Petz: *Quantum Information & Computation* **6**, 507 (2005)
226. M. Ohya, D. Petz: *Quantum Entropy and Its Use*, (Springer, Berlin Heidelberg New York 1993)
227. A. Osterloh, L. Amico, G. Falci et al.: Nature **416**, 608 (2002)
228. E. Ott: *Chaos in dynamical systems*, (Cambridge University Press, Cambridge UK 2002)
229. M. Ozawa, H. Nishimura: Theoret. Informatics and appl. **34**, 379 (2000)
230. P.F. Palmer: J. Math. Phys. **18**, 527 (1977)
231. Y.M. Park: Lett. Math. Phys. **32**, 63 (1994)
232. Y.M. Park, H.H. Shin: Commun. Math. Phys. **144**, 149 (1992)
233. Y.M. Park, H.H. Shin: Commun. Math. Phys. **152**, 497 (1992)
234. V. Paulsen: Cambridge Studies in Advanced Mathematics **78** (2002)
235. P. Pechukas: Phys. Rev. Lett. **73**, 1060 (1994)
236. A. Peres: Phys. Rev. Lett. **77**, 1413 (1996)
237. D. Petz: *An invitation to the Algebra of Canonical Commutation Relations*, (University Press, Leuven 1990)
238. D. Petz: Rev. Math. Phys. **15**, 79 (2003)
239. D. Petz: *Quantum Information Theory and Quantum Statistics*, (Theoretical and Mathematical Physics Series, Springer 2008)
240. D. Petz, M. Mosonyi: J. Math. Phys. **42**, 4857 (2001)
241. M. Popp, F. Verstraete, M. A. Martin-Delgado et al.: Phys. Rev. A **71**, 042306 (2005)
242. J. Preskill: *Lecture Notes: Quantum Computation and Information* www.theory.caltech.edu/ preskill/ph229

243. T.Prosen: J. Phys. A **40**, 7881 (2007)
244. R.T. Powers: *UHF Algebras and Their Applications to Representation of the Anti-Commutation Relations*, (Cargese Lecture Notes in Physics, New York 1970)
245. R.T. Powers, E. Störmer: Commun. Math. Phys. **16**, 1 (1970)
246. R.T. Powers: Can. J. Math. **40**, 86 (1988)
247. G.L. Price: Can. J. Math. **39**, 492 (1987)
248. H. Rauch, S. A. Werner: *Neutron Interferometry*, (Oxford University Press, Oxford 2000)
249. R. Raussendorf, H. J. Briegel: Phys. Rev. Lett. **86**, 5188 (2001)
250. M. Redei, J.S. Summers: *Quantum Probability Theory* `arXiv:quant-ph/0601158`
251. M. Reed, B. Simon: *Methods of Modern Mathematical Physics: Functional Analysis*, (Academic Press, New York 1972)
252. M. Reed, B. Simon: *Methods of Modern Mathematical Physics 4.: Analysis of Operators*, (Academic Press, New York 1978)
253. J.R. Retherford: *Hilbert Space: Compact Operators and the Trace Theorem*, London Mathematical Society Student Text **27**, (Cambridge University Press 1993)
254. J. Rissanen: *Information and Complexity in Statistical Modeling*, (Springer Verlag, 2006)
255. F. Rocca, M. Sirigue, D. Testard: Ann. Inst. H. Poincaré **A10**, 247 (1969)
256. F. Rocca, M. Sirigue, D. Testard: Commun. Math. Phys. **13**, 317 (1969)
257. F. Rocca, M. Sirigue, D. Testard: Commun. Math. Phys. **16**, 119 (1970)
258. W. Rudin: *Real and Complex Analysis*, (McGraw-Hill, New York 1987)
259. W. Rudin: *Fucntional analysis*, (McGraw-Hill, New York 1991)
260. D. Ruelle: *Statistical Mechanics: Rigorous Results*, (Benjamin, New York 1969)
261. M.B. Ruskai: *Lieb's simple proof of concavity of* $\mathrm{Tr}(A^p K^\dagger B^{1-p} K)$ *and remarks on related inequalities*, `quant-ph/0404126`
262. M.B. Ruskai: *Another short and elementary proof of strong sub-additivity of quantum entropy*, `quant-ph/0604206`
263. S. Sakai: *$C^*$ algebras and $W^*$ algebras*, (Springer, Berlin 1971)
264. J.-L. Sauvageot, J.-P. Thouvenot: Commun. Math.Phys. **145**, 411 (1992)
265. F. Scheck: *Mechanics: from Newton's laws to deterministic chaos*, (Springer Verlag, Berlin 1994)
266. P.C. Shields: *The Ergodic Theory of Disrcete Sample Paths*, (AMS, Providence 1996)
267. B. Schumacher: Phys. Rev. A **51**, 2738 (1995)
268. B. Schumacher: *Entropy, Complexity and Computation*, `http://physics.kenyon.edu/coolphys/thrmcmp/newcomp.htm`
269. B. Schumacher, M.D. Westmoreland: Phys. Rev. A **56**, 56 (1997)
270. R. Schatten: *Norm Ideals of Completely Continuous Operators*, (Springer, Berlin 1960)
271. H.G. Schuster: *Deterministic Chaos: an Introduction*, (Weinheim, VHC 1995)
272. M.O. Scully, M.S. Zubairy: *Quantum Optics*, (Cambridge University Press, Cambridge 1997)
273. V.F. Sears: *Neutron Optics*, (Oxford University Press, Oxford 1989)
274. G. L. Sewell: *Quantum Theory of Collective Phenomena*, (Clarendon Press, Oxford 1986)

275. G. L. Sewell: *Quantum Mechanics and its Emergent Macrophysics*, (Princeton University Press, Princeton and Oxford 2002)
276. P.W. Shor: *Equivalence of Additivity Questions in Quantum Information Theory*, quant-ph/0307098
277. B. Simon: *The Statistical Mechanics of Lattice Gases*, (Princeton Univeristy Press, Princeton 1993)
278. R. Simon: Phys. Rev. Lett. **84**, 2726 (2000)
279. Ya.G. Sinai: *Introduction to Ergodic Theory*, (Princeton University Press, Princeton, 1976)
280. C.P. Slichter: *Principle of Magnetic Resonance*, (Springer-Verlag, Berlin 1990)
281. W. Słomczyński: *Dynamical Entropy, Markov Operators, and Iterated Function Systems*, (Wydawnictwo Uniwersytetu JagielloŃskiego, Krakow 2003)
282. W. Słomczyński, K. Życzkowski: J. Math. A **35**, 5674 (1994)
283. R.J. Solomonoff: Inform. Contr. **7**, 1 (1964)
284. R.J. Solomonoff: Inform. Contr. **7**, 224 (1964)
285. H. Spohn: Rev. Mod. Phys. **52** 569 (1980)
286. H. Spohn: J. Math. Phys. **19** 1227 (1980)
287. W.F. Stinespring: Proc. Am. Math. Soc. **6**, 211 (1955)
288. E. Störmer: Proc. Amer. Math. Soc. **86**, 402 (1982)
289. E. Störmer, D. Voiculescu: Commun. Math. Phys. **133**, 521 (1990)
290. F. Strocchi: *Elements of Quantum Mechanics of Infinite Systems*, International School for Advanced Studies Lecture Series **3**, (World Scientific, Singapore 1985)
291. F. Strocchi: *Symmetry Breaking*, LNP **643**, (Springer, Berlin 2005)
292. A. Suarez, R. Silbey, I. Oppenheim: J. Chem. Phys. **97**, 5101 (1992)
293. V.S. Sunder: *An Invitation to von Neumann Algebras*, Universitext, (Springer, New York 1987)
294. K. Svozil: *Randomness and Undecidability in Physics*, (World Scientific 1993)
295. M. Takahashi: *Thermodynamics of One-Dimensional Models*, Cambridge University Press (Cambridge, UK 1999)
296. M. Takesaki: *Theory of Operator Algebras I*, (Springer, New York 1979)
297. B.M. Terhal: Linear Algebra Appl. **323**, 61 (2000)
298. B.M. Terhal, K.G.H. Vollbrecht: Phys. Rev. Lett. **85**, 2625 (2000)
299. W. Thirring: *A course in Mathematical Physics: Classical Dynamical Systems*, (Springer, New York 1978)
300. W. Thirring: *Quantum Mathematical Physics: Atoms, Molecules and Large Systems*, (Springer-Verlag, Berlin, 2002)
301. W. Thirring: *Recent Developments in Mathematical Physics*, (H. Mitter, L. Pittner eds, Springer, 1987)
302. A. Uhlmann: Wiss. Z. Karl-Marx-Univ. Leipzig **21**, 421 (1972)
303. A. Uhlmann: Wiss. Z. Karl-Marx-Univ. Leipzig **22**, 139 (1973)
304. A. Uhlmann: *Optimizing Entropy Relative to a Channel or Subalgebra*, quant-ph/9701014
305. V.A. Uspenskii, A.L. Semenov, A.Kh. Shen: Russian Math. Surveys **45**, 121 (1990)
306. J.J. Vartiainen, M. Möttönen, M.M. Salomaa: Phys. Rev. Lett. **92** (2004)
307. V. Vedral: *Introduction to Quantum Information*, (Oxford University Press, Oxford UK, 2006)

308. R. Verch, R.F. Werner: Rev. Math. Phys. **17**, 545 (2005)
309. P. Vitanyi: IEEE Trans. Inform. Theory **47/6**, 2464 (2001)
310. M. Li, P. Vitanyi: *An Introduction to Kolmogorov Complexity and Its Applications* 2nd ed, (Springer,New York Berlin Heidelberg 1997)
311. D. Voiculescu: Commun. Math. Phys. **144**, 443 (1992)
312. D.F. Walls, G.J. Milburn: *Quantum Optics*, (Springer, Berlin 1994)
313. P. Walters: *An Introduction to Ergodic Theory*, Graduate Texts in Mathematics **79**, (Springer, New York 1982)
314. A. Wehrl: Rev. Mod. Phys. **50**, 221 (1978)
315. U. Weiss: *Quantum Dissipative Systems*, (World Scientific, Singapore 1999)
316. G. Wendin, V.S. Schumeiko: *Superconducting Quantum Circuits, Qubits and Computing*, `arXiv: cond-mat/0508729`
317. R.F. Werner: Phys. Rev. A **40**, 4277 (1989)
318. H. White: Erg. Th. Dyn. Sys. **13**, 807 (1993)
319. J. Wielkie: J. Chem. Phys. **114**, 7736 (2001)
320. W.K. Wootters: Phys. Rev. Lett. **80**, 2245 (1997)
321. S.L. Woronowicz: Rep. Math. Phys. **10**, 165 (1976)
322. G.M. Zaslawski: *Chaos in Dynamic Systems*, (Chur, Harwood Academic Publ. 1985)
323. P. Walther, K.J. Resch, T. Rudolph et al.: Nature **434**, 169 (2005)
324. K. Zhu: *An Introduction to Operator Algebras*, Studies in Mathematics (CRC Press, Boca Raton, Ann Arbor, London, Tokyo 1993)
325. J. Ziv: IEEE Trans. Inform. Theory **18**, 384 (1972)
326. W.H. Zurek: Phys. Rev. A **40**, 4731 (1989)
327. A.K. Zvonkin, L.A. Levin: Russian Mathematical Surveys **25**, 83 (1970)

# Index

**U**



**V**



**W**

FSC
www.fsc.org
MIX
Papier | Fördert
gute Waldnutzung
FSC® C083411

Kolibri 360 GmbH
Ferdinand-Jühlke-Straße 7
99095 Erfurt, Deutschland
produktsicherheit@kolibri360.de